T0340048

MODERN MATHEMATICAL METHODS FOR PHYSICISTS AND ENGINEERS

The advent of powerful desktop computers has revolutionized scientific analysis and engineering design in fields as disparate as particle physics and telecommunications. *Modern Mathematical Methods for Physicists and Engineers* provides an up-to-date mathematical and computational education for students, researchers, and practicing engineers.

The author begins with a review of computation and then deals with a range of key concepts including sets, fields, matrix theory, and vector spaces. He then goes on to cover more advanced subjects such as linear mappings, group theory, and special functions. Throughout, he concentrates exclusively on the most important topics for the working physical scientist or engineer, with the aim of helping them to make intelligent use of the latest computational and analytical methods.

The book contains well over 400 homework problems and covers many topics not dealt with in other textbooks. It will be an ideal textbook for senior undergraduate and graduate students in the physical sciences and engineering, as well as a valuable reference for working engineers.

C. D. Cantrell received his Ph.D. from Princeton University in 1968. He taught at Swarthmore College from 1967 until 1973 and was a staff member at the Los Alamos National Laboratory from 1973 until 1979. Since then he has been at the University of Texas at Dallas, where he is Professor of Physics and Electrical Engineering, and Director of the Photonic Technology and Engineering Center. Professor Cantrell is a consultant for Alcatel USA and Ericsson and is a Fellow of the American Physical Society, the Optical Society of America, and the IEEE.

MODERN MATHEMATICAL METHODS FOR PHYSICISTS AND ENGINEERS

C. D. CANTRELL

CAMBRIDGE
UNIVERSITY PRESS

CAMBRIDGE
UNIVERSITY PRESS

32 Avenue of the Americas, New York NY 10013-2473, USA

Cambridge University Press is part of the University of Cambridge.

It furthers the University's mission by disseminating knowledge in the pursuit of
education, learning and research at the highest international levels of excellence.

www.cambridge.org
Information on this title: www.cambridge.org/9780521598279

First published 2000
Reprinted 2002

A catalogue record for this publication is available from the British Library

Library of Congress Cataloguing in Publication data

Cantrell, C. D. (Cyrus D.), 1940–
    Modern mathematical methods for physicists and engineers / C. D.
Cantrell.
        p.   cm.
    Includes bibliographical references (p.   –   ).
    ISBN 0-521-59180-5 (hb). – ISBN 0-521-59827-3 (pbk.)
    1. Mathematics.   I. Title.
QA37.2.C24   1999
510 – dc21   98-24761
                                                           CIP

ISBN 978-0-521-59827-9 Paperback

To Lynn, Kate, and Sarah

CONTENTS

PREFACE

The purpose of *Modern Mathematical Methods for Physicists and Engineers* is to help graduate and advanced undergraduate students of the physical sciences and engineering acquire a sufficient mathematical background to make intelligent use of modern computational and analytical methods. This book responds to my students' repeated requests for a mathematical methods text with a modern point of view and choice of topics.

For the past fifteen years I have taught graduate courses in computational and mathematical physics. Before introducing the course on which this book is based, I found it necessary, in courses ranging from numerical methods to the applications of group theory in physics, to summarize the rudiments of linear algebra and functional analysis before proceeding to the ostensible subjects of the course. The questions of the students who studied early drafts of this work have helped to shape the presentation. Some students working concurrently in nearby telecommunication, semiconductor, or aerospace, industries have contributed significantly to the substance of portions of the book.

The following is an example of the situations that motivated me to take the time to write a mathematical methods text that breaks significantly with the past: Every semester, students come to my office, puzzled over numerical models in which minor changes in the data produce drastic changes in the outputs. Unfortunately most of these students lack the mathematical background needed to conceptualize some of the most common problems of numerical computation. For an engineer, and for the increasingly large fraction of physics graduates who make careers in numerical modeling or electrical engineering, conceptual understanding of analytical and numerical models is an absolutely essential ingredient of successful designs. A computer can be a tool for understanding, and not merely a means for obtaining a numerical answer of unknown reliability and significance, only in the hands of those who understand the foundations and potential shortcomings of numerical methods. Yet the traditional mathematical methods taught to students in engineering and physics for most of the twentieth century do not provide a sufficient background even for introductory graduate texts on many important contemporary topics, of which numerical computation is only one.

What upper-level undergraduate and first-year graduate students in physics and engineering tend consistently to lack is an understanding of basic mathematical structures – groups, rings, fields, and vector spaces – and of mappings that preserve these structures. In times gone by, students learned mathematical structures though intensive practice with examples. However, in curricula that already are under fire for taking too many years, there simply is no time to learn the language of mathematics by example. Like adults who learn grammar in order to accelerate the acquisition of a foreign language, contemporary students in physics or engineering can more easily acquire a durable understanding of applied and numerical mathematics if they have been exposed to the most essential formal mathematical structures.

The core of *Modern Mathematical Methods for Physicists and Engineers* is linear algebra and basic functional analysis. Computation is the subject of two of the first three chapters because computational examples and exercises occur throughout the book. Chapters on sets and groups, rings and fields provide necessary background for subsequent chapters on vector spaces, inner-product spaces, linear mappings, and matrix representations of finite groups. Group-throry concepts provide an approach to partial differential equations and special functions based on algebra instead of complex analysis. Throughout the book, abstraction is not an end in itself, but a means for students to remember concepts and use them intelligently.

The exercises range in difficulty from simple applications of the definitions in the text to problems that may challenge strong students. In both the text and the exercises, asterisks indicate material that is unusually difficult, and that may be omitted on a first reading.

The manuscript for this book was created in LaTeX on a Macintosh Power Book® using the program Textures®. The illustrations were created using Adobe Illustrator®.

I thank all those who have contributed to this book, especially my students. Special thanks are due to Professors William J. Pervin and Poras Balsara, and to Dawn Hollenbeck, for their valuable comments on portions of the manuscript.

CHAPTER 1

FOUNDATIONS OF COMPUTATION

1.1 INTRODUCTION

This chapter describes representations of integers and fractions that are useful for digital computation and information transmission and introduces the basic concepts of computational accuracy and error propagation. One may ask why it makes sense to begin a book on *mathematical* methods with a computational topic. The answer is that computation is no longer simply a technique for obtaining a numerical answer but has become a tool for research and design that is as important as experimentation and theoretical calculation. The material in this chapter is the foundation for using computation as a tool for understanding.

For example, in many problems the execution time T required for a numerical solution is proportional to some power p of the number N of computational grid points. For such a problem the mean computational speed is

$$s = kN^p/T,$$

where k is a constant that depends on the computer, the programming language, and other factors, and p is a constant that depends on the algorithm. If grid points are placed at points with coordinates $x = lh, y = mh, z = nh$ in a cube with side L, then

$$N = L^3/h^3.$$

The distance between adjacent points, which is a measure of the "resolving power" of the computation, is

$$h = L(k/sT)^{1/(3p)}$$

if one assumes that the time, algorithm, and computational speed are all fixed. It follows that the resolution of a numerical solution, as measured by the grid spacing h, is proportional to $s^{-1/(3p)}$.

If one compares the speed of hand computation with the computational speed of the current generation of high-performance computers, one finds that over the past six decades, s has improved by a factor of 10^9. Therefore the computational resolution for algorithms that are linear in time ($p = 1$) has improved by three orders of magnitude in less than one human lifetime. By way of comparison, the Hubble Space Telescope, the product of more than three centuries of technological development, has a resolution of 0.1 arc-second, which is less than three orders of magnitude better than the resolution of the unaided human eye (0.5–1.5 arc-minute). Moreover, the computational speed s continues to improve by a factor of two every two years, thus improving the computational resolution for a typical three-dimensional problem by a factor of two every eight years.

Until the 1950s, the only feasible way to obtain a numerical result was to use the traditional methods of theoretical physics such as separation of variables, contour integration, and conformal mapping to express the solution of a problem in terms of known functions for which tables of values were available. The inaccessibility, limited execution rate, and small storage capacity of early digital computers, combined with a lack of user-amenable software, kept analytical techniques in the physicist's and engineer's tool kit for another generation. Among the many consequences of the availability of increasingly powerful, inexpensive computers during the 1980s and 1990s is that in high-technology industry computation has become as important as experiment and is much more important than analytical theory.

It is essential for any working physicist or engineer to understand the limitations of what Willis Lamb calls "the numerical approximation," which means computation using a finite set of fixed-precision floating-point numbers (that is, numbers of the form 2.99792458×10^8). An understanding of the consequences of rounding, catastrophic cancellation of significant digits through subtraction, and other basic issues of numerical computation, is indispensable in an age in which physicists and engineers turn to the computer for an answer more often than they turn to analytical techniques. A goal of this chapter, and indeed of the sections on computational methods throughout this book, is to help the reader understand how the mathematical characteristics of a problem, the properties of floating-point arithmetic, the architecture of available computers, and the choice of computational algorithm combine to influence the feasibility and accuracy of a numerical computation.

To profit fully from this chapter the reader should be familiar with the basic principles of computer architecture, as summarized, for example, in Patterson and Hennessy (1994).

1.2 REPRESENTATIONS OF NUMBERS

Like most abstract concepts, numbers can have many different concrete representations. By a **representation** of numbers, we mean a rule that associates a unique string of symbols with each number. It will become clear later that a numerical representation is a mapping, in the sense defined in Section 2.2.2, of a set of abstract numbers onto a set of strings. Historically, the positive integers have been represented by finger signs, by beads on an abacus, by symbols incised on clay tablets with a stylus or written on paper with a brush, pencil, or pen, and by many other means. Positive and negative integers, real numbers, complex numbers, and the values ("true" or "false") of logical variables also can be represented by the states of electronic circuits.

The development and rapid evolution of powerful, inexpensive computers have fundamentally changed the ways in which one models physical systems and designs or uses electronic devices. At the present time, the results of both theoretical models and experimental measurements are almost always expressed in digital form, with a fixed number of significant digits. For this reason, the number representations surveyed in this chapter are fundamental in physics and engineering.

1.2.1 INTEGERS

Base-β Integer Representations

The arabic, or positional, decimal representation of numbers is the product of several thousand years of experimentation with other, less convenient systems. For example, the Babylonians used powers of 60 instead of powers of 10. A consistent base-60 notation would be excessively complex, because it would require 60 different symbols for the "digits" $0, 1, \ldots, 59$. The Roman representation of numbers is decimal (that is, uses powers of 10) but does not use the digits $0, 1, \ldots, 9$ that are used in the Arabic representation. The Roman number MCMXCVI, for example, represents the same integer as 1,996 in Arabic notation. Here, M $= 1000$, C $= 100$, X $= 10$, V $= 5$, XC $= 90$, and so on. Thus a single digit in Arabic notation may correspond to one, two, three, or more letters in the Roman representation. This makes it difficult for computers – and especially for humans – to read Roman numerals. The basic arithmetic operations are also cumbersome in the Roman representation. For example, X – II $=$ VIII, X $\times$ C $=$ M, and so on.

A better idea is to use a *positional* decimal notation in which each digit multiplies a power of 10:

$$1996_{10} = 1 \cdot 10^3 + 9 \cdot 10^2 + 9 \cdot 10^1 + 6 \cdot 10^0. \tag{1.1}$$

Each digit multiplies the power of the base (10) that corresponds to the digit's position, starting with 10^0 at the right and increasing to the left. Among the many advantages of positional notation, multiplication by a power of the base (say, 10^n), one simply shifts left n places. Our grade-school algorithms for the basic arithmetic operations are based on this very simple property. For example, multiplication of two numbers becomes a simple matter of shifting and adding.

One can generalize the decimal positional representation of numbers by using an arbitrary positive integer $\beta \geq 2$ as the base. Every nonzero integer k can be expressed uniquely in a **base-β integer representation** as

$$k = \pm \sum_{j=0}^{n} b_j \, \beta^j, \tag{1.2}$$

where β is called the **base** (or **radix**). The decimal representation uses $\beta = 10$. In Eq. (1.2), each b_j is an integer such that

$$0 \leq b_j < \beta, \tag{1.3}$$

and n is a positive integer such that

$$\beta^n \le k < \beta^{n+1}. \tag{1.4}$$

The nonnegative integers b_j are called **significant digits**. The digit that multiplies the highest power of the base, b_n, is called the **most significant digit**; b_0 is (in this case) the **least significant digit**.

Instead of writing a summation, one normally uses the **positional notation**

$$k = \pm (b_n\, b_{n-1} \cdots b_0)_\beta. \tag{1.5}$$

For example,

$$\begin{aligned}
(255)_{10} &= 2 \cdot 10^2 + 5 \cdot 10^1 + 5 \cdot 10^0 \\
&= (377)_8 = 3 \cdot 8^2 + 7 \cdot 8^1 + 7 \cdot 8^0 \\
&= (11111111)_2 \\
&= 1 \cdot 2^7 + 1 \cdot 2^6 + 1 \cdot 2^5 + 1 \cdot 2^4 \\
&\quad + 1 \cdot 2^3 + 1 \cdot 2^2 + 1 \cdot 2^1 + 1 \cdot 2^0 \\
&= 2^8 - 1,
\end{aligned} \tag{1.6}$$

where we have written most base-10 representations without subscripts in the interest of readability.

To obtain the base-β representation of an integer $k > 0$ one uses a division algorithm: Begin by finding the smallest positive integer n such that

$$k < \beta^{n+1}. \tag{1.7}$$

Then the most significant digit is

$$b_n = \left\lfloor \frac{k}{\beta^n} \right\rfloor. \tag{1.8}$$

In this equation $\lfloor x \rfloor$ is the **integer part** of x, which is defined as the greatest integer that is less than or equal to x. The **fractional part** of x is defined as $x - \lfloor x \rfloor$. Sometimes $\lfloor \cdot \rfloor$ is called the **floor function** because its value is the floor on which the fractional part stands. For example, the integer part of 2.72 is $\lfloor 2.72 \rfloor = 2$, and the fractional part of 2.72 is $2.72 - \lfloor 2.72 \rfloor = 0.72$.

Also, $\lceil x \rceil$ is defined as the smallest integer that is greater than or equal to x, and $\lceil \cdot \rceil$ is called the **ceiling function**.

After one has computed the most significant digit in the base-β representation of k using Eq. (1.8), the equations

$$b_j = \left\lfloor \beta^{-j} \left(k - \sum_{l=j+1}^{n} b_l\, \beta^l \right) \right\rfloor \quad (j = n - 1, \ldots, 0) \tag{1.9}$$

give the remaining digits of k recursively.

To obtain the base-β representation of a negative integer, one can either change the sign to $+$, convert the resulting positive integer, and then change the sign back to $-$, or use the *ceiling* function $\lceil \cdot \rceil$ in Eqs. (1.8) and (1.9) instead of the floor function. For example, to obtain the most significant digit of the base-2 representation of $(-27)_{10} = (-11011)_2$, one calculates $b_4 = \lceil \frac{-27}{2^4} \rceil = -1$.

Commonly Used Bases

A written or printed positional notation of the form of Eq. (1.5) can be read easily by humans and is well adapted to hand computation. Computers can store textual information such as a number written in the positional notation defined in Eq. (1.5) but must convert this information to a different form in order to carry out numerical computations efficiently. In Section 1.2.3 we discuss a commonly used system for storing numerical information as text.

For most computational purposes digital computers represent nonnegative integers with the help of a hardware positional system in which a different state of a dedicated electronic circuit stands for each of the β different possible values of a particular digit b_j. For example, $n + 1$ circuits (with β states each) must be used to represent the $n + 1$ digits in a base-β place-value representation of a positive integer $k = (b_n\, b_{n-1} \cdots b_0)_\beta$.

The electronic components that are most commonly used for hardware positional representations of digits, a capacitor and a flip-flop, have two states (charged or uncharged and on or off, respectively). For this reason most digital computers use one of the bases $\beta = 2^n$ where $n = 1, 3$ or 4. Hand-held calculators, which usually use base $\beta = 10$, represent the ten decimal digits with ten of the sixteen states of a circuit that contains four two-state devices.

The common bases for numerical computation, $\beta = 2$, $\beta = 8$, $\beta = 10$, and $\beta = 16$, give rise to the **binary, octal, decimal**, and **hexadecimal** representations, respectively. The binary, octal, and decimal representations use the familiar digits $0, \ldots, 9$ or a subset thereof. Each of the octal digits,

$$(1)_8 = (001)_2,$$
$$(2)_8 = (010)_2,$$
$$(3)_8 = (011)_2,$$
$$(4)_8 = (100)_2, \tag{1.10}$$
$$(5)_8 = (101)_2,$$
$$(6)_8 = (110)_2,$$
$$(7)_8 = (111)_2,$$

corresponds to a string of three binary digits. For the hexadecimal representation ($\beta = 16$) six additional symbols are necessary to represent the digits that stand for $(10)_{10}$ through $(15)_{10} = \beta - 1$. Because $16 = 2^4$, each hexadecimal digit corresponds to four binary digits.

The standard convention is

$$
\begin{aligned}
(0)_{16} &= (0)_{10} = (0000)_2, \\
(1)_{16} &= (1)_{10} = (0001)_2, \\
(2)_{16} &= (2)_{10} = (0010)_2, \\
(3)_{16} &= (3)_{10} = (0011)_2, \\
(4)_{16} &= (4)_{10} = (0100)_2, \\
(5)_{16} &= (5)_{10} = (0101)_2, \\
(6)_{16} &= (6)_{10} = (0110)_2, \\
(7)_{16} &= (7)_{10} = (0111)_2, \\
(8)_{16} &= (8)_{10} = (1000)_2, \\
(9)_{16} &= (9)_{10} = (1001)_2, \\
(A)_{16} &= (10)_{10} = (1010)_2, \\
(B)_{16} &= (11)_{10} = (1011)_2, \\
(C)_{16} &= (12)_{10} = (1100)_2, \\
(D)_{16} &= (13)_{10} = (1101)_2, \\
(E)_{16} &= (14)_{10} = (1110)_2, \\
(F)_{16} &= (15)_{10} = (1111)_2.
\end{aligned}
\tag{1.11}
$$

Note carefully that $(10)_{16} = (16)_{10}$. Table 1.1 gives the powers of 2 and 16 for the range of exponents that one ordinarily needs for computational purposes.

Bits, Bytes, and Words

The binary digits 0 and 1 often are called **bits**.[1] A string of eight consecutive bits, which is called a **byte**, can be represented compactly by a two-digit string in base 16 – hence the usefulness of hexadecimal notation. For example,

$$
(3F)_{16} = (0011 \quad 1111)_2 = (63)_{10}.
\tag{1.12}
$$

Sometimes a half-byte (four consecutive bits) is called a **nybble**. Evidently the bit pattern in a nybble can be represented uniquely by a single hexadecimal digit.

In modern computers a **word** is always an integral number of bytes (usually 4 or 8). In most currently manufactured computers the random-access memory (RAM) is organized into individually addressable bytes. In a few computer architectures, RAM can be addressed only by words, not by bytes.

There is no limit (other than memory or disk space) to the size of an integer that can be represented in a computer program. However, in any given hardware design only integers of at most a certain maximum length in bits can be represented in fast logic circuits. In

---

[1] *Bit* is a contraction of *BInary digiT*.

TABLE 1.1 Powers of 2 and 16.

Power of 2	Power of 16	Common Name	Decimal Value
2^0	16^0		1
2^1			2
2^2			4
2^3			8
2^4	16^1		16
2^5			32
2^6			64
2^7			128
2^8	16^2		256
2^9			512
2^{10}		1 K	1 024
2^{11}		2 K	2 048
2^{12}	16^3	4 K	4 096
2^{13}		8 K	8 192
2^{14}		16 K	16 384
2^{15}		32 K	32 768
2^{16}	16^4	64 K	65 536
2^{17}		128 K	131 072
2^{18}		256 K	262 144
2^{19}		512 K	524 288
2^{20}	16^5	1 M	1 048 576
2^{21}		2 M	2 097 152
2^{22}		4 M	4 194 304
2^{23}		8 M	8 388 608
2^{24}	16^6	16 M	16 777 216
2^{25}		32 M	33 554 432
2^{26}		64 M	67 108 864
2^{27}		128 M	134 217 728
2^{28}	16^7	256 M	268 435 456
2^{29}		512 M	536 870 912
2^{30}		1 G	1 073 741 824
2^{31}		2 G	2 147 483 648
2^{32}	16^8	4 G	4 294 967 296
2^{33}		8 G	8 589 934 592
2^{34}		16 G	17 179 869 184
2^{35}		32 G	34 359 738 368
2^{36}	16^9	64 G	68 719 476 736
2^{37}		128 G	137 438 953 472
2^{38}		256 G	274 877 906 944
2^{39}		512 G	549 755 813 888
2^{40}	16^{10}	1 T	1 099 511 627 776

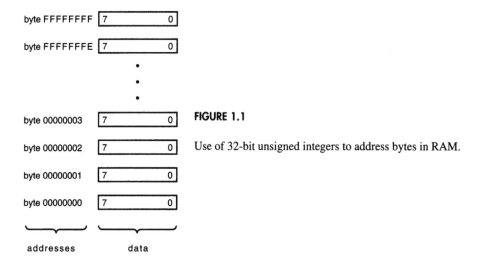

byte FFFFFFFF [7   0]

byte FFFFFFFE [7   0]

byte 00000003 [7   0]

byte 00000002 [7   0]

byte 00000001 [7   0]

byte 00000000 [7   0]

addresses   data

FIGURE 1.1

Use of 32-bit unsigned integers to address bytes in RAM.

most modern computers and general-purpose microprocessors the maximum integer length permitted by the arithmetic and logical unit is either 32 or 64 bits.

Each byte (or, in some cases, each word) that is represented in a computer's RAM is labeled with a unique positive base-β integer with which the byte (or word) can be addressed for the purposes of reading or writing information. For the sake of speed the digits in an address are normally transmitted from the central processing unit (CPU) to the memory unit on parallel conductors. The number of digits in an address is therefore limited by the width of the central processing unit–to–memory address path. In the CRAY® X/MP series of supercomputers, for example, data words were 64 bits long, and each data word in RAM was addressed with a 24-bit integer. Therefore the maximum addressable memory capacity was $2^{24} = 16\,777\,216_{10}$ 8-byte words (the equivalent of 128 megabytes).[2] Many general-purpose microprocessors address memory by bytes using 32-bit addresses (which implies an address space of 2^{32} bytes = 4 gigabytes); see Fig. 1.1.

Unsigned Binary Integer Representation

An engineer or scientist who uses more than one computer family needs to be aware that different computer architectures use different hardware representations of numbers and to know what the major representations are. In a modern computational environment it is common to write and debug a program on a desktop computer, run the program on a supercomputer, and visualize the results on a graphics workstation. These systems may represent integers (for example) in slightly but significantly different ways, ignorance of which can lead to spurious results and unnecessary delays in completing a research project or developing a product.

---

[2] The usual units of memory or mass-storage capacity are 1 kilobyte = 2^{10} bytes = $1\,024$ bytes, 1 megabyte = 2^{20} bytes = $1\,048\,576_{10}$ bytes, 1 gigabyte = 2^{30} bytes = $1\,073\,741\,824_{10}$ bytes, and so forth.

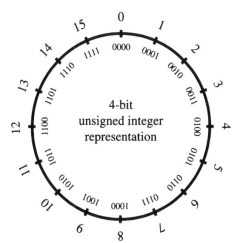

FIGURE 1.2

$n-1$ 0

| b_{n-1} | b_{n-2} | $\cdots$ | b_1 | b_0 |

$$m = b_{n-1} \cdot 2^{n-1} + b_{n-2} \cdot 2^{n-2} + \cdots + b_1 \cdot 2^1 + b_0 \cdot 2^0$$

The bit sequence of the n-bit, binary, unsigned-integer representation.

We begin our survey of the formats in which numbers can be represented in a computer's hardware with **unsigned** (that is, nonnegative) **integers**. Figure 1.2 shows a possible (and common) binary hardware representation of an integer m such that $0 \le m \le 2^n - 1$. For example, the values that can be represented by an unsigned integer if $n = 8$ are $m = 0$ through $m = 255$, inclusive, according to Eq. (1.2). Clearly 2^n nonnegative integers, of which $2^n - 1$ are positive, can be represented with a total of n bits in the unsigned representation shown in Figs. 1.2 and 1.3. The integer 0 is represented by the bit string that consists entirely of zeros. Every nonzero integer within the range of the representation is represented by a nonzero bit string. Figure 1.3 illustrates that one obtains every integer in the range of the representation by starting at 0 and adding 1 repeatedly until one gets to $2^n - 1$. Adding 1 then gives 2^n, which is equivalent to 0 because the n least significant bits of 2^n are all zero. In other words, the marks around the circumference of the circle in Fig. 1.3 are analogous to the hour marks on the face of a clock, except that here one counts by 16s, not by 12s. We show in Sections 4.2.4 and 4.5 that the standard representations of integers with a fixed number (n) of digits are realizations of the group of integers modulo 2^n and the ring of integers modulo 2^n. (For definitions of these basic structures, see Chapter 4.)

Applications of Unsigned Integers

Unsigned integers of a fixed length n can be used if integer data have a fixed maximum magnitude (as is the case for addresses of RAM locations) and if a digitized physical quantity varies between a fixed minimum and a fixed maximum. On a gray-scale computer monitor screen, the shade of gray of a particular pixel can be represented by an 8-bit unsigned integer

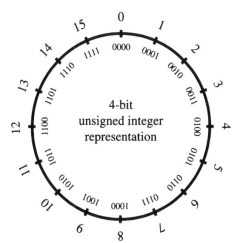

FIGURE 1.3

The 4-bit, binary, unsigned-integer representation.

m ranging from $m = 0$ (black) to $m = (255)_{10}$ (white). For example, $7F_{16} = 127_{10}$ represents 50% gray. A screenful of gray pixels can be regarded as an $M \times N$ matrix

$$\mathbf{P} = \begin{pmatrix} p_{00} & p_{01} & \cdots & p_{0N} \\ \vdots & \vdots & \ddots & \vdots \\ p_{M0} & p_{M1} & \cdots & p_{MN} \end{pmatrix} \tag{1.13}$$

in which the matrix element p_{ij} is an 8-bit unsigned integer that represents the shade of gray of the (i, j) pixel on the screen. This point of view suggests a method of compressing image data so that much fewer than MN bytes are required to transmit an $M \times N$ screenful of data with visually acceptable accuracy; see Section 9.4.5.

On a color computer monitor, the intensity of a red, green, or blue pixel can be specified by an 8-bit unsigned integer ranging from $m = 0$ (zero intensity) to $m = (255)_{10} = FF_{16}$ (maximum intensity). A color can be specified in red-green-blue (RGB) notation by combining the intensities of red, green, and blue in a single 24-bit integer in which the most significant byte is the intensity of red, the middle byte is the intensity of green, and the least significant byte is the intensity of blue. For example, $RGB = FFFF00_{16}$ represents a maximum intensity of yellow ($R = FF_{16}$ = maximum intensity, $G = FF_{16}$ = maximum intensity, $B = 00_{16}$ = zero intensity).

In modern computer systems the number of bits, n, in an unsigned integer is a multiple of 8 in order to permit the hardware representation of an integer to occupy an integral number of bytes (usually 1, 2, 3, 4, or 8). For example, Internet-protocol addresses are unsigned 4-byte integers, the value of which is usually specified by giving the decimal value of each byte, as in 204.179.186.65.

Byte Ordering

In many computers the addresses of bytes in RAM are such that the memory location with the lowest address holds the most significant 8 bits (that is, the **most significant byte**), the memory location with the next lowest address the next most significant byte, and so forth. This byte ordering is called **big-endian** (Fig. 1.4). However, the Intel 80×86 microprocessor family

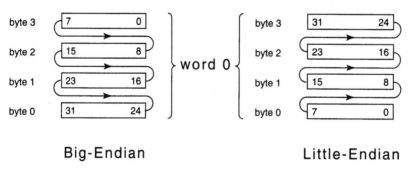

Big-Endian   **Little-Endian**

FIGURE 1.4

Big-endian and little-endian byte orderings. For clarity, only one word of memory is shown. The address of the word is the address of the byte with the lowest address within the word. The arrows point in the direction of increasing significant digits.

uses the opposite byte ordering such that the least significant byte has the lowest address and the most significant byte the highest address. This byte ordering is called **little-endian**. For example, in a 16-bit representation ($n = 16$), the integer $m = (32\,769)_{10} = (80\,01)_{16}$ would be represented in RAM by the consecutive bytes 80 and 01 in the big-endian byte ordering shown in Fig. 1.2, but it would be represented by the consecutive bytes 01 and 80 in a little-endian byte ordering. If, for example, one computes the number $m = (32\,769)_{10}$ on a little-endian computer, which stores the result as 01 80, and then interprets this bit string on a big-endian computer (in order to make a graph, for example), then one obtains the incorrect value $m' = (384)_{10} \neq m$.

It is possible to choose byte ordering when a program is started rather than at the time when the processor is designed. The MIPS® and PowerPC® architectures permit the operating system to set a flag that determines whether big-endian or little-endian byte ordering is to be used.

Fundamental Arithmetic Algorithms

The algorithms for addition, subtraction, multiplication, and division that all of us learned in grade school are valid in any base. For example, in base $\beta = 16$ the subtraction $524610_{10} - 44541_{10} = 7919_{10}$ becomes

$$
\begin{array}{r}
\text{CCEC} \\
- \text{ADFD} \\
\hline
\text{1EEF}
\end{array}
$$

because $1C - D = F$, and so on.

There is no guarantee, of course, that the sum or product of two integers, each of which is smaller than 2^n in magnitude, will have a magnitude smaller than 2^n. Attempting to represent an integer greater in magnitude than 2^n as an unsigned n-bit integer results in **integer overflow**. Overflow is equivalent to going completely around the circumference of the circle in Fig. 1.3 at least once. The interpretation of overflowed integers is system-dependent.

Fast implementations of the fundamental arithmetic algorithms are a subject of current research and, occasionally, public discussion. The notes at the end of this chapter include references to sources of additional information.

Binary Sign-Magnitude Integer Representation

A signed integer $k = (-1)^s m$ that is less than 2^{n-1} in absolute value can be represented in hardware in a **binary sign-magnitude integer format** as a sign bit followed by a string of $n - 1$ bits representing the positive integer $m = (b_{n-2} \cdots b_0)_2$ in Eq. (1.5), as Fig. 1.5 illustrates. The integer 0 is represented twice, as $+0$ and as -0.

FIGURE 1.5

The bit sequence of an n-bit, binary, sign-magnitude integer representation.

In the sign-magnitude representation, dictionary order puts negative numbers after positive numbers if (as usual) one takes $s = 1$ to represent a minus sign. For example, if $n = 2$, then the base-10 integers $-2, -1$, and 1 have the binary sign-magnitude representations 110, 101, and 001, respectively. The dictionary order in this example is 001, 101, 110, which is the reverse of the natural order $-2, -1, 1$ (see Exercise 1.2.4 for the general case).

Quite apart from their uses in numerical computation, signed integers and their sign-magnitude representations find important applications wherever continuously varying physical quantities are digitized and truncated to a finite range of magnitudes.

Biased Integer Representations

In the n-bit biased-B representation, the numerical value assigned to a string of bits is

$$
\begin{aligned}
k &= -B + \sum_{j=0}^{n-1} b_j 2^j \\
&= -B + u,
\end{aligned}
\tag{1.14}
$$

where u is the value that would be assigned to the bit string $b_{n-1} \cdots b_1 b_0$ in an unsigned-integer representation. A biased representation can be used to represent digitized electrical signals and the exponents of floating-point numbers (which we define below). The **bias** B usually is chosen as the integer part of the maximum positive integer that can be represented with n bits,

$$
B = \left\lfloor \frac{2^n - 1}{2} \right\rfloor,
\tag{1.15}
$$

because this choice ensures that $k + B$ is always positive and less than 2^n for integers that lie within the allowed range $-B \le k \le B + 1$.

For example, to represent integers in the range $-7 \le k \le 8$ one can use $B = 7$ and a nybble (a string of four bits), as shown in Fig. 1.6. Requirement (1.15) implies that in a biased representation numerical order is the same as dictionary order, as Fig. 1.6 illustrates.

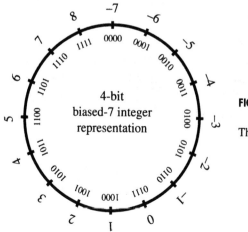

FIGURE 1.6

The 4-bit, binary, biased-7 integer representation.

Probably the best way to understand arithmetic in a biased representation is to regard Eq. (1.15) as a change of the origin of coordinates from 0 to B. Using this picture, one sees at once that subtraction of two biased integers

$$u_1 = k_1 + B$$

and

$$u_2 = k_2 + B \tag{1.16}$$

gives a correct result for the difference of the numerical values $k_1 - k_2$:

$$u_1 - u_2 = k_1 - k_2. \tag{1.17}$$

However, the sum of two unsigned integers u_1, u_2 must be corrected for the fact that the shift of the origin of coordinates has been included twice in $u_1 + u_2$:

$$u_1 + u_2 = k_1 + k_2 + 2B \;\Rightarrow\; k_1 + k_2 = u_1 + u_2 - B. \tag{1.18}$$

Two's-Complement Representations

A **signed integer** is an integer that may be positive, zero, or negative. In nearly all computers, signed integers are represented in an **n-bit two's-complement integer format**:

$$k = -b_n 2^n + \sum_{j=0}^{n-1} b_j 2^j, \tag{1.19}$$

in which the **sign bit** b_n is

$$b_n = \begin{cases} 0, & \text{if } k \geq 0; \\ 1, & \text{if } k < 0. \end{cases} \tag{1.20}$$

For example, the integer $k = -1$ has the 2-bit two's-complement representation 11_2 because $-1 = (-4 + 3)_{10} = (-100 + 11)_2$. The 32-bit two's-complement representation of the integer

$$-2^{20} = -(1048576)_{10} = -(1\,0000\,0000\,0000\,0000\,0000)_2 \tag{1.21}$$

is

$$\begin{aligned}
&1\,0000\,0000\,0000\;0000\,0000\,0000\,0000\,0000 \\
&\qquad\qquad -1\,0000\,0000\,0000\,0000\,0000 \\
&= 1111\,1111\,1111\,0000\,0000\,0000\,0000\,0000.
\end{aligned} \tag{1.22}$$

Readers who are familiar with the concept of congruence defined in Section 4.2.4 should recognize that Eq. (1.19) defines the two's-complement representation of a negative integer k as the smallest positive integer that is congruent to k modulo 2^n. Thus the n-bit

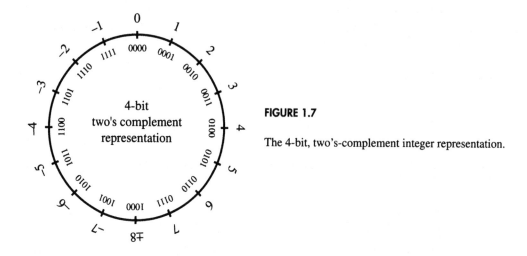

FIGURE 1.7

The 4-bit, two's-complement integer representation.

two's-complement integers are simply a realization of the ring of integers modulo 2^n; see Sections 4.2.4 and 4.5.

Figure 1.7 illustrates the 4-bit two's-complement representation of integers k in the range $-7 \le k \le 7$. Note that in a two's-complement representation, the string such that $b_n = 1$ and $b_{n-1} = \cdots = b_0 = 0$ does not represent a unique integer because its value is both itself and minus itself, yet it is nonzero. Therefore the string $10 \cdots 0$ cannot be considered as a valid two's-complement integer. (This situation arises because the set of integers modulo 2^n contains so-called "proper divisors of zero"; see Section 4.5.1.) Also, for any value of $n \ge 2$, the n-bit string $11 \cdots 1$ is the n-bit two's-complement representative of -1.

1.2.2 RATIONAL NUMBERS AND REAL NUMBERS

Fixed-Point Representations

Generalizing the base-β representation of integers given in Eq. (1.2), one obtains the **fixed-point representation** of a real number r:

$$r = \pm \left[\sum_{j=0}^{n} b_j \beta^j + \sum_{k=1}^{p} \frac{f_k}{\beta^k} \right]. \tag{1.23}$$

(If $p = \infty$, then this infinite series converges by the comparison test using the geometric series; see Section 10.3.2.) The first sum is equal to the integer part of r,

$$\lfloor r \rfloor = \pm (b_n b_{n-1} \cdots b_0)_\beta. \tag{1.24}$$

The second sum gives the fractional part of r. An equivalent form is

$$r = \pm (b_n b_{n-1} \cdots b_0 . f_1 f_2 \cdots f_p)_\beta \tag{1.25}$$

The point is not a *decimal* point unless $\beta = 10$. For a general base (or radix), one speaks instead of the **radix point**.

The decimal ($\beta = 10$) fixed-point representation is familiar because it is widely used for currency. Another fixed-point representation uses the base $\beta = 60$ (thanks to the Babylonians!) for the specialized purpose of representing angles that are less than one degree of arc. For example, an angle of $0.295780°$ can be expressed approximately as 17 minutes, 45 seconds, which means $\frac{17}{60} + \frac{45}{3600}$ degrees.

If p is finite, then r is a **rational number**, that is, r is equal to a quotient of two integers. If $p = \infty$, r may or may not be rational; see the next subsection. If $p = \infty$ the fixed-point representation in Eq. (1.23) is not unique, for the fractional parts of some real numbers can be written in more than one way. In base 10, for example, $1.000 \cdots \times 10^0 = 9.999 \cdots \times 10^{-1}$. If one uses a finite string of digits, this lack of uniqueness is unimportant as long as one rounds an infinite string to the nearest finite string. More information about rounding can be found in Section 1.4.2.

Although fixed-point representations are useful for simple financial computations, they have several disadvantages in simulations and visualizations of fields, waves, and continuous systems in science and engineering:

- In many computations, cancellation of digits as the result of subtraction (see Section 1.4.3) forces one to keep many more digits during a computation than are useful or meaningful in the final result. In a fixed-point representation this is possible only for numbers that are much larger than the minimum positive number that can be represented by Eq. (1.23), which is

$$r_{min} = \beta^{-p}. \tag{1.26}$$

- A fixed-point representation uses digits extravagantly. For example, the fixed-point, decimal representation of Avogadro's number is $N_0 = 602\,000\,000\,000\,000\,000\,000\,000$. In scientific notation, $N_0 = 6.02 \times 10^{23}$, five digits (two for the exponent and three for the significand) take the place of twenty-four digits in the fixed-point representation. Because each digit that is represented in a computer's hardware requires dedicated circuitry in the arithmetic unit and occupies a small but nonzero area on storage devices, there is a strong economic incentive to minimize the number of digits that must be used for scientific and engineering computations.

Nevertheless, the hardware that implements arithmetic operations on numbers in a fixed-point representation is simpler, and significantly faster, than a floating-point arithmetic unit built using the same hardware technology. For this reason, digital signal processors make heavy use of fixed-point representations.

Floating-Point Representations

Several representations are available that require fewer digits than a fixed-point representation. For several centuries, positive real numbers have been represented by their approximate logarithms for the purposes of hand and slide-rule computation. Another, currently more widely used representation of real numbers is the **base-β floating-point represen-tation**

$$r = \pm \beta^n \sum_{j=0}^{\infty} \frac{d_j}{\beta^j} \qquad (1.27)$$

in which the base β is a fixed integer greater than one and each d_j is an integer such that

$$0 \le d_j < \beta. \qquad (1.28)$$

One can see from the more compact notation

$$r = \pm d_0 . d_1 d_2 \cdots \times \beta^n \qquad (1.29)$$

that if the string of nonzero digits is of finite length, as it must be in a practical computation, then Eq. (1.29) is a straightforward generalization of ordinary scientific notation. Like fixed-point representations, floating-point representations are not necessarily unique.

To compute the base-β floating-point expansion of a given real number r, one must first find the exponent $n = \lfloor \log_\beta r \rfloor$ such that

$$\beta^n \le |r| < \beta^{n+1}. \qquad (1.30)$$

Then the equations

$$d_0 = \lfloor \beta^{-n} |r| \rfloor$$

$$d_j = \left\lfloor \beta^{j-n} |r| - \sum_{k=0}^{j-1} d_k \beta^{j-k} \right\rfloor \quad (j > 0) \qquad (1.31)$$

give the digits recursively. (Recall that $\lfloor x \rfloor$ means the integer part of x.)

In general, a string of the form shown in Eq. (1.29) represents a rational number if and only if the fractional part of the string either terminates or repeats indefinitely after some digit (see Exercises 1.2.5 and 1.2.6). Rational numbers that have a terminating representation in one base β, that is, such that $d_j = 0$ for every j greater than some positive integer M, may require an infinite repeating sequence of digits in other bases.

For example, the rational number $\left(\frac{1}{3}\right)_{10} = (3.333 \cdots)_{10} \times 10^{-1}$ has the base-2 (binary) representation

$$\left(\frac{1}{3}\right)_{10} = \frac{1}{4} + \frac{1}{16} + \frac{1}{64} + \frac{1}{256} + \frac{1}{1024} + \frac{1}{4096} \cdots$$

$$= \sum_{k=1}^{\infty} \frac{1}{2^{2k}} \qquad (1.32)$$

$$= (1.0101 \cdots)_2 \times 2^{-2},$$

the base-3 (ternary) representation

$$\left(\frac{1}{3}\right)_{10} = (1.0000 \cdots)_3 \times 3^{-1}, \qquad (1.33)$$

the base-8 (octal) representation

$$\left(\tfrac{1}{3}\right)_{10} = 2 \cdot \tfrac{1}{8} + (4+1)\left(\tfrac{1}{8}\right)^2 + 2\left(\tfrac{1}{8}\right)^3 + (4+1)\left(\tfrac{1}{8}\right)^4 + \cdots$$
$$= (2.5252 \cdots)_8 \times 8^{-1},$$

(1.34)

and the base-16 (hexadecimal) representation

$$\left(\tfrac{1}{3}\right)_{10} = (4+1) \cdot \tfrac{1}{16} + (4+1)\left(\tfrac{1}{16}\right)^2 + (4+1)\left(\tfrac{1}{16}\right)^3 + \cdots$$
$$= (5.5555 \cdots)_{16} \times 16^{-1}.$$

(1.35)

1.2.3 REPRESENTATIONS OF NUMBERS AS TEXT

ASCII Representation of Alphanumeric and Control Characters

It is necessary to define a representation of one's alphabet, punctuation marks, common symbols, and the digits 0 through 9 in order to make it possible for computers to manipulate alphanumeric labels and symbolic names. The most widely used such representation for the Latin alphabet and the symbol set used in North America is the American Standard Code for Information Interchange (ASCII) mapping of characters to 7-bit unsigned integers less than $(128)_{10}$, which is summarized in Table 1.2. Characters that can be printed are called **alphanumeric characters**. Nonprinting characters such as ACK (acknowledge) are called **control characters** because the designers of the ASCII mapping provided them in order to control teletypewriters, obsolete devices that produced output that humans could read. A mapping of letters and other characters to bit strings is a kind of code, although its purpose is not secrecy.

The list given in Table 1.2 includes no unsigned integers greater than $(128)_{10}$. In other words, the most significant bit is always 0. In effect, the standard ASCII mapping is a 7-bit code, not a full 8-bit code. However, one ASCII character is normally represented in hardware by one byte.

A major disadvantage of the ASCII code is that there are not enough 7-bit or 8-bit integers to represent even a small fraction of the set of characters used to write Chinese or Japanese, for example. The usual solution, to use 2-byte codes to represent characters, gives 65,536 possible representations.

The ASCII representatives of characters can be (and are) manipulated within a computer as unsigned 1-byte integers. If the integers that represent letters and digits are arranged in numerical order, the decimal digits come first, followed by the uppercase letters and finally by the lowercase letters. Note that the bit string that represents an uppercase letter (for example, 0100 0001 for A) differs only in a single bit from the string that represents the corresponding lowercase letter (for example, 0110 0001 for a).

The unsigned integers k such that $(128)_{10} \leq k \leq (255)_{10}$ can be mapped to other characters and symbols besides the rather restricted set shown in Table 1.2. Unfortunately no standard character mapping exists for integers greater than $(127)_{10}$.

TABLE 1.2 ASCII Mapping of Characters to the Set of
One-Byte Unsigned Integers Smaller Than $(128)_{10}$

$\beta = 10$	$\beta = 16$	Character	Description
0	00	NUL	null character
1	01	SOH	start of heading
2	02	STX	start of text
3	03	ETX	end of text
4	04	EOT	end of transmit
5	05	ENQ	enquiry
6	06	ACK	acknowledge
7	07	BEL	bell (alert)
8	08	BS	backspace
9	09	HT	horizontal tab
10	0A	LF	line feed
11	0B	VT	vertical tab
12	0C	FF	form feed
13	0D	CR	carriage return
14	0E	SO	shift out
15	0F	SI	shift in
16	10	DLE	data line escape
17	11	DC1	device control 1
18	12	DC2	device control 2
19	13	DC3	device control 3
20	14	DC4	device control 4
21	15	NAK	negative acknowledge
22	16	SYN	synchronous idle
23	17	ETB	end of transmit block
24	18	CAN	cancel
25	19	EM	end of medium
26	1A	SUB	substitute
27	1B	ESC	escape
28	1C	FS	file separator
29	1D	GS	group separator
30	1E	RS	record separator
31	1F	US	unit separator
32	20	SPACE	space
33–47	21–2F	!"#$%&'()*+,-./	punctuation
48–57	30–39	0–9	decimal digits
58–64	3A–3F	:;<=>?	punctuation
65–90	41–5A	A–Z	uppercase letters
91–96	5B–60	[\]^_`	punctuation
97–122	61–7A	a–z	lowercase letters
123–126	7B–7E	{|}~	punctuation
127	7F	DEL	delete

ASCII Representation of Numbers

The ASCII characters (integers) with hexadecimal values 30 through 39 represent the decimal digits. Because there is no ASCII notation for superscripts or subscripts, the exponent in a floating-point representation such as Eq. (1.29) must be represented by some standard symbol. The usual convention for base $\beta = 10$ is to use E (for exponent) or D (for double-precision exponent). For example, one writes the base-10 number 31 415.9 as 3.14159E4 in single precision. (See the next section for a discussion of single-precision and double-precision floating-point representations.)

The ASCII representation has the disadvantage that it wastes computer storage. Every decimal digit is represented by 1 byte. There are only ten decimal digits, but there are 256 different 1-byte integers. A further disadvantage of the ASCII representation is that in order to make use of a number that is stored in ASCII format, a computer must translate from the ASCII representation to a native hardware representation. On a single computer system, storing numeric data in ASCII form and then reading it back imposes a substantial speed penalty in comparison with writing and reading data in a hardware format.

The advantages of the ASCII representation of numbers include the following:

- An ASCII representation can be printed or displayed in a form that humans can read.
- Listings printed on paper, although outmoded from a technological point of view, have a useful life that is much longer than those of the media that are typically used for machine-readable archival storage.
- An unlimited precision and an unlimited range of exponents are available because one can use as many digits as are necessary.
- The ASCII mapping is completely standard except for the representation of the end of a line of text (Table 1.3). There is usually little difficulty in moving alphanumeric data in ASCII form from one kind of computer to another, but there are many ways in which an attempted transfer of data in a hardware format can fail. For example, different computers may use different byte orderings, as was explained in a previous section.

Because of these advantages, the ASCII representation is in wide use for transporting and storing numeric data.

TABLE 1.3 Representations of the End of a Line of Text in the UNIX®, MacOS®, VMS®, and MS-DOS® Operating Systems

Operating System	End of Line (ASCII)	Hexadecimal	Byte Ordering
UNIX	LF	0A	Either
MacOS	CR	0D	Big-Endian
VMS	CR LF	0D0A	Little-Endian
MS-DOS	CR LF	0D0A	Little-Endian

1.2.4 EXERCISES FOR SECTION 1.2

1.2.1 Using the algorithm given in Eqs. (1.8) and (1.9), calculate the representations of the integers 25_{10}, 50_{10}, 250_{10}, 500_{10}, 1025_{10}, 16385_{10}, and 65535_{10} in the bases $\beta = 2$, $\beta = 3$, $\beta = 8$, and $\beta = 16$.

1.2.2 Calculate the following sums without converting to base 10: $(1010101001101011 + 101011100000111)_2$, $(42511 + 47147)_8$, and $(\text{EA4D} + \text{FFF1})_{16}$.

1.2.3 Calculate the following differences without converting to base 10: $(10010001 - 01010000)_2$, $(47747 - 42571)_8$, and $(\text{FAB1} - \text{EC4D})_{16}$.

1.2.4 Calculate the following products without converting to base 10: $(11100100111101 \times 1101)_2$, $(73213 \times 35727)_8$ and $(\text{EA4D} \times \text{FFF1})_{16}$.

1.2.5 Show that every nonterminating, repeating base-β floating-point expansion of a real number, such as

$$r = a_0 . a_1 a_2 \cdots a_m \underbrace{b_1 b_2 \cdots b_n}_{\text{repeats}} \cdots \times \beta^s,$$

in which the string $b_1 b_2 \cdots b_n$ repeats indefinitely, is equal to the quotient of two integers.

Hint: Start with the simplest case in which the preperiodic digits a_j all vanish. Then use the formula for the sum of an infinite geometric series.

*1.2.6 Show that the base-β floating-point representation of the quotient of two integers, m/n, either terminates or repeats indefinitely after some initial string of nonrepeating digits.

1.2.7 Using the algorithm given in Eq. (1.31), find the floating-point representations of $\frac{1}{16}$, $\frac{1}{10}$, $\frac{1}{8}$, $\frac{1}{3}$, $\frac{1}{2}$, $\frac{3}{5}$, and $\frac{2}{3}$ in the bases $\beta = 2$, $\beta = 3$, $\beta = 8$, $\beta = 10$, and $\beta = 16$.

1.2.8 Show that the dictionary order of the integers in a sign-magnitude representation with N binary digits and one sign bit is

$$+0, 1, \cdots, 2^N - 1, -0, -1, \cdots, -(2^N - 1).$$

1.2.9 This problem concerns specific properties of the base-10 representation of integers.

(a) Show that in base $\beta = 10$, an integer $n = d_m d_{m-1} \cdots d_0$ is evenly divisible by 9 if the sum of the digits $\sum_{k=1}^{m} d_k$ is evenly divisible by 9. Also show that if n is evenly divisible by 9, then $\sum_{k=1}^{m} d_k$ is evenly divisible by 9.

(b) Show that in base $\beta = 10$, an integer $n = d_m d_{m-1} \cdots d_0$ is evenly divisible by 3 if the sum of the digits $\sum_{k=1}^{m} d_k$ is evenly divisible by 3. Also show that if n is evenly divisible by 3, then $\sum_{k=1}^{m} d_k$ is evenly divisible by 3.

1.2.10 Prove that if the base-β representation of an integer n is

$$n = \sum_{k=1}^{m} d_k \beta^k, \tag{1.36}$$

then n is a multiple of $\beta - 1$ if and only if $\sum_{k=1}^{m} d_k$ is evenly divisible by $\beta - 1$.

1.2.11 Calculating by hand, find the 32-bit two's-complement representations of the following integers: -1, -4194304_{10}, -8388608_{10}, -16777216_{10}.

1.2.12 Find the base-10 equivalents of all of the possible strings of 4 bits if these strings are interpreted as

(a) 4-bit sign-magnitude representations of integers;

(b) 4-bit two's-complement representations of integers.

1.3 FINITE FLOATING-POINT REPRESENTATIONS

1.3.1 SIMPLE CASES

Normalized Floating-Point Numbers

For computations with real numbers one must use approximate representations such as a **finite base-β floating-point representation**

$$r = (-1)^s \beta^e \sum_{j=0}^{p-1} \frac{d_j}{\beta^j} \tag{1.37}$$

$$= (-1)^s d_0 . d_1 d_2 \cdots d_{p-1} \times \beta^e.$$

The integer e is called the **exponent**. (Please do not confuse e with the base of natural logarithms!). A real number that can be represented exactly by Eq. (1.37) is called a **floating-point number**. As for integers, the most significant digit is d_0, the coefficient of the highest power of the base β. The least significant digit is d_{p-1}. The number of significant digits, p, is called the **precision**. In the hardware floating-point units that are now included with most microprocessors, p is a fixed number.

By convention the digits of the **significand** (or **mantissa**) $d_0 . d_1 d_2 \cdots d_{p-1}$ must be **normalized** to make the first digit d_0 nonzero:

$$d_i = \begin{cases} 0, \ldots, \beta - 1, & \text{if } i = 1, \ldots, p-1; \\ 1, \ldots, \beta - 1, & \text{if } i = 0. \end{cases} \tag{1.38}$$

Some real numbers do not correspond to any normalizable floating-point number; $r = 0$ is the most obvious example. We discuss unnormalizable floating-point numbers in a later section.

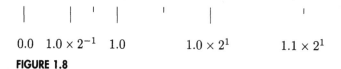

$$0.0 \quad 1.0 \times 2^{-1} \quad 1.0 \qquad 1.0 \times 2^1 \qquad 1.1 \times 2^1$$

FIGURE 1.8

This figure shows all of the nonnegative base-2 floating-point numbers in a toy floating-point representation with two significant digits, a maximum exponent of 1, and a minimum exponent of -1. The negative, nonzero floating-point numbers that exist in this representation are not shown.

The convention that valid floating-point numbers are normalized ensures that there is only one base-β floating-point representative of each floating-point number. To facilitate rounding, which we discuss subsequently, β is always even in practical computations, although odd bases are useful for some theoretical purposes.

In principle the precision p is subject to software control. However, at present all but the simplest numerical computations in physics and engineering require more speed than one can obtain by manipulating strings of digits in software. For this reason only the few values of p that are supported in a computer's hardware are ordinarily available to the scientist-programmer.

Because only a finite number of digits can be used to represent the exponent e in hardware, e is subject to the inequalities

$$e_{\min} \leq e \leq e_{\max}, \tag{1.39}$$

where $e_{\min}$ and $e_{\max}$ are fixed by the designers of the computer's floating-point arithmetic unit. For example, the toy floating-point representation illustrated in Fig. 1.8 can be implemented with $n = 2$ digits and a bias $B = 2$. The only exponents allowed in this representation are $-1, 0$, and 1. Choosing $B > 1$ makes it possible to represent both 0.0 and 1.0×2^{-1}.

In a floating-point representation defined by β, p, $e_{\max}$, and $e_{\min}$, there exist a maximum positive normalized floating-point number

$$r_{\max} = (\beta - 1).(\beta - 1) \cdots (\beta - 1) \times \beta^{e_{\max}} \tag{1.40}$$

and a minimum positive normalized floating-point number

$$r_{\min} = 1.0 \cdots 0 \times \beta^{e_{\min}} = \beta^{e_{\min}}. \tag{1.41}$$

In Fig. 1.8, for example, $r_{\max} = (1 + \frac{1}{2}) \cdot 2 = 3$, and $r_{\min} = 1 \cdot 2^{-1} = \frac{1}{2}$. Attempting to represent a real number that is larger than $r_{\max}$ or smaller than $r_{\min}$ results in **overflow** or **underflow**, respectively. For example, the product of two positive, normalized floating-point numbers that are both greater than $\sqrt{r_{\max}}$ will overflow.

Unnormalizable Floating-Point Numbers

Three important kinds of numbers that result from some floating-point computations are not normalizable and must therefore be represented by special floating-point numbers (if they are represented at all):

- Numbers with exponent $e > e_{max}$. Some floating-point representations use a specific bit string, which is called Inf, to represent any number with $e > e_{max}$ and $d_1 = \cdots = d_{p-1} = 0$. The widely used IEEE-754 floating-point representations use **signed infinity**, that is, $+$Inf and $-$Inf are represented by different bit strings in order to preserve the sign of an overflowed quantity. We have more to say in subsequent sections about the practical uses of floating-point infinity.
- Numbers with exponent $e < e_{min}$. Because underflow is common in scientific computations, sophisticated strategies have been developed for handling it. We discuss the strategy used in the IEEE-754 floating-point representations in a subsequent section.
- Floating-point zero, $0.0\cdots$, which must be represented by at least one special floating-point number. In some floating-point representations zero is **signed**, that is, different floating-point numbers represent $+0.0\cdots$ and $-0.0\cdots$. In other floating-point representations zero is **unsigned**, that is, the real number zero is represented by a unique floating-point number. It makes sense to have signed zero in a floating-point representation that has signed infinity, for then the identity $1/(1/x) = x$ holds if $x = \pm$Inf.

How Many Floating-Point Numbers are There in a Given Representation?

Although the toy floating-point representation illustrated in Fig. 1.8 has too small an exponent range and too few significant digits to be of any practical use, it is helpful in explaining concepts that apply to usable floating-point representations. For example, the toy representation illustrates the fact that the collection of floating-point numbers that are available with given values of β, p, and e is finite (see Fig. 1.8). The number of values of the exponent e is $e_{max} - e_{min} + 1$. The most significant digit of the mantissa must be nonzero; each of the remaining digits can be any one of the base-β digits $0, \ldots, \beta - 1$. Hence, for each value of e, there are $\beta - 1$ values of d_0 and β values of $d_1, \ldots, d_{p-1}$, making $2(\beta - 1)\beta^{p-1}$ normalized floating-point numbers for each value of e, taking the two possible signs into account. Multiplying by the number of values of the exponent, one obtains $2(\beta - 1)\beta^{p-1}(e_{max} - e_{min} + 1)$ valid nonzero floating-point numbers for given values of β, p, e_{max}, and e_{min}. Then there are

$$N(\beta, p, e_{max}, e_{min}) = 2(\beta - 1)\beta^{p-1}(e_{max} - e_{min} + 1)$$

$$+ \begin{cases} 1, & \text{if zero is unsigned, or} \\ 2, & \text{if zero is signed} \end{cases} \tag{1.42}$$

valid floating-point numbers in all. For example, if $\beta = 2$, $p = 2$, $e_{max} = 2$, and $e_{min} = 0$ as in the toy representation of Fig. 1.8, then $N(2, 2, 3, 1) = 2(2 - 1)2^{2-1}(3 - 1 + 1) + 2 = 14$ if zero is signed. (The seven negative floating-point numbers, including -0.0, that exist in this representation are not shown in the figure.)

Figure 1.9 shows the bit sequence of a possible hardware implementation of the toy floating-point representation shown in Fig. 1.8. The highest-order bit is a sign bit s that takes the values 0 or 1 for positive and negative floating-point numbers, respectively. The

FIGURE 1.9

A bit ordering that implements the toy floating-point representation. The significand of a floating-point number represented in this format is equal to $1.f$. The initial 1, which is not stored in order to save space, is called an **implicit 1 bit**. Note that the exponent is biased.

bit string that represents the biased exponent e comes next. The exponent is to be obtained by subtracting the bias $B = 2$ from the unsigned two-digit binary integer e represented by the exponent bit string. For example, the bit string 11 represents the integer $e = 3_{10}$ and corresponds to the exponent $e - B = 3 - 2 = 1$.

Because we have not defined the biased exponent 00 to represent the exponent of any valid floating-point number in the toy format, it is possible to provide for signed zero. We define the bit string 0000 to represent $+0.0\cdots$ and the bit string 1000 to represent $-0.0\cdots$.

Let us consider how to represent the fraction f. In any normalized binary floating-point number, the most significant bit d_0 is always equal to 1. Because it would be a waste of hardware to store something whose value is always the same, it makes sense to use an implicit 1 bit instead of storing d_0. (In the next section we describe the IEEE floating-point representations, in which an implicit 1 bit is used.) A one-bit f represents the remaining digit d_1. The formula for the floating-point number to be represented in hardware by s, e, and f is

$$r = (-1)^s 2^{e-2}(1.f) \tag{1.43}$$

For example, the floating-point number $-1.1 \times 2^{-1} = -(\frac{3}{4})_{10}$ is represented by the bit string 1011. The first digit, 1, is the sign bit s, which is set "on" (i.e., equal to 1) for a negative number. The next two bits, 01, represent the biased exponent $e = (-1) + 2 = 1$. Finally, in this example the fraction bit f is set equal to 1 in order to represent the 1 after the binary point in -1.1×2^{-1}.

Putting the exponent bits before the fraction bit ensures that for positive floating-point numbers, the dictionary order of the bit strings that represent the numbers is the same as the natural order of the numbers themselves. For example, the floating-point number $1.1 \times 2^{-1} = (\frac{3}{4})_{10}$ is represented by the bit string 0011, whereas $1.1 \times 2^0 = (1\frac{1}{2})_{10}$ is represented by the bit string 0101, which follows 0011 in dictionary order.

From now on we shall write as 1. the floating-point number that represents unity and as 0. the floating-point number that represents zero.

Some Consequences of Finite-Precision Arithmetic

Many important physical phenomena are *chaotic*, that is, outputs that correspond to nearby inputs tend to diverge exponentially in time. For a chaotic system, if $d(t)$ is a measure of the distance between two trajectories at a time t, then on average

$$d(t) = 2^{\lambda t/h} d(0), \tag{1.44}$$

where h is the time step size and where λ is called a *Lyapunov exponent*. The formula

$$\lambda = \lim_{N \to \infty} \left\langle \sum_{i=1}^{N} \log_2 \left(\frac{d_{i+1}}{d_i} \right) \right\rangle, \tag{1.45}$$

in which d_i is the distance between two trajectories at the time t_i, d_{i+1} is the distance between the same two trajectories at the time $t_{i+1} = t_i + h$, and angle brackets denote an average over pairs of trajectories, is useful for practical computations of λ.

A major consequence of the fact that hardware-representable floating-point numbers have a fixed number of significant digits is that the results of a simulation of a physical phenomenon with one or more positive Lyapunov exponents become increasingly less reliable as the number of time steps taken by the program increases. For example, suppose that initially the difference between two trajectories is equal to the rounding error, $d(0) = \frac{1}{2}\epsilon_{mach}$ (see Section 1.4.1), and that the trajectories are one-dimensional and are bounded between 0. and 1. After a time t, $d(t) = \frac{1}{2}\epsilon_{mach} 2^{\lambda t/h}$ on average. (This equation cannot hold for all times, because the trajectories are bounded.) If $\lambda t = nh$, and n is a positive integer, the trajectories differ in the n least significant digits. If $n > p$, then on average two trajectories that are initially indistinguishable after rounding have diverged so much that they may no longer have the same most significant digit. Simulations of such a natural phenomenon in a floating-point representation with precision p cannot accurately predict the outcome of a particular initial condition for times longer than ph/λ. This is why weather, for example, cannot accurately be predicted more than a few days in advance.

1.3.2 PRACTICAL FLOATING-POINT REPRESENTATIONS

Two floating-point representations called *single precision* and *double precision* are in widespread use. A 32-bit (4-byte) floating-point representation is called **single precision** by most computer manufacturers, and a 64-bit (8-byte) representation is usually called **double precision**. The single-precision data type is REAL*4 in FORTRAN or float in C; the double-precision data types are REAL*8 and double, respectively. The single-precision and double-precision representations prescribed in IEEE Standard No. 754 (commonly known as "IEEE-754") are worth learning about because every modern microprocessor that is commonly used for general-purpose floating-point calculations conforms (more or less) to the IEEE standards.

The floating-point representations used in some computers do not conform to IEEE-754. The endnotes for this chapter briefly summarize the main features of the CRAY floating-point representation.

Single Precision

IEEE-754 prescribes a base-2 single-precision floating-point representation with $p = 24$ significant digits, a maximum exponent $e_{max} = 127$, and a minimum exponent $e_{min} = -126$. The exponent is stored in biased format with a length of $N = 8$ bits and a bias of

$$B = \left\lfloor \frac{2^8 - 1}{2} \right\rfloor = 127 = 2^7 - 1 \tag{1.46}$$

FIGURE 1.10

The bit sequence of the IEEE-754 single-precision representation.

(see Eq. [1.15]). Figure 1.10 shows the bit ordering in the IEEE-754 single-precision floating-point representation. The use of a common floating-point binary representation facilitates the exchange of data among computers manufactured by different companies.

The formula for the floating-point number with sign bit s, exponent e, and fraction f as illustrated in Fig. 1.10 is

$$r = (-1)^s \, 2^{e-127} \, (1 \, . \, f) \tag{1.47}$$

provided that

$$0 < e < 255 = 2^8 - 1. \tag{1.48}$$

The notation $1 . f$ indicates the use of an implicit 1 bit, as discussed above. The exponent values 0 and 255 designate special reserved operands, which are discussed in a subsequent section. A floating-point number in which both the exponent and the fraction are zero is interpreted as signed zero, $(-1)^s 0$.

For example, the 4-byte hexadecimal word 3F 80 00 00, which is 0011 1111 1000 0000 0000 0000 0000 0000 in binary notation, is the base-2 single-precision representation of the floating-point number 1. Here is the calculation: The sign bit (bit 31) is zero because the binary notation for 3 is 0011. Bits 30 through 23, which are 011 1111 1 in this example, give the biased exponent e, which is equal to 127_{10}. The remaining bits, which are all zero in this example, give the fraction. Then $r = (-1)^0 \, 2^{127-127} \, (1.0) = 1$ in single precision.

To understand in familiar decimal terms the accuracy that can be obtained in the IEEE-754 single-precision floating-point format, let us recall that

$$\beta = 10^{\log_{10} \beta} \tag{1.49}$$

and therefore

$$\beta^e = 10^{e \log_{10} \beta}. \tag{1.50}$$

Thus an exponent e in base β is equivalent to an exponent $e \log_{10} \beta$ in base 10. Likewise, p significant digits in base β are approximately equivalent to $p \log_{10} \beta$ significant digits in base 10. The maximum base-10 exponent in single precision is approximately 38, and the number of significant digits is between 7 and 8. This accuracy is acceptable in many applications, including applications in digital signal processing for which a floating-point representation is desirable.

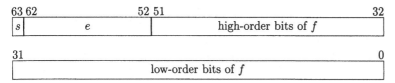

FIGURE 1.11

The bit sequence of the IEEE-754 double-precision representation.

Double Precision

Many scientific users find that the single-precision floating-point representation has too small an exponent range. For example, attempting to evaluate $\hbar^2 = (1.054)^2 \times 10^{-54}$ causes floating-point underflow in single precision. It is less obvious, but nevertheless true, that the single-precision representation also has too few significant digits for many applications. Seven-decimal-digit accuracy would be quite adequate for expressing the results of most physical measurements. However, as we have already pointed out, there are applications in which the output depends sensitively on the input. Computation with a fixed number of significant digits inevitably adds its own input sensitivities to the sensitivities that are characteristic of a particular application. We show in Sections 1.4.3 and 3.2 that subtraction, for example, can reduce the accuracy of a computed result to an effective number of digits far smaller than the precision of the floating-point representation. For these reasons many computer manufacturers make a double-precision representation available in hardware.

Figure 1.11 shows the bit sequence of the IEEE-754 double-precision floating-point representation. The exponent is in biased-integer format with $N = 11$ bits and a bias

$$B = \left\lfloor \frac{2^{11} - 1}{2} \right\rfloor = 1023 = 2^{10} - 1, \tag{1.51}$$

implying a minimum exponent $e_{\min} = -1022$ and a maximum exponent $e_{\max} = 1023$. If

$$0 < e < 2047 = 2^{11} - 1, \tag{1.52}$$

then

$$r = (-1)^s \, 2^{e-1023} \, (1 \cdot f). \tag{1.53}$$

Note, once again, the use of an implicit 1 bit between the exponent and fraction fields to increase the precision from 52 to 53 bits. An exponent value of 0 defines a denormalized number, and an exponent value of 2047 defines an Inf or NaN, as discussed in a subsequent section.

The maximum and minimum decimal exponents obtainable in double precision, approximately -307 and 308, respectively, are adequate for almost all engineering and scientific calculations. The decimal precision is between fifteen and sixteen digits, which seems extravagant until one tries some real computations. (See, for example, the problems at the end of Section 1.4.)

1.3.3 APPROACHING ZERO OR INFINITY GRACEFULLY

Denormalized Floating-Point Numbers

The gap between 0.0 and r_{min} [Eq. (1.41)] that is implied by the representation defined in Eq. (1.37) and illustrated in Fig. 1.8 can cause annoying problems in practical computations. In the example of Fig. 1.8, let $x = 1.0 \times 2^{-1} = (\frac{1}{2})_{10}$ and $y = 1.1 \times 2^{-1} = (\frac{3}{4})_{10}$. The exponent of the difference, $x - y = (\frac{1}{4})_{10} = 1.0 \times 2^{-2}$, is -2, which is smaller than $e_{min} = -1$. If real numbers with exponents that are smaller than e_{min} are "flushed" to zero, then the following FORTRAN fragment, if initialized with the preceding values of x and y, results in underflow to zero and then a divide-by-zero overflow despite the programmer's effort to ensure that division by zero cannot happen:

```
if (x.ne.y) then
  z=1./(x-y)
else
  z=0.
  ierr=1
end if
```

Finding an error of this sort can waste a great deal of one's time unless one is already aware that such an error can occur.

Modern floating-point representations such as IEEE-754 introduce new floating-point numbers between r_{min} and 0.0, called **denormalized numbers,** for the purpose of ensuring that the normalized floating-point number that results after subtracting x from y cannot equal zero unless $x = y$.

In the toy floating-point representation of Fig. 1.8, it makes sense to interpret the floating-point numbers with exponent $e = 0$ and a nonzero fraction $f \neq 0$ as

$$r = (-1)^s 2^{-1}(0.f) = (-1)^s 0.1 \times 2^{-1} \tag{1.54}$$

because the sign and fraction still contain information. This choice interpolates nicely between r_{min} and 0, as shown in Fig. 1.12.

In the IEEE-754 single-precision format, the rule for evaluating a number with exponent $e = 0$ and a nonzero fraction $f \neq 0$ is

$$r = (-1)^s 2^{-126}(0.f), \tag{1.55}$$

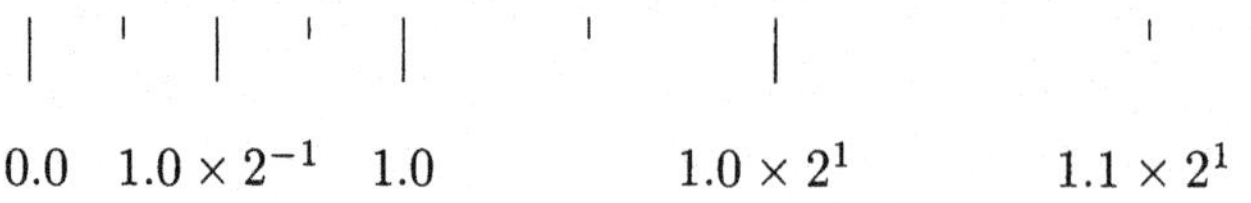

0.0 1.0×2^{-1} 1.0 1.0×2^1 1.1×2^1

FIGURE 1.12

This figure shows an extra (denormalized) floating-point number between $r_{min} = 1.0 \times 2^{-1}$ and 0 in the toy floating-point representation.

which, again, interpolates smoothly between r_{min} and 0. If $e = 0$ and $f \neq 0$ in the IEEE-754 double-precision format, then

$$r = (-1)^s 2^{-1022} (0 . f). \tag{1.56}$$

In both cases the denormalized numbers provide for a gradual loss of significance, depending on the number of zeros between the binary point and the first nonzero bit of f. The price one pays for this graceful behavior may be a substantial performance penalty if the denormalized numbers are implemented in software.

If the denormalized representation is implemented in software rather than in hardware, it is usually possible to instruct a FORTRAN or C compiler not to generate denormalized numbers. Although this practice usually increases execution speed, it may make some common programming practices dangerous, as we illustrate in the FORTRAN code fragment given previously in this section.

Infinities and NaNs

Numbers with exponents that are larger than e_{max} cannot be dealt with elegantly because there is no easy way to approach infinity with a finite number of digits. A floating-point number with exponent $e = e_{max} + 1$ and $f = 0$ is interpreted as infinity, Inf. IEEE-754 requires that a number with $e = e_{max} + 1$ and $f = 0$ must be interpreted as signed infinity, $(-1)^s$ Inf.

All floating-point numbers with exponent $e_{max} + 1$ and nonzero fractions are assigned the value **NaN** (meaning "not a number") regardless of the value of the fraction and regardless of the sign.

Execution may continue even if a program generates NaN or Inf. For example, suppose that one evaluates the function

$$f(x) = \frac{1}{1 + 1/x} \tag{1.57}$$

starting at $x = 0.0$ in steps of 10^{-5}. At the first point the computer divides by zero, obtaining $1/x =$ Inf. Because IEEE-754 prescribes that $y/$Inf $= 0.0$ for every $y \neq 0$, the result is that the computation returns the correct function value, $f(0.0) = 0.0$. Execution then continues with the next value, $x = 1.0 \times 10^{-5}$, unless the user has written a so-called floating-point exception handler that causes some other action to be taken on division by zero. If execution is allowed to continue, then all of the computed function values are correct despite the occurrence of a division by zero.

The following operations generate the value NaN:

- Addition: Inf $+ (-$Inf$)$
- Multiplication: $0 \times$ Inf
- Division: $0/0$ or Inf$/$Inf
- Computation of a remainder (REM): x REM 0 or Inf REM x
- Computation of a square root: $\sqrt{x}$ when $x < 0$

Any operation in which an operand is NaN generates a NaN result.

To see how NaNs can be useful in scientific computations, consider the problem of finding a value of x at which $f(x) = 0.0$ for a given function f. One step in the computation might read, in part,

```
if (f(x).ne.0.0) then
...
else
...
endif
```

If for some x, for example, f were to evaluate $\sqrt{y}$ where $y < 0$, then the subprogram or program segment that evaluates f would return NaN for the value $f(x)$. Any nonzero value of $f(x)$, including an NaN, would cause the program to execute the "then" branch. Instead of stopping, the program would be able to continue until a useful result was produced, provided that the value $f(x)$ is not needed for essential computations.

Some of the references given in the bibliography at the end of this chapter contain additional information on the IEEE-754 representations.

1.3.4 EXERCISES FOR SECTION 1.3

1.3.1   The following numbers are hexadecimal representations of IEEE-754 single-precision floating-point numbers. Note that these representations contain an implicit 1 bit. In each case, give the binary equivalent of the hexadecimal form, the unpacked binary form (that is, the binary form in which the implicit 1 bit is written out explicitly), and the decimal equivalent:

> BF 80 00 00
> 40 00 00 00
> 40 40 00 00

1.3.2   Write and run a program for your computer that
  1. initializes the value of the floating-point variable eps to 1.0, and
  2. divides eps by 2.0 repeatedly, writing the result at each step until the result underflows to zero.

  Explain your output in detail.

1.3.3   Write and run a program for your computer that
  1. initializes the value of the floating-point variable eps to 1.0,
  2. divides eps by 2.0 repeatedly, and
  3. computes $1.0 + eps$, writing the result at each step until the result no longer changes from one step to the next.

  Explain your output in detail.

1.3.4   (From Goldberg [1991].) This problem concerns the evaluation of the function

$$f(x) = \frac{x}{1 + x^2} \tag{1.58}$$

in IEEE-754 floating-point arithmetic.

(a) Show that for all $x > \sqrt{\beta}\,\beta^{e_{max}/2}$, floating-point overflow occurs during the evaluation of Eq. (1.58). Furthermore show that the computed result is equal to 0.0 instead of $1/x$ for all x such that $\sqrt{\beta}\,\beta^{e_{max}/2} < x < r_{max}$ [see Eq. (1.40)].

(b) What result is produced when Eq. (1.58) is evaluated at $x = 0.0$?

(c) Show that the formula

$$f(x) = \frac{1}{x + 1/x}, \tag{1.59}$$

which is mathematically equivalent to Eq. (1.58), produces correct results if $\sqrt{\beta}\,\beta^{e_{max}/2} < x < r_{max}$ and if $x = 0.0$.

1.3.5   The goal of this problem is to understand the result of intentionally or accidentally interpreting an integer stored in two's-complement format as a floating-point number.

(a) Write a C or FORTRAN program that reads a 32-bit integer i4, interprets the bit string that represents i4 as an IEEE-754 single-precision floating-point number r4, and writes r4 as a decimal floating-point number.

(b) Obtain r4 for each of the following values of i4:1, 4194304, 8388608, 16777216, −16777216, −8388608, −4194304, −1.

(c) Explain the results obtained in part (b).

1.4 FLOATING-POINT COMPUTATION

1.4.1 RELATIVE ERROR; MACHINE EPSILON

The **absolute error** of a numerical operation is

absolute error = computed result − exact result.   (1.60)

In physics and engineering one is usually more concerned with the **relative error**,

$$\text{relative error} = \left| \frac{\text{absolute error}}{\text{exact result}} \right|, \tag{1.61}$$

than with the absolute error. Suppose, for example, that a particular numerical operation has led to the floating-point number $1.00 \cdots 00 \times \beta^0$ with an absolute error of $+1$ in the least significant digit (one also says "one unit in the last place" or one **ulp**). Then by Eq. (1.37) (with $r = 1.00 \cdots 01$) the absolute and relative errors are both equal to

$$\epsilon_{\text{mach}} := \beta^{1-p}, \tag{1.62}$$

where ϵ_{mach} is called **machine epsilon**. If the exact result is $1.00 \cdots 00 \times \beta^e$ and the computed result is $1.00 \cdots 01 \times \beta^e$, then the absolute error is β^{e-p+1} but the relative error is still ϵ_{mach}. For this reason machine epsilon is often used as an indicator of the inherent relative error of a floating-point representation.

Machine epsilon also has the significance of being the smallest floating-point number r such that $1.0 + r$ is not equal to 1.0 after rounding; see Exercise 1.4.2.

Machine epsilon does not always give the relative error induced by computing a nearest-neighbor floating-point number instead of the correctly rounded result. For example, if the exact result is $1.00 \cdots 0 \times \beta^e$ and the computed result is $(\beta - 1) . (\beta - 1) \cdots (\beta - 1) \times \beta^{e-1}$, then one computes the relative error as follows:

$$r_{\text{computed}} = \left((\beta - 1) \sum_{j=0}^{p-1} \beta^{-j} \right) \times \beta^{e-1}$$

$$= 1.00 \cdots 0 \times \beta^e - \beta^{e-p}$$

$$r_{\text{exact}} = 1.00 \cdots 0 \times \beta^e \tag{1.63}$$

$$\text{relative error} = \frac{\beta^{e-p}}{\beta^e}$$

$$= \beta^{-p}.$$

However, if the the exact result is $1.00 \cdots 00 \times \beta^e$ and the computed result is $1.00 \cdots 01 \times \beta^e$, then the relative error is

$$\text{relative error} \approx \frac{\beta^{e-p+1}}{\beta^e} \tag{1.64}$$

$$\approx \beta^{1-p}.$$

Thus floating-point numbers are not uniformly distributed. If the value of the exponent increases by 1, the separation between adjacent floating-point numbers increases by a factor of β. One describes this fact by saying that the relative error produced by one ulp "wobbles" by a factor of β. This fact makes it advantageous to choose the smallest possible base, $\beta = 2$.

For example, in Fig. 1.8 adjacent floating-point numbers differ by exactly one ulp. The floating-point number 1.0 differs from the next larger floating-point number by a factor $\beta = 2$ times as much as it differs from the next smaller floating-point number.

1.4.2 ROUNDING

Rounding is the method by which one obtains a base-β floating-point number from a real number r. We shall write the floating-point number obtained by rounding as round (r). One's choice of rounding method may have a major effect on the accuracy of simple computations such as subtraction.

Methods of Rounding

The simplest method of rounding is **truncation**, in which one ends the summation in Eq. (1.37) with the term $j = p - 1$. We show that for truncation,

$$\text{relative error} = \left| \frac{r - \text{round}(r)}{r} \right| \le \epsilon_{\text{mach}}. \tag{1.65}$$

Let

$$r = \beta^e \sum_{j=0}^{\infty} \frac{d_j}{\beta^j}. \tag{1.66}$$

Rounding by truncation gives

$$\text{round}(r) = \beta^e \sum_{j=0}^{p-1} \frac{d_j}{\beta^j}. \tag{1.67}$$

Therefore the relative error is

$$\left| \frac{r - \text{round}(r)}{r} \right| = \left| \frac{\displaystyle\sum_{j=p}^{\infty} \frac{d_j}{\beta^j}}{\displaystyle\sum_{j=0}^{\infty} \frac{d_j}{\beta^j}} \right| \tag{1.68}$$

$$\le (\beta - 1) \sum_{j=p}^{\infty} \frac{1}{\beta^j} = \beta^{1-p}.$$

The inequality follows because every digit in the numerator is not larger than $\beta - 1$ and because the denominator is not smaller than its first term, in which d_0 is not smaller than 1. Because $\epsilon_{\text{mach}} = \beta^{1-p}$, we have established Eq. (1.65).

Another common method is **rounding up**, in which round (r) is computed by truncation if $0 < d_p < \beta/2$. If $d_p \ge \beta/2$, then round (r) is taken as the smallest base-β floating-point number greater than r. It is not difficult to show that the relative error of rounding up is

$$\text{relative error} = \left| \frac{r - \text{round}(r)}{r} \right| \le \frac{1}{2} \epsilon_{\text{mach}}. \tag{1.69}$$

However, continually rounding up may lead the results of some calculations to creep upwards in value, possibly leading to a cumulative error larger than ϵ_{mach}.

All practical methods of rounding other than truncation require one to know more digits than are present in one's floating-point representation. For this reason most modern computers (and all systems that conform to IEEE-754) make use of additional digits in the floating-point unit in order to reduce rounding error. For example, consider the method called **round to even**. In the halfway case in which $d_p = \beta/2$ and $d_j = 0$ for every $j > p$, the least significant digit of round (r) is computed by truncation if d_{p-1} is even, or, if d_{p-1} is odd before rounding, by adding $(\beta/2)\beta^{-p}$ and then truncating. In all other cases one rounds to the nearest base-β floating-point number.

It can be shown that in order to compute an exactly rounded result using round to even, the computer's floating-point unit must keep two additional digits (called *guard digits*) beyond d_{p-1} as well as a "sticky bit," which is set "on" (set to 1) and stays on through the end of the addition if any nonzero bits are shifted past it and thus become inaccessible for the purpose of deciding whether the result is exactly halfway between two valid floating-point numbers.

Rounding Errors in Binary-to-Decimal Conversion

Rounding errors can occur if one converts from decimal to binary or from binary to decimal because a rational number that can be represented exactly by a terminating string of digits in base 10 may be representable only by a repeating binary fraction in base 2.

For example, the base-10 number $\frac{1}{10} = 1 \times 10^{-1}$ can be represented in base 2 only as the infinitely repeating binary fraction $(.00011001100\cdots)_2$. (For simplicity, we shall temporarily work with a fixed binary point.) The nearest (unnormalized) binary fraction that is exactly representable in the IEEE-754 representation is $.00011001100110011001101_2$. Thus, if one sets the value of a single-precision floating-point constant to .1E0 and then prints the value of the constant, one finds that the printed value is approximately .100000001490116E0 instead of .100000000000000E0. Here is the calculation:

$$
\text{round}\,(.1100110011001100110011 00\cdots)
$$

$$
= .11001100110011001100 1101 \times 2^{-3}
$$

$$
= \text{exact} + \text{rounding error}
$$

$$
= .11001100\cdots \times 2^{-3} + (1 - .11001100\cdots) \times 2^{-27}
$$

$$
= \tfrac{1}{10} + \left(2^{-27} - \tfrac{1}{10} \times 2^{-24}\right) \tag{1.70}
$$

$$
= \tfrac{1}{10} + 2^{-24}\left(\tfrac{1}{8} - \tfrac{1}{10}\right)
$$

$$
= \tfrac{1}{10} + \tfrac{2^{-24}}{40}
$$

$$
= \tfrac{1}{10}\left(1 + 2^{-26}\right)_{10}
$$

$$
= (.100000001490116\cdots)_{10}.
$$

Computer users who are unaware that rounding errors can occur because of base conversion are sometimes baffled by small but persistent discrepancies between calculations carried out with and without binary-to-decimal conversion. Moreover, if one writes intermediate data to a disk file and then reads it back in the course of a long computation, the conversion from binary to decimal and back to binary introduces new errors in addition to the errors inherent in floating-point computation.

It may be impossible to recover a unique single-precision binary floating-point number from an output decimal number rounded to eight significant digits or a unique double-precision binary floating-point number from an output decimal number rounded to sixteen significant digits. To recover uniquely the binary floating-point number that existed before base conversion and rounding, one needs nine decimal digits in single precision or seventeen decimal digits in double precision. For proofs of these statements, see Goldberg (1991).

1.4.3 FLOATING-POINT ADDITION AND SUBTRACTION

Fundamental Properties

If x and y are floating-point numbers, then we write the normalized floating-point numbers computed as their sum, difference, product, and quotient as $x \oplus y$, $x \ominus y$, $x \otimes y$, and $x \oslash y$, respectively. (Elsewhere in the book we make use of $\oplus$ for different purposes but in a context that is so different that there should be no confusion.) The IEEE-754 standard requires that the results of floating-point addition, subtraction, multiplication, and division must be equal to the rounded exact results (using round to even).

It is an inescapable consequence of rounding that floating-point addition does not obey one of the basic laws of real numbers, the associative law

$$x + (y + z) = (x + y) + z. \tag{1.71}$$

For example, let x be a floating-point number such that

$$x > \frac{1}{\epsilon_{\text{mach}}}. \tag{1.72}$$

Then

$$x \oplus 1. = x. \tag{1.73}$$

But

$$-x \oplus (x \oplus 1.) = -x \oplus x = 0.$$
$$\neq (-x \oplus x) \oplus 1. = 0. \oplus 1. = 1., \tag{1.74}$$

in which we have used parentheses to force the evaluation of an expression to occur in a specific order.

It is straightforward to show that in floating-point subtraction with p digits and with rounding by truncation, the relative error can be as large as $\beta - 1$. For $\beta = 2$ this means that the relative error can be 1, meaning that the absolute error is as large as the exact result and implying that the most significant digits of the computed and exact results disagree. For $\beta = 10$ the situation is even worse because the relative error can be as large as 9.

The solution to this problem is to increase the number of significant digits used in subtraction. It can be shown that if subtraction is performed with $p + 1$ digits and with rounding by truncation, then the maximum relative error is not greater than ϵ_{mach}.

Catastrophic Cancellation of Significant Digits

From the point of view of a physicist or engineer who merely wants to use a computer to get an accurate numerical answer, one of the most unwelcome features of floating-point computation is **catastrophic cancellation of significant digits**, in which

$$\left| \frac{[\text{round}(x) \ominus \text{round}(y)] - \text{round}(x-y)}{\text{round}(x-y)} \right| \gg \epsilon_{\text{mach}}, \qquad (1.75)$$

where x and y are two real numbers. For example, let $x = 1.005$ and $y = 1.000$. Suppose that x and y are converted to a base-10 floating-point representation with $p = 3$ significant digits, rounded to even, and stored in the computer's memory before being retrieved for the computation of $x \ominus y$. We have round $(x) = 1.00 \times 10^0$ and round $(x - y) = 5.00 \times 10^{-3}$. Therefore

$$\left| \frac{[\text{round}(x) \ominus \text{round}(y)] - \text{round}(x-y)}{\text{round}(x-y)} \right| = \left| \frac{-5.00 \times 10^{-3}}{5.00 \times 10^{-3}} \right| = 1 \gg \epsilon_{\text{mach}}. \qquad (1.76)$$

The most significant digit of the computed result, 0., is different from the most significant digit of the exact result, 5.00×10^{-3}.

One can see beyond the details of these examples to conclude that for every fixed precision p, no matter how large, there exist some subtractions that exhibit catastrophic cancellation. The step of rounding the input data discards essential information carried in the digits that are rounded away. Subtraction may, or may not, remove enough of the more significant digits to create a wildly erroneous result.

1.4.4 EXERCISES FOR SECTION 1.4

1.4.1 Using binary arithmetic ($\beta = 2$), calculating by hand, using truncated arithmetic with $p = 6$, and showing all of your work, evaluate $(1 - x)^2$, $1 - (2x - x^2)$, and $1 - x(2 - x)$ for $x = \frac{28}{32}, \frac{29}{32}, \frac{30}{32}, \frac{31}{32}$.

1.4.2 Write a program that calculates ϵ_{mach} on your computer.

Hint: Begin by normalizing the variable eps to 1.0. Then divide eps by 2.0 and test whether $1.0 + \text{eps}$ is greater than 1.0. Continue until the test fails.

Warning: Read the definition of ϵ_{mach} carefully before you accept your final result.

1.4.3 This problem concerns the roots of the quadratic equation

$$x^2 + \tfrac{17}{8}x + \tfrac{9}{8} = 0 \qquad (1.77)$$

(a) Using binary arithmetic ($\beta = 2$), calculating by hand, and using round to even with $p = 3$, evaluate the roots by substituting the coefficients of the above equation in the quadratic formula.

(b) Compare your answer with the exact solution and explain the origin of any discrepancies.

(c) Propose a strategy for handling certain cases in which the quadratic formula fails and test your strategy on the example of part (a).

1.5 PROPAGATION OF ERRORS

1.5.1 GENERAL FORMULAS

One can estimate the error in a computation either by making a worst-case estimate or by estimating the statistical average of some quantitative measure of error. In this section we summarize basic aspects of the statistical approach to error propagation, which physicists and engineers use to estimate the accuracy of both measurements and computations.

Suppose that the arguments $x_1, \ldots, x_m$ of a function y are experimentally measurable and that their values are subject to random, statistically independent fluctuations from one experiment to another. In more compressed language, $x_1, \ldots, x_m$ are independently distributed random variables. Let the expected value of x_j (that is, the statistical average of x_j over a very large set, or *ensemble*, of experiments) be $E[x_j]$. The expected value has the properties that

$$E[\alpha x_j] = \alpha E[x_j], \tag{1.78}$$

where α is an arbitrary real number, and

$$E[\alpha x_j + \beta x_k] = \alpha E[x_j] + \beta E[x_k], \tag{1.79}$$

where α and β are arbitrary real numbers. Also,

$$E[x_j x_k] = E[x_j] E[x_k] \tag{1.80}$$

if x_j and x_k are independently distributed random variables. This implies nothing about the expected value of a function of a random variable such as $E[x_j^2]$. To compress the notation, we write

$$\mu_j := E[x_j] = \textbf{mean of } x_j \tag{1.81}$$

in the following.

We call the difference between the value of x_j observed in a particular experiment and the ensemble average μ_j the **fluctuation** Δx_j of x_j:

$$\Delta x_j := x_j - \mu_j. \tag{1.82}$$

By this definition, the ensemble average of Δx_j vanishes:

$$E[\Delta x_j] = E[x_j - \mu_j] = E[x_j] - \mu_j = 0. \tag{1.83}$$

The **variance** of the random variable x_j is defined as the quantity

$$\sigma_j^2 := E[(\Delta x_j)^2]. \tag{1.84}$$

Because $E[\Delta x_j] = 0$, σ_j is a quantitative measure of the magnitude of a "typical" fluctuation Δx_j, we call σ_j the **standard deviation** of x_j. The relative error in the value of x_j measured in a single experiment is $|\Delta x_j / \mu_j|$ [see Eq. (1.61)]. Therefore the

$$\textbf{relative standard deviation} := \frac{\sigma_j}{|\mu_j|} \tag{1.85}$$

of the random variable x_j gives the magnitude of a "typical" relative error in a single measurement of x_j.

The fluctuation Δy in the value $y(x_1, \ldots, x_m)$ of a function y caused by fluctuations $\Delta x_1, \ldots, \Delta x_m$ in the values of the arguments $x_1, \ldots, x_m$ is

$$\Delta y := y(x_1 + \Delta x_1, \ldots, x_m + \Delta x_m) - y(x_1, \ldots, x_m)$$
$$\approx \sum_{j=1}^{m} \Delta x_j \frac{\partial y}{\partial x_j} \tag{1.86}$$

by the definition of partial derivatives. Then the square of the fluctuation in the value of y is (approximately) given by the formula

$$(\Delta y)^2 = \sum_{j=1}^{m} \sum_{k=1}^{m} \Delta x_j \Delta x_k \frac{\partial y}{\partial x_j} \frac{\partial y}{\partial x_k}. \tag{1.87}$$

If one assumes that the fluctuations Δx_j are statistically independent, then

$$E[\Delta x_j \Delta x_k] = \delta_{jk} E[(\Delta x_j)^2] \tag{1.88}$$

where the **Kronecker delta** is

$$\delta_{jk} := \begin{cases} 1, & \text{if } j = k; \\ 0, & \text{if } j \neq k. \end{cases} \tag{1.89}$$

Then the statistical average (indicated by $E[\cdot]$) of $(\Delta y)^2$ is

$$\boxed{\begin{aligned} E[(\Delta y)^2] &= \sum_{j=1}^{m} \sum_{k=1}^{m} E[\Delta x_j \Delta x_k] \frac{\partial y}{\partial x_j} \frac{\partial y}{\partial x_k} \\ &= \sum_{j=1}^{m} \left(\frac{\partial y}{\partial x_j} \right)^2 E[(\Delta x_j)^2]. \end{aligned}} \tag{1.90}$$

We shall call Eq. (1.90) the **general error-propagation formula**. Because this formula depends on making a linear approximation to the change in the value of y that results from changes in the values of the arguments, it is not to be trusted if y is a nonlinear function and the relative standard deviations of the x_j are not small compared with unity.

From Eq. (1.90), one can obtain an upper bound that gives a somewhat pessimistic estimate of the variance σ_y^2 in the value of y:

$$
\begin{aligned}
\sigma_y^2 &= \sum_{j=1}^{m} \left(\frac{\partial y}{\partial x_j} \right)^2 \sigma_j^2 \\
&\leq \left(\max_i \left| \frac{\partial y}{\partial x_i} \right| \right)^2 \sum_{j=1}^{m} \sigma_j^2 \\
&\leq R^2 \sum_{j=1}^{m} \sigma_j^2
\end{aligned}
\tag{1.91}
$$

in which

$$
R = \max_i \left| \frac{\partial y}{\partial x_i} \right|
\tag{1.92}
$$

is the maximum rate of change of y with respect to changes in $x_1, \ldots, x_m$. Then the relative standard deviation of y (that is, the standard deviation of y divided by the value of $|y|$) is (approximately) equal to

$$
\frac{\sigma_y}{|E[y]|} = R \left[\sum_{j=1}^{m} \sigma_j^2 \right]^{1/2} .
\tag{1.93}
$$

For example, suppose that $x_1, \ldots, x_m$ are computed, rather than experimentally measured, quantities and that the fluctuations Δx_j result from rounding to the nearest valid floating-point number. Then

$$
\sigma_j = \tfrac{1}{2} \epsilon_{\text{mach}}
\tag{1.94}
$$

and the relative standard deviation of y becomes

$$
\frac{\sigma_y}{|E[y]|} = \frac{1}{2} R \sqrt{m} \, \epsilon_{\text{mach}}.
\tag{1.95}
$$

The right-hand side of this equation is equal to the root-mean-square distance traveled after a random walk of m steps of length $\tfrac{1}{2} R \epsilon_{\text{mach}}$. Our result for the cancellation of significant digits through subtraction, Eq. (1.75), is a special case of Eq. (1.95).

1.5.2 EXAMPLES OF ERROR PROPAGATION

Addition and Subtraction

If

$$
y(x_1, x_2) = x_1 \pm x_2,
\tag{1.96}
$$

then Eq. (1.90) implies that

$$\sigma_y^2 = \sigma_1^2 + \sigma_2^2. \tag{1.97}$$

In other words, the standard deviation of the sum or difference is obtained by adding σ_1 and σ_2 in quadrature, that is, by treating σ_1 and σ_2 as the sides of a right triangle of which σ_y is the hypotenuse.

If y is the sum of m random variables $x_1, \ldots, x_m$,

$$y(x_1, \ldots, x_m) = x_1 + \cdots + x_m, \tag{1.98}$$

then the variance of y is the sum of the variances of the x_i:

$$\sigma_y^2 = \sigma_1^2 + \cdots + \sigma_m^2. \tag{1.99}$$

Multiplication and Division
If

$$y(x_1, x_2) = x_1 x_2, \tag{1.100}$$

then, by Eq. (1.90),

$$\sigma_y^2 = x_2^2 \sigma_1^2 + x_1^2 \sigma_2^2. \tag{1.101}$$

Therefore the relative standard deviation of y is equal to

$$\left(\frac{\sigma_y}{x_1 x_2}\right)^2 = \left(\frac{\sigma_1}{x_1}\right)^2 + \left(\frac{\sigma_2}{x_2}\right)^2. \tag{1.102}$$

For multiplication, the *relative* standard deviations add in quadrature. The same conclusion holds for division.

Standard Deviation of the Mean of N Observations
If y is the mean of $x_1, \ldots, x_N$,

$$y(x_1, \ldots, x_N) = \frac{1}{N} \sum_{j=1}^{N} x_j, \tag{1.103}$$

and if the x_j have identical statistical distributions,

$$\text{for } j = 1, \ldots, N: \mu_j = \mu, \quad \sigma_j = \sigma, \tag{1.104}$$

then the expected value of y is equal to the expected value of any one of the x_j,

$$\mu_y = E[y] = \frac{1}{N} \sum_{j=1}^{N} \mu = \mu. \tag{1.105}$$

However, the standard deviation of y is smaller than the standard deviation of one of the x_j.

Equation (1.90) implies that

$$\sigma_y^2 = \sum_{j=1}^{N} \left(\frac{1}{N}\right)^2 \sigma^2 = \frac{\sigma^2}{N}. \tag{1.106}$$

Therefore the standard deviation of the mean of N independent observations is equal to $1/\sqrt{N}$ times the standard deviation of a single observation:

$$\sigma_y = \frac{\sigma}{\sqrt{N}}. \tag{1.107}$$

In other words, multiple measurements reduce noise.

1.5.3 ESTIMATES OF THE MEAN AND VARIANCE

Up to now we have not taken account of the fact that in a practical situation the values of the mean, μ, and the standard deviation, σ, often are not known. For example, one usually needs to estimate the values of μ and σ of the universe of values of a random variable x, of which the values $x_1, \ldots, x_N$ measured in N independent trials are only a finite subset.

One sees easily that

$$E\left[\sum_{j=1}^{N} x_j\right] = N\mu. \tag{1.108}$$

Therefore

$$\tilde{\mu} := \frac{1}{N} \sum_{j=1}^{N} x_j \tag{1.109}$$

is an estimate of the true mean, μ, because Eq. (1.109) and the definition of the expected value $E[\cdot]$ imply that the mean value of $\tilde{\mu}$ over many observations is equal to μ.

According to the **least-squares criterion**, the best estimate of the true mean μ is the number μ' that minimizes the quantity

$$Q = \sum_{j=1}^{N} (x_j - \mu')^2. \tag{1.110}$$

Exercise 1.5.4 shows that the value of μ' that minimizes Q is

$$\mu' = \tilde{\mu} = \frac{1}{N} \sum_{j=1}^{N} x_j. \tag{1.111}$$

In other words, $\tilde{\mu}$ is the number that best fits the data according to the criterion of least squares. We show in Chapters 8 and 10 that a least-squares criterion can be applied to obtain a best fit in problems involving vectors or functions.

If one recalls that $\sigma^2 = E[(x - \mu)^2]$, reasoning by analogy with Eq. (1.108), it is plausible (but not correct) that the mean of the quantity Q/N may be equal to σ^2 (Q is defined in Eq. (1.110)). Calculating $E[Q]$ for the value $\mu' = \tilde{\mu}$ that minimizes the sum of squares of the differences $x_j - \mu'$, one finds that

$$
\begin{aligned}
E[Q] &= E\left[\sum_{j=1}^{N} \left(x_j - \frac{1}{N} \sum_{k=1}^{N} x_k \right)^2 \right] \\
&= E\left[\sum_{j=1}^{N} \left(\frac{N-1}{N} x_j - \frac{1}{N} \sum_{k \neq j} x_k \right)^2 \right] \\
&= \left(\frac{N-1}{N} \right)^2 E\left[\sum_{j=1}^{N} \left(x_j - \frac{1}{N-1} \sum_{k \neq j} x_k \right)^2 \right].
\end{aligned}
\tag{1.112}
$$

In the second term of the last line of this equation, one recognizes the estimated mean obtained from the $N - 1$ sample values other than x_j. Continuing, one finds that

$$
\begin{aligned}
E[Q] &= \left(\frac{N-1}{N} \right)^2 \sum_{j=1}^{N} E\left[\left(x_j - \frac{1}{N-1} \sum_{k=1}^{N} x_k \right)^2 \right] \\
&= \left(\frac{N-1}{N} \right)^2 \sum_{j=1}^{N} E\left[x_j^2 - \frac{2x_j}{N-1} \sum_{k \neq j} x_k + \frac{1}{(N-1)^2} \left(\sum_{k \neq j} x_k \right)^2 \right].
\end{aligned}
\tag{1.113}
$$

We recall that since each x_j is drawn from the same population,

$$
E[x_j] = E[x] = \mu.
\tag{1.114}
$$

Because we assume that different samples are statistically independent,

$$
E[x_j x_k] = (E[x])^2 = \mu^2
\tag{1.115}
$$

for every $k \neq j$. Because

$$
E[x^2] = \sigma^2 + \mu^2
\tag{1.116}
$$

(see Exercise 1.5.5) and therefore

$$
E[x_j^2] = E[x^2] = \sigma^2 + \mu^2,
\tag{1.117}
$$

one finds that

$$E\left[\left(\sum_{k \neq j} x_k\right)^2\right] = E\left[\sum_{k \neq j} x_k^2 + \sum_{k \neq j} x_k \sum_{l \neq j,k} x_l\right]$$

$$= (N-1)(\sigma^2 + \mu^2) + (N-1)(N-2)\mu^2 \qquad (1.118)$$

$$= (N-1)\sigma^2 + (N-1)^2\mu^2.$$

Substituting this intermediate result into Eq. (1.113) and making use of (1.115), one obtains

$$E[Q] = (N-1)\sigma^2. \qquad (1.119)$$

This result corrects the naive intuition that $E[Q]$ should be equal to $N\sigma^2$. The right factor is $N-1$, not N. One sees from (1.112) that the factor $N-1$ enters because we use the estimated mean of N sample values, $\tilde{\mu}$ (Eq. [1.109]), instead of the true mean μ. As we pointed out above, this introduces the mean of $N-1$ sample values and thereby leads to a factor of $N-1$ instead of N.

It follows from (1.119) that an estimate of the true variance σ^2 using N sampled values $x_1, \ldots, x_N$ is

$$\boxed{\tilde{\sigma}^2 := \frac{1}{N-1} \sum_{j=1}^{N} (x_j - \tilde{\mu})^2.} \qquad (1.120)$$

1.5.4 · EXERCISES FOR SECTION 1.5

1.5.1   This problem concerns the accuracy of computed values of the Cartesian components

$$a_x = a\cos\theta, \quad a_y = a\sin\theta \qquad (1.121)$$

of a vector **a**, given measured or computed values of the magnitude a and the direction θ.

(**a**) Find analytic expressions for the absolute errors Δa_x, Δa_y and the relative errors $(\Delta a_x)/a$, $(\Delta a_y)/a$ in terms of Δa and $\Delta\theta$.

(**b**) Evaluate the relative errors $(\Delta a_x)/a$, $(\Delta a_y)/a$ assuming that

$$\frac{\Delta a}{a} = 0.01 \qquad (1.122)$$

and

$$\Delta\theta = \frac{\pi}{180} \qquad (1.123)$$

for $\theta = \frac{\pi}{180}$, $\theta = \frac{\pi}{6}$, $\theta = \frac{\pi}{3}$ and $\theta = \frac{89\pi}{180}$.

1.5.2   Show that if

$$y(x) = \log x \tag{1.124}$$

then

$$\sigma_y = \frac{\sigma_x}{\mu_x}. \tag{1.125}$$

1.5.3   Show that if

$$z(x) = \alpha x^a y^b \tag{1.126}$$

where α, a, and b are real numbers, then

$$\frac{\sigma_z}{|\mu_z|} = \left[a^2 \left(\frac{\sigma_x}{\mu_x} \right)^2 + b^2 \left(\frac{\sigma_y}{\mu_y} \right)^2 \right]^{1/2}. \tag{1.127}$$

1.5.4   Show that the value of μ' that minimizes

$$Q = \sum_{j=1}^{N} (x_j - \tilde{\mu})^2 \tag{1.128}$$

is

$$\mu' = \tilde{\mu} = \frac{1}{N} \sum_{j=1}^{N} x_j. \tag{1.129}$$

1.5.5   Let x be a random variable with mean μ and standard deviation σ. Show that

$$E[(x - \mu)^2] = E[x^2] - \mu^2. \tag{1.130}$$

1.5.6   Let x be the input, and let y be the output, of a system that is assumed to be linear. The assumption of linearity implies that

$$y = ax + b \tag{1.131}$$

in which a and b are constants to be determined from experiment. Let $x_1, \ldots, x_N$ be the measured values of x, and let $y_1, \ldots, y_N$ be the corresponding measured values of y. The criterion of least squares implies that the best estimates of the slope (or gain) a and intercept (or shift) b are to be found by minimizing the quantity

$$Q = \sum_{i=1}^{N} [y_i - (ax_i + b)]^2. \tag{1.132}$$

Set the partial derivatives of Q with respect to a and b equal to zero and derive the **normal equations**

$$Nb + \left(\sum_{i=1}^{N} x_i \right) a = \sum_{i=1}^{N} y_i$$

$$\left(\sum_{i=1}^{N} x_i \right) b + \left(\sum_{i=1}^{N} (x_i)^2 \right) a = \sum_{i=1}^{N} x_i y_i. \tag{1.133}$$

1.5.7 Solve Eqs. (1.133) and obtain

$$b = \frac{\displaystyle\sum_{i=1}^{N} y_i \sum_{j=1}^{N} (x_j)^2 - \sum_{i=1}^{N} x_i \sum_{j=1}^{N} x_j y_j}{N \displaystyle\sum_{i=1}^{N} (x_i)^2 - \left(\sum_{i=1}^{N} x_i \right)^2}$$

$$a = \frac{N \displaystyle\sum_{i=1}^{N} x_i y_i - \sum_{i=1}^{N} x_i y_i}{N \displaystyle\sum_{i=1}^{N} (x_i)^2 - \left(\sum_{i=1}^{N} x_i \right)^2}. \tag{1.134}$$

1.6 BIBLIOGRAPHY AND ENDNOTES

1.6.1 BIBLIOGRAPHY

Birkhoff, Garrett, and Saunders Mac Lane. *A Survey of Modern Algebra.* 3rd ed. New York: Macmillan, 1965, chapter IV.

Dahlquist, Germund, and Åke Björck. *Numerical Methods* (Ned Anderson, trans.). Englewood Cliffs: Prentice-Hall, 1974, chapter 2.

Goldberg, David. "What Every Computer Scientist Should Know About Floating-Point Arithmetic." *ACM Computing Surveys* 23 (1991): 5–48. This paper contains the proofs of the unproven statements in the section on floating-point computation. Note that Goldberg's ϵ is equal to $\frac{1}{2}\epsilon_{mach}$ in our notation.

Hennessy, John L., and David A. Patterson. *Computer Architecture: A Quantitative Approach.* 2nd ed. San Francisco: Morgan Kaufmann, 1990, appendix A.

Kahan, W. "A Survey of Error Analysis." In C. L. Frieman, editor, *Information Processing 71.* Amsterdam: North-Holland, 1972, pp. 1214–1239.

Patterson, David A., and John L. Hennessy. *Computer Organization and Design: The Hardware–Software Interface.* 2nd ed. San Francisco: Morgan Kaufmann, 1998.

Pentium.ma, a Mathematica Notebook. Wolfram Research, Inc., 1994.

Stevenson, David, et al. "A Proposed Standard for Binary Floating-Point Arithmetic." *IEEE Computer* (March 1981): 51–62.

Swartzlander, Jr., Earl E. *Computer Arithmetic*. Vols. I and II. Los Alamitos: IEEE Computer Society Press, 1990.

Wilkinson, J. H. *Rounding Errors in Algebraic Processes*. New York: Dover, 1994.

1.6.2 ENDNOTES

Section 1.3

The purpose of a floating-point standard is to ensure that the results of the same floating-point operations on different computer systems will be identical if the standard is followed. IEEE stands for the Institute of Electrical and Electronics Engineers. There are actually two IEEE standards for floating-point representations, Standard No. 754 and Standard No. 854. IEEE-754 prescribes the base $\beta = 2$ as well as the arrangement of the exponent and significand bits. IEEE-854 prescribes $\beta = 2$ or $\beta = 10$ (the latter case is for calculators) but does not prescribe a specific format of bits or digits. See References 3–5 for further details.

The CRAY® single-precision floating-point representation uses $\beta = 2$, $p = 48$, $e_{min} = -8192$, and $e_{max} = 8191$. In the CRAY binary single-precision format, bit 63 represents the sign s, bits 62–48 represent the exponent e (biased by 16,384), and bits 47–0 represent the significand f (with no implicit 1 bit). The floating-point number represented by s, e, and f is $r = (-1)^s . f \times 2^{e-16384}$.

The CRAY double-precision floating-point representation, which is implemented only in software, uses $\beta = 2$, $p = 96$, $e_{min} = -8192$, and $e_{max} = 8191$. A double-precision floating-point number is stored in memory as two consecutive single-precision numbers (or *words*), such that the first word contains the sign bit, the exponent, and the 48 most significant bits of the significand. Bits 47–0 of the second word contain the 48 least significant bits of the significand; bits 63–48 are not used.

CHAPTER 2

SETS AND MAPPINGS

2.1 INTRODUCTION

The modern theory of sets was created by Georg Cantor, who published a series of pioneering papers beginning in 1874. Set theory has been extended and refined by many others since Cantor. The vocabulary of elementary set theory has become universal in mathematics. The material selected for this chapter is basic to an understanding of group theory, continuity, convergence, measure, Lebesgue integration, probability theory, fractal sets, and essentially all other topics in modern applied mathematics.

The concept of a **set** of objects (or **elements**) is fundamental in the sense that one usually does not try to define a set in terms of even more basic concepts. In modern set theory one assumes a few plausible axioms, from which one deduces theorems about the properties of sets. Halmos (1960) summarizes the axiomatic approach to set theory. We make no such attempt here. Instead, we assume (naively) that the definitions we make about sets describe objects and relations that can exist.

Many physicists and engineers used to believe that set theory was the epitome of useless abstraction in mathematics. Over the past two decades this dismissive attitude has begun to change to one of grudging acceptance, possibly because the *fractal* sets that one encounters in chaotic systems have properties (such as a nonintegral number of dimensions) that used to be considered "pathological" even by some mathematicians.

Anyone who has watched a curl of smoke ascend, continually stretching out and folding back upon itself, has an intuitive understanding of transformations such as is shown in Fig. 2.1. First let us stretch the unit square to twice its original width and simultaneously compress it to one third of its original height. To model the smoke folding back on itself, we cut the stretched, compressed piece in half vertically and fold the right half into the topmost third of the original square. The equations that describe the transformation of an initial point (x_1, y_1) in the unit square to a final point (x_2, y_2) are

$$\text{for all } x_1 \text{ such that } 0 \leq x_1 \leq \tfrac{1}{2}: \ x_2 = 2x_1, \ y_2 = \tfrac{1}{3}y_1 \tag{2.1}$$

and

$$\text{for all } x_1 \text{ such that } \tfrac{1}{2} < x_1 \leq 1: \ x_2 = 2 - 2x_1, \ y_2 = 1 - \tfrac{1}{3}y_1. \tag{2.2}$$

The next iteration starts with the figure produced by the first step and repeats the transformation in Eqs. (2.1) and (2.2). If one repeats this transformation indefinitely, the y coordinates of

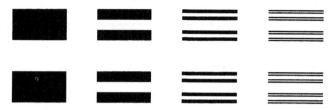

FIGURE 2.1

Illustration of the transformation (2.1) and (2.2).

the points that appear in every figure belong to the **Cantor set** K, which is evidently nothing but a "dust" of points. Thus a commonplace sort of folding–stretching transformation leads directly to one of the supposedly "pathological", or fractal, sets.

Figure 2.2 shows another construction for the Cantor set. One begins with the interval [0, 1]. In the first step, cut out all those points x such that $\frac{1}{3} < x < \frac{2}{3}$. In other words, one cuts out the middle third of the original interval. In the second step, one cuts out the middle thirds of the remaining two intervals, and so on indefinitely. The Cantor set consists of all the points that belong to every figure.

To describe the Cantor set more quantitatively, we make use of the fact that every real number r can be expressed in the base-β floating-point format defined in Eq. (1.27),

$$r = \pm \left[\sum_{j=0}^{\infty} \frac{d_j}{\beta^j} \right] \cdot \beta^n. \tag{2.3}$$

In order to describe the Cantor set, it is convenient to use the base $\beta = 3$. The digits in this base are 0, 1, and 2. The points that survive the first step of the construction shown in Fig. 2.2 clearly have either $d_1 = 0$ (if $0 \leq x \leq \frac{1}{3}$) or $d_1 = 2$ (if $\frac{2}{3} \leq x \leq 1$). Likewise, the points for which $d_1 = 0$ that survive the second step of the construction are of the form $.0d_2d_3\cdots$, in which $d_2 = 0$ (in the left-hand segment) or $d_2 = 2$ (in the right-hand segment). In this way one sees that the Cantor set consists of precisely those points that have a base-3 floating-point expansion in which $n = 0$ and in which only the digits 0 or 2 occur. (Note that the point $x = \frac{1}{3}$, which is one of the points that survives every step and therefore belongs to K, can

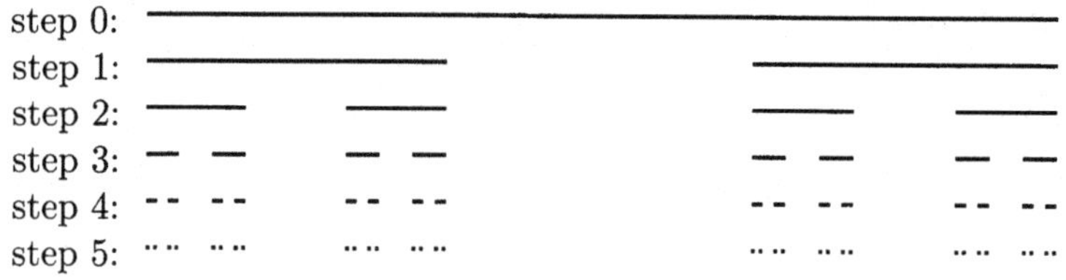

step 0: ——————————————————————————————
step 1: ——————————————   ——————————————
step 2: ——————   ——————   ——————   ——————
step 3: —— ——   —— ——   —— ——   —— ——
step 4: -- --   -- --   -- --   -- --
step 5: ·· ·· ·· ··   ·· ·· ·· ··

FIGURE 2.2

The first several steps of a construction of the Cantor set.

be written in base 3 as either $0.1000\ldots$ or $0.0222\ldots$ before rounding to a finite number of digits. Likewise, there are two base-3 floating-point representations of $x = \frac{2}{3}$ and $x = 1$. All that matters for our characterization of the Cantor set is that, for each point in K, *an expansion exists that contains only the digits 0 and 2*.)

It is natural to ask how "big" the Cantor set K is compared with the interval $[0, 1]$ from which K is constructed. This question has two answers. One answer is, as we show in Section 2.4.5, that every point in $[0, 1]$ can be matched with a unique point in K; therefore K has just as many points as $[0, 1]$ despite the fact that infinitely many points of $[0, 1]$ are deleted in the construction of K. Another answer is that K occupies less space on the real line than the interval $[0, 1]$; therefore K is "smaller" than $[0, 1]$.

2.2 BASIC DEFINITIONS

2.2.1 SETS

If an object belongs to (or is an **element** of) a set S, one writes

$$x \in S. \tag{2.4}$$

One can define a set S by listing its elements, or by giving a rule that defines which elements (out of a defined "universe" of all possible elements) belong to S. For example, one writes

$$X = \{x\} \tag{2.5}$$

to indicate that X is a set with one element, x. One calls a set with one element a **singleton**. Sets need not have a finite number of elements. For example,

$$\mathbb{Z} := \{\ldots, -2, -1, 0, 1, 2, \ldots\}, \tag{2.6}$$

the **set of all integers**, does not have finitely many elements. In fact, neither does the **set of real numbers**, $\mathbb{R}$.

Specifications

A **specification** is a statement that is either true or false for any given object x that belongs to the universe of objects that one has defined. The **axiom of specification** states that one can define a set by giving a criterion that is either true or false for any given element x. More formally, for every universe of elements U and for every valid specification $s(x)$ about an element x, there exists a set

$$S = \{x \in U \mid s(x)\}. \tag{2.7}$$

Equation (2.7) should be read, "S is the set of all elements x of U such that $s(x)$ (is true)." The important words "is true" are understood to be present in this context and therefore are usually omitted. For the rest of this book we enclose the elements of a set in braces. Between the braces that define a set we shall use a vertical bar (|) to mean "such that." Appendix A summarizes our notation.

For example, in the definition of the unit open interval

$$(0, 1) := \{x \in \mathbb{R} \mid 0 < x < 1\}, \tag{2.8}$$

$s(x)$ is the assertion that $0 < x < 1$. One can decompose this assertion into two assertions that must both be true,

- $x > 0$, and
- $x < 1$.

In other words, for every set there is an assertion that is true for every element that belongs to the set and is false for every element that does not belong.

One can also define a set through an assertion that is false for every element of the desired set. If "$s(x)$ is true" is an assertion about an element x, then the **negation** $\neg s(x)$ is the assertion "$s(x)$ is false."

Sets of consecutive integers occur so frequently that it is useful to have a special notation for them. We use the notation

$$(n_1 : n_2) := \{n_1, \ldots, n_2\} \tag{2.9}$$

to denote a set, the elements of which are the integers from n_1 to n_2 (where $n_1 \leq n_2$).

Valid specifications are built out of mathematical operations such as $+$, mathematical relations such as $=$ or $<$, the **universal quantifier** $\forall$ (for every), and the **existential quantifier** $\exists$ (there exists). Except for the bar that ones uses when defining the elements of a set as in Eq. (2.7), one writes "such that" as $\ni$. Colons separate phrases such as "for every x" and "such that" in specifications.

It is useful to express some statements as equations. If the symbol $:=$ appears in an equation, the quantity on the left of the $:=$ is **defined** in terms of the quantity on the right. [The notation $:=$ for "is defined as" is so widespread that it is used even in some computer languages.]

Some statements that we write in the form of equations contain the symbol $\Rightarrow$ ("**implies**"). If a and b are statements, and if $a \Rightarrow b$, then a is called a **sufficient condition** for b because the statement "a is true" is sufficient to guarantee that b is true. If $a \Rightarrow b$, then one says that a is true **only if** b is true, for if b is false then a cannot be true; b is therefore called a **necessary condition** for a. The statement "$b \Rightarrow a$ (is true)" is called the **converse** of the statement "$a \Rightarrow b$ (is true)." The symbol $\Leftrightarrow$ means "**implies and is implied by,**" or "**if and only if,**" or "**is a necessary and sufficient condition for.**"

What we call specifications are, in fact, logical statements built up from the basic logical operations "implies," "and," "or," and negation. Because there is a one-to-one correspondence between specifications and subsets of S, it follows that an assertion about subsets of S is also an assertion about logical statements. In what follows we give additional illustrations of the one-to-one correspondence between set-theoretical statements about the relations among

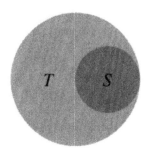

FIGURE 2.3

S is a proper subset of T: $S \subset T$.

unions, intersections, complements, and inclusions of sets, on the one hand, and some of the basic rules of logic, on the other hand.

Subsets

S is called a **subset** (or, more precisely, an **improper subset**) of another set T, written

$$S \subseteq T, \tag{2.10}$$

if and only if

$$\forall x \in S : x \in T \tag{2.11}$$

(for a graphical representation, see Fig. 2.3). If Eq. (2.11) holds, T is called a **superset** of S. For example, the set of valid statement types in FORTRAN 90 is a superset of the set of valid statement types in FORTRAN 77.

The assertion that $S = \{x \in U \mid s(x)\} \subseteq T = \{x \in U \mid t(x)\}$ is equivalent to the assertion that for every x in the universe of elements, if $s(x)$ is true, then $t(x)$ is true:

$$S \subseteq T \Leftrightarrow \forall x \in U : s(x) \Rightarrow t(x). \tag{2.12}$$

Set inclusion ($\subseteq$) therefore corresponds to logical implication ($\Rightarrow$).

It is important to notice that, given two sets S and T, S is not necessarily either a subset or a superset of T because it can be simultaneously true that some elements of S do not belong to T and that some elements of T do not belong to S; hence both $S \subseteq T$ and $T \subseteq S$ are false. In this case S and T are called **incomparable**. For the specification statements s and t that define incomparable sets S and T, neither $s \Rightarrow t$ nor $t \Rightarrow s$ is true.

The **axiom of extension** states that two sets S and T are **equal** (denoted $S = T$) if and only if they have the same elements. Equivalently,

$$S = T \Leftrightarrow [S \subseteq T \text{ and } T \subseteq S]. \tag{2.13}$$

Therefore set equality corresponds to logical implication both ways $s(x) \Leftrightarrow t(x)$.

It follows from the axiom of extension that there is one and only one **empty set** $\emptyset$ with no elements. The empty set is a subset of every subset S of the universe U,

$$\forall S \subseteq U : \emptyset \subseteq S, \tag{2.14}$$

TABLE 2.1 Truth Values of the Propositions $s(x)$ and $t(x)$ for Two Sets S and T Such That $S \subset T$

If	Then	
	$s(x)$ is	$t(x)$ is
$x \in S$	True	True
$x \notin S$ and $x \in T$	False	True
$x \notin T$	False	False

because $s(x)$ is not false (and hence is true) for every $x \in \emptyset$, no matter what specification $s(x)$ defines S.

A set S is properly included in (or is a proper subset of) a set T,

$$S \subset T, \tag{2.15}$$

if and only if $S \subseteq T$ and there exists an element of T that does not belong to S. By this definition,

$$\emptyset \subset S \tag{2.16}$$

for every nonempty set S. If $S \subset T$ and if $s(x)$ and $t(x)$ are the specifications that define S and T, respectively, then for every $x \in S$, $s(x)$ implies $t(x)$, but there exists an $x \in S$ such that $t(x)$ does not imply $s(x)$. Table 2.1 may be helpful in understanding this point.

For example, the **set of positive integers**,

$$\mathbb{Z}^+ := \{1, 2, \ldots\}, \tag{2.17}$$

and the **set of natural numbers**,

$$\mathbb{N} := \{0, 1, 2, \ldots\}, \tag{2.18}$$

are proper subsets of the set of all integers, $\mathbb{Z}$.

Collections of Sets

It is possible for the elements of a set to be sets. In that case one often speaks of a **collection** of sets. Concepts such as "the set of all sets" are not well defined; see Exercise 2.2.1.

Let S be a set. The **power set** of S,

$$\mathfrak{P}(S) := \{X \mid X \subseteq S\}, \tag{2.19}$$

is defined as the collection of all subsets of S. Another notation for the power set is

$$2^S := \mathfrak{P}(S). \tag{2.20}$$

The axiom of extension guarantees the existence of $\mathfrak{P}(S)$. For example, the power set of a two-element set

$$S = \{x, y\} \tag{2.21}$$

is the set

$$\mathfrak{P}(S) = \{\emptyset, \{x\}, \{y\}, \{x, y\}\}. \tag{2.22}$$

We shall show later that, quite apart from its intrinsic interest, the power set of a set S is useful in discussing mappings of S.

2.2.2 MAPPINGS

Domain, Range, and Preimage of a Mapping

As important as sets themselves are mappings of sets. A geographical map represents a particular set, or *domain*, of points on the surface of the Earth by a set of *images* on a piece of paper. In mathematics, a **mapping** or **function**

$$\phi : S \to T \tag{2.23}$$

associates, with every element x of a set S, a unique **image** $y = \phi(x)$ in a (possibly different) set T. One also writes

$$\phi : x \mapsto y \tag{2.24}$$

to show the mapping of $x \in S$ to its image under ϕ.

The set S on which a mapping $\phi : S \to T$ is defined is called the **domain** (i.e., the domain of definition) of ϕ. The **range** (i.e., the range of values) of ϕ is the set

$$\text{range}[\phi] := \{y \in T \mid \exists x \in S :\ni: y = \phi(x)\}, \tag{2.25}$$

which consists of the images of the points of S. The alternative notation

$$\phi(S) := \text{range}[\phi] \tag{2.26}$$

shows the domain S explicitly. Figure 2.4 illustrates the domain and range of a mapping from a subset of the set of real numbers, $\mathbb{R}$, into $\mathbb{R}$.

The set T into which $\phi : S \to T$ maps the elements of S is called the **codomain** of ϕ. Evidently the range of ϕ is a subset of the codomain of ϕ. For example, if $S = \{a, b\}$ and $T = \{c, d\}$, then, under the mapping $\phi : S \to T$ such that

$$\phi(a) = c, \qquad \phi(b) = c, \tag{2.27}$$

ϕ is the mapping under which c is the image of both a and b. Under ϕ, the element d in the codomain T is not the image of any element of S.

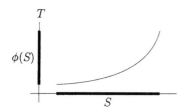

FIGURE 2.4

Illustration of the domain S, range $\phi(S)$, and codomain T of a mapping $\phi : S \rightarrow T$. In this example, ϕ is realized by a real-valued function. The graph of the function is shown as a curve in the x–y plane.

The concept of a mapping is embedded deeply in engineering. For example, one specifies a combinational logic circuit – a circuit that, for each combination of input voltages, produces a unique output voltage that is interpreted as logical "1" or logical "0" – by giving the output that must be produced for each set of inputs. Table 2.2 shows the output (x AND y) that a particular type of circuit, called an AND gate, must produce for each pair of inputs (x, y). In this example, the truth table specifies the input–output mapping ϕ, the domain S is the set of input pairs

$$S = \{(0, 0), (0, 1), (1, 0), (1, 1)\}, \tag{2.28}$$

and the range (here, the set of outputs) is

$$T = \{0, 1\}. \tag{2.29}$$

In this example the codomain is equal to the range.

Another example is the rule for associating letters to decimal digits used on telephone handsets in the United States,

$$\begin{array}{ll} A, B, C \mapsto 2, & D, E, F \mapsto 3, \\ G, H, I \mapsto 4, & J, K, L \mapsto 5, \\ M, N, O \mapsto 6, & P, R, S \mapsto 7, \\ T, U, V \mapsto 8, & W, X, Y \mapsto 9, \end{array} \tag{2.30}$$

TABLE 2.2 Truth Table for an AND Logic Gate Defines a Mapping ϕ from the Set of Input Pairs (x, y) to the Output x AND y

x	y	x AND y
0	0	0
0	1	0
1	0	0
1	1	1

is a mapping for which

$$S = \{A, B, C, D, E, F, G, H, I, J, K, L,$$
$$M, N, O, P, R, S, T, U, V, W, X, Y\}. \quad (2.31)$$

The elements 0 and 1 of the codomain

$$T = \{0, 1, 2, 3, 4, 5, 6, 7, 8, 9\} \quad (2.32)$$

are not the images of any letter under the mapping defined in Eq. (2.30).

The formal statement of the property of **uniqueness**,

$$\forall x, x' \in S : x = x' \Rightarrow \phi(x) = \phi(x'), \quad (2.33)$$

says that one can always apply the same mapping simultaneously to both sides of an equation. A rule that associates more than one image with some s is called **one-to-many** (or **multivalued**). In contemporary mathematics a well-defined function cannot be one-to-many.

We note that a mapping f from one set to another is also called a function. As usual, one writes the **value of the function** f at the point (or for the **argument**) x as

$$f(x). \quad (2.34)$$

If one refers to the function itself, one writes f, *not* $f(x)$. In other words, $f(x)$, $f(t)$, and $f(\omega)$ are the values of the same function at the points x, t, and ω, not different functions. In the few cases in which it is not obvious where to write an argument, one can use the **place-holder notation** $f(\cdot)$ to denote the function itself.

Few concepts in mathematics have undergone more change than that of a function. Until the early twentieth century mathematicians referred to a function by giving the letters that stand for the function and its intended argument, as in "the function $f(x)$." In modern terminology, a function is a mapping f that associates an image $f(x)$ with each point x in some set. To speak of "the function $f(x)$" is to confuse a function (f) with the value of the function for the argument x.

For example, let $\mathbb{R}$ be the set of real numbers. (We have more to say about $\mathbb{R}$ as the book progresses.) The mapping $\phi : \mathbb{R} \to \mathbb{R}$ such that

$$\forall x \in \mathbb{R} : \phi(x) := ax + b \quad (2.35)$$

maps every point x in $\mathbb{R}$ to a unique point $y = ax + b$ in $\mathbb{R}$. From a geometrical point of view, one can think of Eq. (2.35) as representing a stretching transformation (multiplication by a) followed by a translation through b. In fact, the almost trivial mapping defined in Eq. (2.35) is the simplest example of an affine transformation. We shall describe affine transformations in greater generality in Chapter 5.

The **restriction** of a function $f : S \to T$ to a set $S' \subset S$ is the function $g : S' \to T$ with domain S', such that

$$\forall x \in S' : g(x) = f(x). \quad (2.36)$$

The usual notation for the restriction of f to S' is $f \upharpoonright S'$.

A mapping of a set S to a set T is called **onto** (or is a **surjection**) if *every* $y \in T$ is the image of some $x \in S$. For example, if $a \neq 0$, then the mapping defined in Eq. (2.35) carries $\mathbb{R}$ onto $\mathbb{R}$. If ϕ is not onto, then ϕ is called **into**.

If ϕ carries more than one x onto the same image y, then ϕ is called **many-to-one**. For example, the mapping in Eq. (2.27) is many-to-one. To give another example, if $a = 0$ in the mapping in Eq. (2.35), then the entire set of real numbers $\mathbb{R}$ is mapped onto the single point $y = b$. The mapping $\mathbb{R} \to \{b\}$ is infinitely many-to-one.

It is useful to have a notation for the set of all mappings from a set X into a set Y,

$$Y^X := \{\phi \mid \forall x \in X : \phi(x) \in Y\}. \tag{2.37}$$

For example, the set $\mathbb{R}^{\mathbb{N}}$ of all mappings from the set of natural numbers, $\mathbb{N}$, into the set of real numbers, $\mathbb{R}$, is the set of all infinite sequences $(r_0, r_1, \ldots,)$ such that every $r_i \in \mathbb{R}$.

Every mapping $\phi : S \to T$ of a set S into a set T induces a mapping $\hat{\phi} : \mathfrak{P}(S) \to \mathfrak{P}(T)$ of the power set of S into the power set of T. The definition of $\hat{\phi}$ is

$$\forall X \in \mathfrak{P}(S) : \hat{\phi}(X) := \{\phi(x) \mid x \in X\}$$
$$= \phi(X). \tag{2.38}$$

That is, $\hat{\phi}$ carries every subset $X \subseteq S$ onto the set $\phi(X)$, the elements of which are the images of the elements of X under the mapping ϕ.

For example, let

$$S = \{x, y\} \text{ and } T = \{z\} \tag{2.39}$$

and let $\phi : S \to T$ be the mapping such that

$$\phi(x) = \phi(y) = z. \tag{2.40}$$

The mapping ϕ induces a mapping $\hat{\phi} : \mathfrak{P}(S) \to \mathfrak{P}(T)$ such that

$$\hat{\phi}(\{x\}) = \hat{\phi}(\{y\}) = \hat{\phi}(\{x, y\}) = \{z\}. \tag{2.41}$$

Then the image of the power set of S under $\hat{\phi}$ is the set

$$\hat{\phi}(\mathfrak{P}(S)) = \{\emptyset, \{z\}\} = \mathfrak{P}(T). \tag{2.42}$$

One may ask why it is necessary to distinguish $\hat{\phi}$ from ϕ, if the image of X under $\hat{\phi}$ is $\phi(X)$. The answer is that $\hat{\phi}$ and ϕ act on different kinds of objects: ϕ acts on *elements* of S, whereas $\hat{\phi}$ acts on *subsets* of S.

One-to-One Mappings

Geographical mappings (for example) are usually **one-to-one**; that is, every element of the domain corresponds to one, and only one, element of the range. For example, to verify that the mapping defined in Eq. (2.35) is one-to-one if $a \neq 0$, one has only to note that if $\phi(x) = \phi(x')$, then $ax + b = ax' + b$, which implies that $x = x'$ provided that $a \neq 0$.

The formal definition of a one-to-one mapping,

$$[\phi : S \to T \text{ is } (1 - 1)] \Leftrightarrow [\forall x, y \in S : \phi(x) = \phi(y) \Leftrightarrow x = y] \tag{2.43}$$

is used heavily in subsequent chapters. By virtue of the property of uniqueness, one can establish that a given mapping is one-to-one by proving that, for every x and y in the domain of ϕ, the equality of the images $\phi(x)$ and $\phi(y)$ implies the equality of x and y.

A one-to-one mapping $\phi : S \to T$ is also called an **injection**. It is extremely important to notice that a one-to-one mapping is *not* necessarily surjective (onto). For example, the mapping $\phi : S \to T$ of the set $S = \{x\}$ into the set $T = \{y, z\}$ such that $\phi(x) = y$ is one-to-one but is not surjective because there is an element of T that is not the image of any element of S.

A one-to-one mapping ϕ is called a **bijection** if and only if ϕ is both injective and surjective. In other words, a mapping $\phi : S \to T$ is bijective if and only if ϕ is one-to-one and onto. The appropriateness of the term "bijection" is made clear in a subsequent section.

One-to-one mappings are used extensively in computer operating systems. For example, the ASCII mapping of characters to unsigned one-byte integers given in Table 1.2 is one-to-one. Also, the mapping of unsigned n-bit integers to physical memory locations illustrated in Fig. 1.1 is one-to-one.

A set S is called **finite** if and only if there exist a nonnegative integer n and a bijection β such that

$$\forall x \in S : \beta(x) \in (1 : n) \text{ i.e., if and only if } \beta(S) = \{1, \dots, n\}. \tag{2.44}$$

Often one writes the elements of a finite set S as y_i, thereby showing the mapping $\beta : S \to \{1, \dots, n\}$ explicitly:

$$S = \{y_1, \dots, y_n\}. \tag{2.45}$$

The nonnegative integer n is called the **order** of S. We shall use the notation

$$|S| := n \tag{2.46}$$

(n is defined in Eq. (2.44)) for the order of S.

Given the intimate connection between a mapping $\phi : S \to T$ and the power-set mapping $\hat{\phi} : \mathfrak{P}(S) \to \mathfrak{P}(T)$ induced by ϕ, it should come as no surprise that ϕ is one-to-one if and only if $\hat{\phi}$ is one-to-one, and that ϕ is bijective if and only if $\hat{\phi}$ is bijective (see Exercise 2.2.9).

Composition of Mappings

The **product** (or **composition**) of two mappings $\phi : S \to T$ and $\chi : T \to U$ is the mapping $\chi\phi : S \to U$ that results if ϕ is followed by χ:

$$\forall x \in S : \chi\phi(x) := \chi(\phi(x)). \tag{2.47}$$

One also sees the notation $\chi \circ \phi$ for the composition of χ and ϕ. It is an exercise to verify that $\chi \phi$ is a valid mapping if both χ and ϕ are valid mappings.

Note that in the composite mapping $\chi \phi$, it is ϕ that one performs first. The order of mappings goes from right to left on account of the usual convention of writing the element x on which a mapping ϕ acts to the right of the mapping, as in $\phi(x)$.

Often it is useful to represent the composition of $\phi : S \to T$ and $\chi : T \to U$ with the following **commutative diagram**:

$$
\begin{array}{ccc}
 & T & \\
\phi \nearrow & & \searrow \chi \\
S \xrightarrow[\chi \circ \phi]{} & & U
\end{array}
\tag{2.48}
$$

The diagram is called *commutative* because one obtains the same result by going from S to U via the lower branch as one obtains by going via the upper branch $(S \to T \to U)$. The directions of the arrows in the diagram correspond to the directions of the mappings in the sense that the image of S under ϕ is T, and so forth.

For a mapping $\phi : S \to S$ of a set S into itself, we define the **iterates** (or **powers**) of ϕ recursively as

$$
\phi^1 := \phi, \qquad \phi^n := \phi \phi^{n-1}.
\tag{2.49}
$$

Recursion is defined and discussed in Section 2.4.2.

Inverse of a One-to-One Mapping

The **inverse** of a one-to-one mapping $\phi : S \to \phi(S)$ is the mapping $\phi^{-1} : \phi(S) \to S$ that undoes what ϕ does:

$$
\forall x \in S : \phi^{-1}(\phi(x)) = x.
\tag{2.50}
$$

Note that the definition makes sense only because ϕ is one-to-one. If ϕ in Eq. (2.50) were not one-to-one, then ϕ^{-1} would be one-to-many and therefore would not be a legal mapping. (For example, if $\phi(x) = \phi(y) = z$ where $y \neq x$, then there is no valid mapping that takes z into both x and y.)

In other words, *a one-to-one mapping is always invertible on its range*. Obviously ϕ^{-1} is not defined on the elements $y \in T$ (if any exist) such that $y \notin \phi(S)$.

Another way to write the definition of the inverse,

$$
\phi^{-1} \phi = \mathbf{1}_S,
\tag{2.51}
$$

uses the **identity mapping** $\mathbf{1}_S$, which is the mapping that maps every $x \in S$ onto x:

$$
\forall x \in S : \mathbf{1}_S x := x.
\tag{2.52}
$$

Likewise,

$$
\phi \phi^{-1} = \mathbf{1}_{\phi(S)}.
\tag{2.53}
$$

(Challenge: Verify this relation directly from the definitions of the inverse and the identity mapping on $\phi(S)$.) Although each set has its own, unique identity mapping, one usually writes **1** for any identity mapping. If we need to indicate the set on which an identity mapping acts, we shall use a subscript as in Eqs. (2.51) and (2.53).

A bijection $\phi : S \to T$ has an inverse $\phi^{-1} : T \to S$ that is also bijective. In other words, the inverse of a mapping that is one-to-one and onto is also one-to-one and onto. To prove that ϕ^{-1} is onto if $\phi : S \to T$ is one-to-one and onto, it is sufficient to observe that because ϕ is onto, every element $y \in T$ is the image $y = \phi(x)$ of some element $x \in S$. Because the domain of ϕ is S, the definition given in Eq. (2.50) yields a mapping ϕ^{-1} that is onto (and one-to-one).

Let $\phi : S \to T$ and $\chi : T \to U$ be one-to-one mappings. It is easy to show that the product mapping $\chi\phi : S \to U$ is also one-to-one and that the inverse on $\chi(\phi(S))$ is

$$(\chi\phi)^{-1} = \phi^{-1}\chi^{-1} \tag{2.54}$$

(see Exercise 2.2.5). In words: To undo the result of performing ϕ and then χ, one must first undo χ, then ϕ. This result is clear from the commutative diagram (2.48). (If you have trouble visualizing the implications of Eq. [2.54], imagine that ϕ means "put on your socks" and that χ means "put on your shoes.")

If $\phi : S \to T$, $\chi : T \to U$, and $\psi : U \to V$ are all bijections, then a bijective mapping exists between any two of the four sets S, T, U, and V and can be constructed from ϕ, χ, ψ, ϕ^{-1}, χ^{-1}, and ψ^{-1}, as shown in the following commutative diagram:

$$
\begin{array}{ccc}
S & \xrightarrow{\phi} & T \\
\psi\chi\phi \Big\downarrow \;{}^{\xi}\!\diagdown\!\!\!\diagup_{\eta} & & \Big\downarrow \chi \\
V & \xleftarrow[\psi]{} & U
\end{array}
\tag{2.55}
$$

in which

$$\xi = \phi^{-1}\chi^{-1}, \qquad \eta = \chi^{-1}\psi^{-1}. \tag{2.56}$$

Matrix Representation of Bijective Mappings

A bijective mapping of a (small) finite set can be written conveniently as a matrix with two rows. In the first row one writes the elements of the domain. Under each element of the domain one writes its image. The mapping ϕ_1 of $S = \{a, b, c\}$ onto $T = \{d, e, f\}$ in which $a \mapsto d$, $b \mapsto e$ and $c \mapsto f$ can be written as

$$\phi_1 = \begin{pmatrix} a & b & c \\ d & e & f \end{pmatrix} \tag{2.57}$$

Similarly,

$$\phi_2 = \begin{pmatrix} d & e & f \\ g & h & i \end{pmatrix} \tag{2.58}$$

maps T onto $U = \{g, h, i\}$. Both ϕ_1 and ϕ_2 are one-to-one. The product

$$\phi_2\phi_1 = \begin{pmatrix} d & e & f \\ g & h & i \end{pmatrix}\begin{pmatrix} a & b & c \\ d & e & f \end{pmatrix} = \begin{pmatrix} a & b & c \\ g & h & i \end{pmatrix}, \tag{2.59}$$

which maps $a \mapsto d \mapsto g$, $b \mapsto e \mapsto h$, and $c \mapsto f \mapsto i$, is one of $3 \cdot 2 \cdot 1 = 6$ possible bijective mappings of S to U (see Exercise 2.2.2).

In an expression such as Eq. (2.57), the columns can be written in any order as long as the correspondence between the elements of the domain and their images is preserved. In a composition of mappings such as Eq. (2.59), the first matrix conventionally goes on the right, the second goes to the left of the first, and so forth, so that the order of matrices is the same as the order of mappings. In a product of two mappings, therefore, one can make the calculation of $\phi_2\phi_1$ easier by changing the order of the columns in the second mapping ϕ_2 such that the order of the elements in the top row of ϕ_2 is the same as the order of the elements in the bottom row of ϕ_1 (see Exercise 2.2.8).

The inverse mapping ϕ_1^{-1} such that

$$\phi_1^{-1}\phi_1 = 1 \tag{2.60}$$

is

$$\phi_1^{-1} = \begin{pmatrix} d & e & f \\ a & b & c \end{pmatrix}, \tag{2.61}$$

which is simply Eq. (2.57) with the rows interchanged.

Lists

A finite set S, together with a bijective mapping λ of a finite set of integers $(1 : p)$ onto S, is called a **list**. Let

$$\forall i \in (1 : p) : s_i := \lambda(i) \tag{2.62}$$

be the image of the integer i under the mapping λ. Writing λ as

$$\lambda = \begin{pmatrix} 1 & 2 & 3 \cdots \\ s_1 & s_2 & s_3 \cdots \end{pmatrix}, \tag{2.63}$$

one recovers the usual subscripted form of a list, $\{s_1, s_2, s_3, \ldots, s_p\}$. The subscript notation s_i simply reproduces the i-th column of λ.

Lists are used extensively in computer hardware and software. For example, a byte is a list of eight bits. A word is a list of bytes (see Figs. 1.10 and 1.11). RAM is accessed with the help of a mapping of logical addresses $\{0, \ldots, 2^N - 1\}$ to physical locations; in effect, this mapping implements a list.

Manipulating lists is a fundamental task of symbolic computation. General-purpose programs for symbolic calculation such as Mathematica® and Maple® provide powerful tools for using and transforming lists.

Arrays

An **array** is a data structure in which data blocks of the same size are stored adjacent to one another in computer memory. Because they are stored one after the other, it is convenient to refer to array elements using subscripted symbols. For example, the elements of a one-dimensional array named c of length 25 containing the characters of the string "the Schroedinger equation" could be referred to in the C programming language as c[0], c[1], ..., c[24], in which c[0] = 't', c[1] = 'h', c[2] = 'e', and so forth. Computer memory is organized as an array of bytes; see Fig. 1.1.

Although the mapping of subscripts to array elements is transparent to the programmer, the computer must carry out several intermediate mappings in order to achieve the deceptively simple result i $\mapsto$ c[i]. From a strictly logical point of view, the intermediate one-to-one mappings are

$$\text{element of index set} \mapsto \text{logical address of array element}$$
$$\mapsto \text{physical address of array element}$$
$$\mapsto \text{"contents" (or value) of the array element} \tag{2.64}$$

In practice, further intermediate steps are necessary.

Preimage of a Mapping

We show in a previous section that a mapping $\phi : S \to T$ induces a mapping $\hat{\phi}$ of the power sets of S and T defined in Eq. (2.38) such that, for every subset $X \subseteq S$,

$$\hat{\phi}(X) = Y, \tag{2.65}$$

where

$$Y = \{y \in T \mid \exists x \in X :\ni: y = \phi(x)\}. \tag{2.66}$$

For example, if $\phi : \{x, y\} \to \{z\}$ is the mapping defined in Eq. (2.40), and if

$$X = \{x, y\}, \tag{2.67}$$

then one has

$$Y = \{z\}. \tag{2.68}$$

Given a mapping $\phi : S \to T$, for every subset $Y \subseteq T$ of the codomain T there exists a subset $X \subseteq S$ consisting of all of the elements of the domain S that ϕ maps into Y. In terms of the power-set mapping $\hat{\phi}$ defined in Eq. (2.38),

$$Y = \hat{\phi}(X). \tag{2.69}$$

Although $\hat{\phi}$ may not be one-to-one, it is always possible to define a mapping of subsets of T such that the image of Y is X. This mapping is conventionally denoted $\hat{\phi}^{-1}$ or ϕ^{-1}. The formal definition of $\hat{\phi}^{-1}$ is

$$\forall Y \subseteq T : \hat{\phi}^{-1}(Y) := \{x \in S \mid \phi(x) \in Y\}. \tag{2.70}$$

The set $X = \hat{\phi}^{-1}(Y)$ is called the **preimage** (or **inverse image**) of Y. In other words, the preimage of a subset Y of the codomain under a mapping ϕ is the set X whose image is $Y = \hat{\phi}(X)$ under $\hat{\phi}$.

For example, if $\phi : \{x, y\} \to \{z\}$ is the mapping defined in Eq. (2.40), then the preimage of the set $\{z\}$ is the set $\{x, y\}$. Often one writes $\{x, y\} = \phi^{-1}(z)$.

If ϕ is one-to-one (and therefore the inverse mapping ϕ^{-1} is defined on $\phi(S)$ and each of its subsets), the preimage of $T \subseteq \phi(S)$ is the image $S = \phi^{-1}(T)$ under ϕ^{-1}. However, the standard notation for the preimage of a set T is always $\phi^{-1}(T)$, whether or not ϕ is one-to-one.

For example, let $y = \phi(x)$ where $\phi(x) := ax + b$ and a, b, and x are real numbers. If $a \neq 0$, then $\phi^{-1}(y) = (y - b)/a$, which defines a unique real number for each real number y. However, if $a = 0$, then the preimage $\phi^{-1}(y)$ is the empty set unless $y = b$. If $y = b$, then the preimage $\phi^{-1}(y)$ is the entire set of real numbers, because ϕ maps every real number x onto b when $a = 0$.

2.2.3 AXIOM OF CHOICE

A basic postulate of set theory called the **axiom of choice** asserts that it is possible to distinguish elements of a nonempty set from one another by choosing one specific element, then a second, different element, and so on.

Although the axiom of choice is "obvious" for finite sets, it is not always applicable in the context of microscopic physics. A basic postulate of quantum mechanics is that there exist *identical particles* such as electrons that are indistinguishable by any experimental means. Because one cannot build an apparatus that chooses one *specific* electron (for example), one can adopt the point of view that a collection of quantum-mechanical identical particles is not a set in the mathematical sense. An appropriate mathematical construct for identical particles is a multiset, which is defined as follows: A **multiset** is a set M of objects, together with a one-to-one function m from M into $(1 : n)$ (for some positive integer n). The idea is that the integer $m(\psi)$ associated with an element ψ of M is the *multiplicity* of ψ. In more physical terms, we can think of each element ψ of M as a state that a certain kind of particle can occupy, and the associated integer $m(\psi)$ as the *occupancy* of (i.e., the number of identical particles in) the state ψ.

2.2.4 CARTESIAN PRODUCTS

An **ordered pair** (a, b) is a set S, together with the mapping

$$\begin{pmatrix} 1 & 2 \\ a & b \end{pmatrix} \tag{2.71}$$

in which $a \in S$ and $b \in S$. It may happen that $a = b$. (The axiom of choice tells one that the mapping defined in Eq. [2.71] exists.) The mapping defines an ordering (i.e., a comes before b). The elements a and b are called the **components** of the ordered pair (a, b). One can also consider an ordered pair as a list with two elements.

Two ordered pairs are equal if and only if both of their components are equal:

$$(a, b) = (c, d) \Leftrightarrow a = c, \qquad b = d. \tag{2.72}$$

Clearly one can define ordered n-tuples by generalizing the definition of ordered pairs.

Many useful mathematical constructs can be expressed naturally in terms of ordered pairs or ordered n-tuples. For example, a **complex number** $z = x + iy$ (in which $x, y \in \mathbb{R}$) can be thought of as the ordered pair

$$z = (x, y) \tag{2.73}$$

because i merely serves to indicate which of x, y is the imaginary part, that is, the second component of the ordered pair. We write the **set of all complex numbers** as $\mathbb{C}$.

A function f of two variables $x \in S$ and $y \in T$, with values in U, is a function $f : S \times T \to U$ from the Cartesian product $S \times T$ to U.

Another example is the **graph** of a mapping $\phi : S \to T$, which is defined as the set of ordered pairs $(x, \phi(x))$ in which $x \in S$. In the familiar case in which S is an interval of $\mathbb{R}$ and T is a subset of $\mathbb{R}$, the graph of ϕ is a curve such as is illustrated in Fig. 2.4.

Of course, as soon as one defines a new mathematical object, the temptation to define the set of all such objects is irresistible. The **Cartesian product** $A \times B$ of two sets A and B is the set of all ordered pairs (a, b) such that $a \in A$ and $b \in B$:

$$A \times B := \{(a, b) \,|\, a \in A, \ b \in B\}. \tag{2.74}$$

Similarly, the Cartesian product $S_1 \times S_2 \times \cdots \times S_n$ of n sets $S_1, S_2, \ldots, S_n$ is the set of all ordered n-tuples $(x_1, x_2, \ldots, x_n)$ in which $x_1 \in S_1, x_2 \in S_2, \ldots, x_n \in S_n$. If $S_1 = S_2 = \cdots = S_n$, then one writes

$$A^n := \underbrace{A \times A \times \cdots \times A}_{n \text{ times}}. \tag{2.75}$$

For example, the Cartesian coordinates x, y, z of a point in three-dimensional space are the components of an ordered triplet (x, y, z), which is an element of $\mathbb{R}^3$. To give another example, a real-valued function f of three real variables is a mapping $f : \mathbb{R}^3 \to \mathbb{R}$. Yet another example: A complex-valued function f of one complex variable is a mapping $f : \mathbb{R}^2 \to \mathbb{R}^2$.

In principle there is a distinction among ordered triplets such as (x, y, z), $((x, y), z)$, and $(x, (y, z))$. However, it is easy to show that the Cartesian product is associative in the sense that the sets $\{(x, y, z)\}$, $\{((x, y), z)\}$ and $\{(x, (y, z))\}$ are equivalent as defined in Section 2.2.5. Because $(A \times B) \times C$ and $A \times (B \times C)$ are therefore different ways of writing essentially the same set, one simply writes $A \times B \times C$ without worrying about placing parentheses to indicate which Cartesian product is to be taken first.

A **d-dimensional array** is an array, the elements of which are indexed by multiplets with d elements, drawn from $(\mathbb{Z}^+)^d$ or $\mathbb{N}^d$. For example, computer memory is a one-dimensional byte array. The index set is $(0 : 2^N - 1)$, in which N is the number of address bits (see Fig. 1.1). For a two-dimensional array, the index set is the Cartesian product $(0 : m - 1) \times (0 : n - 1)$. As the term "two-dimensional" suggests, one visualizes such an $m \times n$ two-dimensional array as consisting of m horizontal rows and n vertical columns. One writes the elements of a two-dimensional array as c[i][j] (in which $i \in (0 : m - 1)$ and $j \in (0 : n - 1)$) in the C programming language, or c(i,j) (in which $i \in (1 : m)$ and $j \in (1 : n)$) in FORTRAN.

Multidimensional arrays are very widely used in numerical models of physical systems. A four-dimensional array could be used, for example, to store the values of a physical quantity such as electric field sampled at points (x_i, y_j, z_k) and times t_l.

Because of the one-dimensional hardware organization of computer memory, a convention must be adopted in software for the mapping of the index multiplets $(i, j, \ldots)$ of a multidimensional array to one-dimensional memory addresses. The computation of the address of an array element array[i][j] by a C compiler is

$$\text{address of array[i][j]} = \text{ar0} + \text{data\_size} * \text{no\_columns} * \text{i}$$
$$+ \text{data\_size} * \text{j}, \tag{2.76}$$

where ar0 is the address of the first element in the array, array[0][0]. The integer data_size is the number of bytes occupied by one array element. The computation of the address of an array element ar(i,j) by a FORTRAN compiler is

$$\text{address of array(i, j)} = \text{ar0} + \text{data\_size} * \text{no\_rows} * \text{(j - 1)}$$
$$+ \text{data\_size} * \text{(i - 1)}, \tag{2.77}$$

where ar0 is the address of the first element in the array, array(1,1). As a result of these different conventions, array elements are stored and accessed by a C compiler one row after another (as shown in Fig. 2.5); array elements are stored and accessed by a FORTRAN compiler one column after another (as shown in Fig. 2.6).

From the perspective of those who use computers for scientific modeling or engineering design, the most important consequence of the existence of different conventions for storing and accessing the elements of multidimensional arrays is the **stride**, or the difference between the addresses of array elements that are accessed consecutively. Stride is a significant quantity

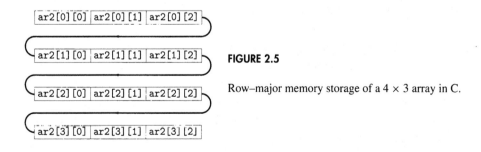

FIGURE 2.5

Row–major memory storage of a 4×3 array in C.

FIGURE 2.6

Column–major memory storage of a 4×3 array in FORTRAN.

because the performance of most hierarchical memory systems deteriorates sharply if the stride becomes so large that fast cache memory is used ineffectively, and as a result the processor has to wait several clock periods while each data word is transferred from main memory.

For example, if the data size of an array element is 4 bytes (as for the IEEE-754 single-precision floating-point representation) and array elements are accessed in the order `array[0][0]`, `array[0][1]`, `array[0][2]`, and so forth from a C program, the stride is equal to 4 bytes because these elements are stored sequentially by C compilers (see Fig. 2.5). However, exactly the same order of access, if performed from a FORTRAN program, results in a stride equal to the data size (4 bytes in this example) *times the number of rows* (see Fig. 2.6). Thus a blunder, such as incrementing the wrong index of a two-dimensional array, or converting a program from FORTRAN to C without changing array indices, can strongly affect performance if either the row dimension (m) or the column dimension (n) is large. We give an example in Section 6.1.3.

2.2.5 EQUIVALENCE AND EQUIVALENCE CLASSES

Two sets are called **equivalent**, written

$$S \sim T, \tag{2.78}$$

if and only if there exists a bijective function ϕ from (all of) S to T. For example, in Eq. (2.57) the mapping ϕ_1 is one-to-one and carries $S = \{a, b, c\}$ onto $T = \{d, e, f\}$. Therefore $S \sim T$.

In general, an **equivalence relation** among elements of a set S is a binary relation (which we denote as $\equiv$) that satisfies the following axioms:

1. *Reflexivity:*

$$\forall a \in S : a \equiv a. \tag{2.79}$$

2. *Symmetry:*

$$\forall a, b \in S : a \equiv b \Rightarrow b \equiv a. \tag{2.80}$$

3. *Transitivity:*

$$\forall a, b, c \in S : [a \equiv b \text{ and } b \equiv c] \Rightarrow a \equiv c. \tag{2.81}$$

Equivalence between sets is an equivalence relation (see Exercise 2.2.3). (*Warning*: The symbol $\sim$ is sometimes used in logic to denote the negation of a statement. In this book, $\neg$ denotes logical negation.)

Let S be a set on which an equivalence relation is defined, and let $a \in S$. The **equivalence class** of a, written $[a]$, is the set of elements that are equivalent to a:

$$[a] := \{b \in S \mid b \equiv a\}. \tag{2.82}$$

Any element that is equivalent to a is called a **representative** of the equivalence class $[a]$.

Two equivalence classes either are identical or have no elements in common. To prove this, suppose that $x \in [a]$ and $x \in [b]$. Then $a \equiv x$ and $x \equiv b$. For every $y \in [a]$, $y \equiv a$; therefore (by transitivity) $y \equiv x$ and $y \equiv b$. Hence $y \in [b]$. It follows that $[a] \subseteq [b]$. One shows similarly that $[b] \subseteq [a]$. Therefore $[a] = [b]$.

The even and odd integers provide one of the simplest examples of equivalence classes. If we make the definition that $a \equiv b$ if and only if a and b are integers that give the same remainder after division by 2, then $[0]$ contains all of the even integers and $[1]$ contains all of the odd integers. We take up this example again as a special case of an equivalence relation called *congruence* in Chapter 4. Other familiar examples of equivalence relations include similarity and congruence in plane and solid geometry.

The real usefulness of an equivalence relation lies in the fact that one can replace equality by equivalence whenever one obtains a mathematical result through a derivation that makes use only of the reflexivity, symmetry, and transitivity of the equality relation. With respect to these three basic properties, equivalence is "just as good as" equality.

The definition of a rational number provides an example of an equivalence relation at work. A **rational number** is an equivalence class (p, q), defined as the set of all ordered pairs of integers (m, n) (in the order [numerator/denominator]) such that the denominator is nonzero, under the equivalence relation

$$(p/q) \equiv (m, n) \Leftrightarrow mq = np. \tag{2.83}$$

Thus, for every nonzero integer q and every integer p,

$$(p/q) = \{(m, n) \in \mathbb{Z}^2 \mid n \neq 0 \text{ and } mq = np\}. \tag{2.84}$$

It is easy to verify that Eq. (2.83) defines an equivalence relation (see Exercise 2.2.4).

The motivation behind Eq. (2.84) is that we want (for example) $2/4$, $3/6$, and $32768/65536$ to define the same rational number, namely, $1/2$. With the definition we have just given, any numerator/denominator pair (m, n) such that $n = 2m$ is a representative of the equivalence class $(1/2)$. Of course, if the only ordered pairs of interest were quotients of integers, one could use numerical equality (which is an equivalence relation!) and forget about the equivalence relation defined in Eq. (2.83). We show in Section 4.5 that Eq. (2.83) permits one to define quotients of elements of a much more general structure than $\mathbb{Z}$ called an *integral domain*. The addition and multiplication of rational numbers defined as equivalence classes are discussed in Section 4.5.

2.2.6 EXERCISES FOR SECTION 2.2

2.2.1 Point out the contradiction that is inherent in the following statement: "Let S be the set of all sets that are not elements of themselves."

[*Hint:* Either $S \in S$ or $S \notin S$, but not both!]

2.2.2 Write down all of the possible one-to-one mappings of $S = \{a, b, c\}$ onto $U = \{g, h, i\}$, using the notation of Eq. (2.57).

2.2.3 Let X be a set whose elements are sets. Prove that the equivalence of sets in X (as defined in Eq. [2.78]) is an equivalence relation as defined in Eqs. (2.79–2.81).

2.2.4 Prove that the definition
$$(m, n) \equiv (p, q) \Leftrightarrow mq = np, \tag{2.85}$$
in which $n \neq 0$ and $q \neq 0$, satisfies the three axioms for an equivalence relation.

2.2.5 Let the mappings $\phi : S \rightarrow T$ and $\chi : T \rightarrow U$ be one-to-one.

1. Prove that the product mapping $\chi \circ \phi : S \rightarrow U$ is also one-to-one.

2. Let $I := \chi(\phi(S))$ be the image of S under $\chi \circ \phi$. Prove that on I, the inverse of $\chi \circ \phi$ is
$$(\chi \circ \phi)^{-1} = \phi^{-1} \circ \chi^{-1}. \tag{2.86}$$

2.2.6 Prove that all lists with n elements are equivalent, where n is a fixed positive integer.

2.2.7 Let each of the mappings $\phi : S \rightarrow T$, $\chi : T \rightarrow U$, and $\psi : U \rightarrow V$ be one-to-one. Prove that each of the mappings ϕ^{-1}, χ^{-1}, ψ^{-1}, $\xi = \phi^{-1}\chi^{-1}$, and $\eta = \chi^{-1}\psi^{-1}$ is one-to-one.

[*Hint:* It may be helpful to refer to the commutative diagram Eq. (2.55).]

2.2.8 Calculate the following products of mappings:

(a)
$$\begin{pmatrix} d & e & f \\ g & h & i \end{pmatrix} \begin{pmatrix} a & b & c \\ f & e & d \end{pmatrix} \tag{2.87}$$

(b)
$$\begin{pmatrix} d & e & f \\ i & h & g \end{pmatrix} \begin{pmatrix} a & b & c \\ f & e & d \end{pmatrix} \tag{2.88}$$

(c)
$$\begin{pmatrix} d & e & f \\ h & g & i \end{pmatrix} \begin{pmatrix} a & b & c \\ f & d & e \end{pmatrix} \tag{2.89}$$

2.2.9   Assume that S and T are sets and that $\phi : S \rightarrow T$ is a mapping.

(a) Prove that ϕ is one-to-one if and only if the power-set mapping $\hat{\phi}$ defined in Eq. (2.38) is one-to-one.

(b) Prove that ϕ is bijective if and only if the power-set mapping $\hat{\phi}$ defined in Eq. (2.38) is bijective.

2.3 UNION, INTERSECTION, AND COMPLEMENT

2.3.1 UNIONS OF SETS

There are several basic ways to make new sets out of already existing sets. The **union**

$$S_1 \cup S_2 := \{x \mid x \in S_1, \text{ or } x \in S_2, \text{ or both}\} \tag{2.90}$$

of two sets S_1 and S_2 is the set of all elements that belong to one or both of the sets S_1 and S_2. If two subsets A and B of S are defined by the specifications $a(x)$ and $b(x)$, respectively, then $A \cup B$ is the subset of $x \in S$ on which either $a(x)$ is true, or $b(x)$ is true, or both. Hence the union of two subsets of S corresponds to the logical inclusive "or" of two propositions.

Obviously the operation of taking the union is commutative,

$$S_1 \cup S_2 = S_2 \cup S_1, \tag{2.91}$$

and associative,

$$S_1 \cup (S_2 \cup S_3) = (S_1 \cup S_2) \cup S_3. \tag{2.92}$$

For this reason the union of sets is sometimes called the **sum** of sets. For example,

$$\{a, b, \} \cup \{c, d\} = \{a, b, c, d\}. \tag{2.93}$$

If $\mathfrak{C}$ is any set whose elements are sets, then the union

$$\bigcup_{S \in \mathfrak{C}} S \tag{2.94}$$

of all the sets in $\mathfrak{C}$ is the set that consists of all elements that belong to at least one set in $\mathfrak{C}$. According to this definition, elements that are common to more than one set of $\mathfrak{C}$ are *not* repeated in the union given in Eq. (2.94). For example,

$$\{a, b, c\} \cup \{b, c, d\} = \{a, b, c, d\}. \tag{2.95}$$

The **Venn diagram** in Fig. 2.7 shows the union of two sets represented as partially overlapping shaded disks.

FIGURE 2.7

Illustration of the union of two sets S_1 and S_2, both represented as shaded disks. Each point that is common to both sets occurs only once, not twice, in $S_1 \cup S_2$.

2.3.2 INTERSECTIONS OF SETS

The **intersection** of two sets,

$$S_1 \cap S_2 := \{x \mid x \in S_1 \text{ and } x \in S_2\}, \tag{2.96}$$

is the set of elements that belong to both S_1 and S_2. Similarly, the intersection

$$\bigcap_{S \in \mathfrak{C}} S \tag{2.97}$$

of the sets in $\mathfrak{C}$ is the set of elements that belong to every set in $\mathfrak{C}$. For example, if $S \subseteq T$, then $S \cap T = S$. In the Venn diagram in Fig. 2.8, the dark area represents the intersection of the shaded disks.

The intersection of two subsets S_1 and S_2 of S corresponds to the logical "and" of two statements. Let S_1 and S_2 be two subsets of S, and let s_1 and s_2 be the specification statements that define S_1 and S_2, respectively. Then $x \in S_1 \cap S_2$ if and only if $s_1(x)$ and $s_2(x)$ are both true.

Sometimes it is useful to have notations other than Eq. (2.94) or Eq. (2.97) for the union and intersection of a family of sets. If there is a one-to-one correspondence between all the elements of an **index set** I and all the elements of a set $\mathfrak{C}$ whose elements are sets, then one can **index** the elements of $\mathfrak{C}$ by the elements of I:

$$\mathfrak{C} = \{S_i \mid i \in I\}. \tag{2.98}$$

The index notation simply means that $\mathfrak{C}$ is the range of a certain bijective function whose domain is I.

Indexing can be used for the elements of any set S. An index set I may be *any* nonempty set that is the domain of a bijective mapping onto S, but most frequently we shall use $I = (1 : n)$

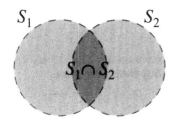

FIGURE 2.8

The dark area is the intersection of the two sets shown as shaded disks.

or $I = \mathbb{Z}^+$, in which $\mathbb{Z}^+$ is the **set of positive integers**. Other index sets such as the set of real numbers, $\mathbb{R}$, occur rather frequently in physics and engineering. For example, a plane wave $\psi_k(x) := e^{ikx}$ can be indexed by the real number k. Sometimes an index mapping $I \to S$ is called a **parametrization** of S.

2.3.3 RELATIVE COMPLEMENT

The **complement of S relative to T**,

$$T \setminus S := \{x \in T \mid x \notin S\}, \tag{2.99}$$

is the set of elements that belong to T but not to S. For example, $\{a, b\} \setminus \{b, c\} = \{a\}$. The relative complement $T \setminus S$ of $S = \{x \mid s(x)\}$ corresponds to the logical negation of $s(x)$ because $T \setminus S$ is the set of elements x in T for which $s(x)$ is not true.

One sees also from the definitions of a subset and of the complement of a set relative to S that if A and B are two nonempty subsets of S then

$$B \subseteq A \Leftrightarrow S \setminus A \subseteq S \setminus B. \tag{2.100}$$

This is equivalent to the familiar rule of logic that "$a(x)$ implies $b(x)$" is true if and only if the **contrapositive** "(not $b(x)$) implies (not $a(x)$)" is true.

The **symmetric difference** of two sets S and T,

$$S \Delta T := (S \setminus T) \cup (T \setminus S), \tag{2.101}$$

corresponds to the logical "exclusive or" of the defining specification statements. For example, $\{a, b\} \Delta \{b, c\} = \{a, c\}$.

From the definitions of the intersection and relative complement it follows (see Exercises 2.3.1 through 2.3.3) that

$$T \setminus (T \setminus S) = S \cap T. \tag{2.102}$$

Therefore

$$S \subseteq T \Leftrightarrow S = T \setminus (T \setminus S) \tag{2.103}$$

and

$$T \setminus S = T \Leftrightarrow T \cap S = \emptyset. \tag{2.104}$$

(See the Venn diagram in Fig. 2.9.)

FIGURE 2.9

In this diagram, the sets S_1 and S_2 are the same as in Figs. 2.7 and 2.8. The complement $S_1 \setminus S_2$ of S_2 relative to S_1 is the shaded area.

2.3.4 DE MORGAN'S LAWS

Suppose that A and B are subsets of a set S. Important relations known as **De Morgan's laws** tell one how to express the complements of the union and intersection of A and B in terms of the complements of A and B relative to S:

$$S \setminus (A \cup B) = (S \setminus A) \cap (S \setminus B) \tag{2.105}$$

$$S \setminus (A \cap B) = (S \setminus A) \cup (S \setminus B). \tag{2.106}$$

The logical relations

$$\neg(a \text{ or } b) = (\neg a) \text{ and } (\neg b) \tag{2.107}$$

and

$$\neg(a \text{ and } b) = (\neg a) \text{ or } (\neg b) \tag{2.108}$$

follow directly from De Morgan's laws. The proofs of Eqs. (2.105) and (2.106) are left as Exercise 2.3.4.

2.3.5 EXERCISES FOR SECTION 2.3

2.3.1 Prove Eq. (2.102).

2.3.2 Prove Eq. (2.103).

2.3.3 Prove Eq. (2.104).

2.3.4 Prove De Morgan's laws, Eqs. (2.105) and (2.106).

2.3.5 Let I be an index set, let A_i be a set (for every $i \in I$), let S be a set, and assume that

$$\forall i \in I : A_i \subseteq S. \tag{2.109}$$

Prove that

$$S \setminus \left(\bigcup_{i \in I} A_i \right) = \bigcap_{i \in I} (S \setminus A_i) \tag{2.110}$$

and

$$S \setminus \left(\bigcap_{i \in I} A_i \right) = \bigcup_{i \in I} (S \setminus A_i). \tag{2.111}$$

2.3.6 Verify that Eq. (2.111) is satisfied if $S = \mathbb{R}$ and

$$A_i = \left(-\frac{1}{i}, \frac{1}{i} \right) \tag{2.112}$$

in which $i \in \mathbb{Z}^+$. By definition, the **open interval** bounded by a and $b > a$ is

$$(a, b) := \{ r \in \mathbb{R} \mid a < r < b \}. \tag{2.113}$$

2.3.7   Express the contrapositive of the assertion

$$(a \text{ and } b) \Rightarrow c \tag{2.114}$$

in terms of the negations $\neg a$ and $\neg b$.

2.4 INFINITE SETS

2.4.1 BASIC PROPERTIES OF INFINITE SETS

If a set S is not finite, it is called **infinite**. We deduce a characteristic property of infinite sets by seeing what happens if a finite set is equivalent to one of its subsets. Let S be finite and let $R \subseteq S$. If $R \sim S$, then there exists a one-to-one mapping χ of S onto R such that

$$\forall x \in R : \exists y \in S : \ni : x = \chi(y). \tag{2.115}$$

Using the notation of Eq. (2.45), we let

$$x_i = \chi(y_i). \tag{2.116}$$

Because $R \subseteq S$, there exists an integer $j \in (1 : n)$ such that

$$y_j = x_i. \tag{2.117}$$

Because domain $[\chi] = S$ and because χ is one-to-one and is defined on every element of S, $\chi(y_i)$ takes on n different values for $1 \leq i \leq n$. If some y_k is not the image of any y_i, then there are at most $n - 1$ distinct images under χ (instead of n), contradicting the assumption that χ is one-to-one. Then every $y_j \in S$ is the image of some y_i under χ, which implies that $y_j \in R$ and therefore that $S \subseteq R$. We have therefore shown (via a slightly formalized commonsense argument) that for a finite set S and one of its subsets R,

$$R \sim S \Rightarrow R = S. \tag{2.118}$$

The converse of Eq. (2.118) is obvious.

Therefore, for a finite set S, a subset R of S is equivalent to S if and only if $R = S$. We show in Eq. (2.100), and it is familiar from elementary logic, that "$a \Rightarrow b$" is equivalent to the contrapositive "$\neg b \Rightarrow \neg a$," in which "$\neg a$" means "not a." We have just shown that "S finite $\Rightarrow \neg(R \subset S \text{ and } R \sim S)$," which is equivalent to "$(R \subset S \text{ and } R \sim S) \Rightarrow \neg(S \text{ finite})$." In other words, a set S is infinite if a proper subset of S is equivalent to S. We show subsequently that if S is infinite, then there exists a proper subset R of S such that $R \sim S$. Therefore *a set S is infinite if and only if a proper subset of S is equivalent to S*. This proposition is sometimes called the **Dedekind–Peirce corollary**.

For example, the mapping

$$\forall m \in \mathbb{Z}^+ : \phi(m) = mn \tag{2.119}$$

(in which n is a fixed nonzero integer) is one-to-one and maps the set of **positive integers**

$$\mathbb{Z}^+ := \{1, 2, \ldots\} \tag{2.120}$$

onto the set of positive multiples of n,

$$n\mathbb{Z}^+ := \{n, 2n, \ldots\}. \tag{2.121}$$

But

$$n\mathbb{Z}^+ \subset \mathbb{Z}^+. \tag{2.122}$$

Indeed, $\mathbb{Z}^+$ is infinite because there is no *one-to-one* mapping of $\mathbb{Z}^+$ onto $(1:n)$ for any $n \in \mathbb{Z}^+$.

2.4.2 INDUCTION AND RECURSION

Sequences

A set S, together with a function from Z to S, in which Z is equivalent to the set of integers $\mathbb{Z}$, is called a **sequence**. Thus, using the notation defined in Eq. (2.37), one can say that a sequence belongs to S^Z. Examples include $S^{\mathbb{Z}}$, $S^{\mathbb{N}}$, and $S^{\mathbb{Z}^+}$. It is usually convenient to write a sequence S in a form that exhibits the mapping from $\mathbb{Z}^+$ (for example) into S, such as

$$(x_1, x_2, \ldots, x_i, \ldots) \quad \text{where} \quad \forall i \in \mathbb{Z}^+ : x_i \in S. \tag{2.123}$$

One calls $x_1, x_2, \ldots$ the **terms** of the sequence. There is no requirement that the terms of a sequence must be distinct; for example, $(1, 1, \ldots)$, every term of which is equal to 1, is a legitimate sequence.

Finite Induction

Sometimes the terms of a sequence are statements (or propositions). Let $x_n = p(n)$, in which p is a proposition that is either true or false given a natural number n. The **principle of finite induction** asserts that

$$\left. \begin{array}{c} p(1) \text{ is true, and} \\ \forall n \in \mathbb{Z}^+ : p(n) \Rightarrow p(n+1) \end{array} \right\} \Rightarrow \forall n \in \mathbb{Z}^+ : p(n) \text{ is true.} \tag{2.124}$$

Note that in order to execute a proof by induction, one must show both that $p(1)$ is true and that $p(n)$ implies $p(n+1)$ for *every* $n \in \mathbb{Z}^+$. The principle of finite induction remains valid if one replaces the set of positive integers $\mathbb{Z}^+$ with the set of natural numbers $\mathbb{N}$. In that case, one begins the inductive sequence with $p(0)$ instead of with $p(1)$, replacing n by $n-1$ everywhere in Eq. (2.124). In general, one can begin the inductive process at any integer n_0, as long as one can prove that $p(n-1) \Rightarrow p(n)$ for every $n > n_0$, and as long as one wants to show that $p(n)$ is true for $n \geq n_0$.

Proof by induction is sometimes an easy way to establish a formula. To prove the standard formula for the sum of a geometric progression, let $p(n)$ be the proposition that

$$x_n = \frac{a - ar^{n+1}}{1 - r} \tag{2.125}$$

in which

$$x_n := a + \sum_{k=1}^{n} ar^k. \tag{2.126}$$

For $n = 1$, this formula reads

$$x_1 = \frac{a - ar^2}{1 - r} = \frac{a(1 - r^2)}{1 - r} = a + ar, \tag{2.127}$$

which is the definition of x_1. Assume that $p(n)$ is true. Then

$$x_{n+1} = x_n + ar^{n+1} = \frac{a - ar^{n+1}}{1 - r} + ar^{n+1} = \frac{a - ar^{n+2}}{1 - r}, \tag{2.128}$$

which is $p(n + 1)$. Because $p(1)$ is true and $p(n)$ implies $p(n + 1)$ for all $n \in \mathbb{Z}^+$, Eq. (2.125) holds for all $n \geq 1$.

Exercises 2.4.1 through 2.4.5 offer further opportunities to practice proof by induction.

Well-Ordering Principle

We illustrate the power of the principle of finite induction by using it to prove the **well-ordering principle** for the set $\mathbb{Z}^+$: *In every nonempty subset R of $\mathbb{Z}^+$, there exists a first (or smallest) element $m \in R$ such that*

$$\forall n \in R : m \leq n. \tag{2.129}$$

Proceeding by contradiction, we assume that R has no first element and let $p(n)$ be the proposition

$$p(n) := [\forall i \in R : n \leq i]. \tag{2.130}$$

The idea is to use induction to show that $p(n)$ is true for all n and then to obtain a contradiction to show that R has a first element after all. Certainly $p(0)$ is true. Suppose that $p(n)$ is true. That means that if $n \in R$, then n is a first element of R, contradicting our assumption that there is no first element of R. Hence $n \notin R$. Therefore there is no $m \in R$ such that $n = m$. With the hypothesis that $p(n)$ is true, it follows that $\forall m \in R : n < m$, and therefore that $\forall m \in R : n + 1 \leq m$. But the last assertion is $p(n + 1)$. Because we have shown that $p(0)$ is true and that $p(n)$ implies $p(n + 1)$, it follows that $p(n)$ is true for all $n \in \mathbb{Z}^+$. Now we close the trap: Because R is nonempty, there exists a natural number $j \in R$. We have just shown that $p(j + 1)$ is true, but then we have the contradiction $j + 1 \leq j$. Therefore R has a first element i. By repeating this argument with the set of elements of R that are not equal to the first element i, we arrange the elements $i, j, k, \ldots$ of any subset of $\mathbb{Z}^+$ such that $i < j < k \ldots$.

Recursive Definitions

Quite often the terms of a sequence are defined **recursively**, meaning that the value of x_{n+1} is defined in terms of the values of x_k, where $k \leq n$. For the sake of simplicity, let us consider sequences belonging to $\mathbb{C}^{\mathbb{Z}^+}$; then the terms of the sequence, x_k, are complex numbers, and $k \geq 1$. (If the notation $\mathbb{C}^{\mathbb{Z}^+}$ is unfamiliar, please see Eq. [2.37].) The principle of finite induction guarantees that if the value of x_1 is known or defined in a way that does not depend on any of the other terms of the sequence, and if a unique value can be computed for x_{n+1} given the values of $x_1, \ldots, x_n$, then the value of x_{n+1} is defined uniquely. We present many important examples of recursive computation, including numerical integration and the solution of ordinary differential equations, in Sections 3.3, 3.4, and 3.5.

For example, the factorial function $x_n = n!$ is defined by the two-term recurrence relation

$$x_{n+1} = (n+1)x_n \tag{2.131}$$

together with the **initial value** or **base value**

$$x_1 = 1. \tag{2.132}$$

To give another example: One gets a two-term recurrence relation,

$$x_{n+1} = x_n + a_{n+1}, \tag{2.133}$$

if x_n is a partial sum of a series,

$$x_n := \sum_{i=1}^{n} a_i. \tag{2.134}$$

The initial value $x_1 = a_1$ is implicit in the definition of x_n as the partial sum of n terms.

Equation (1.9) defines the digits of the base-β representation of an integer k recursively. In that example, Eq. (1.8) defines the initial value.

Very common, and less trivial than two-term recurrence relations, are three-term recurrence relations such as

$$T_{n+1}(\theta) - 2\cos\theta \, T_n(\theta) + T_{n-1}(\theta) = 0, \tag{2.135}$$

the solution of which is defined uniquely if two initial values are given, such as T_0 and T_1. For example, Eq. (2.135) is satisfied by

$$T_n(\theta) = \cos n\theta \tag{2.136}$$

if the initial values are

$$T_0(\theta) = \cos 0 = 1, \qquad T_1(\theta) = \cos\theta. \tag{2.137}$$

However, Eq. (2.135) is also satisfied by the sequence

$$U_n(\theta) = \frac{\sin n\theta}{\sin\theta} \tag{2.138}$$

if the initial values are $U_0(\theta) = 0$, $U_1(\theta) = 1$. In Section 8.2.3 we show that T_n is a Chebyshev polynomial. For now one can simply think of the three-term recurrence relation for T_{n+1} as an inductive algorithm to compute $\cos(n+1)\theta$ in terms of $\{\cos m\theta \mid 0 \le m < n+1\}$.

The computational implementation of a recursive definition such as Eq. (2.133) should be considered rather carefully, as it is easy to lose significant digits through cancellation in evaluating many consecutive recurrence relations. Section 3.3 discusses some of the practical aspects of evaluating the partial sums of a series. Three-term recurrence relations such as Eq. (2.135) can be unstable.

2.4.3 COUNTABLE SETS

Cantor's concept of *countability* is a straightforward generalization of the everyday idea of counting the elements of a finite set by "labeling" them with $1, 2, \ldots, n$. An *infinite* set S such that there exists a bijective function from S to $\mathbb{Z}^+$ is called **countable** (or **denumerable**).

If an infinite set S is not countable, it is called **uncountable**. If a set S is either finite or countable, then S is called **at most countable**. (One frequently sees "countable" used to mean "at most countable.") It follows from these definitions that if a set T is countable, then its elements constitute the terms of a sequence. For the purposes of this book, the terms "finite," "countable," and "uncountable" are mutually exclusive. Despite the normal conventions of the English language, "uncountable" does not mean "not countable" because a set is not countable if it is either finite or uncountable.

For example, the set of all integers, $\mathbb{Z}$, is countable (see Exercise 2.4.6). The set $n\mathbb{Z}^+$ of all positive integral multiples of an integer n defined in Eq. (2.121) is countable. We show subsequently that the set of real numbers, $\mathbb{R}$, is uncountable.

We can now make a statement that partially characterizes infinite subsets of an infinite set: *Every infinite set includes a countable subset.* For example, the set $\mathbb{Z}^+$ of positive integers is a (proper) subset of the set of all real numbers, $\mathbb{R}$.

To begin the proof, let S be an infinite set. We shall show that S includes a countable subset. The axiom of choice tells us that we can choose an element $x_1 \in S$, then $x_2 \in S$ (in which $x_2 \neq x_1$), and so forth, thereby constructing the subset $T := \{x_1, x_2, \ldots\} \subseteq S$. If we were to exhaust S after n steps, where n is some finite positive integer, then S would be finite, contradicting our initial assumption. Then T must be countable, not finite.

The result that every infinite set includes a countable subset implies that countable sets are the "smallest" infinite sets. In order to make this statement rigorous, we would have to discuss the theory of cardinal numbers. (Kronecker criticized this aspect of set theory so strongly that Cantor's contemporaries were reluctant to accept his ideas. Decades later, Hilbert praised Cantor's theory of cardinal numbers as one of his greatest contributions, thereby paving the way for a general acceptance of Cantor's work.)

We are now in a position to finish the proof of the Dedekind–Peirce corollary (see Section 2.4.1). We show that if a set S is infinite, then there exists a proper subset of S that can be put into one-to-one correspondence with S. Because S is infinite, it includes a countable subset $C = \{c_1, c_2, \ldots\}$. Let ϕ be the mapping such that

$$\forall i \in \mathbb{Z}^+ : \phi(c_i) := c_{i+1} \quad \text{and} \quad \forall s \in S : \ni : s \notin C : \phi(s) := s. \tag{2.139}$$

Then ϕ is one-to-one and maps S onto a proper subset of itself because c_1 is not the image of any element of S under ϕ. In other words, there are more elements in S than the elements of the range of ϕ.

If a set S is countable, what about its subsets? Let the infinite set R be a subset of the countable set $S = \{x_1, x_2, \ldots\}$. Using the well-ordering principle, arrange the elements of $R = \{x_i, x_j, x_k, \ldots\}$ so that $i < j < k < \ldots$. Let n_1 be the first integer n such that $x_n \in R$, n_2 the second such integer, and so forth. Then $R = \{x_{n_1}, x_{n_2}, \ldots\}$ is countable because we have constructed a one-to-one mapping from $\mathbb{Z}^+$ onto R. Thus every infinite subset of a countable set is countable. (In particular, if the set S of distinct terms of a sequence is infinite, then S is countable.)

It follows that *if a subset R of a set S is uncountable, then S is uncountable.* (This statement is the contrapositive of the statement that every infinite subset of a countable set is countable; recall Eq. [2.100].) For example, we show in Section 2.4.5 that the **closed interval**

$$[a, b] := \{r \in \mathbb{R} \mid a \le r \le b\} \tag{2.140}$$

is uncountable; clearly $[a, b]$ is a proper subset of the uncountable set of all real numbers, $\mathbb{R}$.

Are there any countable sets other than $\mathbb{Z}^+$, $\mathbb{Z}$, and their infinite subsets? Cantor answered this question in the affirmative by establishing the remarkable result that the set of **rational numbers**, $\mathbb{Q}$, is countable. (Recall that a rational number is an equivalence class of ordered pairs of integers.) Define an array such that the quotient p/q that represents a particular equivalence class in $\mathbb{Q}$ occurs at the intersection of the p-th row and the q-th column:

$$
\begin{array}{llll}
\frac{1}{1} & \frac{1}{2} & \frac{1}{3} & \cdots \\
\frac{2}{1} & \frac{2}{2} & \frac{2}{3} & \cdots \\
\frac{3}{1} & \frac{3}{2} & \frac{3}{3} & \cdots \\
\vdots & & & \ddots
\end{array}
\tag{2.141}
$$

Along the minor diagonal that runs through the p, q intersection, $p+q$ is constant. (Consider the minor diagonal that runs from $1/(p + q - 1)$ to $(p + q - 1)/1$.) To set up a one-to-one correspondence between $\mathbb{Z}^+$ and the elements of the array, we traverse the array along the minor diagonals in the order of increasing values of $p + q$, discarding all fractions that are not in lowest terms (i.e., discarding p/q if q is not relatively prime to p) in order to select a unique representative of each equivalence class. In this way we obtain the sequence $\{1/1 = 1, 1/2, 2/1 = 2, 1/3, 3/1 = 3, \ldots\}$. Every rational number occurs in the array in Eq. (2.141). If we choose a particular rational number, that number is reached in a finite number of steps. This construction results in a one-to-one mapping of the equivalence classes in $\mathbb{Q}$ onto the positive integers $\mathbb{Z}^+$; therefore $\mathbb{Q}$ is countable.

2.4.4 COUNTABLE UNIONS AND INTERSECTIONS

The **union and intersection of a countable collection of sets**,

$$\{S_n \mid n \in \mathbb{Z}^+\}, \tag{2.142}$$

are denoted

$$\bigcup_{n=1}^{\infty} S_n \tag{2.143}$$

and

$$\bigcap_{n=1}^{\infty} S_n, \tag{2.144}$$

respectively. If every set S_n in the collection $\{S_n\}$ is countable, then the **countable union of countable sets**

$$S = \bigcup_{n=1}^{\infty} S_n \tag{2.145}$$

is also countable. To prove this assertion, we recall that because each S_n is countable, its elements can be arranged in a sequence:

$$S_n = (x_{n1}, x_{n2}, \ldots). \tag{2.146}$$

Every element of S occurs at least once in the array

$$
\begin{matrix}
x_{11} & x_{12} & x_{13} & \cdots \\
x_{21} & x_{22} & x_{23} & \cdots \\
x_{31} & x_{32} & x_{33} & \cdots \\
\vdots & \vdots & \vdots & \ddots
\end{matrix}
\tag{2.147}
$$

We can therefore construct a one-to-one mapping of $\mathbb{Z}^+$ onto S in almost exactly the same way in which we proved the countability of the rational numbers. Define the "diagonal" subsets

$$D_k := \{x_{ij} \mid i + j = k, \ k = 2, 3, \ldots\}. \tag{2.148}$$

Each D_k is finite (D_k has $k - 1$ elements). To map $\mathbb{Z}^+$ onto S, we let $1 \mapsto x_{11}$ and then choose elements from D_3 that are different from x_{11} until D_3 is exhausted, then choose elements from D_4 that are different from any of the elements that have already been chosen, and so forth.

2.4.5 UNCOUNTABLE SETS

In view of the countability of $\mathbb{Q}$ and even of a countable union of countable sets, it is important to know that **uncountable sets** exist. Cantor proved that the set of **real numbers**, $\mathbb{R}$, is uncountable. Because every infinite proper subset of a countable set is countable, we can show that $\mathbb{R}$ is uncountable by showing that a proper subset of $\mathbb{R}$ is uncountable. Therefore it is enough for our purposes to show that the set of all real numbers r such that $0 \leq r \leq 1$ (i.e., such that $r \in [0, 1]$) is uncountable. In order to do so, we assume that every infinite decimal fraction that begins with $+0.$ represents a unique real number in $[0, 1]$:

$$0.a_1 a_2 a_3 \cdots \mapsto r \in [0, 1]. \tag{2.149}$$

We also assume that *every* real number in $[0, 1]$ can be represented (not necessarily uniquely) by such a decimal fraction. Let D be the set of all decimal fractions that begin with $+0$. In order to establish a one-to-one correspondence between D and $[0, 1]$, we must eliminate all but one of the decimal fractions (such as $0.500\cdots$ and $0.499\cdots$) that represent the same real number. Therefore we define the mapping from $[0, 1]$ to D in such a way that every real number that can be exactly represented by a decimal fraction of the form $0.a_1a_2\cdots a_{n-1}a_n000\cdots$ is represented instead by a decimal fraction of the form $0.a_1a_2\cdots a_{n-1}(a_n - 1)999\cdots$. Now suppose (for the sake of obtaining a contradiction) that D is countable:

$$D = \{r_1, r_2, \ldots\}. \tag{2.150}$$

Define an array that consists of the decimal fractions that belong to D listed in the order in which they occur in the above sequence:

$$\begin{aligned}
r_1 &= 0.d_{11}d_{12}d_{13}\cdots \\
r_2 &= 0.d_{21}d_{22}d_{23}\cdots \\
r_3 &= 0.d_{31}d_{32}d_{33}\cdots \\
&\vdots \qquad\qquad \ddots
\end{aligned} \tag{2.151}$$

Then construct a decimal fraction

$$t = 0.b_1b_2b_3\cdots \tag{2.152}$$

by defining the digit b_n as follows:

$$b_n = \begin{cases} d_{nn} - 1, & \text{if } d_{nn} \neq 0; \\ 1, & \text{if } d_{nn} = 0. \end{cases} \tag{2.153}$$

By assumption, t represents a unique real number in $[0, 1]$. By construction, t differs from every element of D in at least one decimal digit, and no digit of t is equal to 9. Hence the real number that t represents is not equal to any of the real numbers represented by elements of D. The observation that $0 < t < 1$ completes the contradiction with our assumption that the countable set D contains a representative of every real number in $[0, 1]$.

We use the result that the set of real numbers in the interval $[0, 1]$ is uncountable to show that the Cantor set

$$K := \left\{ r \in \mathbb{R} \,\middle|\, r = \sum_{j=1}^{\infty} \frac{d_j}{3^j} : \ni: \forall\, j : d_j = 0 \text{ or } d_j = 2 \right\} \tag{2.154}$$

is uncountable. (Recall that a real number r belongs to K if and only if $r \in [0, 1]$ and the base-3 expansion of r contains only 0s and 2s.) Now we define a one-to-one mapping $\phi : K \to [0, 1]$:

$$\forall\, r \in K : \phi(r) := \sum_{j=1}^{\infty} \frac{d'_j}{2^j}, \qquad d'_j := \begin{cases} 0 & \text{if } d_j = 0 \\ 1 & \text{if } d_j = 2. \end{cases} \tag{2.155}$$

All we had to do to define ϕ was to change the base from 3 to 2 and change all the 2s to 1s! Clearly ϕ is one-to-one. The images of ϕ represent every point in $[0, 1]$ through a binary (base-2) expansion. Hence K is equivalent to $[0, 1]$ and is therefore uncountable.

Now that we have seen Cantor's potent idea of "taking apart" the floating-point expansion of a real number digit-by-digit in an invertible way, we can prove the highly nonintuitive theorem that a unit square is equivalent to a unit line segment:

$$[0, 1] \times [0, 1] \sim [0, 1]. \tag{2.156}$$

Consider any element (r, s) of $[0, 1] \times [0, 1]$ in which

$$r = 0.a_1 a_2 a_3 \cdots, \qquad s = 0.b_1 b_2 b_3 \cdots \tag{2.157}$$

(in some base β) and define a new real number t by interleaving the digits of r and s (again in base β):

$$t := 0.a_1 b_1 a_2 b_2 a_3 b_3 \cdots. \tag{2.158}$$

Every real number t in $[0, 1]$ is the image of some ordered pair (r, s) because we can distribute alternate digits of t to r and s to obtain the (unique) preimage of t. Moreover, two ordered pairs are mapped onto the same t if and only if they are equal (so long as we resolve the possible nonuniqueness of the floating-point expansion as we have already done above). Therefore $[0, 1] \times [0, 1] \sim [0, 1]$.

2.4.6 EXERCISES FOR SECTION 2.4

2.4.1   Prove by induction:

(a)

$$\sum_{i=1}^{n} i = \frac{n(n+1)}{2}. \tag{2.159}$$

(b)

$$\sum_{i=1}^{n} i^2 = \frac{n(n+1)(2n+1)}{6}. \tag{2.160}$$

(c)

$$\sum_{i=1}^{n} i^3 = \left[\frac{n(n+1)}{2}\right]^2. \tag{2.161}$$

2.4.2   The symbol

$$\binom{n}{r} := \frac{n!}{r!\,(n-r)!} \tag{2.162}$$

is called a **binomial coefficient**.

(a) Prove the following addition theorem for binomial coefficients:

$$\forall r = 1, \ldots, n : \binom{n+1}{r} = \binom{n}{r-1} + \binom{n}{r}. \tag{2.163}$$

What construction for $\binom{n}{r}$ does Eq. (2.163) imply if the binomial coefficients are arranged in an array such that $\binom{n}{r}$ falls in (horizontal) row n and (vertical) column r?

(b) Using the principle of finite induction, prove the **binomial theorem,**

$$(x+y)^n = \sum_{r=0}^{n} \binom{n}{r} x^r y^{n-r}. \tag{2.164}$$

2.4.3   Using the principle of finite induction, prove **Leibniz's formula for the** n-th **derivative of a product** of two n-times-differentiable functions u and v,

$$\frac{d^n(uv)}{dx^n} = \sum_{r=0}^{n} \binom{n}{r} \frac{d^r u}{dx^r} \frac{d^{n-r} v}{dx^{n-r}}. \tag{2.165}$$

2.4.4   Let $\phi_1 : S_1 \to S_2, \ldots, \phi_n : S_n \to S_{n+1}, \ldots$ be mappings. Using the principle of finite induction, prove that

$$\forall n \in \mathbb{Z}^+ : (\phi_1 \cdots \phi_n)^{-1} = \phi_n^{-1} \cdots \phi_1^{-1}. \tag{2.166}$$

2.4.5   Using the principle of finite induction, prove that

$$\sum_{k=0}^{K} \cos[(2k+1)\theta] = \frac{\sin[2(K+1)\theta]}{2\sin\theta}. \tag{2.167}$$

Hint: Use trigonometric identities to turn $\sin\theta \sum_{k=0}^{K} \cos[(2k+1)\theta]$ into a telescoping sum.

2.4.6   Prove that the set $\mathbb{Z} = \{\ldots, -2, -1, 0, 1, 2, \ldots\}$ of *all* integers is countable.

2.4.7   Hilbert's hotel has a countable collection of rooms. Assume that a countable set of guests has already checked into Hilbert's hotel, making the hotel "full" in the sense that every room is occupied.

(a) Show that the hotel can accommodate another guest who arrives later.

(b) Show that the hotel can even accommodate a countable set of later arrivals.

These results constitute the **Hilbert hotel paradox**.

2.4.8   Prove that $\mathbb{Q}^n$ is countable, where $\mathbb{Q}$ is the set of rational numbers and n is a positive integer.

2.4.9   Use the principle of finite induction to prove the following theorem about the base-2 (binary) representation of an integer $m \in \mathbb{N}$: If n bits (not counting leading zeroes) are necessary to represent $m \in \mathbb{N}$ in the base $\beta = 2$, then

$$n \geq \log_2(m+1). \tag{2.168}$$

2.5 ORDERED AND PARTIALLY ORDERED SETS

2.5.1 PARTIAL ORDERINGS

In some sets it makes sense to talk about an *ordering* or *partial ordering* of the elements of the set. If S is a set, a **partial ordering** on S is a binary (meaning two-element) relation (conventionally denoted $\prec$ and pronounced "precedes") that obeys the following axioms:

1. *Antisymmetry:* The statements

$$x \prec y, \qquad x = y, \qquad y \prec x \tag{2.169}$$

   are mutually exclusive (i.e., at most one of these statements can be true). Note that it is *not* assumed that any of these statements must apply to any given pair x, y. One speaks of a *partial* ordering because the definition does not require that an ordering relation must exist between every two elements x and y.

2. *Transitivity:*

$$x \prec y \text{ and } y \prec z \Rightarrow x \prec z. \tag{2.170}$$

If we compare these axioms with the axioms for an equivalence relation, we see that here the axiom of antisymmetry replaces the axioms of reflexivity and symmetry for an equivalence relation. Evidently the relation of proper inclusion is a partial ordering of the subsets of a set T. **A partially ordered set** (or **poset**) is one on which a partial ordering is defined.

For example, the set of real numbers, $\mathbb{R}$, is partially ordered if $\prec$ is defined as numerical inequality, $<$. For another example, consider the set

$$S = \{\emptyset, \{x\}, \{y\}, \{x, y\}\} \quad \text{in which } x \neq y. \tag{2.171}$$

If one takes $\prec$ to be set inclusion ($\subset$), it is easy to see with the help of a graph such as Fig. 2.10 that S is partially ordered. Note that the order relation $\subset$ is undefined between the two sets $\{x\}$ and $\{y\}$.

The symbol $\preceq$ means "precedes or is equal to."

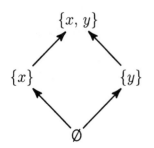

FIGURE 2.10

The set $\{\emptyset, \{x\}, \{y\}, \{x, y\}\}$ is partially ordered by inclusion. Arrows represent the direction of inclusion; for example, $\emptyset \subset \{x\}$. Inclusions that follow from transitivity, such as $\emptyset \subset \{x, y\}$, are not shown.

2.5.2 ORDERINGS; UPPER AND LOWER BOUNDS

A **chain** is a partially ordered set in which any two elements obey the following axiom:

3. *Comparability:* For every pair of elements x, y of S, one and only one of the statements Eq. (2.169) is true.

A partial ordering that obeys this axiom is called an **ordering**. A chain is also called an **ordered** (or **totally ordered**) **set**.

For example, $\mathbb{Z}$, $\mathbb{Q}$, and $\mathbb{R}$ are chains (i.e., ordered sets) if the order relation is defined to be numerical inequality, $<$.

In an ordered set it is possible to define a *bounded* subset. Let S be a subset of a partially ordered set T. If there exists an element $u \in T$ such that

$$x \in S \Rightarrow x \preceq u, \tag{2.172}$$

then u is an **upper bound** of S. One says equivalently that S is **bounded above**. If it is also true that

$$\forall z \in T : z \text{ is an upper bound of } S \Rightarrow u \preceq z, \tag{2.173}$$

then u is called the **least upper bound** (or **supremum**) of S, and is written

$$u = \text{lub } S = \text{sup } S. \tag{2.174}$$

If a least upper bound exists, it is unique. Similarly, if T is an ordered set, $l \in T$, $S \subseteq T$, and

$$x \in S \Rightarrow l \preceq x, \tag{2.175}$$

then l is called a **lower bound** of S. If, in addition,

$$\forall z \in T : z \text{ is a lower bound} \Rightarrow z \preceq l, \tag{2.176}$$

then l is called the **greatest lower bound** (or **infimum**) of S, and is denoted

$$l = \text{glb } S = \text{inf } S. \tag{2.177}$$

If S has a greatest lower bound, then the definition implies that inf S is unique. Finally, a subset S of an ordered set T is called **bounded** if and only if there exist (in T) an upper bound and a lower bound of S.

For example, if $T = \mathbb{R}$, if $\prec$ is numerical inequality $(<)$ and if

$$S := \left\{ \frac{1}{2^n} \,\middle|\, n \in \mathbb{Z}^+ \right\}, \tag{2.178}$$

then the supremum of S belongs to S,

$$\text{sup } S = \tfrac{1}{2}, \tag{2.179}$$

but the infimum of S belongs to T but not to S:

$$\inf S = 0. \tag{2.180}$$

To show that $\inf S = 0$, we observe that, for every positive integer n, $0 < 2^{-n}$ (therefore 0 is a lower bound of S), and that for every $\epsilon > 0$, no matter how small, there exists a positive integer n such that $2^{-n} < \epsilon$ (therefore no lower bound of S is greater than 0).

The discussion of metric spaces in Section 10.1 shows that it is possible to define bounded subsets of a set in which there is no natural ordering. However, constructs such as upper and lower bounds, which depend on the existence of an ordering, cannot be usefully generalized to an arbitrary metric space.

2.5.3 MAXIMAL CHAINS

A chain is called **maximal** if and only if it is not properly included in any other chain. For example, the set of ordered pairs with identical real elements,

$$X = \{(x, x) | x \in \mathbb{R}\}, \tag{2.181}$$

is a maximal chain in $\mathbb{R}^2$ if one defines the ordering

$$(x, x) \prec (x', x') \Leftrightarrow x < x'. \tag{2.182}$$

In Fig. 2.10, $\{\emptyset, \{x\}, \{x, y\}\}$ and $\{\emptyset, \{y\}, \{x, y\}\}$ are both maximal chains, but $\{\emptyset, \{x\}\}$ is a chain which is not maximal. **Zorn's lemma** asserts that in every nonempty partially ordered set in which every chain has an upper bound there exists at least one maximal chain. (See Halmos's book [1960] for a proof.)

Zorn's lemma implies that the **ordered set of real numbers** (in which the order $\prec$ is numerical inequality, $<$) has the following fundamental property: *If $S \subseteq \mathbb{R}$ is nonempty and is bounded above, then S has a least upper bound (supremum).* It follows directly by letting $x \mapsto -x$ that *if $S \subseteq \mathbb{R}$ is nonempty and is bounded below, then S has a greatest lower bound (infimum).*

2.5.4 EXERCISES FOR SECTION 2.5

2.5.1   Find the infimum x_i and the supremum x_s of each of the following sets S (assuming that $\prec$ means $<$) and say whether x_i or x_s (or both) belong to S. Explain your answers carefully.

(a) $S = \{e^x \mid x \in \mathbb{R}\}$.

(b) $S = \{e^x \mid 0 < x < \infty\}$.

(c) $S = \{\sin x \mid x \in \mathbb{R}\}$.

(d) $S = \bigcup_{n=1}^{\infty} \left(-\frac{1}{n}, \frac{1}{n}\right)$

(e) $S = \bigcap_{n=1}^{\infty} \left(-\frac{1}{n}, \frac{1}{n}\right)$

2.5.2   Determine whether the following set of real numbers has an infimum x_i or a supremum x_s:

$$S = \{|\exp(1/z)| \mid z \in \mathbb{C}\}. \tag{2.183}$$

If a supremum or an infimum exists, explain whether it belongs to S.

2.5.3   Find the infimum x_i and supremum x_s of the Cantor set, K. Explain whether x_i and x_s belong to K.

2.6 BIBLIOGRAPHY

Birkhoff, Garrett, and Saunders Mac Lane. *A Survey of Modern Algebra*. 3 ed. Macmillan, New York, 1965, chapters I, XI, XII.

Halmos, Paul R. *Naive Set Theory*. Van Nostrand, Princeton, 1960.

Kolmogorov, Andrei Nikolaevich, and Sergei Vasilevich Fomin. *Introductory Real Analysis*. Dover, New York, 1975, chapter 1.

Loomis, Lynn H., and Shlomo Sternberg. *Advanced Calculus*. Addison-Wesley, Reading, 1968.

Solow, Daniel. *How to Read and Do Proofs*. 2 ed. Wiley, New York, 1990.

Weiss, Mark Allen. *Data Structures and Algorithm Analysis*. 2 ed. Benjamin/Cummings, Reading, 1995.

CHAPTER 3

EVALUATION OF FUNCTIONS

3.1 INTRODUCTION

The ready availability of calculators and computers that are preprogrammed to compute the most important transcendental functions, such as *sin*, *cos*, and *exp*, has rendered most tabulations of the values of these functions obsolete for practical purposes other than checking the accuracy of one's computed results. However, one must still know how to compute accurate values of functions that one defines for one's own purposes, as well as some of the less-common transcendental functions. The purpose of this chapter is to survey some of the more important methods of function evaluation, as well as their pitfalls.

3.2 SENSITIVITY AND CONDITION NUMBER

3.2.1 DEFINITIONS

The numerical phenomenon of catastrophic cancellation of significant digits (see Section 1.4.3) illustrates the fact that the values of some functions are sensitive to small changes in the values of the arguments. The **sensitivity** $S(f; x, y)$ of the value $f(x, y)$ to the input data y is the relative change in $f(x, y)$ per unit relative change in y,

$$
\begin{aligned}
S(f; x, y) &:= \lim_{x' \to x, y' \to y} \left| \frac{y[f(x, y') - f(x, y)]}{(y' - y) f(x, y)} \right| \\
&\approx \left| \frac{y}{f(x, y)} \frac{\partial f(x, y)}{\partial y} \right|.
\end{aligned}
\tag{3.1}
$$

The sensitivity $S(f; x, y)$ is also called the **condition number** for the computation of $f(x, y)$. The condition number (or sensitivity) is dimensionless and is independent of the absolute magnitudes of x, y, and $f(x, y)$. In many cases it is correct to interpret the sensitivity as the ratio

$$
\frac{\Delta f / f}{\Delta y / y}
\tag{3.2}
$$

of the relative error in the computed value of a function f to the relative error in the input value, y.

For example, if $f(x, y) = x - y$, one has

$$S(f; x, y) = \frac{|y|}{|x - y|}, \tag{3.3}$$

which becomes large without bound as y approaches x. The subtraction of two nearly equal numbers is an example of an **ill-conditioned problem**, meaning that the computation of $f(x, y) = x - y$ is extremely sensitive to the input value y if $y \approx x$.

Many important practical problems can be ill-conditioned, including many for which software is available even on hand-held calculators, such as solving systems of linear equations (see Section 9.5.3).

3.2.2 EVALUATION OF POLYNOMIALS

Polynomials are among the simplest functions for user-written programs to evaluate; yet, even here, catastrophic cancellation can cause significant problems. For example, let us evaluate the polynomial $(1 - x)^2$ if x is close to 1. One can evaluate this expression in the power form $(1 - x)^2$, the grouped form $1 - (2x - x^2)$, or the nested form $1 - x(2 - x)$. We evaluate the expanded form in a hypothetical floating-point representation in which $\beta = 2$ and $p = 6$. For the sake of making the results visible on a graph, at each step we truncate to 8 significant bits and then round to 6 significant bits. If

$$\begin{aligned}
x &= 1.00000 \times 2^0 - 1.10000 \times 2^{-4} \\
&= 1.00000 - 0.00011 \\
&= 0.11101 \\
&= 1.11010 \times 2^{-1},
\end{aligned} \tag{3.4}$$

then one finds straightforwardly that $x^2 = 1.101001001 \times 2^{-1}$ (exactly) and that round $(x^2) = 1.10100 \times 2^{-1}$ (after truncation to 8 bits). Also, $2x = 1.11010 \times 2^0$ (both exactly and after rounding). Then $2x - \text{round}(x^2) = 1.00000 \times 2^0$, leading to the erroneous computed result $1 - (2x - \text{round}(x^2)) = 0$. The exact result is $(1 - x)^2 = 1.00100 \times 2^{-7}$. Figure 3.1 shows the results for $x \in [.75, 1.25]$.

In this example, the relative change in the computed value of the function is equal to 1, which is 32 times $\epsilon_{\text{mach}} = 2^{-5}$. The relative change in the input value of x is $(1.10000 \times 2^{-4})_2 = (3/32)_{10}$; hence the sensitivity of evaluating the polynomial $1 - 2x + x^2$ in this particular example is equal to $32/3 \approx 11$. The root cause of the high sensitivity of polynomial evaluation near repeated roots, as in this example, is the high condition number of subtraction of two nearly equal quantities. In the next example we see that the exact value of the condition number may depend strongly on the point at which the function is evaluated; hence one should compute a condition number for a range of input values. For a conservative estimate, one should use the maximum of the absolute values of the computed sensitivities.

Figure 3.2 shows a somewhat more realistic example, the evaluation of a polynomial in the power form $(1 - x)^6$ and in the expanded form $x^6 - 6x^5 + 15x^4 - 20x^3 + 15x^2 - 6x + 1$

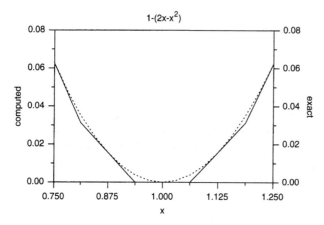

FIGURE 3.1

Evaluation of $1 - (2x - x^2)$ exactly (*dashed curve*) and in a floating-point representation with $\beta = 2$ and $p = 6$ (*solid line*).

using IEEE single-precision arithmetic. In this case machine epsilon is $\epsilon_{mach} \approx 1.192 \times 10^{-7}$, or roughly one-eighth of the distance between the 0 and the 10 on the vertical scale in Fig. 3.2. The large amplitude of the numerical "noise" in this example would cause the failure of simple root-finding algorithms that test whether a function has opposite signs at the end points of an interval, and if so, conclude that a root of the function lies between those points.

One finds by inspecting the numerical data from which Fig. 3.2 was prepared that the condition number for evaluating the polynomial $x^6 - 6x^5 + 15x^4 - 20x^3 + 15x^2 - 6x + 1$

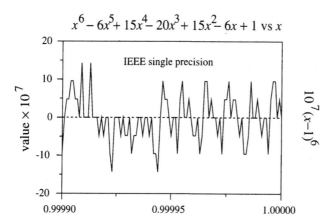

FIGURE 3.2

Numerical evaluation of a polynomial in the power form $(1 - x)^6$ (*dashed line* and *right vertical axis*) and in the expanded form $x^6 - 6x^5 + 15x^4 - 20x^3 + 15x^2 - 6x + 1$ (*solid line* and *left vertical axis*), using IEEE single-precision floating-point arithmetic in both cases. The computed values plotted on the vertical axes have been multiplied by 10^7. The value of ϵ_{mach} is 1.192×10^{-7}; the value of one ulp is 5.96×10^{-8} in the range shown.

in the range shown fluctuates between (approximately) 0 and 15. The conclusion is that one cannot assume that as ordinary an operation as polynomial evaluation is always well-conditioned if one is using fixed-precision floating-point arithmetic.

3.2.3 MULTIPLE ROOTS OF POLYNOMIALS

If r of the roots $z_1, \ldots, z_n$ of an n-th-degree polynomial equation

$$z^n + a_1 z^{n-1} + \cdots + a_n = 0 \tag{3.5}$$

are equal, for example, if $z_1 = \cdots = z_r$, then one says that the **multiplicity** of the root z_1 is equal to r. A root is called **multiple** if its multiplicity is greater than 1. The numerical problem of finding the value of a multiple root is ill-conditioned. To take an extreme case, let

$$p(z, \alpha) := \sum_{k=1}^{n} \binom{n}{k} z^k (-1)^{n-k} + \alpha. \tag{3.6}$$

The fundamental theorem of algebra implies that for every value of the constant term α, the n-th-degree polynomial equation

$$p(z, \alpha) = 0 \tag{3.7}$$

has n roots $z_1(\alpha), \ldots, z_n(\alpha)$, which are generally complex numbers. If

$$\alpha = (-1)^n, \tag{3.8}$$

then the root 1 is repeated n times:

$$p(z, \alpha)\big|_{\alpha=(-1)^n} = (z - 1)^n. \tag{3.9}$$

If α is changed by one ulp (unit in the last place, u),

$$\alpha = (-1)^n - u, \tag{3.10}$$

then Eq. (3.7) becomes

$$(z - 1)^n = u, \tag{3.11}$$

the roots of which are

$$z_m - 1 = u^{1/n} e^{2\pi i m/n}, \qquad m = 0, 1, \ldots, n - 1. \tag{3.12}$$

If plotted in the complex plane, the z_m are represented by points that are equally spaced around the circumference of a circle of radius $u^{1/n}$ centered at $z = 1$. If $n \gg 1$, then $u^{1/n} \gg u$.

For example, if $p = 24$ and $n = 8$, then $u = 2^{-24}$; the absolute and relative errors are $u^{1/8} = 2^{-3} = 0.125_{10}$. For Eq. (3.6) with $\alpha = (-1)^n$, changing the constant term by one digit in the least-significant place causes *all* of the roots to differ from the correct value by

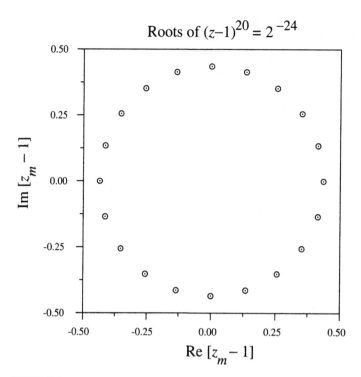

FIGURE 3.3

The points represent the error in the m-th root, $z_m - 1$, of the polynomial equation $(z - 1)^{20} = 2^{-24}$, for $m \in (0:19)$.

so large an amount that the real root $1 + u^{1/n}$ differs from the correct value by one unit in the third place after the binary point! Figure 3.3 shows an even more extreme case.

The sensitivity of the repeated root z_1 to the value of α in Eq. (3.7) follows from the definition of computational sensitivity, Eq. (3.2):

$$S(z_1; \alpha) = \left| \frac{\alpha[z_1(\alpha') - z_1(\alpha)]}{(\alpha' - \alpha)z_1(\alpha)} \right| \approx \left| \frac{u^{1/n}}{u} \right| = \frac{1}{u^{(n-1)/n}} \gg 1. \tag{3.13}$$

If $n \gg 1$, the sensitivity is approximately equal to $1/u$, the reciprocal of one unit in the last (least-significant) place. According to Eq. (1.63), in the worst case this implies a sensitivity of β^p. In IEEE-754 single precision, then, the sensitivity can be as large as $2^{24} \approx 1.7 \times 10^7$. As we show in the example in the preceding paragraph, a change in the least-significant bit of α can cause a relative change of the order of magnitude of 1 in the computed values of the roots (see Fig. 3.3).

Practical algorithms for finding the roots of polynomials must be designed in such a way that repeated roots can be computed accurately. One technique is **deflation**, in which one divides the polynomial p whose roots one wants to know by a monomial $x - x_0$, in which x_0 is a known root. The quotient is then a polynomial of lower degree. In principle, one can continue to deflate a polynomial until all roots have been found. However, in order to deflate a polynomial one must already have computed a root x_0 accurately. In practice, the difficulty of computing x_0 with high accuracy may limit the usefulness of deflation. Press et al. (1992) summarize methods for finding roots of polynomials.

Analytical formulas for the roots of quadratic, cubic, or quartic equations give the roots of polynomials of degrees two, three, or four as algebraic functions of the coefficients. For example, one learns in school that the roots of the quadratic equation

$$ax^2 + bx + c = 0 \tag{3.14}$$

are

$$x_1 = \frac{-b + \sqrt{D}}{2a}, \qquad x_2 = \frac{-b - \sqrt{D}}{2a} \tag{3.15}$$

in which

$$D = b^2 - 4ac \tag{3.16}$$

is called the **discriminant**. Unfortunately, if $4ac \ll b^2$ and the discriminant $D \approx b^2$ are computed using a finite, fixed number of significant digits, catastrophic cancellation occurs in the computation of x_1 (if $b > 0$) or x_2 (if $b < 0$) because $4ac$ contributes few, if any, significant digits to $\sqrt{D}$. Similar difficulties occur in the formulas for the roots of cubic and quartic equations.

Of course, if one has the foresight to recognize catastrophic cancellation, then one is aware that the computation of the expression

$$A = b + \text{sign}(b)\sqrt{D} \tag{3.17}$$

is always well conditioned. This gives one root, $-A/2a$. The product of the roots is c/a; hence the other root is $-2c/A$. The computation of $\sqrt{D}$ may result in a loss of significant digits; see Exercise 3.2.3.

Another technique for computing the roots of a polynomial p (with real coefficients) is to find a balanced matrix $\mathbf{M}$ of which p is the characteristic polynomial. Then the roots of p are the eigenvalues of $\mathbf{M}$, which can be computed numerically by an algorithm that is always well conditioned.

The first step in finding $\mathbf{M}$ is to compute the so-called companion matrix of the polynomial p, as in Exercise 6.5.3. The subsequent steps of balancing and root-finding are well discussed in Press et al.

3.2.4 EXERCISES FOR SECTION 3.2

3.2.1 Use the "Microscope" package of *Mathematica* to analyze the errors in evaluating the right-hand side of the binomial theorem (2.164) for $y = 1$ and for values of x near 1 for $n = 1, 5, 10, 15$, and 20.

3.2.2   Write a computer program in C, C++, or FORTRAN to analyze the errors in evaluating the right-hand side of the binomial theorem (2.164) for $y = 1$ and for values of x near 1 for $n = 1, 5, 10, 15$, and 20.

3.2.3   Explore one of the limitations of the square-root function in your computer's floating-point processor by writing and running a program in C, C++, or FORTRAN to extract the square root of 2.0_{10} n times and then square the result n times to obtain a floating-point number s_n. If the hardware square-root function were perfectly accurate, s_n would always be 2.0. Use your program to make a table showing n and s_n for values of n from 1 to 25, in IEEE-754 single precision.

3.3 RECURSION AND ITERATION

In Section 2.4.2 we discussed the principle of finite induction and gave examples of recursive definitions, including partial sums of series. Many problems in engineering and mathematical physics can usefully be solved by breaking them up into steps, such that the result for step n depends on the results obtained for steps before n. Optimization, finding roots of polynomials and other functions, numerical integration, and solution of initial-value problems by applying finite-difference approximations to ordinary differential equations are only a few of the important applications of recursive methods.

3.3.1 FINDING ROOTS BY BISECTION

If a continuously differentiable function $f : \mathbb{R} \to \mathbb{R}$ has opposite signs at $x = b$ and $x = c$, then Rolle's theorem asserts that a zero of f lies between b and c. If one bisects the interval $[b, c]$ and evaluates f at the bisection point, then one can decide in which of the two subintervals the zero lies. After k steps, one has an interval of length $2^{-k}|b - c|$ in which a root is sure to lie. Putting it a little differently, each step of the bisection method improves the accuracy of the answer by one bit. The method of bisection fails if catastrophic cancellation of significant digits occurs, as in Fig. 3.2, and if the zero is of even order, that is, if $f(x) = (x - a)^{2k}g(x)$, where $g(a) \neq 0$. For a zero of even order, $f(a + \epsilon)$ and $f(a - \epsilon)$ have the same sign if ϵ is small enough that $g(x - \epsilon)$ and $g(x + \epsilon)$ have the same sign.

3.3.2 NEWTON–RAPHSON METHOD

Derivation

Let f be a function of one real variable x, such that the first derivative of f is continuous for all x. For example, f may be a polynomial in x. Let a be a zero of $f : f(a) = 0$. If x is sufficiently close to a one has

$$f(x) \approx f(a) + (x - a)f'(x)$$
$$\approx (x - a)f'(x). \tag{3.18}$$

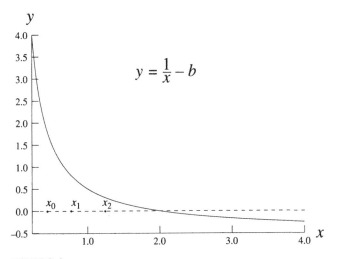

FIGURE 3.4

Illustration of the first two steps of Newton–Raphson iteration for the function $f(x) = 1/x - b$, with $b = 0.5$.

Then

$$a \approx x - \frac{f(x)}{f'(x)}. \tag{3.19}$$

The **Newton–Raphson method** consists of iterating the difference equation

$$x_{k+1} = x_k - \frac{f(x_k)}{f'(x_k)} \tag{3.20}$$

to convergence. Figure 3.4 illustrates the resulting recursive method for finding a root.

Rate of Convergence

The best feature of the Newton–Raphson method is that if $f'(a) \neq 0$ and if one is able to choose a starting value x_0 that is close to a (in a sense which we shall make precise in a moment), then each iteration approximately doubles the number of accurately known significant digits in one's approximation to a. For any function that vanishes at $x = a$, but that has a nonzero derivative there, one can write approximately

$$f(x) = B(x - a) + C(x - a)^2 \tag{3.21}$$

in which $B = df/dx|_a$ and $C = \frac{1}{2} d^2 f/dx^2|_a$. The Newton–Raphson difference equation for this function is

$$x_{k+1} = x_k - \frac{(x_k - a) + \dfrac{C}{B}(x_k - a)^2}{1 + 2\dfrac{C}{B}(x_k - a)}, \tag{3.22}$$

which becomes

$$x_{k+1} = a + \frac{C}{B}(x_k - a)^2 \tag{3.23}$$

if x is sufficiently close to a that

$$2\frac{C}{B}|x_k - a| \ll 1. \tag{3.24}$$

Equation (3.23) predicts that if inequality (3.24) is already obeyed when $k = 0$, then

$$x_{k+1} - a \approx \left(\frac{C}{B}\right)^k (x_0 - a)^{2k}, \tag{3.25}$$

according to which the number of accurate significant digits approximately doubles with each iteration.

Newton–Raphson iteration is much less effective in finding a multiple zero than it is in finding a simple zero. If a function f has a zero of multiplicity m at $x = a$, then $f(x) \approx (x - a)^m f^{(m)}(a)/m!$ for x sufficiently close to a. One finds that in this case

$$x_{k+1} - a \approx \frac{m - 1}{m}(x_k - a), \tag{3.26}$$

which predicts that when x_k is close to the multiple root at $x = a$, the distance to the true root diminishes linearly by a factor of $(m - 1)/m$ with each iteration rather than diminishing quadratically as in Eq. (3.25).

Clearly one mode of failure of the Newton–Raphson method is that one may happen on a value x_k such that $f'(x_k) \approx 0$, sending x_{k+1} far away from the desired zero. Therefore any practical implementation must include a "sanity check" at each step of the iteration to make certain that x_{k+1} does not lie outside some predefined interval. For this reason one often uses the Newton–Raphson method for improving the accuracy of (or "polishing") a zero that has been found approximately by some other method.

Reciprocal Approximation Instead of Floating-Point Division

Another computational problem for which the Newton–Raphson method is well adapted is – surprisingly – floating-point division. Some microprocessor and computer vendors use a Newton–Raphson difference equation, plus a lookup table for the starting value, instead of true floating-point division. To find the number a such that

$$a = \frac{1}{b} \tag{3.27}$$

in which b is real and nonzero, one can find the zero of the function

$$f(x) = \frac{1}{x} - b. \tag{3.28}$$

Because $f'(x) = -1/x^2$, the Newton–Raphson difference equation for f is

$$x_{k+1} = x_k - \left(-x_k + x_k^2 b\right) = 2x_k - x_k^2 b. \tag{3.29}$$

Here, too, convergence is quadratic if the starting value x_0 is close to $1/b$ (see Exercise 3.3.3). The value of x_k for some predetermined value of k is called the **reciprocal approximation** to $1/b$.

3.3.3 EVALUATION OF SERIES

Many of the special functions of engineering and mathematical physics are defined in terms of infinite series. Even though recurrence relations often suffer less than series from catastrophic cancellation of significant digits, there may be ranges of the argument for which evaluating a series is faster than using recurrence relations.

Let

$$s = \sum_{i=1}^{\infty} x_i \tag{3.30}$$

be a convergent infinite series in the sense that

$$s = \lim_{N \to \infty} s_N \tag{3.31}$$

in which s_N is the N-th partial sum,

$$s_N = \sum_{i=1}^{N} x_i. \tag{3.32}$$

Because we already know that subtraction of two quantities of nearly the same magnitude is ill-conditioned, we assume that the terms x_i are all of one sign.

For the sake of convenience in the following discussion we assume that the magnitudes $|x_i|$ decrease monotonically as i increases. We make no use of this assumption in our derivations.

Natural Order

Given the definition of convergence of a series, the most natural method for approximating the value of s is to use the difference equation

$$s_{k+1} = s_k + x_{k+1} \tag{3.33}$$

until the value of s_{k+1} no longer changes, that is, until $|x_{k+1}| < \epsilon_{\text{mach}}|s_k|$. This is almost certainly the least accurate way of obtaining an approximation to the true value, s. Adding a very small term x_{k+1} to a much larger number s_k is guaranteed to lose the maximum possible number of significant digits of x_{k+1} because x_k must be shifted many places in order to make its exponent equal to that of s_k.

Kahan (1972) pointed out that a so-called inverse error analysis of Eq. (3.33) answers the question, To what problem have we computed the correct answer? rather than the question, What is the error in our computation? According to the discussion of rounding errors in Section 1.4.2, the floating-point difference equation

$$s_{k+1} = s_k \oplus x_{k+1} \tag{3.34}$$

implies that

$$s_{k+1} = (1 + \delta_{k+1})(s_k + x_{k+1}), \tag{3.35}$$

where

$$|\delta_{k+1}| \leq \tfrac{1}{2}\epsilon_{\text{mach}} \tag{3.36}$$

if one rounds the result to the nearest valid floating-point number. Then the solution of Eq. (3.34) is

$$s_N \approx \sum_{j=1}^{N} \left(1 + \sum_{k=j}^{N} \delta_k\right) x_j \tag{3.37}$$

in which we have ignored all products of the small numbers δ_k because they are of order ϵ_{mach}^2. In view of Eq. (3.36), one has

$$s_N \approx \sum_{j=1}^{N} (1 + \xi_j) x_j \tag{3.38}$$

in which

$$|\xi_j| \leq (1 + \epsilon_{\text{mach}}/2)^{N-j+1} - 1. \tag{3.39}$$

Continuing to ignore numbers of order ϵ_{mach}^m for $m > 1$, one finds that the error incurred by summing in natural order may be larger than $\tfrac{1}{2}N\epsilon_{\text{mach}}|x_1|$, which is N times the roundoff error $\tfrac{1}{2}\epsilon_{\text{mach}}|x_1|$.

If one sums in natural order, but in double precision, then the double-precision value of machine epsilon, $\epsilon_{\text{mach,dp}}$, enters the error estimate. If the number of terms obeys the inequality $N > \epsilon_{\text{mach,sp}}/\epsilon_{\text{mach,dp}}$, then a double-precision computation produces a result that has single-precision accuracy.

Reverse Order

The most obvious improvement in summation strategy is to add the terms in the inverse order, starting with x_N and ending with x_1. An error analysis similar to the one that we carried out for summation in natural order (Kahan 1972) shows that Eq. (3.38) still holds but that Eq. (3.39) is replaced by

$$|\xi_j| \leq (1 + \epsilon_{\text{mach}}/2)^j - 1. \tag{3.40}$$

Reversing the order has shifted the largest relative error to the term of smallest magnitude, x_N, which now may have a relative error as great as N times round-off error. The largest term, x_1, now suffers no more than the usual round-off error of $\frac{1}{2}\epsilon_{mach}|x_1|$. If the computation is carried out in double precision, then of course one should use the double-precision value of ϵ_{mach}.

However, in order to test the convergence of the partial sums one must compute several, perhaps many, such sums for different values of N. For this reason summation in reverse order involves much more computational effort than summation in natural order unless one happens to know in advance the value of N that ensures the desired accuracy.

Kahan's Summation Method

A method invented by Kahan cleverly recovers the bits that are lost in the process of adding a small and a large number and preserves this information in the form of an accumulated correction. Kahan's summation algorithm, and many more interesting examples of error analysis, are described (Kahan 1972). For a detailed error analysis of Kahan's method, the reader should consult Goldberg (1991).

The following FORTRAN segment, due to Kahan, implements his summation algorithm:

```
    s = 0.
    c = 0.
    do 100 j = 1,N
      y = c + x(j)
      t = s + y
      c = (s - t) + y
100 s = t
    s = s + c
```

The method works because the variable c contains the information that was lost as the result of adding x(j) to s.

A detailed error analysis shows that if one uses Kahan's method to compute the sum of N terms of a series, the result is of the same form as Eq. (3.38) with

$$|\xi_j| \leq \epsilon_{mach} + O(\epsilon_{mach}^2). \tag{3.41}$$

If the number of terms, N, is sufficiently small that summing in natural order using double precision gives more accuracy than rounding the exact sum to single precision would give, then there is no point in using Kahan's method. One uses Kahan's method either if one is already using double precision and needs improved accuracy or if hardware double precision is not available.

3.3.4 EXERCISES FOR SECTION 3.3

3.3.1   The infinite series for the sine function,

$$\sin x = \sum_{m=0}^{\infty} \frac{(-1)^m x^{2m+1}}{(2m+1)!}, \tag{3.42}$$

is mathematically correct in the sense that it can be shown to converge to the sine of x for any finite real argument x. The goal of this problem is to determine experimentally the range of values of x for which a truncated series is computationally useful.

(a) Write a computer program that evaluates the sum

$$\sin x \approx \sum_{m=0}^{M} \frac{(-1)^m x^{2m+1}}{(2m + 1)!} \qquad (3.43)$$

using single-precision arithmetic. Choose M by requiring that the magnitude of the first neglected term shall be less than machine epsilon.

(b) Use your program for $x = 1.0 \times 10^{-2}$ and 1.0. Compare your results with the values tabulated in Abramowitz and Stegun (1964), Table 4.6.

(c) Use your program for $x = 10.0$, 20.0, 30.0, 40.0, and 50.0. Compare your results with the values tabulated in Abramowitz and Stegun (1964), Table 4.8, and with the values obtained by calling the sine routine that is built into your compiler's math library.

*(d) Explain any discrepancies between your computed values of $\sin x$ and the tabulated values.

Which results can you improve by using double precision instead of single precision? Which results can you improve by summing in an order other than natural order? Which results, if any, can you improve by using Kahan's summation formula?

3.3.2   The n-th partial sum of the power series for the irrational number e (the base of natural logarithms) is

$$e = \sum_{k=0}^{n} \frac{1}{k!}. \qquad (3.44)$$

Find the relative error in the value of e computed using

(a) $n = 3$,

(b) $n = 5$,

using single precision and double precision in natural order and in an order other than natural order. Which results, if any, can you improve by using Kahan's summation formula?

3.3.3   The n-th partial sum of the power series for the irrational number e^{-1} (where e is the base of natural logarithms) is

$$e^{-1} = \sum_{k=0}^{n} \frac{(-1)^k}{k!}. \qquad (3.45)$$

Find the relative error in the value of e computed using

(a) $n = 3$,

(b) $n = 5$,

using single precision and double precision in natural order and in an order other than natural order. Which results, if any, can you improve by using Kahan's summation formula? Comment on the differences between your answers for this exercise and your answers for Exercise 3.3.2.

3.3.4 The purpose of this problem is to investigate an iterative method of computing the square root of a number.

(a) Show that the equation

$$x^2 = a \tag{3.46}$$

is equivalent to the equation

$$x = g(x), \tag{3.47}$$

where

$$g(x) := \tfrac{1}{2}\left(x + \tfrac{a}{x}\right). \tag{3.48}$$

Plot graphs of the two sides of Eq. (3.47) for selected values of a.

(b) Show that if the numbers $x_0, x_1, \ldots, x_k, \ldots$ obey the difference equation

$$x_{k+1} = g(x_k), \tag{3.49}$$

then

$$\lim_{k \to \infty} x_k = \sqrt{a}. \tag{3.50}$$

(c) Show that if x_0 is sufficiently close to a, the number of significant bits approximately doubles with each iteration.

3.3.5 Show that the reciprocal approximation (3.29) converges quadratically by showing that if

$$x_k = \frac{1}{b} + \epsilon, \tag{3.51}$$

then

$$x_{k+1} = -b\epsilon^2 + \frac{1}{b}. \tag{3.52}$$

3.4 INTRODUCTION TO NUMERICAL INTEGRATION

In addition to the rounding errors surveyed above, computations may suffer from errors that are inherent in the algorithm that is used to solve a scientific or engineering problem. This is especially true for recursive algorithms, in which the value to be computed depends on previously computed values. The error after many recursive steps may greatly exceed the error in a single step.

The classic approach to numerical evaluation of the one-dimensional Riemann integral

$$I = \int_a^b f(x)\,dx \tag{3.53}$$

(in which $f : [a, b] \to \mathbb{R}$ is a real-valued function) is to sample f at $N + 1$ points

$$x_0 = a, x_1 = a + h, \ldots, x_N = b \tag{3.54}$$

and to approximate I as a weighted sum of sampled values of f:

$$I \approx \sum_{i=0}^N w_i\, f(x_i). \tag{3.55}$$

Algorithms for numerical integration (often called "rules") fall into several categories. In the simplest category are integration rules that, first, break the integral I into a sum,

$$\int_a^b f(x)\,dx = \sum_{i=1}^N I_i, \tag{3.56}$$

in which each term I_i,

$$I_i := \int_{x_{i-1}}^{x_i} f(x)\,dx, \tag{3.57}$$

is the value of the integral of f over the interval (x_{i-1}, x_i), and, second, approximate the value of each I_i using a numerical integration method. One calls the interval (x_{i-1}, x_i) a **panel**.

In the category of Gaussian quadrature methods, the sum (3.55) is a global approximation to the value of the integral, but the individual terms in the sum do not approximate the integral of f on (x_{i-1}, x_i). In yet another category, the so-called Monte Carlo methods, the integral is approximated by using a generator of pseudorandom numbers to choose the points x_i.

In this section we describe only selected examples from the simplest category of one-dimensional numerical integration methods: the right-handed rectangle rule, the centered rectangle rule, the trapezoidal rule, and Simpson's rule.

Given a set of sample points (3.54), the k-th **step size** is defined as the distance between x_k and the preceding point:

$$h_k := x_k - x_{k-1}. \tag{3.58}$$

In the common case of uniform sampling, all step sizes are equal to

$$h := \frac{b - a}{N}, \tag{3.59}$$

and $x_k = kh + x_0$.

3.4.1 RECTANGLE RULES

The simplest approximations to a function $f : [a, b] \to \mathbb{R}$ for the purposes of numerical integration are the **staircase approximations**, in which the approximating function is piecewise constant on each panel. For the **right-hand rectangle rule**, one chooses the value of the approximating staircase function on the i-th panel as the value of f at the right end of the panel:

$$\int_a^b f(x)\, dx \approx R_h[f], \tag{3.60}$$

in which the right-hand rectangle functional is defined as

$$R_h[f] := \sum_{i=1}^{N} r_i[f], \tag{3.61}$$

and the approximation to the value of the integral on one panel is

$$r_i[f] := h_i f(x_i). \tag{3.62}$$

If the sampling grid is uniform ($h_i = h = (b-a)/N$ for every i), then the right-hand rectangle rule for approximating the integral I is

$$R_h[f] = h \sum_{i=1}^{N} f(x_i). \tag{3.63}$$

One can see instantly from Fig. 3.5 that the right-hand rectangle rule errs by neglecting the areas of the roughly triangular regions between the rectangles and the curve $\{x, f(x)\}$.

Of course, there is also a left-hand rectangle rule, according to which the value of the integral on one panel is approximated as the product of the width of the panel and the value of the function at the left edge of the panel:

$$l_i[f] := h_i f(x_{i-1}). \tag{3.64}$$

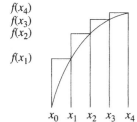

FIGURE 3.5

Approximate evaluation of the integral $\int_{x_0}^{x_4} f(x)\, dx$ using the right-hand rectangle rule.

The left-hand rectangle rule for integration is equivalent to Euler's method for solving a first-order ordinary differential equation; see Section 3.5.1.

A different choice of points at which the function f is sampled leads to an improved approximation for the integral $I = \int_a^b f(x)\,dx$. The **centered rectangle rule** for approximating I is

$$\int_a^b f(x)\,dx \approx C_h[f] \tag{3.65}$$

in which the approximate value of the integral over one panel is taken as

$$c_i[f] := h_i f\left(x_{i-\frac{1}{2}}\right), \tag{3.66}$$

and $x_{i-\frac{1}{2}} := (x_i - x_{i-1})/2$.

Thus the centered rectangle rule leads to the integration formula

$$\begin{aligned}
C_h[f] &= \sum_{i=1}^{N} c_i[f] \\
&= h \sum_{i=1}^{N} f\left(x_{i-\frac{1}{2}}\right).
\end{aligned} \tag{3.67}$$

3.4.2 TRAPEZOIDAL RULE

Another way to try to improve on the right-hand rectangle rule is to take the curve between successive sample points as a straight line instead of a step function. In this case the region that approximates the area under the curve in one panel is a trapezoid, the area of which is

$$t_i[f] := \tfrac{1}{2}h_i[f(x_{i-1}) + f(x_i)]. \tag{3.68}$$

The resulting **trapezoidal rule** is

$$\begin{aligned}
T_h[f] &= h \sum_{i=1}^{N} t_i[f] \\
&= \frac{h}{2}f(x_0) + h \sum_{i=1}^{N-1} f(x_i) + \frac{h}{2}f(x_N).
\end{aligned} \tag{3.69}$$

3.4.3 LOCAL AND GLOBAL ERRORS

The error in a numerical computation is the difference between the computed and the exact values. The classic approach to analyzing the error in a numerical algorithm that accepts a function as the input and produces a number as the output is to expand the input function in a Taylor series about one of the sampling points with the goal of expressing the error as a power series in the step size, h. If an integral is expressed approximately as a sum over panels, then the **global error** (the error in the sum) is equal to the sum of the **local errors** (the errors on individual panels).

For example, integration over one panel of the Taylor series of f about x_i,

$$f(x) = f(x_i) + \frac{df}{dx}\bigg|_{x_i} (x - x_i) + \cdots \tag{3.70}$$

gives

$$\int_{x_{i-1}}^{x_i} f(x)\,dx = h_i f(x_i) - \frac{1}{2}h_i^2 \frac{df}{dx}\bigg|_{x_i} + \cdots \tag{3.71}$$

The first term in this expansion is the one-panel right-hand rectangle rule. The second term, which is proportional to h^2, is the leading contribution to the local (single-panel) error in the right-hand rectangle rule. Summing the one-panel results for the integral (assuming uniform sampling) gives

$$\int_a^b f(x)\,dx \approx R_h[f] - \frac{h^2}{2} \sum_{i=1}^N \frac{df}{dx}\bigg|_{x_i}$$

$$\approx R_h[f] - h\left(\frac{b-a}{2}\right)\left(\frac{1}{N}\sum_{i=1}^N \frac{df}{dx}\bigg|_{x_i}\right). \tag{3.72}$$

The second line of the preceding equation gives the global (multiple-panel) error, which is proportional to h. Note that the global error is of lower degree in the step size than the local error (h versus h^2) and that the coefficient of h is the mean of the values of the first derivative at the sample points.

For the centered rectangle rule, the input function is sampled at the intermediate points $x_{i-\frac{1}{2}}$. Hence a useful Taylor expansion is

$$f(x) = f\left(x_{i-\frac{1}{2}}\right) + \frac{df}{dx}\bigg|_{x_{i-\frac{1}{2}}} \left(x - x_{i-\frac{1}{2}}\right) + \frac{1}{2}\frac{d^2 f}{dx^2}\bigg|_{x_{i-\frac{1}{2}}} \left(x - x_{i-\frac{1}{2}}\right)^2$$

$$+ \frac{1}{6}\frac{d^3 f}{dx^3}\bigg|_{x_{i-\frac{1}{2}}} \left(x - x_{i-\frac{1}{2}}\right)^3 + \frac{1}{24}\frac{d^4 f}{dx^4}\bigg|_{x_{i-\frac{1}{2}}} \left(x - x_{i-\frac{1}{2}}\right)^4 + \cdots \tag{3.73}$$

Integration of a typical term of this expansion over one panel gives

$$\int_{x_{i-1}}^{x_i} \left(x - x_{i-\frac{1}{2}}\right)^p dx = \begin{cases} h_i & \text{if } p = 0; \\ 0 & \text{if } p = 1; \\ \frac{1}{12}h_i^3 & \text{if } p = 2; \\ 0 & \text{if } p = 3; \\ \frac{1}{80}h_i^5 & \text{if } p = 4. \end{cases} \tag{3.74}$$

The resulting expression for the integral of f over a single panel is

$$
\begin{aligned}
\int_{x_{i-1}}^{x_i} f(x)\,dx &= c_i[f] + p_i[f] + q_i[f] + \cdots \\
&= t_i[f] - 2p_i[f] - 4q_i[f] + \cdots
\end{aligned}
\tag{3.75}
$$

in which the error terms,

$$
p_i[f] := \frac{1}{24} h_i^3 \left.\frac{d^2 f}{dx^2}\right|_{x_{i-\frac{1}{2}}}
\tag{3.76}
$$

and

$$
q_i[f] := \frac{1}{1920} h_i^5 \left.\frac{d^4 f}{dx^4}\right|_{x_{i-\frac{1}{2}}},
\tag{3.77}
$$

are of degrees three and five, respectively, in the step size h.

The result of this power-series attack on the problem of estimating the error in numerical integration is a power series in the step size, the first term of which is a known integration "rule." The low-order terms can be arranged so that the first term is either the centered rectangle rule,

$$
\begin{aligned}
\int_a^b f(x)\,dx &\approx C_h[f] + \frac{h^3}{24} \sum_{i=1}^{N} \left.\frac{d^2 f}{dx^2}\right|_{x_{i-\frac{1}{2}}} \\
&\approx C_h[f] + h^2 \left(\frac{b-a}{24}\right) \left(\frac{1}{N} \sum_{i=1}^{N} \left.\frac{d^2 f}{dx^2}\right|_{x_{i-\frac{1}{2}}}\right),
\end{aligned}
\tag{3.78}
$$

or the trapezoidal rule,

$$
\begin{aligned}
\int_a^b f(x)\,dx &\approx T_h[f] - \frac{h^3}{12} \sum_{i=1}^{N} \left.\frac{d^2 f}{dx^2}\right|_{x_{i-\frac{1}{2}}} \\
&\approx T_h[f] - h^2 \left(\frac{b-a}{12}\right) \left(\frac{1}{N} \sum_{i=1}^{N} \left.\frac{d^2 f}{dx^2}\right|_{x_{i-\frac{1}{2}}}\right)
\end{aligned}
\tag{3.79}
$$

Two important points can be extracted from the two preceding formulas. One is that, as expected, the global error is one order lower in the step size h than the local error. The second point is that the leading error terms of the trapezoidal and centered rectangle rules are opposite in sign and differ in magnitude by a factor of two. The latter point suggests combining the one-panel formulas for the trapezoidal and centered-rectangle rules in such a way that the order-h^2 error of the combination vanishes.

With this idea in mind, let

$$s_i[f] := \tfrac{2}{3}c_i[f] + \tfrac{1}{3}t_i[f]$$
$$= \tfrac{1}{6}h_i\left[f(x_{i-1}) + 4f\left(x_{i-\frac{1}{2}}\right) + f(x_i)\right]. \tag{3.80}$$

Equation (3.75) implies then that

$$\int_{x_{i-1}}^{x_i} f(x)\,dx = s_i[f] + \tfrac{2}{3}q_i[f] + \cdots \tag{3.81}$$

Approximating the integral of f over one panel as $s_i[f]$ is significantly more accurate than using either $c_i[f]$ or $t_i[f]$.

Summing the single-panel formulas for $s_i[f]$,

$$S_h[f] := \sum_{i=1}^{N} s_i[f], \tag{3.82}$$

leads to a significantly more accurate integration rule than either the centered-rectangle rule or the trapezoidal rule, at least according to the criterion that a numerical approximation is more accurate if its error is of higher order in the step size. The result is **Simpson's rule**,

$$
\begin{aligned}
\int_a^b f(x)\,dx &\approx S_h[f] - \frac{2}{3}\sum_{i=1}^{N} q_i \\
&\approx S_h[f] - \frac{1}{2880}h^5 \sum_{i=1}^{N} \left.\frac{d^4 f}{dx^4}\right|_{x_{i-\frac{1}{2}}} \\
&\approx S_h[f] - \frac{1}{90}\left(\frac{h}{2}\right)^5 \sum_{i=1}^{N} \left.\frac{d^4 f}{dx^4}\right|_{x_{i-\frac{1}{2}}} \\
&\approx S_h[f] - \frac{b-a}{180}\left(\frac{h}{2}\right)^4 \left(\frac{1}{N}\sum_{i=1}^{N} \left.\frac{d^4 f}{dx^4}\right|_{x_{i-\frac{1}{2}}}\right).
\end{aligned}
\tag{3.83}
$$

The real sampling interval, in terms of which we have expressed Simpson's rule in the last line, is the distance between $x_{i-\frac{1}{2}}$ and x_i, which is $h/2$, not h.

3.4.4 EXERCISES FOR SECTION 3.4

3.4.1 Use Eq. (3.74) to show that

  (a) The single-panel, right-hand rectangle rule is exact if f is a constant.

  (b) The single-panel, centered rectangle rule is exact if f is a polynomial of degree 1, $f(x) = a_1 x + a_0$.

  (c) The trapezoidal rule is also exact if f is a polynomial of degree 1.

(d) Simpson's rule is exact if f is a polynomial of degree 3.

(e) Exhibit a polynomial of degree 2 for which the centered rectangle rule fails.

3.4.2   The purpose of this problem is to analyze the behavior of the simple integration rules in this chapter for sinusoidal, instead of polynomial, inputs.

(a) Show that if the integral of

$$f(x) = \sin\left(\frac{\pi x}{h}\right) \tag{3.84}$$

is evaluated using the rectangle rules or the trapezoidal rule, then the relative error of the numerical estimate of $\int_a^b f(x)\,dx$ approaches zero as $(b - a)/h \to \infty$.

(b) Show that if the same integral as in part (a) is evaluated using Simpson's rule, then the relative error of the numerical estimate approaches unity as $(b - a)/h \to \infty$. Discuss the reason for this behavior.

3.5 SOLUTION OF DIFFERENTIAL EQUATIONS

One of the most powerful, general ways to evaluate a function is to solve a differential equation. The basic laws of continuum mechanics, electromagnetics, and quantum mechanics are expressed in terms of partial differential equations. Solving these equations is one of the chief goals of theoretical and computational physics and is one of the most important practical problems in the design of high-speed digital and communication systems. The methods presented here are the building blocks from which one constructs useful computational methods.

To solve a differential equation means to find a real-valued function $u : \mathbb{R}^n \to \mathbb{R}$ (or a complex-valued function $v : \mathbb{R}^n \to \mathbb{C}$) that satisfies the given differential equation and that takes prescribed values on some subset of its domain. If the highest-order derivative of u in the differential equation is of order m, then one says that the differential equation is of **order** m.

If the values of u and its derivatives up to order $m - 1$ are given on some set S that divides the domain of u into two disjoint subsets, then the differential equation and the prescription for the values of u on S define an **initial-value problem,** which one solves to obtain u on one of the disjoint subsets. For example, if $n = 4$ and $S = \{(x, y, z, t_0) \mid x, y, z \in \mathbb{R}\}$, then all four-tuples (x, y, z, t) such that $t \neq t_0$ fall into one of two disjoint subsets $R_<$, $R_>$ of $\mathbb{R}^4$, defined as follows: If $t < t_0$, then $(x, y, z, t) \in R_<$; if $t > t_0$, then $(x, y, z, t) \in R_>$. An initial-value problem provides a physically reasonable description of the evolution for times $t > t_0$ of a field (temperature, for example), the value of which is given everywhere at an initial time $t = t_0$.

In contrast, a **boundary-value problem** consists of a differential equation and prescribed values of the unknown function u on the boundary of the domain of u. For example, Laplace's equation, $\nabla^2 \phi = 0$, and boundary conditions that specify the value of the potential ϕ on all conducting surfaces constitute a boundary-value problem.

In this section we concentrate on initial-value problems involving first-order differential equations of the form

$$\frac{dy}{dx} = f(x, y) \tag{3.85}$$

and an initial value $y(x_0) = y_0$. The restriction to first-order equations turns out not to be particularly important. A second-order ordinary differential equation of the form

$$\frac{d^2 y}{dx^2} = g(x, y) \tag{3.86}$$

is equivalent to a system of two first-order differential equations. To demonstrate this, let us consider the definition

$$\frac{dy}{dx} = y' \tag{3.87}$$

as an additional differential equation. The original second-order equation is first-order in y':

$$\frac{dy'}{dx} = g(x, y). \tag{3.88}$$

These two equations, taken together, constitute a system of two first-order equations. Initial or boundary conditions must be transferred from the original second-order equation. For example, the solution of the second-order equation

$$\frac{d^2 y}{dx^2} = -y \tag{3.89}$$

subject to the initial conditions $y(0) = 0$ and $y'(0) = 1$ is $y(x) = \sin x$. This same solution can be found by solving the system

$$\frac{dy}{dx} = y'$$
$$\frac{dy'}{dx} = -y \tag{3.90}$$

with the same initial conditions. Although no computation is necessary to obtain the solutions of either the second-order differential equation or the equivalent first-order system in this example, the method of reducing a differential equation of second or higher order to an equivalent system of first-order equations permits one to use the same suite of computational methods for all orders and for general ordinary differential equations that do not have an analytic solution.

3.5.1 EULER'S METHOD

The elementary definition of the derivative of a function f,

$$\frac{dy}{dx}(x) = \lim_{h \to 0} \frac{y(x + h) - y(x)}{h}, \tag{3.91}$$

suggests that in order to obtain a numerical solution of the ordinary differential equation (3.85) on the interval $[0, L]$, all one has to do is to select a finite value of h,

$$h = \frac{L}{N},$$ (3.92)

write

$$y(x + h) \approx y(x) + h \left. \frac{dy}{dx} \right|_x,$$ (3.93)

and then apply this relation recursively to get $y(x + 2h)$, and so forth.

It turns out that the approximation (3.93) is nothing more than a version of the rectangle rule of numerical integration. One can verify by differentiation with respect to h that the differential equation (3.85) is equivalent to the integral equation

$$y(x + h) = y(x) + \int_x^{x+h} f(x', y(x')) \, dx'.$$ (3.94)

Equation (3.93) follows if one evaluates this integral using a left-hand rectangle rule. If one applies other integration rules to this same integral, one obtains different methods for solving ordinary differential equations. We discuss a few of these methods below.

To find the numerical values $c_0, \ldots, c_N$ of a **computed solution** c at the points

$$x_0, \ldots, x_N = x_0 + L = x_0 + Nh,$$ (3.95)

one approximates the differential equation by the difference equation

$$\frac{c_{n+1} - c_n}{h} = f(x_n, c_n).$$ (3.96)

Then one can compute the value of the computed solution on the next sampling point, c_{n+1} in terms of c_n and f_n from the equation

$$\boxed{c_{n+1} = c_n + hf(x_n, c_n).}$$ (3.97)

Given the initial value

$$c_0 = y_0,$$ (3.98)

which must be known in order to solve either the differential equation or the difference equation, one can compute f_0, then c_1, then f_1, then c_2, and so on, eventually obtaining the value of c_N. The algorithm (3.97) for solving the differential equation (3.85) is called **Euler's method**.

Euler's method is **explicit**, meaning that the unknown value c_{n+1} does not occur as an argument of the "right-hand side" f. Methods in which c_{n+1} does occur as an argument of

f are called **implicit**. An explicit method usually is computationally faster than an implicit method of comparable accuracy because an implicit equation must be solved iteratively in most cases.

3.5.2 TRUNCATION ERROR OF EULER'S METHOD

In this section we analyze the accuracy of Euler's method for solving first-order ordinary differential equations of the form $dy/dx = f(x, y)$ and compare the error of the algorithm – the so-called truncation error – with the rounding error that results from applying the method for many steps.

If the function f in Eq. (3.85) is sufficiently differentiable, one can write

$$f(x, y) = f(x, y(x_0)) + (y - y_0)\frac{\partial f}{\partial y} + O((y - y_0)^2) \tag{3.99}$$

in which $y_0 := y(x_0)$ is an approximation to $y(x)$. For the purpose of analyzing *local* accuracy and stability, one needs only an estimate of the effects of small changes in the value of f. For a first approximation, then, it is enough to use a differential equation that is linear in y. We choose the linearized test problem

$$f(x, y) = ay(x) \Rightarrow f(x_n, c_n) = ac_n \tag{3.100}$$

in which a is a constant. Then the difference equation reduces to a two-term recurrence relation,

$$c_{n+1} = (1 + ha)c_n, \tag{3.101}$$

the general solution of which is

$$c_N = \xi^N c_0 \tag{3.102}$$

in which

$$\xi = 1 + ha. \tag{3.103}$$

The analytical solution of the differential equation

$$\frac{dy}{dx} = ay \tag{3.104}$$

on the interval $[x_0, x]$ with the initial condition

$$y(x_0) = y_0 = c_0 = K \tag{3.105}$$

(in which K is a constant) is

$$y(x) = Ke^{ax}. \tag{3.106}$$

We compare these two expressions on an interval of one step, $[x_0, x_1]$ and on the larger interval $[x_0, x_N]$.

On an interval whose length is one step size, such as $[x_0, x_1]$, one has

$$y_1 = y(x_1) = Ke^{ha}. \tag{3.107}$$

The solution of the difference equation (3.101) for one step is

$$c_1 = K\xi \tag{3.108}$$

in which $\xi = 1 + ha$. The difference between the computed and exact solutions for one step (from x_0 to x_1) is called the **local truncation error**,

$$T_1 := c_1 - y_1. \tag{3.109}$$

Substituting c_1 from Eq. (3.108) and expanding the exact one-step solution ($y = e^{ax}$) in a Maclaurin series through the quadratic term, one obtains

$$\begin{aligned}
T_1 &= c_1 - y_1 \\
&= K[\xi - e^{ha}] \\
&\approx -\frac{K}{2}(ha)^2,
\end{aligned} \tag{3.110}$$

provided that $ha \ll 1$. Because the local truncation error is $O(h^2)$, Euler's method is locally accurate to first order in the step size, h.

The **global truncation error** T_N is the difference between the computed and exact solutions at $x_N = L$, given the initial condition (3.105):

$$\begin{aligned}
T_N &= c_N - y_N \\
&= K[\xi^N - e^{Nha}] \\
&= K[\xi^{L/h} - e^{aL}] \\
&\approx -\frac{K}{2}a^2Lh.
\end{aligned} \tag{3.111}$$

Note that the global truncation error of Euler's method is $O(h)$, not $O(h^2)$, because $T_N \approx NT_1$ and because $Nh = L$ is fixed.

One is tempted to conclude that all that is necessary to obtain an arbitrarily small global truncation error in solving the differential equation (3.104) is to make the step size h sufficiently small. However, given a fixed point $x = L$ at which the solution is needed, the number of steps N increases as the step size decreases. It follows that there is an optimum step size h_{opt} at which the sum of the truncation and rounding errors is a minimum.

Figure 3.6 shows the global error in Euler's method, as computed using IEEE-754 single-precision arithmetic. The global error plotted here includes both rounding error and truncation error because it is not possible to "turn off" rounding error. As a function of the step size h, the global error fluctuates (because of rounding error) but decreases by roughly an order of

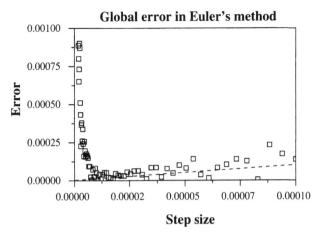

FIGURE 3.6

Absolute value of the global error in Euler's method for $L = 1$ and $a = 1$. The computation was performed using IEEE-754 single-precision arithmetic. The global truncation error predicted by Eq. (3.111) is shown as a dashed line. For step sizes larger than approximately $h = 0.00001$, the error decreases roughly linearly in h with considerable scatter due to rounding error. For step sizes much smaller than $h = 0.00002$, the error increases rapidly because of an accumulation of rounding errors.

magnitude as the step size decreases from $h = 0.0001$ to the vicinity of $h = 0.00001$ in qualitative agreement with the estimate of global truncation error in Eq. (3.111). For step sizes that are significantly less than $h = 0.00001$, the global error increases because rounding error increases more rapidly than the global truncation error decreases. For this particular example, then, $h_{\text{opt}} \approx 0.00001$.

3.5.3 STABILITY ANALYSIS OF EULER'S METHOD

In fields as far apart as engineering and economics, a system is called **stable** if and only if a bounded input produces a bounded output. Because Eq. (3.102) implies that the computed solution at x_n is proportional to ξ^n, the approximate solution of the differential equation $y' = f(x, y)$ computed by Euler's method is stable if and only if ξ^n does not grow geometrically:

$$|\xi| \leq 1 \implies |1 + ha| \leq 1. \tag{3.112}$$

Because a can be either real or complex, the set of values of ha for which the computed solution is stable lies within the disk of radius 1 centered at $ha = -1$ in the complex plane (Fig. 3.7). For a stable system, $\text{Re}(a) < 0$. In computing a numerical solution of a differential equation (or a system of differential equations) describing a stable system, it is essential to choose the step size, h, so that ha lies within the region of stability of the numerical method.

Figure 3.8 illustrates the consequences of misusing a finite-difference method by choosing a step size that is so large that ha lies outside the region of stability. In this example, $a = -1$

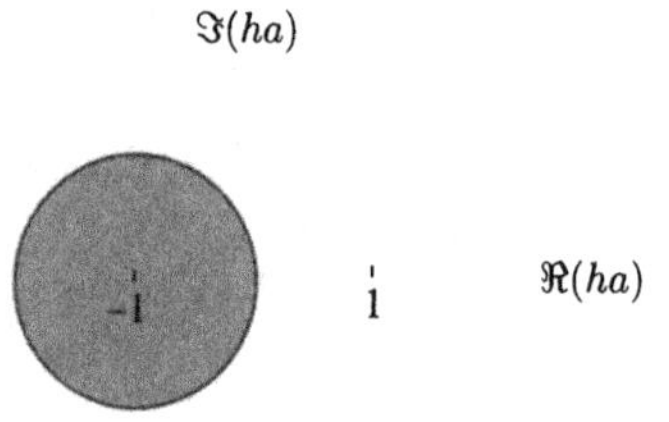

$\Im(ha)$

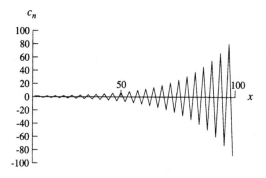

$\Re(ha)$

FIGURE 3.7

The region of stability of Euler's method, computed using the test equation $y' = ay$, is a disk of radius 1 centered at $z = -1$, for which $z = ha$.

and $h = 2.1$, giving $\xi = 1 + ha = -1.1$, which violates the stability criterion for Euler's method, Eq. (3.112). The computed solution is wrong in every detail. The magnitude of the computed solution is greater than 1 at each x_n; the analytical solution, $y(x) = e^{ax}$, is less than 1 for every $x > 0$. The analytical solution approaches zero monotonically as $x \to \infty$, but the computed solution oscillates and its magnitude becomes large without limit as $x \to \infty$.

3.5.4 SELECTED FINITE-DIFFERENCE METHODS

Midpoint Method

In order to avoid the large $(O(h^2))$ local truncation error of Euler's method, one naturally looks for a more accurate integration formula that one can turn into a recursive algorithm for solving a differential equation. The integral equation

$$y(x + h) = y(x - h) + \int_{x-h}^{x+h} f(x', y(x')) \, dx' \tag{3.113}$$

is completely equivalent to the original differential equation (3.85). If one uses the centered rectangle rule (on a panel of width $2h$) to obtain an approximation for $y(x_{n+1})$, one obtains the **midpoint method**

$$\boxed{c_{n+1} = 2hf(x_n, c_n) + c_{n-1}} \tag{3.114}$$

FIGURE 3.8

The solution computed to the equation $y' = y$ by Euler's method with a step size $h = 2.1$, which is outside of the region of stability. The initial condition is $y(0) = 1$. The analytical solution is so heavily damped, and is of such a small magnitude, that on the scale of this graph it nearly coincides with the horizontal axis.

In order to analyze the properties of the solutions of the midpoint difference equation (3.114), let us assume that $f(x, y) = ay$ and substitute the trial solution

$$c_n = c_0 \xi^n \tag{3.115}$$

in Eq. (3.114). The resulting equation is

$$\xi^{n+1} = 2ha\xi^n + \xi^{n-1}. \tag{3.116}$$

After canceling the common factor ξ^{n-1}, one obtains the **characteristic equation** of the midpoint method,

$$\xi^2 - 2ha\xi - 1 = 0. \tag{3.117}$$

The midpoint method is an example of a **two-step method**, meaning that the difference equation that defines the method, Eq. (3.114), includes the values of the computed solution at three different points $(x_{n-1}, x_n$ and $x_{n+1})$. Hence the midpoint method's characteristic equation involves ξ^2, ξ and $\xi^0 = 1$, and therefore is quadratic. The Euler method is a one-step method, and therefore its characteristic equation, Eq. (3.103), is linear.

The quadratic formula gives the solution of the midpoint characteristic equation as

$$\xi = ha \pm \sqrt{(ha)^2 + 1}. \tag{3.118}$$

One root is approximately equal to the exact one-step solution, e^{ha}, of the original differential equation $y' = ay$:

$$\begin{aligned} \xi_1 &= 1 + ha + \frac{(ha)^2}{2} + O((ha)^3) \\ &= e^{ha} + O((ha)^3). \end{aligned} \tag{3.119}$$

However, there is an additional root,

$$\xi_2 = -\frac{1}{\xi_1}. \tag{3.120}$$

As we shall see momentarily, the existence of ξ_2 renders the midpoint method useless for all but a restricted class of differential equations.

In a multistep method for solving first-order differential equations, the root of the characteristic equation that is approximately equal to e^{ha} is called the **principal characteristic root**. The other roots, if there are any, are called **parasitic roots**. For the midpoint method, ξ_1 is the principal root, and ξ_2 is the parasitic root.

The general solution of the midpoint difference equation (3.114) is a linear combination of powers of the two characteristic roots, provided that $\xi_2 \neq \xi_1$:

$$c_n = \alpha\xi_1^n + \beta\xi_2^n. \tag{3.121}$$

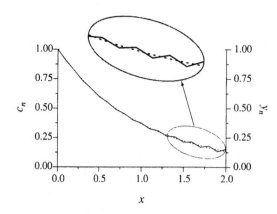

c_n

y_n

x

FIGURE 3.9

The computed (*solid line*) and exact (*dashed line*) solutions of the differential equation $y' = -y$, illustrating a weak instability. The first step of the computed solution was taken using Euler's method; the remaining steps were taken with the midpoint method.

(The underlying mathematics that makes this assertion rigorously true is covered in Section 9.6.2). Because there are two constants (α and β) in Eq. (3.121), two initial conditions are necessary to start the midpoint method. Given the original initial condition $c_0 = y_0$, one uses a one-step method (such as one of the Runge–Kutta methods) to obtain c_1. The equations

$$c_0 = \alpha + \beta$$
$$c_1 = \alpha\xi_1 + \beta\xi_2 \tag{3.122}$$

then determine α and β. For a k-step method, k initial points are required.

Because the computed value c_1 "almost never" matches the exact value $y(x_1)$, one usually has $\beta \neq 0$. Therefore one cannot eliminate the parasitic root ξ_2 or its contribution to c_n, $\beta\xi_2^n$. Figure 3.9 shows the consequences of attempting to use the midpoint method to compute the solution of the differential equation $y' = -y$ with the initial condition $y(0) = 1$ and with a step size $h = 0.1$. In this example,

$$\xi_1 = ha + \sqrt{1 + (ha)^2} = -0.1 + \sqrt{1.01} \approx 0.9050, \tag{3.123}$$
$$\xi_2 = -\frac{1}{\xi_1} \approx -1.1050, \tag{3.124}$$

and

$$\alpha \approx 0.9975, \qquad \beta \approx 0.0025. \tag{3.125}$$

Because $\beta \neq 0$, the geometrically growing parasitic contribution $\beta\xi_2^n$ becomes large enough to make the sum of the parasitic contribution and the decaying principal contribution oscillate noticeably, as shown in Fig. 3.9. For still larger values of x, the parasitic contribution overwhelms the principal contribution. In contrast to our example of instability in Euler's method, in which the modulus of the principal (and only) characteristic root is greater than 1, here the modulus of the principal root is less than 1. This is an example of a **weak instability** in which the modulus of the principal characteristic root is not greater than 1 but the modulus of at least one of the parasitic roots is greater than 1.

Evidently the concept of stability needs to be refined in the light of the computational fact that parasitic roots exist in all multistep methods. The usual definition is that a finite-difference method is **absolutely stable** if and only if

- The modulus of the principal characteristic root is not greater than 1, and
- No parasitic root has a greater modulus than the principal characteristic root.

One also calls a finite-difference method **relatively stable** if and only if no parasitic root has a greater modulus than the principal characteristic root. Both definitions exclude finite-difference methods that exhibit weak instability, such as the midpoint method.

The set of values of ha for which a finite-difference method for the differential equation $y' = ay$ is absolutely stable is called the **region of absolute stability** of the method. For Euler's method, the region of absolute stability is the disk of unit radius centered at $ha = -1$, as shown in Fig. 3.7. For the midpoint method, the region of absolute stability is the line segment from $-i$ to i in the complex plane (see Exercise 3.5.2).

A finite-difference method must be absolutely stable in order to provide accurate solutions of differential equations that describe stable systems. If the system being described is unstable, then at least one of the functions that describes the system grows exponentially. In this case one cannot require the finite-difference method to be absolutely stable, but one must still require relative stability in order to compute an accurate solution. Even if one uses a "canned" program written by "experts" to solve a differential equation, there can be no excuse for using a finite-difference method outside of its region of absolute or relative stability, whichever is applicable.

Backward Euler Method

The **backward Euler method** is similar to Euler's method, except that the derivative term is evaluated at the point at which the solution is to be computed rather than the point at which the solution has been computed:

$$c_{n+1} = hf(x_{n+1}, c_{n+1}) + c_n. \tag{3.126}$$

This results in an **implicit** method in which the unknown value of the computed solution, c_{n+1}, occurs both on the left-hand side and as an argument of the possibly nonlinear function f. If one uses an implicit finite-difference method, then one must supply an algorithm that can solve an implicit difference equation such as Eq. (3.126) quickly and accurately.

For the linearized test problem $dy/dx = ay$, the backward Euler difference equation becomes

$$c_{n+1} = hac_{n+1} + c_n$$
$$\Rightarrow c_{n+1} = \xi c_n \tag{3.127}$$

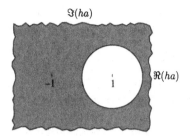

$\Im(ha)$

$\Re(ha)$

FIGURE 3.10

The region of stability of the backward Euler method, computed using the test equation $y' = ay$, is the set of points outside of, and on the circumference of, a disk of radius 1 centered at $z = 1$, for which $z = ha$.

in which the principal characteristic root is

$$\xi = \frac{1}{1 - ha}.$$ (3.128)

Because this is a one-step method, there are no parasitic roots. The solution of the difference equation for the test problem is

$$c_n = c_0 \xi^n,$$ (3.129)

provided that one neglects roundoff error.

The region of absolute stability of the backward Euler method is the set of points $z = ha$ such that

$$\frac{1}{|1 - z|} \leq 1 \Rightarrow |1 - z| \geq 1.$$ (3.130)

This set consists of the points that lie *outside* of, or on the circumference of, the disk in Fig. 3.10.

Trapezoidal Method

If one applies the one-panel trapezoidal rule to the integral (3.94), one obtains the difference equation for the **trapezoidal method**,

$$c_{n+1} = c_n + \frac{h}{2}(f(x_{n+1}, c_{n+1}) + f(x_n, c_n)).$$ (3.131)

Evidently this is a one-step, implicit method. Because c_{n+1} occurs both on the right-hand side and as an argument of f, this equation must be solved iteratively in general.

For the linearized test problem (3.100), the trapezoidal difference equation becomes

$$c_{n+1} = c_n + \frac{ha}{2}(c_{n+1} + c_n).$$ (3.132)

Solving for c_{n+1}, one gets

$$c_{n+1} = \xi c_n$$ (3.133)

in which the principal (and only) characteristic root is

$$\xi = \frac{1 + \frac{1}{2}ha}{1 - \frac{1}{2}ha}. \tag{3.134}$$

The region of absolute stability in the plane of complex $z = ha$ is the set such that

$$\left| \frac{1 + z/2}{1 - z/2} \right| \leq 1. \tag{3.135}$$

This set is the left half-plane (see Exercise 3.5.3).

Midpoint-Trapezoidal Predictor–Corrector Method

Reviewing the choices for finite-difference methods with local truncation errors of order h^3, one sees that the midpoint method is explicit, hence computationally fast, but is weakly unstable, and the trapezoidal method is stable but implicit and hence computationally slow. In order to obtain a finite-difference method that combines the stability of the trapezoidal method with the computational speed of the midpoint method, one can use the midpoint difference equation to obtain a prediction p_{n+1} of the value of the computed function at x_{n+1}:

$$p_{n+1} = 2hf(x_n, c_n) + c_{n-1}. \tag{3.136}$$

One then uses the predicted value of c_{n+1} in the trapezoidal difference equation:

$$c_{n+1} = \frac{h}{2}[f(x_{n+1}, p_{n+1}) + f(x_n, c_n)] + c_n. \tag{3.137}$$

These two difference equations constitute the **midpoint-trapezoidal predictor–corrector method**.

The solution of the midpoint-trapezoidal difference equations for the usual linearized test problem $dy/dx = ay$ is

$$c_n = \alpha \xi_1^n + \beta \xi_2^n, \tag{3.138}$$

for which the characteristic roots are

$$\begin{aligned}
\xi_1 &= \frac{1}{2}\left[1 + \frac{ha}{2} + (ha)^2\right] \\
&\quad + \frac{1}{2}\left\{\left[1 + \frac{ha}{2} + (ha)^2\right]^2 + 2ha\right\}^{1/2} \\
&= e^{ha} + O((ha)^3) \quad \text{(principal root)} \\
\xi_2 &= \frac{1}{2}\left[1 + \frac{1}{2}ha + (ha)^2\right] \\
&\quad - \frac{1}{2}\left\{\left[1 + \frac{1}{2}ha + (ha)^2\right]^2 + 2ha\right\}^{1/2} \\
&\quad \text{(parasitic root)}.
\end{aligned} \tag{3.139}$$

Figure 3.11 shows the region of absolute stability that these roots imply.

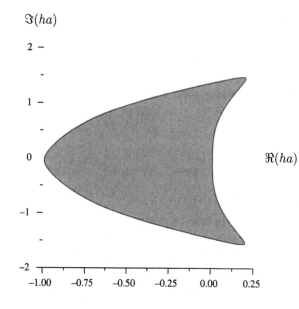

FIGURE 3.11

The region of absolute stability of the midpoint-trapezoidal predictor–corrector method, computed using the test equation $y' = ay$, in the plane of complex $z = ha$.

3.5.5 EXERCISES FOR SECTION 3.5

3.5.1   Compute the values of $\sin x$ for $x = .01, .1, 1.0, 10.0$, and 100.0 by using Euler's method, the midpoint method, and the midpoint-trapezoidal predictor–corrector method to solve the system of differential equations

$$\frac{dy}{dx} = y'$$
$$\frac{dy'}{dx} = -y \tag{3.140}$$

with the initial conditions $y(0) = 0$, $y'(0) = 1$. Compare your results with the values returned by the sine function in your compiler's library of mathematical functions and with the results of Exercise 3.3.1.

3.5.2   Show that the region of absolute stability of the midpoint method is the line segment joining $-i$ and i in the complex ha-plane.

3.5.3   Show that if a complex number $z = x + iy$ satisfies Eq. (3.135), then $x \le 0$. Conclude that the region of absolute stability of the trapezoidal method is the left half-plane.

3.5.4   Derive Eq. (3.139) for the characteristic roots of the midpoint-trapezoidal predictor–corrector finite-difference method.

3.5.5   Graph e^{ha}, $|\xi_1|$ (and $|\xi_2|$, if applicable), for *real* values of ha, for the following finite-difference methods:

1. Euler's method

2. The midpoint method

3. The backward Euler method

4. The trapezoidal method

5. The midpoint-trapezoidal predictor–corrector method

3.5.6 Graph e^{ha}, $|\xi_1|$ (and $|\xi_2|$, if applicable), for *purely imaginary* values of ha, for the following finite-difference methods:

1. Euler's method

2. The midpoint method

3. The backward Euler method

4. The trapezoidal method

5. The midpoint-trapezoidal predictor–corrector method

3.5.7 In some applications, such as phase-sensitive nonlinear processes, it is necessary to choose a finite-difference method with high phase accuracy. The appropriate linearized test problem for an undamped, phase-sensitive process is

$$\frac{dy}{dx} = ia''y, \tag{3.141}$$

where a'' is real. Show that if one defines the **local phase error** of a finite-difference method as

$$\Phi := \theta_1 - ha'' \tag{3.142}$$

in which θ_1 is the phase of the principal characteristic root,

$$\xi_1 = |\xi_1|\, e^{i\theta_1}, \tag{3.143}$$

and if $|\Phi| \ll 1$, then the local phase error is approximately equal to the modulus of the local truncation error:

$$\Phi \approx |\xi_1 - e^{ha}|. \tag{3.144}$$

3.5.8 Write a C, C++, or FORTRAN program to compute a numerical solution of the following time-dependent Schrödinger equation for a two-level quantum system, using the midpoint-trapezoidal predictor-corrector method:

$$\frac{dc_0}{dt} = ic_1$$
$$\frac{dc_1}{dt} = i\Delta c_1 + ic_0. \tag{3.145}$$

Assume that Δ is real, and that the initial conditions are

$$c_0 = 1$$
$$c_1 = 0. \tag{3.146}$$

1. Compute the solution, and graph $|c_0|$ and $|c_1|$ versus step number for at least twenty periods of oscillation for each of the following conditions:

    (a) $\Delta = 0.1$ and for the following values of h: $h = .01, 0.1, 1.0, 2.0, 3.0, 4.0$

(b) $\Delta = 1.0$ and for the following values of h: $h = .01, 0.1, 1.0, 2.0, 3.0, 4.0$

(c) $\Delta = 10.0$ and for the following values of h: $h = .01, 0.1, 1.0, 2.0, 3.0, 4.0$

2. Plot the following analytical solution for each of the preceding sets of conditions and compare with the numerical solution,

$$c_0(t) = e^{i\Delta t/2}\left[\cos\omega_R t - i\frac{\Delta}{2\omega_R}\sin\omega_R t\right]$$
$$c_1(t) = \frac{ie^{i\Delta t/2}}{\omega_R}\sin\omega_R t, \tag{3.147}$$

where

$$\omega_R = \left[1 + \left(\frac{\Delta}{2}\right)^2\right]^{\frac{1}{2}}. \tag{3.148}$$

3.6 BIBLIOGRAPHY

Milton, Abramowitz, and Irene A. Stegun, "Handbook of Mathematical Functions with Formulas, Graphs, and Mathematical Tables." Washington: National Bureau of Standards, Applied Mathematics Series, Vol. 55, 1964.

Dahlquist, Germund, and Åke Björck. *Numerical Methods* (Ned Anderson, trans.). Englewood Cliffs: Prentice-Hall, 1974, chapter 2.

Goldberg, David. "What Every Computer Scientist Should Know About Floating-Point Arithmetic." *ACM Computing Surveys* 23(1991):5–48. This paper contains the proofs of the unproven statements in the section on floating-point computation. Note that Goldberg's ϵ is equal to $\frac{1}{2}\epsilon_{mach}$ in our notation.

Kahan, W. "A Survey of Error Analysis." In C. L. Frieman editor, *Information Processing 71*. Amsterdam: North-Holland, 1972, pp. 1214–1239.

Press, William H., Saul A. Teukolsky, William T. Vetterling, and Brian P. Flannery. *Numerical Recipes*. 2nd ed. New York: Cambridge, 1992.

Wilkinson, J. H. *Rounding Errors in Algebraic Processes*. New York: Dover, 1994.

CHAPTER 4

GROUPS, RINGS, AND FIELDS

4.1 INTRODUCTION

Group theory began with Lagrange's work in 1770 and 1771 on groups of substitutions (which we would call permutations) of the roots of polynomial equations. Many fundamental group-theoretical concepts were introduced by Évariste Galois, an eccentric genius who was killed in a duel in 1832 at the age of twenty. After several decades it was finally recognized that Galois had succeeded in creating a theory that explains why polynomial equations of degree five or greater cannot generally be solved in radicals. Arthur Cayley introduced the concept of an abstract group in 1849 through 1854, but his work was ignored for several more decades. In 1894, Sophus Lie (pronounced "Lee") published a pioneering paper on continuous groups. His objective was to create a comprehensive theory of the solvability of differential equations in quadratures, much as Galois had done for the solvability of polynomial equations in radicals. During the twentieth century nearly every area of mathematics has been simplified and generalized by introducing abstract algebraic structures such as groups.

Group theory can be applied usefully in almost any situation in which a physical system possesses a geometrical symmetry. Calculations of the transmission and reflection coefficients for microwaves or light waves at waveguide junctions, for example, become significantly simpler if the waveguide geometry is unchanged under reflection in a plane or under a rotation through some angle. Also, the symmetry of a crystal's unit cell determines important relations among the coefficients that relate an input, such as a strain tensor, to an output such as the magnitude and direction of a piezoelectrically generated electric field.

Significant applications of group theory to quantum physics began in 1928 with the publication of Weyl (1928). Eugene P. Wigner and Giulio Racah systematically applied the theory of the matrix representations of groups to simplify calculations in atomic, molecular, and solid-state physics by taking advantage of geometrical symmetries (such as the invariance of the Hamiltonian of an isolated atom under rotations) and of the symmetry or antisymmetry of the wave function under permutations of identical particles. Wigner (1959) made it clear that all selection rules involving angular momentum or parity can be derived from geometrical symmetry principles. However, group theory was not universally appreciated by physicists. Some referred derisively to "die Gruppenpest," meaning "the plague of group (theory)." Many physicists preferred to use simpler but often more cumbersome techniques such as commutation relations to derive results that could have been obtained from the theory of Lie groups.

Group theory's importance in the quantum physics of crystalline solids derives from the rotational–reflectional symmetry of the crystal's unit cell and from the symmetry of the entire

121

crystal lattice under a group of rotations, reflections, and translations. The number of energy bands of conduction electrons for a given wave vector, for example, can be determined directly by symmetry considerations with the help of group theory. Chemists also make extensive use of group theory in classifying the electronic, vibrational, and rotational energy levels of polyatomic molecules.

The most fundamental applications of group theory in physics involve *dynamical*, not geometrical, symmetries. According to the "Eightfold Way" [now known as flavor $SU(3)$], which was created in 1961 by Murray Gell-Mann and Yuval Ne'eman, the strong interaction that binds atomic nuclei is symmetric under transformations of the Lie group $SU(3)$. This approach began to be taken seriously after the discovery in 1964 of the Ω^- particle, the existence of which is predicted by flavor $SU(3)$ symmetry. Even then, some physicists who were committed to other approaches to understanding the strong interaction regarded group theory as "the sheerest garbage," in the words of one.

The primacy of group theory in elementary-particle physics is now established beyond a reasonable doubt. The current Standard Model of the elementary particles ascribes $SU(3)$ symmetry to the strong interaction and $U(1) \times SU(2)$ symmetry to the weak and electromagnetic interactions. Essential features of the Standard Model have been confirmed by the experimental discovery of quarks and the W and Z particles.

The theory of matrix representations of groups has become so fundamental in physics that elementary particles and the energy levels of atomic, molecular, and crystalline systems are routinely identified with abstract vectors that transform among one another according to the irreducible representations of certain groups. Over the past two centuries group theory has evolved from being a curiosity to being a major part of the foundation on which modern mathematics and physics are built.

From a student's point of view, the theory of groups and their representations is useful because it provides a few organizing principles that substantially simplify the assimilation of difficult, detailed material, even in "classical" areas such as the special functions of mathematical physics. Groups are the building blocks from which more complex and familiar structures such as vector spaces are assembled.

4.2 GROUPS

4.2.1 AXIOMS

A **group** $(G, \circ)$ is a set G, together with a function $\circ$ from the Cartesian product $G \times G$ to G (recall Section 2.2.4) that obeys the following axioms:

1. *Closure:G* is closed under $\circ$:

$$\forall a, b \in G : \exists c \in G : \ni: \quad c = a \circ b. \tag{4.1}$$

It is equally correct (and perhaps a little clearer) to write

$$c = \circ(a, b). \tag{4.2}$$

One can think of ∘ as standing for "operation" (or "op"). Sometimes one calls ∘ the **group operation** and $a \circ b$ the **group product** of b and a. (In writing the product $a \circ b$ for a group of symmetry operations, one thinks of carrying out b, then a.) Instead of writing $(G, \circ)$, one can say that G **is a group under** ∘.

2. *Associativity*:

$$\forall a, b, c \in G : a \circ (b \circ c) = (a \circ b) \circ c. \tag{4.3}$$

One can equally well write

$$\circ(a, \circ(b, c)) = \circ(\circ(a, b), c). \tag{4.4}$$

3. *Unit element (identity)*: There exists a left identity $e \in G$:

$$\exists e \in G : \ni : \forall a \in G : \quad e \circ a = a. \tag{4.5}$$

The conventional use of the letter e stems from the German word *Einheit*, meaning unity (as in the number 1).

4. *Inverse*: For each $a \in G$, there exists a left inverse (written a^{-1}, read "a inverse"):

$$\forall a \in G : \exists a^{-1} \in G : \ni : \quad a^{-1} \circ a = e. \tag{4.6}$$

To define a group operation ∘ on a set G, one must specify the image $c = a \circ b$ for all possible pairs a, b in such a way that axioms 1 through 4 are satisfied. If the definition of ∘ is clear from the context, we shall refer to "the group G" rather than to "the group $(G, \circ)$" and shall write $a^{-1}a$ instead of $a^{-1} \circ a$.

Do any groups exist? Yes: The set $\{1\}$ that consists of the real number 1 is a group under ordinary multiplication, because $(\{1\}, \cdot)$ satisfies axioms 1 through 4 trivially. We give several less trivial examples in subsequent sections.

Finite and Infinite Groups

If G is a finite set, then one calls $(G, \circ)$ a **finite group**. The number of elements of G is called the **order** of the group $(G, \circ)$ and is usually written $|G|$. For example, the set of rotational symmetries of an isosceles triangle is a group C_2 of order $|C_2| = 2$, which we describe in Section 4.2.2. Also, the rotations and reflections that leave a cube invariant are the elements of the *full octahedral group*, O_h. There are forty-eight such rotations and reflections; hence the order of O_h is 48.

If the set G is infinite, then $(G, \circ)$ is an **infinite group**. Among the most familiar infinite groups are **groups of numbers** such as $(\mathbb{Z}, +)$, $(\mathbb{Q}, +)$, $(\mathbb{Q}, \cdot)$, $(\mathbb{R}, +)$, $(\mathbb{R}, \cdot)$, $(\mathbb{C}, +)$, and $(\mathbb{C}, \cdot)$; $\mathbb{Z}, \mathbb{Q}, \mathbb{R}$, and $\mathbb{C}$ are the integers, rational numbers, real numbers, and complex numbers, respectively (see Exercises 4.2.1–4.2.7).

Another class of frequently encountered infinite groups is the set

$$n\mathbb{Z} := \{nj \mid j \in \mathbb{Z}\} \tag{4.7}$$

of **multiples of an integer n** under addition. The sum of two multiples of n is also a multiple of n; therefore the closure axiom is satisfied. Associativity follows from the associativity of integer addition. The identity element is 0, and the additive inverse of nj is $-nj$. Therefore $(n\mathbb{Z}, +)$ is a group.

Another class of infinite groups, additive groups of vectors, is fundamentally important in engineering and physics. For example, the set $\mathbb{R}^2$ of ordered pairs of real numbers

$$\mathbf{r} = \begin{pmatrix} x \\ y \end{pmatrix} \tag{4.8}$$

(better known as the set of two-dimensional real vectors) is a group under the operation of vector addition,

$$\mathbf{r} + \mathbf{r}' = \begin{pmatrix} x \\ y \end{pmatrix} + \begin{pmatrix} x' \\ y' \end{pmatrix} = \begin{pmatrix} x + x' \\ y + y' \end{pmatrix}. \tag{4.9}$$

(We show in Chapter 6 that in order to maintain compatibility with the usual convention that matrices act on vectors from the left, one must write ordered pairs that one interprets as vectors as columns rather than rows.) In $(\mathbb{R}^2, +)$, the identity element is the zero vector,

$$\mathbf{0} = \begin{pmatrix} 0 \\ 0 \end{pmatrix} \tag{4.10}$$

and the additive inverse of the vector $\mathbf{r}$ is

$$-\mathbf{r} = \begin{pmatrix} -x \\ -y \end{pmatrix}. \tag{4.11}$$

The elements of **matrix groups** are the matrix realizations of linear mappings (see Chapter 6) or affine mappings (see Appendix B), instead of numbers. Infinite groups of linear mappings that are important in physics, such as the translation and orthogonal groups, are discussed in Chapters 11 and 12.

For example, the set $GL(2, \mathbb{R})$ of 2×2 matrices

$$\mathbf{A} = \begin{pmatrix} a & b \\ c & d \end{pmatrix} \tag{4.12}$$

with real elements a, b, c, d and with a nonzero determinant

$$\det[\mathbf{A}] := ad - bc \tag{4.13}$$

is an infinite group. (The notation $GL(2, \mathbb{R})$ means "the general linear group of a real vector space with dimension 2;" see Chapter 6.) It is straightforward to verify that the identity element in $GL(2, \mathbb{R})$ is the 2×2 unit matrix

$$\mathbf{1}_{2 \times 2} = \begin{pmatrix} 1 & 0 \\ 0 & 1 \end{pmatrix} \tag{4.14}$$

and that the inverse of the matrix **A** above is

$$\mathbf{A}^{-1} = \frac{1}{\det[\mathbf{A}]} \begin{pmatrix} d & -b \\ -c & a \end{pmatrix}. \tag{4.15}$$

The product **AB** of two matrices

$$\mathbf{A} = \begin{pmatrix} a & b \\ c & d \end{pmatrix}, \quad \mathbf{B} = \begin{pmatrix} a' & b' \\ c' & d' \end{pmatrix} \tag{4.16}$$

in $GL(2, \mathbb{R})$ has the determinant

$$\det[\mathbf{AB}] = a'd'ad + b'c'bc - (a'd'bc + b'c'ad) = \det[\mathbf{A}]\det[\mathbf{B}], \tag{4.17}$$

which vanishes if and only if the determinant of **A** or **B** vanishes. Hence $GL(2, \mathbb{R})$ is closed under matrix multiplication. The proof that the multiplication of 2×2 matrices is associative is Exercise 4.2.20. It follows that $GL(2, \mathbb{R})$ is a group under the operation of matrix multiplication.

We discuss matrices and determinants in greater detail in Chapter 6.

Abelian and Non-Abelian Groups
Axiom 1 does not require the **commutative law**

$$\forall a, b \in G : a \circ b = b \circ a \tag{4.18}$$

to hold. A group in which Eq. (4.18) holds is called **Abelian**. From the equivalent form $\circ(a, b) = \circ(b, a)$, one sees that Eq. (4.18) strongly restricts the function $\circ$. For example, if a and b are real numbers and $f(a, b) = ab^2$, then $f(b, a) = ba^2 \neq f(a, b)$.

The commutative law, Eq. (4.18), fails in matrix groups. For example, let G be the group $GL(2, \mathbb{R})$ of 2×2 matrices with real elements and nonzero determinants (see Eqs. [4.12]–[4.17]). The matrix

$$\sigma_x = \begin{pmatrix} 0 & 1 \\ 1 & 0 \end{pmatrix} \tag{4.19}$$

belongs to $GL(2, \mathbb{R})$ because

$$\det[\sigma_x] = -1 \neq 0. \tag{4.20}$$

Let

$$\mathbf{A} = \begin{pmatrix} a & b \\ c & d \end{pmatrix} \tag{4.21}$$

belong to $GL(2, \mathbb{R})$. Then

$$\sigma_x \mathbf{A} = \begin{pmatrix} c & d \\ a & b \end{pmatrix} \tag{4.22}$$

but

$$\mathbf{A}\sigma_x = \begin{pmatrix} b & a \\ d & c \end{pmatrix},$$ (4.23)

which implies that $\sigma_x \mathbf{A} = \mathbf{A}\sigma_x$ if and only if $c = b$ and $d = a$. Therefore $GL(2, \mathbb{R})$ is not Abelian.

If a group G is Abelian, one often writes the group operation $\circ$ as addition ($a \circ b = a + b$) because of the commutativity of the addition of numbers. In many important cases, group elements are mappings whose compositions (or products) do not obey the commutative law. In most of our discussion of groups we shall therefore write $\circ$ as composition with the understanding that if we discuss Abelian groups we may change our notation to addition.

Nonassociative Function

Axiom 2 imposes an important constraint on the function $\circ$. One cannot assume, just because exact numerical addition and multiplication are associative, that any function from $G \times G$ to G obeys Eq. (4.4). For example, let $G = \mathbb{R} \setminus \{0\}$ (the nonzero real numbers) and $f(a, b) = ab^2$. Then $f(f(a, b), c) = ab^2c^2$, but $f(a, f(b, c)) = ab^2c^4$. Hence (G, f) is not a group.

Axiom 2 is also violated if $\circ$ is the addition of floating-point numbers rounded to a finite, fixed precision because the order in which successive additions and roundings are carried out affects the final result (see Eqs. [1.71]–[1.74]). Unlike the set of rational numbers $\mathbb{Q}$, the set of floating-point numbers in a given finite floating-point representation is not a group under addition.

Right Identity and Right Inverse

Axioms 3 and 4 imply the existence of a right inverse and a right identity. We begin the proof by observing that if $a \in G$, then (by axiom 4) there exists a left inverse $a^{-1} \in G$, and therefore (by axiom 3)

$$a^{-1}aa^{-1} = ea^{-1} = a^{-1}.$$ (4.24)

In turn, a^{-1} has a left inverse $(a^{-1})^{-1} \in G$:

$$(a^{-1})^{-1}a^{-1} = e.$$ (4.25)

Then

$$aa^{-1} = e(aa^{-1}) = ((a^{-1})^{-1}a^{-1})(aa^{-1}).$$ (4.26)

By the associative law and Eq. (4.24), the right-hand side of Eq. (4.26) is equal to

$$(a^{-1})^{-1}a^{-1}aa^{-1} = (a^{-1})^{-1}ea^{-1} = (a^{-1})^{-1}a^{-1} = e.$$ (4.27)

By Eqs. (4.25) and (4.27),

$$(a^{-1})^{-1}a^{-1}aa^{-1} = e = eaa^{-1} \Rightarrow aa^{-1} = e.$$ (4.28)

One can therefore replace Eq. (4.6) with the more general requirement

$$a^{-1}a = aa^{-1} = e. \tag{4.29}$$

It follows immediately from Eq. (4.29) that

$$ae = aa^{-1}a = ea. \tag{4.30}$$

Therefore a left identity is also a right identity, and axiom 3 becomes the more general law

$$ae = ea = a. \tag{4.31}$$

We have not shown yet that there is *only one* identity element in a group G, or that for any given element of G there is *only one* inverse. The uniqueness of the group identity and the uniqueness of the inverse a^{-1} are the subjects of Exercises 4.2.8 and 4.2.9.

One can show easily that the inverse of the product ab is

$$(ab)^{-1} = b^{-1}a^{-1}, \tag{4.32}$$

as is true generally for mappings (see Eq. [2.54] and Exercise 4.2.19).

An important property that follows immediately from the basic group axioms is the **cancellation law**:

$$ab = ac \Rightarrow b = c \quad \text{and} \quad ba = ca \Rightarrow b = c. \tag{4.33}$$

The first statement follows if one multiplies both sides of the equality $ab = ac$ by a^{-1} from the left: $a^{-1}ab = a^{-1}ac$. To establish the second statement in Eq. (4.33), multiply with a^{-1} from the right.

4.2.2 TWO-ELEMENT GROUP

Perhaps the most important finite group is the **two-element group** $\{e, a\}$. In order to see whether it is possible to define a group operation on a two-element set $\{e, a\}$, one must investigate the four possible images $d = b \circ c$ where $b \in \{e, a\}$ and $c \in \{e, a\}$. One element of $\{e, a\}$ must be the group identity; we choose e. Then the products $e \circ e = e, e \circ a = a \circ e = a$ are determined by group axiom 3. The product

$$a^2 := aa \tag{4.34}$$

must be equal to e because $aa = a = ae$ would imply $a = e$ by the cancellation law. The equation $a^2 = e$ implies that $a^{-1} = a$. Then group axiom 4 is satisfied. It is straightforward to verify that the function $\circ$ such that $e \circ e = e$ and $e \circ a = a \circ e = a$ satisfies the associative law (group axiom 2).

Therefore we have defined a group, which we write as

$$C_2 := \{e, a \mid a^2 = e\}. \tag{4.35}$$

TABLE 4.1 Multiplication
Table of the Group C_2

	First Operation	
C_2	e	a
e	e	a
a	a	e

Table 4.1 gives the multiplication table of C_2. (In the table, *first operation* refers to the group element on the right in a two-element product.) One realization of C_2 is the set $\{1, -1\}$ under the operation of multiplication. Another realization is the set

$$\left\{ \mathbf{1}_{2\times 2}, \begin{pmatrix} 1 & 0 \\ 0 & -1 \end{pmatrix} \right\} \tag{4.36}$$

under the operation of matrix multiplication.

Generators of a Group

If every element of a group $(G, \circ)$ can be obtained by applying the group operation $\circ$ to the elements of some subset S finitely many times, then one says that S **generates** G. The elements of S are called **generators** of $(G, \circ)$. For example, the set $\{a \mid a^2 = e\}$ generates C_2.

Physical Realizations of the Two-Element Group

Two-fold symmetries occur whenever an object or an interaction is symmetric under a geometrical transformation that is its own inverse ($a^2 = e$). For example, C_2 is the symmetry group of an isosceles, nonequilateral triangle (see Fig. 4.1). Many animals are externally bilaterally symmetric, meaning that their appearance does not change under reflection through a certain plane running down the middle of the animal. The equilibrium configurations of some molecules such as hydrogen peroxide (HOOH) are symmetric under a 180° rotation about a certain axis (in this case, an axis that bisects the O–O bond).

More fundamental examples of C_2 symmetry include symmetry under **parity, time inversion**, and **charge conjugation**. The **parity group** P is a two-element group

$$P = \{\{\mathsf{P}_e, \mathsf{P}_i\}, \circ\}, \tag{4.37}$$

the elements of which are mappings P_e and P_i that act as follows on functions of $\mathbf{r} = (x, y, z)$:

$$(\mathsf{P}_e f)(\mathbf{r}) := f(\mathbf{r}), \quad (\mathsf{P}_i f)(\mathbf{r}) := f(-\mathbf{r}). \tag{4.38}$$

Evidently P_e is the identity mapping. The mapping P_i is called **inversion**. The group operation $\circ$ is the composition of mappings as defined in Eq. (2.47).

One says that a function $f : \mathbb{R}^3 \to \mathbb{R}$ has **even parity** if

$$(\mathsf{P}_i f)(\mathbf{r}) = f(\mathbf{r}) \tag{4.39}$$

and odd parity if

$$(\mathsf{P}_i f)(\mathbf{r}) = -f(\mathbf{r}). \tag{4.40}$$

Note that $\mathsf{P}_i f = (+1)f$ if f has even parity and that $\mathsf{P}_i f = (-1)f$ if f has odd parity. It is no accident that the numbers $\{1, -1\}$ realize the two-element group (under the operation of numerical multiplication). This is an example of a representation of a group, a topic that we discuss in greater detail in Section 4.3.1 and Chapter 11.

Every function $f : \mathbb{R}^3 \to \mathbb{R}$ can be expressed as the sum of an even-parity function and an odd-parity function. Let

$$f_+(\mathbf{r}) := \tfrac{1}{2}[f(\mathbf{r}) + f(-\mathbf{r})] \tag{4.41}$$

$$f_-(\mathbf{r}) := \tfrac{1}{2}[f(\mathbf{r}) - f(-\mathbf{r})]. \tag{4.42}$$

One sees immediately that f_+ has even parity, that f_- has odd parity, and that the function f is equal to the sum of its even-parity and odd-parity parts:

$$f = f_+ + f_-. \tag{4.43}$$

Subgroups

If $(G, \circ)$ is a group, H is a subset of G and $(H, \circ)$ is a group, then $(H, \circ)$ is called a **subgroup** of $(G, \circ)$. Usually one just says, "H is a subgroup of G." The essential point of the definition is that H is a sub*set* of G *and* H is a group under the same operation $\circ$ as G. A group G is always a subgroup of itself. If H is a subgroup and a proper subset of G, then H is called a **proper subgroup** of G. For example, $\{e\}$ and $\{e, a\}$ are both subgroups of C_2, but only $\{e\}$ is a proper subgroup. To give another example, the additive group $(n\mathbb{Z}, +)$ of integral multiples of an integer n is a proper subgroup of the additive group of integers, $(\mathbb{Z}, +)$.

The additive group $(\mathbb{R}^2, +)$ affords easily visualized examples of subgroups. It is easy to see that the set of vectors that point along the x-axis,

$$X = \left\{ \begin{pmatrix} x \\ 0 \end{pmatrix} \right\}, \tag{4.44}$$

is a subgroup of $(\mathbb{R}^2, +)$. X is closed under vector addition and inherits the property of associativity from $(\mathbb{R}^2, +)$. Obviously the zero vector $\mathbf{0}$ belongs to X. The additive inverse $-\mathbf{r}$ of a vector $\mathbf{x} \in X$ also points along the x-axis (the y-component is still zero; the x-component is $-x$); therefore $-\mathbf{r} \in X$.

Not every subset S of G is a subgroup of G because the axiom of closure, Eq. (4.1), requires that a subgroup H contain the resultants of all possible group products of elements of H (see Exercise 4.2.10). For example, $S = \{a\}$ is not a subgroup of C_2 because $a^2 = e \notin S$; therefore S is not closed under the group operation.

Because a subgroup H is a group, it has an identity element e'. Certainly $e' \in G$ because H is a subset of G. Exercise 4.2.11 shows that e' is equal to e, the identity element of G. Clearly the subgroup $\{e\}$ that consists only of the identity element is a proper subgroup of any group G that contains an element $a \neq e$.

4.2.3 ORBITS AND COSETS

Group Operations on a Set

In many practical applications one has not only a symmetry group but also a set on which mappings can be defined in a natural way, such that the mappings are compatible with the group product and leave the set unchanged. Given a group G and a set S, one calls S a **G-set** (meaning a set on which G operates) if and only if there exists a mapping t from $G \times S$ to S such that

$$\forall g \in G : \forall s \in S : t(g, s) \in S, \tag{4.45}$$

$$\forall s \in S : t(e, s) = s, \tag{4.46}$$

and such that the composition of mappings is **compatible** with the group product in the sense that

$$\forall g, g' \in G : \forall s \in S : t(g', t(g, s)) = t(g'g, s). \tag{4.47}$$

By an abuse of notation, one sometimes writes $t(g, s) = gs$.

For example, the set of points in an isosceles, nonequilateral triangle is invariant under the operation of rotation through $180°$ about the axis of the triangle (Fig. 4.1). (One has to regard the triangle as a flat three-dimensional figure because the rotation temporarily takes the triangle out of the plane.) Two such rotations return every point of the triangle to its original location; hence the $180°$ rotation is its own inverse. The identity operation is to do nothing. Associativity is easy to verify in this case. Therefore the points of the triangle in

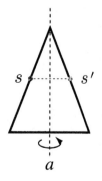

FIGURE 4.1

Rotational symmetry of an isosceles, nonequilateral triangle. The operation a is rotation through $180°$ about the indicated axis. The point s' belongs to the orbit of s.

Fig. 4.1 are the elements of a G-set, in which G is a group with two elements (rotation through $0°$ and $180°$).

It is possible for S to be G itself. For example, let s be any element of G. The mapping

$$\forall g \in G : \ t_r(g, s) := gs \tag{4.48}$$

carries $s \in G$ to $s' = gs \in G$ and obviously obeys Eq. (4.46). In order to verify compatibility, we let

$$s' = t_r(g, s) = gs \quad \text{and} \quad s'' = t_r(g', s') = g's', \tag{4.49}$$

which implies that

$$s'' = g'gs = t_r(g'g, s), \tag{4.50}$$

thereby demonstrating compatibility.

Likewise, the mapping

$$\forall g \in G : \ t_l(g, s) := sg^{-1} \tag{4.51}$$

maps S to S and obeys Eq. (4.46). To verify that Eq. (4.51) is compatible with the group product, let

$$s' = t_l(g, s) = sg^{-1} \quad \text{and} \quad s'' = t_l(g', s') = s'g'^{-1}. \tag{4.52}$$

Then

$$s'' = sg^{-1}g'^{-1} = s(g'g)^{-1} = t_l(g'g, s). \tag{4.53}$$

Therefore t_l is compatible with the group product.

For future reference we shall need to know that $t_l(g, \cdot)$ and $t_r(g, \cdot)$ are bijective. For example, let b be any element of G, and let

$$c = bg. \tag{4.54}$$

Then

$$b = cg^{-1} = t_l(g, c), \tag{4.55}$$

that is, b is the image of c under $t_l(g, \cdot)$. Because every element of G is the image of some other element of G under $t_l(g, \cdot)$, it follows that $t_l(g, \cdot)$ is surjective. To show that $t_l(g, \cdot)$ is bijective, suppose that the images of two elements of G under $t_l(g, \cdot)$ are equal:

$$t_l(g, s) = t_l(g, s') \Rightarrow sg^{-1} = s'g^{-1}. \tag{4.56}$$

By the cancellation law, it follows that the preimages are equal: $s' = s$. Because $t_l(g, \cdot)$ is both surjective and injective, it is bijective. The proof that $t_r(g, \cdot)$ is bijective is similar (Exercise 4.2.21).

Again let s be any element of G. The **conjugacy mapping**

$$\forall g \in G : \ t_c(g, s) := gsg^{-1}, \tag{4.57}$$

which (in a loose sense) combines the mappings t_r and t_l, also obeys the compatibility equation Eq. (4.46). To show that Eq. (4.57) is compatible with the group product (Eq. [4.47]), one uses the relation $(g_2 g_1)^{-1} = g_1^{-1} g_2^{-1}$ for the inverse of a product (see Exercise 4.2.19):

$$\begin{aligned} t_c(g', t_c(g, s)) &= t_c(g', gsg^{-1}) \\ &= g'gsg^{-1}g'^{-1} \\ &= t_c(g'g, s). \end{aligned} \tag{4.58}$$

It is straightforward to show that $t_c(g, \cdot)$ is bijective (Exercise 4.2.22). In fact, in an Abelian group the conjugacy mapping is just the identity mapping.

In this section we have exhibited three different mappings under which a group G is a G-set. Therefore a G-set exists for every group G. We show in Section 4.2.8 that the mappings t_r, t_l, and t_c are of fundamental importance for understanding the structure of a group.

Orbits

If S is a G-set, then the set that consists of all of the images of a given, fixed point $s \in S$ under the operations of G,

$$\text{orbit}[s] := \{s' \in S \mid \exists g \in G :\ni: s' = t(g, s)\}, \tag{4.59}$$

is called the **orbit** of s. (If one needs to be precise, one can say that orbit$[s]$ is the orbit of s under the mapping t and the group G.) For example, the orbit of the right-hand vertex of the base of the triangle in Fig. 4.1 is the set that consists of both base vertices.

It is easy to show that the relation of belonging to the same orbit is an equivalence relation:

$$s \equiv s' \Leftrightarrow s' \in \text{orbit}[s]. \tag{4.60}$$

Reflexivity is trivial. If $s' \equiv s$, then s' belongs to the orbit of s, and there exists a group element g such that $s' = t(g, s)$. Then Eq. (4.47) implies that $s = t(g^{-1}, s')$, whence $s \equiv s'$. To check transitivity, note that if s' is in the orbit of s and if s'' is in the orbit of s', then

$$\exists g, g' \in G :\ni: \ s' = t(g, s), \quad s'' = t(g', s') \tag{4.61}$$

whence

$$s'' = t(g', t(g, s)) = t(g'g, s) \tag{4.62}$$

by Eq. (4.47). Therefore s'' lies in the orbit of s.

Because membership in the same orbit is an equivalence relation, it follows at once that different orbits are disjoint (see Section 2.2.5). Therefore, because every point in S belongs to an orbit (its own), the distinct orbits partition S into disjoint subsets. For example, every point on the perimeter of the isosceles triangle in Fig. 4.1 belongs to a two-element orbit, except for the two points at which the axis a intersects the triangle, which are in orbits by themselves.

Because distinct orbits are disjoint, each orbit is a G-set. Exercise 4.2.24 shows that the smallest possible G-set is the orbit of a single point. For example, let $\mathbf{r}$ be a point in the Euclidean plane ($\mathbb{R}^2$), and let G be the two-element group that consists of the identity and the operation of reflection in the origin. Thus $G = \{e, i \mid \forall \mathbf{r} \in \mathbb{R}^2 : i\mathbf{r} = -\mathbf{r}\}$. The smallest possible G-set containing $\mathbf{r}$ is the set $\{\mathbf{r}, -\mathbf{r}\}$, consisting of the point $\mathbf{r}$ and its image under inversion in the origin. The origin is the smallest G-set, consisting of one point.

Cosets

Let H be a subgroup of a group G and let $a \in G$. The **left coset** of H generated by a is the set

$$aH := \{ah \mid h \in H\}. \tag{4.63}$$

For example, if $G = C_2$ and $H = \{e\}$ then the left cosets are the sets $eH = \{e\}$ and $aH = \{a\}$. Likewise, the **right coset** of H generated by $a \in G$ is the set

$$Ha := \{ha \mid h \in H\}. \tag{4.64}$$

For example, in the additive group of real two-dimensional vectors, $(\mathbb{R}^2, +)$, let $\mathbf{a}$ be a vector with a nonzero y-component a_y:

$$\mathbf{a} = \begin{pmatrix} a_x \\ a_y \end{pmatrix}, \quad a_y \neq 0. \tag{4.65}$$

Also, let X be the subgroup of vectors that lie along the x-axis (see Eq. [4.44]). The coset

$$\mathbf{a} + X = \left\{ \begin{pmatrix} a_x + x \\ a_y \end{pmatrix} \middle| x \in \mathbb{R} \right\} \tag{4.66}$$

of X generated by $\mathbf{a}$ is the set of vectors whose tips lie on the line that is parallel to the x-axis and that passes through the point $x = 0$, $y = a_y$. Every vector that belongs to this coset has a y-component equal to a_y.

A coset should not be confused with a subgroup. A left (or right) coset of a subgroup H generated by $a \in G$ is a subgroup of G if and only if $a \in H$ (see Exercise 4.2.12). For example, in $G = C_2$ the only left coset that is a subgroup of C_2 is $eH = H$. In the previous example of $(\mathbb{R}^2, +)$, the coset $\mathbf{a} + X$ is not a subgroup, for it is not closed under vector addition. The sum of two vectors, each of which has a y-component equal to a_y, is a vector with a y-component of $2a_y$.

Evidently the left coset generated by a is simply the orbit of a under the mapping $t_l : G \to G$ defined in Eq. (4.51). Because orbits are equivalence classes, it follows that, for every subgroup H of G, the cosets of H are equivalence classes:

$$a \equiv b \Leftrightarrow aH = bH \tag{4.67}$$

Moreover, two left cosets aH and bH are either identical or disjoint (i.e., have no elements in common). The same assertion is true for right cosets.

Similarly, the right cosets of any subgroup H of a group G are equivalence classes and therefore are either disjoint or identical.

Lagrange's Theorem

Our definitions and results so far apply to any subgroup of any group. Now suppose that G is a finite group and that H is a subgroup of G; then H is also a finite group. We show that every left coset of H has the same number of elements by showing that aH and bH are equivalent as sets:

$$\forall a, b \in G : aH \sim bH. \tag{4.68}$$

Define the mapping

$$\phi : ah \mapsto bh \tag{4.69}$$

for every $h \in H$. Therefore

$$\forall c \in aH : \quad \phi(c) = ba^{-1}c \tag{4.70}$$

because ϕ must map a to b. To see whether ϕ is injective, assume that two images $\phi(c)$, $\phi(c')$ are equal and check whether the preimages c, c' must be equal. One has

$$\phi(c) = \phi(c') \Rightarrow ba^{-1}c = ba^{-1}c' \Rightarrow c = c', \tag{4.71}$$

the last step following from the cancellation law. Therefore ϕ is injective. Also, ϕ is surjective, for if $d = bh \in bH$, then d is the image of $ah = a(b^{-1}d)$ under ϕ. Because aH and bH are finite and equivalent, they have the same number of elements. (For infinite groups, the same argument shows that aH and bH have the same cardinal number of elements.)

Let $(G, \circ)$ be a finite group of order g, and let $(H, \circ)$ be a subgroup of order h. Let the disjoint left cosets of H be $a_1H, a_2H, \ldots, a_kH$. Because G is a finite set, k is a finite, positive integer. Each coset a_iH has h elements. Because each of the g elements of G occurs in one and only one coset a_iH,

$$g = hk \Rightarrow k = \frac{g}{h}. \tag{4.72}$$

This is **Lagrange's theorem**: *The order of a subgroup of a finite group is an integral divisor of the order of the group.*

An immediate consequence of Lagrange's theorem is that if G is a finite group whose order is a prime number, then G has no proper subgroups with more than one element. Because the set $\{e\}$ that consists only of the group identity element is a subset of every subgroup, it follows that $\{e\}$ and G are the only subgroups of a finite group G of prime order. (For the converse, see Exercise 4.2.13.)

Three-Element Group

Lagrange's theorem and its corollary for a group of prime order suggest that it may be rather easy to derive the possible structures of groups with only a few elements. In a **three-element group** $\{e, a, b\}$ one has $a^2 \neq e$, for if a^2 were equal to e then the two-element group $\{e, a \mid a^2 = e\}$ would be a subgroup of a three-element group, a conclusion which Lagrange's theorem rules out. Clearly, either $a^2 \neq a$ or $a = e$. Therefore $b = a^2$. If

$$a^3 = a^2 a = aaa = ab = ba \qquad (4.73)$$

were equal to a or b, then one would have either $b = e$ or $a = e$. Therefore $a^3 = e$, from which it follows that $a^{-1} = a^2$, because $aa^2 = e$. We have shown that the only possible three-element group is

$$C_3 := \{e, a, a^2 \mid a^3 = e\}. \qquad (4.74)$$

Table 4.2 gives the multiplication table.

One realization of C_3 is the group $\{R_{\pi/3}, R_{2\pi/3}, R_0 = e\}$ consisting of rotations through $120°$ and $240°$ (a and a^2, respectively), plus the identity rotation through $0°$. In this realization, C_3 is a subgroup of the symmetry group of an equilateral triangle. Symmetry under the group C_3 is widespread in nature, from the macroscopic scale (e.g., the flowers of plants in the lily family) to the molecular scale (e.g., the ammonia molecule NH_3).

Let us work out a minimal G-set for C_3 in the realization $\{R_{\pi/3}, R_{2\pi/3}, R_0 = e\}$. If $\mathbf{r}$ is a point on the x-axis in the Euclidean plane, then

$$\text{orbit}[\mathbf{r}] = \{\mathbf{r}, R_{\pi/3}\mathbf{r}, R_{2\pi/3}\mathbf{r}\}. \qquad (4.75)$$

The three points in orbit[$\mathbf{r}$] lie at the vertices of an equilateral triangle, as illustrated in Fig. 4.2.

TABLE 4.2 Multiplication Table of the Group C_3, in which $b = a^2$

	First Operation		
C_3	e	a	b
e	e	a	b
a	a	b	e
b	b	e	a

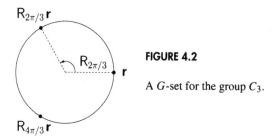

FIGURE 4.2

A G-set for the group C_3.

4.2.4 CYCLIC GROUPS

Definition of a Cyclic Group

The two-element group C_2 and the three-element group C_3 illustrate a certain kind of group structure. As we have already done for a^2 and a^3, we define the positive integral **powers** of an element a of a group G recursively:

$$a^1 := a \tag{4.76}$$

$$\forall n \in \mathbb{Z}^+ : a^{n+1} := aa^n. \tag{4.77}$$

We define negative integral powers as

$$a^{-n} := (a^{-1})^n. \tag{4.78}$$

Exercise 4.2.15 shows that the usual exponent laws $a^m a^n = a^{m+n}$ and $(a^m)^n = a^{mn}$ hold.

A group G is called **cyclic** if and only if every element of G is an integral power of a single element of G:

$$G \text{ is cyclic} \Leftrightarrow \exists a \in G :\ni: \forall b \in G : \exists m \in \mathbb{Z} :\ni: b = a^m. \tag{4.79}$$

Two cases are possible for an element a of a cyclic group G: Either all the powers of a are distinct (meaning that $a^m = a^n \Rightarrow m = n$), or for some m and n

$$a^m = a^n. \tag{4.80}$$

In the first case G is the infinite cyclic group

$$\{\ldots, a^{-m}, \ldots, a^{-1}, e, a, \ldots, a^m, \ldots\}. \tag{4.81}$$

Suppose that in the second case $n > m$; then repeated multiplication by a^{-1} (if $m > 0$) or a (if $m < 0$) gives

$$a^{n-m} = e. \tag{4.82}$$

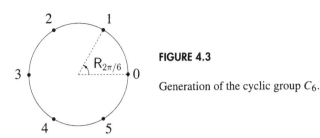

FIGURE 4.3

Generation of the cyclic group C_6.

Let P be the set of all positive integers k such that $a^k = e$. By the well-ordering principle, P has a least (first) element r. Then we have the structure of G:

$$G = C_r := \{e, a, \ldots, a^{r-1} \mid a^r = e\}. \tag{4.83}$$

One calls r the **order** of the group element a because r is the order of the cyclic group C_r that a generates. G contains an inverse for every element, for $a^s a^{r-s} = a^r = e$, implying that $(a^s)^{-1} = a^{r-s}$ for every s such that $1 \le s \le r - 1$. Every group whose order is a prime number is cyclic (see Exercise 4.2.16).

For example, if n is a positive integer, then the rotations $R_{2m\pi/n}$ of a plane figure (about an axis that is perpendicular to the plane) through the angles $2\pi/n, 2 \cdot 2\pi/n, \ldots, (n-1)2\pi/n$ are the elements of a cyclic group that realizes C_n. The generator is the operation of rotation through $2\pi/n$, $R_{2\pi/n}$. A minimal G-set consists of an initial point and its images under the rotations $R_{2\pi m/n}$ for $m \in (0 : n - 1)$. Figure 4.3 shows the rotation that generates the cyclic group C_6 and the points of the G-set (labeled "0" through "5").

The term *cyclic* comes from the Greek word $\kappa\nu\kappa\lambda o\varsigma$, meaning "circle." The relation $a^n = e$, which is satisfied by the generator of a cyclic group, implies that the group can be realized as the group of rotational transformations of a set of n beads equally spaced around the circumference of a circle, as shown in Fig. 4.3.

Certain cyclic groups whose order is a power of a prime number are important in some modern methods of generating pseudorandom numbers for computations.

One-Dimensional Crystal Lattices

Perhaps the most physically important realization of the general cyclic group C_n is a **one-dimensional crystal lattice** with periodic boundary conditions. One can imagine such a "crystal" as a lattice of n identical atoms (or n identical potential wells) equally spaced around the perimeter of a circle. The spacing a is called the **lattice period**. In order to describe the lattice mathematically, we define (for each $a \in \mathbb{R}$) a **lattice translation operator** P_a, which maps functions defined on $\mathbb{R}$ in the following way:

$$\forall x \in [0, L] : (P_a f)(x) := f(x - a). \tag{4.84}$$

The m-th power of P_a is $(P_a)^m$, the translation of a function through m lattice periods:

$$((P_a)^m f)(x) := (P_{ma} f)(x) = f(x - ma). \tag{4.85}$$

If one assumes periodic boundary conditions, then all physically meaningful functions (such as waves, for example) must be periodic with the period $L = na$, in which L is the length of the crystal:

$$\forall x \in [0, L]: \ f(x \pm L) = f(x). \tag{4.86}$$

By applying this condition recursively we can extend a function f defined on $[0, L]$ such that $f(L) = f(0)$ to a function defined on $(-\infty, \infty)$. By Eq. (4.86) and the definition of L,

$$((P_a)^n f)(x) = f(x - na) = f(x - L) = f(x). \tag{4.87}$$

Hence

$$(P_a)^n = \mathbf{1}, \tag{4.88}$$

where $\mathbf{1}$ is the **identity operator** such that

$$(\mathbf{1} f)(x) := f(x). \tag{4.89}$$

The inverse of a lattice translation is $((P_a)^m)^{-1} = (P_a)^{n-m}$ in view of Eq. (4.88). The set

$$T_n := \{\mathbf{1}, P_a, (P_a)^2, \ldots, (P_a)^{n-1}\} \tag{4.90}$$

therefore is a realization of the cyclic group C_n.

Cyclic Groups of Nonprime Order

If the order of a cyclic group C_n is not prime, then not all of the elements of C_n have the same order. For example, in

$$C_4 := \{e, a, a^2, a^3 \mid a^4 = e\} \tag{4.91}$$

the elements a and a^3 have order 4, because $(a^3)^2 = a^2$ and $(a^3)^3 = a$, implying that $k = 4$ is the smallest integer such that $(a^3)^k = e$. However, the element a^2 has order 2 because $(a^2)^2 = e$. Now consider C_n and suppose that $n = kl$ and $k, l \in \mathbb{Z}^+, k \neq 1, l \neq 1$ (i.e., k and l are nontrivial divisors of n). By the exponent law, $(a^k)^l = (a^l)^k = a^n = e$. Therefore a^k has order l and a^l has order k. We conclude that if n is not prime, then for every nontrivial divisor k of n there exists an element $b = a^{n/k}$ of the cyclic group C_n such that the order of b is equal to k.

Four-Element Groups

To illustrate the usefulness of the concept of the order of an element of a group, we work out the possible structures of the groups with four elements. Clearly one structure is the four-element cyclic group C_4. Let $V = \{e, a, b, c\}$ be a different four-element group (if one exists), and let k be the order of a. If $k = 4$, then V is just C_4. If $k = 3$, then $\{e, a, a^2\}$ is a subgroup of V; but this is forbidden by Lagrange's theorem. If $k = 1$, then V does not have four distinct elements. Therefore in a four-element group other than C_4 (if one exists), the order of a must be $k = 2$. Similar comments apply to b and c. Therefore the only possible structure of a four-element group other than C_4 is

$$V = \{e, a, b, c \mid a^2 = b^2 = c^2 = e\}. \tag{4.92}$$

(By the way, V stands for *Vierergruppe*, meaning "group of four.") Of course, we have not yet shown that V is a group. To do so, one must come up with a multiplication table that obeys the group axioms (see Exercise 4.2.17).

Complex n-th Roots of Unity

The n-th **complex roots of unity** are an important realization of the cyclic group C_n. Each of the complex numbers

$$z_1 = 1, z_2 = \omega, z_3 = \omega^2, \ldots, z_n = \omega^{n-1}, \tag{4.93}$$

in which

$$\omega = e^{2\pi i/n}, \tag{4.94}$$

is an n-th root of unity, that is, $z_1, \ldots, z_n$ are the n roots of the n-th-degree polynomial equation

$$z^n - 1 = 0. \tag{4.95}$$

Complex multiplication by ω rotates an arbitrary complex number $z = \rho e^{i\theta}$ through $2\pi/n$:

$$\omega z = \rho e^{i(\theta + 2\pi/n)}. \tag{4.96}$$

Therefore

$$K_n := \{1, \omega, \ldots, \omega^{n-1} \mid \omega = e^{2\pi i/n}\} \tag{4.97}$$

is a realization of C_n if the group operation is complex multiplication.

Integers Modulo n

Another important cyclic group is $(\mathbb{Z}_n, +)$, in which $\mathbb{Z}_n$ is the group of **integers modulo** n. Two integers $k, l \in \mathbb{Z}$ are called **congruent (modulo n)**, written

$$k \equiv l \pmod{n} \tag{4.98}$$

if and only if they leave the same remainder r such that $0 \le r < n$ if divided by n:

$$k = pn + r, \quad l = qn + r. \tag{4.99}$$

It is clearly equivalent to make the definition that $k \equiv l \pmod{n}$ if and only if k and l differ by an integral multiple of n:

$$k \equiv l \pmod{n} \Leftrightarrow \exists m \in \mathbb{Z} : \ni: \quad k - l = mn. \tag{4.100}$$

Because r is an integer, its possible values are $0, 1, \ldots, n - 1$.

Congruence is clearly an equivalence relation. The **equivalence class of** r **modulo** n is

$$r_n := \{k \in \mathbb{Z} \mid k \equiv r \pmod{n}\}. \tag{4.101}$$

$\mathbb{Z}_n$ is the set of all such equivalence classes:

$$\mathbb{Z}_n := \{r_n \mid 0 \leq r \leq n - 1\}. \tag{4.102}$$

For example, $\mathbb{Z}_2 = \{0_2, 1_2\}$ is the set whose elements are the equivalence classes of the even and odd integers.

There exists an operation of addition such that $(\mathbb{Z}_n, +)$ is a group. The simplest definition of **addition modulo** n is

$$r_n + s_n := (r + s)_n. \tag{4.103}$$

Note that we are adding equivalence classes, not integers! The definition says that the sum (modulo n) of equivalence classes represented by the integers r and s is the equivalence class represented by $r + s$. For example,

$$1_2 + 1_2 = 0_2 \tag{4.104}$$

restates the familiar rule that "odd plus odd is even." Similarly,

$$0_2 + 0_2 = 0_2 \tag{4.105}$$

implies the rule "even plus even is even," and

$$1_2 + 0_2 = 1_2 \tag{4.106}$$

implies that "odd plus even is odd."

The definition (4.103) makes the addition of equivalence classes commutative. The associativity of addition modulo n follows from the associativity of addition in $\mathbb{Z}$. The equivalence class

$$0_n = n\mathbb{Z} := \{mn \mid m \in \mathbb{Z}\} \tag{4.107}$$

is the additive identity modulo n because adding a multiple of n to each element of the equivalence class r_n leaves r_n invariant. The additive inverse (modulo n) of r_n is

$$-(r_n) = (n - r)_n, \tag{4.108}$$

for

$$r_n + (n - r)_n = n_n = 0_n. \tag{4.109}$$

Therefore $(\mathbb{Z}_n, +)$ is a group. Because the group operation $+$ is commutative, $\mathbb{Z}_n$ is Abelian. Moreover, Eqs. (4.76) and (4.77) imply that the m-th power of the equivalence class $1_n \in \mathbb{Z}_n$ is

$$(1_n)^m = \underbrace{(1 + 1 + \cdots + 1)}_{m \text{ terms}}_n = m_n. \qquad (4.110)$$

This implies that $\mathbb{Z}_n$ consists only of powers of the element 1_n and therefore is a cyclic group.

For example, the set of n-bit two's-complement integers is simply the set of integers modulo 2^n, $\mathbb{Z}_{2^n}$. According to Eq. (4.108), the additive inverse of k (modulo 2^n) is

$$-\left(k_{2^n}\right) = (2^n - k)_{2^n}, \qquad (4.111)$$

which is the n-bit two's complement representative of $-k$. In Fig. 1.7, for example, the additive inverse of 5 is $11 = 2^4 - 5$, the additive inverse of 7 is $9 = 2^4 - 7$, and so on.

Difference Equations and n-Cycles

Concepts derived from cyclic groups are important in several fields of engineering and applied physics. For example, in the study of so-called bifurcations in nonlinear systems one encounters mappings $\phi : S \to S$ and points $x_i \in S$ obtained by recursively applying a mapping ϕ,

$$x_k = \phi(x_{k-1}), \qquad (4.112)$$

which returns to the starting point after a finite number of steps:

$$\exists n \in \mathbb{Z}^+ : \ni: \quad x_n = x_1. \qquad (4.113)$$

Equation (4.112) is an example of a **difference equation**. A subset $X = \{x_1, \ldots, x_{n-1}\}$ that obeys Eqs. (4.112) and (4.113) is called an $\boldsymbol{n}$**-cycle**. An n-cycle X has the important property of being invariant under ϕ:

$$\phi(X) = X. \qquad (4.114)$$

For example, if one defines ϕ to mean multiplication by ω (see Eq. [4.94]), then K_n (see Eq. [4.97]) is an n-cycle.

We give an important example from the field of solid-state physics in Chapter 7.

4.2.5 DIHEDRAL GROUPS

The group that is generated by one rotation through $2\pi/n$ and one C_2 rotation about a perpendicular axis is called the **dihedral group** D_n. The dihedral group D_2, which has the same multiplication table as the Viergruppe V, is a special case. D_2 is the symmetry group of a rectangle or a rectangular parallelepiped; see Fig. 4.4 and Exercise 4.2.17.

If $n > 2$, D_n is the symmetry group of the regular polygon with n sides. Except for D_2, the dihedral groups are non-Abelian. For example, the full rotational symmetry group of an

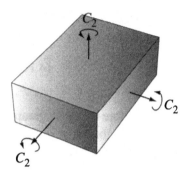

FIGURE 4.4

Rotational symmetries of a rectangular parallelepiped.

FIGURE 4.5

Rotational symmetries of an equilateral triangle or a triangular prism.

equilateral triangle is a six-element non-Abelian group called D_3, of which the three-element cyclic group C_3 is a subset. The C_3 rotations through $+2\pi/3$ and $-2\pi/3$ radians are labeled d and f, respectively, in Fig. 4.5. The identity rotation e is not shown. The three elements of D_3 that do not belong to C_3 are rotations through π radians (which we call C_2 rotations) about each symmetry axis of the triangle. In Fig. 4.5, the axes of the C_2 rotations are shown as dashed lines, and the C_2 rotations are labeled a, b, and c. It is easy to see that a group that contains the C_3 rotation d through $2\pi/3$, the C_3^2 rotation $d^2 = f$ through $-2\pi/3$, and one C_2 rotation about an axis perpendicular to the axis of the rotations d and f must contain two additional C_2 rotations about axes that are the images under d and f of the axis of the original C_2 rotation.

Each entry in Table 4.3 gives the result $a_2 a_1$ of performing first a_1, then a_2 on the triangle shown in Fig. 4.5. The rotations in the table are understood to be performed about axes that

TABLE 4.3 Multiplication
Table of the Group D_3

D_3	First Operation					
	e	a	b	c	d	f
e	e	a	b	c	d	f
a	a	e	d	f	b	c
b	b	f	e	d	c	a
c	c	d	f	e	a	b
d	d	c	a	b	f	e
f	f	b	c	a	e	d

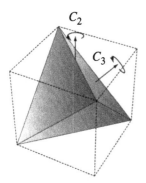

FIGURE 4.6

Rotational symmetries of a regular tetrahedron.

are fixed in space, *not* attached to the triangle. For example, in space-fixed axes one finds that the result of ad is the same as if b alone had been performed. In axes fixed in the triangle one would find $ad = c$ instead of $ad = b$. Because D_3 is non-Abelian, the table is not symmetric about the main diagonal e, e, e, e, f, d.

The subjects of Exercises 4.2.14 and 4.2.18 are the multiplication table and subgroups of D_3. Exercises 4.2.25 through 4.2.27 concern another dihedral group, D_4.

4.2.6 CUBIC GROUPS

The groups of symmetries of the regular tetrahedron and the regular octahedron, which are called the **cubic groups**, are especially important in semiconductor physics. The tetrahedral group T has twelve elements (e, three C_2 rotations, four C_3 rotations, and four C_3^2 rotations), two of which are illustrated in Fig. 4.6. The octahedral group O has twenty-four elements (e, eight C_3 rotations, three C_4^2 rotations, six C_2 rotations, and six C_4 rotations); see Fig. 4.7.

4.2.7 CONTINUOUS GROUPS

We have already described several infinite groups, including the additive group of integers, $(\mathbb{Z}, +)$; the infinite cyclic group, Eq. (4.81); and the group $GL(2, \mathbb{R})$ of invertible 2×2 matrices, Eq. (4.12). To understand some of the differences among these infinite groups, recall from Eq. (2.98) that a parametrization of a group G is a bijective mapping $P \to G$

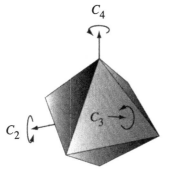

FIGURE 4.7

Rotational symmetries of a regular octahedron.

from a set P to G. The elements of a finite group G can be parametrized with the set $P = (1 : |G|)$. For example, one can write the elements of G as $a_1, \ldots, a_{|G|}$. An infinite cyclic group can be parametrized with the set of all integers, $\mathbb{Z}$. However, to parametrize $GL(2, \mathbb{R})$ requires four parameters, the matrix elements a, b, c, and d, each of which can range over the entire real line. If one lets N be the set of points (a, b, c, d) in $\mathbb{R}^4$ for which the determinant of the matrix $\mathbf{A}$ in Eq. (4.12) vanishes, then the parameter set for $GL(2, \mathbb{R})$ is $\mathbb{R}^4 \setminus N$, which is considerably "bigger" than $\mathbb{Z}$ (see Section 10.2.2).

Some of the most important groups in engineering and physics are infinite and have parameter sets that are subsets of $\mathbb{R}^p$, in which p is a nonnegative integer. The parameter set for the group of all proper rotations in two-dimensional Euclidean space, $SO(2, \mathbb{R})$, is the half-open interval $[0, 2\pi) = \{\phi \in \mathbb{R} \mid 0 \leq \phi < 2\pi\}$. The group elements are matrices that depend on a single parameter, ϕ:

$$\mathbf{R}_\phi = \begin{pmatrix} \cos\phi & -\sin\phi \\ \sin\phi & \cos\phi \end{pmatrix}. \tag{4.115}$$

The notation $SO(2, \mathbb{R})$ means "the special orthogonal group on $\mathbb{R}^2$"; the group is "special" because its elements must have a determinant equal to $+1$. "Improper rotations" such as reflection in the y axis,

$$\begin{pmatrix} 1 & 0 \\ 0 & -1 \end{pmatrix}, \tag{4.116}$$

are thereby excluded.

To specify an element of the group of proper rotations in three-dimensional Euclidean space, $SO(3, \mathbb{R})$, one must specify both the angle of rotation, ϕ, and the axis of rotation, which is defined by a unit vector $\mathbf{n}$. Any two real numbers n_x and n_y such that $n_x^2 + n_y^2 \leq 1$ specify a unit vector such that $n_x^2 + n_y^2 + n_z^2 = 1$; hence the parameter set for $SO(3, \mathbb{R})$ is a subset of $\mathbb{R}^3$. Because a rotation through a negative angle about an axis $\mathbf{n}$ is the same as a positive rotation about the axis $-\mathbf{n}$, the angle of rotation is restricted to lie in the interval $[0, \pi)$. Therefore an axis and angle of rotation are specified uniquely by $\langle n_x, n_y, \phi \rangle$. Because the axis of rotation can point in any direction, the parameter space for $SO(3, \mathbb{R})$ is an open ball (a ball without "skin"; see Section 10.1.4) with a radius equal to π. An element of $SO(3, \mathbb{R})$ is a 3×3 matrix, which we shall write as $\mathbf{R}_{\mathbf{n},\phi}$.

If there exists a parameter set P for a group G such that P is a subset of $\mathbb{R}^p$ (in which p is a nonnegative integer, $p \in \mathbb{Z}^+$) and such that the elements of G are continuous functions of the parameters $(\alpha_1, \ldots, \alpha_p) \in \mathbb{R}^p$, then G is called a **continuous group**. One might well ask, How can one talk about continuity of group elements? To define continuity, one must first have a definition of distance. Let us define the distance between two 2×2 matrices

$$\mathbf{A} = \begin{pmatrix} a & b \\ c & d \end{pmatrix}, \quad \mathbf{B} = \begin{pmatrix} a' & b' \\ c' & d' \end{pmatrix} \tag{4.117}$$

as

$$\|\mathbf{A}' - \mathbf{A}\|_F = [(a' - a)^2 + (b' - b)^2 + (c' - c)^2 + (d' - d)^2]^{\frac{1}{2}} = |\mathbf{r}' - \mathbf{r}|. \tag{4.118}$$

The parameters are $\alpha_1 = a$, $\alpha_2 = b$, $\alpha_3 = c$, and $\alpha_4 = d$, $\mathbf{r}$ is the element (a, b, c, d) of $\mathbb{R}^4$, and $|\mathbf{r}' - \mathbf{r}|$ is the usual Euclidean distance between two points. In other words, we define the distance between two elements of $GL(2, \mathbb{R})$ as the distance between the points in parameter space that label the group elements. In Chapter 8 we call $\|\mathbf{A}' - \mathbf{A}\|_F$ the "Frobenius norm" of the matrix difference $\mathbf{A}' - \mathbf{A}$. In freshman calculus, one learns that a function $f : \mathbb{R} \to \mathbb{R}$ is continuous at x if and only if for every $\epsilon > 0$, there exists a real number $\delta > 0$ such that $|f(x') - f(x)| < \epsilon$ for all x' such that $|x' - x| < \delta$. In this example, $\|\mathbf{A}' - \mathbf{A}\|_F$ is analogous to $|f(x') - f(x)|$, and $[(a' - a)^2 + (b' - b)^2 + (c' - c)^2 + (d' - d)^2]^{\frac{1}{2}}$ is analogous to $|x' - x|$. Thus our definition of the distance between two 2×2 matrices automatically makes the matrix a continuous function of its elements, which are the parameters of the group in this case.

If the parameter set $P \subseteq \mathbb{R}^p$ of a continuous group G is bounded and closed, or is a subset of a bounded, closed subset of $\mathbb{R}^p$, then the group G is called **compact**. The rotation groups $SO(2, \mathbb{R})$ and $SO(3, \mathbb{R})$ are compact. If the parameter set P is not bounded or closed, then G is called **noncompact**. For example, the continuous translation groups are noncompact.

If the elements of a continuous group G are differentiable infinitely many times with respect to the parameters, then G is called a **Lie group**. Again as in freshman calculus, we define the derivative of a matrix as the limit of a difference quotient:

$$\frac{\partial \mathbf{A}}{\partial \alpha} := \lim_{\Delta\alpha \to 0} \left(\frac{1}{\Delta\alpha} [\mathbf{A}(\alpha + \Delta\alpha) - \mathbf{A}(\alpha)] \right). \tag{4.119}$$

For example,

$$\frac{\partial}{\partial a} \begin{pmatrix} a & b \\ c & d \end{pmatrix} = \begin{pmatrix} 1 & 0 \\ 0 & 0 \end{pmatrix}. \tag{4.120}$$

The elements of the group $GL(2, \mathbb{R})$ are infinitely differentiable functions of the group parameters (a, b, c, d). Likewise, the elements of $SO(2, \mathbb{R})$ are infinitely differentiable functions of ϕ, and the elements of $SO(3, \mathbb{R})$ are infinitely differentiable functions of n_x, n_y, and ϕ.

Differentiability, continuity of derivatives, and analyticity (expansion of a function in a power series) are the mathematical properties that make it possible to get correct results by working with "infinitesimal" changes in the arguments of functions. We (heuristically) use infinitesimals to explain some of the basic ideas of Lie groups. Readers who need to review matrix algebra are encouraged to read Section 6.1.2 before proceeding further.

In $SO(2, \mathbb{R})$, for example, one can write

$$\mathbf{R}_{\phi+\Delta\phi} \approx \mathbf{R}_\phi + \Delta\phi \frac{\partial}{\partial\phi} \mathbf{R}_\phi$$

$$\approx \left[1 + \Delta\phi \left(\frac{\partial}{\partial\phi} \mathbf{R}_\phi \right) \mathbf{R}_\phi^{-1} \right] \mathbf{R}_\phi. \tag{4.121}$$

Using the trigonometric identity $\cos^2 \phi + \sin^2 \phi = 1$, one finds that

$$\left(\frac{\partial}{\partial \phi}\mathbf{R}_\phi\right)\mathbf{R}_\phi^{-1} = \begin{pmatrix} -\sin \phi & -\cos \phi \\ \cos \phi & -\sin \phi \end{pmatrix}\begin{pmatrix} \cos \phi & \sin \phi \\ -\sin \phi & \cos \phi \end{pmatrix}$$

$$= \begin{pmatrix} 0 & -1 \\ 1 & 0 \end{pmatrix} = -i\sigma_2 \tag{4.122}$$

in which

$$\sigma_2 := \begin{pmatrix} 0 & -i \\ i & 0 \end{pmatrix} \tag{4.123}$$

is one of the Pauli matrices used in quantum physics. Equation (4.115) implies that the matrix of an infinitesimal rotation is the sum of the unit matrix and a matrix that is proportional to σ_2:

$$\mathbf{R}_{\Delta\phi} \approx \begin{pmatrix} 1 & -\Delta\phi \\ \Delta\phi & 1 \end{pmatrix} = \mathbf{1} - i\Delta\phi\sigma_2. \tag{4.124}$$

Therefore, from Eqs. (4.121) and (4.122), one has the approximate expression

$$\boxed{\mathbf{R}_{\phi+\Delta\phi} \approx [\mathbf{1} - i\Delta\phi\sigma_2]\mathbf{R}_\phi} \tag{4.125}$$

for the effect of a finite rotation (implemented by $\mathbf{R}_\phi$) followed by an infinitesimal rotation. Because of the property expressed in Eq. (4.125), $(\partial\mathbf{R}_\phi/\partial\phi)\mathbf{R}_\phi^{-1}$ is called the **infinitesimal generator** of rotations in the plane.

Pushing further, let us try to approximate a finite rotation through an angle ϕ as the product of a large number of small rotations, each through an angle $\Delta\phi = \phi/N$:

$$\mathbf{R}_\phi \approx \left[\mathbf{1} - i\frac{\phi}{N}\sigma_2\right]^N$$

$$\xrightarrow[N \to \infty]{} \exp[-i\phi\sigma_2]. \tag{4.126}$$

The second line follows heuristically from the freshman calculus definition of the exponential function. In order to define the exponential of a matrix, we use a power series:

$$\exp[-i\phi\sigma_2] := \sum_{n=0}^\infty \frac{(-i\phi)^n}{n!}\sigma_2^n. \tag{4.127}$$

This definition makes sense because the product and sum of 2×2 matrices are again 2×2 matrices and because one guesses (correctly) that it is possible to define convergence for such a series; see Section 10.3 for the details.

We now show how to reduce the apparently formidable matrix power series to two simple terms. The observation that

$$\sigma_2^2 = -\begin{pmatrix} 1 & 0 \\ 0 & 1 \end{pmatrix} = -\mathbf{1} \tag{4.128}$$

implies that

$$\sigma_2^n = \begin{cases} (-1)^{\frac{n}{2}} \mathbf{1}, & \text{if } n \text{ is even,} \\ (-i)^{\frac{n+1}{2}} \sigma_2, & \text{if } n \text{ is odd.} \end{cases} \tag{4.129}$$

This property, and the absolute convergence of the series in Eq. (4.127), permit one to rearrange the terms in the series into a series in ϕ times σ_2, plus a different series in ϕ times the unit matrix $\mathbf{1}$:

$$\begin{aligned} \exp[-i\phi\sigma_2] &= \left(1 - \frac{\phi^2}{2!} + \cdots\right)\mathbf{1} - i\left(\phi - \frac{\phi^3}{3!} + \cdots\right)\sigma_2 \\ &= \cos\phi\, \mathbf{1} - i\sin\phi\, \sigma_2. \end{aligned} \tag{4.130}$$

One verifies easily, using the definition of σ_2, that this equation gives the correct formula for a finite rotation, in agreement with Eq. (4.115).

We have shown that every matrix in the group $SO(2, \mathbb{R})$ is equal to the exponential of a constant times the infinitesimal generator, $-i\sigma_2$:

$$\mathbf{R}_\phi = \exp[-i\phi\sigma_2]. \tag{4.131}$$

In quantum mechanics, one defines the **angular momentum operator** as i times the infinitesimal generator of rotations. For vectors in the two-dimensional plane, then, the angular momentum operator is σ_2.

The general point that one should remember is that, in a Lie group, all of the group elements that are sufficiently close to the identity element (which, in a matrix group, is the unit matrix) can be obtained by exponentiating (a linear combination of) the infinitesimal generator(s).

4.2.8 CLASSES OF CONJUGATE ELEMENTS

Conjugacy
The mapping $t_c : G \to G$ such that

$$\forall x \in G : t_c(a, x) = axa^{-1} \tag{4.132}$$

is already familiar [see Eq. (4.57)]. An element $y \in G$ that belongs to the orbit of x under t_c, that is, such that

$$y = axa^{-1} = t_c(a, x) \tag{4.133}$$

for some $a \in G$, is called **conjugate** to x. For example, b is conjugate to a in the group D_3, because $f^{-1}af = daf = dc = b$. To understand intuitively what conjugacy means in this case, note that, in Fig. 4.5, f rotates the axis of b into the axis of a. Thus the mapping $f^{-1}af$ first rotates the b axis into the a axis, then rotates through π about a, then rotates the a axis into the b axis.

Conjugacy is an equivalence relation, for the relation of belonging to the same orbit is an equivalence relation (see Sections 2.2.5 and 4.2.3).

The orbit of an element $x \in G$ under the conjugacy mapping t_c [see Eq. (4.57)],

$$\mathcal{C}_x := \{y \in G \mid \exists a \in G : \ni : y = axa^{-1}\}, \tag{4.134}$$

is called the **conjugacy class of**, the **class of**, or the **class represented by**, the element x. Because conjugacy is an equivalence relation, the classes of a given group G are equivalence classes and therefore are disjoint. Clearly the identity e of any group is in a class by itself. In an Abelian group, every element is in a class by itself. For example, because D_2 is Abelian, D_2 has four classes, each containing one element.

In a non-Abelian group, some classes contain more than one element. For example, in D_3 the classes are

$$\mathcal{C}_e = \{e\}, \quad \mathcal{C}_a = \{a, b, c\}, \quad \mathcal{C}_d = \{d, f\} \tag{4.135}$$

(see Exercise 4.3.8). Note that $\mathcal{C}_a$ contains all of the C_2 rotations, and $\mathcal{C}_d$ contains all of the C_3 rotations, that belong to the group D_3.

In the tetrahedral group T there are four classes: the class of the identity, a class of three C_2 rotations, a class of four C_3 rotations, and a class of four C_3^2 rotations. The five classes of the octahedral group O are the class of the identity, a class of eight C_3 rotations, a class of three C_4^2 rotations, a class of six C_2 rotations, and a class of six C_4 rotations.

Let S be a subset of a group G. Let us write

$$t_c(a, S) := \{axa^{-1} \mid x \in S\} \tag{4.136}$$

for the image of S under a conjugacy mapping $t_c(a, \cdot)$ (see Eq. [4.57]). We show that S is equal to a union of conjugacy classes of G if and only if

$$\forall a \in G : t_c(a, S) = S, \tag{4.137}$$

that is, if and only if S is invariant under all conjugacy mappings of G. To prove the "if" part ($\forall a \in G : t_c(a, S) = S$ implies that S is a union of classes), note that if $t_c(a, S) = S$ for all $a \in G$, then for every $x \in S$ the conjugacy class of x is a subset of S: $\mathcal{C}_x \subseteq S$. To start the proof of the "only if" part, note that the assumption that S is a union of classes implies that for every $x \in S$, $\mathcal{C}_x \subseteq S$. Then for every $a \in G$, $t_c(a, x) \in S$, from which it follows that $t_c(a, S) \subseteq S$. To show that $t_c(a, S) = S$, one must also show that $S \subseteq t_c(a, S)$. Let $x \in S$. To demonstrate that $x \in t_c(a, S)$, observe that $x = t_c(a, y)$ in which $y = a^{-1}xa$. But $y \in \mathcal{C}_x$. Because $\mathcal{C}_x \subseteq S$, it follows that $y \in S$. Hence $x = t_c(a, y) \in t_c(a, S)$, which completes the proof.

Classes in $SO(3, \mathbb{R})$

For a more interesting example, let us turn to the group $SO(3, \mathbb{R})$ of **proper rotations in three-dimensional space**, which is one of the most important groups in physics. We state a few plausible facts without giving a derivation. The product of any two rotations is a rotation. An element $R_{n,\phi}$ of $SO(3, \mathbb{R})$ is uniquely specified by a unit vector $\mathbf{n}$ that points in the direction of the axis of rotation and an angle of rotation ϕ in the interval $[0, \pi)$. The sense of positive rotation is determined by the right-hand rule. The inverse of $R_{n,\phi}$ can be written either as $R_{-n,\phi}$ or as $R_{n,-\phi}$. Rotations $R_{m,\theta}$ and $R_{n,\phi}$ about different axes m and n do not commute, that is,

$$R_{m,\theta} R_{n,\phi} \neq R_{n,\phi} R_{m,\theta}. \tag{4.138}$$

Let us derive the class of an element $x \in SO(3, \mathbb{R})$. Define

$$x = R_{n,\phi} \tag{4.139}$$

and

$$a = R_{m,\theta} \Rightarrow a^{-1} = R_{m,-\theta}. \tag{4.140}$$

Next, let n' be the unit vector obtained by rotating n about the axis m through the angle θ:

$$n' := R_{m,\theta} n. \tag{4.141}$$

The rotation

$$y = axa^{-1} = R_{m,\theta} R_{n,\phi} R_{m,-\theta} \tag{4.142}$$

is conjugate to x. Equation (4.142) instructs one to perform first the rotation that takes n' into n, then a rotation through the angle ϕ about the axis n, and finally the rotation that takes n back into n'. The net result is a rotation about n' through the angle ϕ. Therefore the elements of the class $\mathcal{C}_x$, for which x is defined in Eq. (4.139), are rotations through the same angle, ϕ, about different axes. Each angle of rotation gives rise to a different class in the group $SO(3, \mathbb{R})$. This statement obviously holds in the finite subgroups of $SO(3, \mathbb{R})$ such as C_3, D_3, and the other crystallographic point groups.

It can be shown that every three-dimensional rotation through the angle ϕ is conjugate to $R_{n,\phi}$, with n an arbitrary unit vector. This statement is not necessarily true in a finite subgroup G of $SO(3, \mathbb{R})$ because the rotation $R_{m,\theta}$ that rotates n into n' may not belong to G. For example, in the octahedral group O the C_4^2 rotations about the axes through the vertexes of the octahedron in Fig. 4.7 and the C_2 rotations about the axes that bisect the edges of the octahedron are in different classes.

4.2.9 EXERCISES FOR SECTION 4.2

4.2.1   Prove that the integers under addition, $(\mathbb{Z}, +)$, satisfy all the axioms for an Abelian group.

   Hint: [Note that the identity element is 0, $(m)^{-1} = -m$, and so forth.]

4.2.2   Prove that the rational numbers under addition, $(\mathbb{Q}, +)$, satisfy all of the group axioms.

   Hint: [Recall from Section 2.1 that one defines a rational number (p/q) as the equivalence class of all fractions np/nq.]

4.2.3   Show that if one represents a real number $r \in \mathbb{R}$ in the base-β floating-point format

$$r = \pm \left[\sum_{j=0}^{\infty} \frac{d_j}{\beta^j} \right] \cdot \beta^n, \tag{4.143}$$

in which $\beta > 1$, $\beta \in \mathbb{Z}^+$ and

$$\forall j \in \mathbb{Z}^+ : 0 \le d_j < \beta, \tag{4.144}$$

then $(\mathbb{R}, +)$ satisfies all of the group axioms.

4.2.4   Show that if one represents a complex number $x + iy = z \in \mathbb{C}$ as an ordered pair,

$$z := \langle x, y \rangle, \tag{4.145}$$

in which $x, y \in \mathbb{R}$, and if one defines addition componentwise,

$$z_1 + z_2 = \langle x_1 + x_2, y_1 + y_2 \rangle, \tag{4.146}$$

then $(\mathbb{C}, +)$ satisfies all of the group axioms.

4.2.5   Prove that the nonzero rational numbers under multiplication, $(\mathbb{Q} \setminus \{0\}, \cdot)$, satisfy all of the group axioms if one defines

$$(p_1/q_1)(p_2/q_2) := (p_1 p_2 / q_1 q_2). \tag{4.147}$$

4.2.6   Prove that the nonzero real numbers under multiplication, $(\mathbb{R} \setminus \{0\}, \cdot)$, satisfy all of the group axioms.

4.2.7   Prove that the set of nonzero complex numbers under multiplication, $(\mathbb{C} \setminus \{(0, 0)\}, \cdot)$ satisfies all of the group axioms if one defines the product $z_1 z_2$ as

$$z_1 z_2 := (x_1, y_1)(x_2, y_2) := (x_1 x_2 - y_1 y_2, x_1 y_2 + x_2 y_1). \tag{4.148}$$

4.2.8   Let G be a group. Prove that if for some $f \in G$

$$\forall a \in G : af = a, \tag{4.149}$$

then $f = e$. Thus the identity element is unique.

4.2.9   Prove that

$$\forall a \in G : ab = e \Rightarrow b = a^{-1}. \tag{4.150}$$

Therefore the inverse of every $a \in G$ is unique.

4.2.10   Show that if G is a group and H is a subset of G such that

$$\forall h, h' \in H : hh' \in H \tag{4.151}$$

and

$$\forall h \in H : \exists h^{-1} \in H, \tag{4.152}$$

then H is a subgroup of G.

4.2.11 Prove that if H is a subgroup of a group G, the identity element e' of H is equal to the identity element e of G.

4.2.12 Prove that if a is an element of a group G and if H is a subgroup of G, then aH is a subgroup of G if and only if a belongs to H.

Hint: An easy way to prove the "only if" part is to show that if a does not belong to H, then $a^{-1} \notin H$ and $e \notin aH$, which shows that aH is not a subgroup of G.

4.2.13* Prove that if G and $\{e\}$ are the only subgroups of a finite group G, then the order of G is a prime number.

4.2.14 This exercise pertains to the dihedral group D_3.

(a) Find all of the subgroups of D_3.

(b) Find all of the cosets of each of the subgroups found in part (a).

4.2.15 Prove that if $(G, \circ,)$ is a group, if $a \in G$, and if one defines positive powers of a recursively through the equations

$$a \circ a = a^2 \tag{4.153}$$

$$a \circ a^{n-1} = a^n, \tag{4.154}$$

then

$$a^m \circ a^n = a^{m+n} \tag{4.155}$$

and

$$(a^m)^n = a^{mn}. \tag{4.156}$$

4.2.16 Prove that every group whose order (number of elements) is a prime integer p is cyclic.

4.2.17 Work out a multiplication table that makes V (see Eq. [4.91]) into a group. Is V Abelian?

Hint: A physical realization of this group is the symmetry group D_2 of a rectangle. Choose x and y axes parallel to the sides of the rectangle. Then identify b, a, c with rotations through $180°$ about the x, y, and z axes, respectively, as in Fig. 4.8. With these choices, you should find that $ab = c$.

4.2.18 Verify the multiplication table of D_3 (Table 4.3).

4.2.19 Verify that the group inverse of the product ab is $(ab)^{-1} = b^{-1}a^{-1}$, as claimed in Eq. (4.32).

4.2.20 Verify that the multiplication of 2×2 matrices is associative.

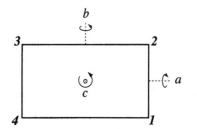

FIGURE 4.8

Rotational symmetry operations of a rectangle.

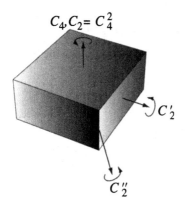

FIGURE 4.9

Rotational symmetry of a square prism. The operations shown generate the group D_4.

4.2.21   Verify that the mapping $t_r : G \rightarrow G$ defined in Eq. (4.48) is bijective.

4.2.22   Verify that the mapping $t_c : G \rightarrow G$ defined in Eq. (4.57) is bijective.

4.2.23   Let G be a group, and let S and T be G-sets. Prove that $S \cap T$ is a G-set.

4.2.24   Let G be a group, let S be a G-set, and let $s \in S$. Prove that the smallest G-set that contains s is orbit[s].

4.2.25   Obtain the multiplication table of the group D_4. The group operations are shown in Fig. 4.9.

4.2.26   Find all of the subgroups of the group D_4.

4.3 GROUP HOMOMORPHISMS

4.3.1 DEFINITIONS AND BASIC PROPERTIES

In modern mathematics one not only defines abstract structures such as groups; one also studies mappings that preserve these structures. This section describes one of the most basic examples of a structure-preserving mapping, the homomorphism of one group into another. Group homomorphisms are a fundamental tool for revealing the structure of groups on both an abstract and a practical level.

Special kinds of group homomorphisms, known as group representations, give rise to the so-called special functions of physics and engineering, as well as to selection rules for atomic transitions and interactions of elementary particles.

Group Homomorphisms

Let G and G' be groups, let $\circ$ be the group operation in G, and let $\circ'$ be the group operation in G'. A **group homomorphism** of G to G' is a mapping ϕ from G into or onto G' such that

$$\forall a, b \in G : \quad \phi(a \circ b) = \phi(a) \circ' \phi(b). \tag{4.157}$$

Note that a and b belong to G; $\phi(a)$ and $\phi(b)$ belong to G'. If one does not write the group operations explicitly and if one defines

$$a' := \phi(a), \quad b' := \phi(b), \tag{4.158}$$

then the definition of a homomorphism looks simpler:

$$\forall a, b \in G : \quad \phi(ab) = a'b'. \tag{4.159}$$

However, when one constructs a homomorphism it is always a good idea to write down what the group operations $\circ$ and $\circ'$ in G and G' are, even if one does not write the group operation in every equation.

Later we shall define homomorphisms of structures other than groups. For now, we shall call a group homomorphism a **homomorphism** because the context is clear.

For example, every group G can be mapped homomorphically onto the one-element group $C_1 = \{e' \mid e'^2 = e'\}$ by the mapping

$$\forall a \in G : \quad \phi(a) = e'. \tag{4.160}$$

The mapping ϕ of this example is called the **identity homomorphism**.

The three-to-one mapping ψ such that

$$\begin{aligned} \psi(e) = \psi(d) = \psi(f) = e' \\ \psi(a) = \psi(b) = \psi(c) = a' \end{aligned} \tag{4.161}$$

is a homomorphism of $G = D_3$ onto $G' = C_2 = \{e', a'\}$. For example,

$$\begin{aligned} \psi(ab) = \psi(d) = e' \\ \psi(a)\psi(b) = a'a' = e'. \end{aligned} \tag{4.162}$$

The reader should verify other cases using Table 4.3 (see Exercise 4.3.9).

By Eqs. (4.103) and (4.157), the mapping $\phi_n : (\mathbb{Z}, +) \to (\mathbb{Z}_n, +)$ from the integers to the integers modulo n such that

$$\forall r \in \mathbb{Z} : \quad \phi_n(r) = r_n \in \mathbb{Z}_n \tag{4.163}$$

is also a homomorphism. Because ϕ_n maps r and every integer that differs from r by a multiple of n onto the same (countably infinite) equivalence class r_n, ϕ_n definitely is not injective.

As in the preceding examples, a homomorphism $\phi : G \to G'$ always maps the identity $e \in G$ onto the identity $e' \in G'$:

$$\phi(e) = e'. \tag{4.164}$$

For, suppose that $f := \phi(e) \in G'$. Then

$$e = e^2 \Rightarrow f = f^2. \tag{4.165}$$

By the definition of the identity in G',

$$f = fe'. \tag{4.166}$$

Now apply the cancellation law in G' to the equation

$$f^2 = fe' \tag{4.167}$$

to obtain

$$f = e', \tag{4.168}$$

which is what we set out to prove.

Another basic property of a group homomorphism ϕ is that inverses map to inverses:

$$\forall a \in G : [\phi(a)]^{-1} = \phi(a^{-1}). \tag{4.169}$$

From Eq. (4.164) and the fact that ϕ is a homomorphism one gets

$$\phi(aa^{-1}) = \phi(e) \Rightarrow \phi(a)\phi(a^{-1}) = e' \tag{4.170}$$

for every $a \in G$, which proves Eq. (4.169).

Matrix Representations of Groups

If a matrix group G' is a homomorphic image of a group G under a mapping Γ, then one says that Γ is a **matrix representation** of G and that G' **represents** G. If the matrices that belong to G' have n rows and columns, then one says that the representation Γ is *n*-**dimensional**. To give a trivial example, the **identity representation** $\Gamma^{(A_1)}$ such that

$$\forall a \in G : \Gamma^{(A_1)}(a) = 1 \in \mathbb{R} \tag{4.171}$$

maps G onto the multiplicative group $G' = (\{1\}, \cdot)$. Because the real number 1 can be considered as a 1×1 matrix, this is a one-dimensional representation of the one-element group $C_1 = \{e \mid e^2 = e\}$.

A less trivial example is the injective one-dimensional representation

$$e \mapsto 1, \quad a \mapsto -1 \tag{4.172}$$

of the two-element group $C_2 = \{e, a\}$. This representation is important in physics because of its close relationship with the parity group.

Another example is the mapping $\Gamma^{(A_2)} : D_3 \to \{1, -1\}$ such that

$$\begin{aligned}
\Gamma^{(A_2)}(e) &= \Gamma^{(A_2)}(d) = \Gamma^{(A_2)}(f) = 1 \\
\Gamma^{(A_2)}(a) &= \Gamma^{(A_2)}(b) = \Gamma^{(A_2)}(c) = -1,
\end{aligned} \tag{4.173}$$

which, evidently, is a 1×1 matrix realization of Eq. (4.161).

Both $\Gamma^{(A_1)}$ and $\Gamma^{(A_2)}$ are many-to-one, one-dimensional matrix representations. We show subsequently that D_3 has an injective representation by a group of two-dimensional matrices.

In Chapter 11 we show that for every finite group G, a function on a G-set can always be expanded in terms of functions that transform according to certain special representations of G (the so-called irreducible representations). For example, any function $f : \mathbb{R}^3 \to \mathbb{R}$ can be expressed as the sum of its even-parity and odd-parity parts (see Eq. [4.43]). The invariance of basic physical interactions (such as the Coulomb interaction $V(\mathbf{r}) = Ze^2/|\mathbf{r}|$) under the two-element inversion group leads to fundamental selection rules for observable physical processes such as the absorption and emission of radiation.

Operator Representations of Groups

Although the term "operator representation of a group" sounds formidable, operator representations are both familiar and useful in physics and engineering. For example, the mappings P_e and P_i defined in Eq. (4.38) constitute an operator representation of the two-element group C_2. Operator representations occur naturally whenever one defines functions on a symmetric structure.

Although operators are defined formally in Chapter 10, we adopt for now the informal definition that an operator is a mapping that maps functions defined on some subset of $\mathbb{R}^n$ to other functions defined on the same subset. Operators that represent group elements must be linear mappings as defined in Section 6.1.1. Because we make no use here of the property of linearity, it is not necessary to study Chapter 6 before reading farther.

We define a class of operator representations that is large enough to include most of the operator representations of interest in physics and engineering. Let G be a group such that every element $\mathsf{R} \in G$ maps a point $\mathbf{x} \in \mathbb{R}^n$ to another point $\mathsf{R}\mathbf{x} \in \mathbb{R}^n$. For example, G might be the group of rigid translations and rotations that leave a face-centered cubic lattice invariant. Let $f : \mathbb{R}^n \to \mathbb{C}$ be a complex-valued function on $\mathbb{R}^n$. For example, if $n = 3$, f might be the wave function of a conduction-band electron in a semiconductor. For every $\mathsf{R} \in G$ we define an operator $P_{\mathsf{R}} : f \mapsto P_{\mathsf{R}}f$ as follows:

$$\forall f : \ \mathbb{R}^n \to \mathbb{C} : \quad \forall \mathbf{x} \in \mathbb{R}^n : \ \forall \mathsf{R} \in G : \ (P_{\mathsf{R}}f)(\mathsf{R}\mathbf{x}) := f(\mathbf{x}). \tag{4.174}$$

In other words, the transformed function $P_{\mathsf{R}}f$ is defined to have the same value at the transformed point $\mathsf{R}\mathbf{x}$ as the original function f has at the original point $\mathbf{x}$. The definition implies that the value of the transformed function $P_{\mathsf{R}}f$ at a point $\mathbf{x} \in \mathbb{R}^n$ is

$$(P_{\mathsf{R}}f)(\mathbf{x}) = f(\mathsf{R}^{-1}\mathbf{x}). \tag{4.175}$$

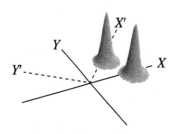

FIGURE 4.10

Transformation of a function on $\mathbb{R}^2$ through rotation, illustrating Eq. (4.175).

To understand why the inverse of R *must* appear, note that the commutative diagram

$$\begin{array}{ccc}
 & \mathbb{R}^n & \\
 R \nearrow & & \searrow P_R f \\
\mathbb{R}^n & \xrightarrow{\quad f \quad} & \mathbb{C}
\end{array}$$

(4.176)

is equivalent to Eq. (4.174), which in turn is equivalent to Eq. (4.175). Figure 4.10 gives further insight into Eq. (4.174).

To show that the set of operators $\{P_R \mid R \in G\}$ is a group and that the mapping

$$\phi : R \mapsto P_R$$

(4.177)

is a homomorphism, we calculate the product mapping $P_S P_R$:

$$
\begin{aligned}
(P_S P_R f)(\mathbf{x}) &= (P_S(P_R f))(\mathbf{x}) \\
&= (P_S f)(R^{-1}\mathbf{x}) \\
&= f(R^{-1}S^{-1}\mathbf{x}) \\
&= f((SR)^{-1}\mathbf{x})) \\
&= (P_{SR} f)(\mathbf{x}).
\end{aligned}
$$

(4.178)

Note that in passing from the second to the third line, $\mathbf{x}$ is mapped to $S^{-1}\mathbf{x}$. (If one were [incorrectly] to write $(P_S f)(R^{-1}\mathbf{x}) = f(S^{-1}R^{-1}\mathbf{x})$, then one would wrongly regard f as a function of $R^{-1}\mathbf{x}$ and $P_S f$ as a function of $\mathbf{x}$.)

The operators P_R clearly satisfy the group closure axiom by virtue of the relation

$$P_S P_R = P_{SR},$$

(4.179)

which we just proved. Associativity follows from the associativity of composition of functions. The existence of an inverse and identity element also follows from Eq. (4.179). It follows that the set $\{P_R \mid R \in G\}$ is a group. Then the mapping

$$\phi : R \mapsto P_R$$

(4.180)

is an operator representation of G. One says that the set of functions $\{P_R \mid R \in G\}$ **carries the representation** ϕ of G.

Occasionally we need the following generalization: Let G be a group, let S be a G-set, and let $f : S \to \mathbb{C}$ be a complex-valued function that is defined on S. For every $c \in G$, define

$$\forall x \in S : (P_c f)(cx) := f(x). \tag{4.181}$$

An equivalent definition is

$$\forall x \in S : (P_c f)(x) := f(c^{-1}x). \tag{4.182}$$

Then the mapping

$$c \mapsto P_c \tag{4.183}$$

is a representation of G. The proof is the same as the proof of Eq. (4.179).

Group Isomorphisms

If a group homomorphism $\phi : G \to G'$ is injective and surjective, then ϕ is called a **group isomorphism**. We indicate the relation of isomorphism with the same symbol ($\cong$) that one would use for geometrical congruence:

$$G' \text{ is isomorphic to } G \Leftrightarrow G' \cong G. \tag{4.184}$$

One can see immediately that $\cong$ is an equivalence relation. Moreover, any injective group homomorphism ϕ carries G onto $\phi(G)$ and therefore defines an isomorphism of G and $\phi(G)$.

Not every injective mapping of the elements of a group is an isomorphism. For example, Exercise 4.2.21 shows that the mapping $t_r(a, x) = ax$ defined in Eq. (4.48) (in which a is a fixed element of G) is bijective, considered as a mapping of sets. However, $t_r(a, \cdot)$ is not a group isomorphism, for in general the homomorphism law $\phi(x)\phi(y) = \phi(xy)$ does not hold (one has $t_r(a, x)t_r(a, y) = axay$, but $t_r(a, xy) = axy$).

If $G' \cong G$ under a mapping ϕ, then one says that the mapping ϕ (or, by an abuse of terminology, the group G') is a **faithful representation** of G. For example, the matrix-valued mapping $\Gamma^{(E)}$ such that

$$\Gamma^{(E)}(e) = \begin{pmatrix} 1 & 0 \\ 0 & 1 \end{pmatrix}, \quad \Gamma^{(E)}(a) = \begin{pmatrix} -1 & 0 \\ 0 & 1 \end{pmatrix},$$

$$\Gamma^{(E)}(b) = \frac{1}{2}\begin{pmatrix} 1 & \sqrt{3} \\ \sqrt{3} & -1 \end{pmatrix}, \quad \Gamma^{(E)}(c) = \frac{1}{2}\begin{pmatrix} 1 & -\sqrt{3} \\ -\sqrt{3} & -1 \end{pmatrix}, \tag{4.185}$$

$$\Gamma^{(E)}(d) = \frac{1}{2}\begin{pmatrix} -1 & -\sqrt{3} \\ \sqrt{3} & -1 \end{pmatrix}, \quad \Gamma^{(E)}(f) = \frac{1}{2}\begin{pmatrix} -1 & \sqrt{3} \\ -\sqrt{3} & -1 \end{pmatrix}$$

is a faithful representation of D_3. (It is straightforward to see that the set of matrices shown is a group under the operation of matrix multiplication and that $\Gamma^{(E)}$ is an isomorphism; see Exercise 4.3.10.)

The cyclic groups $(C_n, \circ)$ and $(K_n, \cdot)$ furnish another example of isomorphism. One sees at once that $C_n \cong K_n$ under the mapping

$$\forall r \in (0 : n - 1) : \quad \phi(a^r) = \omega^r. \tag{4.186}$$

The multiplicative group $(K_n, \cdot)$ and the additive group $(\mathbb{Z}_n, +)$ are isomorphic under the mapping

$$\forall r \in (0 : n - 1) : \quad \psi(\omega^r) = r_n. \tag{4.187}$$

Lastly, $(\mathbb{Z}_n, +)$ and $(C_n, \circ)$ are isomorphic under the mapping

$$\forall r \in (0 : n - 1) : \quad \chi(r_n) = a^r. \tag{4.188}$$

Because we have shown that

$$(C_n, \circ) \cong (K_n, \cdot) \cong (\mathbb{Z}_n, +) \cong (C_n, \circ), \tag{4.189}$$

it follows (by various compositions of $\phi, \psi, \chi, \phi^{-1}, \psi^{-1}$, and χ^{-1}) that any two of the groups $(C_n, \circ), (K_n, \cdot)$, and $(\mathbb{Z}_n, +)$ are isomorphic. Note that the group operation in $(K_n, \cdot)$ is complex multiplication, but the group operation in $(\mathbb{Z}_n, +)$ is addition modulo n. Nevertheless $(K_n, \cdot)$ and $(\mathbb{Z}_n, +)$ are both realizations of the same abstract group, $(C_n, \circ)$.

4.3.2 NORMAL SUBGROUPS

Kernel of a Homomorphism

Let us study more closely the properties of a group homomorphism ϕ that is not injective. First, we assume for the sake of simplicity (and without losing any useful generality) that ϕ is surjective:

$$G' = \phi(G). \tag{4.190}$$

Because ϕ is not injective, there exist elements $a, b \in G$ such that

$$\phi(a) = \phi(b). \tag{4.191}$$

Then, by Eqs. (4.157) and (4.169),

$$\phi(a^{-1})\phi(b) = e' = \phi(a)\phi(b^{-1}). \tag{4.192}$$

It follows that

$$\forall a, b \in G : \quad \phi(a) = \phi(b) \Leftrightarrow \phi(a^{-1}b) = e'. \tag{4.193}$$

Therefore one can study all the elements of G that are mapped onto a single element of G' by studying the preimage of a single element, the identity e' of G'.

The set ker[ϕ] of elements of a group G that are mapped onto the identity $e' \in G'$ by a group homomorphism ϕ is called the **kernel** of ϕ:

$$\ker[\phi] := \{a \in G \mid \phi(a) = e'\}. \tag{4.194}$$

It is straightforward to show, one by one, that ker[ϕ] satisfies each of the group axioms and therefore is a subgroup of G.

- Group axiom 1: Because $\phi(ab) = \phi(a)\phi(b) = e'$ for every pair of elements a, b in ker[ϕ], ker[ϕ] is closed under the group operation in G.
- Group axiom 2: The composition of elements of ker[ϕ] is associative by virtue of the associativity of the group operation in G.
- Group axiom 3: By Eq. (4.164), e belongs to ker[ϕ].
- Group axiom 4: By Eq. (4.169), a^{-1} belongs to ker[ϕ] for every $a \in$ ker[ϕ].

For example, the kernel ker[ϕ_n] of the homomorphism $\phi_n : \mathbb{Z} \to \mathbb{Z}_n$ (Eq. [4.163]) is the set $n\mathbb{Z}$, which consists of the integral multiples of n (see Eq. [4.107]). In accordance with the theorem that we just proved, $(n\mathbb{Z}, +)$ is a group. (The identity element is 0, the inverse of mn is $-mn$, and so forth.) To give another example, the kernel of the homomorphism Eq. (4.173) is $\{e, d, f\} \cong C_3$.

From Eqs. (4.193) and (4.194) one gets an intuitively appealing characterization of the preimage $\phi^{-1}(a')$ of an element $a' \in G'$. (Recall that $\phi^{-1}(a')$ denotes the set of elements of G that are mapped onto a', even if ϕ is not invertible; see Section 2.1). According to Eq. (4.193),

$$\phi(a) = \phi(b) \Leftrightarrow a^{-1}b \in \ker[\phi]. \tag{4.195}$$

Equivalently,

$$\phi(a) = \phi(b) = a' \Leftrightarrow \exists k \in \ker[\phi] : \ni: \ b = ak. \tag{4.196}$$

This means that $b \in G$ is mapped onto the same image in G' as $a \in G$ if and only if b is a "multiple" of a by some element k of the kernel of ϕ, exactly as in the example of $\phi_n : \mathbb{Z} \to \mathbb{Z}_n$.

An immediate consequence of Eqs. (4.195–4.196) is the important result that if $\phi : G \to G'$ is a homomorphism of G *onto* G', then G and G' are isomorphic if and only if ker[ϕ] contains only the identity element of G:

$$G' \cong G \Leftrightarrow \ker[\phi : G \to G'] = \{e \in G\}. \tag{4.197}$$

To prove the "if" assertion, suppose that ker[ϕ] = $\{e\}$. Then, if two images under ϕ are equal, $\phi(b) = \phi(a)$, it follows that $a^{-1}b = e$; therefore the preimages are equal, $b = a$. To prove the "only if" part, assume that $G' \cong G$. Then

$$[\forall a, b \in G : \ \phi(b) = \phi(a) \Rightarrow b = a \Rightarrow a^{-1}b = e] \Rightarrow \ker[\phi] = \{e\}, \tag{4.198}$$

which completes the proof.

Quotient Groups

Although the kernel of every homomorphism is a subgroup, not every subgroup can be the kernel of a homomorphism. If $k \in \ker[\phi]$, then $aka^{-1} \in \ker[\phi]$ for every $a \in G$:

$$\forall a \in G : \; \phi(aka^{-1}) = \phi(a)e'[\phi(a)]^{-1} = e'. \tag{4.199}$$

Conversely, if $aka^{-1} \in \ker[\phi]$ for every $a \in G$, it follows that $k \in \ker[\phi]$ [take $a = e$]. Then

$$k \in \ker[\phi] \Leftrightarrow \forall a \in G : aka^{-1} \in \ker[\phi]. \tag{4.200}$$

Therefore the kernel of a homomorphism is a special kind of subgroup.

Évariste Galois introduced the structure of which $\ker[\phi]$ appears, at least so far, to be a special case. A subgroup H of a group G is called **normal in** G (or **invariant in** G) if and only if

$$\forall x \in H : \; \forall a \in G : \; axa^{-1} \in H. \tag{4.201}$$

If G is Abelian, then every subgroup is normal. Beware: If H is normal in a non-Abelian group G, it does not necessarily follow that H is normal in a group of which G is a subgroup. For example, it can be shown that there is only one normal subgroup of the symmetric (i.e., permutation) group S_n on n objects. The symmetric group S_3 is isomorphic to D_3 and therefore has a normal subgroup of order 3. This subgroup of S_3 is also a subgroup of every S_n (for which $n > 3$) but is normal only in S_3.

The definition of a normal subgroup, Eq. (4.201), implies that if H is normal in G and $x \in H$, then $\mathcal{C}_x \subset H$. Therefore another definition of a normal subgroup is that H is normal in G if and only if H is a subgroup and is the union of a collection of classes of conjugate elements of G.

For example, in D_3 the normal subgroup $\{e, d, f\}$ is the union of two classes,

$$\{e, d, f\} = \mathcal{C}_e \cup \mathcal{C}_d, \tag{4.202}$$

but the nonnormal subgroup $\{e, a\}$ is not the union of any two classes of D_3. The only class that contains a is $\mathcal{C}_a = \{a, b, c\}$. Therefore the only union of classes that contains a and is a subgroup is $\mathcal{C}_e \cup \mathcal{C}_a \cup \mathcal{C}_d = D_3$.

Not every union of classes is a subgroup; therefore one cannot construct normal subgroups by randomly taking unions of classes. For example, in D_3 the union $\mathcal{C}_e \cup \mathcal{C}_a$ is not closed under the group operation (recall that $ab = d$) and therefore is not a subgroup. The union $\mathcal{C}_a \cup \mathcal{C}_d$ does not contain the identity and therefore is not a subgroup.

We have shown that the kernel of every homomorphism of a group G onto a group G' is a normal subgroup of G. We show now that every normal subgroup of G is the kernel of some homomorphism of G. Anticipating this result, we summarize: *A subgroup H of a group G is the kernel of a homomorphism of G if and only if H is normal in G.*

For example, the kernel of the homomorphism ϕ defined in Eq. (4.161) is $\{e, d, f\}$, which is a normal subgroup of D_3. On the other hand, the subgroup $\{e, a\}$ of D_3 is not the kernel of any homomorphism ψ of D_3 because $\{e, a\}$ is not normal in D_3. For instance, one finds that $dad^{-1} = daf = dc = b \notin \{e, a\}$. It follows that none of the C_2 subgroups is normal in D_3;

hence there is one and only one nontrivial homomorphism of D_3, namely, the one defined by Eq. (4.161).

The proof proceeds via a direct (although, at first sight, somewhat implausible) construction of a group G' such that $\phi : (G, \circ) \to (G', \circ')$ is a homomorphism and $H = \ker[\phi]$. It turns out that one can take G' to be the set of (distinct) cosets of H:

$$G' = \{aH \mid a \in G\}. \tag{4.203}$$

It is not obvious that G' is a group. Recalling the definition of a coset (Eq. [4.63]), we define **coset multiplication** as follows:

$$aH \circ' bH := \{ahbh' \mid h, h' \in H\}. \tag{4.204}$$

Because H is normal,

$$\forall b \in G : \forall h \in H : \exists h'' \in H : \ni: h = bh''b^{-1}. \tag{4.205}$$

Of course, $h'' = b^{-1}hb$; one knows from Eq. (4.201) that $h'' \in H$. One verifies by inspection that the mapping $t_c(b, \cdot) : H \to H$ defined by the equation $t_c(b, h) = b^{-1}hb$ is such that

$$t_c(b, h)t_c(b, h') = t_c(b, hh'). \tag{4.206}$$

Exercise 4.2.22 shows that $t_c(b, \cdot)$ is bijective. It follows that $t_c(b, \cdot)$ is an isomorphism. From Eqs. (4.204) and (4.205), one gets

$$aH \circ' bH = \{abh''h' \mid h', h'' \in H\} = \{abh''' \mid h''' \in H\} = abH \tag{4.207}$$

in which $h''' = h''h'$. (The set of products $h''h'$ contains every element of H; see Exercise 4.2.21.) This result establishes the first group axiom (closure). The third axiom is also satisfied, for

$$e' = eH = H \tag{4.208}$$

is the **identity coset**. The relation

$$(aH)^{-1} = a^{-1}H \tag{4.209}$$

(which follows from Eq. [4.207]) establishes the existence of an inverse for every element of G'. The proof that coset multiplication is associative is left as an exercise (Exercise 4.3.7).

The coset group G' is usually called the **quotient group** of G with respect to H and is written

$$G' = G/H \tag{4.210}$$

(read "G mod H"). The justification for this terminology is that elements a, b of G that are mapped onto the same coset aH in G/H differ only by an element of H (see Eq. [4.196]), and therefore the quotient group is G modulo H in the same sense in which $\mathbb{Z}_n$ is $\mathbb{Z}$ modulo (the multiples of) n.

4.3.3 DIRECT PRODUCT GROUPS

Let $(G_1, \circ_1), \ldots, (G_n, \circ_n)$ be groups, and let G be the Cartesian product

$$G := G_1 \times \cdots \times G_n. \tag{4.211}$$

Let $\langle a_1, \ldots, a_n \rangle \in G$, where $\forall i \in (1 : n) : a_i \in G_i$. If one defines the binary operation

$$\forall a = \langle a_1, \ldots, a_n \rangle \in G : \forall b = \langle b_1, \ldots, b_n \rangle \in G :$$
$$a \circ b := \langle a_1 \circ_1 b_1, \ldots, a_n \circ_n b_n \rangle, \tag{4.212}$$

then $(G, \circ)$ is a group. G is closed under the operation $\circ$, and $\circ$ is associative, because every G_i is closed under $\circ_i$ and $\circ_i$ is associative. The identity element of G is $\langle e_1, \ldots, e_n \rangle$, in which e_i is the identity element of G_i. The inverse of $\langle a_1, \ldots, a_n \rangle \in G$ is $\langle a_1^{-1}, \ldots, a_n^{-1} \rangle$. G is called the **direct product** of $(G_1, \circ_1), \ldots, (G_n, \circ_n)$, and the groups $(G_i, \circ_i)$ are called the **direct factors** of G. The order of G is the product of the orders $|G_i|$ of the direct factors:

$$|G| = \prod_{i=1}^{n} |G_i|. \tag{4.213}$$

For example, the elements of the direct product group $C_2 \times C_2$ are $e = \langle e_1, e_2 \rangle$, $a = \langle a_1, e_2 \rangle$, $b = \langle e_1, a_2 \rangle$ and $c = \langle a_1, a_2 \rangle$. The order of $C_2 \times C_2$ is $2 \times 2 = 4$. Clearly $ab = c$, etc.; one finds that $V \cong C_2 \times C_2$.

Another important example of a direct product group is

$$O_h = C_i \times O, \tag{4.214}$$

in which $C_i \cong C_2$ is the inversion group. The order of O_h is $24 \times 2 = 48$.

Let

$$\hat{G}_i := \{ \langle e_1, \ldots, e_{i-1}, x_i, e_{i+1}, \ldots, e_n \rangle \mid x_i \in G_i \}. \tag{4.215}$$

Obviously $\hat{G}_i$ is a group and $\hat{G}_i \cong G_i$ under the mapping

$$\forall x_i \in G_i : x_i \mapsto \langle e_1, \ldots, e_{i-1}, x_i, e_{i+1}, \ldots, e_n \rangle. \tag{4.216}$$

We show that $\hat{G}_1, \ldots, \hat{G}_n$ are normal subgroups of G. Choose any $x \in \hat{G}_i$, and let $a \in G$. Then there exist elements $a_1 \in G_1, \ldots, a_n \in G_n$ and $x_i \in \hat{G}_i$ such that $a = \langle a_1, \ldots, a_n \rangle$ and $x = \langle e_1, \ldots, e_{i-1}, x_i, e_{i+1}, \ldots, e_n \rangle$. One sees at once that

$$axa^{-1} = \langle e_1, \ldots, e_{i-1}, a_i x_i a_i^{-1}, e_{i+1}, \ldots, e_n \rangle \in \hat{G}_i. \tag{4.217}$$

Therefore $\hat{G}_i$ is normal in G.

Another important fact about direct product groups is that elements from different direct factors commute. Let $x \in \hat{G}_i$ and $y \in \hat{G}_j$ with $i < j$. Then one has

$$xy = \langle e_1, \ldots, e_{i-1}, x_i, e_{i+1}, \ldots, e_{j-1}, y_j, e_{j+1}, \ldots, e_n \rangle$$
$$= yx. \tag{4.218}$$

We show now that the preceding properties uniquely characterize a direct product of groups. Let $H_1, \ldots, H_n$ be subgroups of a group G such that

1. The elements of different subgroups H_i commute:

$$\forall a_i \in H_i : \forall a_j \in H_j : \ni: j \neq i : a_i a_j = a_j a_i. \tag{4.219}$$

2. Every element of G can be expressed uniquely as a product of elements of the subgroups H_i:

$$\forall a \in G : \exists! a_1 \in H_1 : \ldots : \exists! a_n \in H_n : \ni: a = a_1 a_2 \cdots a_n. \tag{4.220}$$

3. The only element common to any two subgroups is the identity of G:

$$a \in H_i \quad \text{and} \quad a \in H_j \ (i \neq j) \Rightarrow a = e. \tag{4.221}$$

Then G is isomorphic to $H_1 \times \cdots \times H_n$.

To prove this theorem, note first that condition (2) implies that the mapping

$$\forall a = a_1 a_2 \cdots a_n \in G : \langle a_1, \ldots, a_n \rangle \mapsto a \tag{4.222}$$

is bijective and that condition (1) implies that the subgroups H_i are normal in G. To show that $G \cong H_1 \times \cdots \times H_n$, it is enough to observe that

$$\forall a = \langle a_1, \ldots, a_n \rangle \in G : \forall b = \langle b_1, \ldots, b_n \rangle \in G :$$
$$a \circ b = \langle a_1 b_1, \ldots, a_n b_n \rangle \mapsto a_1 b_1 \cdots a_n b_n = a_1 \cdots a_n b_1 \cdots b_n. \tag{4.223}$$

Note that condition (2) guarantees that one can assume that factorizations into products of elements of the H_i exist for any elements a and b of G. The step $a_1 b_1 \cdots a_n b_n = a_1 \cdots a_n b_1 \cdots b_n$ follows from condition (1).

For example, in $G = D_2$ the two subgroups $H_1 = \{e, a\}$, $H_2 = \{e, b\}$ satisfy conditions (1) through (3), for the only element of D_2 that belongs neither to H_1 nor to H_2 is $c = ab = ba$, and the only common element of H_1 and H_2 is e. Also, $a_1 b_1 a_2 b_2 = a_1 a_2 b_1 b_2$ because $b_1 \in H_1$ and $a_2 \in H_2$; then condition (1) implies that $b_1 a_2 = a_2 b_1$. It follows that $D_2 \cong H_1 \times H_2 \cong C_2 \times C_2$.

A direct product of Abelian groups is usually called a **direct sum** because one usually writes the group operation as addition if a group is Abelian. The direct sum of H_1 and H_2 is written $H_1 \oplus H_2$. For example, the additive group of $\mathbb{C}$ is the direct sum of the additive groups of two copies of $\mathbb{R}$,

$$(\mathbb{C}, +) = (\mathbb{R}, +) \oplus (\mathbb{R}, +), \tag{4.224}$$

because $\langle x, y \rangle = \langle x, 0 \rangle + \langle 0, y \rangle$ uniquely. Of course, in $(\mathbb{C}, +)$ the group operation reads $\langle x, y \rangle + \langle x', y' \rangle = \langle x + x', y + y' \rangle$.

Another important property of a direct product of two groups, $G = G_1 \times G_2$, is that

$$G_1 \cong G/\hat{G}_2 \quad \text{and} \quad G_2 \cong G/\hat{G}_1; \tag{4.225}$$

$\hat{G}_1 \cong G_1$ and $\hat{G}_2 \cong G_2$ are defined in Eq. (4.215). To construct a proof, let us peel away a layer of abstraction from the definition of an element of the factor group $G/\hat{G}_2$. An element of $G/\hat{G}_2$ is a coset $a\hat{G}_2$ of $\hat{G}_2$. Let $a = \langle a_1, a_2 \rangle$. Then

$$
\begin{aligned}
a\hat{G}_2 &= \{\langle a_1, a_2 \rangle \circ \langle e_1, g_2 \rangle \mid g_2 \in G_2\} \\
&= \{\langle a_1, a_2 g_2 \rangle \mid g_2 \in G_2\} \\
&= \{\langle a_1, h_2 \rangle \mid h_2 \in G_2\} \\
&= \langle a_1, e_2 \rangle \hat{G}_2.
\end{aligned}
\tag{4.226}
$$

But the mapping

$$
\forall a_1 \in G_1 : \langle a_1, e_2 \rangle \hat{G}_2 \mapsto \langle a_1, e_2 \rangle \in \hat{G}_1
\tag{4.227}
$$

is injective. This mapping is also a homomorphism, for the kernel, $\langle e_1, e_2 \rangle \hat{G}_2 = \hat{G}_2$, is normal in G. It follows that $G_1 \cong G/\hat{G}_2$.

4.3.4 EXERCISES FOR SECTION 4.3

4.3.1 Let k, l, n be positive integers such that $k \neq 1$ and $l \neq 1$, and let $n = kl$. Construct a homomorphism ϕ_{nk} of $\mathbb{Z}_n$ onto $\mathbb{Z}_k$. Is ϕ_{nk} an isomorphism?

4.3.2 Construct a homomorphism of the group V (Eq. [4.91]) onto C_2.

4.3.3 Find all the normal subgroups of the group D_3.

4.3.4 Show that the subset $\{e, a, b, c\}$ of the group D_3 is not the kernel of any homomorphism of D_3.

4.3.5 Show that in the group D_3, the cosets aH, bH, and cH are equal.

4.3.6 Establish the multiplication table for the cosets eH, aH of the subgroup $\{e, d, f\}$ in D_3.

4.3.7 Prove the associativity of the coset multiplication defined in Eq. (4.207) in which H is a normal subgroup of G.

4.3.8 Show that for the group D_3 and the subgroup $H = \{e, d, f\}$, the distinct classes are

$$
\mathcal{C}_e = \{e\}, \quad \mathcal{C}_a = \{a, b, c\}, \quad \mathcal{C}_d = \{d, f\}.
\tag{4.228}
$$

4.3.9 Verify Eq. (4.159) for the mapping given in Eq. (4.161) for every entry in Table 4.3.

4.3.10 Verify that the set of 2×2 matrices shown in Eq. (4.185) is a group under matrix multiplication and that Γ_E is an isomorphism.

4.3.11 Prove that C_6 is isomorphic to the direct product of the groups $\{e, a^2, a^4\}$ and $\{e, a^3\}$.

4.4 *SYMMETRIC GROUPS

4.4.1 PERMUTATIONS

Definitions

A **permutation** ϕ is a injective mapping of a finite set T onto itself. We can show right away that one does not need to study all finite sets because one can set up an injective correspondence between every permutation of an arbitrary finite set T and a permutation of a particular finite set of integers. Let the number of elements of T be $n \in \mathbb{Z}^+$. Then there exists an injective mapping ψ of

$$(1 : n) := \{1, \ldots, n\} \tag{4.229}$$

onto T,

$$\forall t \in T : \exists j \in (1 : n) : \ni: \ t = t_j := \psi(j), \tag{4.230}$$

because all finite sets with the same number of elements are equivalent. Apply a permutation ϕ to the elements of T such that

$$t' = \phi(t) = \phi\psi(j). \tag{4.231}$$

Now map t' back into $(1 : n)$:

$$j' := \psi^{-1}(t') = \psi^{-1}\phi\psi(j). \tag{4.232}$$

Therefore every permutation ϕ of T induces a unique permutation

$$\pi := \psi^{-1}\phi\psi \tag{4.233}$$

of $(1 : n)$, and vice versa. (One can see that π is injective and onto because it is the composition of three mappings, each of which is injective and onto.) The commutative diagram

$$
\begin{array}{ccc}
T & \xrightarrow{\phi} & T \\
\psi \uparrow & & \downarrow \psi^{-1} \\
(1 : n) & \xrightarrow[\pi]{} & (1 : n)
\end{array}
\tag{4.234}
$$

summarizes the relations among π, ϕ, and ψ.

One can write permutations with the same array notation as one uses for any mapping of a finite set (see Eq. [2.57]):

$$\pi = \begin{pmatrix} 1 & 2 & \cdots & n \\ \pi(1) & \pi(2) & \cdots & \pi(n) \end{pmatrix} \tag{4.235}$$

For example, the permutation π that takes $(1, 2, 3)$ onto $(2, 3, 1)$ is

$$\pi = \begin{pmatrix} 1 & 2 & 3 \\ 2 & 3 & 1 \end{pmatrix} \tag{4.236}$$

in which $2 = \pi(1)$ and so forth.

Groups of Permutations

If a permutation π_1 is followed by another permutation π_2, then by definition the product $\pi_2\pi_1$ is such that for every i

$$\pi_2\pi_1(i) = \pi_2(\pi_1(i)). \tag{4.237}$$

As in Eq. (2.59), it follows that

$$
\begin{aligned}
\pi_2\pi_1 &= \begin{pmatrix} \pi_1(1) & \pi_1(2) & \cdots & \pi_1(n) \\ \pi_2(\pi_1(1)) & \pi_2(\pi_1(2)) & \cdots & \pi_2(\pi_1(n)) \end{pmatrix} \\
&\quad\cdot \begin{pmatrix} 1 & 2 & \cdots & n \\ \pi_1(1) & \pi_1(2) & \cdots & \pi_1(n) \end{pmatrix} \\
&= \begin{pmatrix} 1 & 2 & \cdots & n \\ \pi_2\pi_1(1) & \pi_2\pi_1(2) & \cdots & \pi_2\pi_1(n) \end{pmatrix}.
\end{aligned} \tag{4.238}
$$

Therefore the composition of two permutations is another permutation, implying that the set

$$S_n := \{\pi \mid \pi(1 : n) = (1 : n)\} \tag{4.239}$$

satisfies the axiom of closure. The associativity of products of three mappings follows from the definition of composition. The identity permutation is

$$\begin{pmatrix} 1 & 2 & \cdots & n \\ 1 & 2 & \cdots & n \end{pmatrix}. \tag{4.240}$$

For every permutation $\pi \in S_n$, there exists the inverse permutation

$$\pi^{-1} = \begin{pmatrix} \pi(1) & \pi(2) & \cdots & \pi(n) \\ 1 & 2 & \cdots & n \end{pmatrix}. \tag{4.241}$$

Therefore S_n is a group, the **symmetric group** (or **permutation group**) on n objects. The order of S_n is equal to the number of distinct permutations of n objects. One can choose $\pi(1)$ in any one of n ways, $\pi(2)$ in $n - 1$ ways, and so forth, and finally $\pi(n)$ in only one way, because only one element of $(1 : n)$ has not yet been chosen. Then

$$n! := n(n - 1) \cdots 2 \cdot 1 \tag{4.242}$$

is the order of the symmetric group S_n.

4.4.2 CAYLEY'S THEOREM

The matrix representations of S_n are useful in many applications in atomic, nuclear, and elementary-particle physics. **Cayley's theorem** provides a fundamental reason for interest in S_n: *Every finite group G of order n is isomorphic to a subgroup of S_n*. At least in principle, if one studies the symmetric groups, one studies all finite groups. The fact that a normal subgroup of G need not be normal in S_n limits the usefulness of Cayley's theorem.

The proof of Cayley's theorem depends on the observation that left multiplication by a fixed element of G defines a bijective mapping of G onto G, hence a permutation of the elements of G. Write the elements of G as $a_1, \ldots, a_n$. For any $b \in G$, let

$$\pi_{l(b)} := \begin{pmatrix} a_1 & \cdots & a_n \\ ba_1 & \cdots & ba_n \end{pmatrix}. \tag{4.243}$$

Exercise 4.4.8 shows that $\pi_{l(b)}$ is a bijective mapping of G onto G and therefore is a permutation of the elements of G. This result is known as the **rearrangement theorem** because it implies that left multiplication simply rearranges (permutes) the elements of G.

Let

$$P := \{\pi_{l(b)} \mid b \in G\} \tag{4.244}$$

be the set that consists of all of the permutations of G that are induced by left multiplication. To establish that P is a group, observe that the product $\pi_{l(c)}\pi_{l(b)}$ can be written as

$$\begin{aligned} \pi_{l(c)}\pi_{l(b)} &= \begin{pmatrix} a_1 & \cdots & a_n \\ ca_1 & \cdots & ca_n \end{pmatrix} \begin{pmatrix} a_1 & \cdots & a_n \\ ba_1 & \cdots & ba_n \end{pmatrix} \\ &= \begin{pmatrix} ba_1 & \cdots & ba_n \\ cba_1 & \cdots & cba_n \end{pmatrix} \begin{pmatrix} a_1 & \cdots & a_n \\ ba_1 & \cdots & ba_n \end{pmatrix}. \end{aligned} \tag{4.245}$$

Therefore

$$\pi_{l(c)}\pi_{l(b)} = \begin{pmatrix} a_1 & \cdots & a_n \\ cba_1 & \cdots & cba_n \end{pmatrix} = \pi_{l(cb)}. \tag{4.246}$$

Because Eq. (4.246) establishes closure, and the demonstration that the other group axioms are satisfied is trivial, we have shown both that P is a group and that the mapping $\phi : G \to P$ such that

$$\phi(a) = \pi_{l(a)} \tag{4.247}$$

is a homomorphism of G onto P. Moreover, ϕ is injective because $\pi_{l(b)} = \pi_{l(a)}$ if and only if $b = a$ ($\pi_{l(b)}(e) = \pi_{l(a)}(e) \Rightarrow b = a$). Therefore G and P are isomorphic:

$$G \cong P. \tag{4.248}$$

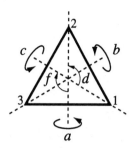

FIGURE 4.11

Rotational and permutational symmetries of an equilateral triangle. The vertexes have been labeled *1*, *2*, and *3*.

Let G be the image of $(1 : n)$ under the injective mapping ψ such that

$$\forall a_j \in G: \quad a_j = \psi(j).$$

(4.249)

Then, for every $\pi_a \in P$, $\psi^{-1}\pi_{l(a)}\psi$ is a permutation of $(1 : n)$ and hence an element of S_n. Because $\psi^{-1}\pi_{l(a)}\psi$ is injective, it follows that P (and therefore G) is isomorphic to a subgroup of S_n. The diagram (4.234) applies to Cayley's theorem if one makes the identifications $T = G$ and $\phi = \pi_a$.

For example, one can see from Fig. 4.11 that D_3 is isomorphic to S_3 because every rotation in D_3 permutes the vertexes and their labels (1, 2, or 3) and every permutation in S_3 corresponds to one and only one rotation. The permutations that correspond to the elements of D_3 are

$$\pi_{l(e)} = \begin{pmatrix} 1 & 2 & 3 \\ 1 & 2 & 3 \end{pmatrix}, \quad \pi_{l(a)} = \begin{pmatrix} 1 & 2 & 3 \\ 3 & 2 & 1 \end{pmatrix}, \quad \pi_{l(b)} = \begin{pmatrix} 1 & 2 & 3 \\ 2 & 1 & 3 \end{pmatrix},$$

$$\pi_{l(c)} = \begin{pmatrix} 1 & 2 & 3 \\ 1 & 3 & 2 \end{pmatrix}, \quad \pi_{l(d)} = \begin{pmatrix} 1 & 2 & 3 \\ 2 & 3 & 1 \end{pmatrix}, \quad \pi_{l(f)} = \begin{pmatrix} 1 & 2 & 3 \\ 3 & 1 & 2 \end{pmatrix}.$$

(4.250)

The group $D_2 \cong V$, the elements of which are illustrated in Fig. 4.12, furnishes another example of Cayley's theorem. One finds that

$$\pi_{l(e)} = \begin{pmatrix} 1 & 2 & 3 & 4 \\ 1 & 2 & 3 & 4 \end{pmatrix}, \quad \pi_{l(a)} = \begin{pmatrix} 1 & 2 & 3 & 4 \\ 2 & 1 & 4 & 3 \end{pmatrix},$$

$$\pi_{l(b)} = \begin{pmatrix} 1 & 2 & 3 & 4 \\ 4 & 3 & 2 & 1 \end{pmatrix}, \quad \pi_{l(c)} = \begin{pmatrix} 1 & 2 & 3 & 4 \\ 3 & 4 & 1 & 2 \end{pmatrix}.$$

(4.251)

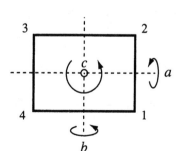

FIGURE 4.12

Rotational and permutational symmetries of a rectangle. The vertexes have been labeled *1*, *2*, *3*, and *4*.

Therefore the four-element group D_2 is isomorphic to a proper subgroup of the symmetric group S_4, the order of which is $4! = 24$.

4.4.3 CYCLIC PERMUTATIONS

Transpositions and *m*-Cycles

Although the notation used in Eq. (4.235) is accurate, it is cumbersome for permutations in which (for example) only two symbols are changed. The notation defined in Eq. (4.235) also fails to make obvious some important properties of a permutation π. A **transposition** (i, j) is a permutation in which i and j are interchanged and all other elements of $(1 : n)$ are left alone. Clearly (i, j) generates a cyclic subgroup of order two because $(i, j)^2$ is the identity permutation. If $j = i + 1$ or $i = j + 1$, then (i, j) is called an **adjacent transposition**. A **three-cycle** is a permutation

$$\pi = (i, j, k) = \begin{pmatrix} i & j & k \\ j & k & i \end{pmatrix} \tag{4.252}$$

such that

$$\pi(i) = j, \quad \pi(j) = k, \quad \pi(k) = i, \tag{4.253}$$

and all other elements of $(1 : n)$ are left alone. A three-cycle generates a cyclic subgroup of order three because

$$\pi^2(i) = k, \quad \pi^2(j) = i, \quad \pi^2(k) = j \tag{4.254}$$

and therefore

$$\pi^3(i) = i, \quad \pi^3(j) = j, \quad \pi^3(k) = k. \tag{4.255}$$

For example, in Eq. (4.250)

$$\pi_d = (1, 3, 2), \quad \pi_f = (1, 2, 3). \tag{4.256}$$

In general, an ***m*-cycle** is a permutation $\pi = (i_1, \ldots, i_m)$ such that

$$\pi(i_1) = i_2, \ldots, \pi(i_{m-1}) = i_m, \pi(i_m) = i_1. \tag{4.257}$$

An m-cycle generates a cyclic subgroup of order m. In order to understand Eq. (4.257), it is helpful to visualize $i_1, \ldots, i_m$ as the labels of points spaced equally around the circumference of a circle, as in Fig. 4.3.

By definition, any **one-cycle** (i) leaves i (and all other elements of $(1 : n)$) unchanged. If, for some m, a permutation π is an m-cycle, then π is called a **cyclic permutation** or just a **cycle**. Two cyclic permutations are called **disjoint** if and only if they are permutations of disjoint subsets of $(1 : n)$.

Expression of a Permutation as a Product of Disjoint m-Cycles

We use the principle of finite induction to show that any permutation π of $(1 : n)$ can be expressed uniquely (up to order) as a product of disjoint cyclic permutations. (See, for example, Exercise 4.4.7.) For example,

$$\pi_1 = \begin{pmatrix} 1 & 2 & 3 \\ 1 & 3 & 2 \end{pmatrix} = (1)(23),$$

$$\pi_2 = \begin{pmatrix} 1 & 2 & 3 & 4 & 5 \\ 4 & 3 & 2 & 5 & 1 \end{pmatrix} = (145)(23). \tag{4.258}$$

To start the proof, we observe that the theorem is trivially true for $n = 1$. Assume that it holds for $(1 : n)$. Let π be any permutation of $(1 : n + 1)$. Consider the set

$$P_1 := \{1, \pi(1), \pi^2(1), \ldots, \pi^{n+1}(1)\}, \tag{4.259}$$

the elements of which are 1 and the images of 1 under $n + 1$ successive iterations of π. For example, for π_2 above one has $\pi_2(1) = 4$, $\pi_2^2(1) = 5$, $\pi_2^3(1) = 1$; hence $P_1 = \{1, 4, 5\}$. If all of the images of 1 under π were distinct, then P_1 would have $n + 2$ elements. But $P_1 \subseteq (1 : n + 1)$. Therefore there exists an integer $m \in (1 : n + 1) :\ni: \pi^m(1) = 1$. Let l be the smallest such m. Certainly $l \geq 1$ and

$$P_1 = \{1, \pi(1), \ldots, \pi^{l-1}(1)\}. \tag{4.260}$$

(Recall that by definition all of the elements of a set are distinct.) Therefore the number of elements of $(1 : n + 1) \backslash P_1$ is less than or equal to n.

By construction, $\pi P_1 = P_1$, that is, P_1 is invariant under π. Because π is injective, it follows that $\pi(1 : n + 1) \backslash P_1 = (1 : n + 1) \backslash P_1$. Restricted to P_1, π is the same mapping as the permutation

$$(1, \pi(1), \ldots, \pi^{l-1}(1)) = \begin{pmatrix} 1 & \pi(1) & \cdots & \pi^{l-1}(1) \\ \pi(1) & \pi^2(1) & \cdots & 1 \end{pmatrix}. \tag{4.261}$$

Therefore π is equal to the product of the cyclic permutation $(1, \pi(1), \ldots, \pi^{l-1}(1))$ of P_1 and a permutation π' that acts on $(1 : n + 1) \backslash P_1$:

$$\pi = \pi' \cdot (1, \pi(1), \ldots, \pi^{l-1}(1)). \tag{4.262}$$

Because $(1 : n + 1) \backslash P_1$ has at most n elements, the inductive hypothesis implies that π' can be expressed as a product of disjoint cyclic permutations. The expression Eq. (4.262) is unique because $\pi(1), \ldots, \pi^{l-1}(1)$ are uniquely determined once π is defined. Therefore the theorem is true for $n + 1$ and hence for all n. This proof provides a construction for expressing any given permutation as a product of disjoint cycles. Exercise 4.4.5 gives an example.

4.4.4 EVEN AND ODD PERMUTATIONS

Permutations of the Arguments of a Polynomial

If p is a polynomial in $x_1, \ldots, x_n$, then one can use Eq. (4.182) to define the action of a permutation $\pi \in S_n$ on p:

$$(P_\pi p)(x_1, \ldots, x_n) = p\left(x_{\pi^{-1}(1)}, \ldots, x_{\pi^{-1}(n)}\right). \tag{4.263}$$

For example, let $n = 3$ and $\pi = (123)$. Then

$$\begin{aligned}(P_{(123)}p)(x_1, x_2, x_3) &= p\left(x_{(123)^{-1}(1)}, x_{(123)^{-1}(2)}, x_{(123)^{-1}(3)}\right) \\ &= p(x_3, x_1, x_2)\end{aligned} \tag{4.264}$$

because $(123)^{-1} = (132)$.

In order to understand the meaning of the definition (4.263), it may help to think of the arguments of p as boxes labeled $1, 2, \ldots, n$. Then Eq. (4.263) can be read, "For each $i \in (1 : n)$, replace the contents of box i with the contents of the box labeled $\pi^{-1}(i)$." For example, if one applies $P_{(12)}$ to the right-hand side of Eq. (4.264), the box prescription gives

$$\begin{aligned}\left(P_{(12)}P_{(123)}p\right)(x_1, x_2, x_3) &= \left(P_{(12)}p\right)(x_3, x_1, x_2) \\ &= p(x_1, x_3, x_2) \\ &= \left(P_{(23)}p\right)(x_1, x_2, x_3),\end{aligned} \tag{4.265}$$

which satisfies the homomorphism law because $(12)(123) = (23)$. However, if one mistakenly applies the permutation (12) to x_1 and x_2 rather than to the first and second arguments of p, one obtains the incorrect result $p(x_3, x_2, x_1) = (P_{(13)}p)(x_1, x_2, x_3)$ instead of Eq. (4.265).

With the definition (4.263), the operators P_π are a group, called the **group of the polynomial p**, and the mapping

$$\pi \mapsto P_\pi \tag{4.266}$$

is an operator representation of S_n (see Section 4.3.1 for a proof). For example, the group of the polynomial

$$p(x_1, x_2, x_3, x_4) = x_1 x_2 + x_3 x_4 \tag{4.267}$$

contains the operators $P_{(1)(2)(3)(4)}$, $P_{(12)}$, $P_{(34)}$, and $P_{(12)(34)}$ and therefore is isomorphic to D_2.

Sign Representation of S_n

The polynomial

$$v_n(x_1, \ldots, x_n) = \prod_{j=1}^{n-1} \prod_{i=j+1}^{n} (x_i - x_j), \tag{4.268}$$

which we shall usually write in the more symmetrical form

$$v_n(x_1, \ldots, x_n) = \prod_{j<i}(x_i - x_j), \tag{4.269}$$

carries an important representation of S_n.

For S_3, which is isomorphic to D_3, one has

$$v_3(x_1, x_2, x_3) = (x_2 - x_1)(x_3 - x_1)(x_3 - x_2), \tag{4.270}$$

and, according to Eq. (4.264),

$$\begin{aligned}
\left(P_{(123)}v_3\right)(x_1, x_2, x_3) &= v_3(x_3, x_1, x_2) \\
&= (x_1 - x_3)(x_2 - x_3)(x_2 - x_1) \\
&= v_3(x_1, x_2, x_3).
\end{aligned} \tag{4.271}$$

For $\pi = (132) = (123)^{-1}$ one obtains also

$$\left(P_{(132)}v_3\right)(x_1, x_2, x_3) = v_3(x_1, x_2, x_3). \tag{4.272}$$

However, v_3 changes sign under any of the transpositions (12), (13), and (23). For example,

$$\begin{aligned}
\left(P_{(12)}v_3\right)(x_1, x_2, x_3) &= v_3(x_2, x_1, x_3) \\
&= (x_1 - x_2)(x_3 - x_2)(x_3 - x_1) \\
&= -v_3(x_1, x_2, x_3).
\end{aligned} \tag{4.273}$$

Therefore for every permutation $\pi \in S_3$ one has

$$(P_\pi v_3)(x_1, x_2, x_3) = \sigma(\pi)v_3(x_1, x_2, x_3) \tag{4.274}$$

in which

$$\sigma(\pi) = \begin{cases} 1, & \text{if } \pi = (1)(2)(3), (123) \text{ or } (132); \\ -1, & \text{if } \pi = (12), (13) \text{ or } (23). \end{cases} \tag{4.275}$$

We have shown that $\sigma(\pi) = 1$ if π belongs to the normal subgroup $A_3 = \{(1)(2)(3),$ $(123), (132)\}$ of S_3, which (according to Eq. [4.250]) is isomorphic to the normal subgroup $\{e, d, f\}$ of D_3. Also, $\sigma(\pi) = -1$ if π belongs to the coset of A_3 generated by (12), which is $\{(12), (13), (23)\}$. Therefore the mapping

$$\pi \mapsto \sigma(\pi) \tag{4.276}$$

defines a representation of S_3 that is equivalent to the A_2 representation of D_3 defined in Eq. (4.173).

In the general case, v_n contains one factor for every pair of integers (i, j) in which $i, j \in (1 : n)$ and $i > j$. Therefore $P_\pi v_n$ contains the same factors, generally with reversals of the order of some pairs (i, j). It follows that

$$(P_\pi v_n)(x_1, \ldots, x_n) = \sigma(\pi) v_n(x_1, \ldots, x_n) \tag{4.277}$$

in which $\sigma(\pi) = \pm 1$. From the definition (4.263) it follows that

$$(P_{\pi_2} P_{\pi_1} v_n)(x_1, \ldots, x_n) = (P_{\pi_2 \pi_1} v_n)(x_1, \ldots, x_n) \tag{4.278}$$

(see Section 4.3.1). From Eq. (4.277) one has then

$$\sigma(\pi_2 \pi_1) = \sigma(\pi_2) \sigma(\pi_1), \tag{4.279}$$

which implies that the mapping Eq. (4.276) is a representation of S_n. One calls $\sigma(\pi)$ the **sign** of the permutation π and σ the **sign representation** of S_n.

Alternating Group
Permutations $\pi \in S_n$ such that $\sigma(\pi) = +1$ are called **even permutations**, and permutations $\pi \in S_n$ such that $\sigma(\pi) = -1$ are called **odd permutations**. Clearly the following rules apply to products of even and odd permutations:

$$\begin{aligned} \text{even} \times \text{even} = \text{odd} \times \text{odd} = \text{even}, \\ \text{even} \times \text{odd} = \text{odd} \times \text{even} = \text{odd}. \end{aligned} \tag{4.280}$$

Let A_n be the set of even permutations in S_n. Because any product of even permutations is even, it follows that A_n is closed under composition of permutations. Because $A_n \subset S_n$, products of permutations in A_n are associative. The identity permutation belongs to A_n. Finally, $\sigma(\pi^{-1}) = \sigma(\pi)$ (see Exercise 4.4.9). Therefore A_n is a group. One calls A_n the **alternating group** on n objects.

From Eq. (4.279) and Exercise 4.4.9 it follows that

$$\sigma(\pi \pi' \pi^{-1}) = \sigma(\pi'). \tag{4.281}$$

Therefore A_n is a normal subgroup of S_n.

If π is a fixed odd permutation and π' is any odd permutation, then $\pi^{-1} \pi'$ is even and therefore belongs to A_n. It follows that $\pi' = \pi(\pi^{-1} \pi')$ belongs to the left coset πA_n. Therefore there are only two cosets of A_n in S_n, namely, $e A_n = A_n$ and πA_n (in which π is any odd permutation). By Lagrange's theorem, the order of A_n therefore is equal to one half of the order of S_n. It follows that the order of the alternating group A_n is $n!/2$.

4.4.5 EXERCISES FOR SECTION 4.4

4.4.1   Find the single permutation that is the same mapping as the product

$$\begin{pmatrix} 3 & 2 & 1 \\ 3 & 1 & 2 \end{pmatrix} \begin{pmatrix} 1 & 2 & 3 \\ 3 & 2 & 1 \end{pmatrix}. \tag{4.282}$$

4.4.2   Evaluate the following products of permutations, starting on the right and working towards the left in each case:

$$(12)(123), \quad (12)(132), \quad (23)(123), (23)(132), \quad (13)(123), \quad (13)(132). \tag{4.283}$$

4.4.3   Show that

$$(123)^{-1} = (132). \tag{4.284}$$

4.4.4   Show that

$$(12)(23)(12) = (13). \tag{4.285}$$

4.4.5   Show that

$$\begin{pmatrix} 1 & 2 & 3 & 4 & 5 \\ 2 & 1 & 4 & 3 & 5 \end{pmatrix} = (12)(345). \tag{4.286}$$

4.4.6   Show that any transposition is conjugate to an adjacent transposition:

$$(i, i + v) = (i + 1, i + v)(i, i + 1)(i + 1, i + v). \tag{4.287}$$

Conclude that all transpositions (i, j) are in the same class.

4.4.7   Show that

(a)

$$(12345) = (12)(23)(34)(45) \tag{4.288}$$

(b)

$$(12345) = (15)(14)(13)(12) \tag{4.289}$$

Does the order of the transpositions in (a) or (b) matter?

4.4.8   Let $G = \{a_1, \ldots, a_n\}$ be a finite group, let $b \in G$, and let

$$\pi_{l(b)} := \begin{pmatrix} a_1 & \cdots & a_n \\ ba_1 & \cdots & ba_n \end{pmatrix}. \tag{4.290}$$

Prove that $\pi_{l(b)} : G \to G$ is bijective and therefore is a permutation of G.

Hint: As noted in the section, this result implies that left multiplication simply rearranges (permutes) the elements of G. For this reason the result that $\pi_{l(b)}$ is a permutation of G is called the **rearrangement theorem**.

4.4.9   Prove that for every $\pi \in S_n$,

$$\sigma(\pi^{-1}) = \sigma(\pi). \tag{4.291}$$

4.4.10 Let $G = \{a_1, , \ldots, a_n\}$ be a finite group, let $b \in G$, and let

$$\pi_{c(b)} := \begin{pmatrix} a_1 & \cdots & a_n \\ ba_1b^{-1} & \cdots & ba_nb^{-1} \end{pmatrix}. \tag{4.292}$$

Evidently $\pi_{c(b)}$ is a conjugacy mapping as defined in Eq. (4.57). Prove that $\pi_{c(b)} : G \to G$ is bijective, and therefore is a permutation of G. Also prove that, for any group element $a \in G$, the restriction of $\pi_{c(b)}$ to the class of a, $\pi_{c(b)} \upharpoonright \mathcal{C}_a$, is a permutation of $\mathcal{C}_a$.

4.4.11 Eqs. (4.259–4.262) establish that every permutation $\pi \in S_n$ can be expressed uniquely as a product of cyclic permutations. Assume that π has v_1 1-cycles, v_2 2-cycles, $\ldots$, v_n n-cycles; obviously,

$$\sum_{k=1}^{n} k v_k = n. \tag{4.293}$$

Prove that a permutation $\pi' \in S_n$ belongs to the class of π if and only if π' has the same cycle structure as π. Conclude that there exists a one-to-one correspondence between classes $\mathcal{C}_v \subset S_n$ and n-tuples $v = \langle v_1, \ldots, v_n \rangle$ that satisfy Eq. (4.293).

4.5 RINGS AND INTEGRAL DOMAINS

4.5.1 AXIOMS AND EXAMPLES

Axioms for a Ring

When the law of composition of pairs of elements is multiplication instead of addition, the integers $\mathbb{Z}$ satisfy the first three group axioms. The product of two integers is an integer (closure), and the product of three integers is associative. There exists a multiplicative identity element, 1. However, under multiplication $\mathbb{Z}$ does not satisfy the fourth group axiom (existence of a multiplicative inverse). Therefore $(\mathbb{Z}, \cdot)$ is not a group. In fact, $\mathbb{Z}$ is the prototypical example of a different algebraic structure. A **ring** is a set R, together with two functions $+$ and $\cdot$ from $R \times R$ to R, that obeys the following axioms:

1. $(R, +)$ *is an Abelian group.* One writes the additive identity as 0 and the additive inverse of a as $-a$.
2. *Closure under multiplication:*

   $$\forall a, b \in R : \exists c \in R : \ni: \ c = ab. \tag{4.294}$$

3. *Associativity of multiplication:*

   $$\forall a, b, c \in R : a(bc) = (ab)c. \tag{4.295}$$

4. *Existence of a multiplicative identity (unit element):*

   $$\exists 1 \in R : \ni: 1 \neq 0: \quad \text{and} \quad \forall a \in R : 1a = a1 = a. \tag{4.296}$$

The multiplicative unit element 1 is unique.

5. *Right and left distributive laws:*

$$\forall a, b, c \in R : \quad a(b + c) = ab + ac. \tag{4.297}$$

$$\forall a, b, c \in R : \quad (a + b)c = ac + bc. \tag{4.298}$$

Ring of Integers Modulo *n*

We showed in Section 4.2.4 that the n-bit two's-complement integers, which are used in integer computation, realize the additive group $\mathbb{Z}_{2^m}$ of the integers modulo 2^m. The set of integers modulo n, $\mathbb{Z}_n$, is also a ring if one makes a useful definition of the product of two equivalence classes modulo n, say, k_n and l_n. (It turns out that the multiplication of integers in the m-bit two's-complement representation is simply multiplication in $\mathbb{Z}_{2^m}$.) The most natural definition is that the product $k_n l_n$ in $\mathbb{Z}_n$ is the equivalence class $(kl)_n$ defined by the product kl (calculated in $\mathbb{Z}$):

$$\forall k, l \in \mathbb{Z} : \quad k_n l_n := (kl)_n. \tag{4.299}$$

For example, in $\mathbb{Z}_8$ one has $3_8 \cdot 5_8 = 15_8 = 7_8$.

To show that this definition makes sense, we must show first that the product $k_n l_n$ depends only on the equivalence classes k_n and l_n and not on the representatives k and l. For example, in $\mathbb{Z}_8$ one can verify that $7_8 = 3_8 \cdot 5_8 = 11_8 \cdot 5_8$. Let

$$a \equiv k \pmod{n}, \qquad b \equiv l \pmod{n}. \tag{4.300}$$

Then there exist integers p and q such that

$$a = k + pn, \quad b = l + qn, \tag{4.301}$$

and

$$ab = (k + pn)(l + qn) = kl + n(kq + lp) + n^2 pq. \tag{4.302}$$

The terms that have n as a factor drop out if one takes the remainder after division by n. Hence

$$ab \equiv kl \pmod{n}, \tag{4.303}$$

which establishes that if $a \in k_n$ and $b \in l_n$, then

$$k_n l_n = (kl)_n = (ab)_n = a_n b_n. \tag{4.304}$$

Therefore the product $k_n l_n$ depends only on the equivalence classes, not on the specific representatives k and l.

Equation (4.304) is important in integer computation because the numerical value of a product of m-bit integers in $\mathbb{Z}_{2^m}$ is always reduced to a representative that lies between 0 and

$2^m - 1$ by discarding all of the high-order "overflow" bits that do not fit into the m-bit format. Keeping only bits 0 through $m - 1$ and discarding bits m, $m + 1$, and so forth corresponds to throwing away the integral multiples of $2^m = n$ in Eq. (4.302).

The associativity of multiplication in $\mathbb{Z}_n$ follows from the associativity of multiplication in $\mathbb{Z}$ and the definition Eq. (4.299). Exercise 4.5.6 proves the distributivity of multiplication in $\mathbb{Z}_n$.

Ring of Real $m \times m$ Matrices

Another example of a ring is the set $\mathbb{R}^{m \times m}$ of all $m \times m$ matrices with real elements. We have already pointed out that $\mathbb{R}^{m \times m}$ is an Abelian group under the operation of matrix addition. The additional features that make it a ring are its multiplicative identity, the $m \times m$ unit matrix

$$1 = \begin{pmatrix} 1 & 0 & \cdots & 0 \\ 0 & 1 & \cdots & 0 \\ \vdots & \vdots & \ddots & \vdots \\ 0 & 0 & \cdots & 1 \end{pmatrix}, \tag{4.305}$$

and the associativity of the multiplication of square matrices, which follows from the associativity of the composition of mappings (see Section 6.1.1). The proof of the distributive laws in this example is straightforward but tedious.

Special Kinds of Rings

If multiplication is commutative in a ring R, that is, if

$$\forall a, b \in R: \quad ab = ba, \tag{4.306}$$

then R is called a **commutative ring**. Examples of commutative rings include the integers modulo n, $\mathbb{Z}_n$; the ring of polynomials in x with integer coefficients, $\mathbb{Z}[x]$; and the ring of polynomials in x with real coefficients, $\mathbb{R}[x]$. The set $\mathbb{Z}[\sqrt{2}]$ of linear combinations $m + n\sqrt{2}$, in which m and n belong to $\mathbb{Z}$, is also a commutative ring. However, matrix rings such as $\mathbb{R}^{m \times m}$ are not commutative because matrix multiplication is not commutative in general (see Eq. [4.23]).

If a subset S of a ring R is itself a ring under the same operations of addition and multiplication as R, then S is called a **subring** of R.

A ring is a group under addition, but not necessarily under multiplication. The ring axioms do not imply the existence of a multiplicative inverse for every element a (that is, an element a^{-1} such that $aa^{-1} = 1$). For example, the matrix

$$\sigma_+ = \begin{pmatrix} 0 & 1 \\ 0 & 0 \end{pmatrix} \tag{4.307}$$

has no inverse in $\mathbb{R}^{2\times 2}$. One can verify that a 2×2 matrix $\mathbf{A}$ such that

$$\sigma_+ \mathbf{A} = \begin{pmatrix} 1 & 0 \\ 0 & 1 \end{pmatrix} \tag{4.308}$$

cannot exist by noting that the matrix product

$$\begin{pmatrix} 0 & 1 \\ 0 & 0 \end{pmatrix}\begin{pmatrix} a & b \\ c & d \end{pmatrix} = \begin{pmatrix} c & d \\ 0 & 0 \end{pmatrix} \tag{4.309}$$

cannot be equal to the 2×2 identity matrix.

Integral Domains

Axioms 1 through 5 do not forbid the existence of **proper divisors of zero**, that is, nonzero elements a and b such that $ab = 0$. If a ring R contains proper divisors of zero, then it is not correct to cancel common factors in an equation such as $ab = ac$ because $a(b - c) = 0$ does not imply that either $a = 0$ or $b - c = 0$.

For example, the product of the nonzero matrix σ_+ with itself is the 2×2 null matrix:

$$\begin{pmatrix} 0 & 1 \\ 0 & 0 \end{pmatrix}\begin{pmatrix} 0 & 1 \\ 0 & 0 \end{pmatrix} = \begin{pmatrix} 0 & 0 \\ 0 & 0 \end{pmatrix}. \tag{4.310}$$

Thus the ring $\mathbb{R}^{2\times 2}$ contains zero divisors.

For an example using the finite ring $\mathbb{Z}_n$ of integers modulo n, let the positive integer n be the product of the positive integers k and l, for which $1 < k < n$ and $1 < l < n$. Then $k_n \neq 0$ and $l_n \neq 0$ in $\mathbb{Z}_n$, but

$$k_n l_n \equiv 0 \pmod{n}. \tag{4.311}$$

In the specific case $n = 6$, one has $2_6 \neq 0$ and $3_6 \neq 0$, but $2_6 \cdot 3_6 = 6_6 = 0_6$.

In the ring $\mathbb{Z}_{2^m}$ (the m-bit two's-complement representation of signed integers), any two elements $(2^k)_{2^m}$ and $(2^l)_{2^m}$ such that $k + l = m$ are proper divisors of zero. For example, in $\mathbb{Z}_{16}$, the equivalence classes 2_{16}, 4_{16}, and 8_{16} are proper divisors of zero because $2 \cdot 8 = 16$ and $4 \cdot 4 = 16$. This is a striking instance in which the computer arithmetic of integers obeys quite different rules than arithmetic in $\mathbb{Z}$.

An **integral domain** D is a commutative ring in which the **multiplicative cancellation law**

$$\forall a \in D : \exists : a \neq 0 : \forall b, c \in D : \quad ab = ac \Leftrightarrow b = c \tag{4.312}$$

holds. Equation (4.312) forbids the existence of zero divisors. The ring of integers $\mathbb{Z}$ and the rings $\mathbb{Z}_p$ (for which p is a prime number) are examples of integral domains. Other examples include $\mathbb{Z}[x]$, $\mathbb{R}[x]$, and $\mathbb{Z}[\sqrt{2}]$.

4.5.2 BASIC PROPERTIES OF RINGS

Multiplicative Properties

Certain properties that are familiar from $\mathbb{Z}$ follow for a general ring R directly from axioms 1 through 5. For example,

$$\forall a \in R: \quad a0 = 0a = 0. \tag{4.313}$$

The proof is that from the existence of an additive identity 0 and the right distributive law one has

$$aa = a(a + 0) = aa + a0 \tag{4.314}$$

and

$$aa = aa + 0. \tag{4.315}$$

Then the additive cancellation law implies that $a0 = 0$. The proof that $0a = 0$ is entirely similar.

Another familiar result,

$$\forall a, b \in R: \quad (-a)(-b) = ab, \tag{4.316}$$

can be established by evaluating $[ab + a(-b)] + (-a)(-b)$ in two different ways. First,

$$\begin{aligned}
[ab + a(-b)] + (-a)(-b) &= ab + [a(-b) + (-a)(-b)] \\
&= ab + [a + (-a)](-b) = ab + 0(-b) \\
&= ab + 0 = ab.
\end{aligned} \tag{4.317}$$

Second,

$$\begin{aligned}
[ab + a(-b)] + (-a)(-b) &= a[b + (-b)] + (-a)(-b) \\
&= a0 + (-a)(-b) = (-a)(-b).
\end{aligned} \tag{4.318}$$

A comparison of Eqs. (4.317) and (4.318) establishes Eq. (4.316).

Characteristic of a Ring

The **characteristic** of a ring R is the additive order of its unit element, which is defined as the smallest positive integer $m \in \mathbb{Z}^+$ such that

$$m \cdot 1 = 0, \tag{4.319}$$

where $m \cdot 1$ is the m-th "power" under addition of the ring's multiplicative unit element:

$$m \cdot 1 := \underbrace{1 + 1 + \cdots + 1}_{m \text{ terms}}. \tag{4.320}$$

The simplest example occurs in the ring of integers modulo n: The characteristic of $\mathbb{Z}_n$ is n. Another example occurs in the ring $(\mathbb{Z}_n)^{m \times m}$ of $m \times m$ matrices whose elements are integers modulo n. In this ring, the unit element is the $m \times m$ unit matrix (modulo n),

$$\mathbf{1}_n := \begin{pmatrix} 1_n & 0_n & \cdots & 0_n \\ 0_n & 1_n & \cdots & 0_n \\ \vdots & \vdots & \ddots & \vdots \\ 0_n & 0_n & \cdots & 1_n \end{pmatrix}, \tag{4.321}$$

for which one has

$$n \cdot \mathbf{1}_n = \begin{pmatrix} n \cdot 1_n & n \cdot 0_n & \cdots & n \cdot 0_n \\ n \cdot 0_n & n \cdot 1_n & \cdots & n \cdot 0_n \\ \vdots & \vdots & \ddots & \vdots \\ n \cdot 0_n & n \cdot 0_n & \cdots & n \cdot 1_n \end{pmatrix} = \begin{pmatrix} 0_n & 0_n & \cdots & 0_n \\ 0_n & 0_n & \cdots & 0_n \\ \vdots & \vdots & \ddots & \vdots \\ 0_n & 0_n & \cdots & 0_n \end{pmatrix} \tag{4.322}$$

because $n \cdot 1_n = n_n = 0_n$ in the ring $\mathbb{Z}_n$ to which the elements of $\mathbf{1}_n$ belong.

We define the characteristic of $\mathbb{Z}$ to be ∞; another logical choice is 0.

4.5.3 RATIONAL NUMBERS

The set of **rational numbers** $\mathbb{Q}$ defined in Eq. (2.84) is an integral domain of great importance in both abstract and numerical mathematics. In showing that $\mathbb{Q}$ is an integral domain, we shall make use only of the fact that $\mathbb{Q}$ is a set of equivalence classes of ordered numerator/denominator pairs of elements of an integral domain D (see Eq. [2.85] for the equivalence relation). Because we do not make use of the fact that $D = \mathbb{Z}$, our derivation shows that it is possible to define an integral domain that consists of equivalence classes of ordered pairs whose components are drawn from an arbitrary integral domain such as $\mathbb{Z}[x]$ or $\mathbb{R}[x]$.

We define addition in $\mathbb{Q}$ as follows:

$$(a/b) + (c/d) = ((ad + bc)/bd). \tag{4.323}$$

Recall that (a/b) is a valid rational number if and only if $a, b \in \mathbb{Z}$, and $b \neq 0$. To reduce the sum of two rational numbers to lowest terms (such that the numerator and denominator are relatively prime), one needs to know that

$$\forall c \in \mathbb{Z} : \ni: c \neq 0 : \ (a/b) = (ac/bc). \tag{4.324}$$

This follows from the facts that $(a, b) \equiv (ac, bc)$ (see Exercise 2.2.4) and that two equivalence classes are equal if and only if their representatives are equivalent (see Section 2.2.5). Then, for all $d \in \mathbb{Z}$ such that $d \neq 0$,

$$0 := (0/d) \tag{4.325}$$

is the additive identity element (take $c = 0$ in Eq. [4.323] and apply Eq. [4.324]). The additive inverse of (a/b) is

$$-(a/b) = (-a/b), \tag{4.326}$$

for

$$(a/b) + (-a/b) = ((ab + (-a)b)/b^2) = (0/b^2) = 0. \tag{4.327}$$

The commutative law for addition in $\mathbb{Q}$ follows from the fact that addition in $\mathbb{Z}$ is commutative. We define the product of two elements of $\mathbb{Q}$ as

$$(a/b)(c/d) := (ac/bd). \tag{4.328}$$

The commutativity of this product follows from the commutativity of multiplication in $\mathbb{Z}$. Exercises 4.5.4 and 4.5.5 provide proofs of associativity and distributivity. One can see immediately from Eq. (4.328) that the multiplicative unit element is $(1/1)$. The final point in showing that $\mathbb{Q}$ is an integral domain is to prove the multiplicative cancellation law. From Eq. (4.328) one sees that if $a \neq 0$, then

$$(a/b)(c/d) = (a/b)(f/g) \Leftrightarrow (ac/bd) = (af/bg) \Leftrightarrow abcg = abdf. \tag{4.329}$$

Because $b \neq 0$ (because it is a denominator) and $a \neq 0$ (by assumption), the multiplicative cancellation law in $\mathbb{Z}$ implies that

$$abcg = abdf \Leftrightarrow cg = df. \tag{4.330}$$

But the second equality in Eq. (4.330) holds if and only if $(c/d) = (f/g)$ according to Eq. (4.328). Then

$$(a/b)(c/d) = (a/b)(f/g) \Leftrightarrow (c/d) = (f/g), \tag{4.331}$$

which establishes the multiplicative cancellation law in $\mathbb{Q}$. Therefore $\mathbb{Q}$ is an integral domain.

4.5.4 *RING HOMOMORPHISMS

Definition and Examples

For every properly defined algebraic structure, there is a structure-preserving transformation. A **ring homomorphism** is a mapping ϕ of a ring R to a ring R' such that

$$\forall x, y \in R : \quad \phi(x + y) = \phi(x) + \phi(y), \tag{4.332}$$

$$\forall x, y \in R : \quad \phi(xy) = \phi(x)\phi(y), \tag{4.333}$$

and

$$\phi(1) = 1' \tag{4.334}$$

in which $1'$ is the unit element of R'. For example, the mapping $\phi : \mathbb{Z} \rightarrow \mathbb{Z}_n$ such that

$$\forall k \in \mathbb{Z} : \phi(k) = k_n \tag{4.335}$$

is a ring homomorphism of $\mathbb{Z}$ onto $\mathbb{Z}_n$.

From Eq. (4.332), ϕ is a homomorphism of the additive group of R to the additive group of R'. Therefore

$$\phi(0) = 0' \tag{4.336}$$

(in which $0'$ is the additive identity element of R'), and

$$\forall x \in R : \phi(-x) = -\phi(x). \tag{4.337}$$

In particular,

$$\forall x, y \in R : \phi(x - y) = \phi(x) - \phi(y). \tag{4.338}$$

If a ring homomorphism ϕ is injective and onto, then ϕ is called a **ring isomorphism**. For example, let R be any ring of characteristic n. We show that the additive powers $m \cdot 1$ of the unit element of R constitute a subring of R that is isomorphic to the ring of integers modulo n, $\mathbb{Z}_n$. We begin by showing that $m \cdot 1$ has an additive inverse. In order to ensure consistency with Eq. (4.313) one must define $m \cdot 1$ for $m < 0$ as

$$\forall m < 0 : m \cdot 1 := -((-m) \cdot 1). \tag{4.339}$$

Then $(m + (-m)) \cdot 1 = 0 \cdot 1 = 0$. (Note that in Eq. (4.339) the leftmost minus sign indicates the additive inverse in R; the other minus sign indicates the additive inverse in $\mathbb{Z}$.) Now it follows by definition that for all integers m, q one has a distributive law for the "multiples" of ring elements by integers

$$m \cdot 1 + q \cdot 1 = (m + q) \cdot 1. \tag{4.340}$$

The remaining question concerns the product $(m \cdot 1)(q \cdot 1)$. It is easy to show from the distributive laws, Eqs. (4.297) and (4.298), that

$$(m \cdot 1)(q \cdot 1) = (mq) \cdot 1 \tag{4.341}$$

for all integers m, q. (For example, $(2 \cdot 1)(2 \cdot 1) = (1+1)(1+1) = (1+1+1+1) \cdot 1 = (2 \cdot 2) \cdot 1$.) From Eqs. (4.340) and (4.341) it follows that the mapping

$$m \cdot 1 \mapsto m_n \tag{4.342}$$

is a ring homomorphism.

It is easy to see that the above mapping is a ring isomorphism, for

$$m_n = r_n \Rightarrow \exists q \in \mathbb{Z} : \ni : m = qn + r. \tag{4.343}$$

Because the characteristic of R is n,

$$m \cdot 1 = (qn) \cdot 1 + r \cdot 1 \Rightarrow m \cdot 1 = r \cdot 1. \tag{4.344}$$

Therefore $m \cdot 1 \mapsto m_n$ is a ring isomorphism. Every ring R of characteristic $n < \infty$ includes an isomorphic copy of the n-element ring of integers modulo n, $\mathbb{Z}_n$.

Ideals

As with groups, one can characterize a ring homomorphism completely by looking at a particular substructure. Suppose that

$$\phi(x) = \phi(y) \tag{4.345}$$

for some $x, y \in R$. Then, by Eq. (4.338),

$$\phi(x - y) = 0'. \tag{4.346}$$

Therefore it makes sense to define the set $J \subseteq R$ that consists of the preimage of the additive identity element $0' \in R'$. J is called the **kernel** of the ring homomorphism ϕ. For example, the set $n\mathbb{Z}$ of the positive and negative multiples (including 0) of the positive integer n is the kernel of the ring homomorphism Eq. (4.335).

We deduce a few of the properties of J. From now on we exclude the trivial ring that consists only of the zero element; then $1 \neq 0$ and $1' \neq 0'$. Because $\phi(1) = 1' \neq 0$, it follows that

$$1 \notin J. \tag{4.347}$$

Also,

$$\forall x, y \in J : \quad \phi(x - y) = 0' - 0' = 0' \Rightarrow x - y \in J \tag{4.348}$$

and

$$\forall r \in R : \quad \forall x \in J : \quad \phi(rx) = \phi(r)\phi(x) = \phi(r)0' = 0' \Rightarrow rx \in J. \tag{4.349}$$

The next step is to formalize these properties by defining a new substructure.

A subset I of a ring R is called a **left ideal** if and only if

$$\forall x, y \in I : \quad x - y \in I \tag{4.350}$$

and

$$\forall r \in R : \forall x \in I : \quad rx \in I. \tag{4.351}$$

If (instead of Eq. [4.351]) $xr \in I$ for every $r \in R$, then I is called a **right ideal**. In a commutative ring, every left ideal is a right ideal and vice versa.

For example, $n\mathbb{Z}$ is an ideal of $\mathbb{Z}$. The polynomials of which $1 - x$ is a factor are an ideal in $\mathbb{R}[x]$. The kernel of every homomorphism of a ring R is an ideal of R.

4.5.5 EXERCISES FOR SECTION 4.5

4.5.1   Let R be a ring. Prove that $R[x]$, the set of **polynomials** in an indeterminate x with coefficients in R, is a ring if one defines the addition and multiplication of polynomials as in high-school algebra.

4.5.2*   Let $\mathbb{Z}[\sqrt{2}]$ be the set

$$\mathbb{Z}[\sqrt{2}] := \{m + n\sqrt{2} \mid m, n \in \mathbb{Z}\}. \tag{4.352}$$

Prove that $\mathbb{Z}[\sqrt{2}]$ is an integral domain.

4.5.3   Prove that if R is a ring, then the unit element of R is unique. That is, prove that if for some $e \in R$,

$$\forall a \in R : ae = ea = a, \tag{4.353}$$

then $e = 1$.

4.5.4   Show that the product defined in Eq. (4.327) is associative.

4.5.5   Show that the definitions given in Eqs. (4.323), (4.324), and (4.327) obey the distributive laws given in Eqs. (4.297) and (4.298).

4.5.6   Show that the definition of multiplication in $\mathbb{Z}_n$ given in Eq. (4.299) obeys the distributive laws, Eqs. (4.297) and (4.298).

4.5.7   Establish Eq. (4.341).

4.5.8   Let p be a prime number. Prove that $\mathbb{Z}_p$ is an integral domain.

4.6 FIELDS

4.6.1 AXIOMS AND EXAMPLES

A **field** $\mathbb{F}$ is an integral domain in which there exists at least one nonzero element and in which the set of nonzero elements, $\mathbb{F}\backslash\{0\}$, is an Abelian group under multiplication. Therefore, in addition to the ring axioms Eqs. (4.294) through (4.298), the commutative law for multiplication Eq. (4.306), and the multiplicative cancellation law Eq. (4.299), the elements of a field $\mathbb{F}$ have the property that

$$\forall x \in \mathbb{F}\backslash\{0\} : \exists \frac{1}{x} := x^{-1} \in \mathbb{F} : \ni: \ x \cdot \frac{1}{x} = \frac{1}{x} \cdot x = 1. \tag{4.354}$$

The rational numbers $\mathbb{Q}$, real numbers $\mathbb{R}$, and complex numbers $\mathbb{C}$ are fields that are fundamentally important in physics and mathematics. One can see immediately from Eq. (4.328) that for any two nonzero integers p, q,

$$x = (p/q) \Rightarrow x^{-1} = (q/p). \tag{4.355}$$

Therefore the rational numbers $\mathbb{Q}$ are a field.

In Eq. (2.73) we defined a complex number z as any element of $\mathbb{R}^2 : z = (x, y)$. Exercise 4.2.4 shows that $\mathbb{C}$, the set of complex numbers, is an Abelian group under addition, and Exercise 4.2.7 shows that the set of nonzero complex numbers under multiplication, $(\mathbb{C}\backslash\{(0, 0)\}, \cdot)$, is an Abelian group. Therefore $\mathbb{C}$ is a field.

Two fields $\mathbb{F}$ and $\mathbb{F}'$ are called **isomorphic** if and only if the additive groups of $\mathbb{F}$ and $\mathbb{F}'$ are isomorphic and the multiplicative groups of the nonzero elements of $\mathbb{F}$ and $\mathbb{F}'$ are isomorphic.

A field $\mathbb{F}'$ is called a **subfield** of a field $\mathbb{F}$ if and only if $\mathbb{F}'$ is a subset of $\mathbb{F}$ and $\mathbb{F}'$ is a field. For example, the complex field $\mathbb{C} = \{(x, y) \,|\, x, y \in \mathbb{R}\}$ includes the subset $\mathbb{F}' = \{(x, 0) \,|\, x \in \mathbb{R}\}$, which is isomorphic to $\mathbb{R}$ and is a subfield of $\mathbb{C}$.

In Chapters 1 and 2 we make use of the fact that every real number r can be expressed in the base-β floating-point format

$$r = \pm \left[\sum_{j=0}^{\infty} \frac{d_j}{\beta^j} \right] \cdot \beta^t, \tag{4.356}$$

in which

$$\forall j \in \mathbb{Z}^+ : \quad 0 \le d_j < \beta. \tag{4.357}$$

This representation is not unique for reasons that we have already discussed. Therefore, one defines two series in powers of β^{-1} as equivalent if and only if the difference of the two series converges to zero. Exercise 4.2.3 can easily be generalized to show that if one defines $\mathbb{R}$ as the set of all equivalence classes of series of the form Eq. (4.356), then $(\mathbb{R}, +)$ is an Abelian group. Exercise 4.2.6 shows that $(\mathbb{R}\backslash\{0\}, \cdot)$ is also an Abelian group. Therefore $\mathbb{R}$ is a field, the **field of real numbers**.

It is clear that every r for which the series Eq. (4.356) terminates,

$$\exists n \in \mathbb{Z}^+ : \ni: \ \forall j > n: \ d_j = 0, \tag{4.358}$$

or for which the digits are periodic,

$$\exists n, k \in \mathbb{Z}^+ : \ni: \ \forall j > n: \ d_j + k = d_j, \tag{4.359}$$

r can be expressed as the quotient of two integers and therefore is the representative of an equivalence class in $\mathbb{Q}$ (see Exercise 1.2.5). The converse is also true: Every quotient p/q of nonzero integers can be expressed as either a terminating series such as Eq. (4.358) or a repeating series such as Eq. (4.359). Moreover, these quotients obey the laws of addition and multiplication for rational numbers, Eqs. (4.323) and (4.328). Therefore $\mathbb{R}$ contains an isomorphic copy of $\mathbb{Q}$; one says that $\mathbb{Q}$ can be **embedded** in $\mathbb{R}$. The elements of $\mathbb{R}$ that are not in the copy of $\mathbb{Q}$ are called **irrational numbers**.

4.6.2 *GALOIS FIELDS

Fermat's Theorem

Finite fields are called **Galois fields** after their discoverer. It turns out that there is a unique Galois field $\mathbb{GF}(q)$ with q elements. If one searches for finite fields among the finite rings $\mathbb{Z}_n$, one sees at once that the existence of proper divisors of zero if n is not prime (see Eq. [4.311]) rules out all candidates except for the integral domains $\mathbb{Z}_p$, for which p is prime.

To show that an integral domain $\mathbb{Z}_p$ is a field, we must prove that every $k_p \in \{1_p, \ldots,$ $(p-1)_p\}$ has a multiplicative inverse $1/k_p$. The existence of a multiplicative inverse (modulo p) is a direct consequence of **Fermat's theorem** (not to be confused with Fermat's *last* theorem!): If $n \in \mathbb{Z}^+$ and if p is prime, then

$$n^p \equiv n \quad (\text{mod } p). \tag{4.360}$$

It is convenient to prove Fermat's theorem by induction. The theorem is true for $n = 1$ for any p. For a given p, assume that the theorem holds for n. By the binomial theorem,

$$(n + 1)^p = \sum_{r=0}^{p} \binom{p}{r} n^{p-r}. \tag{4.361}$$

For all $r > 0$, the binomial coefficient is

$$\binom{p}{r} = \frac{p!}{r!(p-r)!} = \frac{p(p-1)\cdots(p-r+1)}{1 \cdot 2 \cdots r}. \tag{4.362}$$

Because p is prime, it follows that for any r such that $0 < r < p$ the factor p in the numerator of the last term in Eq. (4.362) cannot be canceled by any factors in the denominator. Therefore every binomial coefficient in $(n+1)^p$ except for $r = 0$ and $r = p$ is divisible by p, implying that

$$(n + 1)^p \equiv n^p + 1 \quad (\text{mod } p). \tag{4.363}$$

Now the inductive hypothesis Eq. (4.360) implies that

$$(n + 1)^p \equiv n + 1 \quad (\text{mod } p). \tag{4.364}$$

This establishes the theorem.

It follows from Eq. (4.360) that

$$(n^p - n)_p = (n(n^{p-1} - 1))_p = n_p(n^{p-1} - 1)_p = 0_p. \tag{4.365}$$

Because $\mathbb{Z}_p$ has no zero divisors, either $n_p = 0_p$ or $(n^{p-1} - 1)_p = 0_p$. It follows that if $n_p \neq 0_p$, then

$$n^{p-1} \equiv 1 \pmod{p}. \tag{4.366}$$

Therefore

$$\forall k_p \in \mathbb{Z}_p : \frac{1}{k_p} = (k_p)^{p-2} \tag{4.367}$$

because Eq. (4.366) implies that

$$k_p \cdot (k_p)^{p-2} = 1. \tag{4.368}$$

Therefore $\mathbb{Z}_p$ is a field for every prime number p.

To show that $\mathbb{Z}_p$ is the only p-element field, let K_p be any field with p elements. Because $1 \in K_p$, the additive powers $m \cdot 1$ defined in Eq. (4.320) belong to K_p, for every integer m. Then the set $\{1\}$ generates an additive cyclic subgroup A. The elements $0 \cdot 1, 1 \cdot 1, 2 \cdot 1, \ldots, (p - 1) \cdot 1 \in A$ are distinct unless $m \cdot 1 = 0$ for some $m < p$. If $m < p$ and $m \cdot 1 = 0$, then the p-element additive group K_n includes an m-element additive subgroup, which contradicts Lagrange's theorem. Then $A = K_p$. It follows from Eq. (4.340) that K_p and $\mathbb{Z}_p$ are isomorphic as additive groups under the mapping

$$m \cdot 1 \mapsto m_p. \tag{4.369}$$

Then Eq. (4.341) implies that K_p and $\mathbb{Z}_p$ are isomorphic as fields. We can therefore write

$$\mathbb{Z}_p \cong \mathrm{GF}(p), \tag{4.370}$$

in which *the* Galois field $\mathrm{GF}(p)$ is uniquely determined up to isomorphism.

For example, in the two-element field $\mathrm{GF}(2) = \mathbb{Z}_2$, $1_2/1_2 = 1$. In the three-element field $\mathbb{Z}_3 = \{0, 1_3, 2_3\}$, one has

$$2_3^2 \equiv 1 \pmod{3}. \tag{4.371}$$

Therefore $1/2_3 = 2_3$. In $\mathrm{GF}(5)$, $(1/2)_5 = (2_5)^{5-2}(= 8_5) = 3_5$.

Construction of $\mathrm{GF}(4)$

There are Galois fields other than $\mathbb{Z}_p$. To give an example, we construct the four-element field $\mathrm{GF}(4)$. One knows that $\mathrm{GF}(4) \neq \mathbb{Z}_4$ because $\mathbb{Z}_4$ is not a field (there are zero divisors: $2 \cdot 2 \equiv 0 \pmod{4}$). Certainly $\mathrm{GF}(4)$ must contain 0 and 1; let

$$\mathrm{GF}(4) = \{0, 1, \omega, \omega'\}. \tag{4.372}$$

The nonzero elements constitute an Abelian multiplicative group of order three. Up to isomorphism, there is only one group of order three, the cyclic group C_3. Therefore

$$\omega' = \omega^2, \quad \omega^3 = 1. \tag{4.373}$$

It follows that

$$\frac{1}{\omega} = \omega^2, \quad \frac{1}{\omega^2} = \omega. \tag{4.374}$$

We turn next to the additive group structure of $\mathbb{GF}(4)$. First, let us find the additive inverse of 1. If $-1 = \omega$, then (by Eq. [4.316]) $\omega^2 = 1$, which is impossible. If $-1 = \omega^2$, then $(\omega^2)^2 = 1 \Rightarrow \omega^4 = 1$. With Eq. (4.373), this implies that $\omega = 1$ – another contradiction. Therefore $-1 = 1$ in $\mathbb{GF}(4)$, that is,

$$1 \in \mathbb{GF}(4) \Rightarrow 1 + 1 = 0. \tag{4.375}$$

By Eq. (4.296), $1 \cdot 1 = 1$. By Eq. (4.313), $1 \cdot 0 = 0$. Therefore $\mathbb{GF}(4)$ includes an isomorphic copy of the two-element Galois field $\mathbb{GF}(2) = \mathbb{Z}_2$.

The remaining unknown quantities are $-\omega$, $-\omega^2$, $1 + \omega$, $1 + \omega^2$, and $\omega + \omega^2$. One can find $-\omega$ immediately:

$$\omega + \omega = (1 + 1)\omega = 0\omega = 0 \Rightarrow -\omega = \omega \tag{4.376}$$

by Eq. (4.313). Likewise, $\omega^2 + \omega^2 = 0$. One can see that $1 + \omega \neq 0$ (because $\omega + \omega = 0$ and $1 \neq \omega$), $1 + \omega \neq 1$ (because $\omega \neq 0$), and $1 + \omega \neq \omega$ (because $1 \neq 0$). Therefore

$$1 + \omega = \omega^2 \Rightarrow 1 + \omega + \omega^2 = 0. \tag{4.377}$$

But Eq. (4.377) implies that ω is one of the primitive cube roots of unity,

$$\omega = e^{\pm 2\pi i/3}. \tag{4.378}$$

Many of the properties we have established for $\mathbb{GF}(4)$ can be generalized to arbitrary Galois fields.

Order of a Galois Field

Consider a Galois field $\mathbb{GF}(q)$ in which q is not prime. Because $1 \in \mathbb{GF}(q)$, the additive powers $m \cdot 1$ defined in Eq. (4.320) belong to $\mathbb{GF}(q)$. For some $m \leq q$, $m \cdot 1 = 0$ because the set $\{m \cdot 1 \mid 1 \leq m \leq q + 1\}$ is a subset of the q-element set $\mathbb{GF}(q)$. Therefore $\mathbb{GF}(q)$ contains an isomorphic copy of $\mathbb{Z}_m$. Unless m is equal to a prime number p, $\mathbb{Z}_m$ (and therefore $\mathbb{GF}(q)$) contains zero divisors (see Section 4.5.1). Therefore m is a prime number: $m = p$. The distributive laws (4.297) and (4.298) imply that the law of multiplication of the multiples of unity $m \cdot 1$ is Eq. (4.299). It follows that every Galois field $\mathbb{GF}(q)$ includes unique isomorphic copy Π of some $\mathbb{Z}_p$. Π is called the **prime field** of $\mathbb{GF}(q)$.

The $q - 1$ nonzero elements of the field $\mathbb{GF}(q)$ constitute an Abelian multiplicative group. The multiplicative order l of any $\alpha \neq 0$ in $\mathbb{GF}(q)$ must be a divisor of $q - 1$ because the cyclic group $\{1, \alpha, \ldots, \alpha^{l-1}\}$ is a subgroup of $\mathbb{GF}(q)$. Therefore $\alpha^{q-1} = (\alpha^l)^{(q-1)/l} = 1^{(q-1)/l} = 1$, which implies that

$$\forall \alpha \in \mathbb{GF}(q): \quad \alpha^{q-1} = 1. \tag{4.379}$$

It follows that every nonzero element of $\mathbb{GF}(q)$ is a $(q-1)$-th root of unity.

At this point the concept of a vector space makes a decisive contribution (see Section 5.2). It turns out that $\mathbb{GF}(q)$ is a finite-dimensional vector space over its prime field Π. Therefore (see Exercises 4.2.10 and 4.2.11) one has

$$q = p^n \tag{4.380}$$

for some positive integer n. Therefore every Galois field is $\mathbb{GF}(p^n)$ for some prime number p and some $n \in \mathbb{Z}^+$.

4.6.3 EXERCISES FOR SECTION 4.6

4.6.1   Let D be an integral domain. Prove that if the set of elements of D is finite, then D is a field.

Hint: Let a be any nonzero element of D and show that the mapping $\phi_a : D \to D$ defined by

$$\forall b \in D: \quad \phi_a(b) = ab \tag{4.381}$$

is bijective.

4.6.2   Prove that if a, b, c, and d are any four *nonzero* elements of any field $\mathbb{F}$, then

$$\frac{a}{b}(:= ab^{-1} = b^{-1}a) = \frac{c}{d} \Leftrightarrow ad = bc \tag{4.382}$$

$$\frac{a}{b}\frac{c}{d} = \frac{ac}{bd} \tag{4.383}$$

$$\frac{a}{b}\frac{b}{a} = 1. \tag{4.384}$$

4.7 BIBLIOGRAPHY

Birkhoff, Garrett, and Saunders Mac Lane. *A Survey of Modern Algebra*. 3rd ed. Macmillan, 1965, chapters I–VI.

Hall, Marshall. *The Theory of Groups*. Macmillan, 1959, chapters 1, 2.

Hamermesh, Morton. *Group Theory and Its Application to Physical Problems*. Dover, 1990, chapter 1.

Mac Lane, Saunders, and Garrett Birkhoff. *Algebra*. Macmillan, 1967, chapters III–V.

van der Waerden, Bartel Leendert. *Modern Algebra*. Vol. I. Frederick Ungar, 1953, chapters II–IV.

Weyl, Hermann. *The Theory of Groups and Quantum Mechanics* (H. P. Robertson trans.). Dover, 1950 (first published 1928).

Wigner, Eugene P. *Group Theory and Its Applications to the Quantum Mechanics of Atomic Spectra*. Academic Press, 1959, chapters 7 and 8.

CHAPTER 5

VECTOR SPACES

5.1 INTRODUCTION

The mathematics of linear spaces is essential for physicists and engineers because it is the foundation on which many theories and models of important materials, devices, and systems are built. The linearity of a fundamental physical theory, quantum mechanics, has been established experimentally to very high precision; see Bollinger et al. (1989) and Weinberg (1989). The equations of classical electromagnetic theory, which describe everything from electrical networks to laser beams, are linear to a good approximation. For these reasons, if for no others, physicists and engineers must take linear algebra and linear functional analysis seriously. The fact that even nonlinear structures such as curved space-time may have useful local linear approximations has led to many linearized models of systems in fields ranging from physics to economics. The high degree of development of numerical linear algebra, its importance in methods for solving ordinary and partial differential equations, and the availability of powerful, inexpensive computers have made a good knowledge of linear algebra indispensable for a working physicist or engineer.

The fundamental structure in linear algebra and linear functional analysis is that of a vector space. In this introduction we use one of the most important partial differential equations in quantum mechanics and electromagnetic theory to illustrate why one needs to study vector spaces. The generalization to two or three (spatial) dimensions of the ordinary (temporal) differential equation for the displacement $y(t)$ of a particle undergoing simple harmonic motion,

$$\left(\frac{d^2}{dt^2} + \omega^2\right)y = 0, \tag{5.1}$$

is the **homogeneous Helmholtz equation** for a function ψ,

$$(\nabla^2 + k^2)\psi = 0. \tag{5.2}$$

Therefore one would expect the homogeneous Helmholtz equation to occur (almost) everywhere spatial waves are found.

In quantum mechanics, the Schrödinger equation for a single particle subject to a potential V is

$$\mathsf{H}\psi(\mathbf{r}) = E\psi(\mathbf{r}), \tag{5.3}$$

in which $\mathbf{r}$ is a point in ordinary three-dimensional space, E is the energy (which must be a real number), and H is the single-particle Hamiltonian operator,

$$\mathsf{H} = -\frac{\hbar^2}{2m}\nabla^2 + V(\mathbf{r}). \tag{5.4}$$

After a little algebra one sees that the complex-valued wave function ψ, which is assumed to provide a complete description of the state of the particle, satisfies the homogeneous Helmholtz equation (5.2) with

$$k^2 := \frac{2m}{\hbar^2}[E - V(\mathbf{r})]. \tag{5.5}$$

In electrodynamics, the source-free Maxwell equations lead to Eq. (5.2) for each Cartesian component of the electric and magnetic fields if one assumes a harmonic time dependence (i.e., if one assumes that each field component is proportional to $\cos(\omega t + \phi)$) and if one discards a term proportional to $\nabla(\nabla \cdot \mathbf{E})$. (Note that $\nabla(\nabla \cdot \mathbf{E})$ vanishes on any open subset of $\mathbb{R}^3$ on which the dielectric permittivity ϵ is constant.) In this case $k^2 = \epsilon\omega^2/c^2$ is piecewise constant.

Vector spaces play an essential rôle in the implementation of **symmetry principles in physics** by "carrying" the irreducible representations of symmetry groups such as $SU(2)$ and $SU(3)$. In fact, a special case of the homogeneous Helmholtz equation can be derived from a symmetry principle. If k^2 is independent of position, then it can be shown that Eq. (5.2) is the most general second-order partial differential equation that is invariant under the group $E(3)$ of translations and rotations in three-dimensional Euclidean space. Far from being an esoteric fact, the invariance of the Helmholtz operator $\nabla^2 + k^2$ under $E(3)$ motivates the definition of important special functions (spherical Bessel functions and spherical harmonics) and also helps one to derive many of the properties of these functions in a simple manner.

For our current purposes, the first interesting property of the Helmholtz operator $\nabla^2 + k^2$ is that it is **additive**:

$$(\nabla^2 + k^2)(\psi + \phi) = (\nabla^2 + k^2)\psi + (\nabla^2 + k^2)\phi. \tag{5.6}$$

It follows that if ψ and ϕ are solutions of the homogeneous Helmholtz equation,

$$(\nabla^2 + k^2)\psi = 0 = (\nabla^2 + k^2)\phi, \tag{5.7}$$

then $\psi + \phi$ is another solution:

$$(\nabla^2 + k^2)(\psi + \phi) = 0. \tag{5.8}$$

Clearly $-\psi$ is a solution if ψ is, and the sum $\psi + (-\psi) = 0$ solves Eq. (5.2) trivially. Therefore the solutions of the homogeneous Helmholtz equation satisfy all the axioms of an additive group.

The second property of the set of solutions of Eq. (5.2) that interests us now is **closure under scalar multiplication**. If ψ is a solution, then so is $\alpha\psi$, for any real or complex number

α. An equation (and, by extension, the physical system that it describes) for which $\alpha\psi + \beta\phi$ is a solution if ψ and ϕ are solutions is called **linear** and, in physics and engineering, is said to obey the **principle of superposition**. An extremely important consequence is that a system that obeys the superposition principle can be broken up conceptually into simple parts, and solutions for the parts can then be combined into a solution for the whole system.

The principle of superposition applies to all objects that are mapped onto zero by a linear operator, of which $\nabla^2 + k^2$ is only one example. The evaluation of a function at a certain point is also a linear operation. Functions that all satisfy the same **homogeneous boundary conditions** on an interval $[a, b]$, such as **Dirichlet conditions**

$$f(a) = f(b) = 0 \tag{5.9}$$

or **Neumann conditions**

$$f'(a) = f'(b) = 0 \tag{5.10}$$

or **mixed conditions**

$$f'(a) + \eta f(a) = f'(b) + \eta f(b) = 0, \tag{5.11}$$

also satisfy the principle of superposition. From a physicist's or an engineer's point of view, it is reasonable to regard the superposition principle as what motivates the axioms that define a vector space.

Given that that many different realizations of vector spaces occur in physics and engineering, it is most efficient to study vector spaces in a general setting before studying the vector spaces that occur in specific methods or applications. One goal of our study is to understand how to expand any solution of an equation such as Eq. (5.2) in terms of a *basis* of functions that are products of solutions of ordinary differential equations, as in the method of separation of variables. We take a first step toward that goal in this chapter and another step in Chapter 10. Another, closely related goal is to develop a geometrical interpretation of vector spaces so that we can make useful mental pictures of the vector spaces that occur in various applications.

5.2 BASIC DEFINITIONS AND EXAMPLES

5.2.1 AXIOMS FOR A VECTOR SPACE

A **vector space** $\mathcal{V}$ is a set of elements $\bar{x}$ (called **vectors**) that satisfy the following two sets of axioms:

- $\mathcal{V}$ is a commutative group under addition:

  1. $\mathcal{V}$ *is closed under a commutative binary operation, called* **addition**:

  $$\forall \bar{x} \in \mathcal{V} : \forall \bar{y} \in \mathcal{V} : \bar{x} + \bar{y} = \bar{y} + \bar{x} \in \mathcal{V}. \tag{5.12}$$

2. *Vector addition is associative:*

$$\overline{x} + (\overline{y} + \overline{z}) = (\overline{x} + \overline{y}) + \overline{z}. \tag{5.13}$$

3. *There exists an additive identity, the* **zero vector** $\overline{0}$:

$$\exists \overline{0} \in \mathcal{V} :\ni: \forall \overline{x} \in \mathcal{V} : \overline{x} + \overline{0} = \overline{x}. \tag{5.14}$$

4. *There exists an additive inverse* $-\overline{x}$ *for every vector* $\overline{x}$:

$$\forall \overline{x} \in \mathcal{V} : \exists - \overline{x} \in \mathcal{V} :\ni: \overline{x} + (-\overline{x}) = \overline{0}. \tag{5.15}$$

- $\mathcal{V}$ satisfies the following additional axioms with respect to a number field $\mathbb{F}$, whose elements α are called **scalars**:

5. $\mathcal{V}$ *is closed under scalar multiplication:*

$$\forall \overline{x} \in \mathcal{V} : \forall \alpha \in \mathbb{F} : \alpha \overline{x} \in \mathcal{V}. \tag{5.16}$$

By definition, $\overline{x}\alpha := \alpha\overline{x}$.

6. *Scalar multiplication is distributive with respect to elements of both* $\mathcal{V}$ *and* $\mathbb{F}$:

$$\forall \overline{x}, \ \overline{y} \in \mathcal{V} : \forall \alpha \in \mathbb{F} : \alpha(\overline{x} + \overline{y}) = \alpha \overline{x} + \alpha \overline{y} \tag{5.17}$$

$$\forall \overline{x} \in \mathcal{V} : \forall \alpha, \ \ \beta \in \mathbb{F} : (\alpha + \beta)\overline{x} = \alpha \overline{x} + \beta \overline{x}. \tag{5.18}$$

7. *Scalar multiplication is associative:*

$$\forall \overline{x} \in \mathcal{V} : \forall \alpha, \ \ \beta \in \mathbb{F} : \alpha(\beta \overline{x}) = (\alpha \beta)\overline{x} := \alpha \beta \overline{x}. \tag{5.19}$$

8. *Multiplication with the zero scalar* $0 \in \mathbb{F}$ *gives the zero vector,* $\overline{0} \in \mathcal{V}$:

$$\forall \overline{x} \in \mathcal{V} : \ \ 0\overline{x} = \overline{0}. \tag{5.20}$$

9. *The unit scalar* $1 \in \mathbb{F}$ *has the property that*

$$\forall \overline{x} \in \mathcal{V} : \ \ 1\overline{x} = \overline{x}. \tag{5.21}$$

Ordinarily $\mathbb{F}$ is either the field of real numbers, $\mathbb{R}$, or the field of complex numbers, $\mathbb{C}$. A vector space over $\mathbb{R}$ is called a **real vector space**. If $\mathbb{F} = \mathbb{C}$, then $\mathcal{V}$ is a **complex vector space**. Every real vector space $\mathcal{V}$ can be embedded in a natural way in a complex vector space, which is called the **complex extension** or **complexification** of $\mathcal{V}$.

5.2.2 SELECTED REALIZATIONS OF THE VECTOR-SPACE AXIOMS

A Field is a Vector Space

Every number field is a vector space over itself. The complex numbers, $\mathbb{C}$, are a vector space over the field of real numbers, $\mathbb{R}$.

Spaces of Column Vectors

Definitions   In Section 2.2.4 we defined $\mathbb{F}^n$ as the set of n-tuples $\langle x^1, \ldots, x^n \rangle$ in which every $x^i \in \mathbb{F}$. For the sake of compatibility with standard matrix notation, from now on we shall write elements of $\mathbb{F}^n$ as **column vectors**

$$\mathbf{x} := \begin{pmatrix} x^1 \\ x^2 \\ \vdots \\ x^n \end{pmatrix}. \tag{5.22}$$

With the definitions

$$\mathbf{x} + \mathbf{y} := \begin{pmatrix} x^1 + y^1 \\ x^2 + y^2 \\ \vdots \\ x^n + y^n \end{pmatrix}, \quad \forall \alpha \in \mathbb{F} : \alpha \mathbf{x} := \begin{pmatrix} \alpha x^1 \\ \alpha x^2 \\ \vdots \\ \alpha x^n \end{pmatrix} \tag{5.23}$$

$\mathbb{F}^n$ is a vector space over the number field $\mathbb{F}$. For instance, the x- and y-components of a displacement vector in two-dimensional space can be regarded as the entries of a column vector

$$\mathbf{r} = \begin{pmatrix} x \\ y \end{pmatrix} \tag{5.24}$$

in $\mathbb{R}^2$; see Fig. 5.1. The reason for using superscripts to label the components of a column vector $\mathbf{x}$ is that x^i is the i-th contravariant component of $\mathbf{x}$ relative to a certain "canonical" basis; see Section 5.4.1.

Storage of vectors in computer memory   For computational purposes, a vector belonging to $\mathbb{R}^n$ is approximated by a vector in $\mathbb{Q}^n$ and stored as an array (see Section 2.2.2) in RAM, as shown in Figs. 2.5 and 2.6. The array elements are represented by "words" of 4 bytes each

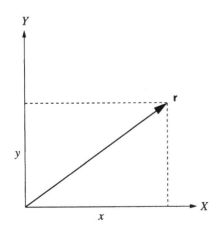

FIGURE 5.1

Representation of a vector $\mathbf{r}$ belonging to $\mathbb{R}^2$ as a displacement in the x–y plane.

(in a single-precision representation) or 8 bytes each (in a double-precision representation). (Floating-point representations are discussed in Section 1.2.2.)

Vectors of sampled values   One of the industrially significant applications of finite-dimensional vector spaces occurs in signal processing in which one applies various transformations to vectors of data values. For example, suppose that a voltage $x(t)$ is sampled at the times $t_0, \ldots, t_{n-1}$. Because voltages obey the superposition principle, it is useful to consider the sampled values $x(t_0), \ldots, x(t_{n-1})$ as the contravariant components of a vector $\mathbf{x} \in \mathbb{R}^n$ with the identification

$$x^i := x(t_i). \tag{5.25}$$

Common operations such as finding the discrete Fourier transform of the signal, which we discuss in Section 8.5, then become linear mappings on the vector space $\mathbb{R}^n$, which contains all possible sample vectors.

Many important applications of vectors of sampled values occur in numerical computation. For example, the computed solution of an ordinary differential equation is a list of numbers $\langle c_0, \ldots, c_{n-1} \rangle$ that approximate the values $y(t_0), \ldots, y(t_{n-1})$ of the exact solution at the sampling times $t_0, \ldots, t_{n-1}$ (see Section 3.5). Frequently it is useful to consider the list $\langle c_0, \ldots, c_{n-1} \rangle$ as belonging to $\mathbb{R}^n$ for the same reasons that we just gave for sampled voltages.

Discrete quantum systems   A **state vector** of a finite, discrete quantum system is an element ψ of $\mathbb{C}^n$ such that the **normalization condition**

$$\sum_{i=1}^{n} |\psi^i|^2 = 1 \tag{5.26}$$

is obeyed. The integer n is usually equal to the number of energy levels of the system. For example, the magnetic sublevels of an atom with a total angular momentum equal to $(n-1)/2$ (in units of $\hbar$) are a physical realization of a quantum system with n states. Computational models of quantum systems with an infinite set of energy levels must assume a finite set of energy levels.

The contravariant components ψ^i are called **probability amplitudes** in quantum mechanics because the probability of finding the system in state i is equal to $|\psi^i|^2$.

Matrix Spaces

Generalizing the definitions for column vectors, one sees that the set $\mathbb{F}^{m \times n}$ of $m \times n$ arrays, or **matrices**,

$$\mathbf{A} := \begin{pmatrix} a_1^1 & a_2^1 & \cdots & a_n^1 \\ a_1^2 & a_2^2 & \cdots & a_n^2 \\ \vdots & \vdots & \ddots & \vdots \\ a_1^m & a_2^m & \cdots & a_n^m \end{pmatrix}, \tag{5.27}$$

with elements $a^i_j \in \mathbb{F}$, is a vector space over $\mathbb{F}$. In the notation used in this book, the number i of the horizontal row to which an element a^i_j belongs is a superscript; the number j of the vertical column to which a^i_j belongs is a subscript.

The scalar multiple $\alpha\mathbf{A}$ of a matrix $\mathbf{A} \in \mathbb{F}^{m \times n}$ is defined to be the matrix obtained by multiplying every element of $\mathbf{A}$ by α. Addition is defined componentwise. That is, the matrix equation $\mathbf{C} = \mathbf{A} + \mathbf{B}$ means that

$$c^i_j = a^i_j + b^i_j \tag{5.28}$$

for all i and j such that $1 \le i \le m$ and $1 \le j \le n$.

Sequence Spaces

Definitions Let $\phi(\mathbb{F})$ be the set of **finite sequences** with elements $x_n \in \mathbb{F}$:

$$\overline{x} = (x_1, x_2, \ldots, x_n, \ldots) \in \phi(\mathbb{F}) \Leftrightarrow \exists N_{\overline{x}} :\ni: \forall n > N_{\overline{x}} : x_n = 0. \tag{5.29}$$

We define addition componentwise:

$$\forall \overline{x}, \overline{y} \in \phi(\mathbb{F}) : \overline{x} + \overline{y} := (x_1 + y_1, x_2 + y_2, \ldots, x_n + y_n, \ldots). \tag{5.30}$$

The scalar multiple $\alpha\overline{x}$ (where $\alpha \in \mathbb{F}$) is defined as

$$\alpha\overline{x} := (\alpha x_1, \alpha x_2, \ldots, \alpha x_n, \ldots). \tag{5.31}$$

With these definitions, $\phi(\mathbb{F})$ is a vector space over $\mathbb{F}$ (see Exercise 5.2.15). The **space of all sequences** with elements x_i drawn from $\mathbb{F} = \mathbb{R}$ or $\mathbb{C}$,

$$\mathbb{F}^{\mathbb{Z}^+} := \{\overline{x} = (x_1, x_2, \ldots, x_n, \ldots) \mid x_i \in \mathbb{F}\}, \tag{5.32}$$

the **space of bounded sequences**

$$m(\mathbb{F}) := \{\overline{x} \in \mathbb{F}^{\mathbb{Z}^+} \mid \exists K_{\overline{x}} \in \mathbb{R} :\ni: \forall i \in \mathbb{Z}^+ : |x_i| \le K_{\overline{x}}\}, \tag{5.33}$$

the **space $c(\mathbb{F})$ of convergent sequences**

$$c(\mathbb{F}) := \{\overline{x} \in \mathbb{F}^{\mathbb{Z}^+} \mid \exists x \in \mathbb{F} :\ni: x_n \to x\}, \tag{5.34}$$

and the **space of null sequences,**

$$c_0(\mathbb{F}) := \{\overline{x} \in c(\mathbb{F}) \mid x_n \to 0\}, \tag{5.35}$$

are vector spaces over $\mathbb{F}$ if addition and scalar multiplication are defined as for $\phi(\mathbb{F})$ (see Exercises 5.2.16 and 5.2.17). Because $\mathbb{N} \sim \mathbb{Z}^+$, one can also write the space of all sequences with terms in $\mathbb{F}$ as

$$\mathbb{F}^{\mathbb{N}} := \{\overline{x} = (x_0, x_1, \ldots, x_n, \ldots) \mid x_i \in \mathbb{F}\}, \tag{5.36}$$

the only difference with respect to $\mathbb{F}^{\mathbb{Z}^+}$ being that one starts labeling the terms of the sequence at 0 instead of 1.

Sometimes it is inconvenient to label a sequence starting with any finite integer. Because $\mathbb{Z} \sim \mathbb{Z}^+$, one can also write a sequence in the form

$$(\ldots, x_{-m}, \ldots, x_{-1}, x_0, x_1 \ldots, x_n, \ldots), \tag{5.37}$$

in which the labels belong to $\mathbb{Z}$ rather than to $\mathbb{Z}^+$ or $\mathbb{N}$. The standard set-theoretic notation for the space of sequences of the form Eq. (5.37) is $\mathbb{F}^{\mathbb{Z}}$ (see Section 2.2.2).

Data streams   For the purposes of signal processing, it is often convenient to discretize a signal whose value at time t is $x(t)$ by sampling at times

$$t_n = nh, \tag{5.38}$$

where h is the sampling interval and $n \in \mathbb{Z}$. In order to make use of properties such as invariance under shifts of the origin of time, one regards a vector of sampled values

$$x_n = x(t_n) \tag{5.39}$$

as belonging to $\mathbb{R}^{\mathbb{Z}}$ or $\mathbb{C}^{\mathbb{Z}}$.

ℓ^2 Spaces

The set $\ell^2(\mathbb{Z}^+, \mathbb{F})$ of (possibly infinite) sequences

$$\overline{x} = (x_1, x_2, \ldots, x_n, \ldots), \quad x_i \in \mathbb{F} = \mathbb{R} \text{ or } \mathbb{C} \tag{5.40}$$

such that

$$\sum_{n=1}^{\infty} |x_n|^2 < \infty \tag{5.41}$$

is a vector space over $\mathbb{F}$ if we define addition and scalar multiplication as for $\phi(\mathbb{Z}^+, \mathbb{F})$. Note that $\phi(\mathbb{Z}^+, \mathbb{F})$ is a proper subspace of $\ell^2(\mathbb{Z}^+, \mathbb{F})$.

The vector space $\ell^2(\mathbb{Z}^+, \mathbb{R})$ was first introduced by David Hilbert in order to study integral equations. Erhard Schmidt and Maurice Fréchet generalized Hilbert's work and gave a geometrical interpretation of $\ell^2(\mathbb{Z}^+, \mathbb{R})$ and $\ell^2(\mathbb{Z}^+, \mathbb{C})$.

The space $\ell^2(\mathbb{Z}^+, \mathbb{C})$ is important in physics because any element of $\ell^2(\mathbb{Z}^+, \mathbb{C})$ such that

$$\sum_{n=1}^{\infty} |x_n|^2 = 1 \tag{5.42}$$

is a possible **state vector** of a **discrete quantum system**. It is these state vectors on which the (infinite-dimensional) matrices that correspond to physical observables act in Heisenberg's formulation of quantum mechanics. We study $\ell^2(\mathbb{Z}^+, \mathbb{C})$ in Chapter 10.

Spaces of Continuously Differentiable Functions

Let $\mathcal{C}^n(X; \mathbb{F})$ be the set of n-times continuously differentiable functions from the set X into the number field $\mathbb{F} = \mathbb{R}$ or $\mathbb{C}$. $\mathcal{C}^0(X; \mathbb{F})$ is the set of continuous functions $f : X \to \mathbb{F}$. If the context makes it clear which field is meant, we write $\mathcal{C}^n(X)$. We assume that X is a closed interval in $\mathbb{R}$. The definition implies that

$$f \in \mathcal{C}^n(X; \mathbb{F}) \Leftrightarrow f' \in \mathcal{C}^{n-1}(X; \mathbb{F}), \ldots, f^{(n)} \in \mathcal{C}^0(X; \mathbb{F}). \tag{5.43}$$

Let $f, g \in \mathcal{C}^n(X; \mathbb{F})$. The obvious definition of the sum $f + g$ is

$$\forall x \in X : \quad (f + g)(x) := f(x) + g(x), \tag{5.44}$$

and the obvious definition of the scalar multiple αf is

$$\forall x \in X : \quad (\alpha f)(x) := \alpha f(x). \tag{5.45}$$

With these definitions, $\mathcal{C}^n(X; \mathbb{F})$ is a vector space over $\mathbb{F}$ (see Exercise 5.2.8).

Spaces of Square-Integrable Functions

Let $\mathcal{L}^2_w([a, b]; \mathbb{F})$ be the set of **square-integrable functions** f with respect to the weight function $w : \mathbb{F} \to \mathbb{R}^+$ from the interval $[a, b]$ into $\mathbb{F} = \mathbb{R}$ or $\mathbb{C}$. In other words, a function $f : [a, b] \to \mathbb{F}$ belongs to $\mathcal{L}^2_w([a, b]; \mathbb{F})$ if and only if

$$\int_a^b |f(x)|^2 w(x) \, dx < \infty. \tag{5.46}$$

If the weight function is equal to 1 on the region of integration, we shall omit the subscript w. $\mathcal{L}^2_w([a, b]; \mathbb{F})$ is a vector space over $\mathbb{F}$. The weight function $w(x)$ must be real and nonnegative and can vanish only at isolated points in $[a, b]$.

In equations such as Eq. (5.46), one really should use a Lebesgue integral in which $w(x) \, dx$ is replaced by $d\mu(x)$ in which μ is a measure (see Bartle [1966]). However, in many applications the function f is Riemann-integrable, and

$$\int_a^x f(x') \, d\mu(x') = \int_a^x f(x')w(x') \, dx' \tag{5.47}$$

in which the integral on the left is a Lebesgue integral and the integral on the right is a Riemann integral.

The space $\mathcal{L}^2_w([a, b]; \mathbb{F})$ and its generalizations $\mathcal{L}^2_w([0, \infty); \mathbb{F})$, $\mathcal{L}^2_w(\mathbb{R}; \mathbb{F})$, and $\mathcal{L}^2_w(\mathbb{R}^n; \mathbb{F})$ are fundamental in physics and engineering because they are complete (see Chapter 10) and because of the many situations in which a quadratic functional such as Eq. (5.46) has a physical meaning such as total probability or total energy. The vector space $\mathcal{C}^0([a, b]; \mathbb{F})$ of continuous, $\mathbb{F}$-valued functions on a finite interval $[a, b]$ is a proper subspace of $\mathcal{L}^2_w([a, b]; \mathbb{F})$. Every element ψ of $\mathcal{L}^2(\mathbb{R}^3; \mathbb{C})$ such that

$$\int |\psi(\mathbf{r})|^2 d^3 r = 1 \tag{5.48}$$

is a possible **wave function** of a quantum-mechanical particle in "free space" (the term often used in physics for $\mathbb{R}^3$). The wave function of a system of n particles in free space belongs to $\mathcal{L}^2(\mathbb{R}^{3n}; \mathbb{C})$ and obeys a normalization condition of the same form as Eq. (5.48) except that the region of integration is $\mathbb{R}^{3n}$.

Sturm–Liouville Solution Spaces

The set of functions $f \in \mathcal{C}^2([a, b]; \mathbb{F})$ (in which $\mathbb{F} = \mathbb{R}$ or $\mathbb{C}$) that are solutions of the (**homogeneous**) **Sturm–Liouville differential equation**

$$\frac{d}{dx}\left[p(x)\frac{df}{dx} \right] + q(x)f = 0 \tag{5.49}$$

and that satisfy homogeneous boundary conditions (such as the Dirichlet conditions, Eq. [5.9]) at a and b, is a vector space over $\mathbb{F}$ and is called the **solution space** of the differential equation. The solution space is a proper subspace of $\mathcal{C}^2([a, b]; \mathbb{F})$. For now we simply assume that such solutions exist. In Chapter 12 we construct solutions of Bessel's equation, which is of the form of Eq. (5.49) (see Exercise 5.2.21).

It turns out that one must require p to be strictly positive in (a, b):

$$\forall x \in (a, b): \quad p(x) > 0. \tag{5.50}$$

However, in some cases of great practical interest p vanishes or is singular at a, or b, or both (see Exercises 5.2.19–5.2.21).

Often one sees the Sturm–Liouville differential equation written in the form

$$\frac{d^2 f}{dx^2} + P\frac{df}{dx} + Qf = 0 \tag{5.51}$$

in which

$$P = \frac{1}{p}\frac{dp}{dx}, \quad Q = \frac{q}{p}. \tag{5.52}$$

We use Eq. (5.49) instead of Eq. (5.51) because p turns out to be inversely proportional to the weight function in a useful inner product (see Chapter 8 for definitions and further details). The Sturm–Liouville differential equation can also be put into the form

$$\frac{d^2 z}{dx^2} + gz = 0, \tag{5.53}$$

which is the equation of motion of a simple harmonic oscillator with a variable frequency (see Exercise 5.2.18).

Ordinary differential equations of Sturm-Liouville form arise naturally if one tries to separate partial differential equations such as Laplace's equation, the Helmholtz equation, or the time-independent Schrödinger equation in curvilinear coordinates. Almost every one of the special functions of mathematical physics (such as Bessel, Legendre, Hermite, Laguerre, and hypergeometric functions, to mention a few) satisfies an equation of the form of Eq. (5.49). We shall study some properties of the solutions of Sturm–Liouville equations in conjunction with the special functions.

The **inhomogeneous Sturm–Liouville differential equation** is

$$\frac{d}{dx}\left[p(x)\frac{df}{dx}\right] + q(x)f = s(x) \tag{5.54}$$

in which the source function s is nonnull. The set of solutions of this equation is *not* a vector space. However, solving this equation and closely related inhomogeneous partial differential equations, subject to prescribed boundary conditions, is a fundamental problem in electromagnetics, quantum mechanics, and many other areas. We discuss the matrix version of the Green-function technique for solving inhomogeneous linear equations in Section 6.7.4.

5.2.3 VECTOR SUBSPACES

As in other areas of mathematics, the existence of subsets that obey the same structural axioms as a parent set gives the theory of vector spaces richness, interest, and utility. A **vector subspace** (or, if the context is clear, often simply a **subspace**) of a vector space $\mathcal{V}$ is a subset $\mathcal{W} \subseteq \mathcal{V}$ that is itself a vector space. $\mathcal{W}$ is a **proper subspace** of $\mathcal{V}$ if and only if there is at least one element of $\mathcal{V}$ that does not belong to $\mathcal{W}$, and $\mathcal{W} \neq \{\overline{0}\}$.

Every vector space contains the subspace that consists only of the zero vector,

$$\mathcal{O} := \{\overline{0}\}. \tag{5.55}$$

If $\mathcal{V}$ contains any nonzero vectors, then $\mathcal{O}$ is a proper subspace of $\mathcal{V}$. At the other extreme, $\mathcal{V}$ is always a subspace of itself. If $\mathcal{V}$ contains a nonzero vector $\overline{x}$, then the set of all scalar multiples of $\overline{x}$,

$$\{\alpha\,\overline{x}\,|\,\alpha \in \mathbb{F}\}, \tag{5.56}$$

is a subspace of $\mathcal{V}$ that is called the **subspace generated by** $\overline{x}$. In Fig. 5.2, **0** indicates the zero vector. The nonzero vector **w** generates a subspace whose elements correspond to the points of the line through the origin parallel to **w**. (We write vectors that are equal to n-tuples of real or complex scalars with boldface sansserif letters. Because **w** belongs to two-dimensional Euclidean space, one can write it as a column vector with two components x and y, as in Eq. [5.23].)

More generally, the subspace $\mathcal{W}$ that consists of all of the finite linear combinations of a set of vectors, $S = \{\overline{v}_1, \ldots, \overline{v}_m\}$, is called the **subspace generated by** S. The vectors $\overline{v}_1, \ldots, \overline{v}_m$ are called **generators** of $\mathcal{W}$. For example, the x–y plane in three-dimensional Cartesian coordinates is generated by the unit vectors **i** and **j** along the x- and y-axes, respectively.

There are several methods by which one can uniquely characterize a proper vector subspace $\mathcal{W}$ for either computational or theoretical purposes:

- Give a list of (possibly linearly dependent) vectors that span $\mathcal{W}$ (see Section 5.3).
- Give a basis of $\mathcal{W}$ (see Section 5.4).

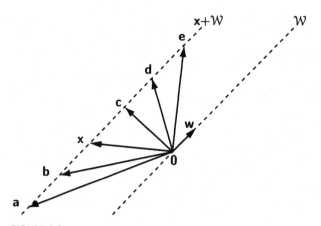

FIGURE 5.2

The vectors **a**, **b**, **c**, **d**, and **e** belong to the affine subspace **x** + $\mathcal{W}$.

- Specify the annihilator of $\mathcal{W}$ by either of the first two methods (see Section 7.4).
- Specify the orthogonal complement of $\mathcal{W}$ by either of the first two methods (see Section 8.3.3).
- Give a set of linear equations that specify either the annihilator of $\mathcal{W}$ or the orthogonal complement of $\mathcal{W}$ (see Section 8.3.3).

Translations and Affine Subspaces

A subspace $\mathcal{W}$ of a vector space $\mathcal{V}$ obeys the axioms for an additive group and is therefore a subgroup of the additive group $\mathcal{V}$. We know from Chapter 4 that for every subgroup there exists a family of cosets. A coset of a subspace $\mathcal{W}$ of a vector space $\mathcal{V}$ is the set

$$\overline{x} + \mathcal{W} = \mathcal{W} + \overline{x} := \{\overline{x} + \overline{w} \mid \overline{w} \in \mathcal{W}\}. \tag{5.57}$$

The coset $\overline{x} + \mathcal{W}$ is also known as the **affine subspace** (or **affine linear variety**) of $\mathcal{V}$ through $\overline{x}$ and parallel to $\mathcal{W}$. Figure 5.3 shows examples of vectors that belong to a particular affine subspace in $\mathbb{R}^2$.

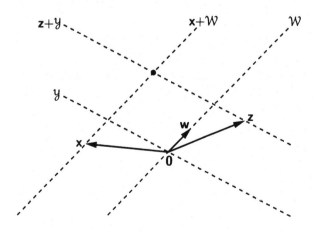

FIGURE 5.3

Vector and affine subspaces in the plane.

For example, let $\mathcal{V} = \mathbb{R}^2$ and

$$W = \left\{ \alpha \begin{pmatrix} 1 \\ 1 \end{pmatrix} \,\middle|\, \alpha \in \mathbb{R} \right\}. \tag{5.58}$$

If $\mathbf{x} = \begin{pmatrix} -1 \\ 0 \end{pmatrix}$, the affine subspace $\mathbf{x} + W$ is

$$\begin{pmatrix} -1 \\ 0 \end{pmatrix} + W = \left\{ \begin{pmatrix} \alpha - 1 \\ \alpha \end{pmatrix} \,\middle|\, \alpha \in \mathbb{R} \right\}. \tag{5.59}$$

Every vector subspace generates a family of affine subspaces through the operation of **translation**. By definition, the **translate** $\mathsf{T}_{\bar{x}}\bar{u}$ of a vector $\bar{u} \in \mathcal{V}$ by the vector $\bar{x}$ is

$$\mathsf{T}_{\bar{x}}\bar{u} := \bar{x} + \bar{u}. \tag{5.60}$$

In a vector space $\mathcal{V}$, one can map a subset $U \subseteq \mathcal{V}$ onto another subset of $\mathcal{V}$ by translating U by a fixed vector $\bar{x} \in \mathcal{V}$:

$$U \mapsto U' := \mathsf{T}_{\bar{x}}U := \bar{x} + U := \{\bar{x} + \bar{u} \,|\, \bar{u} \in U\}. \tag{5.61}$$

Two subsets U and U' of $\mathcal{V}$ are called **parallel** if and only if there exists a translation that maps U onto U'. For example, the subset U' in Fig. 5.4 is parallel to U because U' is derived from U by a translation through $\mathbf{x}$.

Clearly any affine subspace $\bar{x} + W$ is a translation of W and is therefore parallel to W. In fact, any translation of an affine subspace of $\mathcal{V}$ is a parallel affine subspace of $\mathcal{V}$.

Every translation $\mathsf{T}_{\bar{x}}$ (where $\bar{x} \in \mathcal{V}$) carries $\mathcal{V}$ onto itself, for $\bar{y} \in \mathcal{V}$ is the image of $\bar{y} - \bar{x}$ under $\mathsf{T}_{\bar{x}}$. Let

$$T(\mathcal{V}) := \{\mathsf{T}_{\bar{x}} \,|\, \bar{x} \in \mathcal{V}\} \tag{5.62}$$

be the set of translations of $\mathcal{V}$. The law of composition of translations is

$$\mathsf{T}_{\bar{y}}\mathsf{T}_{\bar{x}} = \mathsf{T}_{\bar{y}+\bar{x}} = \mathsf{T}_{\bar{x}+\bar{y}} = \mathsf{T}_{\bar{x}}\mathsf{T}_{\bar{y}}, \tag{5.63}$$

the identity translation is $\mathsf{T}_{\bar{0}}$, and the inverse of $\mathsf{T}_{\bar{x}}$ is $\mathsf{T}_{-\bar{x}}$. The associativity of the law of composition follows from the associativity of addition in $\mathcal{V}$. Therefore one calls $T(\mathcal{V})$ the **group of translations of** $\mathcal{V}$.

FIGURE 5.4

U and U' are parallel subsets of $\mathbb{R}^3$; $\mathbf{x}$ is the translation vector by which $U \mapsto U'$.

An **affine space** is a set $\mathcal{A}$ of **points** a such that $T(\mathcal{V})$ acts on $\mathcal{A}$ as follows: For every $a \in \mathcal{A}$, and for every pair of vectors $\overline{x}$, $\overline{y} \in \mathcal{V}$, $T_{\overline{x}+\overline{y}}a = T_{\overline{x}}(T_{\overline{y}}a)$, and for every pair of points a, $b \in \mathcal{A}$, there exists a unique vector $\overline{x} \in \mathcal{V}$ such that $T_{\overline{x}}b = a$. The vector $\overline{x}$ is called the **displacement** from b to a and is written $\overline{x} = a - b$. Clearly $\mathcal{V}$ itself can be looked upon as an affine space. Whether one considers $\mathcal{V}$ as a vector space or as an affine space depends on one's point of view. An **affine subspace** $\mathcal{B}$ of $\mathcal{A}$ is a subset of $\mathcal{A}$ that is an affine space.

A fundamental difference between affine spaces and vector spaces is that in an affine space the addition and scalar multiplication of points are undefined. Because an affine space $\mathcal{A}$ therefore has no additive identity, there is no preferred "zero" point in $\mathcal{A}$. If we designate a point $b \in \mathcal{A}$ as the **origin of coordinates**, then every affine point $a \in \mathcal{A}$ corresponds to a unique vector $\overline{x} = a - b \in \mathcal{V}$ and vice versa. This correspondence defines the **attachment of $\mathcal{V}$ to $\mathcal{A}$ at** b. The concept of attaching a vector space to an affine space at a point restates the concept of choosing an origin of coordinates without making any reference to coordinate axes.

In an affine space it makes sense to define translations and parallelism as in a vector space. The intersection of two affine subspaces is another affine subspace (see Exercise 5.2.14). If one considers a vector space $\mathcal{V}$ as an affine space, then two vectors $\overline{x}$ and $\overline{y}$ are points that belong to the same affine subspace of $\mathcal{V}$ parallel to the vector subspace $\mathcal{W}$ if and only if $\overline{x} - \overline{y}$ belongs to $\mathcal{W}$ (see Exercise 5.2.13). It follows that an affine subspace is itself an affine space. Affine subspaces are therefore the natural substructures of an affine space.

For example, let $\mathcal{W}$ be the subspace of the plane generated by the vector displacement (relative to a chosen origin) $\mathbf{w}$, as shown in Fig. 5.3. Then the coset $\mathbf{x} + \mathcal{W}$ is the line through the point $\mathbf{x}$ that is parallel to the line $\mathcal{W}$. Clearly $\mathbf{x} + \mathcal{W}$ is the image of $\mathcal{W}$ under a translation through $\mathbf{x}$. Both $\mathcal{W}$ and $\mathbf{x} + \mathcal{W}$ are affine subspaces of the plane. The intersection of the affine subspaces $\mathbf{x} + \mathcal{W}$ and $\mathbf{z} + \mathcal{Y}$ is another affine subspace consisting of a single point. Therefore the affine substructures of the plane are points and lines. If we choose a different origin of coordinates (that is, if we assign a different point as the zero vector $\mathbf{0}$), then $\mathcal{W}$ and $\mathbf{x} + \mathcal{W}$ are still parallel affine subspaces, although their labels as cosets of a vector subspace are different because of the displacement of the origin. Similarly, any line through the origin is a vector subspace of three-dimensional space and so is any plane through the origin. If $\mathcal{W}$ is a plane through the origin, then the affine subspace $\mathbf{x} + \mathcal{W}$ is the plane parallel to $\mathcal{W}$ through the point $\mathbf{x}$. The intersection of a line and a plane is an affine subspace that consists of a single point, and the intersection of two planes is an affine subspace consisting of a line.

Translations and Conservation Laws

The group of translations of three-dimensional space, $T(\mathbb{R}^3)$, is significant in physics because the invariance of the Hamiltonian or Lagrangian under $T(\mathbb{R}^3)$ implies the **conservation of momentum**. For example, if the classical single-particle Hamiltonian

$$H = \frac{\mathbf{p}^2}{2m} + V(\mathbf{r}) \tag{5.64}$$

is invariant under $T(\mathbb{R}^3)$, then the potential energy V is a constant, for $V(\mathbf{r} + \mathbf{a}) = V(\mathbf{r})$ for all $\mathbf{a}$. (The translation does not act on $\mathbf{p}$ because the components of $\mathbf{p}$ and the components of $\mathbf{r}$ are independent variables in the Hamiltonian formulation of classical mechanics.) Then Hamilton's equation for the momentum $\mathbf{p}$ becomes

$$\frac{d\mathbf{p}}{dt} = -\nabla_r H = 0, \tag{5.65}$$

implying that $\mathbf{p}$ is constant in time.

In quantum mechanics, the momentum operator $\mathbf{p}$ is constant in time (in the Heisenberg picture) if and only if

$$i\hbar \frac{d\mathbf{p}}{dt} = [\mathbf{p}, \mathsf{H}] := \mathbf{p}\mathsf{H} - \mathsf{H}\mathbf{p} = 0. \tag{5.66}$$

But the operator $\mathbf{p}$ commutes with the quantum-mechanical single-particle Hamiltonian operator H, which has exactly the same form as the classical Hamiltonian Eq. (5.64), if and only if

$$[\mathbf{p}, V]\psi(\mathbf{r}) = -i\hbar(\nabla_r V)\psi(\mathbf{r}) = \mathbf{0} \tag{5.67}$$

for all continuously differentiable, square-integrable wave functions ψ. As in classical mechanics, if V is continuously differentiable and if

$$\forall \mathbf{a} : V(\mathbf{r} + \mathbf{a}) = V(\mathbf{r}) \Rightarrow \nabla_r V = 0, \tag{5.68}$$

then the momentum operator is constant in time $(d\mathbf{p}/dt = \mathbf{0})$.

5.2.4 *COMMENTS ON VECTOR-SPACE AXIOMS

*Zero Vector Versus Zero Scalar

The zero vector $\bar{0}$ is not the same object as the zero scalar 0. For example, in three-dimensional space the zero vector is the *displacement* $\bar{0} = 0\hat{\imath} + 0\hat{\jmath} + 0\hat{k}$, but the zero scalar is the *real number* 0.

Finite Versus Infinite Linear Combinations

Axioms 1 and 5 imply that all **finite linear combinations**

$$\sum_{j=1}^{N} \alpha^j \bar{x}_j \quad \text{(in which } N \text{ is a finite positive integer)} \tag{5.69}$$

of elements of $\mathcal{V}$ belong to $\mathcal{V}$. (For the rest of this chapter, letter superscripts are indices, not powers.) However, *infinite* linear combinations of functions arise in the course of solving the partial differential equations of electromagnetic theory, quantum mechanics, and many other fields of physics. In order to attach a useful meaning to infinite sums we need to define additional, nonalgebraic concepts such as the *limit* of a sequence of sums of n vectors as n increases without bound. Limits and infinite linear combinations are discussed in Chapter 10.

*Number Field

We assume that the number field $\mathbb{F}$ has *characteristic* ∞, that is, that the sum $1 + 1 + \cdots + 1$ of any finite number of 1s never vanishes. This assumption rules out finite fields. For example, if V is a vector space over the two-element field $\mathbb{GF}(2)$, in which $1 + 1 = 0$, Eqs. (5.18) and (5.20) imply that

$$\forall \overline{x} \in V : \overline{x} + \overline{x} = 1\overline{x} + 1\overline{x} = (1 + 1)\overline{x} = 0\overline{x} = \overline{0}. \tag{5.70}$$

Then $\overline{x} = -\overline{x}$ even if $\overline{x} \neq \overline{0}$. Vector spaces over finite fields occur in Galois theory and have important applications to pseudorandom-number generators and coding theory.

*Distance

Axioms 1 through 9 do not provide a way to answer the question, How close is one vector to another? Many of the vector spaces used in physics and engineering have an additional structural property, the existence of a **metric** or **distance**. In three-dimensional space, if x, y, z and x', y', z' are the components of $\mathbf{r}$ and $\mathbf{r}'$ along mutually perpendicular axes, then Pythagoras's theorem tells us that the distance between $\mathbf{r}$ and $\mathbf{r}'$ is

$$d(\mathbf{r}, \mathbf{r}') = [(x - x')^2 + (y - y')^2 + (z - z')^2]^{1/2}. \tag{5.71}$$

Not all vector spaces have a physically or mathematically meaningful distance function. We discuss some of the mathematical consequences of the existence of a distance in Chapter 8.

*Why Are Real Numbers Necessary?

One is inclined to ask why one needs the full set of real numbers, $\mathbb{R}$, if all physical measurements and numerical computations use only the field of rational numbers $\mathbb{Q}$. (In fact, a digital computer's floating-point hardware uses a *finite* subset of $\mathbb{Q}$!)

A fundamental reason why we need the full set of real numbers is that it would be inconvenient to use a number field that did not contain all the (finite) distances between vectors. The values of the distance function Eq. (5.71) are generally irrational (i.e., in $\mathbb{R}$ but not in $\mathbb{Q}$), even if the components of $\mathbf{r}$ and $\mathbf{r}'$ are rational.

Another fundamental reason is that $\mathbb{Q}$ lacks the property of *completeness*. In Chapter 10 we see that a metric space is called *complete* if it contains the limits of all of its convergent sequences. There are convergent sequences of rational numbers whose limits do *not* lie in $\mathbb{Q}$, such as a sequence of increasingly accurate decimal approximations to the irrational number π. Completeness is essential in order to be able to define important analytical properties, such as continuity, at all points. For example, if $\{\alpha_i\}$ is a sequence of rational numbers that converges to an *irrational* limit α and if f is a function that is defined only on the rational numbers $\mathbb{Q}$, then it would make no sense to say that $f(\alpha_i) \to f(\alpha)$ as $\alpha_i \to \alpha$ because α

lies outside of the number field on which f is defined. We discuss completeness in greater detail in Chapter 10. Both $\mathbb{R}$ and $\mathbb{C}$ are complete.

*Modules over Rings

The vector-space axioms given in Section 5.2.1 make no reference to the commutativity of scalar multiplication ($\beta\alpha = \alpha\beta$) or to the existence of a multiplicative inverse $1/\alpha$ for every nonzero scalar α. It follows that one could weaken the definition of a vector space, and thereby define a more general structure, by replacing the field $\mathbb{F}$ with a collection R of scalars α that have only those properties that are necessary for consistency with axioms 5 through 9. In detail, R must be a commutative group under addition, must be closed under a multiplication that is distributive over addition $[\alpha(\beta + \gamma) = \alpha\beta + \alpha\gamma]$, and must have a unit element 1 such that $1\alpha = \alpha 1 = \alpha$. In other words, R must be a ring, as defined in Section 4.5. An additive group $\mathcal{V}$ that satisfies axioms 5 through 9 and in which the scalars belong to a ring R instead of to a field $\mathbb{F}$ is called an **R-module**. (An additive group is sometimes called a *module*.) If R is the ring of integers (i.e., the set $\mathbb{Z}$ of all integers under the elementary operations of addition and multiplication), then one speaks of a $\mathbb{Z}$-**module**. A vector space is a special case of an R-module because every field is also a ring. We do not follow the usual approach in modern algebra of establishing the basic properties of R-modules before specializing to vector spaces because R-modules that are not vector spaces and that are neither supersets nor subsets of vector spaces are of limited usefulness in physics and engineering.

*Notation

Different notations for vectors and their components have evolved in different areas of physics and engineering. Although some of the reasons for using the notation $\overline{x}$ for an abstract vector depend on concepts that have not been introduced yet in this book, it is important to understand from the beginning that the notation used here has not been chosen arbitrarily. In quantum mechanics one often uses a Dirac "bra vector":

$$|x\rangle := \overline{x}. \tag{5.72}$$

Although Dirac's notation is perfectly adapted for many quantum-mechanical calculations, it is awkward for expressing some fundamental vector-space concepts such as the dual space (the space of linear functionals on $\mathcal{V}$, which we study in Chapter 7). The concept of the dual space is essential for understanding infinite-dimensional vector spaces such as are used in quantum theory, communications theory, and many other areas.

The notation $\overline{x}$ for a vector has significant advantages with respect to Dirac's notation and to the usual mathematical notation, an italic lower-case letter such as x. Writing a vector with a bar makes it easy to distinguish a vector from a scalar, or a vector that has been labeled with an index ($\overline{x}_i$), from a covariant vector component (x_i). Also, $\overline{x}$ recalls the widespread notation $\vec{r}$ for a vector in three-dimensional Euclidean space. We use the bar notation for abstract vectors throughout this book. The only exception is that we write elements of the vector spaces $\mathbb{F}^n$ (defined in a subsequent section) with bold sansserif letters, in conformity with a widespread convention.

5.2.5 EXERCISES FOR SECTION 5.2

5.2.1. Let W and W' be vector subspaces of a vector space V. This problem concerns the set-theoretic intersection

$$W \cap W' := \{\overline{w} \in V \mid \overline{w} \in W \text{ and } \overline{w} \in W'\}. \tag{5.73}$$

(a) Show that $W \cap W'$ is not the empty set $\emptyset$.

(b) Prove that $W \cap W'$ is a subspace of each of the vector spaces W, W', and V.

(c) Let I be a set. Assume that for every $i \in I$, W_i is a subspace of V. Prove that the intersection

$$W := \bigcap_{i \in I} W_i \tag{5.74}$$

is a subspace of V.

5.2.2. Give an example of two proper subspaces W, W' of $V = \mathbb{R}^3$ such that

$$0 \subset W \cap W' \subset V \tag{5.75}$$

(in which $\subset$ means proper inclusion). Your example should illustrate a basic theorem of solid geometry.

5.2.3. Prove in general (not by example) that if V is a vector space and W is a proper subspace of V, then the set complement

$$V \setminus W := \{\overline{x} \in V \mid \overline{x} \notin W\} \tag{5.76}$$

does not satisfy all of the vector-space axioms (and is therefore not a subspace of V).

Hint: Prove that

$$\exists \overline{x} \in V \setminus W :\ni: \overline{x} \neq \overline{0}. \tag{5.77}$$

Then establish a contradiction.

5.2.4. Prove in general (not by example) that if W and W' are proper subspaces of a vector space V, then the union $W \cup W'$ is not a subspace of V.

Hint: Use the result of the preceding problem and De Morgan's laws.

5.2.5. Let a_i and b_i be finite real numbers for $i = 1, 2, 3$. Show that the set of vectors in $\mathbb{R}^3$ such that

$$x^i \in [a_i, b_i] \text{ for } i = 1, 2, 3 \tag{5.78}$$

(i.e., the set of vectors that lie in a bounded rectangular region) does not obey all of the vector-space axioms. Then show that the set of vectors that lie in any bounded region of $\mathbb{R}^3$ is not a vector space.

5.2.6. Let S be the set of n-tuples whose elements are all integers (i.e., every $x^i \in \mathbb{Z}$), and let addition and scalar multiplication be defined as in Eq. (5.23). Prove that S is not a vector space over $\mathbb{Q}$ or $\mathbb{R}$.

5.2.7. Let S be the set of column vectors whose components are all nonnegative real numbers {i.e., every $x^i \in [0, \infty)$}, and let addition and scalar multiplication be defined as in Eq. (5.23). Prove that S is not a vector space over $\mathbb{R}$.

5.2.8. Prove that $\mathcal{C}^n([a, b]; \mathbb{R})$ is a real vector space.

Hint: Recall the definition of continuity:

$$f \in \mathcal{C}^0([a,b]; \mathbb{R}) \Leftrightarrow$$
$$\forall x \in [a, b] : \forall \epsilon > 0 : \exists \delta > 0 :\ni: \tag{5.79}$$
$$\forall x' \in [a, b] :\ni: |x' - x| < \delta : |f(x') - f(x)| < \epsilon.$$

To prove closure under addition, you may find the inequality

$$|a + b - (c + d)| \le |a - c| + |b - d| \tag{5.80}$$

useful.

5.2.9. Prove that the set of finite Fourier sums

$$\left\{ \sum_{m=-M}^{N} c_m e^{imx} \,\middle|\, c_m \in \mathbb{C} \right\}, \tag{5.81}$$

in which $M, N \in \mathbb{Z}^+$, and $x \in \mathbb{R}$ are fixed is a complex vector space.

5.2.10. Prove that the elements $\{0, 1, \omega, \omega^2\}$ of the four-element field $\mathbb{GF}(4)$ are a vector space over the prime field $\mathbb{Z}_2$ of integers modulo 2.

5.2.11. Let $\mathbb{GF}(q)$ be a Galois field.

(a) Prove that $\mathbb{GF}(q)$ is a vector space over its prime field Π. (See Section 4.6.2.)

(b) Prove that the set of n-element column vectors Eq. (5.23) such that every component x^i belongs to Π has exactly p^n elements, where p is a prime.

5.2.12. Let $\mathcal{W}$ be a subspace of a vector space $\mathcal{V}$. Prove that the zero vector of $\mathcal{V}$, $\bar{0}_V$, is the zero vector of $\mathcal{W}$. (It is not enough to observe that $\mathcal{W}$ contains a zero vector $\bar{0}_W$. You need to show that $\bar{0}_W = \bar{0}_V$.)

5.2.13. Let $\mathcal{V}$ be a real or complex vector space and let $\mathcal{W}$ be a vector subspace of $\mathcal{V}$. Prove that any two vectors $\bar{y}$ and $\bar{z}$ of $\mathcal{V}$ belong to the same affine subspace $\bar{x} + \mathcal{W}$ if and only if

$$\bar{y} - \bar{z} \in \mathcal{W}. \tag{5.82}$$

5.2.14. Let $\mathcal{W}$ and $\mathcal{W}'$ be proper vector subspaces of the real or complex vector space $\mathcal{V}$. Assume that the intersection $(\bar{x} + \mathcal{W}) \cap (\bar{x}' + \mathcal{W}')$ of the affine subspaces $\bar{x} + \mathcal{W}$ and $\bar{x}' + \mathcal{W}'$ contains a vector $\bar{y}$.

(a) Prove that

$$\bar{x} + \mathcal{W} = \bar{y} + \mathcal{W} \tag{5.83}$$

and

$$\bar{x}' + \mathcal{W}' = \bar{y} + \mathcal{W}'. \tag{5.84}$$

(b) Prove that the intersection of two affine subspaces is an affine subspace:

$$(\bar{x} + \mathcal{W}) \cap (\bar{x}' + \mathcal{W}') = \bar{y} + (\mathcal{W} \cap \mathcal{W}'). \tag{5.85}$$

(In Fig. 5.3, a black dot indicates the intersection of $\mathbf{z} + \mathcal{Y}$ and $\mathbf{x} + \mathcal{W}$.)

(c) Give an example in three-dimensional space in which $\bar{x}$ and $\bar{x}'$ are both nonzero.

5.2.15. Verify that $\phi(\mathbb{F})$ satisfies all of the axioms for a vector space.

5.2.16. Verify that $c(\mathbb{F})$ satisfies all of the axioms for a vector space.

Hint: You may find Eq. (5.80) useful.

5.2.17. Verify that $c_0(\mathbb{F})$ satisfies all of the axioms for a vector space.

Hint: You may find Eq. (5.80) useful.

5.2.18. Using the substitution

$$z = \frac{f}{\sqrt{p}}, \tag{5.86}$$

put the Sturm–Liouville equation (5.49) into the form of Eq. (5.53).

5.2.19. The standard form of the differential equation satisfied by the Legendre polynomials is

$$(1 - x^2)\frac{d^2 f}{dx^2} - 2x\frac{df}{dx} + n(n + 1)f = 0. \tag{5.87}$$

Put this equation into the Sturm–Liouville form Eq. (5.49), identifying the function p in the process.

5.2.20. The standard form of **Chebyshev's differential equation** is

$$(1 - x^2)\frac{d^2 f}{dx^2} - x\frac{df}{dx} + n^2 f = 0 \tag{5.88}$$

in which n is a nonnegative integer. Put this equation into the Sturm–Liouville form Eq. (5.49), identifying the function p in the process.

5.2.21. The standard form of **Bessel's differential equation** is

$$x^2\frac{d^2 f}{dx^2} + x\frac{df}{dx} + (x^2 - \nu^2)f = 0 \tag{5.89}$$

in which ν is a real constant. Put Bessel's equation into the Sturm–Liouville form Eq. (5.49), identifying the function p in the process.

5.2.22. Given two vectors $\bar{x}, \bar{y}$ in a real or complex vector space $\mathcal{V}$ such that $\bar{x} - \bar{y} \in \mathcal{W}$, in which $\mathcal{W}$ is a vector subspace of $\mathcal{V}$, let

$$Z := \{\bar{z}_\lambda = \lambda\bar{x} + (1 - \lambda)\bar{y} \,|\, 0 \leq \lambda \leq 1\}. \tag{5.90}$$

Show that for all λ such that $0 < \lambda < 1$

$$\bar{x} - \bar{z}_\lambda \in \mathcal{W} \tag{5.91}$$

and therefore that Z is a proper subset of the affine subspace parallel to $\mathcal{W}$ through $\bar{x}$ (or $\bar{y}$). Z is called the **chord** between $\bar{x}$ and $\bar{y}$.

5.2.23. This problem concerns two-element column vectors

$$\mathbf{r} = \begin{pmatrix} x \\ y \end{pmatrix} \in \mathbb{R}^2 \tag{5.92}$$

(a) The equation that defines an element of the vector subspace $\mathcal{W}$ generated by a fixed vector $\mathbf{w} \in \mathbb{R}^2$ is

$$\mathbf{r} = \alpha\mathbf{w}. \tag{5.93}$$

From Eq. (5.93), which gives the parametric equations

$$x = f_x(\alpha), \quad y = f_y(\alpha) \tag{5.94}$$

of the line defined by $\mathcal{W}$, obtain a linear relation between x and y.

(b) The equation that defines an element of an affine subspace $\mathbf{r} + \mathcal{W}$ that is parallel to $\mathcal{W}$ is

$$\mathbf{r}' = \mathbf{r} + \alpha\mathbf{w} \in \mathbf{r} + \mathcal{W}. \tag{5.95}$$

From this equation, obtain a linear relation between x' and y'.

Hint: This is the equation of the line defined by $\mathbf{r} + \mathcal{W}$.

(c) Compare the equations of the lines in parts (a) and (b). State your observations as a formula for the coefficients of the equation of the line in $\mathbb{R}^2$ defined by an affine subspace that is parallel to the vector subspace that is generated by a given vector.

5.3 LINEAR INDEPENDENCE AND LINEAR DEPENDENCE

5.3.1 DEFINITIONS

A set $S = \{\bar{x}_i \,|\, i = 1, \ldots, p \text{ or } i \in \mathbb{Z}^+\}$ of *nonzero* vectors is called **linearly independent** if and only if for every finite m,

$$\sum_{i=1}^{m} \lambda^i \bar{x}_i = \bar{0}, \ \bar{x}_i \in S, \ \lambda^i \in \mathbb{F} \Rightarrow \forall i \in (1 : m) : \lambda^i = 0, \tag{5.96}$$

that is, the coefficients in any finite linear combination of elements of S that gives the zero vector must all be equal to the zero scalar. For example, the set

$$\left\{ \begin{pmatrix} 1 \\ 0 \end{pmatrix}, \begin{pmatrix} 0 \\ 1 \end{pmatrix} \right\} \tag{5.97}$$

is linearly independent, for

$$\lambda^1 \begin{pmatrix} 1 \\ 0 \end{pmatrix} + \lambda^2 \begin{pmatrix} 0 \\ 1 \end{pmatrix} = \begin{pmatrix} \lambda^1 \\ \lambda^2 \end{pmatrix} = \begin{pmatrix} 0 \\ 0 \end{pmatrix} \tag{5.98}$$

if and only if $\lambda^1 = \lambda^2 = 0$. The vector space $\mathbb{C}$ over the real field $\mathbb{R}$ furnishes another example: $S = \{1, i\}$ is linearly independent.

If S is not linearly independent, then it is called (or the vectors that belong to S are called) **linearly dependent**. For example, the vectors

$$\mathbf{x}_1 = \begin{pmatrix} 0 \\ 1 \\ -1 \end{pmatrix}, \quad \mathbf{x}_2 = \begin{pmatrix} 2 \\ 1 \\ 1 \end{pmatrix}, \quad \mathbf{x}_3 = \begin{pmatrix} 3 \\ 1 \\ 2 \end{pmatrix} \tag{5.99}$$

in $\mathbb{R}^3$ are linearly dependent because

$$\mathbf{x}_1 - 3\mathbf{x}_2 + 2\mathbf{x}_3 = \mathbf{0}. \tag{5.100}$$

5.3.2 BASIC RESULTS ON LINEAR DEPENDENCE

In any vector space, two vectors $\overline{x}$ and $\overline{y}$ are linearly dependent if and only if there exist nonzero scalars α and β such that

$$\alpha\overline{x} + \beta\overline{y} = \overline{0}, \tag{5.101}$$

that is, if and only if $\overline{y}$ is a scalar multiple of $\overline{x}$:

$$\overline{y} = -\frac{\alpha}{\beta}\overline{x}. \tag{5.102}$$

In $\mathbb{R}^3$, three vectors $\mathbf{x}, \mathbf{y}, \mathbf{z}$ are linearly dependent if and only if there exist scalars α, β, γ such that at least two are nonzero (say, α and β) and

$$\mathbf{x} = -\frac{\beta}{\alpha}\mathbf{y} - \frac{\gamma}{\alpha}\mathbf{z}. \tag{5.103}$$

This equation states that $\mathbf{x}$ belongs to the vector subspace generated by $\mathbf{y}$ and $\mathbf{z}$. In $\mathbb{R}^3$, that subspace is a plane unless $\mathbf{y}$ belongs to the vector subspace (the line!) generated by $\mathbf{z}$. Equivalently, $\mathbf{x}, \mathbf{y}, \mathbf{z} \in \mathbb{R}^3$ are linearly dependent if and only if

$$\mathbf{x} \cdot (\mathbf{y} \times \mathbf{z}) = 0 \tag{5.104}$$

(see Exercise 5.3.4).

The notations $\{\overline{x}_1, \ldots, \overline{x}_p\}$ or $\{\overline{x}_i \,|\, i \in \mathbb{Z}^+\}$ for S indicate that S is a set of vectors, each of which is labeled with a unique integer. We recall that a finite set (of vectors) S, together with a one-to-one mapping of a finite subset $(1 : p)$ of $\mathbb{Z}^+$ onto S, is called a **list**. If $T = \{\overline{x}_{i_1}, \ldots, \overline{x}_{i_j}\}$, where $i_1 \leq i_2 \leq \cdots \leq i_j$ and where $\{i_1, \ldots, i_j\}$ is a subset of $\{1, \ldots, p\}$, then one calls T a **sublist** of S.

If a list of vectors $\overline{x}_1, \ldots, \overline{x}_p$ is linearly independent, then the coefficients of each $\overline{x}_i$ $(i \in (1 : p))$ on each side of a homogeneous linear equation are equal:

$$\sum_{i=1}^{p} \lambda^i \overline{x}_i = \sum_{i=1}^{p} \mu^i \overline{x}_i \Rightarrow \forall i \in (1 : p) : \lambda^i = \mu^i, \tag{5.105}$$

as one can see by transposing the right-hand sum, using the associative law, and applying the definition of linear independence. In practice, this simple result is one of the more frequently used applications of the concept of linear independence. For example, Exercise 5.3.8 shows that $\sin mx$ and $\cos mx$ are linearly independent. Therefore separate equalities hold for the coefficients of $\sin mx$ and $\cos mx$ in a linear trigonometric equation involving sines and cosines of integer multiples of x.

Criterion for Linear Dependence

Tests for linear dependence are important for both abstract and practical purposes. We show that Eq. (5.96) implies, and is implied by, the following abstract criterion for linear dependence: The vectors $\overline{x}_1, \ldots, \overline{x}_p$ are linearly dependent if and only if at least one vector, say $\overline{x}_k$, is a linear combination of the vectors $\overline{x}_1, \ldots, \overline{x}_{k-1}$ that *precede* $\overline{x}_k$ in the list $S = \{\overline{x}_1, \ldots, \overline{x}_p\}$. It is equivalent to say that S is linearly dependent if and only if $\overline{x}_p$ belongs to the vector subspace generated by its predecessors. For example, if the elements of S are the vectors $\mathbf{x}_1$, $\mathbf{x}_2$ and $\mathbf{x}_3$ of Eq. (5.99), then $\mathbf{x}_3$ belongs to the vector subspace generated by $\mathbf{x}_1$ and $\mathbf{x}_2$, for $\mathbf{x}_3 = -\frac{1}{2}\mathbf{x}_1 + \frac{3}{2}\mathbf{x}_2$.

To prove the criterion, we observe that if S is linearly dependent, then there exist scalars λ^i (at least two of which are nonzero) such that

$$\sum_{i=1}^{p} \lambda^i \overline{x}_i = \overline{0}. \tag{5.106}$$

Let k be the greatest value of i such that $\lambda^i \neq 0$. Then

$$\overline{x}_k = \left(-\frac{\lambda^1}{\lambda^k}\right) \overline{x}_1 + \cdots + \left(-\frac{\lambda^{k-1}}{\lambda^k}\right) \overline{x}_{k-1}. \tag{5.107}$$

Conversely, if

$$\overline{x}_k = \alpha^1 \overline{x}_1 + \cdots + \alpha^{k-1} \overline{x}_{k-1} \tag{5.108}$$

then

$$\alpha^1 \overline{x}_1 + \cdots + \alpha^{k-1} \overline{x}_{k-1} + (-1)\overline{x}_k = \overline{0}, \tag{5.109}$$

implying that S is linearly dependent. (We can never encounter $k = 1$, for then $\lambda^1 \bar{x}_1 = \bar{0}$ and $\lambda^1 \neq 0$ imply that $\bar{x}_1 = \bar{0}$, which contradicts the assumption that every vector in S is nonzero.)

In Section 5.4.4 we develop a practical test for linear dependence based on the relation between linear dependence and the properties of systems of linear equations. We show that to every linearly dependent list of n vectors there corresponds an inconsistent system of n linear equations. Practical tests for linear dependence are directly related to the problem of estimating the so-called condition number of the matrix of coefficients in a system of linear equations. A list that is mathematically linearly independent may be linearly dependent for some practical purposes such as floating-point computation. Exercise 5.3.5 gives an example.

Finitely Spanned Vector Spaces

A list T containing n vectors **spans** a vector space $\mathcal{V}$, written

$$\mathcal{V} = \text{span}[T], \tag{5.110}$$

if and only if every vector in $\mathcal{V}$ can be expressed as a *finite* linear combination of vectors in T,

$$\bar{x} \in \mathcal{V} \Rightarrow \exists \alpha^1, \ldots, \alpha^n \in \mathbb{F} : \exists \bar{t}_{i_1}, \ldots, \bar{t}_{i_n} \in T :\ni:$$

$$\bar{x} = \sum_{j=1}^{n} \alpha^j \bar{t}_{i_j}. \tag{5.111}$$

For example,

$$\mathbb{R}^2 = \text{span}\left[\left\{\begin{pmatrix} 1 \\ 0 \end{pmatrix}, \begin{pmatrix} 0 \\ 1 \end{pmatrix}\right\}\right]. \tag{5.112}$$

The list T need not be linearly independent. For example, the vector space of polynomials $p(x) = a_0 + a_1 x$ of degree not greater than 1 over $\mathbb{R}$ is spanned by the list $T = \{1, \pi, x\}$.

Any list S of nonzero vectors spans the vector space $\mathcal{V}$ that consists of the finite linear combinations of the elements of S. A vector space is called **finitely spanned** if and only if there exists a finite list S such that

$$\mathcal{V} = \text{span}[S]. \tag{5.113}$$

Evidently the vectors of S are generators of $\mathcal{V}$. Not every vector space is finitely spanned. For example, $\mathbb{F}^{\mathbb{Z}^+}$ is not finitely spanned: Why?

In a finitely spanned vector space, every finite list of nonzero vectors such as $S = \{\bar{x}_1, \ldots, \bar{x}_p\}$ contains a *linearly independent* sublist T such that

$$\text{span}[T] = \text{span}[S]. \tag{5.114}$$

If S is linearly independent, then we can take $T = S$. In order to see how to construct T given a linearly dependent S, we observe that there must be a vector $\bar{x}_j \in S$ that is linearly dependent on its predecessors. Define a new list

$$S_1 := \{\overline{x}_1, \ldots, \overline{x}_{j-1}, \overline{x}_{j+1}, \ldots, \overline{x}_p\}. \tag{5.115}$$

In other words, S_1 is S with the element $\overline{x}_j$ deleted. Clearly span $[S_1] =$ span $[S]$. If S_1 is linearly dependent, then we again find and delete a vector $\overline{x}_k$ that is linearly dependent on its predecessors, and so on until after l steps we arrive at a list $T := S_l$ in which no vector is linearly dependent on its predecessors. Therefore T is linearly independent. This construction ensures that Eq. (5.114) is satisfied.

The basic result on linear dependence is that, if n vectors $\overline{x}_1, \ldots, \overline{x}_n$ span a vector space $\mathcal{V}$,

$$\mathcal{V} = \text{span}[\overline{x}_1, \ldots, \overline{x}_n], \tag{5.116}$$

and if $\mathcal{V}$ contains r linearly independent vectors $\overline{y}_1, \ldots, \overline{y}_r$, then

$$n \geq r. \tag{5.117}$$

To derive this result, we begin by considering the $(n+1)$-element list

$$B_1 := \{\overline{y}_1, \overline{x}_1, \ldots, \overline{x}_n\}. \tag{5.118}$$

Certainly B_1 spans $\mathcal{V}$. Also, B_1 is linearly dependent, for $\{\overline{x}_i\}$ spans $\mathcal{V}$ and therefore $\overline{y}_1$ can be expressed as a linear combination of the $\{\overline{x}_i\}$. Then some element of B_1 is linearly dependent on its predecessors. Because $\overline{y}_1$ has no predecessors, it follows that some $\overline{x}_i$ is linearly dependent on $\overline{y}_1, \ldots, \overline{x}_{i-1}$. We define a new $(n+1)$-element list with $\overline{x}_i$ deleted and with $\overline{y}_2$ as the first element,

$$B_2 := \{\overline{y}_2, \overline{y}_1, \overline{x}_1, \ldots, \overline{x}_{i-1}, \overline{x}_{i+1}, \ldots, \overline{x}_n\}. \tag{5.119}$$

Clearly span$[B_2] =$ span$[B_1] = \mathcal{V}$. Again, B_2 is linearly dependent, for $\overline{y}_2$ can be expressed as a linear combination of $\overline{x}_1, \ldots, \overline{x}_n$ and hence as a linear combination of the vectors (other than $\overline{y}_1$) in B_2. We continue the construction, at each step adding another $\overline{y}_j$ and deleting an $\overline{x}_k$ that is linearly dependent on its predecessors. Because the $\{\overline{y}_j\}$ have been assumed to be linearly independent, no $\overline{y}_j$ can be deleted. After r steps every $\overline{y}_j$ will have been added and r of the $\overline{x}_k$s will have been deleted, resulting in the $(n+1)$-element, linearly dependent list $B_r = \{\overline{y}_1, \ldots, \overline{y}_r, \overline{x}_{i_1} \ldots, \overline{x}_{i_k}\}$. By assumption, $\{\overline{y}_1, \ldots, \overline{y}_r\}$ is linearly independent. Therefore

$$r < n + 1 \Rightarrow r \leq n, \tag{5.120}$$

establishing Eq. (5.118).

We have not yet discussed what is probably the most powerful tool for ensuring the linear independence of a list. That tool, the concept of orthogonality, is one of the main topics of Chapter 8.

5.3.3 EXAMPLES OF LINEAR INDEPENDENCE

Space of Complex Numbers

If $\mathbb{C}$ is considered as a vector space over $\mathbb{R}$, then 1 and i are linearly independent. Therefore one can separately equate the real and imaginary parts of a complex equation:

$$a + bi = c + di \Rightarrow a = c \text{ and } b = d. \tag{5.121}$$

The list $\{1, i\}$ spans $\mathbb{C}$.

Spaces of Column Vectors

The column vectors

$$\mathbf{e}_1 = \begin{pmatrix} 1 \\ 0 \\ 0 \\ \vdots \\ 0 \\ 0 \end{pmatrix}, \quad \mathbf{e}_2 = \begin{pmatrix} 0 \\ 1 \\ 0 \\ \vdots \\ 0 \\ 0 \end{pmatrix}, \ldots, \quad \mathbf{e}_n = \begin{pmatrix} 0 \\ 0 \\ 0 \\ \vdots \\ 0 \\ 1 \end{pmatrix} \tag{5.122}$$

are linearly independent, for

$$\sum_{i=1}^{n} \lambda^i \, \mathbf{e}_i = \begin{pmatrix} \lambda^1 \\ \lambda^2 \\ \lambda^3 \\ \vdots \\ \lambda^n \end{pmatrix} = \begin{pmatrix} 0 \\ 0 \\ 0 \\ \vdots \\ 0 \end{pmatrix} = \mathbf{0} \tag{5.123}$$

if and only if $\lambda^1 = \cdots = \lambda^n = 0$. The list $\{\mathbf{e}_1, \ldots, \mathbf{e}_n\}$ spans $\mathbb{F}^n$.

In $\mathbb{F}^n$, linear dependence and linear independence can be expressed in terms of the solutions of a system of linear equations. If applied to a list $X = \{\mathbf{x}_1, \ldots, \mathbf{x}_m\}$, the criterion for linear dependence, Eq. (5.96), implies that

$$\forall j \in (1:n): \sum_{i=1}^{m} \lambda^i x_i^j = 0. \tag{5.124}$$

This is a set of n linear equations in the m unknowns $\lambda^1, \ldots, \lambda^m$. If $n > m$, then the linear system Eq. (5.124) is overdetermined. The only general solution of an overdetermined system is $\lambda^i = 0$ for all i, which implies that X is linearly independent. For some sets of coefficients x_i^j, however, a solution exists in which some coefficients λ^i are nonzero. In these special cases, X is linearly dependent. If $n = m$, then the trivial solution ($\lambda^i = 0$ for all i) is the only solution, and the list X is linearly independent if and only if the determinant of the coefficients x_i^j is nonzero. If $n < m$, then there exist infinitely many nonzero solution sets

$\{\lambda^1, \ldots, \lambda^m\}$; the list X is therefore linearly dependent. We show that these familiar properties generalize easily to abstract vector spaces.

Matrix Spaces

The $m \times n$ matrices $\mathbf{E}_i^j$ that have $1 \in \mathbb{F}$ at the intersection of the ith row and jth column, and 0 elsewhere, are linearly independent in $\mathbb{F}^{m \times n}$, for

$$\sum_{i=1}^{m} \sum_{j=1}^{n} \lambda_j^i \, \mathbf{E}_i^j = \begin{pmatrix} \lambda_1^1 & \cdots & \lambda_n^1 \\ \vdots & \ddots & \vdots \\ \lambda_1^m & \cdots & \lambda_n^m \end{pmatrix} = \begin{pmatrix} 0 & \cdots & 0 \\ \vdots & \ddots & \vdots \\ 0 & \cdots & 0 \end{pmatrix} \tag{5.125}$$

if and only if $\lambda_1^1 = \cdots = \lambda_n^m = 0$.

For a general list of matrices $\{\mathbf{A}_{(1)}, \ldots, \mathbf{A}_{(m)}\}$, the criterion for linear dependence, Eq. (5.96), implies that

$$\forall j \in (1:n): \ \forall k \in (1:n): \ \sum_{i=1}^{m} \lambda^i a_{(i)k}^j = 0 \tag{5.126}$$

in which $a_{(i)k}^j$ is the element of $\mathbf{A}_i$ in row j and column k. Apart from trivial differences of notation, Eq. (5.126) is the same as Eq. (5.96).

Sequence Spaces

The sequences

$$\bar{e}_n = (0, \ldots, 0, 1, 0, \ldots) \tag{5.127}$$

with 1 in the n-th place and zero elsewhere, are linearly independent in the vector spaces $\phi(\mathbb{F})$, $\mathbb{F}^{\mathbb{N}}$, $\mathbb{F}^{\mathbb{Z}^+}$, $\mathbb{F}^{\mathbb{Z}}$, $m(\mathbb{F})$, $c(\mathbb{F})$, $c_0(\mathbb{F})$, and $\ell^2(\mathbb{F})$ in which $\mathbb{F}$ is either $\mathbb{R}$ or $\mathbb{C}$.

For example, if one regards the elements of $\mathbb{F}^{\mathbb{Z}}$ as discrete-time (sampled) signals, then $\bar{e}_n$ represents a unit impulse occurring at time $t_n = nh$.

Spaces of Differentiable Functions

Let y_1 and y_2 belong to $\mathcal{C}^1([a, b]; \mathbb{R})$. Then the results of Section 6.5.2 imply that y_1 and y_2 are linearly independent on $[a, b]$ if and only if the **Wronskian**

$$W[y_1, y_2] := y_1 y_2' - y_1' y_2 \tag{5.128}$$

does not vanish identically on $[a, b]$. (Here $y_i' := dy_i/dx$.) For, if there exist real numbers α^1, α^2 such that

$$\forall x \in [a, b]: \ \alpha^1 y_1(x) + \alpha^2 y_2(x) = 0, \tag{5.129}$$

then one can differentiate to obtain another relation,

$$\forall x \in [a, b]: \ \alpha^1 y_1'(x) + \alpha^2 y_2'(x) = 0. \tag{5.130}$$

For any $x \in [a, b]$, the only solution of Eqs. (5.129) and (5.130) is $\alpha^1 = \alpha^2 = 0$ unless the determinant of the 2×2 matrix whose elements are the coefficients of α^1 and α^2 vanishes:

$$\det \begin{pmatrix} y_1(x) & y_2(x) \\ y_1'(x) & y_2'(x) \end{pmatrix} = W[y_1, y_2] = 0. \tag{5.131}$$

Conversely, if $\forall x \in [a, b] : W[y_1, y_2] = 0$, then y_1 and y_2 are linearly dependent on the interval $[a, b]$. For example, $W[1, x] = 1$; therefore $\{1, x\}$ is linearly independent. An evaluation of the Wronskian shows that for every $m \in \mathbb{R} \setminus \{0\}$, $\sin mx$ and $\cos mx$ are linearly independent on $[-\pi, \pi]$ (see Exercise 5.3.8).

Linear dependence should be distinguished carefully from other functional relationships that might loosely be called "dependence." For example, e^x and e^{-x} are "dependent" because $e^{-x} = 1/e^x$, but e^x and e^{-x} are *linearly* independent on all of $\mathbb{R}$ because $W[e^x, e^{-x}] = 2$.

Similarly, if the functions $y_1, \ldots, y_n$ belong to $\mathcal{C}^{n-1}([a, b]; \mathbb{R})$, then the list $\{y_1, \ldots, y_n\}$ is linearly independent on $[a, b]$ if and only if the linear combinations

$$\alpha^1 y_1 + \cdots + \alpha^n y_n, \ldots, \alpha^1 \frac{d^{n-1} y_1}{dx^{n-1}} + \cdots + \alpha^n \frac{d^{n-1} y_n}{dx^{n-1}} \tag{5.132}$$

all vanish. By the theory of linear equations, these n equations have a nonzero solution for $\alpha^1, \ldots, \alpha^n$ (and therefore the list $\{y_1, \ldots, y_n\}$ is linearly dependent) if and only if the **Wronskian of n functions**

$$W[y_1, \ldots, y_n] = \det \begin{pmatrix} y_1 & y_2 & \cdots & y_n \\ \dfrac{dy_1}{dx} & \dfrac{dy_2}{dx} & \cdots & \dfrac{dy_n}{dx} \\ \vdots & \vdots & \ddots & \vdots \\ \dfrac{d^{n-1} y_1}{dx^{n-1}} & \dfrac{d^{n-1} y_2}{dx^{n-1}} & \cdots & \dfrac{d^{n-1} y_n}{dx^{n-1}} \end{pmatrix} \tag{5.133}$$

vanishes on $[a, b]$. Conversely, if the preceding Wronskian is nonzero on $[a, b]$, then $y_1, \ldots, y_n$ are linearly independent. For example, the $n + 1$ functions $\{1, x, \ldots, x^n\}$ are linearly independent on $\mathbb{R}$ (see Exercise 5.3.9).

Sturm–Liouville Solution Spaces

An evaluation of the Wronskian of any three solutions y_1, y_2, y_3 of the homogeneous Sturm–Liouville differential equation Eq. (5.49) shows that the list $\{y_1, y_2, y_3\}$ is linearly dependent because the differential equation provides a linear relation between the last row and the first two rows of the Wronskian determinant. For every value of x for which $p(x) \neq 0$, Eq. (5.49) $[(pf')' + qf = 0]$ implies that for $i = 1, 2, 3$

$$y_i'' = \alpha y_i' + \beta y_i \tag{5.134}$$

in which

$$\alpha = -\frac{p'}{p}, \quad \beta = -\frac{q}{p}. \tag{5.135}$$

Then

$$W[y_1, y_2, y_3] = \det \begin{pmatrix} y_1 & y_2 & y_3 \\ y_1' & y_2' & y_3' \\ y_1'' & y_2'' & y_3'' \end{pmatrix}$$

$$= \det \begin{pmatrix} y_1 & y_2 & y_3 \\ y_1' & y_2' & y_3' \\ \alpha y_1' + \beta y_1 & \alpha y_2' + \beta y_2 & \alpha y_3' + \beta y_3 \end{pmatrix} = 0. \tag{5.136}$$

5.3.4 EXERCISES FOR SECTION 5.3

5.3.1. Test the list

$$\left\{ \mathbf{a} = \begin{pmatrix} 1 \\ 1 \\ 0 \end{pmatrix}, \mathbf{b} = \begin{pmatrix} 1 \\ -1 \\ 0 \end{pmatrix}, \mathbf{c} = \begin{pmatrix} \sqrt{2} \\ \pi \\ 0 \end{pmatrix} \right\} \tag{5.137}$$

for linear dependence.

5.3.2. The unit vectors $\hat{\imath}, \hat{\jmath}, \hat{k}$ along the x-, y-, and z-axes in three-dimensional space satisfy the following **cross-product relations**:

$$\hat{\imath} \times \hat{\imath} = 0, \quad \hat{\jmath} \times \hat{\jmath} = 0, \quad \hat{k} \times \hat{k} = 0 \tag{5.138}$$

$$\hat{\imath} \times \hat{\jmath} = -\hat{\jmath} \times \hat{\imath} = \hat{k}, \quad \hat{\jmath} \times \hat{k} = -\hat{k} \times \hat{\jmath} = \hat{\imath}, \quad \hat{k} \times \hat{\imath} = -\hat{\imath} \times \hat{k} = \hat{\jmath} \tag{5.139}$$

By definition, $(x\hat{\imath}) \times (y\hat{\jmath}) = xy\,\hat{\imath} \times \hat{\jmath} = xy\,\hat{k}$, and $\mathbf{0} \times \hat{\imath} = \mathbf{0}$, and so forth. Assuming that these equations are correct, use them to prove that the list $\{\hat{\imath}, \hat{\jmath}, \hat{k}\}$ is linearly independent.

Hint: Assume that $\{\hat{\imath}, \hat{\jmath}, \hat{k}\}$ is linearly dependent and develop a contradiction.

5.3.3. Use the hypotheses and results of the preceding problem to prove that two vectors **a, b** in three-dimensional space are linearly dependent if and only if

$$\mathbf{a} \times \mathbf{b} = \mathbf{0}. \tag{5.140}$$

5.3.4. This problem also concerns linear dependence in three-dimensional space.

(a) Prove that any three vectors in $\mathbb{R}^3$ that lie in the same plane are linearly dependent.

(b) Prove that three nonzero vectors **a**, **b**, **c** are linearly dependent if and only if

$$\mathbf{a} \cdot (\mathbf{b} \times \mathbf{c}) = 0. \tag{5.141}$$

Give a geometrical interpretation of this equation.

Hint: You do not need to use any material from subsequent chapters. Use the elementary geometrical definitions of dot and cross product.

5.3.5. Computers usually perform numerical computations with a fixed number of significant digits. In order to see what can happen to a list that is mathematically linearly independent in this "real-world" situation, consider the list

$$S = \left\{ \begin{pmatrix} 1/3 \\ 2/3 \end{pmatrix}, \begin{pmatrix} -3 \\ -7 \end{pmatrix} \right\}. \tag{5.142}$$

(a) Show that S is linearly independent.

(b) Show that if all the components of the vectors in the list are rounded to the nearest floating-point number in the format $\pm d \times 10^e$, where d and e are one-digit *integers* (such that $0 \le d \le 9$ and $-9 \le e \le 9$), then S becomes linearly dependent after rounding.

Comment: This example is a trivial illustration of a real problem. The obvious solution here – to use two significant digits d_1, d_2 – will not solve all of the problems of this sort that occur in practice. In fact, there is *no* finite number of significant digits that will solve all such problems. We show in Chapters 6 and 9 that it is possible to define quantities that measure the closeness of a list of vectors to linear dependence and to use one of these quantities, together with other information such as the number of significant digits in one's floating-point representation, to detect lists that are "dangerously" close to linear dependence.

5.3.6. You have already shown in Exercise 5.2.10 that the finite field $\mathbb{GF}(4)$ is a vector space over its prime field, the field $\mathbb{Z}_2$ of integers modulo 2. Give an example of elements of $\mathbb{GF}(4)$ that are linearly dependent over $\mathbb{Z}_2$ but linearly independent over $\mathbb{R}$.

5.3.7. Prove that if S is a list of vectors,

$$S = \{\overline{x}_1, \dots, \overline{x}_p\}, \tag{5.143}$$

then span$[S]$ is equal to the intersection of all the subspaces of $\mathcal{V}$ that include S.

5.3.8. Prove that for all $m \in \mathbb{Z}^+$, $\sin mx$ and $\cos mx$ are linearly independent on $[-\pi, \pi]$.

5.3.9. Prove that for every $n > 1$, the list

$$S = \{1, x, \dots, x^n\} \tag{5.144}$$

is linearly independent on $\mathbb{R}$.

Answer: $W[1, x, \dots, x^n] = 1!\, 2! \cdots n!.$

5.3.10. A **quaternion** is a linear combination

$$x = x_0 1 + x_1 i + x_2 j + x_3 k \tag{5.145}$$

in which every x_i is a real number, **1** is the unit quaternion such that

$$1x = x1 = x \tag{5.146}$$

for all quaternions **x**, and the quaternions **i**, **j**, **k** have the following multiplication properties:

$$i^2 = j^2 = k^2 = -1, \tag{5.147}$$

$$ij = -ji = k, \quad jk = -kj = i, \quad ki = -ik = j. \tag{5.148}$$

By definition, a product such as $(a\mathbf{x})(b\mathbf{y})$ equals $ab\mathbf{xy}$, and $0\mathbf{x}$ equals the zero quaternion **0**. Prove that the list $\{1, i, j, k\}$ is linearly independent.

5.3.11. Test the following for linear dependence:

(a) The list $\{1, \sigma_1, \sigma_2, \sigma_3\}$, in which

$$1 := \begin{pmatrix} 1 & 0 \\ 0 & 1 \end{pmatrix}, \quad \sigma_1 := \begin{pmatrix} 0 & 1 \\ 1 & 0 \end{pmatrix},$$

$$\sigma_2 := \begin{pmatrix} 0 & -i \\ i & 0 \end{pmatrix}, \quad \sigma_3 := \begin{pmatrix} 1 & 0 \\ 0 & -1 \end{pmatrix}. \tag{5.149}$$

The matrices $\sigma_1, \sigma_2, \sigma_3$ are called the **Pauli spin matrices**.

(b) The list $\{A_{(1)}, A_{(2)}, A_{(3)}\}$ in which

$$A_{(1)} := \begin{pmatrix} 1 & 2 \\ 3 & 4 \end{pmatrix}, \quad A_{(2)} := \begin{pmatrix} 2 & 3 \\ 4 & 5 \end{pmatrix}, \quad A_{(3)} := \begin{pmatrix} 3 & 4 \\ 5 & 6 \end{pmatrix}. \tag{5.150}$$

(c) The list $\{A_{(1)}, A_{(2)}, A_{(3)}, A_{(4)}, A_{(5)}\}$ in which $A_{(1)}, A_{(2)}$, and $A_{(3)}$ are defined in Eq. (5.150) and

$$A_{(4)} := \begin{pmatrix} 5 & 6 \\ 7 & 8 \end{pmatrix}, \quad A_{(5)} := \begin{pmatrix} 6 & 7 \\ 8 & 9 \end{pmatrix}. \tag{5.151}$$

5.4 BASES AND DIMENSION

5.4.1 DIMENSION OF A VECTOR SPACE

A **basis** of a vector space $\mathcal{V}$ is a set

$$B = \{\bar{e}_i \in \mathcal{V} \mid i \in I\} \tag{5.152}$$

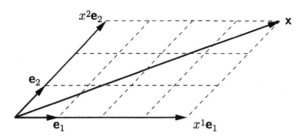

FIGURE 5.5

The contravariant components of a vector $\mathbf{x} = x^1\mathbf{e}_1 + x^2\mathbf{e}_2$ in $\mathbb{R}^2$ are the coordinates of $\mathbf{x}$ relative to axes defined by $\mathbf{e}_1$ and $\mathbf{e}_2$.

such that the index set I is either finite or countable and such that every $\overline{x} \in \mathcal{V}$ has a *unique* expansion in terms of the elements of B,

$$\overline{x} = \sum_{i \in I} x^i \, \overline{e}_i. \tag{5.153}$$

In this equation, the numbers x^i are called the (**contravariant**) **components** (or **coordinates**) of the vector $\overline{x}$ **relative to the basis** $\{\overline{e}_i\}$. We abbreviate the latter phrase as rel $\{\overline{e}_i\}$. We shall discuss the meaning of the term *contravariant* in the next section. Eq. (5.153) is called the **coordinate expansion** of $\overline{x}$ (rel $\{\overline{e}_i\}$). The subspace $\mathcal{W}_i$ generated by a basis vector $\overline{e}_i$ is the i-th **coordinate axis**. (The plural of *basis* is *bases*, pronounced with a long e.)

If the index set I is finite, then B is a list. Therefore $\mathcal{V}$ is finitely spanned, and the list T in Eq. (5.114) is a basis of $\mathcal{V}$. The uniqueness of the contravariant components x^i in the expansion in Eq. (5.153) is guaranteed by the linear independence of the elements of T (see Exercise 5.4.3).

From Eq. (5.153) and from the example shown in Fig. 5.5, one can see that the contravariant components of $\overline{x}$ (rel $\{\overline{e}_i\}$) are the **parallel projections** of $\overline{x}$ along the directions of the basis vectors. In our two-dimensional example, the contravariant components x^1 and x^2 of $\mathbf{x} = x^1\mathbf{e}_1 + x^2\mathbf{e}_2$ are the lengths of adjacent sides of a parallelogram, the included diagonal of which corresponds to $\mathbf{x}$. In order to compute the parallel projections of a given vector $\overline{x}$ relative to a basis $\{\overline{e}'_i\}$ given the expansion of $\overline{x}$ relative to a different basis $\{\overline{e}_i\}$, one must, in effect, solve a system of linear equations (see Chapters 7 and 9).

Powerful, practical methods exist for constructing bases, such as successive orthogonalization (see Section 8.2) and the determination of the eigenvectors of certain kinds of linear transformations, or the eigenfunctions of certain kinds of linear operators. If the basis vectors $\overline{e}_i$ are eigenvectors of a physically important linear transformation, and if the corresponding eigenvalues are real and are oscillation frequencies, then the $\overline{e}_i$ are called **normal modes**. If the basis vectors $\{\overline{e}_i\}$ are eigenvectors of the Hamiltonian of a discrete quantum system, then the $\overline{e}_i$ are called **energy eigenstates** because the eigenvalues of the Hamiltonian are quantized energies.

*Infinite Bases

If the index set I in Eq. (5.152) is countable, then the sequence B is sometimes called a **Schauder basis** of $\mathcal{V}$. If $\mathcal{V}$ is not finitely spanned, the summation in Eq. (5.153) may contain an infinite set of nonzero terms and may therefore be meaningless unless one has additional information about $\mathcal{V}$. For example, the summation makes sense if it makes sense to talk about the convergence of a sequence of vectors in $\mathcal{V}$ and if the sequence of partial sums

$$\bar{x}_n = \sum_{i=1}^{n} x^i \, \bar{e}_i \tag{5.154}$$

converges in some norm to $\bar{x}$ as n becomes large without bound. We see in Chapter 10 that this approach is possible for many of the infinite-dimensional vector spaces that are important in physics and engineering.

One can show, using Zorn's lemma (see Section 2.5.3), that in every vector space $\mathcal{V}$ it is possible to define a set of vectors, called a **Hamel basis** of $\mathcal{V}$, such that every vector $\bar{x} \in \mathcal{V}$ is a *finite* linear combination of elements of the Hamel basis and such that every finite subset of the Hamel basis is linearly independent (see Exercise 5.4.10). If $\mathcal{V}$ is finitely spanned, then a Hamel basis is a basis as defined here, and vice versa. If $\mathcal{V}$ has a countable basis B in the sense that the sequence of partial sums Eq. (5.154) converges to a limit, then B need not be a Hamel basis because the coordinate expansions of some elements of $\mathcal{V}$ may require an infinite subset of B. For vector spaces in which a (finite or countable) basis as defined above does not exist, Hamel bases turn out not to be very useful in practice because they are uncountable and because Zorn's lemma does not provide a practical construction.

Finite-Dimensional Vector Spaces

By definition, two bases of a vector space $\mathcal{V}$ are the same if and only if they have the same elements and the same order of elements. We show that if a basis of a vector space $\mathcal{V}$ has a finite number of elements, n, then all bases of $\mathcal{V}$ have n elements. Anticipating this result, we define the (finite) number of basis elements as the **dimension** ($\dim[\mathcal{V}]$) of the vector space $\mathcal{V}$, which is said to be **finite-dimensional**. If $\mathcal{V}$ has a basis that must be indexed by an infinite set, then $\mathcal{V}$ is called **infinite-dimensional**. Because a basis of the vector space $\mathcal{O} = \{\bar{0}\}$ is the empty set, it follows that $\dim[\mathcal{O}] = 0$.

Let $\mathcal{V}$ be a finite-dimensional vector space. To see that all bases of $\mathcal{V}$ have the same number of elements, suppose that

$$B := \{\bar{x}_1, \ldots, \bar{x}_n\} \tag{5.155}$$

is a basis of $\mathcal{V}$, and let

$$C := \{\bar{y}_1, \ldots, \bar{y}_r\} \tag{5.156}$$

be another basis of $\mathcal{V}$. Because $\text{span}[B] = \mathcal{V}$ and because C is linearly independent by the definition of a basis, we have $n \geq r$. Interchanging B and C, we have $r \geq n$. Then $n = r$.

In summary: *The dimension of $\mathcal{V}$ is a property of $\mathcal{V}$, not of a particular basis of $\mathcal{V}$. Moreover, any $(n + 1)$ nonzero vectors of an n-dimensional vector space $\mathcal{V}$ are linearly dependent, and no set of $(n - 1)$ vectors can span $\mathcal{V}$.*

Summation Convention

In physics, if dim[$\mathcal{V}$] is clear from the context one often writes Eq. (5.153) in the form

$$\overline{x} = x^i \overline{e}_i, \tag{5.157}$$

where $x^i \in \mathbb{F}$. Equation (5.157) uses the **Einstein summation convention**:

1. If an index appears twice in an equation, once as a superscript and once as a subscript, then that index is summed from 1 through $n = \dim[\mathcal{V}]$.
2. Every equation in which an index appears once on each side as a superscript or as a subscript is valid for all values of the index ranging from 1 through n. (The ordinary meaning of a superscript as a power is suspended unless the superscript is applied to an expression that is enclosed in delimiters such as parentheses, bars, or square brackets.)

For the rest of this book, we use the Einstein summation convention for coordinate expansions in a finite-dimensional vector space unless a summation symbol or the abbreviation no Σ (no summation) appears in the equation.

Extension of a Linearly Independent List to a Basis

One can think of a vector in an n-dimensional vector space as representing a point in an abstract space and of a set of basis vectors as defining a set of coordinate axes in this abstract space. An n-dimensional vector space such that $0 < n < \infty$ has infinitely many bases, each with n elements. Every physicist or engineer knows that one's choice of a set of coordinate axes can make a problem either easy or hard to solve. Therefore one needs to know how to construct bases that are adapted to a particular purpose.

We show that every linearly independent list

$$C := \{\overline{y}_1, \ldots, \overline{y}_r\} \tag{5.158}$$

in a finite-dimensional vector space $\mathcal{V}$ is a sublist of a basis of $\mathcal{V}$. Equivalently, every linearly independent list C in $\mathcal{V}$ either is a basis of $\mathcal{V}$ or can be **extended to a basis of** $\mathcal{V}$. To see why, let $\{\overline{x}_1, \ldots, \overline{x}_n\}$ be a basis of $\mathcal{V}$, and define the list

$$D := \{\overline{y}_1, \ldots, \overline{y}_r, \overline{x}_1, \ldots, \overline{x}_n\}. \tag{5.159}$$

Certainly D is linearly dependent and spans $\mathcal{V}$. One can extract a linearly independent sublist $E \subset D$ such that $\text{span}[E] = \text{span}[D] = \mathcal{V}$ by deleting every element of D that is linearly dependent on its predecessors, starting with $\overline{x}_n$. None of the $\{\overline{y}_i\}$ can be deleted, for we have assumed that C is linearly independent. By construction, E is a basis of $\mathcal{V}$ that includes C.

A fundamental consequence is that *the dimension of a proper subspace $\mathcal{W} \subset \mathcal{V}$ is less than the dimension of $\mathcal{V}$*:

$$\mathcal{W} \subset \mathcal{V} \Rightarrow \dim[\mathcal{W}] < \dim[\mathcal{V}]. \tag{5.160}$$

For, if there exists a vector $\overline{x}$ in $\mathcal{V}$ that does not belong to $\mathcal{W}$, and if $\{\overline{y}_1, \ldots, \overline{y}_m\}$ is a basis of $\mathcal{W}$, then the list $\{\overline{y}_1, \ldots, \overline{y}_m, \overline{x}\}$ is linearly independent and can therefore be extended to a basis of $\mathcal{V}$. Then $\dim[\mathcal{W}] + 1 \leq \dim[\mathcal{V}]$.

*R-Modules

If one tries to generalize the concepts of basis and dimension from a vector space to a general R-module, one discovers quickly that the existence of a basis of a vector space is a nontrivial property. The construction of a basis for a finitely spanned vector space depends on the existence of an inverse for every nonzero scalar. Therefore one cannot conclude that a basis exists for every finitely spanned R-module.

In at least one important case one can say a bit more because of the direct relationship that we show in Section 4.5 between the integers $\mathbb{Z}$ and the rational numbers (quotients) $\mathbb{Q}$. The set $\mathbb{Z}^n$ of n-element column vectors with integer components is a $\mathbb{Z}$-module, that is, an R-module over the ring of integers. If $\bar{z} \in \mathbb{Z}^n$ and if m is any nonzero integer, then $(1/m)\bar{z} \in \mathbb{Q}^n$. It follows that if a list $\{\bar{z}_1, \ldots, \bar{z}_p\}$ is linearly dependent in $\mathbb{Z}^n$, then the list $\{(1/m)\bar{z}_1, \ldots, (1/m)\bar{z}_p\}$ is linearly dependent in $\mathbb{Q}^n$. Also, if a list $\{\bar{r}_1, \ldots, \bar{r}_p\}$ is linearly dependent in $\mathbb{Q}^n$, then the list $\{ql\,\bar{r}_1, \ldots, ql\,\bar{r}_p\}$ is linearly dependent in $\mathbb{Z}^n$, where q is the least common multiple of the denominators of the components of $\{\bar{r}_1, \ldots, \bar{r}_p\}$ and l is the least common multiple of the denominators of the coefficients λ^i in Eq. (5.106). Certainly there exist bases of $\mathbb{Z}^n$; the list $\{\mathbf{e}_1, \ldots, \mathbf{e}_n\}$ is an example (see Eq. [5.122]). Every basis of $\mathbb{Z}^n$ is a basis of $\mathbb{Q}^n$. (The converse is false.) Therefore every basis of $\mathbb{Z}^n$ has the same number of elements (n), which we call the *dimension* of $\mathbb{Z}^n$.

Even though $\mathbb{Z}^n$ has a definite dimension, its dimension and the dimensions of its proper submodules generally do not satisfy Eq. (5.160). For example, consider $\mathbb{Z}^2$, the set of column vectors with two integer components. The linearly independent two-element list

$$\left\{\mathbf{e}_1 = \begin{pmatrix} 1 \\ 0 \end{pmatrix}, \mathbf{e}_2 = \begin{pmatrix} 0 \\ 1 \end{pmatrix}\right\} \tag{5.161}$$

clearly spans $\mathbb{Z}^2$ (and $\mathbb{Q}^2$). However, another linearly independent two-element list,

$$\left\{\begin{pmatrix} 2 \\ 0 \end{pmatrix}, \begin{pmatrix} 0 \\ 2 \end{pmatrix}\right\}, \tag{5.162}$$

spans all of $\mathbb{Q}^2$ but spans only the proper submodule of $\mathbb{Z}^2$ that consists of column vectors whose components are *even* integers. In this example, the dimension of a proper submodule is *equal* to the dimension of its parent $\mathbb{Z}$-module, in violation of the inequality (5.160). It follows that even for a $\mathbb{Z}$-module one cannot define a dimension that has all the properties of the dimension of a vector space.

5.4.2 SELECTED REALIZATIONS OF VECTOR-SPACE BASES

Complex Numbers

A basis of $\mathbb{C}$ over $\mathbb{R}$ is $\{1, i\}$:

$$\mathbb{C} = \mathrm{span}[\{1, i\}]. \tag{5.163}$$

The dimension is two.

Spaces of Column Vectors

Canonical basis in $\mathbb{R}^n$ and $\mathbb{C}^n$ The list of column vectors $\{\mathbf{e}_i \mid i = 1, 2, \ldots, n\}$ defined in Eq. (5.122) spans $\mathbb{R}^n$ and $\mathbb{C}^n$ and is linearly independent according to Eq. (5.123). Therefore $\{\mathbf{e}_i\}$ is a basis of $\mathbb{F}^n$. The basis $\{\mathbf{e}_i\}$ is usually called the **canonical basis** (or the **standard basis**). The coordinate expansion of $\mathbf{x} \in \mathbb{F}^n$ relative to the canonical basis is

$$\mathbf{x} = x^i \, \mathbf{e}_i \tag{5.164}$$

(compare this equation with Eq. [5.23]). The **Kronecker delta** is the function on $\mathbb{Z} \times \mathbb{Z}$ such that

$$\delta_j^i = \begin{cases} 0 & \text{if } i \neq j; \\ 1 & \text{if } i = j. \end{cases} \tag{5.165}$$

In terms of the Kronecker delta, the coordinate expansion of the basis vector $\mathbf{e}_i$ is

$$\mathbf{e}_i = \delta_i^j \mathbf{e}_j. \tag{5.166}$$

Therefore the contravariant components of $\mathbf{e}_i$ are $\delta_i^1, \ldots, \delta_i^n$.

Hyperplanes in $\mathbb{R}^n$ and $\mathbb{C}^n$ In $\mathbb{F}^n$, it is natural to call a translate of a one-dimensional subspace a **line** and a translate of an $(n - 1)$-dimensional subspace a **hyperplane** (recall Eq. [5.61]). In view of the fact that any n-dimensional vector space $\mathcal{V}$ over $\mathbb{F}$ is isomorphic to $\mathbb{F}^n$ (see Section 5.4.3), it is equally natural to extend these definitions to $\mathcal{V}$.

Matrix Spaces

The set of matrices $\mathbf{E}_i^j$ defined in Eq. (5.125) is the canonical basis of $\mathbb{F}^{m \times n}$. Indeed, if $\mathbf{A}$ is the matrix defined in Eq. (5.27), then

$$\mathbf{A} = a_j^i \mathbf{E}_i^j. \tag{5.167}$$

Therefore the matrix elements a_j^i are the coordinates of $\mathbf{A}$ (rel$\{\mathbf{E}_i^j\}$). The element of the matrix $\mathbf{E}_i^j$ at the intersection of the k-th row and l-th column is

$$E_{il}^{jk} = \delta_i^k \delta_l^j. \tag{5.168}$$

The dimension of $\mathbb{F}^{m \times n}$ is mn.

Sequence Spaces

The set of sequences $\bar{e}_i$ such that the i-th entry is 1 and all other entries are 0,

$$\bar{e}_i := (\delta_i^1, \delta_i^2, \ldots, \delta_i^n, \ldots), \tag{5.169}$$

is a basis of the infinite-dimensional vector spaces $\phi(\mathbb{F})$, $\mathbb{F}^{\mathbb{Z}^+}$, $\mathbb{F}^{\mathbb{N}}$, $\mathbb{F}^{\mathbb{Z}}$, $c_0(\mathbb{F})$, $c(\mathbb{F})$, and $\ell^2(\mathbb{F})$. The proof for $\phi(\mathbb{F})$ is immediate; the proof for $c(\mathbb{F})$ and $c_0(\mathbb{F})$ is straightforward. We prove that $\{\bar{e}_i \mid i \in \mathbb{Z}^+\}$ is a Schauder basis of $\ell^2(\mathbb{F})$ in Chapter 10.

Spaces of Continuous Functions

We show in Appendix F that if a and b are finite, then the set of monomials $\{1, x, x^2, \ldots,$ $x^n, \ldots\}$ is a basis of $\mathcal{C}^0([a, b]; \mathbb{F})$ and that therefore the dimension is countably infinite.

Spaces of Square-Integrable Functions

Chapters 10 and 12 give several examples of polynomial bases of $\mathcal{L}_w^2([a, b]; \mathbb{F})$ as well as examples of other special-function bases such as the Fourier basis $\{e^{imx}\}$ (for a constant weight function on $[-\pi, \pi]$) and the Fourier–Bessel basis $\{J_m(j_{m,n}x/a) \mid \forall n \in \mathbb{Z}^+ : J_m(j_{m,n}) = 0\}$ (for a weight function proportional to x on the interval $[0, a]$).

Sturm–Liouville Solution Spaces

We show in Section 5.3.3 that any three distinct solutions y_1, y_2, y_3 of the homogeneous Sturm–Liouville differential equation are linearly dependent. Therefore the dimension of the solution space of Eq. (5.49) is less than three. One way to establish that the dimension is equal to two is to show that the Wronskian equation (5.127) is nonzero on (a, b). Assume that y_1 satisfies Eq. (5.49) and homogeneous boundary conditions at a and b, that $y_1 \in \mathcal{C}^2([a, b]; \mathbb{R})$, and that p is nonzero on (a, b). Write the Sturm–Liouville differential equation in the form

$$py_i'' + p'y_i' + qy_i = 0 \tag{5.170}$$

for $i = 1, 2$. We proceed heuristically, trying a solution of the form

$$y_2(x) = c(x)y_1(x). \tag{5.171}$$

(This is an example of the method of variation of parameters.) From Eq. (5.170) one finds the differential equation

$$c'' + \left(2\frac{y_1'}{y_1} + \frac{p'}{p}\right)c' = 0 \Rightarrow \frac{d}{dx}\log c' = -\frac{d}{dx}\log p - 2\frac{d}{dx}\log y_1 \tag{5.172}$$

for the unknown function c. Then

$$c' = \frac{K}{py_1^2} \tag{5.173}$$

in which K is a real constant. Now we can find the Wronskian, which involves only y_1, c, and c':

$$W[y_1, cy_1](x) = c'(x)y_1^2(x) = \frac{K}{p(x)}. \tag{5.174}$$

Because p is finite, the Wronskian cannot vanish. Therefore y_1 and cy_1 are linearly independent on (a, b).

It remains for us to see whether $y_2 = cy_1$ is a solution of the differential equation Eq. (5.170). The solution of Eq. (5.173) for c is

$$c(x) = K \int_a^x \frac{p(x')}{[y_1(x')]^2} \, dx', \tag{5.175}$$

and therefore our tentative solution y_2 is

$$y_2(x) = K y_1(x) \int_a^x \frac{p(x')}{[y_1(x')]^2} \, dx'. \tag{5.176}$$

Equation (5.176) is known as **Liouville's formula**. It is straightforward to verify that Eq. (5.176) satisfies the Sturm–Liouville differential equation Eq. (5.170) (see Exercise 5.4.8). Equation (5.176) shows that a formula for a second solution y_2 exists that is valid on some open interval of $[a, b]$, even though Eq. (5.176) is formally singular at points at which y_1 vanishes.

Therefore, if a solution of a Sturm–Liouville equation exists that belongs to $\mathcal{C}^2([a, b]; \mathbb{R})$, then the dimension of the solution space is two.

5.4.3 VECTOR-SPACE ISOMORPHISMS

Definitions

In order to show that two vector spaces $\mathcal{V}$ and $\mathcal{V}'$ have the same structure, one must be able to make a *structure-preserving* correspondence between the elements of $\mathcal{V}$ and the elements of $\mathcal{V}'$. Two vector spaces $\mathcal{V}$ and $\mathcal{V}'$ are called (**vector**)-**isomorphic**, written

$$\mathcal{V}' \cong \mathcal{V}, \tag{5.177}$$

if and only if there exists a one-to-one mapping $\mathsf{M} : \overline{x} \mapsto \overline{x}' = \mathsf{M}\overline{x}$ from $\mathcal{V}$ *onto* $\mathcal{V}'$ that preserves the vector-space structure. Formally, a mapping M is called a (**vector**) **isomorphism** of $\mathcal{V}$ and $\mathcal{V}'$ if and only if M is defined on all of $\mathcal{V}$ and the following conditions are met:

1. M *is one-to-one:*

$$\forall \overline{x}, \overline{y} \in \mathcal{V} : \mathsf{M}\overline{x} = \mathsf{M}\overline{y} \Rightarrow \overline{x} = \overline{y} \tag{5.178}$$

2. M *maps $\mathcal{V}$ onto $\mathcal{V}'$:*

$$\forall \overline{x}' \in \mathcal{V}' : \exists \overline{x} \in \mathcal{V} :\ni: \ \overline{x}' = \mathsf{M}\overline{x} \tag{5.179}$$

3. M *is* **additive***:*

$$\forall \overline{x}, \overline{y} \in \mathcal{V} : \ \mathsf{M}(\overline{x} + \overline{y}) = \mathsf{M}\overline{x} + \mathsf{M}\overline{y}. \tag{5.180}$$

4. M *is* **homogeneous***:*

$$\forall \alpha \in \mathbb{F} : \forall \overline{x} \in \mathcal{V} : \ \mathsf{M}(\alpha \overline{x}) = \alpha(\mathsf{M}\overline{x}). \tag{5.181}$$

A mapping that is additive and homogeneous, but that is not required to be one-to-one or onto, is called **linear**. We discuss linear mappings in some detail in Chapter 6.

For example, let $\mathcal{V} = \mathbb{R}^{(2)}[x]$ and $\mathcal{V}' = \mathbb{R}^3$, in which $\mathbb{R}^{(m)}[x]$ is the space of polynomials of degree m or less over $\mathbb{R}$ in the unknown x. An element of $\mathbb{R}^{(2)}[x]$ can be written as

$$\overline{p} = a_0 + a_1 x + a_2 x^2 \tag{5.182}$$

in which every $a_i \in \mathbb{R}$. Define the mapping $M : \mathbb{R}^{(2)}[x] \to \mathbb{R}^3$ such that

$$
\begin{aligned}
M(1) &= \mathbf{e}_1 \\
M(x) &= \mathbf{e}_2 \\
M(x^2) &= \mathbf{e}_3.
\end{aligned}
\tag{5.183}
$$

Then

$$
M(a_0 + a_1 x + a_2 x^2) = \begin{pmatrix} a_0 \\ a_1 \\ a_2 \end{pmatrix}.
\tag{5.184}
$$

Clearly M is one-to-one, onto, additive, and homogeneous and is therefore a vector isomorphism. However, M does not preserve properties that are unrelated to the axioms for a vector isomorphism such as inner products or the possibility of dividing a quadratic polynomial by a linear polynomial.

Another example of a vector isomorphism is a **permutation of the basis elements,**

$$
\bar{e}_i \mapsto A_\pi \bar{e}_i := \bar{e}_{\pi(i)}
\tag{5.185}
$$

in which π is a permutation of $(1:n)$. The image of a vector $\bar{x} = x^i \bar{e}_i$ under A_π is

$$
\begin{aligned}
A_\pi \bar{x} &= x^i A_\pi \bar{e}_i \\
&= x^i \bar{e}_{\pi(i)} \\
&= x^{\pi^{-1}(i)} \bar{e}_i.
\end{aligned}
\tag{5.186}
$$

It is easy to show that A_π is a vector isomorphism.

For example, let $\mathcal{V} = \mathbb{R}^3$ and $\pi = (123)$. Then

$$
\begin{aligned}
A_{(123)}\mathbf{e}_1 &= \mathbf{e}_2 \\
A_{(123)}\mathbf{e}_2 &= \mathbf{e}_3 \\
A_{(123)}\mathbf{e}_3 &= \mathbf{e}_1
\end{aligned}
\tag{5.187}
$$

and therefore

$$
\begin{aligned}
A_{(123)} \begin{pmatrix} x^1 \\ x^2 \\ x^3 \end{pmatrix} &= A_{(123)}(x^1 \mathbf{e}_1 + x^2 \mathbf{e}_2 + x^3 \mathbf{e}_3) \\
&= x^1 \mathbf{e}_2 + x^2 \mathbf{e}_3 + x^3 \mathbf{e}_1 \\
&= \begin{pmatrix} x^2 \\ x^3 \\ x^1 \end{pmatrix}.
\end{aligned}
\tag{5.188}
$$

This result is consistent with Eq. (5.186), for $(123)^{-1} = (231)$.

Basis permutations are of considerable practical importance in the numerical solution of systems of linear equations by the method of Gaussian elimination (see Section 5.4.4).

Yet another example of a vector isomorphism is the mapping $C : \mathbb{F}^{n \times m} \to \mathbb{F}^{mn}$ such that

$$\forall \mathbf{A} \in \mathbb{F}^{n \times m} : C\mathbf{A} := \mathbf{x} \tag{5.189}$$

in which

$$x^{(i,j)} := a_j^i. \tag{5.190}$$

The notation $x^{(i,j)}$ indicates that the contravariant components of the vector $\mathbf{x}$ are to be indexed with ordered pairs (i, j) formed from the row and column indices that label elements of the matrix $\mathbf{A}$. What this mapping accomplishes is to stretch out the matrix $\mathbf{A}$ into a single column vector $\mathbf{x}$ with mn elements by writing the rows of $\mathbf{A}$ as column vectors, one on top of another. In the C programming language, C is the mapping by which two-dimensional arrays are mapped onto vectors for the purpose of sequential storage in memory. The mapping

$$F : \mathbf{A} \mapsto \mathbf{y} \tag{5.191}$$

in which

$$y^{(j,i)} := a_j^i \tag{5.192}$$

produces a similar result, except that the elements of $\mathbf{y}$ are obtained by writing the columns of $\mathbf{A}$ one on top of another. F is the matrix-to-vector mapping employed in the FORTRAN programming language.

Products of Vector Isomorphisms

The **product** or **composition** of two vector isomorphisms $M_1 : \mathcal{V}_{(1)} \to \mathcal{V}_{(2)}$ and $M_2 : \mathcal{V}_{(2)} \to \mathcal{V}_{(3)}$ is the mapping $M_{21} : \mathcal{V}_{(1)} \to \mathcal{V}_{(3)}$ such that for every $\overline{x}_{(1)} \in \mathcal{V}_{(1)}$,

$$M_{21}\overline{x}_{(1)} := M_2(M_1\overline{x}_{(1)}). \tag{5.193}$$

It is straightforward to check that M_{21} is a vector isomorphism and that the product of any three vector isomorphisms is associative (see Exercise 5.4.12).

It follows from Eqs. (5.178) through (5.181) that the relation of vector isomorphism is reflexive, symmetric, and transitive:

$$\mathcal{V} \cong \mathcal{V}, \tag{5.194}$$

$$\mathcal{V}' \cong \mathcal{V} \Rightarrow \mathcal{V} \cong \mathcal{V}', \tag{5.195}$$

$$\mathcal{V}'' \cong \mathcal{V}' \text{ and } \mathcal{V}' \cong \mathcal{V} \Rightarrow \mathcal{V}'' \cong \mathcal{V}. \tag{5.196}$$

Therefore vector isomorphism is an equivalence relation.

Characterization of a Vector Space by its Dimension

A finite-dimensional vector space is completely characterized, up to a vector isomorphism, by its field $\mathbb{F}$ and its dimension n. We demonstrate this proposition in two steps: First, we show that the dimension of a vector space that is isomorphic to $\mathcal{V}$ is equal to the dimension of $\mathcal{V}$. In the second step, we show that any two vector spaces over the same field that have the same dimension are isomorphic.

Let $\mathsf{M} : \mathcal{V} \to \mathcal{V}'$ be a vector isomorphism. From Eq. (5.178) it follows that

$$\mathsf{M}(\overline{x} - \overline{y}) = \overline{0}' \in \mathcal{V}' \Leftrightarrow \overline{x} = \overline{y}. \tag{5.197}$$

Then the only vector in $\mathcal{V}$ that M carries onto the zero vector $\overline{0}'$ of $\mathcal{V}'$ is the zero vector $\overline{0}$ of $\mathcal{V}$:

$$\forall \overline{x} \in \mathcal{V}: \ \mathsf{M}\overline{x} = \overline{0}' \Rightarrow \overline{x} = \overline{0}. \tag{5.198}$$

From this equation it follows that a list $\{\overline{x}_1, \ldots, \overline{x}_n\}$ is linearly independent in $\mathcal{V}$ if and only if the list $\{\mathsf{M}\overline{x}_1, \ldots, \mathsf{M}\overline{x}_n\}$ is linearly independent in $\mathcal{V}'$. (The same statement is true if "independent" is replaced by "dependent" everywhere.) Then $\{\mathsf{M}\overline{x}_1, \ldots, \mathsf{M}\overline{x}_n\}$ is a basis of $\mathcal{V}'$ if and only if $\{\overline{x}_1, \ldots, \overline{x}_n\}$ is a basis of $\mathcal{V}$. We have shown that

$$\mathcal{V}' \cong \mathcal{V} \Rightarrow \dim[\mathcal{V}'] = \dim[\mathcal{V}]. \tag{5.199}$$

Conversely, if $\dim[\mathcal{V}'] = \dim[\mathcal{V}] = n$, then the mapping

$$\forall i \in (1:n): \ \mathsf{M}\overline{e}_i = \overline{e}'_i \tag{5.200}$$

in which $\{\overline{e}_i\}$ and $\{\overline{e}'_i\}$ are bases of $\mathcal{V}$ and $\mathcal{V}'$, respectively, is an isomorphism of $\mathcal{V}$ and $\mathcal{V}'$. Therefore *two finite-dimensional vector spaces over the same field $\mathbb{F}$ are isomorphic if and only if*

$$\dim[\mathcal{V}] = \dim[\mathcal{V}']. \tag{5.201}$$

Equivalently, every finite-dimensional vector space over a field $\mathbb{F}$ is uniquely characterized (up to isomorphism) by its dimension.

In particular, *every n-dimensional vector space over a field $\mathbb{F}$ is isomorphic to the space of column vectors $\mathbb{F}^n$ under the mapping* $\mathsf{M} : \mathcal{V} \to \mathbb{F}^n$ *defined by*

$$\forall i \in (1:n): \ \mathbf{e}_i = \mathsf{M}\overline{e}_i, \tag{5.202}$$

where $\{\overline{e}_i\}$ is a basis of $\mathcal{V}$ and $\{\mathbf{e}_i\}$ is the canonical basis of $\mathbb{F}^n$ defined in Eq. (5.122).

It is important to recognize that vector isomorphisms such as the one defined in Eq. (5.202) are not required to preserve important properties of the bases $\{\overline{e}_i\}$ or $\{\mathbf{e}_i\}$, such as orthogonality, that do not pertain to Eqs. (5.178) through (5.181). For example, $\{\mathbf{e}_1, \mathbf{e}_2, \mathbf{e}_3\}$ might be the list $\{\hat{\imath}, \hat{\jmath}, \hat{k}\}$ of mutually perpendicular unit vectors in $\mathbb{R}^3$, and $\{\overline{e}_1, \overline{e}_2, \overline{e}_3\}$ might be *any* list of three linearly independent vectors in any real vector space.

Any list of three noncoplanar vectors $\mathbf{a}$, $\mathbf{b}$, $\mathbf{c}$ is a basis of $\mathbb{R}^3$. If $\mathbf{a}$, $\mathbf{b}$, and $\mathbf{c}$ are not mutually perpendicular, then the contravariant components x^i (rel$\{\mathbf{a}, \mathbf{b}, \mathbf{c}\}$) are called **oblique coordinates**. Oblique coordinates arise frequently in areas such as general relativity in which differential geometry is applied to physics.

If $M : V \to V'$ is a vector isomorphism and if $V' = V$, then M is called a (**vector**) **automorphism** of V. Let aut $[V]$ be the set of all automorphisms of V. We already know that aut $[V]$ is closed under composition and that the product of three automorphisms is associative. The identity automorphism **1** maps every vector $x \in V$ onto itself. Finally, because every $A \in$ aut $[V]$ is an isomorphism, it has an inverse. Therefore we can call aut $[V]$ the **group of vector automorphisms of** V.

We see in Chapter 6 that if V is a finite-dimensional complex vector space, then aut $[V]$ has a faithful matrix representation that is called the **full linear group** of V and is denoted $GL(n, \mathbb{C})$. Several of the subgroups of $GL(n, \mathbb{C})$ have important applications in physics such as the special linear groups $SL(n, \mathbb{C})$ and $SL(n, \mathbb{R})$, the unitary group $U(n)$, the special unitary group $SU(n)$, the full orthogonal group $O(n)$, the special orthogonal group $SO(n)$, and (if n is even) the symplectic group $Sp(n)$.

Physical Examples of Vector Isomorphisms

The isomorphism of two vector spaces is a necessary, but not a sufficient, condition for the vector spaces to be physically equivalent. For example, the complex numbers, $\mathbb{C}$, are a two-dimensional real vector space (see Eq. [5.163]). Therefore

$$\mathbb{C} \cong \mathbb{R}^2. \tag{5.203}$$

This vector isomorphism permits one to visualize complex numbers as vectors in a two-dimensional plane. Thus one can easily draw pictures that show phase relationships of time-varying quantities in application areas ranging from alternating-current circuits to optics.

Quantum mechanics provides a famous example of isomorphic, physically equivalent vector spaces: If one formulates quantum mechanics in terms of $\ell^2(\mathbb{Z}^+, \mathbb{C})$ as the underlying space, then one arrives naturally at Heisenberg's matrix mechanics. If the underlying space is taken to be $\mathcal{L}^2([a, b]; \mathbb{C})$, then one arrives at Schrödinger's wave mechanics. A necessary condition for the correctness of the physical statement that matrix mechanics and wave mechanics are equivalent is the mathematical statement that $\ell^2(\mathbb{Z}^+, \mathbb{C})$ is isomorphic to $\mathcal{L}^2([a, b]; \mathbb{C})$.

Elementary-particle physics provides an example of two vector spaces that are isomorphic but not physically equivalent. The vector spaces of the internal angular momentum (spin) of an electron and of the isotopic spin (isospin) of a nucleon are both isomorphic to $\mathbb{C}^2$. In the former case, every element of $\mathbb{C}^2$ represents a physically realizable state of the electron's spin. In the latter case, only the elements $e^{i\theta} \mathbf{e}_1$ and $e^{i\theta} \mathbf{e}_2$ of $\mathbb{C}^2$ represent physically realizable states corresponding to the proton and neutron, respectively. (The vectors $\mathbf{e}_1$ and $\mathbf{e}_2$ are defined in Eq. [5.122].)

5.4.4 GAUSSIAN ELIMINATION AND LINEAR DEPENDENCE

Gaussian Elimination Applied to Rows

The concept of a vector isomorphism defined in Section 5.4.3 gives one a tool for developing useful algorithms. Consider the problem of determining whether a list

$$X = \{\overline{x}_1, \ldots, \overline{x}_m\} \tag{5.204}$$

is linearly independent or linearly dependent in a vector space $\mathcal{V}$. We solve this problem by constructing consecutive vector isomorphisms that finally permit us to see at a glance whether $\dim[\mathrm{span}[X]]$ is equal to or less than m. If equality holds, then X is linearly independent; if not, then X is linearly dependent.

We begin by introducing a basis in $\mathcal{V}$:

$$\bar{x}_i = x_i^j \, \bar{e}_j. \tag{5.205}$$

If one writes the contravariant components of all the vectors $\bar{x}_i$ in an array, one obtains the matrix

$$\mathbf{X} := \begin{pmatrix} x_1^1 & x_2^1 & \cdots & x_m^1 \\ x_1^2 & x_2^2 & \cdots & x_m^2 \\ \vdots & \vdots & \ddots & \vdots \\ x_1^n & x_2^n & \cdots & x_m^n \end{pmatrix}. \tag{5.206}$$

If necessary, one must permute the basis elements (as described in Section 5.4.3), or the vectors $\bar{x}_i$, or both, in order to make

$$x_1^1 \neq 0 \tag{5.207}$$

after relabeling. Permuting the basis elements or vectors, which evidently permutes the columns or rows of $\mathbf{X}$, respectively, is called **pivoting**. The elements x_i^i (no Σ) are called **pivots** (or **pivotal elements**).

If one cannot make the upper-left element of $\mathbf{X}$ nonzero by any row or column permutation, then all of the elements of $\mathbf{X}$ are zero and X is linearly dependent. Assuming the contrary, we define the vector-space isomorphism

$$\bar{e}_k \mapsto \bar{e}_k^{(2)} = \mathsf{M}^{(1)} \bar{e}_k := \begin{cases} \bar{e}_1 + \sum_{j=2}^n m_1^j \, \bar{e}_j, & \text{if } k = 1; \\ \bar{e}_k, & \text{if } k \in (2:n), \end{cases} \tag{5.208}$$

where

$$m_1^i := -\frac{x_1^i}{x_1^1}. \tag{5.209}$$

$\mathsf{M}^{(1)}$ does not change the dimension of $\mathrm{span}[X]$. (Why?) $\mathsf{M}^{(1)}$ is an example of a **Gauss transformation**.

The effect of $\mathsf{M}^{(1)}$ is to add m_1^i times the first row of $\mathbf{X}$ to the the i-th row. The inverse mapping is therefore

$$\bar{e}_k^{(2)} \mapsto \bar{e}_k = (\mathsf{M}^{(1)})^{-1} \bar{e}_k^{(2)} := \begin{cases} \bar{e}_1^{(2)} - \sum_{j=2}^n m_1^j \, \bar{e}_j^{(2)}, & \text{if } k = 1; \\ \bar{e}_k^{(2)}, & \text{if } k \in (2:n). \end{cases} \tag{5.210}$$

Under $M^{(1)}$,

$$\bar{x}_i \mapsto \bar{x}_i^{(2)} = x_i^j \bar{e}_j^{(2)}$$

$$= x_i^1 \left(\bar{e}_1 + \sum_{j=2}^{n} m_1^j \bar{e}_j \right) + \sum_{j=2}^{n} x_i^j \bar{e}_j$$

$$= x_i^1 \bar{e}_1 + \sum_{j=2}^{n} \left(x_i^j + m_1^j x_i^1 \right) \bar{e}_j \tag{5.211}$$

$$= x_i^{(2)j} \bar{e}_j.$$

Note that $\bar{x}_i^{(2)}$ is defined to have the same contravariant components $(\text{rel}\{\bar{e}_j^{(2)}\})$ as $\bar{x}_i$ has $(\text{rel}\{\bar{e}_j\})$.

If $i = 1$, then $x_1^j + m_1^j x_1^1 = 0$ for $j \in (2 : n)$. It follows that relative to the basis $\{\bar{e}_1, \ldots, \bar{e}_n\}$,

$$\bar{x}_1^{(2)} = \begin{pmatrix} x_1^{(2)1} \\ 0 \\ \vdots \\ 0 \end{pmatrix}, \quad \bar{x}_2^{(2)} = \begin{pmatrix} x_2^{(2)1} \\ x_2^{(2)2} \\ \vdots \\ x_2^{(2)n} \end{pmatrix}, \quad \ldots, \quad \bar{x}_m^{(2)} = \begin{pmatrix} x_m^{(2)1} \\ x_m^{(2)2} \\ \vdots \\ x_m^{(2)n} \end{pmatrix}. \tag{5.212}$$

The effect of the Gauss transformation $M^{(1)}$ is to zero out the elements of the first column of $\mathbf{X}^{(2)}$ below $x_1^1 = x_1^{(2)1}$.

The next step is to make another Gauss transformation $M^{(2)}$ that leaves the first column and row of $\mathbf{X}^{(2)}$ alone and acts only on the block below the first row and to the right of the first column. If $x_2^{(2)2} = 0$, then one must permute the rows or columns (or both) of the block below the first row and to the right of the first column until one brings a nonzero element to the second row, second column position. If this is impossible, then $x_i^{(2)2} = \cdots = x_i^{(2)n} = 0$ for $i \in (1 : m)$, which means that $\bar{x}_i^{(2)} = x_i^{(2)1} \bar{e}_1$. Therefore span $[X]$ is one-dimensional, which implies that X is linearly dependent if $m > 1$. If, on the contrary, a nonzero pivot $x_2^{(2)2}$ can be found, then one can proceed to zero out the elements of $\bar{x}_2^{(3)}$ in the third through n-th rows with a second Gauss transformation, and so on.

If nonzero pivots can be found at every step, then by a succession of Gauss transformations (possibly accompanied by permutations of the labels of the vectors or the basis elements) one can bring $\mathbf{X}$ to the upper-triangular form

$$\mathbf{X}^{(m)} = M^{(m-1)} \cdots M^{(1)} \mathbf{X}$$

$$= \begin{pmatrix} x_1^1 & x_2^{(m)1} & x_3^{(m)1} & \cdots & x_m^{(m)1} \\ 0 & x_2^{(m)2} & x_3^{(m)2} & \cdots & x_m^{(m)2} \\ 0 & 0 & x_3^{(m)3} & \cdots & x_m^{(m)3} \\ \vdots & \vdots & \vdots & \ddots & \vdots \\ 0 & 0 & 0 & \cdots & x_m^{(m)m} \end{pmatrix} \tag{5.213}$$

in which every diagonal component $x_i^{(m)i}$ is nonzero. If this is the case, and $m \leq n$, then the columns of $\mathbf{X}^{(m)}$ give the contravariant components of a linearly independent, m-element list $\{\overline{x}_1^{(m)}, \ldots, \overline{x}_m^{(m)}\}$ that is the image of X under the vector isomorphism $\mathsf{M}^{(m-1)} \cdots \mathsf{M}^{(1)}$. It follows from Exercise 5.4.16 that the list X is linearly independent. Therefore

$$\dim[\text{span}[X]] = \begin{cases} m, & \text{if } m \leq n; \\ n, & \text{otherwise,} \end{cases} \tag{5.214}$$

provided that a nonzero pivot can be found at every step of Gaussian elimination.

If a nonzero pivot *cannot* be found at step l, then the components $x_l^{(l)i}$ vanish for all $i \geq l$. It follows that $\overline{x}_l^{(l)}$ is linearly dependent on its predecessors $\overline{x}_1^{(l)}, \ldots, \overline{x}_{l-1}^{(l)}$. Therefore the list X is linearly dependent in this case. The dimension of span $[X]$ is equal to the number of vectors $\overline{x}_i^{(n)}$ in which the last nonzero component (i.e., the component $x_i^{(n)k}$ such that $x_i^{(n)l} = 0$ for every $l > k$) is a pivot. For example, if

$$\mathbf{x}_1 = \begin{pmatrix} 1 \\ 2 \\ 3 \end{pmatrix}, \quad \mathbf{x}_2 = \begin{pmatrix} 3 \\ 6 \\ 9 \end{pmatrix}, \quad \mathbf{x}_3 = \begin{pmatrix} 5 \\ 17 \\ 36 \end{pmatrix}, \tag{5.215}$$

then

$$\mathbf{X}^{(2)} = \begin{pmatrix} 1 & 3 & 5 \\ 0 & 0 & 7 \\ 0 & 0 & 21 \end{pmatrix} \tag{5.216}$$

and

$$\mathbf{X}^{(3)} = \begin{pmatrix} 1 & 3 & 5 \\ 0 & 0 & 7 \\ 0 & 0 & 0 \end{pmatrix}. \tag{5.217}$$

The nonzero pivots are 1 and 7. Because the first and second columns of $\mathbf{X}^{(2)}$ are linearly dependent, the dimension of span$[\mathbf{x}_1, \mathbf{x}_2, \mathbf{x}_3]$ is 2.

Gaussian Elimination Applied to Columns

In order to find a basis, and not merely the dimension, of span$[X]$, one must form linear combinations of the *columns* of $\mathbf{X}$ such that the new columns are linearly independent. Gaussian elimination, applied to the columns of $\mathbf{X}$, is a convenient procedure for computing a basis of the vector subspace spanned by the list X. For example, consider the linear combinations

$$\mathbf{y}_2 := \mathbf{x}_2 - 3\mathbf{x}_1 = \begin{pmatrix} 0 \\ 0 \\ 0 \end{pmatrix}, \quad \mathbf{y}_3 := \mathbf{x}_3 - 5\mathbf{x}_1 = \begin{pmatrix} 0 \\ 7 \\ 21 \end{pmatrix} \tag{5.218}$$

of the vectors in the list given in Eq. (5.215). In the new vectors $\mathbf{y}_2$ and $\mathbf{y}_3$, the first component is zero. The list $\{\mathbf{x}_1, \mathbf{y}_3\}$ is linearly independent and has the same span as $\{\mathbf{x}_1, \mathbf{x}_2, \mathbf{x}_3\}$ because $\mathbf{x}_1$, $\mathbf{x}_2$, and $\mathbf{x}_3$ are linear combinations of $\mathbf{x}_1$ and $\mathbf{y}_3$.

If one is given a general list of vectors X, the first step in computing a basis of span$[X]$ is to choose linear combinations $\mathbf{y}_2^{(2)}, \ldots, \mathbf{y}_m^{(2)}$ such that

$$y_2^{(2)1} = \cdots = y_m^{(2)1} = 0. \tag{5.219}$$

One continues by finding (if possible) a vector $\mathbf{y}_i^{(2)}$ such that $y_i^{(2)2} \neq 0$, permuting $\mathbf{y}_i^{(2)}$ to be the second vector in the list, and using the component $y_i^{(2)2}$ of $\mathbf{y}_i^{(2)}$ as a pivot to obtain a new list in which the first two contravariant components of the third, fourth, and so on through m-th vectors are zero. The nonzero vectors remaining after m steps (or after l steps, if only l nonzero pivots can be found) are a basis of span$[X]$.

Computation of Contravariant Components

Gaussian elimination provides a method for computing the contravariant components of vectors in $\mathbb{F}^n$ relative to an arbitrary basis $\{\mathbf{b}_1, \ldots, \mathbf{b}_n\}$. Let

$$\mathbf{x} = x^i \mathbf{e}_i = x'^j \mathbf{b}_j; \tag{5.220}$$

then one must solve the system of linear equations

$$x'^1 b_1^1 + \cdots + x'^n b_n^1 = x^1$$
$$\vdots \tag{5.221}$$
$$x'^1 b_1^n + \cdots + x'^n b_n^n = x^n$$

in order to obtain the contravariant components $x'^1, \ldots, x'^n$ of $\mathbf{x}$ (rel$\{\mathbf{b}_i\}$). To obtain the solution, one applies Gaussian elimination to the list $\{\mathbf{b}_1, \ldots, \mathbf{b}_n, \mathbf{x}\}$ in which the contravariant components are given relative to the canonical basis $\{\mathbf{e}_i\}$.

For example, in order to find the contravariant components of

$$\mathbf{x} = \begin{pmatrix} x \\ y \\ z \end{pmatrix} = x' \mathbf{b}_1 + y' \mathbf{b}_2 + z' \mathbf{b}_3 \tag{5.222}$$

relative to the basis

$$\left\{ \mathbf{b}_1 = \begin{pmatrix} 1 \\ 1 \\ 0 \end{pmatrix}, \quad \mathbf{b}_2 = \begin{pmatrix} 1 \\ 0 \\ 1 \end{pmatrix}, \quad \mathbf{b}_3 = \begin{pmatrix} 1 \\ 2 \\ 1 \end{pmatrix} \right\} \tag{5.223}$$

of $\mathbb{R}^3$, one must solve the linear system

$$\begin{aligned} x' + y' + z' &= x \\ x' + 0 + 2z' &= y \\ 0 + y' + z' &= z \end{aligned} \tag{5.224}$$

for x', y' and z'. The first step of Gaussian elimination gives

$$\begin{aligned}
x' + y' + z' &= x \\
-y' + z' &= y - x \\
y' + z' &= z,
\end{aligned}$$

(5.225)

and the second step gives

$$\begin{aligned}
x' + y' + z' &= x \\
-y' + 2z' &= y - x \\
2z' &= y - x + z.
\end{aligned}$$

(5.226)

Back-substituting

$$z' = \tfrac{1}{2}(y - x + z)$$

(5.227)

into the second equation of Eq. (5.227), one gets

$$y' = \tfrac{1}{2}(x - y + z).$$

(5.228)

Then the first equation of Eq. (5.227) yields

$$x' = x - z.$$

(5.229)

Every system of n linear equations in n unknowns with a nonzero right-hand side defines a set of contravariant components relative to a basis, the contravariant components of which relative to the canonical basis are given by the coefficients of the unknowns.

We discuss the solution of systems of linear equations in greater depth in Sections 6.6 and 9.5.

5.4.5 EXERCISES FOR SECTION 5.4

5.4.1. Show that the list $\{1, \sigma_1, \sigma_2, \sigma_3\}$ defined in Eq. (5.149) is a basis of the vector space $\mathbb{C}^{2\times 2}$.

5.4.2. In the vector space $\mathbb{R}^2$ (two-dimensional Euclidean space), choose the basis

$$\mathbf{b}_1 = \begin{pmatrix} 1 \\ 0 \end{pmatrix}, \quad \mathbf{b}_2 = \begin{pmatrix} 1 \\ 1 \end{pmatrix}.$$

(5.230)

For an arbitrary vector

$$\mathbf{r} = \begin{pmatrix} x \\ y \end{pmatrix}$$

(5.231)

in $\mathbb{R}^2$, find the contravariant components of $\mathbf{r}$ relative to the basis $\{\mathbf{b}_1, \mathbf{b}_2\}$.

5.4.3.   Let

$$B := \{\overline{e}_1, \ldots, \overline{e}_n\} \tag{5.232}$$

be a linearly independent set that spans a vector space $\mathcal{V}$. Prove that every $\overline{x} \in \mathcal{V}$ can be expressed *uniquely* in the form

$$\overline{x} = \sum_{i=1}^{n} x^i \, \overline{e}_i. \tag{5.233}$$

5.4.4.   You have already shown in Exercises 5.2.10 and 5.2.11 that a finite field $\mathbb{GF}(q)$ is a vector space $\mathcal{V}$ over its prime field Π. Prove that $\dim[\mathcal{V}]$ is a finite positive integer n and therefore that $\mathcal{V}$ is isomorphic to Π^n. (This completes the proof, begun in Exercise 5.2.11, that $q = p^n$.)

5.4.5.   Find the dimension of the vector space of quaternions over the field of real numbers $\mathbb{R}$. (See Exercise 5.2.13)

5.4.6.   Prove that the vector space of quaternions is isomorphic to the *real* vector space spanned by the list $\{1, i\sigma_1, i\sigma_2, i\sigma_3\}$, in which 1 is the 2×2 unit matrix and $\{\sigma_i\}$ are the Pauli spin matrices (see Exercise 5.4.1 and Eq. (5.149)).

5.4.7.   Let M be a mapping of a vector space $\mathcal{V}$ onto a vector space $\mathcal{V}'$ (both over the same field $\mathbb{F}$) that satisfies Eqs. (5.178) through (5.181). Prove that M is an isomorphism of $\mathcal{V}$ and $\mathcal{V}'$ if and only if the following statement is true: *Every proper subspace $\mathcal{W}$ of $\mathcal{V}$ is isomorphic to some proper subspace $\mathcal{W}'$ of $\mathcal{V}'$, and every proper subspace $\mathcal{W}'$ of $\mathcal{V}'$ is isomorphic to some proper subspace $\mathcal{W}$ of $\mathcal{V}$.*

5.4.8.   Verify that Eq. (5.176) satisfies the Sturm–Liouville differential equation, Eq. (5.170).

5.4.9.   The equation of motion of the undamped simple harmonic oscillator, Eq. (5.1), is already in Sturm–Liouville form. Verify that if one takes $y_1(t) = \sin \omega t$, then Eq. (5.176) gives a second, linearly independent solution of Eq. (5.1).

Hint: You may invoke the result of Exercise 5.3.8 to establish the linear independence of your second solution.

*5.4.10.   Prove that every vector space $\mathcal{V}$ has a Hamel basis.

Hint: Let S be the set of all linearly independent subsets of $\mathcal{V}$. Partially order S by inclusion. Then apply Zorn's lemma.

5.4.11.   Let an additive, homogeneous mapping $A : \mathbb{R}^2 \to \mathbb{R}^2$ map the basis vectors as follows:

$$\mathbf{e}_1 \mapsto \mathbf{e}_1' = \cos\theta \mathbf{e}_1 + \sin\theta \mathbf{e}_2,$$
$$\mathbf{e}_2 \mapsto \mathbf{e}_2' = -\sin\theta \mathbf{e}_1 + \cos\theta \mathbf{e}_2 \tag{5.234}$$

in which $\theta \in [0, 2\pi)$. Prove that A is an automorphism of $\mathbb{R}^2$. Sketch the initial and final basis vectors for $\theta = \pi/3$. (A is an example of a **rotation**.)

Hint: Obtain the new contravariant coordinates x'^1, x'^2 such that

$$\mathbf{x} \mapsto \mathbf{x}' := x^1 \mathbf{e}_1' + x^2 \mathbf{e}_2' = x'^1 \mathbf{e}_1 + x'^2 \mathbf{e}_2. \tag{5.235}$$

5.4.12. This problem concerns successive vector isomorphisms.

(a) Prove that the mapping M_{12} defined in Eq. (5.193) is a vector isomorphism.

(b) Prove that the product of any three vector isomorphisms is associative.

5.4.13. Verify that if $\mathbb{F} = \mathbb{R}$ or $\mathbb{C}$, then

$$\dim[\mathbb{F}^n] = n. \tag{5.236}$$

5.4.14 Test for linear dependence and find the dimension of the span of each of the following lists:

(a)

$$\mathbf{x}_1 = \begin{pmatrix} 1 \\ 3 \\ 5 \end{pmatrix}, \quad \mathbf{x}_2 = \begin{pmatrix} 1 \\ 2 \\ 3 \end{pmatrix}, \quad \mathbf{x}_3 = \begin{pmatrix} 0 \\ -1 \\ -2 \end{pmatrix} \tag{5.237}$$

(b)

$$\mathbf{x}_1 = \begin{pmatrix} 1 \\ 1 \\ -1 \end{pmatrix}, \quad \mathbf{x}_2 = \begin{pmatrix} -1 \\ 1 \\ 1 \end{pmatrix}, \quad \mathbf{x}_3 = \begin{pmatrix} 1 \\ 1 \\ 1 \end{pmatrix}. \tag{5.238}$$

5.4.15 Find bases of the spans of the lists in the preceding exercise.

5.4.16 Prove that if $M : \mathcal{V}_{(1)} \rightarrow \mathcal{V}_{(2)}$ is a vector isomorphism, then the list $X = \{\overline{x}_1, \ldots, \overline{x}_m\}$ is linearly independent in $\mathcal{V}_{(1)}$ if and only if the list $MX = \{M\overline{x}_1, \ldots, M\overline{x}_m\}$ is linearly independent in $\mathcal{V}_{(2)}$.

5.5 COMPLEMENTARY SUBSPACES

5.5.1 VECTOR COMPLEMENTS AND DIRECT SUMS

Until now we have regarded an affine subspace $\overline{x} + \mathcal{W}$ of a vector space $\mathcal{V}$ simply as a subset of $\mathcal{V}$ that is derived from a vector subspace $\mathcal{W}$ by a translation through $\overline{x}$. However, one knows from Chapter 4 that $\overline{x} + \mathcal{W}$ is also a coset of the normal, additive subgroup $\mathcal{W}$. Therefore the set $\mathcal{V}/\mathcal{W}$ of distinct cosets $\overline{x} + \mathcal{W}$ of $\mathcal{W}$ is a group in its own right. $\mathcal{V}/\mathcal{W}$ is called the **quotient space** of $\mathcal{W}$ in $\mathcal{V}$.

We see in this section that for each vector subspace $\mathcal{W}$, the quotient space $\mathcal{V}/\mathcal{W}$ induces a family of so-called complementary subspaces to $\mathcal{W}$ in $\mathcal{V}$. This turns out to be the natural abstract setting for several problems of practical importance, such as how to construct a solution of the inhomogeneous version of Eq. (5.49) subject to prescribed boundary conditions. We discuss this application in discussion of Green functions.

5.5.2 DEFINITION OF COMPLEMENTARY SUBSPACES

The quotient space V/W is a vector space over the same field $\mathbb{F}$ over which V is a vector space (see Exercise 5.5.1). The dimension of the quotient space is called the **codimension** of W in V and is written

$$\text{codim}_V[W] := \dim[V/W]. \tag{5.239}$$

For the applications studied in this book, $\text{codim}_V[W]$ is finite. (We do not assume, unless we say otherwise, that $\dim[W]$ is finite.) Let

$$m' := \text{codim}_V[W], \tag{5.240}$$

and let

$$T := \{\overline{f}_1 + W, \ldots, \overline{f}_{m'} + W\} \tag{5.241}$$

be a basis of V/W. The list $\{\overline{f}_1, \ldots, \overline{f}_{m'}\}$ is linearly independent (otherwise T would be linearly dependent). Because every $\overline{f}_j \in V \backslash W$ (why?), one sees that $\overline{f}_j$ is linearly independent of any list of vectors drawn from W. Define the vector subspace

$$W' := \text{span}[\overline{f}_1, \ldots, \overline{f}_{m'}]. \tag{5.242}$$

W and W' are called **complementary subspaces**. We shall also call W' a **vector complement** of W. (One also hears W' called a **supplement** of W.)

The mapping $\overline{f}_i + W \mapsto \overline{f}_i$ is one-to-one and onto. Therefore V/W is isomorphic to W':

$$V/W \cong W'. \tag{5.243}$$

Because no nonzero vector of W' can belong to W,

$$W \cap W' = 0. \tag{5.244}$$

(Recall that 0 is $\{\overline{0}\}$, not the empty set!)

One should take care not to confuse a vector complement W' of a vector subspace W with the set-theoretical complement $V \backslash W$. In Fig. 5.6, for example, the subspace W_2 consists

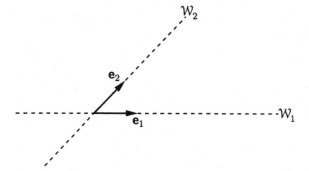

FIGURE 5.6

W_1 and W_2 are complementary subspaces in $\mathbb{R}^2$.

of the scalar multiples of e_2 and therefore corresponds to the coordinate axis defined by e_2. Because there is a one-to-one correspondence between vectors and points of the plane, the set-theoretical complement $\mathcal{V} \backslash \mathcal{W}_1$ contains all the points of the plane that do not lie on the coordinate axis defined by e_1. That is, $\mathcal{V} \backslash \mathcal{W}_1$ contains all the vectors that do not belong to the subspace $\mathcal{W}_1$. The set of nonzero vectors that belong to $\mathcal{W}_2 := \text{span}[e_2]$ is a proper subset of $\mathcal{V} \backslash \mathcal{W}_1$.

Equation (5.255) shows that two vector subspaces $\mathcal{W}$ and $\mathcal{W}'$ of a finite-dimensional vector space $\mathcal{V}$ are complementary if $\mathcal{W}'$ has the *dimensions* of $\mathcal{V}$ that are missing from $\mathcal{W}$, and the set-theoretical complement $\mathcal{V} \backslash \mathcal{W}$ contains the *elements* of $\mathcal{V}$ that are missing from $\mathcal{W}$.

5.5.3 DIMENSIONS OF COMPLEMENTARY SUBSPACES

Given a vector subspace $\mathcal{W}$ and a vector complement $\mathcal{W}'$ of $\mathcal{W}$, there is a unique decomposition $\overline{x} = \overline{w} + \overline{w}'$ of every vector $\overline{x} \in \mathcal{V}$ such that $\overline{w} \in \mathcal{W}$ and $\overline{w}' \in \mathcal{W}'$. We begin the proof of this proposition by noting that for every $\overline{x} \in \mathcal{V}$, $\overline{x} + \mathcal{W}$ has a *unique* coordinate expansion (rel T), by Eq. (5.241):

$$\overline{x} + \mathcal{W} = \sum_{i=1}^{m'} \xi^i (\overline{f}_i + \mathcal{W}) = \left(\sum_{i=1}^{m'} \xi^i \overline{f}_i \right) + \mathcal{W}. \tag{5.245}$$

If $\overline{x}$ does not belong to $\mathcal{W}$, then some ξ^i is nonzero. Consider the vector

$$\overline{w} = \overline{x} - \overline{w}' \tag{5.246}$$

in which

$$\overline{w}' = \sum_{i=1}^{m'} \xi^i \overline{f}_i. \tag{5.247}$$

Clearly $\overline{w}' \in \mathcal{W}'$. Because $\overline{w} \in \mathcal{W}$ (Exercise 5.5.3), one sees that every $\overline{x} \in \mathcal{V}$ has the decomposition

$$\overline{x} = \overline{w} + \overline{w}' \tag{5.248}$$

in which $\overline{w} \in \mathcal{W}$ and $\overline{w}' \in \mathcal{W}'$. This decomposition is unique, given $\mathcal{W}$ and $\mathcal{W}'$ (see Exercise 5.5.3). The vector $\overline{w}$ is called the **parallel projection** of $\overline{x}$ on the subspace $\mathcal{W}$.

We show now that complementarity of vector spaces is symmetric. That is, if $\mathcal{W}'$ is complementary to $\mathcal{W}$, then $\mathcal{W}$ is complementary to $\mathcal{W}'$. For,

$$\mathcal{V}/\mathcal{W}' = \{ \overline{x} + \mathcal{W}' \mid \overline{x} \in \mathcal{V} \}. \tag{5.249}$$

By Eq. (5.248), $\overline{x} = \overline{w} + \overline{w}'$. Then $\overline{x} + \mathcal{W}' = \overline{w} + \mathcal{W}'$. Why? Therefore

$$\mathcal{V}/\mathcal{W}' = \{ \overline{w} + \mathcal{W}' \mid \overline{w} \in \mathcal{W} \} \cong \mathcal{W}, \tag{5.250}$$

as claimed.

Every proper subspace $\mathcal{W}$ of $\mathcal{V}$ except for $\mathcal{O}$ has uncountably many complementary subspaces $\mathcal{W}'$. To see this, let

$$\overline{g}_i = \overline{f}_i + \overline{w}_i \quad \text{in which} \quad i \in (1:m') \tag{5.251}$$

and every $\overline{w}_i \in \mathcal{W}$. Then the basis T of $\mathcal{V}/\mathcal{W}$ is unchanged, but

$$\mathcal{U}' := \mathrm{span}[\overline{g}_1, \ldots, \overline{g}_{m'}] \tag{5.252}$$

is different from $\mathcal{W}'$ if any $\overline{w}_i$ is nonzero. For, if $\overline{w}_1 \neq \overline{0}$ (for example) and if $\overline{g}_1 \in \mathcal{W}'$, then $\overline{g}_1 - \overline{f}_1 = \overline{w}_1 \in \mathcal{W}'$. By Eq. (5.244), $\overline{w}_1 \in \mathcal{W}$ and $\overline{w}_1 \in \mathcal{W}'$ imply $\overline{w}_1 = \overline{0}$, a contradiction.

If $\dim[\mathcal{W}]$ is finite, let $\{\overline{e}_1, \ldots, \overline{e}_m\}$ be a basis of $\mathcal{W}$. Because the decomposition Eq. (5.248) is unique,

$$\{\overline{e}_1, \ldots, \overline{e}_m, \overline{f}_1, \ldots, \overline{f}_{m'}\} \tag{5.253}$$

is a basis of $\mathcal{V}$. Then

$$m + m' = \dim[\mathcal{V}]. \tag{5.254}$$

We have shown that if $\mathcal{W}$ and $\mathcal{W}'$ are complementary subspaces of a finite-dimensional vector space $\mathcal{V}$, then

$$\dim[\mathcal{W}] + \mathrm{codim}_{\mathcal{V}}[\mathcal{W}] = \dim[\mathcal{W}] + \dim[\mathcal{W}'] = \dim[\mathcal{V}]. \tag{5.255}$$

5.5.4 DIRECT SUMS OF VECTOR SPACES

A vector space $\mathcal{V}$ is called the **direct sum** of two vector spaces $\mathcal{W}$ and $\mathcal{W}'$, written

$$\mathcal{V} = \mathcal{W} \oplus \mathcal{W}', \tag{5.256}$$

if and only if every vector $\overline{x} \in \mathcal{V}$ can be expressed uniquely as a sum

$$\overline{x} = \overline{y} + \overline{y}' \tag{5.257}$$

of vectors

$$\overline{y} \in \mathcal{W} \quad \text{and} \quad \overline{y}' \in \mathcal{W}' \tag{5.258}$$

and if the only element that belongs to both $\mathcal{W}$ and $\mathcal{W}'$ is the additive identity element $\overline{0}$. The vector spaces $\mathcal{W}$ and $\mathcal{W}'$ are called **direct summands** of $\mathcal{V}$. The (Abelian) additive group of a direct sum $\mathcal{W} \oplus \mathcal{W}'$ is the group direct product of the additive groups of $\mathcal{W}$ and $\mathcal{W}'$ (see Section 4.3.3).

Evidently complementary subspaces are direct summands of $\mathcal{V}$. We show subsequently that, conversely, direct summands $\mathcal{W}$ and $\mathcal{W}'$ of $\mathcal{V}$ are complementary subspaces.

The decomposition of a vector space of functions that carry a representation of a physically important group into a direct sum of vector spaces that carry irreducible representations of the group, which we discuss in Chapter 11, is of fundamental importance in modern physics as well as in the mathematical theory of group representations.

5.5.5 BASES OF COMPLEMENTARY SUBSPACES

In practical computations one usually defines a vector subspace by giving a basis. We therefore restate the major results concerning complementary subspaces in terms of bases.

Let $\mathcal{W}$ and $\mathcal{W}'$ be complementary subspaces of a vector space $\mathcal{V}$ with dimensions m and m', respectively. By Eq. (5.254), $m + m' = n := \dim[\mathcal{V}]$. Let

$$B := \{\bar{e}_1, \ldots, \bar{e}_m\} \tag{5.259}$$

be a basis of $\mathcal{W}$, and let

$$B' := \{\bar{f}_1, \ldots, \bar{f}_{m'}\} \tag{5.260}$$

be a basis of $\mathcal{W}'$. By Eq. (5.253), $B \cup B'$ is a basis of $\mathcal{V}$. In other words, if $\mathcal{W}$ and $\mathcal{W}'$ are complementary subspaces of a vector space $\mathcal{V}$, then the union of the bases of $\mathcal{W}$ and $\mathcal{W}'$ is a basis of $\mathcal{V}$.

Conversely, let B and B' be linearly independent sets of linearly independent vectors belonging to a vector space $\mathcal{V}$, and let

$$B \cup B' = \{\bar{e}_1, \ldots, \bar{e}_m, \bar{f}_1, \ldots, \bar{f}_{m'}\} \tag{5.261}$$

span $\mathcal{V}$. Define

$$\mathcal{W} = \text{span}[B] \quad \text{and} \quad \mathcal{W}' = \text{span}[B']. \tag{5.262}$$

We claim that $\mathcal{W}$ and $\mathcal{W}'$ are complementary subspaces of $\mathcal{V}$.

To prove this assertion, let us begin by noting that $B \cup B'$ is linearly independent and spans $\mathcal{V}$ and therefore is a basis of $\mathcal{V}$. Relative to the basis $B \cup B'$, every vector $\bar{x} \in \mathcal{V}$ has a unique coordinate expansion

$$\bar{x} = \bar{w} + \bar{w}' \tag{5.263}$$

in which

$$\bar{w} = \sum_{j=1}^{m} x^j \bar{e}_j \in \mathcal{W} \tag{5.264}$$

and

$$\bar{w}' = \sum_{j=m+1}^{m'} x^j \bar{f}_{j-m} \in \mathcal{W}'. \tag{5.265}$$

Therefore every element of $\mathcal{V}$ can be expressed uniquely as the sum of an element of $\mathcal{W}$ and an element of $\mathcal{W}'$. Moreover, only the zero vector belongs to both $\mathcal{W}$ and $\mathcal{W}'$, for

$$\bar{z} = \sum_{j=1}^{m} z^j \, \bar{e}_j = \sum_{j=m+1}^{m'} z^j \, \bar{f}_{j-m} \tag{5.266}$$

implies that

$$\sum_{j=1}^{m} z^j \, \bar{e}_j = - \sum_{j=m+1}^{m'} z^j \, \bar{f}_{j-m} = \bar{0}. \tag{5.267}$$

Hence

$$\forall j \in (1 : m + m') : \ z^j = 0 \tag{5.268}$$

in view of the linear independence of B and B'. It follows that $\mathcal{V}$ is the direct sum of $\mathcal{W}$ and $\mathcal{W}'$:

$$\mathcal{V} = \mathcal{W} \oplus \mathcal{W}'. \tag{5.269}$$

Then $\mathcal{W}$ and $\mathcal{W}'$ are complementary, as we claim.

5.5.6 EXAMPLES OF DIRECT SUMS OF VECTOR SPACES

Space of Complex Numbers

The vector space $\mathbb{C}$ over $\mathbb{R}$ is isomorphic to the direct sum of two one-dimensional vector spaces:

$$\mathbb{C} \cong \mathrm{span}[1] \oplus \mathrm{span}[i]. \tag{5.270}$$

Spaces of Column Vectors

The canonical form of the direct sum of two finite-dimensional vector spaces over a field $\mathbb{F}$ follows from the isomorphism (5.202). Suppose that $\mathcal{W}$ is (isomorphic to) $\mathbb{F}^m$ and $\mathcal{W}'$ is (isomorphic to) $\mathbb{F}^n$. Then $\mathcal{W} \oplus \mathcal{W}'$ is (isomorphic to) $\mathbb{F}^{m+n}$ under the mapping that identifies the first m contravariant components (rel$\{e_1, \ldots, e_{m+n}\}$) of a vector $z = Mx + Mx' \in \mathbb{F}^{m+n}$ with the contravariant components $x^1, \ldots, x^m$ of $x \in \mathcal{W}$ and identifies the last n contravariant components of z with the contravariant components $x^{m+1}, \ldots, x^{m+n}$ of $x' \in \mathcal{W}'$:

$$x \mapsto \begin{pmatrix} x^1 \\ \vdots \\ x^m \\ 0 \\ \vdots \\ 0 \end{pmatrix}, \quad x' \mapsto \begin{pmatrix} 0 \\ \vdots \\ 0 \\ x^{m+1} \\ \vdots \\ x^{m+n} \end{pmatrix}, \quad x + x' \mapsto \begin{pmatrix} x^1 \\ \vdots \\ x^m \\ x^{m+1} \\ \vdots \\ x^n \end{pmatrix}. \tag{5.271}$$

Matrix Spaces

If W and W' are isomorphic to the finite-dimensional vector spaces $\mathbb{F}^{m \times m}$ and $\mathbb{F}^{n \times n}$, respectively, then the canonical realization of the direct sum $W \oplus W'$ is the vector space $\mathbb{F}^{m \times m} \oplus \mathbb{F}^{n \times n}$. The representative of $\bar{z} = \bar{x} + \bar{x}'$ is the block-diagonal matrix

$$\begin{pmatrix} \mathbf{X} & \mathbf{0} \\ \mathbf{0} & \mathbf{X}' \end{pmatrix} \tag{5.272}$$

in which the upper-left $m \times m$ block $\mathbf{X}$ corresponds to $\bar{x} \in W$ and the lower-right $n \times n$ block $\mathbf{X}'$ corresponds to $\bar{x}' \in W'$.

The primary use of matrices is to represent linear mappings. A block-diagonal matrix such as Eq. (5.272) is the matrix of the direct sum of two linear mappings, which is defined in Section 6.1.4.

5.5.7 EXERCISES FOR SECTION 5.5

5.5.1. Let V be a vector space over a field $\mathbb{F}$, and let W be a subspace of V. Prove that V/W is a vector space over $\mathbb{F}$.

5.5.2. Show that a subspace W that is complementary to a given subspace W cannot be parallel to W.

5.5.3. Let W and W' be complementary subspaces of a vector space V over $\mathbb{F}$. Prove that if

$$\bar{x} = \bar{v} + \bar{v}' \tag{5.273}$$

and

$$\bar{x} = \bar{w} + \bar{w}' \tag{5.274}$$

in which

$$\bar{v} \in W, \quad \bar{w} \in W, \quad \bar{v}' \in W', \quad \bar{w}' \in W' \tag{5.275}$$

then

$$\bar{v} = \bar{w} \tag{5.276}$$

and

$$\bar{v}' = \bar{w}'. \tag{5.277}$$

Conclude that the decomposition given in Eq. (5.248) is unique.

5.5.4. Let W be a vector subspace of a vector space V, and let W' be a complementary subspace of W. For every $\bar{x} \in V$, let $\bar{x} = \bar{w} + \bar{w}'$ and define the **parallel**

projector $\Pi_{\mathcal{W}}$ **on** $\mathcal{W}$ as follows:

$$\Pi_{\mathcal{W}}\overline{x} := \overline{w}. \tag{5.278}$$

Prove that

$$\Pi_{\mathcal{W}}^2 := \Pi_{\mathcal{W}}\Pi_{\mathcal{W}} = \Pi_{\mathcal{W}}. \tag{5.279}$$

Hint: An operator that is equal to its own square is called **idempotent**.

5.6 BIBLIOGRAPHY AND ENDNOTES

5.6.1 BIBLIOGRAPHY

Bartle, Robert G. *The Elements of Integration.* New York: Wiley, 1966.

Garrett, Birkhoff, and Saunders Mac Lane. *A Survey of Modern Algebra.* 3rd ed. New York: Macmillan, 1965, chapters VII, IX.

Bollinger, J. J., D. J. Heinzen, Wayne M. Itano, S. L. Gilbert, and D. J. Wineland. "Test of the Linearity of Quantum Mechanics by rf Spectroscopy of the ^{9}Be$^+$ Ground State." *Physical Review Letters* 63, (1989): 1031–1034.

Crampin, M., and F. A. E. Pirani. *Applicable Differential Geometry.* New York: Cambridge, 1986, chapter 1.

Halmos, Paul R. *Finite-Dimensional Vector Spaces.* 2nd ed. Princeton: Van Nostrand, 1958, chapter I.

Kolmogorov, Andrei Nikolaevich, and Sergei Vasilevich Fomin. *Introductory Real Analysis.* New York: Dover, 1975, chapter 4.

Loomis, Lynn H., and Shlomo Sternberg. *Advanced Calculus.* Reading: Addison-Wesley, 1968, chapters 1 and 2

Messiah, Albert. *Quantum Mechanics.* Vols. I, II. Amsterdam: North-Holland, 1962.

Prugovečki, Eduard. *Quantum Mechanics in Hilbert Space.* 2nd ed. Academic Press, 1981.

Weinberg, Steven. "Precision Tests of Quantum Mechanics." *Physical Review Letters* 62 (1989): 485–488 .

Wilansky, Albert. *Modern Methods in Topological Vector Spaces.* New York: McGraw-Hill, 1978, §§1-5, 5-4.

5.6.2 ENDNOTES

Section 5.1

Tests of the linearity of quantum mechanics have been proposed by several authors and most recently by [Weinberg 1989]. The time-dependent Schrödinger equation can be written as

$$i\hbar\frac{\partial\psi}{\partial t} = \frac{\partial h(\psi, \psi^*)}{\partial \psi^*}. \tag{5.280}$$

In standard quantum mechanics the function h is bilinear in ψ and ψ^* and is homogeneous in ψ

[i.e., $\forall \lambda \in \mathbb{C}: h(\lambda\psi, \psi^*) = \lambda h(\psi, \psi^*)$]. Weinberg considered possible modifications in which h contains a small nonbilinear term but is still homogeneous. (Homogeneity turns out to permit a proper treatment of physically separated quantum systems.) Weinberg proposed an experimental test, which was carried out by [Bollinger 1989]. The latter authors found an upper limit of 2.4×10^{-20} eV (electron volts) for the nonlinear contribution to the ^{9}Be$^+$ nuclear Hamiltonian, which is less than 4 parts in 10^{27} of the average binding energy per nucleon in ^{9}Be$^+$.

CHAPTER 6

LINEAR MAPPINGS I

6.1 LINEAR MAPPINGS AND THEIR MATRICES

6.1.1 BASIC PROPERTIES

Linear mappings make vector spaces useful in engineering and the physical sciences. The properties of linear mappings of finite-dimensional vector spaces are the subject of linear algebra, which is probably the most important part of mathematics from the point of view of a working scientist or engineer. The properties of linear mappings of infinite-dimensional vector spaces, which are usually called *linear operators*, are the subject of a branch of twentieth-century mathematics called *functional analysis*.

Axioms for Linear Mappings

Let $\mathcal{V}_{(1)}$ and $\mathcal{V}_{(2)}$ be vector spaces over the same field $\mathbb{F}$, and let $\mathcal{D}$ be a vector subspace of $\mathcal{V}_{(1)}$. A **linear mapping** (or **linear transformation**) is a mapping $A : \mathcal{D} \to \mathcal{V}_{(2)}$ from $\mathcal{D}$ *into* $\mathcal{V}_{(2)}$ that obeys the following axioms:

1. *Additivity:*

$$\forall \overline{x}, \overline{y} \in \mathcal{D} : \; A(\overline{x} + \overline{y}) = A\overline{x} + A\overline{y}. \tag{6.1}$$

The property of additivity implies that A is a homomorphism of the additive group $(\mathcal{V}_{(1)}, +)$ into the additive group $(\mathcal{V}_{(2)}, +)$.

2. *Homogeneity:*

$$\forall \overline{x} \in \mathcal{D} : \forall \alpha \in \mathbb{F} : \; A(\alpha\overline{x}) = \alpha A(\overline{x}). \tag{6.2}$$

This axiom requires that the mapping A must commute with scalar multiplication. An additive mapping of a *complex* vector space $\mathcal{V}_{(1)}$ into a complex vector space $\mathcal{V}_{(2)}$ such that

$$\forall \overline{x} \in \mathcal{D} : \forall \alpha \in \mathbb{F} : \; A(\alpha\overline{x}) = \alpha^* A(\overline{x}) \tag{6.3}$$

is called **antilinear**.

The field $\mathbb{F}$ is either the field of real numbers, $\mathbb{R}$, or the field of complex numbers, $\mathbb{C}$. Linear mappings usually are denoted by sansserif capital letters such as A. (The notation A comes from the French word *application*, which means "mapping" in a mathematical context.)

An obvious example of a linear mapping is the **identity mapping** $1_V : V \to V$ such that

$$\forall \overline{x} \in V : \quad 1_V \overline{x} := \overline{x}. \tag{6.4}$$

Clearly 1_V is additive and homogeneous and therefore satisfies the axioms for a linear mapping. Usually we drop the subscript and refer only to 1.

Another example of a linear mapping is a uniform dilatation of the real line,

$$\forall x \in \mathbb{R} : x \mapsto ax, \tag{6.5}$$

in which $a \neq 0$. In other sections of this chapter we study linear mappings of the form $x \mapsto \mathbf{A}x$, where $\mathbf{A}$ is a matrix. In a uniform dilatation, the constant a is the simplest possible matrix (1×1).

Every vector isomorphism is also a one-to-one linear mapping from $V_{(1)}$ onto $V_{(2)}$. Physically important isomorphic linear mappings include rotations and Lorentz transformations.

The mapping $\mathsf{F} : \mathbb{R} \to \mathbb{R}$ such that

$$\mathsf{F}x := ax + b \tag{6.6}$$

fails to satisfy the axioms for a linear mapping because of the presence of the inhomogeneous term b. In fact, $x \mapsto ax + b$ is an example of an *affine mapping*, that is, a mapping $\mathsf{F} : V_{(1)} \to V_{(2)}$ such that for all $\overline{x} \in V_{(1)}$,

$$\mathsf{F}\overline{x} = \mathsf{A}\overline{x} + \overline{a} \tag{6.7}$$

in which $\mathsf{A} : V_{(1)} \to V_{(2)}$ is a linear mapping and $\overline{a}$ is a fixed vector in $V_{(1)}$ (see Appendix B). Affine mappings are also called *inhomogeneous linear mappings*. For example, inhomogeneous Lorentz transformations (i.e., elements of the Poincaré group) are affine mappings. A mapping that is neither linear nor affine is called **nonlinear**.

A linear mapping on an infinite-dimensional vector space is called a **linear operator**.

Product of Two Linear Mappings

The **product of two linear mappings** $\mathsf{A} : V_{(1)} \to V_{(2)}$ and $\mathsf{B} : V_{(2)} \to V_{(3)}$ is the linear mapping $\mathsf{BA} : V_{(1)} \to V_{(3)}$ that results from applying B after A:

$$\mathsf{BA}\overline{x} := \mathsf{B}(\mathsf{A}\overline{x}). \tag{6.8}$$

It is straightforward to verify that BA is a linear mapping if B and A are linear mappings. The definition of a mapping (see Section 2.2.2) implies that the composition of three or more linear mappings is associative.

Vector Spaces of Linear Mappings

The **set of all linear mappings** from $V_{(1)}$ into $V_{(2)}$ is denoted $\hom[V_{(1)}, V_{(2)}]$. The **sum of two linear mappings** A and B is the linear mapping $\mathsf{A} + \mathsf{B}$ such that

$$\forall \overline{x} \in V_{(1)} : (\mathsf{A} + \mathsf{B})\overline{x} := \mathsf{A}\overline{x} + \mathsf{B}\overline{x}. \tag{6.9}$$

The **scalar multiple** αA is the linear mapping such that

$$\forall \overline{x} \in \mathcal{V}_{(1)} : \forall \alpha \in \mathbb{F} : (\alpha A)\overline{x} := \alpha A\overline{x}. \tag{6.10}$$

With these definitions it is easy to show that $\hom[\mathcal{V}_{(1)}, \mathcal{V}_{(2)}]$ is a vector space over $\mathbb{F}$ (see Exercise 6.1.1). We show in a subsequent section that if we write $\dim[\mathcal{V}_{(1)}]$ as n_1 and $\dim[\mathcal{V}_{(2)}]$ as n_2, then $\dim[\hom[\mathcal{V}_{(1)}, \mathcal{V}_{(2)}]]$ is equal to the product $n_1 n_2$.

Domain and Range of a Linear Mapping

As is true for any mapping, the **domain** of a linear mapping A is the set of vectors of $\mathcal{V}_{(1)}$ on which A is defined,

$$\text{domain}[A] := \left\{ \overline{x} \in \mathcal{V}_{(1)} \mid A\overline{x} \in \mathcal{V}_{(2)} \right\}. \tag{6.11}$$

It follows from the additivity and homogeneity of A that $\text{domain}[A]$ is a vector subspace of $\mathcal{V}_{(1)}$. Hence the vector subspace in the definition of a linear mapping, Eq. (6.1), is the domain of the mapping:

$$\mathcal{D} = \text{domain}[A]. \tag{6.12}$$

The **range** of A is the image of the domain $\mathcal{D}$ under A,

$$\text{range}[A] := \{A\overline{x} \mid \overline{x} \in \mathcal{D}\}. \tag{6.13}$$

If $\text{range}[A] = \mathcal{V}_{(2)}$, then A is called **onto**. In any event, $\text{range}[A]$ is a vector subspace of $\mathcal{V}_{(2)}$ (see Exercises 6.1.2 and 6.1.3). For the remainder of this chapter we usually assume for the sake of simplicity that a linear mapping A is defined on every element of $\mathcal{V}_{(1)}$:

$$\text{domain}[A] = \mathcal{V}_{(1)}. \tag{6.14}$$

This assumption does no harm if $\mathcal{V}$ is finite-dimensional. However, if $\mathcal{V}_{(1)}$ is infinite-dimensional, the domain of a linear mapping need not be all of $\mathcal{V}_{(1)}$.

6.1.2 MATRIX OF A LINEAR MAPPING

So far we have defined some of the basic properties of linear mappings in a generic or basis-free manner. A basis-free approach helps one to understand that many properties of a linear mapping are independent of the bases that one chooses for $\mathcal{V}_{(1)}$ and $\mathcal{V}_{(2)}$. A basis-free approach may be especially useful for understanding abstract properties if one, or both, of the vector spaces $\mathcal{V}_{(1)}$ and $\mathcal{V}_{(2)}$ are infinite-dimensional. For the purposes of numerical computation with linear mappings one is restricted to finite-dimensional vector spaces. Also, one is usually forced to introduce bases of $\mathcal{V}_{(1)}$ and $\mathcal{V}_{(2)}$ because explicit bases make it possible to realize abstract linear mappings as arrays (i.e., matrices), which can conveniently be stored and manipulated.

Let $\{\bar{e}_1, \ldots, \bar{e}_{n_1}\}$ and $\{\bar{e}'_1, \ldots, \bar{e}'_{n_2}\}$ be bases of the finite-dimensional vector spaces $\mathcal{V}_{(1)}$ and $\mathcal{V}_{(2)}$, respectively. Because $\mathcal{V}_{(1)}$ and $\mathcal{V}_{(2)}$ may have different dimensions, we need to extend the Einstein summation convention as follows for the purposes of this chapter: *Every index attached to a vector drawn from a vector space $\mathcal{V}_{(i)}$ will run from 1 to $n_i = \dim[\mathcal{V}_{(i)}]$.*

Matrix Elements of a Linear Mapping

Let the image under A of a basis vector $\bar{e}_i$ of $\mathcal{V}_{(1)}$ be

$$\bar{a}_i := A\bar{e}_i. \tag{6.15}$$

The **matrix elements** a_i^j of A ($\mathrm{rel}\{\bar{e}_i\}$ and $\{\bar{e}'_j\}$) are defined as the contravariant components of $\bar{a}_i$ ($\mathrm{rel}\{\bar{e}'_j\}$):

$$\bar{a}_i = a_i^j \bar{e}'_j. \tag{6.16}$$

(The summation convention is in force; hence $\sum_{j=1}^{n_2}$ is understood!) The matrix elements a_i^j depend on both bases $\{\bar{e}_i\}$ and $\{\bar{e}'_j\}$. Changing a single basis element, or even changing the order of basis elements, changes the matrix elements a_i^j. Elsewhere in this chapter we study how the matrix of a linear mapping is altered by changes of bases in the underlying vector spaces.

The image under A of an arbitrary vector $\bar{x} \in \mathcal{V}_{(1)}$ is

$$\bar{x}' = A\bar{x} = A(x^i \bar{e}_i)$$
$$= x^i A\bar{e}_i = x^i \bar{a}_i. \tag{6.17}$$

But

$$\bar{x}' = x'^j \bar{e}'_j; \tag{6.18}$$

therefore the contravariant components of $\bar{x}'$ ($\mathrm{rel}\{\bar{e}'_j\}$) are

$$\forall j \in (1:n_2): \quad x'^j = a_i^j x^i. \tag{6.19}$$

Relative to fixed bases $\{\bar{e}_i\}$ and $\{\bar{e}'_j\}$, the array of matrix elements $\{a_i^j\}$ completely specifies the linear mapping A.

Column Rank of a Linear Mapping

The **column rank** of a linear mapping A is defined as the dimension of range[A]:

$$\mathrm{rank}[A] := \dim[\mathrm{range}[A]]. \tag{6.20}$$

According to Eq. (6.17), every image vector $\bar{x}' = A\bar{x}$ under A is a linear combination of the vectors $\bar{a}_i$. Because the set $\{\bar{a}_i\}$ spans range[A],

$$\mathrm{rank}[A] = \dim[\mathrm{span}\{\bar{a}_i\}]. \tag{6.21}$$

The reason why rank[A] is called the *column* rank becomes obvious in Eq. (6.34).

The definition of column rank does not presuppose any constraint on the dimensions n_1 of $\mathcal{V}_{(1)}$ and n_2 of $\mathcal{V}_{(2)}$ except that n_1 and n_2 are finite; n_1 may be less than, equal to, or greater than n_2. The dimension of range[A] is independent of the basis we choose for $\mathcal{V}_{(2)}$. We shall see later that rank[A] is also independent of the basis we choose in $\mathcal{V}_{(1)}$. Because

$$\text{range}[A] \subseteq \mathcal{V}_{(2)}, \tag{6.22}$$

obviously

$$\text{rank}[A] \leq \dim[\mathcal{V}_{(2)}] := n_2. \tag{6.23}$$

It is also clear that the dimension of range[A] cannot exceed the dimension of $\mathcal{V}_{(1)} = $ domain [A],

$$\text{rank}[A] \leq \dim[\text{domain}[A]] = \dim[\mathcal{V}_{(1)}] = n_1, \tag{6.24}$$

because the n_1 vectors $\bar{a}_1, \ldots, \bar{a}_{n_1}$ span the range of A.

Matrices, Viewed as Lists of Column Vectors

Because the matrix elements a_i^j specify a linear mapping A uniquely once the bases are given, it makes sense to represent A by an array of its matrix elements. We show that there is only one reasonable way to arrange the array of matrix elements given our convention of representing elements of the space $\mathbb{F}^n$ as column vectors.

According to Eq. (5.202), there exist vector isomorphisms $M_{(1)} : \mathcal{V}_{(1)} \to \mathbb{F}^{n_1}$ and $M_{(2)} : \mathcal{V}_{(2)} \to \mathbb{F}^{n_2}$ such that

$$\begin{aligned} \forall i \in (1:n_1) : \ & M_{(1)}\bar{e}_i = \mathbf{e}_i \in \mathbb{F}^{n_1} \\ \forall j \in (1:n_2) : \ & M_{(2)}\bar{e}'_j = \mathbf{e}'_j \in \mathbb{F}^{n_2} \end{aligned} \tag{6.25}$$

in which $\{\mathbf{e}_i\}$ and $\{\mathbf{e}'_j\}$ are the canonical bases of $\mathbb{F}^{n_1}$ and $\mathbb{F}^{n_2}$ (see Eq. [5.122]). Then an n_1-component vector $\mathbf{x}$ is the image of an abstract vector $\bar{x}$ under the mapping $M_{(1)}$,

$$\mathbf{x} = x^i \mathbf{e}_i = x^i M_{(1)}\bar{e}_i = M_{(1)} x^i \bar{e}_i = M_{(1)}\bar{x}. \tag{6.26}$$

Likewise,

$$\mathbf{x}' = x'^j \mathbf{e}'_j = x'^j M_{(2)}\bar{e}'_j = M_{(2)}\bar{x}'. \tag{6.27}$$

Applying $M_{(2)}$ to both sides of the equation $\bar{x}' = A\bar{x}$, one obtains

$$\mathbf{x}' = M_{(2)}\bar{x}' = M_{(2)}A\bar{x} = \left(M_{(2)}AM_{(1)}^{-1}\right)\left(M_{(1)}\bar{x}\right) = \mathbf{A}\mathbf{x} \tag{6.28}$$

in which

$$\mathbf{A} := M_{(2)}AM_{(1)}^{-1} \tag{6.29}$$

is a linear mapping of $\mathbb{F}^{n_1}$ into $\mathbb{F}^{n_2}$. It follows from Eq. (6.28) that

$$\mathbf{a}_i := \mathsf{M}_{(2)}\bar{a}_i \tag{6.30}$$

is the image of $\mathbf{e}_i$ under $\mathbf{A}$,

$$\mathbf{a}_i = a_i^j \mathbf{e}_j' = \begin{pmatrix} a_i^1 \\ a_i^2 \\ \vdots \\ a_i^{n_2} \end{pmatrix} = \mathbf{A}\mathbf{e}_i. \tag{6.31}$$

In other words, $\mathbf{A}$ is the linear mapping of $\mathbb{F}^{n_1}$ into $\mathbb{F}^{n_2}$ whose matrix elements relative to the canonical column-vector bases in $\mathbb{F}^{n_1}$ and $\mathbb{F}^{n_2}$ are the same as the matrix elements of the linear mapping A relative to the bases $\{\bar{e}_i\}$ and $\{\bar{e}_j'\}$ of $\mathcal{V}_{(1)}$ and $\mathcal{V}_{(2)}$. The commutative diagram

$$\begin{array}{ccc} \mathbb{F}^{n_1} & \xrightarrow{\;\mathbf{A}\;} & \mathbb{F}^{n_2} \\ \mathsf{M}_{(1)} \Big\uparrow & & \Big\uparrow \mathsf{M}_{(2)} \\ \mathcal{V}_{(1)} & \xrightarrow[\;A\;]{} & \mathcal{V}_{(2)} \end{array} \tag{6.32}$$

summarizes the relationships among the mappings A, $\mathbf{A}$, $\mathsf{M}_{(1)}$, and $\mathsf{M}_{(2)}$. One sees at once that $\mathbf{A}$ is certainly a linear mapping because it is the product of three linear mappings.

From Eqs. (6.33) and (6.19) one sees that every vector $\mathbf{x}'$ that is the image of a vector $\mathbf{x}$ under $\mathbf{A}$ is a linear combination of the vectors $\{\mathbf{a}_i\}$ defined previously:

$$\mathbf{x}' = x^i a_i^j \mathbf{e}_j' = x^i \mathbf{a}_i. \tag{6.33}$$

Then Eq. (6.26) implies that the column vectors $\mathbf{a}_i$ span the range of $\mathbf{A}$:

$$\text{range}[\mathbf{A}] = \text{span}\{\mathbf{a}_1, \ldots, \mathbf{a}_{n_1}\}. \tag{6.34}$$

For this reason range[$\mathbf{A}$] is often called the **column space of $\mathbf{A}$**.

In view of the fundamental importance of the column vectors $\mathbf{a}_i$, it makes sense to represent $\mathbf{A}$ as a **matrix** whose i-th column is $\mathbf{a}_i$:

$$\mathbf{A} = \begin{pmatrix} a_1^1 & a_2^1 & \cdots & a_{n_1}^1 \\ a_1^2 & a_2^2 & \cdots & a_{n_1}^2 \\ \vdots & \vdots & \ddots & \vdots \\ a_1^{n_2} & a_2^{n_2} & \cdots & a_{n_1}^{n_2} \end{pmatrix}. \tag{6.35}$$

One can also write $\mathbf{A}$ in the **column-partitioned** form

$$\mathbf{A} = (\mathbf{a}_1, \ldots, \mathbf{a}_{n_1}), \tag{6.36}$$

which is a list of the images $\mathbf{a}_i = \mathbf{A}\mathbf{e}_i$ of the canonical basis under $\mathbf{A}$.

This book uses a superscript–subscript notation (a_i^j) for matrix elements. The superscript (here, j) labels the row of the matrix to which a_i^j belongs. The subscript (here, i) labels the column.

If one needs to refer to a matrix in terms of its matrix elements, one surrounds the matrix elements with some sort of special brackets. Often one sees the notation $\|a_i^j\|$ used to mean "the matrix whose elements are a_i^j." Because $\| \cdot \|$ is also a common notation for the norm of a vector or matrix, it seems best to use a different notation to indicate the matrix itself. We shall write

$$\mathbf{A} = \left[a_i^j\right] \tag{6.37}$$

to mean that $\mathbf{A}$ is the matrix with elements a_i^j.

The **dimensions** of a matrix are the number of its rows and the number of its columns. Matrix dimensions are usually quoted in the format "rows × columns." Thus $\mathbf{A}$ in Eq. (6.35) is an $n_2 \times n_1$ matrix. **Square matrices**, in which $n_2 = n_1$, realize linear mappings between vector spaces that have equal dimensions and are therefore isomorphic.

An important square matrix is the **identity matrix** (or **unit matrix**) $\mathbf{1}$, which realizes the identity mapping defined in Eq. (6.4). The elements of the identity matrix are

$$\mathbf{1}_i^j = \delta_i^j \tag{6.38}$$

(see Eq. [5.165]). We use the same symbol for the identity matrix and the identity mapping.

The **transpose** of the matrix $\mathbf{A}$ defined in Eq. (6.35) is the matrix $\mathbf{A}^T$ obtained by interchanging the columns and rows of $\mathbf{A}$:

$$\mathbf{A}^T = \begin{pmatrix} a_1^1 & a_1^2 & \cdots & a_1^{n_1} \\ a_2^1 & a_2^2 & \cdots & a_2^{n_1} \\ \vdots & \vdots & \ddots & \vdots \\ a_{n_2}^1 & a_{n_2}^2 & \cdots & a_{n_2}^{n_1} \end{pmatrix}. \tag{6.39}$$

The row i, column j element t_j^i of $\mathbf{T} = \mathbf{A}^T$ is equal to the column i, row j element of $\mathbf{A}$:

$$t_j^i = a_i^j. \tag{6.40}$$

The picture

$$\tag{6.41}$$

shows the relation between the dimensions of $\mathbf{A}$ and the dimensions of $\mathbf{A}^T$.

The transpose $\mathbf{A}^T$ of a matrix $\mathbf{A}$ is not directly useful in describing the range of $\mathbf{A}$. For now, $\mathbf{A}^T$ is simply a handy notation. For example, $\mathbf{x}^T = (x^1, x^2, \ldots, x^n)^T$ is a space-saving way to write a column vector. The fundamental significance of the transpose of a matrix is discussed in Chapter 9.

Matrix–Vector Product

The equation $x'^j = a_i^j x^i$ (Eq. [6.19]) tells us that we can picture the action of a linear mapping A upon a vector $\overline{x}$ in terms of **matrix–vector multiplication**:

The element x'^j in the j-th row of the image vector $\mathbf{x}' = \mathbf{A}\mathbf{x}$ is equal to the sum of the element-by-element products of the j-th row of the matrix $\mathbf{A}$ with the column vector $\mathbf{x}$:

$$x'^j = \left(a_1^j, \ldots, a_{n_1}^j\right) \begin{pmatrix} x^1 \\ \vdots \\ x^{n_1} \end{pmatrix} := a_1^j x^1 + \cdots + a_{n_1}^j x^{n_1}. \tag{6.42}$$

The computation of matrix products is the subject of Section 6.1.3.

One can look upon the matrix–vector equation

$$\mathbf{A}\mathbf{x} = \mathbf{b} \tag{6.43}$$

as a system of linear equations, the j-th of which is

$$a_1^j x^1 + \cdots + a_{n_1}^j x^{n_1} = b_j. \tag{6.44}$$

The number of unknowns is equal to the number of components of the vector $\mathbf{x}$, n_1. The number of equations is equal to the number of components of $\mathbf{b}$, n_2.

One needs to have a clear understanding of the relations among the dimensions of the matrix $\mathbf{A}$, the vector $\mathbf{x}$ on which it acts, and the image vector $\mathbf{b}$ in Eq. (6.43). The following picture applies to the case in which $\mathbf{x}$, $\mathbf{b}$, and $\mathbf{A}$ belong to $\mathbb{F}^{n_1}$, $\mathbb{F}^{n_2}$, and $\mathbb{F}^{n_2 \times n_1}$, respectively, and in which $n_1 < n_2$:

$$\tag{6.45}$$

In the case shown here, rank$[\mathbf{A}]$ is not greater than n_1 because the dimension of range$[\mathbf{A}]$ cannot exceed the number of columns of $\mathbf{A}$.

If $n_1 > n_2$, the block picture of the equation $\mathbf{A}\mathbf{x} = \mathbf{b}$ becomes

$$\begin{array}{c} \overset{n_1}{\boxed{\mathbf{A}}} \end{array} \; \boxed{\mathbf{x}} = \boxed{\mathbf{b}}$$

$$(6.46)$$

In this case rank[$\mathbf{A}$] cannot exceed n_2 because the dimension of range[$\mathbf{A}$] can be no greater than the dimension of $\mathcal{V}_{(2)}$. We conclude that

$$\text{rank}[\mathbf{A}] \leq \min\{n_1, n_2\}. \tag{6.47}$$

Visualization of Linear Mappings

According to Eq. (6.31), the column vector $\mathbf{a}_i$ of a matrix $\mathbf{A}$ is the image of $\mathbf{e}_i$ under $\mathbf{A}$. If the dimension of range[$\mathbf{A}$] is not greater than 3, then one can learn much by plotting the columns of $\mathbf{A}$ as vectors, lines, or points in three-dimensional space. For example, consider the matrix

$$\mathbf{A} = \begin{pmatrix} 3 & 2 \\ 2 & 6 \end{pmatrix}. \tag{6.48}$$

Figure 6.1(a) shows the basis vectors $\mathbf{e}_1$ and $\mathbf{e}_2$ as mutually perpendicular vectors of unit length, as usual. Figure 6.1(b) shows the images $\mathbf{a}_1 = (3, 2)^T$ and $\mathbf{a}_2 = (2, 6)^T$ of the basis vectors under $\mathbf{A}$. Clearly the image of a vector $\mathbf{x}$ under $\mathbf{A}$ is generally not parallel to $\mathbf{x}$.

Physics and engineering furnish many examples of relations in which a physically important vector quantity is the image of another under a linear mapping. For example, the relationship between the electric displacement $\mathbf{D}$ and the electric field $\mathbf{E}$ in a homogeneous, anisotropic dielectric is

$$\mathbf{D} = \epsilon\mathbf{E} \tag{6.49}$$

in which ϵ is the **dielectric–permittivity matrix** (also called the *dielectric–permittivity tensor*). In this example, $\mathbf{D}$ need not be (and, in general, is not) parallel to $\mathbf{E}$, and the

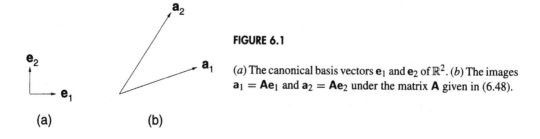

FIGURE 6.1

(a) The canonical basis vectors $\mathbf{e}_1$ and $\mathbf{e}_2$ of $\mathbb{R}^2$. (b) The images $\mathbf{a}_1 = \mathbf{A}\mathbf{e}_1$ and $\mathbf{a}_2 = \mathbf{A}\mathbf{e}_2$ under the matrix $\mathbf{A}$ given in (6.48).

(a)   (b)

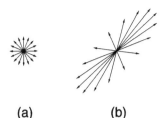

FIGURE 6.2

(a) A set of vectors of unit length; angular interval $= \pi/8$. (b) The images of the vectors in (a) under the matrix $\mathbf{A}$ given in (6.48).

(a)   (b)

magnitude of $\mathbf{D}$ is greater in some directions than in others given a constant magnitude of $\mathbf{E}$. In anisotropic materials, the latter phenomenon is the cause of birefringence.

Figure 6.2 shows a set of vectors rotated through equal angles with respect to one another, as well as the image of this set under the matrix $\mathbf{A}$ defined in Eq. (6.48). If the vectors in the left-hand part of Fig. 6.2 are electric-field vectors and if $\mathbf{A}$ is equal to the dielectric-permittivity matrix ϵ, then Fig. 6.2(b) shows the electric-displacement vectors induced by the electric fields shown in Fig. 6.2(a).

We see in a subsequent section why the application of the matrix $\mathbf{A}$ appears to rotate and stretch the initial set of vectors shown on the left side of Fig. 6.2.

Computation of the Column Rank

The method of Gaussian elimination developed in Section 5.4.4 for the purpose of determining the dimension of the span of a list X of vectors is a practical technique for systematically computing the rank of an arbitrary linear mapping of a finite dimensional vector space. To use this method, one performs Gaussian elimination on the rows of the matrix $\mathbf{A}$ relative to chosen bases.

For example, applying Gaussian elimination to the rows of the matrix $\mathbf{A}$ in Eq. (6.48) results in the matrix

$$\mathbf{A}^{(2)} = \begin{pmatrix} 3 & 2 \\ 0 & \frac{14}{3} \end{pmatrix}. \tag{6.50}$$

Therefore rank$[\mathbf{A}] = 2$.

Computation of a Basis of the Range

Gaussian elimination, applied to the columns of a matrix $\mathbf{A}$, gives a basis of the span of the columns and therefore a basis of range$[\mathbf{A}]$. For example, to find a basis of the range of a linear mapping with the matrix $\mathbf{A}$ defined in Eq. (6.48), one applies Gaussian elimination to the columns, obtaining

$$\mathbf{A}' = \begin{pmatrix} 3 & 0 \\ 2 & \frac{14}{3} \end{pmatrix}. \tag{6.51}$$

Therefore the range of $\mathbf{A}$ is spanned by

$$\mathbf{a}_1 = \begin{pmatrix} 3 \\ 2 \end{pmatrix}, \quad \mathbf{e}_2 = \begin{pmatrix} 0 \\ 1 \end{pmatrix}. \tag{6.52}$$

Many other bases can be found for the range of $\mathbf{A}$; later we shall make use of the basis

$$\left\{ \mathbf{f}_1 = \begin{pmatrix} 1 \\ 2 \end{pmatrix}, \mathbf{f}_2 = \begin{pmatrix} -2 \\ 1 \end{pmatrix} \right\}. \tag{6.53}$$

The first step of Gaussian elimination of the columns of the matrix

$$\mathbf{B} = \begin{pmatrix} 1 & 2 & 1 \\ 2 & 3 & 3 \\ 1 & 1 & 2 \end{pmatrix} \tag{6.54}$$

gives the matrix

$$\begin{pmatrix} 1 & 0 & 0 \\ 2 & -1 & 1 \\ 1 & -1 & 1 \end{pmatrix}, \tag{6.55}$$

and the second step gives

$$\begin{pmatrix} 1 & 0 & 0 \\ 2 & -1 & 0 \\ 1 & -1 & 0 \end{pmatrix}. \tag{6.56}$$

Obviously the vectors

$$\left\{ \mathbf{r}_1 = \begin{pmatrix} 1 \\ 2 \\ 1 \end{pmatrix}, \mathbf{r}_2 = \begin{pmatrix} 0 \\ -1 \\ -1 \end{pmatrix} \right\} \tag{6.57}$$

span range[$\mathbf{B}$]. A somewhat simpler basis of range[$\mathbf{B}$] is

$$\left\{ \mathbf{g}_1 = \begin{pmatrix} 1 \\ 1 \\ 0 \end{pmatrix}, \mathbf{g}_2 = \begin{pmatrix} 0 \\ 1 \\ 1 \end{pmatrix} \right\}, \tag{6.58}$$

where $\mathbf{g}_1 = \mathbf{r}_1 + \mathbf{r}_2$ and $\mathbf{g}_2 = -\mathbf{r}_2$. Note that in this example the dimension of the range is less than the dimension of the domain.

For another example, consider the matrix

$$\mathbf{C} = \begin{pmatrix} 1 & 5 & 9 & 13 & 17 \\ 2 & 6 & 10 & 14 & 18 \\ 3 & 7 & 11 & 15 & 19 \\ 4 & 8 & 12 & 16 & 20 \end{pmatrix}. \tag{6.59}$$

After the first step of Gaussian elimination on the columns of $\mathbf{C}$, one has

$$\mathbf{C}' = \begin{pmatrix} 1 & 0 & 0 & 0 & 0 \\ 2 & -4 & -8 & -12 & -16 \\ 3 & -8 & -16 & -24 & -32 \\ 4 & -12 & -24 & -36 & -48 \end{pmatrix}. \tag{6.60}$$

The second step results in

$$\mathbf{C}'' = \begin{pmatrix} 1 & 0 & 0 & 0 & 0 \\ 2 & -4 & 0 & 0 & 0 \\ 3 & -8 & 0 & 0 & 0 \\ 4 & -12 & 0 & 0 & 0 \end{pmatrix}. \tag{6.61}$$

Therefore

$$\left\{ \begin{pmatrix} 1 \\ 2 \\ 3 \\ 4 \end{pmatrix}, \begin{pmatrix} 0 \\ 1 \\ 2 \\ 3 \end{pmatrix} \right\} \tag{6.62}$$

is a basis of the range of $\mathbf{C}$; clearly rank$[\mathbf{C}] = 2$.

Matrix of the Product of Two Linear Mappings

Let B be a linear mapping from $\mathcal{V}_{(2)}$ into $\mathcal{V}_{(3)}$. Choose a basis $\{\bar{e}''_1, \ldots, \bar{e}''_{n_3}\}$ of $\mathcal{V}_{(3)}$. Let

$$\begin{aligned} \bar{b}_j &:= B\bar{e}'_j \\ &= b^k_j \bar{e}''_k. \end{aligned} \tag{6.63}$$

From Eqs. (6.15), (6.17), and (6.63), the image of the vector $\bar{x}$ under the product mapping

$$C := BA \tag{6.64}$$

is

$$\begin{aligned} \bar{x}'' = C\bar{x} &= B(A\bar{x}) \\ &= B(x^i \bar{a}_i) \\ &= B\big(x^i a^j_i \bar{e}'_j\big) \\ &= x^i \bar{c}_i, \end{aligned} \tag{6.65}$$

in which

$$\bar{c}_i = a^j_i \bar{b}_j, \quad \bar{b}_j = B\bar{e}'_j. \tag{6.66}$$

We see not only that the vectors $\bar{c}_i$ span the range of C but also that every vector $\bar{x}''$ in range$[C]$ is a linear combination of the vectors $\bar{b}_j$ that span range$[B]$. Therefore

$$\text{range}[BA] \subseteq \text{range}[B]. \tag{6.67}$$

It follows that

$$\text{rank}[BA] \le \text{rank}[B], \tag{6.68}$$

that is, the rank of BA cannot exceed (and may be less than) the rank of B.

From Eqs. (6.65) and (6.66) one sees at once that

$$\bar{x}'' = x''^k \bar{e}''_k,$$ (6.69)

where the contravariant components of $\bar{x}''$ or $\mathbf{x}''$ are

$$x''^k = c_i^k x^i = b_j^k a_i^j x^i.$$ (6.70)

Therefore the matrix elements of C, relative to the bases $\{\bar{e}_1, \ldots, \bar{e}_{n_1}\}$ of $\mathcal{V}_{(1)}$, $\{\bar{e}'_1, \ldots, \bar{e}'_{n_2}\}$ of $\mathcal{V}_{(2)}$ and $\{\bar{e}''_1, \ldots, \bar{e}''_{n_3}\}$ of $\mathcal{V}_{(3)}$, are

$$c_k^i = b_k^j a_j^i.$$ (6.71)

This equation follows directly from the definition of the composition of two mappings and the coordinate expansion of a vector relative to a basis.

Let us verify that the mapping $\mathbf{C} : \mathbb{F}^{n_1} \to \mathbb{F}^{n_3}$ such that

$$\mathbf{C} = \mathsf{M}_{(3)} C \mathsf{M}_{(1)}^{-1}$$ (6.72)

is the product of the mappings $\mathbf{A}$ and $\mathbf{B}$ in which $\mathsf{M}_{(3)} : \mathcal{V}_{(3)} \to \mathbb{F}^{n_3}$ is the canonical mapping such that

$$\forall i \in (1 : n_3): \quad \mathsf{M}_{(3)} \bar{e}_i = \mathbf{e}_i \in \mathbb{F}^{n_1}.$$ (6.73)

The mapping $\mathbf{C} : \mathbb{F}^{n_1} \to \mathbb{F}^{n_3}$ that has the same matrix elements (relative to $\{\mathbf{e}_1, \ldots, \mathbf{e}_{n_1}\}$ and $\{\mathbf{e}''_1, \ldots, \mathbf{e}''_{n_3}\}$) as $C : \mathcal{V}_{(1)} \to \mathcal{V}_{(3)}$ has (relative to $\{\bar{e}_1, \ldots, \bar{e}_{n_1}\}$ and $\{\bar{e}''_1, \ldots, \bar{e}''_{n_3}\}$) is

$$\begin{aligned}
\mathbf{C} &= \mathsf{M}_{(3)} BA \mathsf{M}_{(1)}^{-1} \\
&= \mathsf{M}_{(3)} B \mathsf{M}_{(2)}^{-1} \mathsf{M}_{(2)} A \mathsf{M}_{(1)}^{-1} \\
&= \mathbf{BA}.
\end{aligned}$$ (6.74)

The commutative diagram

(6.75)

summarizes the relations among the mappings in Eq. (6.74).

Because one usually identifies **A** and **B** with their matrices, **C** is called the **matrix product** of **B** and **A**. Equation (6.71) tells one how to compute the $n_3 \times n_1$ matrix **C** associated with the abstract linear mapping C, given the matrices **B** and **A**: *The row k, column i element of the matrix* **C** = **BA** *that realizes the linear-mapping product* C = BA *is equal to the dot product of the k-th row of the matrix* **B** *with the i-th column of the matrix* **A**. (Recall that the dot product of two vectors $(a^1, a^2, a^3)^T$ and $(b^1, b^2, b^3)^T \epsilon \mathbb{R}^3$ is $a^1b^1 + a^2b^2 + a^3b^3$.)

In the matrix product **BA**, the dimensions of **B** and **A** must be **compatible**, that is, the number of columns of **B** must equal the number of rows of **A**, as in the following picture:

$$(6.76)$$

If the dimensions of two matrices **B** and **A** are **incompatible**, that is, if the number of columns of **B** is not equal to the number of rows of **A**, then one cannot form a matrix product **BA**. Incompatibility comes about if a linear mapping A maps $\mathcal{V}_{(1)}$ into a vector space $\mathcal{V}_{(2)}$ of a dimension that is different from the dimension of the space on which a linear mapping B acts.

It is straightforward to show that the formula for the transpose $(\mathbf{BA})^T$ of a matrix product follows from Eqs. (6.71) and (6.40):

$$(\mathbf{BA})^T = \mathbf{A}^T \mathbf{B}^T \tag{6.77}$$

(see Exercise 6.1.15).

6.1.3 COMPUTATION OF MATRIX PRODUCTS

Two different problems can occur in the numerical evaluation of matrix products, a task for which modern computers would seem to be ideally suited:

- Loss of significance through cancellation (see Section 8.1.3)
- Strong dependence of computational speed for large matrices on program details

We discuss the second problem in this section.

A numerical computation of the matrix product **C** = **BA**, in which the dimensions are as shown in Eq. (6.76), requires n_2 floating-point multiplications for each row of **B** and for each column of **A**, making a total of $n_1 n_2 n_3$ floating-point multiplications. It is possible to do slightly better than this estimate suggests by using Strassen multiplication and similar techniques (Golub and Van Loan 1989).

MATRIX MULTIPLICATION

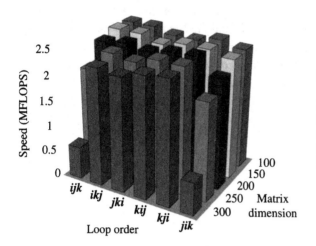

FIGURE 6.3

Performance of IEEE-754 double-precision FORTRAN matrix multiplication versus matrix size and loop ordering. Cache sizes: Level 1: 32 kB; Level 2: 512 kB.

Computation of a matrix product using Eq. (6.71), $c_k^i = b_k^j a_j^i$, requires three nested loops, one for each of the indices i, j, and k. The following program fragment is a possible FORTRAN implementation of these three loops:[1]

```
      do 150 i = 1,n
        do 150 j = 1,n
          do 150 k = 1,n
            c(i,k) = c(i,k)+a(i,j)*b(j,k)
150     continue
```

The order of the nested loops in this example, ijk, is one of six possible loop orderings corresponding to the six permutations of the induction variables i, j, and k. Each loop ordering is computationally correct, but, as Fig. 6.3 illustrates, some loop orderings show a much slower rate of execution than others for matrices larger than a certain size, which depends on the number of bytes of cache memory. One can see from the figure that for the largest matrix dimension (300 × 300) and in FORTRAN, the ijk loop ordering is a factor of five slower than the best ordering, kji.

In order to understand the dependence of the speed of matrix multiplication on loop ordering, it is useful to refer to Figs. 2.5 and 2.6 and Eqs. (2.76) and (2.77). In the ijk loop ordering, and in FORTRAN, the innermost loop (summation on k) accesses the elements of the matrices **B** and **C** column-by-column within any given row. Therefore the stride is

---

[1] An implementation of the same three loops in C would use the syntax for(i=1;i<=n;i++) instead of do 150 i = 1,n, etc.

equal to the number of bytes occupied by one matrix element *times the number of rows*. In the fastest FORTRAN loop ordering, kji, the innermost loop (load/store on i) accesses the elements of the matrices **A** and **C** row-by-row within any given column. Therefore the stride is equal to the number of bytes occupied by one matrix element, which is as small as possible.

Although an accurate estimate of the matrix size at which performance degrades by 10% or more depends on system-specific details, it should be clear from this discussion that, in a large matrix computation, a little attention to the details of loop ordering and memory access can be rewarded with significant improvements in performance.

6.1.4 INVARIANT SUBSPACES AND DIRECT SUMS

A subspace W of a vector space V is called **invariant under a linear mapping** $A : V \to V$ if and only if

$$\overline{w} \in W \Rightarrow A\overline{w} \in W, \tag{6.78}$$

or, in other words, if and only if

$$AW \subseteq W. \tag{6.79}$$

Let $\{\overline{e}_1, \ldots, \overline{e}_m\}$ be a basis of W. Extend this basis of W to a basis $\{\overline{e}_1, \ldots, \overline{e}_m, \overline{e}_{m+1}, \ldots, \overline{e}_n\}$ of V. Because $AW \subseteq W$, it follows that for every basis element $\overline{e}_i$ of W, $A\overline{e}_i$ is equal to a linear combination of $\{\overline{e}_1, \ldots, \overline{e}_m\}$:

$$\forall i \in (1 : m) : A\overline{e}_i = \sum_{j=1}^{m} a_i^j \overline{e}_j. \tag{6.80}$$

Then the matrix elements of A in the first m columns must vanish below row m:

$$\forall i \in (1 : m) : \forall j \in (m + 1 : n) : a_i^j = 0. \tag{6.81}$$

The form of the matrix of a linear mapping that leaves a subspace W invariant therefore is

$$
\mathbf{A} =
\begin{array}{c}
\\ 1 \\ \vdots \\ m \\ m+1 \\ \vdots \\ n
\end{array}
\begin{array}{c}
\begin{array}{cccccc}
1 & \cdots & m & m+1 & \cdots & n
\end{array} \\
\left(
\begin{array}{cccccc}
a_1^1 & \cdots & a_m^1 & a_{m+1}^1 & \cdots & a_n^1 \\
\vdots & \ddots & \vdots & \vdots & \ddots & \vdots \\
a_1^m & \cdots & a_m^m & a_{m+1}^m & \cdots & a_n^m \\
0 & \cdots & 0 & a_{m+1}^{m+1} & \cdots & a_n^{m+1} \\
\vdots & \ddots & \vdots & \vdots & \ddots & \vdots \\
0 & \cdots & 0 & a_{m+1}^n & \cdots & a_n^n
\end{array}
\right)
\end{array}
\tag{6.82}
$$

relative to any basis of V that extends a basis of W.

Suppose that a vector space V is the direct sum of two complementary subspaces W and W', each of which is invariant under a linear mapping A:

$$V = W \oplus W', \quad AW \subseteq W, \quad AW' \subseteq W'. \tag{6.83}$$

This situation often arises in the problem of finding matrix eigenvalues and eigenvectors and also in group representation theory. Let $\{\bar{e}_i \mid i \in (1 : m)\}$ be a basis of W, and let $\{\bar{e}_j \mid j \in (m + 1 : n)\}$ be a basis of W'. The union of these two bases is a basis of V relative to which the matrix of A takes on the block-diagonal form

$$
\mathbf{A} = \begin{array}{c}
\begin{array}{c} 1 \\ \vdots \\ m \\ m+1 \\ \vdots \\ n \end{array}
\begin{pmatrix}
a_1^1 & \cdots & a_m^1 & 0 & \cdots & 0 \\
\vdots & \ddots & \vdots & \vdots & \ddots & \vdots \\
a_1^m & \cdots & a_m^m & 0 & \cdots & 0 \\
0 & \cdots & 0 & a_{m+1}^{m+1} & \cdots & a_n^{m+1} \\
\vdots & \ddots & \vdots & \vdots & \ddots & \vdots \\
0 & \cdots & 0 & a_{m+1}^n & \cdots & a_n^n
\end{pmatrix}
\end{array}
\tag{6.84}
$$

because, for every vector $\bar{w} \in W$, the image $A\bar{w}$ has no contravariant components along any of the basis vectors of W'.

Equivalently, if V is the direct sum of subspaces that are invariant under a linear mapping A, and if $\mathbf{A}$ is the matrix of A relative to *any* basis, then there exists a nonsingular matrix $\mathbf{T}$ such that $\mathbf{TAT}^{-1}$ takes on the block-diagonal form of Eq. (6.84). The matrix $\mathbf{T}^{-1}$ is the matrix that transforms the original basis into a basis with respect to which the matrix of A is block-diagonal; see Section 6.2.2.

If Eqs. (6.83) and (6.84) apply, the upper-left-hand block is the matrix of the restriction

$$A_W = A \upharpoonright W \tag{6.85}$$

of A to W, and the lower-right-hand block is the matrix of the restriction

$$A_{W'} = A \upharpoonright W' \tag{6.86}$$

of A to W'. (See Eq. [2.36] for the definition of the restriction of a mapping.) In this case one says that the linear mapping A is the **direct sum** of the linear mappings A_W and $A_{W'}$ and writes

$$A = A_W \oplus A_{W'}. \tag{6.87}$$

Conversely, whenever one encounters a vector space V such that $V = W \oplus W'$, then one can uniquely decompose any $\bar{x} \in V$ into the sum of vectors $\bar{w}$ and $\bar{w}'$ in W and W', respectively. If there exist linear mappings B : $W \to W$ of W into W and C : $W' \to W'$ of W' into W', then the linear mapping A such that

$$A\bar{x} := B\bar{w} + C\bar{w}' \tag{6.88}$$

is the direct sum A = B $\oplus$ C.

6.1.5 OTHER EXAMPLES OF LINEAR MAPPINGS

Complex Numbers

Considering the complex numbers $\mathbb{C}$ as a vector space over $\mathbb{R}$, we define the linear mapping

$$\forall z \in \mathbb{C} : \mathsf{A}z := \alpha z \tag{6.89}$$

in which $\alpha \in \mathbb{C}$ is any nonzero complex number. Certainly A is linear. From the polar decompositions $z = |z|e^{i\theta}$ and $\alpha = |\alpha|e^{i\phi}$ one sees that A rotates z through the angle ϕ and uniformly dilates z by the factor $|\alpha|$:

$$\alpha z = |\alpha| \cdot |z|e^{i(\theta + \phi)}. \tag{6.90}$$

To make contact with the theory developed in this section, let us map $\mathbb{C}$ onto $\mathbb{R}^2$ with the mapping $\mathsf{M} : 1 \mapsto \mathbf{e}_1, i \mapsto \mathbf{e}_2$. Then

$$\mathsf{M}z = \mathbf{z} = \begin{pmatrix} z' \\ z'' \end{pmatrix} \tag{6.91}$$

in which we write the real and imaginary parts of z and α as $z = z' + iz''$ and $\alpha = \alpha' + i\alpha''$ as usual. From the definition of a product in $\mathbb{C}$,

$$\mathsf{A}z = \alpha z = (\alpha'z' - \alpha''z'') + i(\alpha''z' + \alpha'z'') \tag{6.92}$$

$$\Rightarrow \mathbf{A}\mathbf{z} := \mathsf{M}(\alpha z) = \begin{pmatrix} \alpha'z' - \alpha''z'' \\ \alpha''z' + \alpha'z'' \end{pmatrix}. \tag{6.93}$$

Then the matrix in $\mathbb{R}^{2 \times 2}$ that realizes A is

$$\mathbf{A} = \mathsf{M}\mathsf{A}\mathsf{M}^{-1} = \begin{pmatrix} \alpha' & -\alpha'' \\ \alpha'' & \alpha' \end{pmatrix} = \begin{pmatrix} |\alpha| & 0 \\ 0 & |\alpha| \end{pmatrix} \begin{pmatrix} \cos\phi & -\sin\phi \\ \sin\phi & \cos\phi \end{pmatrix} \tag{6.94}$$

in which $\cos\phi = \alpha'/|\alpha|$, $\sin\phi = \alpha''/|\alpha|$, and $|\alpha| = [(\alpha')^2 + (\alpha'')^2]^{1/2}$. The first matrix on the right-hand side represents a uniform dilatation that stretches (if $|\alpha| > 1$) or compresses (if $|\alpha| < 1$) all vectors by the same factor, $|\alpha|$ (see Exercise 6.1.11). The second matrix on the right-hand side represents a rotation in $\mathbb{R}^2$ (see Exercise 6.1.13).

ℓ^2 Spaces

The sequence space $\ell^2(\mathbb{N}, \mathbb{C})$ is isomorphic to the state space of any discrete quantum system (see Section 5.2.2). For example, let the sequence $\bar{e}_n$ defined in Eq. (5.127) be the representative in $\ell^2(\mathbb{N}, \mathbb{C})$ of the state of a quantum-mechanical simple harmonic oscillator with n quanta $\hbar\omega$ in which $n \in \mathbb{N}$. The **harmonic-oscillator Hamiltonian** H such that

$$\forall n \in \mathbb{N} : \mathsf{H}\bar{e}_n := \left(n + \tfrac{1}{2}\right)\hbar\omega\bar{e}_n, \tag{6.95}$$

the **harmonic-oscillator annihilation operator** a such that

$$\begin{cases} a\bar{e}_n := \bar{0}, & \text{if } n = 1 \\ a\bar{e}_n := \sqrt{n}\bar{e}_{n-1}, & \text{if } n > 1, \end{cases} \tag{6.96}$$

and the **harmonic-oscillator creation operator** $a^\dagger$ such that

$$\forall n \in \mathbb{N} : a^\dagger \bar{e}_n := \sqrt{n+1}\bar{e}_{n+1} \tag{6.97}$$

are also linear operators whose domains are properly included in $\ell^2(\mathbb{Z}^+, \mathbb{F})$ (see Exercise 6.1.8). (Chapter 9 clarifies the notation $a^\dagger$. For now, one can regard $a^\dagger$ as just another linear mapping.)

It is straightforward to verify that

$$H = \hbar\omega\left(a^\dagger a + \frac{1}{2}\mathbf{1}\right) \tag{6.98}$$

(see Exercise 6.1.9). Note, however, that because n is unbounded and

$$a^\dagger a\bar{e}_n = n\bar{e}_n, \tag{6.99}$$

H is unbounded in the sense that there is no limit to the norm of the image $H\bar{u}$ of a unit vector $\bar{u}$.

Spaces of Differentiable Functions

Most of the fundamental laws of physics are expressed through differential equations of the form

$$L\bar{f} = \bar{b} \tag{6.100}$$

in which L is a linear differential operator (and therefore is a linear mapping) and $\bar{f}$ and $\bar{b}$ are functions belonging to some infinite-dimensional vector space $\mathcal{V}$ such as the intersection of $\mathbb{C}^\infty(\mathbb{R}^m; \mathbb{C})$ and the space of square-integrable real-valued functions with m real arguments, $\mathcal{L}^2(\mathbb{R}^m; \mathbb{R})$.

The equation

$$L\bar{f} = \bar{b}, \tag{6.101}$$

in which L is a linear differential operator, is the prototype of many of the inhomogeneous ordinary and partial differential equations of mathematical physics such as **Poisson's equation**

$$\nabla^2 \phi(\mathbf{r}) = -4\pi \rho(\mathbf{r}) \tag{6.102}$$

and the **inhomogeneous wave equation**

$$\left(\nabla^2 - \frac{1}{c^2}\frac{\partial^2}{\partial t^2}\right)\mathbf{A}(\mathbf{r}, t) = -\frac{4\pi}{c}\mathbf{J}(\mathbf{r}, t). \tag{6.103}$$

The **inverse problem** of obtaining $\overline{f}$ from Eq. (6.100) if L is a known linear differential operator and $\overline{b}$ is a known function is one of the key problems of mathematical physics.

Discretization and Interpolation Mappings

In order to perform numerical computations using equations that are expressed in terms of continuous functions, a physicist or engineer must somehow map $\mathcal{V}$ into $\mathbb{R}^n$ or $\mathbb{C}^n$. If one then approximates the operator L in Eq. (6.100) as a linear mapping $L_n : \mathbb{R}^n \to \mathbb{R}^n$ or $L_n : \mathbb{C}^n \to \mathbb{C}^n$ (for example), one obtains a system of linear equations that can be solved by Gaussian elimination or other methods. Because an understanding of this and closely related approaches is an essential tool for all physicists and for many engineers, it is important to understand the basic properties of mappings from infinite-dimensional to finite-dimensional vector space.

For the sake of simplicity and utility the only infinite-dimensional vector spaces in our discussion of discretization and interpolation mappings are spaces of continuously differentiable real-valued functions. Let $\mathcal{V}$ be the infinite-dimensional vector space of m-times continuously differentiable real-valued functions with r real arguments, $\mathcal{C}^m(\mathbb{R}^r; \mathbb{R})$. The **sampling mapping** is the mapping $S_n : \mathcal{V} \to \mathbb{R}^n$ such that

$$\forall f \in \mathcal{V} : S_n f := \begin{pmatrix} f(\mathbf{y}_1) \\ \vdots \\ f(\mathbf{y}_n) \end{pmatrix} \tag{6.104}$$

for a fixed set of **sampling points** points $\mathbf{y}_1, \ldots, \mathbf{y}_n \in \mathbb{R}^r$. The vector

$$\mathbf{f}_n := \begin{pmatrix} f(\mathbf{y}_1) \\ \vdots \\ f(\mathbf{y}_n) \end{pmatrix} \in \mathbb{R}^n \tag{6.105}$$

is the **vector of sampled values**. The mapping $S_n : f \mapsto \mathbf{f}_n$ is linear (see Exercise 6.1.17). S_n is an example of a **discretization mapping**, which is defined as a linear mapping from $\mathcal{C}^m(\mathbb{R}^r; \mathbb{R})$ into $\mathbb{R}^n$.

The goal of discretization is usually either to convert a function or an analog signal to a form that can conveniently be stored in a computer or to approximate the infinite-dimensional inverse problem defined in Eq. (6.100) by the finite-dimensional inverse problem

$$L_n \mathbf{f}_n = \mathbf{b}_n \tag{6.106}$$

in which $L_n \in \mathbb{F}^{n \times n}$. We discuss approximations under which one maps a differential operator L to a finite-dimensional matrix L_n in subsequent parts of the book.

Suppose now that one has obtained a solution $\mathbf{f}_n$ of the approximate inverse problem Eq. (6.106). In order to recover a continuously differentiable function that (one hopes) approximates the solution of Eq. (6.100), one must carry out an **approximation mapping**

$$A_n : \mathbb{R}^n \to \mathcal{V}, \tag{6.107}$$

which is defined as a linear mapping such that the function $A_n f_n$ approximates the true solution f according to some criterion of closeness. For example, for functions of a single real variable $(r = 1)$ the simplest choice is **linear interpolation**,

$$\forall i = 2, \ldots, n : \forall x \in [x_{i-1}, x_i] :$$

$$(A_n f_n)(x) := \frac{(x - x_{i-1}) f(x_i) + (x_i - x) f(x_{i-1})}{x_i - x_{i-1}}, \qquad (6.108)$$

which approximates the graph of f on the interval $[x_{i-1}, x_i]$ by a line segment drawn from $(x_{i-1}, f(x_{i-1}))$ to $(x_i, f(x_i))$. Other important approximation mappings include Lagrangian interpolation (see Section 7.6.1), spline interpolation, and approximation by a linear combination of orthogonal functions (see Section 8.4).

6.1.6 EXERCISES FOR SECTION 6.1

6.1.1   Let $\mathcal{V}_{(1)}$ and $\mathcal{V}_{(2)}$ be real or complex vector spaces over the same field $\mathbb{F} = \mathbb{R}$ or $\mathbb{C}$. Prove that the set of all linear mappings from $\mathcal{V}_{(1)}$ into $\mathcal{V}_{(2)}$, $\mathrm{hom}[\mathcal{V}_{(1)}, \mathcal{V}_{(2)}]$ is a vector space over $\mathbb{F}$.

6.1.2   Prove that if $A : \mathcal{V}_{(1)} \to \mathcal{V}_{(2)}$ is a linear mapping, then domain$[A]$ is a vector subspace of $\mathcal{V}_{(1)}$.

6.1.3   Prove that if $A : \mathcal{V}_{(1)} \to \mathcal{V}_{(2)}$ is a linear mapping, then range$[A]$ is a vector subspace of $\mathcal{V}_{(2)}$.

6.1.4   Find the rank of each of the following matrices by finding a basis of the column space:

(a)
$$\begin{pmatrix} 1 & 1 & 1 \\ 1 & 1 & 1 \\ 1 & 1 & 1 \end{pmatrix} \qquad (6.109)$$

(b)
$$\begin{pmatrix} 0 & 1 & 1 \\ 0 & 0 & 1 \\ 1 & 1 & 0 \end{pmatrix} \qquad (6.110)$$

(c)
$$\begin{pmatrix} 1 & 1 & 2 \\ 0 & 2 & 1 \\ 2 & 1 & 2 \end{pmatrix} \qquad (6.111)$$

(d)
$$\begin{pmatrix} 1 & 2 & 3 \\ 0 & 0 & 0 \\ 7 & 8 & 9 \end{pmatrix} \qquad (6.112)$$

6.1.5   The following questions concern the mapping $\mathbf{A} : \mathbb{R}^3 \to \mathbb{R}^2$ such that

$$\mathbf{A} \begin{pmatrix} x \\ y \\ z \end{pmatrix} = \begin{pmatrix} x + 3y + 2z \\ y + z \end{pmatrix}. \qquad (6.113)$$

(a) Show that **A** is a linear mapping.

(b) Find the matrix of **A** relative to the canonical basis $\{e_i\}$, Eq. (5.122).

(c) Find range[**A**] and rank[**A**].

6.1.6 This problem concerns the mapping $\mathbf{A} : \mathbb{R}^2 \to \mathbb{R}^2$ such that

$$\mathbf{Ax} = \mathbf{x}' \tag{6.114}$$

in which

$$\mathbf{x} = \begin{pmatrix} x \\ y \end{pmatrix}, \quad \mathbf{x}' = \begin{pmatrix} x + y \\ 3x + 2y \end{pmatrix}. \tag{6.115}$$

(a) Show that **A** is a linear mapping.

(b) Make coordinate expansions of **x** and **x**′ relative to the basis

$$\left\{ \mathbf{b}_1 := \begin{pmatrix} 1 \\ 2 \end{pmatrix}, \quad \mathbf{b}_2 := \begin{pmatrix} 3 \\ 1 \end{pmatrix} \right\}. \tag{6.116}$$

Hint: Set up and solve systems of linear equations to find the unknown contravariant components.

(c) Find the matrix of **A** relative to the basis of part (b).

6.1.7 Give at least one example of a sequence $\overline{x} \in \ell^2(\mathbb{Z}^+, \mathbb{C})$ such that

$$A\overline{x} \notin \ell^2(\mathbb{Z}^+, \mathbb{C}) \tag{6.117}$$

for each of the three cases $A = H$, $A = a$, and $A = a^\dagger$ (see Eqs. [6.95]–[6.97]).

6.1.8 Show that the harmonic-oscillator Hamiltonian is equal to

$$H = \hbar\omega \left(a^\dagger a + \tfrac{1}{2}\mathbf{1} \right) \tag{6.118}$$

in which H, the annihilation operator a, and the creation operator $a^\dagger$ are defined in Eqs. (6.95) through (6.97).

6.1.9 Show that

$$\forall n \in \mathbb{N} : [a, a^\dagger]\overline{e}_n = \overline{e}_n \tag{6.119}$$

in which

$$[a, a^\dagger] := aa^\dagger - a^\dagger a \tag{6.120}$$

is called the **commutator** of a and $a^\dagger$.

6.1.10 Find a 3×3 matrix **A** such that the vector equation in $\mathbb{R}^3$

$$\mathbf{c} = \mathbf{a} \times \mathbf{b} \tag{6.121}$$

can be written as

$$\mathbf{c} = \mathbf{Ab}. \tag{6.122}$$

6.1.11 Let $D : \mathbb{R}^2 \to \mathbb{R}^2$ map the contravariant components of a vector relative to the canonical basis $\{e_1, e_2\}$ as follows:

$$\begin{aligned} x^1 &\mapsto x'^1 = \alpha x^1 \\ x^2 &\mapsto x'^2 = \beta x^2 \end{aligned} \qquad (6.123)$$

in which α, $\beta \in [0, \infty)$.

(a) Demonstrate that D is a linear mapping in $\mathbb{R}^2$.

(b) Sketch the final shape of a unit square, the lower-left corner of which is located at the origin, for $\alpha = 2$, $\beta = 3$.

(c) Find the matrix of D relative to the basis $\{e_1, e_2\}$.

D is an example of a **deformation**.

6.1.12 Let $S : \mathbb{R}^2 \to \mathbb{R}^2$ map the contravariant components of a vector relative to the canonical basis $\{e_1, e_2\}$ as follows:

$$\begin{aligned} x^1 &\mapsto x'^1 = x^1 \\ x^2 &\mapsto x'^2 = x^2 + \alpha x^1 \end{aligned} \qquad (6.124)$$

in which α is real and nonzero.

(a) Demonstrate that S is a linear mapping in $\mathbb{R}^2$.

(b) Sketch the final shape of a unit square, the lower-left corner of which is located at the origin, for $\alpha = 1$.

(c) Find the matrix of S relative to the basis $\{e_1, e_2\}$.

S is an example of a **shear**. You may find it helpful to refer to Exercise 5.2.23.

6.1.13 Let $R : \mathbb{R}^2 \to \mathbb{R}^2$ map the contravariant components of a vector, relative to the canonical basis $\{e_1, e_2\}$, as follows:

$$\begin{aligned} x^1 &\mapsto x'^1 = (\cos \theta) x^1 + (\sin \theta) x^2 \\ x^2 &\mapsto x'^2 = -(\sin \theta) x^1 + (\cos \theta) x^2, \end{aligned} \qquad (6.125)$$

in which $\theta \in [0, 2\pi)$.

(a) Demonstrate that R is a linear mapping in $\mathbb{R}^2$.

(b) Sketch the final orientation of a unit square, the lower-left corner of which is located at the origin, for $\theta = \pi/3$.

(c) Find the matrix of R relative to the basis $\{e_1, e_2\}$.

R is an example of a **rotation**.

6.1.14 Let $\Lambda : \mathbb{R}^2 \to \mathbb{R}^2$ map the contravariant components of a vector $\mathbf{x}$, relative to the canonical basis $\{e_0, e_1\}$, as follows:

$$\begin{aligned} x^0 &\mapsto x'^0 = -(\sinh \chi) x^1 + (\cosh \chi) x^0 \\ x^1 &\mapsto x'^1 = (\cosh \chi) x^1 - (\sinh \chi) x^0 \end{aligned} \qquad (6.126)$$

in which χ is real and positive.

    (a) Show that Λ is a Lorentz transformation (in two-dimensional space-time) if one makes the following identifications:

$$x^0 = ct$$
$$x^1 = x$$
$$\chi = \tanh^{-1}\beta \tag{6.127}$$
$$\beta = \frac{v}{c}$$

    (in which v is velocity and c is the speed of light).

    (b) Demonstrate that Λ is a linear mapping in $\mathbb{R}^2$.

    (c) Find the matrix of Λ relative to the basis $\{\mathbf{e}_1, \mathbf{e}_2\}$.

6.1.15 Establish Eq. (6.77).

6.1.16 Derive the 3×3 matrix $\mathbf{P}$ (relative to the Cartesian basis $\{\mathbf{i}, \mathbf{j}, \mathbf{k}\}$) such that for every vector $\mathbf{a} \in \mathbb{R}^3$,

$$\mathbf{Pa} = \mathbf{n}(\mathbf{n} \cdot \mathbf{a}) \tag{6.128}$$

in which $\mathbf{n}$ is an arbitrary unit vector.

6.1.17 Demonstrate that the mappings S_n defined in Eq. (6.104) and A_n defined in Eq. (6.108) are linear mappings.

6.2 NONSINGULAR LINEAR MAPPINGS

6.2.1 DEFINITIONS AND BASIC PROPERTIES

A linear mapping $\mathsf{A} : \mathcal{V}_{(1)} \to \mathcal{V}_{(2)}$ is called **nonsingular** if and only if it is one-to-one, that is, if and only if

$$\mathsf{A}\overline{x} = \mathsf{A}\overline{y} \Rightarrow \overline{x} = \overline{y}. \tag{6.129}$$

Because a linear mapping A is additive, an equivalent definition is that A is nonsingular if and only if

$$\ker[\mathbf{A}] = \mathcal{O} \tag{6.130}$$

(in other words, $\mathsf{A}(\overline{x} - \overline{y}) = \overline{0}$ if and only if $\overline{x} - \overline{y} = \overline{0}$). Therefore a linear mapping A is nonsingular if and only if

$$\mathsf{A}\overline{x} = \overline{0} \Rightarrow \overline{x} = \overline{0}. \tag{6.131}$$

For example, the linear mapping

$$A : \mathbf{x} = \begin{pmatrix} x^1 \\ x^2 \end{pmatrix} \mapsto \mathbf{x}' = \begin{pmatrix} 3x^1 + 2x^2 \\ \frac{14}{3}x^2 \end{pmatrix} \tag{6.132}$$

is nonsingular because $A\mathbf{x} = \mathbf{0}$ if and only if $\frac{14}{3}x^2 = 0$ and $3x^1 + 2x^2 = 0$ and hence if and only if $\mathbf{x} = \mathbf{0}$.

A nonsingular linear mapping $A : \text{domain}[A] \to \text{range}[A]$ satisfies the definition of a vector isomorphism, Eqs. (5.178) through (5.181). It follows that if a linear mapping A is nonsingular, then its range and domain are isomorphic,

$$A \text{ nonsingular} \Rightarrow \text{range}[A] \cong \text{domain}[A], \tag{6.133}$$

regardless of the dimension of $\mathcal{V}_{(1)}$. For example, the domain and range of the nonsingular linear mapping Eq. (6.132) both have dimension 2 over the real field and therefore are isomorphic. The converse of Eq. (6.133) holds only in finite-dimensional spaces; see below.

It is not necessary for a linear mapping $A : \mathcal{V}_{(1)} \to \mathcal{V}_{(2)}$ to map $\mathcal{V}_{(1)}$ onto $\mathcal{V}_{(2)}$ (as in Eq. [6.132]) in order for A to be nonsingular. The range of A may be a proper subspace of $\mathcal{V}_{(2)}$. For example, let $\mathcal{V}_{(1)} = \mathbb{R}$ and $\mathcal{V}_{(2)} = \mathbb{R}^2$. Define the mapping

$$\forall x \in \mathbb{R} : x \mapsto Ax = \begin{pmatrix} 2 \\ 1 \end{pmatrix} x = \begin{pmatrix} 2x \\ x \end{pmatrix} \in \mathbb{R}^2. \tag{6.134}$$

The domain of A, $\mathbb{R}$, is one-dimensional. The range of A is the one-dimensional subspace of $\mathbb{R}^2$ spanned by the vector $\mathbf{w} = (2, 1)^T$. Clearly $Ax = \mathbf{0}$ if and only if $x = 0$. Therefore A is nonsingular and $\text{range}[A] \cong \text{domain}[A]$, but $\text{range}[A] \subset \mathcal{V}_{(2)}$.

If $\mathcal{V}_{(1)}$ is finite-dimensional, then the converse of Eq. (6.133) holds. That is, if $\text{range}[A] \cong \text{domain}[A]$ and if $\mathcal{V}_{(1)}$ is finite-dimensional, then $A : \mathcal{V}_{(1)} \to \mathcal{V}_{(2)}$ is nonsingular. The proof is a matter of applying the definition of a vector isomorphism: Since $\mathcal{V}_{(1)}$ is of finite dimension and $\text{range}[A] \cong \text{domain}[A]$, $\text{domain}[A] = \mathcal{V}_{(1)}$. Then the mapping $\bar{e}_i \mapsto \bar{a}_i$ for $i \in (1 : \dim[\mathcal{V}_{(1)}])$ is a vector isomorphism and hence is one-to-one by definition. Then Eq. (6.129) holds, implying that A is nonsingular.

For example, the matrix $\mathbf{A}$ defined in Eq. (6.48) is nonsingular because $\dim[\text{range}[\mathbf{A}]] = \text{rank}[\mathbf{A}] = 2 = \dim[\text{domain}[\mathbf{A}]]$. Because the dimensions of $\text{domain}[\mathbf{A}]$ and $\text{range}[\mathbf{A}]$ are equal, it follows that $\text{range}[A] \cong \text{domain}[A]$. Then $\mathbf{A}$ is nonsingular.

In a finite-dimensional vector space $\mathcal{V}_{(1)}$, the simplest criterion that ensures that a linear mapping A is nonsingular is that A must have **full rank**, that is,

$$\text{rank}[A] = \dim[\mathcal{V}_{(1)}]. \tag{6.135}$$

A linear mapping, or a matrix, that does not have full rank is called **rank-deficient**.

Let $\mathbf{A}$ be an $n_2 \times n_1$ matrix. The dimension of $\mathcal{V}_{(1)}$ is n_1 and the dimension of $\mathcal{V}_{(2)}$ is n_2. Because the dimension of the column space of $\mathbf{A}$ cannot exceed n_2, it follows that a necessary condition for $\mathbf{A}$ to have full rank is that

$$n_1 \leq n_2. \tag{6.136}$$

That is, **A** cannot be nonsingular unless the number of rows is at least as great as the number of columns. A matrix that has more columns than rows violates Eq. (6.136) and therefore is singular (see Section 6.3).

Although it is true in both finite-dimensional and infinite-dimensional vector spaces that if a linear mapping A is nonsingular, then range[A] $\cong$ domain[A], the converse ("if range[A] $\cong$ domain[A], then A is nonsingular") does *not* hold in infinite-dimensional spaces. Section 6.3.4 provides a counterexample involving shift operators on sequence spaces.

A nonsingular linear mapping A is also called **invertible** because it is possible to define a unique **inverse linear mapping** A^{-1} : range[A] $\rightarrow$ domain[A] such that

$$\forall \bar{x} \in \text{domain}[A] : \forall \bar{x}' \in \text{range}[A] :\ni: A\bar{x} = \bar{x}' :\ A^{-1}\bar{x}' := \bar{x}. \tag{6.137}$$

If A is nonsingular, then by the definition of an inverse

$$AA^{-1} = 1_{\text{range}[A]}, \qquad A^{-1}A = 1_{\text{domain}[A]}. \tag{6.138}$$

Note that A^{-1} goes from range[A] to domain[A] and not from $\mathcal{V}_{(2)}$ to domain[A]. The distinction is important if range[A] is a proper subset of $\mathcal{V}_{(2)}$. For example, the inverse of the linear mapping Eq. (6.134) maps a two-dimensional vector to the real number that is equal to the vector's first component,

$$\forall \mathbf{w} \in \mathcal{W} : \mathbf{w} \mapsto x = w^2 \tag{6.139}$$

in which $\mathcal{W}$ is the subspace of $\mathbb{R}^2$ generated by the vector $(2, 1)^T$.

An $n_2 \times n_1$ matrix **A** that realizes a nonsingular linear mapping is called a **nonsingular matrix**. If the range of **A** is all of $\mathbb{F}^{n_2}$, then $\dim[\text{range}[\mathbf{A}]] = n_2 = \dim[\text{domain}[\mathbf{A}]] = n_1$, which implies that **A** is a square matrix. A rectangular matrix such as $(Z, 1)^T$ is nonsingular, but its range is not all of $\mathbb{F}^{n_2} = \mathbb{R}^2$ in this example.

Every nonsingular $n_1 \times n_1$ matrix possesses a unique $n_1 \times n_1$ **inverse matrix** $\mathbf{A}^{-1}$, which is defined as A^{-1} is defined in Eq. (6.137). The matrix $\mathbf{A}^{-1}$ that realizes A^{-1} has the properties that

$$\mathbf{A}\mathbf{A}^{-1} = \mathbf{A}^{-1}\mathbf{A} = 1_{n_1 \times n_1} \tag{6.140}$$

in which $1_{n_1 \times n_1}$ means the $n_1 \times n_1$ identity matrix.

For example, the inverse of the matrix **A** defined in Eq. (6.48) is

$$\mathbf{A}^{-1} = \frac{1}{14}\begin{pmatrix} 6 & -2 \\ -2 & 3 \end{pmatrix} \Rightarrow \mathbf{A}\mathbf{A}^{-1} = \mathbf{A}^{-1}\mathbf{A} = \begin{pmatrix} 1 & 0 \\ 0 & 1 \end{pmatrix}. \tag{6.141}$$

The inverse of a 2×2 matrix can be computed conveniently by using Cramer's rule for the solution of a system of linear equations $\mathbf{Ax} = \mathbf{b}$, which gives an explicit formula in terms of determinants for $\mathbf{x} = \mathbf{A}^{-1}\mathbf{b}$. Cramer's formula for $\mathbf{x}$ is, in effect, a formula for the inverse of the coefficient matrix **A**. For the reasons stated in Section 6.5.2, Cramer's formula is not computationally useful unless the dimensions of **A** are 2×2.

As is true for mappings in general, the inverse of the product $\mathbf{BA}$ of two nonsingular square matrices $\mathbf{B}$ and $\mathbf{A}$ is the product of the inverses taken in the opposite order,

$$(\mathbf{BA})^{-1} = \mathbf{A}^{-1}\mathbf{B}^{-1}. \tag{6.142}$$

Taking the transpose of both sides of Eq. (6.142) and using Eq. (6.77), one obtains

$$(\mathbf{A}^{-1})^T \mathbf{A}^T = \mathbf{A}^T (\mathbf{A}^{-1})^T = \mathbf{1}, \tag{6.143}$$

which says that for nonsingular square matrices, the inverse of the transpose is the transpose of the inverse:

$$(\mathbf{A}^T)^{-1} = (\mathbf{A}^{-1})^T. \tag{6.144}$$

A nonsingular rectangular matrix such as

$$\mathbf{A} = \begin{pmatrix} 2 \\ 1 \end{pmatrix}, \tag{6.145}$$

which realizes the linear mapping Eq. (6.134) relative to the canonical basis in $\mathbb{R}^2$, has no inverse matrix in the sense of Eq. (6.140). Instead, a nonsingular rectangular matrix possesses a *left inverse* such that

$$\mathbf{A}^L \mathbf{A} = \mathbf{1}_{n_1 \times n_1}. \tag{6.146}$$

For example,

$$\mathbf{A}^L = \tfrac{1}{5}(2, 1) \tag{6.147}$$

is a left inverse of the matrix Eq. (6.145), for

$$\mathbf{A}^L \mathbf{A} = \frac{1}{5}(2, 1) \begin{pmatrix} 2 \\ 1 \end{pmatrix} = 1 \in \mathbb{R}. \tag{6.148}$$

The matrix

$$\mathbf{A}'^L = (0, 1) \tag{6.149}$$

is another left inverse of Eq. (6.145). Unlike the inverse of a nonsingular square matrix, a left inverse of a nonsingular rectangular matrix cannot be unique because it must map a space of dimension n_2 onto a space of a smaller dimension. The mapping Eq. (6.139) of the one-dimensional subspace $\mathcal{W}$ onto the real line is one-to-one, but there are infinitely many ways to go in the other direction, mapping a vector $\mathbf{r} \in \mathbb{R}^2$ into $\mathcal{W}$. It can be shown that Eq. (6.147) performs an orthogonal projection of $\mathbf{r}$ onto $\mathcal{W}$ (as described in Sections 8.2.1 and 8.3) and then performs the inverse mapping from $\mathcal{W}$ to $\mathbb{R}$ defined in Eq. (6.139). Equation (6.149) makes a parallel projection of $\mathbf{r}$ onto $\mathcal{W}$ (as described in Section 5.4.1) before performing the mapping Eq. (6.139).

6.2.2 CHANGE OF BASIS

One can consider a nonsingular linear mapping $\mathsf{T} : \mathcal{V} \to \mathcal{V}$ of a vector space $\mathcal{V}$ into itself as a change of basis. In other words, one adopts a passive point of view in which vectors remain the same but the bases change (see Appendix B). Let the coordinate expansions of a vector $\bar{x}$ relative to the bases $\{\bar{e}_i\}$ and $\{\bar{e}'_j\}$ be

$$x^i \bar{e}_i = x'^j \bar{e}'_j \tag{6.150}$$

in which (by Eq. [B.29])

$$\bar{e}'_j = \mathsf{B}\bar{e}_j \tag{6.151}$$

and B is a vector automorphism. Let

$$\mathsf{T} := \mathsf{B}^{-1}. \tag{6.152}$$

Therefore

$$\bar{e}_i = \mathsf{T}\bar{e}'_i. \tag{6.153}$$

In terms of the matrix elements t_i^j of T, Eq. (6.153) reads

$$\bar{e}_i = t_i^j \bar{e}'_j. \tag{6.154}$$

If we substitute this equation into Eq. (6.150), we obtain a formula for the contravariant components of $\bar{x}$ relative to the new basis:

$$x'^j = t_i^j x^i. \tag{6.155}$$

The matrix realization of Eq. (6.155) is

$$\mathbf{x}' = \mathbf{T}\mathbf{x} \tag{6.156}$$

in which the column vectors $\mathbf{x}$ and $\mathbf{x}'$ represent the same abstract vector $\bar{x}$ relative to different bases and $\mathbf{T}$ is a nonsingular matrix.

Let $\mathsf{A} : \mathcal{V} \to \mathcal{V}$ be any linear mapping in $\mathcal{V}$, let $\mathbf{A}$ be the matrix of A relative to the basis $\{\bar{e}_i\}$, and let $\bar{y}$ be the image of $\bar{x}$ under A:

$$\bar{y} = \mathsf{A}\bar{x}. \tag{6.157}$$

The matrix realization of Eq. (6.157) is

$$\mathbf{y} = \mathbf{A}\mathbf{x}. \tag{6.158}$$

To calculate the matrix of the linear mapping A relative to the new basis $\{\bar{e}'_j\}$ (Eq. [6.151]), note that (by Eq. [6.156]) the column vector that represents $\bar{y}$ relative to the new basis is

$$\mathbf{y}' = \mathbf{T}\mathbf{y} = \mathbf{T}\mathbf{A}\mathbf{x} = \mathbf{T}\mathbf{A}\underbrace{\mathbf{T}^{-1}\mathbf{T}}_{\mathbf{1}}\mathbf{x}. \tag{6.159}$$

Then

$$\mathbf{y}' = \mathbf{A}'\mathbf{x}' \tag{6.160}$$

in which the matrix

$$\mathbf{A}' = \mathbf{T}\mathbf{A}\mathbf{T}^{-1} \tag{6.161}$$

realizes the same abstract linear mapping A as the matrix $\mathbf{A}$ relative to a different basis.

The matrix $\mathbf{A}' = \mathbf{T}\mathbf{A}\mathbf{T}^{-1}$ is called the **similarity transform** of $\mathbf{A}$ by $\mathbf{T}$. $\mathbf{A}$ and $\mathbf{A}'$ are called **similar matrices**. Because the similarity transform of a matrix represents the same abstract linear mapping as the original matrix (or by Exercise 6.1.10), all matrices that are similar to one another have the same rank:

$$\text{rank}[\mathbf{A}'] = \text{rank}[\mathbf{T}\mathbf{A}\mathbf{T}^{-1}] = \text{rank}[\mathbf{A}] = \text{rank}[A]. \tag{6.162}$$

As we have defined it here, a similarity transformation can be applied only to square matrices.

More generally, let $\mathbf{A}$ be an $n_2 \times n_1$ rectangular matrix that realizes a linear mapping $A : \mathcal{V}_{(1)} \to \mathcal{V}_{(2)}$, relative to bases $\{\bar{e}_{(1)i}\}$ and $\{\bar{e}_{(2)j}\}$, respectively. Make the basis changes

$$\begin{aligned} \bar{e}'_{(1)i} &= \mathbf{T}^{-1}_{(1)}\bar{e}_{(1)i} \\ \bar{e}'_{(2)j} &= \mathbf{T}^{-1}_{(2)}\bar{e}_{(2)j}. \end{aligned} \tag{6.163}$$

In $\mathcal{V}_{(1)}$, the column vectors $\mathbf{x}_{(1)}$ and $\mathbf{x}'_{(1)}$, in which

$$\mathbf{x}'_{(1)} = \mathbf{T}_{(1)}\mathbf{x}_{(1)}, \tag{6.164}$$

represent the same abstract vector $\bar{x}_{(1)}$ relative to the bases $\{\bar{e}_{(1)i}\}$ and $\{\bar{e}'_{(1)i}\}$, respectively. Likewise,

$$\mathbf{x}'_{(2)} = \mathbf{T}_{(2)}\mathbf{x}_{(2)} \tag{6.165}$$

holds for an abstract vector $\bar{x}_{(2)} \in \mathcal{V}_{(2)}$. If

$$\mathbf{y}_{(2)} = \mathbf{A}\mathbf{x}_{(1)}, \tag{6.166}$$

relative to the bases $\{\bar{e}_{(1)i}\}$ and $\{\bar{e}_{(2)j}\}$, then relative to the bases $\{\bar{e}'_{(1)i}\}$ and $\{\bar{e}'_{(2)j}\}$ one has

$$\mathbf{y}'_{(2)} = \mathbf{A}'\mathbf{x}'_{(1)} \tag{6.167}$$

in which

$$\mathbf{A}' = \mathbf{T}_{(2)}\mathbf{A}\mathbf{T}_{(1)}^{-1}. \tag{6.168}$$

We shall apply basis changes to general rectangular matrices in deriving the singular-value decomposition (see Section 9.4.1).

6.2.3 PERMUTATION MATRICES

If the elements of a basis are permuted as in Eq. (5.185),

$$A_\pi \bar{e}_i = \bar{e}_{\pi(i)}, \tag{6.169}$$

the contravariant components of a vector $\bar{x} = x^i \bar{e}_i$ relative to the new basis are

$$x'^j = x^{\pi^{-1}(j)} \tag{6.170}$$

according to Eq. (5.186). The transformation from $\{x^i\}$ to $\{x'^j\}$ can be written as a linear transformation $x'^j = a_i^j x^i$ if the elements a_i^j of a matrix $\mathbf{A}_\pi$ are defined as follows:

$$a_i^j = \delta_i^{\pi^{-1}(j)}. \tag{6.171}$$

A matrix whose elements obey this equation for some permutation π is called a **permutation matrix**. Because there exists a unique integer i such that $i = \pi(j)$ for $j = 1, 2, \ldots, n$, only one matrix element can be nonzero in each row and in each column of a permutation matrix, and that nonzero element is equal to 1. Conversely, a matrix in which all nonzero elements are equal to 1 and in which only one element in each row and column is nonzero is the matrix of some permutation.

Every permutation matrix $\mathbf{A}_\pi$ is nonsingular. It follows at once from the defining Eq. (6.169) that the inverse of $\mathbf{A}_\pi$ is $\mathbf{A}_{\pi^{-1}}$.

For example, the matrix of the permutation (123) constructed according to Eq. (6.171) is

$$\mathbf{A}_{(123)} = \begin{pmatrix} 0 & 0 & 1 \\ 1 & 0 & 0 \\ 0 & 1 & 0 \end{pmatrix}. \tag{6.172}$$

Indeed,

$$\mathbf{A}_{(123)}\mathbf{e}_1 = \mathbf{e}_2, \quad \mathbf{A}_{(123)}\mathbf{e}_2 = \mathbf{e}_3, \quad \mathbf{A}_{(123)}\mathbf{e}_3 = \mathbf{e}_1 \tag{6.173}$$

as one would expect. If a permutation matrix $\mathbf{A}_\pi$ acts on a column vector $\mathbf{x}$, the effect is to apply the permutation π^{-1} to the rows of $\mathbf{x}$; see Eqs. (5.187) and (5.188).

6.2.4 GENERAL LINEAR GROUP OF A VECTOR SPACE

The nonsingular linear mappings of a vector space $\mathcal{V}$ onto itself and the operation of composition defined in Eq. (6.8) satisfy the group axioms: The composition of two nonsingular linear mappings of $\mathcal{V}$ is another nonsingular linear mapping of $\mathcal{V}$, the composition of mappings is associative, and there exists an identity linear mapping. We have already shown that if A is nonsingular, then A^{-1} exists. The group of nonsingular linear mappings of an n-dimensional vector space $\mathcal{V}$ over a field $\mathbb{F}$ is called the **general linear group** of $\mathcal{V}$ over $\mathbb{F}$ and is denoted $GL(\mathcal{V})$. Because all vector spaces over $\mathbb{F}$ with the same dimension n are isomorphic, their general linear groups are isomorphic too (see Exercise 6.2.1). Therefore one writes $GL(n, \mathbb{F})$ instead of $GL(\mathcal{V})$ unless one wishes to emphasize the underlying vector space $\mathcal{V}$. The set of nonsingular matrices with elements in the field $\mathbb{F} = \mathbb{R}$ or $\mathbb{C}$ is a group that is isomorphic to $GL(n, \mathbb{F})$ (see Exercise 6.2.2).

Many matrix groups that are important in physics, including $SU(n)$, $U(n)$, $O(n)$, and $SO(n)$, to name a few, are subgroups of $GL(n, \mathbb{C})$.

6.2.5 EXERCISES FOR SECTION 6.2

6.2.1 Prove that if $\mathcal{V}_{(1)}$ and $\mathcal{V}_{(2)}$ are n-dimensional vector spaces over the same field $\mathbb{F}$, then their general linear groups are isomorphic:

$$GL\big(\mathcal{V}_{(1)}\big) \cong GL\big(\mathcal{V}_{(2)}\big). \tag{6.174}$$

6.2.2 Prove that if $\mathcal{V}$ is an n-dimensional vector space over a field $\mathbb{F}$, then the general linear group of $\mathcal{V}$ is isomorphic to the group of nonsingular matrices with elements in $\mathbb{F}$.

6.2.3 A translation mapping $T_{\bar{a}}$ can be realized by a matrix by the following method (or trick, depending on one's point of view):

(a) Let $\mathbf{x} \in \mathbb{F}^n$. Prove that the mapping $\mathsf{M} : \mathbb{F}^n \to \mathbb{F}^{n+1}$ such that

$$\mathsf{M}\mathbf{x} = \mathsf{M} \begin{pmatrix} x^1 \\ x^2 \\ \vdots \\ x^n \end{pmatrix} := \begin{pmatrix} x^1 \\ x^2 \\ \vdots \\ x^n \\ 1 \end{pmatrix} \tag{6.175}$$

is one-to-one. Is M a *linear* mapping? Show how to define M^{-1}.

(b) Prove that the $(n + 1) \times (n + 1)$ matrix

$$
\mathsf{T_a} := \begin{pmatrix}
1 & 0 & 0 & \cdots & 0 & 0 & a^1 \\
0 & 1 & 0 & \cdots & 0 & 0 & a^2 \\
0 & 0 & 1 & \cdots & 0 & 0 & a^3 \\
\vdots & \vdots & \vdots & \ddots & \vdots & \vdots & \vdots \\
0 & 0 & 0 & \cdots & 1 & 0 & a^{n-1} \\
0 & 0 & 0 & \cdots & 0 & 1 & a^n \\
0 & 0 & 0 & \cdots & 0 & 0 & 1
\end{pmatrix}
\tag{6.176}
$$

is nonsingular.

(c) Prove that

$$
\mathsf{M}^{-1}\mathsf{T_a}\begin{pmatrix} x^1 \\ x^2 \\ \vdots \\ x^n \\ 1 \end{pmatrix} = \begin{pmatrix} x^1 + a^1 \\ x^2 + a^2 \\ \vdots \\ x^n + a^n \end{pmatrix}.
\tag{6.177}
$$

If $n = 2$, the transpose of $\mathsf{T_a}$ is a special case of the **current transformation matrix** in the PostScript™ page description language, which is used widely for computer graphics.

6.2.4   Show that

$$
\mathsf{T_a}\mathsf{T_b} = \mathsf{T_{a+b}}
\tag{6.178}
$$

in which $\mathsf{T_a}$ is defined in Eq. (6.176). Is the set of matrices $\{\mathsf{T_a} \,|\, \mathbf{a} \in \mathbb{R}^n\}$ a representation of $T(\mathbb{R}^n)$?

6.2.5   Let $\mathbb{R}^2$ be realized as the x–y plane, and let $\mathbf{n}$ be a unit vector along the z-axis.

(a) Show that if $\mathbf{a} \in \mathbb{R}^2$ is a nonzero vector in the plane, then $\mathbf{a}$ and $\mathbf{n} \times \mathbf{a}$ span $\mathbb{R}^2$.

(b) Let $\mathbf{a}'$ be the image of $\mathbf{a}$ under an active rotation in the positive sense about the axis $\mathbf{n}$ through the angle θ. Show that

$$
\mathbf{a}' = (\cos\theta)\mathbf{a} + (\sin\theta)\mathbf{n} \times \mathbf{a}.
\tag{6.179}
$$

(c) From the result of (b), obtain the matrix $\mathbf{R}$ such that

$$
\mathbf{a}' = \mathbf{R}\mathbf{a}.
\tag{6.180}
$$

6.2.6   The goal of this problem is to construct an explicit expression for the matrix of a three-dimensional rotation in terms of the axis and angle of rotation and to show that the resulting matrix is nonsingular. In Fig. 6.4, $\mathbf{n}$ is a unit vector that points along the axis of rotation and θ is the angle of rotation. In order to derive the form of a rotation matrix in $\mathbb{R}^3$, it is convenient to reduce the problem to one in $\mathbb{R}^2$ (see the previous exercise).

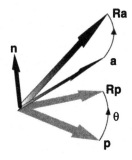

FIGURE 6.4

Finite rotation of a vector **a** about an axis **n**. The angle of rotation is θ. The image of **a** under the rotation is **Ra**. The orthogonal projections **p** and **Rp** of **a** and **Ra** on the plane perpendicular to **n** are shown as shadows.

(a) Let $\mathbf{a} \in \mathbb{R}^3$. Show that the component of **a** perpendicular to **n** is

$$\mathbf{p} = \mathbf{n} \times (\mathbf{a} \times \mathbf{n}). \tag{6.181}$$

Check that $\mathbf{p} \cdot \mathbf{n} = 0$; hence **p** is in the plane perpendicular to **n**.

(b) Show that the component of **a** parallel to **n** is

$$\mathbf{z} = \mathbf{a} - \mathbf{p} = \mathbf{n}(\mathbf{n} \cdot \mathbf{a}). \tag{6.182}$$

(c) Let $R = R_{\mathbf{n},\theta}$ be the linear mapping that rotates the vector **a** about the axis specified by the unit vector **n** through the angle θ in the positive (that is, right-handed) sense of rotation. Let **R** be the matrix of R relative to the Cartesian basis $\{\mathbf{i}, \mathbf{j}, \mathbf{k}\}$. Show that

$$\mathbf{Rz} = R_{\mathbf{n},\theta}\mathbf{z} = \mathbf{z} \tag{6.183}$$

and

$$\mathbf{Rp} = R_{\mathbf{n},\theta}\mathbf{p} = \mathbf{b} + \mathbf{c} \tag{6.184}$$

in which

$$\begin{aligned} \mathbf{b} &= \cos\theta\,\mathbf{p}, \\ \mathbf{c} &= \sin\theta\,\mathbf{n} \times \mathbf{p}. \end{aligned} \tag{6.185}$$

(d) Use the preceding results to show that

$$\mathbf{Ra} = R_{\mathbf{n},\theta}\mathbf{a} = \cos\theta\,\mathbf{a} + (1 - \cos\theta)\,\mathbf{n}(\mathbf{n} \cdot \mathbf{a}) + \sin\theta\,\mathbf{n} \times \mathbf{a} \tag{6.186}$$

and that $R_{\mathbf{n},\theta}$ is a linear mapping.

(e) Use the properties of the vector dot and cross products to show that $R_{\mathbf{n},\theta}$ is nonsingular.

(f) Obtain the matrix **R** relative to the basis $\{\mathbf{i}, \mathbf{j}, \mathbf{k}\}$.

6.2.7 Prove that the conjugacy class of $\mathbf{A} \in GL(n, \mathbb{F})$ is the set of matrices over $\mathbb{F}$ that are similar to **A**.

6.2.8   Prove that the similarity relation between matrices is an equivalence relation. In other words, show that **A** is similar to itself, that if **B** is similar to **A** then **A** is similar to **B**, and that if **B** is similar to **A** and **C** is similar to **B** then **C** is similar to **A**.

6.2.9   Construct the inverse of the permutation matrix $\mathbf{A}_{(123)}$ using Eq. (6.171).

6.2.10   Prove that if **A** and **B** are similar matrices, then

$$\text{rank}[\mathbf{B}] = \text{rank}[\mathbf{A}]. \tag{6.187}$$

6.3 SINGULAR LINEAR MAPPINGS

6.3.1 SINGULARITY AND LINEAR DEPENDENCE

A linear mapping A that is *not* one-to-one is called **singular**. A finite-dimensional matrix **A** that realizes a singular linear mapping A is called a **singular matrix**.

We show that a matrix **A** is singular if and only if its column vectors are linearly dependent. Assume that **A** is singular. If the columns of **A** were linearly independent, then by Eq. (6.21) the dimension of range[**A**] would be equal to n_1, the dimension of $\mathcal{V}_{(1)}$. We would therefore have the contradiction that **A** is nonsingular. This establishes the "only if" part of the proposition. If we assume that the columns of **A** are linearly dependent, then rank[**A**] $< n_1$. Therefore **A** is singular, by the contrapositive of Eq. (6.133). This establishes the "if" part and completes the proof.

For example, the columns of the matrix **B** defined in Eq. (6.54) are linearly dependent, according to Eqs. (6.55) through (6.58). Therefore **B** is singular. To see that **B** is not one-to-one, let $\mathbf{B} = (\mathbf{b}_1, \mathbf{b}_2, \mathbf{b}_3)$ be the column partition of **B**. Express the columns of **B**, which are the images of the canonical basis, in terms of the basis $\{\mathbf{g}_1, \mathbf{g}_2\}$ of the range of **B** defined in Eq. (6.58):

$$\begin{aligned}
\mathbf{b}_1 &= \mathbf{B}\mathbf{e}_1 = \mathbf{g}_1 + \mathbf{g}_2 \\
\mathbf{b}_2 &= \mathbf{B}\mathbf{e}_2 = 2\mathbf{g}_1 + \mathbf{g}_2 \\
\mathbf{b}_3 &= \mathbf{B}\mathbf{e}_3 = \mathbf{g}_1 + 2\mathbf{g}_2.
\end{aligned} \tag{6.188}$$

It follows that

$$\mathbf{B}(-2\mathbf{e}_1 + \mathbf{e}_2 + \mathbf{e}_3) = \mathbf{g}_1 + \mathbf{g}_2 = \mathbf{B}\mathbf{e}_1. \tag{6.189}$$

Because **B** carries the two different vectors

$$\mathbf{e}_1, \quad -2\mathbf{e}_1 + \mathbf{e}_2 + \mathbf{e}_3 \tag{6.190}$$

onto the same image, $\mathbf{g}_1 + \mathbf{g}_2$, **B** is not one-to-one and therefore is singular.

One of the most important practical mathematical problems that a working engineer or physicist can face is the detection of singularity or nearness to singularity of a matrix. Unfortunately there is usually no way to tell at a glance whether a given matrix **A** is singular. We see in Section 6.5.2 that a matrix is singular if and only if its determinant vanishes. But testing whether $\det[\mathbf{A}] = 0$ by traditional methods is computationally infeasible in most practical cases. Practical tests for nearness to singularity depend on the singular-value decomposition derived in Section 9.4.1

6.3.2 VISUALIZATION OF SINGULAR LINEAR MAPPINGS

The key feature of a singular linear mapping A, the reduced dimensionality of its range as compared with its domain, can be strikingly evident in visualizations of A. In Figs. 6.1 and 6.2 we saw an example of visualization of a nonsingular linear mapping in two dimensions. For an example of visualization of a singular linear mapping, consider the matrix $\mathbf{B}$ defined in Eq. (6.54). The basis vectors $\mathbf{e}_1$, $\mathbf{e}_2$, $\mathbf{e}_3$ should be imagined as mutually perpendicular vectors of unit length pointing along the X, Y, and Z axes, as usual. Figure 6.5 shows the basis-vector images $\mathbf{b}_1$, $\mathbf{b}_2$, $\mathbf{b}_3$ as lines seen from two different viewpoints. The unit vector $\mathbf{e}_1$ along the X axis is mapped onto the shortest of the three lines shown inside the bounding box in the left-hand part of Fig. 6.5. The unit vector along the Y axis is mapped onto the line nearest to the X axis. From the viewpoint shown in the right-hand part of the figure, it is clear that $\mathbf{b}_1$, $\mathbf{b}_2$, $\mathbf{b}_3$ all lie in the same plane. Therefore the dimension of range[$\mathbf{B}$] is 2, in agreement with our computation of a basis of range[$\mathbf{B}$] in Eqs. (6.55) through (6.58).

In this example one could have noticed without making a plot that the columns are linearly dependent because

$$\mathbf{b}_3 = 3\mathbf{b}_1 - \mathbf{b}_2. \tag{6.191}$$

The real usefulness of plotting the columns of a matrix results from the facts that making a plot may help one to identify column vectors that are nearly linearly dependent and that even if the columns are linearly independent a plot shows immediately how the matrix transforms the coordinate axes.

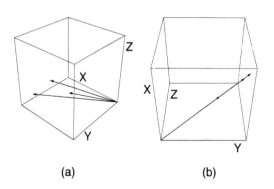

(a)   (b)

FIGURE 6.5

(*a*) Perspective view showing the three column vectors of the matrix **B** defined in Eq. (6.54). (*b*) Rotated view showing that the vectors lie in a common plane.

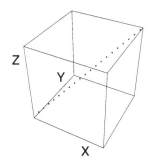

FIGURE 6.6

The column vectors $\mathbf{a}_1, \mathbf{a}_2, \ldots, \mathbf{a}_{20}$ of the matrix defined in (6.192) plotted as points in three-dimensional space. The point closest to the lower left-hand corner of the bounding box corresponds to the vector $\mathbf{a}_1$.

For an example with higher dimensionality, consider the 3×20 matrix

$$\mathbf{A} = (\mathbf{a}_1, \mathbf{a}_2, \ldots, \mathbf{a}_{20}) = \begin{pmatrix} 1 & 4 & \cdots & 3n-2 & \cdots & 58 \\ 2 & 5 & \cdots & 3n-1 & \cdots & 59 \\ 3 & 6 & \cdots & 3n & \cdots & 60 \end{pmatrix}. \tag{6.192}$$

Figure 6.6 shows the result of plotting $\mathbf{a}_1, \mathbf{a}_2, \ldots, \mathbf{a}_{20}$ as a set of points in three-dimensional space. The distance from the origin of coordinates of the point that represents $\mathbf{a}_i$ increases with i.

According to the figure, the points that represent the vectors $\mathbf{a}_i$ lie on a straight line. If the line passed through the origin (which it does not), then $\dim[\mathrm{range}[\mathbf{A}]] = \mathrm{rank}[\mathbf{A}]$ would be equal to 1. It is straightforward to verify that all of the column vectors of the matrix $\mathbf{A}$ in Eq. (6.192) lie on the line

$$\mathbf{x} = \mathbf{x}_0 + \alpha \mathbf{x}_1 \tag{6.193}$$

in which

$$\mathbf{x}_0 = \begin{pmatrix} -1 \\ 0 \\ 1 \end{pmatrix}, \quad \mathbf{x}_1 = \begin{pmatrix} 1 \\ 1 \\ 1 \end{pmatrix}. \tag{6.194}$$

For example, $\mathbf{a}_1 = \mathbf{x}_0 + 2\mathbf{x}_1$. (Observe that Eq. [6.193] defines an affine subspace $\mathbf{x}_0 + \mathcal{W}$ in which $\mathcal{W} = \mathrm{span}[\mathbf{x}_1]$.) Because the linearly independent vectors $\mathbf{x}_0$ and $\mathbf{x}_1$ span range$[\mathbf{A}]$, it follows that $\mathrm{rank}[\mathbf{A}] = 2$.

6.3.3 NULL SPACE OF A LINEAR MAPPING

A linear mapping $\mathsf{A} : \mathcal{V}_{(1)} \to \mathcal{V}_{(2)}$ is a homomorphism of the additive group $(\mathcal{V}_{(1)}, +)$ into the additive group $(\mathcal{V}_{(2)}, +)$. We know from Chapter 4 that a homomorphism is characterized by its kernel. The kernel of a linear mapping A, which is usually called the **null space** of A, is the set of elements of $\mathcal{V}_{(1)}$ that A maps to the additive identity $\overline{0}' \in \mathcal{V}_{(2)}$:

$$\mathrm{null}[\mathsf{A}] := \left\{ \overline{z} \in \mathcal{V}_{(1)} \,\middle|\, \mathsf{A}\overline{z} = \overline{0}' \in \mathcal{V}_{(2)} \right\}. \tag{6.195}$$

Because null[A] is closed under addition and scalar multiplication, it is a vector subspace of $\mathcal{V}_{(1)}$ (see Exercise 6.3.6). Because every subspace of $\mathcal{V}_{(1)}$ is a normal subgroup of the group $(\mathcal{V}_{(1)}, +)$, any vector subspace $\mathcal{W} \subseteq \mathcal{V}_{(1)}$ can be the null space of some linear mapping.

For example, the vector

$$\mathbf{z} = \begin{pmatrix} 3 \\ -1 \\ -1 \end{pmatrix} \tag{6.196}$$

spans the null space of the matrix $\mathbf{B}$ defined in Eq. (6.54). The reader should verify that $\mathbf{Bz} = \mathbf{0}$ and observe that the vectors $\mathbf{e}_1, -2\mathbf{e}_1 + \mathbf{e}_2 + \mathbf{e}_3$, which are carried onto the same image vector according to Eq. (6.190), differ by a scalar multiple of $\mathbf{z}$. Subsequently we see how to compute a basis of the null space of a linear mapping systematically. In the simplest cases one can find a basis of null[$\mathbf{A}$] by trial and error (see Exercise 6.3.4).

If null[A] is finite-dimensional, then its dimension is called the **nullity** of A:

$$\text{nullity}[A] := \dim[\text{null}[A]]. \tag{6.197}$$

Because

$$\text{null}[A] \subseteq \mathcal{V}_{(1)}, \tag{6.198}$$

we see from Eq. (5.160) that

$$\text{nullity}[A] \leq n_1 := \dim\left[\mathcal{V}_{(1)}\right]. \tag{6.199}$$

For example, according to Eq. (6.196), nullity[$\mathbf{B}$] $= 1$ in which $\mathbf{B}$ is the matrix defined in Eq. (6.54).

From an intuitive point of view, the nullity of a linear mapping measures, in terms of vector-space dimensions, the information that is lost as a result of applying the mapping. We show that if the nullity is nonzero, then the linear mapping is singular. A null space that contains more than the null vector corresponds, in the theory of group homomorphisms, to a kernel that contains more than the group identity element; see Eq. (4.198). Therefore a linear mapping A is singular if and only if null[A] contains a nonzero vector:

$$\boxed{\text{A is singular} \Leftrightarrow \text{null}[A] \neq \mathcal{O} \Leftrightarrow \text{nullity}[A] > 0} \tag{6.200}$$

in which $\mathcal{O} := \{\bar{0} \in \mathcal{V}_{(1)}\}$ is the subspace that consists of the zero vector alone. For, if A is not one-to-one, then there exist vectors $\bar{x}, \bar{y} \in \mathcal{V}_{(1)}$ such that

$$\bar{z} := \bar{x} - \bar{y} \neq \bar{0} \tag{6.201}$$

and such that

$$A\bar{x} = A\bar{y}. \tag{6.202}$$

From the additivity of A it follows that

$$A(\bar{x} - \bar{y}) = A\bar{x} - A\bar{y} = A\bar{z} = \bar{0}' \in \mathcal{V}_{(2)} \tag{6.203}$$

and therefore that

$$\bar{z} \in \text{null}[A] \tag{6.204}$$

in which $\bar{z}$ is nonzero. Conversely, if null$[A] \neq \mathcal{O}$, then Eq. (6.203) implies Eq. (6.202) for all vectors $\bar{x}$ and $\bar{y}$ that satisfy Eq. (6.201); therefore A is not one-to-one. If null$[A] \neq \mathcal{O}$, then the dimension of null$[A]$ is greater than zero, and the converse. This establishes Eq. (6.200). Clearly we have also shown that A is *non*singular if and only if null$[A] = \mathcal{O}$, which is true if and only if nullity$[A] = 0$.

To find the null space and nullity of the product BA : $\mathcal{V}_{(1)} \to \mathcal{V}_{(3)}$ of two linear mappings A : $\mathcal{V}_{(1)} \to \mathcal{V}_{(2)}$ and B : $\mathcal{V}_{(2)} \to \mathcal{V}_{(3)}$, one must look not only for the vectors in $\mathcal{V}_{(1)}$ that A maps to $\bar{0}'$ but also for the vectors in $\mathcal{V}_{(1)}$ whose images in $\mathcal{V}_{(2)}$ are mapped by B to $\bar{0}'' \in \mathcal{V}_{(3)}$. Exercise 6.3.5 summarizes the results.

6.3.4 OTHER EXAMPLES OF A SINGULAR LINEAR MAPPINGS

ℓ^2 Spaces

In the infinite-dimensional vector space $\ell^2(\mathbb{Z}^+, \mathbb{F})$, the **right-shift operator** S such that

$$\begin{aligned} &\forall n \in \mathbb{Z}^+ : S\bar{e}_n := \bar{e}_{n+1} \\ &\Rightarrow S(x_1, x_2, x_3, \ldots) = (0, x_1, x_2, x_3, \ldots) \end{aligned} \tag{6.205}$$

is a linear operator, and so is the **left-shift operator** $S^\dagger$ such that

$$S^\dagger \bar{e}_n := \begin{cases} \bar{0} & \text{if } n = 1; \\ \bar{e}_{n-1} & \text{if } n > 1. \end{cases} \tag{6.206}$$

The action of $S^\dagger$ on a sequence is

$$S^\dagger(x_1, x_2, x_3, \ldots) = (x_2, x_3, \ldots). \tag{6.207}$$

The right-shift operator S defined in Eq. (6.205) is nonsingular, but the left-shift operator $S^\dagger$ is singular because $S^\dagger \bar{e}_1 = \bar{0}$. If one left-shifts a sequence in $\ell^2(\mathbb{Z}^+, \mathbb{F})$, one loses information irrecoverably. Hence nullity$[S^\dagger] = 1$. The domain of $S^\dagger$ is

$$\text{domain}[S^\dagger] = \ell^2(\mathbb{Z}^+, \mathbb{F}) \tag{6.208}$$

because the normalization sum Eq. (5.41) for the image sequence $S^\dagger \bar{x}$ is equal to $\sum_2^\infty |x^n|^2$, which is finite if the normalization condition given in Eq. (5.41) holds for $\bar{x}$. The range of $S^\dagger$ is also $\ell^2(\mathbb{Z}^+, \mathbb{F})$ because every sequence $\bar{x}$ in $\ell^2(\mathbb{Z}^+, \mathbb{F})$ is the image under $S^\dagger$ of an element of $\ell^2(\mathbb{Z}^+, \mathbb{F})$:

$$\overline{x} = (x_1, x_2, x_3, \ldots) = S^\dagger(0, x_1, x_2, x_3, \ldots) = S^\dagger S \overline{x}. \tag{6.209}$$

Because $S^\dagger$ is singular and

$$\text{range}[S^\dagger] = \text{domain}[S^\dagger], \tag{6.210}$$

it follows that the proposition "range[A] $\cong$ domain[A] implies that A is nonsingular" does *not* hold for the infinite-dimensional vector space $\ell^2(\mathbb{Z}^+, \mathbb{F})$.

Moreover, one sees at once from the definitions that $S^\dagger$ is a left inverse but not a right inverse of S:

$$S^\dagger S = 1, \quad SS^\dagger \neq 1. \tag{6.211}$$

Exactly as in the Hilbert hotel "paradox" (Exercise 2.4.7), $S^\dagger$ shifts every component of $\overline{x}$ to make room for an additional component equal to zero.

The shift operators are defined also on the more general sequence spaces $\mathbb{F}^{\mathbb{Z}^+}, \mathbb{F}^{\mathbb{N}}, \ell^2(\mathbb{Z}, \mathbb{F})$, and $\mathbb{F}^{\mathbb{Z}}$. It is important to notice that $S^\dagger$ is *not* singular on the spaces $\ell^2(\mathbb{Z}, \mathbb{F})$ or $\mathbb{F}^{\mathbb{Z}}$, because the sequences

$$\overline{u} = (\ldots, u_{-|m|}, \ldots, u_{-1}, u_0, u_1 \ldots, u_n, \ldots) \tag{6.212}$$

that belong to these spaces do not begin or end at any finite index. For example, one shows easily that

$$\forall \overline{u} \in \mathbb{F}^{\mathbb{Z}} : SS^\dagger \overline{u} = S^\dagger S \overline{u} = \overline{u} \tag{6.213}$$

(see Exercises 6.3.7 and 6.3.8). A similar result holds for $\ell^2(\mathbb{Z}, \mathbb{F})$. Therefore, on $\ell^2(\mathbb{Z}, \mathbb{F})$ and $\mathbb{F}^{\mathbb{Z}}$, S and $S^\dagger$ are both nonsingular and are inverses of one another.

Null Space of a Differential Operator

The null space of a linear differential operator A is the solution space of the homogeneous linear differential equation $A\overline{x} = \overline{0}$. For example, let $\mathcal{V} = \mathcal{C}^2([a, b]; \mathbb{R})$ and

$$A = \frac{d^2}{dx^2}. \tag{6.214}$$

Then

$$\text{null}[A] = \text{span}\{1, x\} \tag{6.215}$$

because $y(x) = 1$ and $y(x) = x$ are linearly independent solutions of the homogeneous equation

$$\frac{d^2 y}{dx^2}(x) = 0. \tag{6.216}$$

In agreement with Eq. (5.136), nullity[A] = 2.

6.3.5 EXERCISES FOR SECTION 6.3

6.3.1   Let **a** and **b** be the columns of a 2×2 matrix **A**. Prove that **A** is singular if and only if

$$\mathbf{a} \times \mathbf{b} = \mathbf{0}. \tag{6.217}$$

Interpret this result geometrically.

6.3.2   Let **a**, **b**, and **c** be the columns of a 3×3 matrix **A**. Prove that **A** is singular if and only if

$$\mathbf{a} \cdot (\mathbf{b} \times \mathbf{c}) = \mathbf{0}. \tag{6.218}$$

Interpret this result geometrically.

6.3.3   Prove that a linear mapping A is singular if and only if

$$\text{rank}[A] < \dim[\text{domain}[A]]. \tag{6.219}$$

6.3.4   You are given the matrix of a linear mapping A,

$$\mathbf{A} = \begin{pmatrix} 1 & 1 & 0 \\ 0 & 1 & 1 \\ 1 & 1 & 0 \end{pmatrix}, \tag{6.220}$$

relative to some (not necessarily orthonormal) basis of a three-dimensional vector space over $\mathbb{R}$.

(a) Write down a column-vector basis of range[**A**].

(b) Write down a column-vector basis of null[**A**].

(c) Find rank[**A**] and nullity[**A**].

6.3.5   This problem concerns the null space of the product BA : $\mathcal{V}_{(1)} \to \mathcal{V}_{(3)}$ of two linear mappings A : $\mathcal{V}_{(1)} \to \mathcal{V}_{(2)}$ and B : $\mathcal{V}_{(2)} \to \mathcal{V}_{(3)}$.

(a) Prove that

$$\text{nullity}[BA] = \text{nullity}[A] + \dim[\text{range}[A] \cap \text{null}[B]] \tag{6.221}$$

Hint: Construct a basis of range[A] ∩ null[B]. Then extend this to a basis of range[A].

(b) Using the result of part (a), prove **Sylvester's law of nullity**:

$$\text{nullity}[A] \leq \text{nullity}[BA] \leq \text{nullity}[A] + \text{nullity}[B]. \tag{6.222}$$

(c) Construct an example in which $\mathcal{V}_{(1)} = \mathcal{V}_{(2)} = \mathbb{R}^3$ and in which nullity[BA] = nullity[A] + nullity[B].

6.3.6   A : $\mathcal{V}_{(1)} \to \mathcal{V}_{(2)}$ is a linear mapping. Prove that null[A] is a vector subspace of $\mathcal{V}_{(1)}$.

6.3.7 Prove that the right-shift operator S and the left-shift operator $S^\dagger$ are *non*singular on the space $\mathbb{F}^{\mathbb{Z}}$.

6.3.8 Prove that Eq. (6.213) holds.

6.4 INTRODUCTION TO DIGITAL FILTERS

Apart from their applications in signal processing, digital filters are of fundamental importance to working physicists and engineers because many numerical methods, especially those for the solution of ordinary and partial differential equations, can be regarded as digital filters. Methods that have already been developed for digital filters provide a useful new set of tools with which to analyze and evaluate other numerical algorithms. Digital filtering may also be necessary in order to assure both the stability and the accuracy of a numerical method.

This section defines nonrecursive and recursive digital filters, derives a simple expression for noise amplification or quenching by a digital filter, and examines the basic properties of difference operators.

6.4.1 DEFINITIONS

Let $\bar{u} \in \mathbb{F}^{\mathbb{Z}}$ be a "double-ended" sequence $(\ldots, u_{-m}, \ldots, u_0, u_1, \ldots, u_n, \ldots)$ of real or complex numbers u_k. For example, $\bar{u}$ may be an idealized stream of time-dependent data values sampled at discrete times t_n,

$$u_n = u(t_n), \tag{6.223}$$

from which one wishes to extract information (such as the recent mean price of a share of stock or the coordinates of an aircraft). Often, although not always, the sampling times t_n are uniformly spaced. Each canonical basis element $\bar{e}_n$, in which the n-th term is equal to 1 and all other terms are equal to zero, is the sequence that corresponds to a **unit impulse** at time $t = t_n$. Because $\{\bar{e}_n\}$ is a basis of $\mathbb{F}^{\mathbb{Z}}$, one can regard every sequence (or sampled data stream) as resulting from a superposition of unit impulses.

A **nonrecursive digital filter** with constant coefficients is a linear mapping $C : \mathbb{F}^{\mathbb{Z}} \to \mathbb{F}^{\mathbb{Z}}$ such that

$$\begin{aligned} y_n &= (C\bar{u})_n \\ &= \sum_{k=-N}^{N} c_k u_{n-k} \end{aligned} \tag{6.224}$$

in which N is a finite positive integer. We use the notation

$$\bar{y} = C\bar{u} \tag{6.225}$$

for nonrecursive digital filters if there is no need to draw attention to individual terms. We use the notation

$$\overline{y} = \Phi \overline{u} \tag{6.226}$$

for more general linear digital filters.

For example, if $N = 2$, then the calculation of y_n can be represented schematically as

$$
\begin{array}{c}
\vdots \\
+ u_{n-3} \cdot 0 \\
+ u_{n-2} \cdot c_2 \\
+ u_{n-1} \cdot c_1 \\
y_n = + u_n \cdot c_0 \\
+ u_{n+1} \cdot c_{-1} \\
+ u_{n+2} \cdot c_{-2} \\
+ u_{n+3} \cdot 0 \\
\vdots
\end{array}
\tag{6.227}
$$

The coefficients c_k are then moved down one step, and the process is repeated to obtain y_{n+1}, and so on. Because one index is k and the other is $n - k$, one calls Eq. (6.224) a **discrete convolution** of the data stream $\overline{u}$ with the **window** specified by the coefficients $c_{-N}, \ldots, c_N$. We consider only filters with constant coefficients (in which c_k is independent of n). However, filters with variable coefficients (in which c_k depends on n) are also useful.

In order to form an intuitive picture of a nonrecursive digital filter, recall that $\overline{e}_n$ can be interpreted as the sequence that corresponds to a unit impulse at time $t = t_n$. The response $\overline{y}$ to a unit impulse is

$$y_m = \sum_{k=-N}^{N} c_k \delta_{m-k,n} = c_{m-n}. \tag{6.228}$$

Therefore the sequence $(\ldots, 0, c_{-N}, \ldots, c_N, 0, \ldots,)$ is the **impulse response**, that is, the response to a unit impulse.

To measure the coefficients of a filter, then, one simply records the filter's response to a unit impulse $\overline{e}_n$ at a convenient time t_n. It is not necessary to know the details that may determine the coefficients c_k. One needs to know only the output that results when the input is a unit impulse. This point of view is especially useful for analyzing the stability and accuracy of numerical methods for solving ordinary and partial differential equations.

The **centered moving average**

$$y_n = \frac{1}{2N + 1} \sum_{k=-N}^{N} u_{n-k} \tag{6.229}$$

in which every coefficient is equal to the reciprocal of the number of points, is an important nonrecursive digital filter, as is the $(2N + 1)$-point **straight-line differentiator**

$$y_n = -\frac{\sum_{k=-N}^{N} k u_{n-k}}{\sum_{m=-N}^{N} m^2}. \tag{6.230}$$

It is easy to verify that if the sampled values u_n increase linearly, $u_{n-k} = \alpha(n - k)$, then $y_n = \alpha$ (see Exercise 6.4.5). Differentiators are singular linear mappings because they must map to zero any sequence $\bar{u}$ in which all terms are equal.

With the definitions

$$s_k = \tfrac{1}{2}(c_k + c_{-k}), \qquad a_k = \tfrac{1}{2}(c_k - c_{-k}), \tag{6.231}$$

the coefficients of any nonrecursive digital filter can be expressed as a sum

$$c_k = a_k + s_k \tag{6.232}$$

in which the coefficients s_k are symmetric,

$$s_{-k} = s_k, \tag{6.233}$$

and the coefficients a_k are antisymmetric,

$$a_{-k} = -a_k. \tag{6.234}$$

Symmetric filters smooth the data; antisymmetric filters detect trends. The moving average is the simplest symmetric filter; the straight-line differentiator is the simplest antisymmetric filter.

A **causal, recursive digital filter** with constant coefficients is a linear mapping of $\mathbb{F}^{\mathbb{Z}}$ into $\mathbb{F}^{\mathbb{Z}}$ such that

$$y_n = \sum_{k=-N}^{N} c_k u_{n-k} + \sum_{k=1}^{M} d_k y_{n-k} \tag{6.235}$$

in which M and N are positive integers. (The reader should check that the mapping Eq. [6.235] is additive and homogeneous; see Exercise 6.4.6). One can write equivalently

$$\bar{y} = C\bar{u} + D\bar{y} \tag{6.236}$$

in which the definitions of the linear mappings $C : \mathbb{F}^{\mathbb{Z}} \to \mathbb{F}^{\mathbb{Z}}$ and $D : \mathbb{F}^{\mathbb{Z}} \to \mathbb{F}^{\mathbb{Z}}$ are obvious from Eq. (6.235). The recursive digital filter defined in Eqs. (6.235) and (6.236) is called "causal" because y_n does not depend on y_m where $m > n$. The terms y_{n+1}, y_{n+2}, and so forth represent output values at times t_{n+1}, t_{n+2}, and so forth, which are "in the future," that is, later than t_n.

Numerical methods for computing the values of definite integrals or finding solutions of differential equations are recursive digital filters. For example, the trapezoidal algorithm for numerical integration or for the solution of a first-order linear differential equation,

$$y_n = y_{n-1} + \tfrac{1}{2}h(u_n + u_{n-1}), \tag{6.237}$$

is of the recursive form Eq. (6.235) with $d_1 = 1$ and $c_0 = c_1 = h/2$; all other coefficients are equal to zero. Euler's method Eq. (3.97) for solving a first-order ordinary differential equation is also of the form Eq. (6.235), the only nonzero coefficients being $c_1 = h$ and $d_1 = 1$.

Another example of a recursive digital filter is the **first-order autoregressive process**

$$y_n = \rho y_{n-1} + a_n \tag{6.238}$$

in which $0 < \rho < 1$ is fixed and a_n is a random variable such that

$$E[a_m a_n] = \sigma_a^2 \delta_n^m \tag{6.239}$$

in which $E[a_m a_n]$ is the mean (expected) value of $a_m a_n$ and σ_a^2 is the variance of the random variable a. One can show (see Exercise 6.4.1) that Eq. (6.238) produces a sequence $\{y_n\}$ that has a mean value of zero and decays approximately exponentially with a time constant that depends on the value of ρ. The variance of y_n depends on σ_a and ρ.

6.4.2 NOISE AMPLIFICATION BY DIGITAL FILTERS

The user of digital filters must be aware of the possibility that a filter may amplify roundoff errors and statistical uncertainties in the data. Suppose that the value recorded for u_k is equal to $u_k + e_k$, in which e_k is the error. For the sake of simplicity we assume that the errors of different sampled values in the data stream are uncorrelated and that all of the errors have the same variance σ (see Eq. [1.84]):

$$\forall k \in \mathbb{Z} : \begin{cases} E[e_k] = 0, \\ E[e_k e_l] = \sigma^2 \delta_k^l \end{cases} \tag{6.240}$$

in which δ_l^l is the Kronecker delta. In this model, the expected value of y_n is equal to the "true" value:

$$E[y_n] = E\left[\sum_{-N}^{N} c_k(u_{n-k} + e_{n-k})\right] = \sum_{-N}^{N} c_k u_{n-k}. \tag{6.241}$$

The variance of y_n is equal to

$$E[(y_n - E[y_n])^2] = E\left[\left(\sum_{-N}^{N} c_k e_{n-k}\right)^2\right]$$

$$= E\left[\sum_{k=-N}^{N} \sum_{m=-N}^{N} c_k c_m e_{n-k} e_{n-m}\right] \tag{6.242}$$

$$= \sigma^2 \sum_{k=-N}^{N} c_k^2.$$

Because the expected value of e_n^2 is equal to σ^2, it follows that the variance of y_n is larger (or smaller) than that of the error by a factor equal to $\sum c_k^2$ and that the standard deviation of y_n is larger (or smaller) than the standard deviation of the error by a factor equal to $\left[\sum c_k^2\right]^{1/2}$.

For example, the noise amplification factor of the $(2N+1)$-point centered moving average is $\left[\sum c_k^2\right]^{1/2} = (2N+1)^{-1/2} < 1$. As one should expect of a smoothing filter, the moving average quenches noise.

6.4.3 DIFFERENCE OPERATORS

Forward Differences

Let

$$\overline{y} = (y_1, y_2, \ldots) \in \mathbb{F}^{\mathbb{Z}^+}. \tag{6.243}$$

In applications, the terms $y_1, y_2, \ldots$ of $\overline{y}$ often are the values of a function at sample points $x_1, x_2, \ldots$. For purposes of approximating derivatives it is useful to consider the sequence $\Delta \overline{y}$ of **forward differences**,

$$\Delta \overline{y} := (y_2 - y_1, y_3 - y_2, \ldots), \tag{6.244}$$

the n-th term of which is

$$\begin{aligned}
\Delta y_n &:= (\Delta \overline{y})_n \\
&= y_{n+1} - y_n.
\end{aligned} \tag{6.245}$$

The symbol Δ is called the **forward difference operator**. It is singular and has a nullity of one because Δ maps any sequence $\overline{y}$ in which all terms are equal, $\overline{y} = (c, c, \ldots)$, to $\overline{0} = (0, 0, \ldots)$.

The sequence obtained by applying Δ to the sequence of forward differences,

$$\Delta^2 \overline{y} := \Delta[\Delta \overline{y}], \tag{6.246}$$

is called the sequence of **second forward differences**. One has

$$\begin{aligned}
\Delta^2 y_n &= \Delta(y_2 - y_1, y_3 - y_2, y_4 - y_3, \ldots) \\
&= (y_3 - 2y_2 + y_1, y_4 - 2y_3 + y_2, \ldots),
\end{aligned} \tag{6.247}$$

in which

$$\begin{aligned}
\Delta^2 y_n &:= (\Delta^2 \overline{y})_n \\
&= y_{n+2} - 2y_{n+1} + y_n
\end{aligned} \tag{6.248}$$

is the general term.

We shall derive a formula for the sequence of **k-th forward differences**,

$$\Delta^k \overline{y} := \Delta[\Delta^{k-1} \overline{y}]. \tag{6.249}$$

From the definitions of Δ and $S^\dagger$ (Eqs. [6.244] and [6.206]) it follows that

$$\Delta = S^\dagger - 1. \tag{6.250}$$

One sees immediately from the binomial theorem that

$$\Delta^k \bar{y} = (S^\dagger - 1)^k \bar{y} = \sum_{r=0}^{k} \binom{k}{r} (-1)^r (S^\dagger)^{k-r} \bar{y}. \tag{6.251}$$

Then the n-th term of the sequence of k-th forward differences is

$$\Delta^k y_n = \sum_{r=0}^{k} \binom{k}{r} (-1)^r y_{n-k+r}. \tag{6.252}$$

Considered as a digital filter, Δ^k has a noise amplification factor equal to the square root of

$$\sum c_i^2 = \sum_{r=0}^{k} \binom{k}{r}^2 = \binom{2k}{k} \tag{6.253}$$

in which the second step follows from an identity involving binomial coefficients (see Exercise 6.4.3). For example, for $k = 5$ one obtains a noise amplification factor of $\sqrt{252} \approx 15.9$.

Sometimes one needs a formula for y_{n+k} in terms of forward differences. From the equation

$$S^\dagger = 1 + \Delta \tag{6.254}$$

it follows that

$$(S^\dagger)^k \bar{y} = (1 + \Delta)^k \bar{y} = \sum_{r=0}^{k} \binom{k}{r} \Delta^r \bar{y}. \tag{6.255}$$

Then

$$\begin{aligned} y_{n+k} &= \sum_{r=0}^{k} \binom{k}{r} \Delta^r y_n \\ &= y_n + \sum_{r=1}^{k} \frac{k \cdots (k-r+1)}{r!} \Delta^r y_n. \end{aligned} \tag{6.256}$$

Null Space of the Forward Difference Operator

The equation

$$\Delta y_n = 0 \tag{6.257}$$

implies the recurrence relation $y_{n+1} - y_n = 0$. Therefore the null space of the forward difference operator is one-dimensional and is spanned by the sequence $(1, 1, \ldots, 1, \ldots)$.

To find the null space of Δ^2, note that a sequence $\overline{y}$ belongs to the null space of the second-difference operator, null[Δ^2], if and only if the sequence of first differences,

$$
\begin{aligned}
y_n' &:= \Delta y_n \\
&= y_{n+1} - y_n,
\end{aligned}
\tag{6.258}
$$

belongs to the null space of the first-difference operator, null[Δ]. Then $y_n' = a_1$, in which a_1 is a constant. Therefore

$$
y_n = a_0 + a_1 n
\tag{6.259}
$$

in which a_0 is a constant. In other words, the null space of Δ^2 consists of all linear combinations of the sequence $(1, 1, \ldots, 1, \ldots)$ and the sequence $(1, 2, \ldots, n, \ldots)$, the terms of which are the values of the polynomial $p(x) = x$ sampled at the points $x_n = n$. It follows that the dimension of null[Δ^2] is equal to 2.

We use the principle of finite induction to show that if a sequence $\overline{y}$ belongs to null[Δ^k], then there exist scalars $a_0, \ldots, a_{k-1}$ such that the general term of $\overline{y}$ is

$$
y_n = a_0 + a_1 n + \cdots + a_{k-1} n^{k-1}.
\tag{6.260}
$$

We have shown that the theorem holds for $k = 1$ and $k = 2$. Assume that it holds for k, and let

$$
\Delta^{k+1} y_n = 0.
\tag{6.261}
$$

Then

$$
\Delta^k y_n' = 0
\tag{6.262}
$$

in which

$$
y_n' = y_{n+1} - y_n.
\tag{6.263}
$$

By the inductive hypothesis, there exist scalars $a_0', \ldots, a_{k-1}'$ such that

$$
y_n' = a_0' + a_1' n + \cdots + a_{k-1}' n^{k-1}.
\tag{6.264}
$$

Then the terms of $\overline{y}$ satisfy the two-term recurrence relation

$$
y_{n+1} = y_n + a_0' + a_1' n + \cdots + a_{k-1}' n^{k-1},
\tag{6.265}
$$

the solution of which is a polynomial in n,

$$
y_{n+1} = y_1 + \sum_{r=1}^{n} [a_0' + a_1' r + \cdots + a_{k-1}' r^{k-1}].
\tag{6.266}
$$

The term of highest degree is a polynomial of degree k in n,

$$\sum_{r=1}^{n} r^{k-1} = \frac{B_k(n+1) - B_k(0)}{k}, \tag{6.267}$$

in which B_k is the Bernoulli polynomial of degree k. This shows that Eq. (6.260) holds for $k+1$. An explicit calculation of the coefficients a_l is not needed for the inductive proof. Thus the k-th difference operator, Δ^k, is a digital filter that annihilates all polynomials of degree less than or equal to $k-1$.

By Exercise 6.5.10, the sequences

$$\bar{v}_1 = (1, 1, \ldots, 1, \ldots), \ldots, \bar{v}_k = (1, 2^{k-1}, \ldots, n^{k-1}, \ldots) \tag{6.268}$$

are linearly independent. We have just shown that $\{\bar{u}_1, \ldots, \bar{u}_k\}$ span null$[\Delta^k]$. Therefore the dimension of null$[\Delta^k]$ is equal to k.

Newton's Forward Formula

Newton's method for interpolating data values $y_n, \ldots, y_{n+k}$, which represent the values of a function sampled at uniformly spaced points

$$x_l = lh + x_0, \tag{6.269}$$

is to form forward differences of the data sequence and to replace $n + k \in \mathbb{Z}^+$ with the real number

$$\xi = \frac{x - x_0}{h} \tag{6.270}$$

in the binomial coefficient on the right-hand side of Eq. (6.256). The result is a polynomial in ξ,

$$p_k(\xi) = y_n + \sum_{r=1}^{k} \frac{(\xi - n) \cdots (\xi - n - r + 1)}{r!} \Delta^r y_n, \tag{6.271}$$

which interpolates the points $y_n, \ldots, y_{n+k}$:

$$\forall l \in (n : n + k) : p_k(x_l) = y_l. \tag{6.272}$$

Equation (6.271) is known as **Newton's forward formula**. To establish Eq. (6.272), note that if $\xi = l$, then all terms in the sum

$$p_k(l) = y_n + \sum_{r=1}^{k} \frac{(l - n) \cdots (l - n - r + 1)}{r!} \Delta^r y_n \tag{6.273}$$

such that $n + r > l$ vanish. Then the right-hand side of Eq. (6.273) becomes

$$y_n + \sum_{r=1}^{l-n} \frac{(l-n)\cdots(l-n-r+1)}{r!} \Delta^r y_n = y_l \tag{6.274}$$

according to Eq. (6.256). This verifies Eq. (6.272).

Newton's forward formula has the useful feature that it is relatively easy to add a new data point y_{n+k+1} (representing a sampled value $y(x_{n+k+1})$) to an already interpolated dataset. One has only to add the single term $((\xi - n)(\xi - n - 1)\cdots(\xi - n - k)/(k + 1)!)\Delta^{k+1}y_n$ to $p_k(\xi)$ in order to obtain $p_{k+1}(\xi)$.

Apart from convenience, however, Newton's forward formula has limited practical usefulness for interpolation unless the function being interpolated varies slowly between $x = x_n$ and $x = x_{n+k}$. It turns out that that polynomial interpolation using uniformly spaced points causes the interpolated curve to overshoot or undershoot if $|\Delta y_l| \gg |y_l|$ for any $l \in (n : n+k)$. Also, the property of reproducing the given values y_l, Eq. (6.272), depends on the accurate cancellation of many terms. Given that forward difference operators amplify noise, one may not be able to attain high accuracy in interpolating with a single, globally fit polynomial such as Eq. (6.271).

Newton–Cotes Integration Formulas

The classical way to obtain formulas for numerical integration is to fit a polynomial to the integrand at a set of sample points and then integrate the polynomial exactly. The celebrated Newton–Cotes integration formulas follow by direct integration of Newton's forward formula, Eq. (6.271), from y_n to y_{n+k}. For example, let

$$z_n = \int_{x_0}^{x_n} y(x)\,dx \tag{6.275}$$

be the n-th term in the sequence of sampled values of the integral of a function y. To derive the trapezoidal approximation to z_n, take $k = 1$ in Eq. (6.271) and integrate from x_n to x_{n+1} with respect to x (or from n to $n + 1$ with respect to ξ):

$$\int_{x_n}^{x_{n+1}} y(x)\,dx \approx h \int_n^{n+1} p_1(\xi)\,d\xi = \frac{1}{2}h(y_n + y_{n+1}). \tag{6.276}$$

Then

$$z_{n+1} \approx z_n + \tfrac{1}{2}h(y_n + y_{n+1}), \tag{6.277}$$

which is the **trapezoidal rule** of numerical integration for a single "panel," that is, for the interval between one pair of sample points. Similarly, for $k = 2$ one obtains **Simpson's rule** for one panel,

$$z_{n+2} \approx z_n + \tfrac{1}{3}h(y_n + 4y_{n+1} + y_{n+2}). \tag{6.278}$$

There are two approaches to quantifying the errors in these approximations: the time-honored method of Taylor series expansions used in Section 3.4.3 and the method of transfer-function analysis used in Section 11.6.5.

Backward Differences

The sequence $\nabla \overline{y}$ of **backward differences** of a sequence $\overline{y}$ is

$$\nabla \overline{y} := (y_1, y_2 - y_1, \ldots), \quad \text{if } \overline{y} \in \mathbb{F}^{\mathbb{Z}^+}, \quad \text{or } (y_0, y_1 - y_0, \ldots) \text{ if } \overline{y} \in \mathbb{F}^{\mathbb{Z}}. \tag{6.279}$$

The n-th term is

$$\begin{aligned}
\nabla y_n &:= (\nabla \overline{y})_n \\
&= y_n - y_{n-1}.
\end{aligned} \tag{6.280}$$

The **backward difference operator** ∇ is singular on $\mathbb{F}^{\mathbb{Z}}$ but is nonsingular on $\mathbb{F}^{\mathbb{Z}^+}$ because $\nabla y_1 = y_1$ (and therefore the entire sequence can be reconstructed from the values $\nabla y_2 = y_2 - y_1, \ldots$).

From the definition of the right-shift operator, Eq. (6.205), it follows that

$$S = 1 - \nabla. \tag{6.281}$$

With the binomial theorem one obtains the expression

$$\nabla^k y_n = \sum_{r=0}^{k} \binom{k}{r} (-1)^r y_{n-r} \tag{6.282}$$

for the terms of the sequence of k-th backward differences.

Newton's Backward Formula

Newton's interpolation formula using backward differences follows by generalizing Eq. (6.282) in a manner similar to that in which we derived Eq. (6.272). The result is a polynomial

$$q_k(\eta) = y_n + \sum_{r=1}^{k} \frac{\eta \cdots (\eta + r - 1)}{r!} \nabla^r y_n, \tag{6.283}$$

in the variable

$$\eta = \frac{x - x_n}{h}, \tag{6.284}$$

such that q_k interpolates the points $y_n, \ldots, y_{n-k}$. Equation (6.283), which is known as **Newton's backward formula**, is useful in deriving methods for solving stiff systems of ordinary differential equations.

6.4.4 EXERCISES FOR SECTION 6.4

6.4.1 Show for the first-order autoregressive process Eq. (6.238) that if $0 < m < n$, then

$$E[y_n] = \rho^n E[y_0], \tag{6.285}$$

$$E[y_m y_n] = \rho^{n-m} E[y_m^2] \tag{6.286}$$

and

$$E[y_m^2] = \rho^{2m} E[y_0^2] + \sum_{r=0}^{m-1} \rho^{2r} \sigma_a^2. \tag{6.287}$$

Show that as $m \to \infty$ with $n - m$ fixed,

$$E[y_n] \to 0 \tag{6.288}$$

and

$$E[y_m^2] \to \frac{\sigma_a^2}{1 - \rho^2}. \tag{6.289}$$

Therefore, after the memory of the initial value y_0 has decayed away, y_n has a mean of zero. The constants σ_a and ρ can be chosen to produce a desired steady-state variance and decay rate.

6.4.2 Demonstrate that the interpolating polynomial defined in Eq. (6.271) is

$$p_3(x) = y_1 + \frac{(x - x_1)}{h} \Delta y_1 + \frac{(x - x_1)(x - x_2)}{2h^2} \Delta^2 y_1 \tag{6.290}$$

for $k = 3$ and for equally spaced sampling points x_1, $x_2 = x_1 + h$, and $x_3 = x_1 + 2h$. Also demonstrate that p_3 can be written in the form

$$p_3(x) = y_1 \frac{(x - x_2)(x - x_3)}{(x_1 - x_2)(x_1 - x_3)} + y_2 \frac{(x - x_1)(x - x_3)}{(x_2 - x_1)(x_2 - x_3)}$$
$$+ y_3 \frac{(x - x_1)(x - x_2)}{(x_3 - x_1)(x_3 - x_2)}. \tag{6.291}$$

6.4.3 Use the identity

$$\sum_{m=0}^{n} \binom{r}{m} \binom{s}{n - m} = \binom{r + s}{n} \tag{6.292}$$

(in which $r + s \geq n$) to establish Eq. (6.253).

6.4.4 Use Eqs. (6.211), (6.250), and (6.281) to establish the following relations:

$$\Delta = \nabla + \Delta \nabla, \tag{6.293}$$

$$S^{\dagger} = 1 + \nabla + \Delta \nabla. \tag{6.294}$$

6.4.5 Verify for the differentiator defined in Eq. (6.230) that if the sampled values increase linearly, $u_{n-k} = \alpha(n - k)$, then $y_n = \alpha$.

6.4.6 Verify that the mapping defined in Eq. (6.235) is additive and homogeneous.

6.4.7 Find a basis of the null space of the straight-line differentiator defined in Eq. (6.230).

6.5 TRACE AND DETERMINANT

6.5.1 TRACE OF A LINEAR MAPPING

Let $A : V \to V$ be a linear mapping of a vector space V into itself, and let $\mathbf{A}$ be the matrix of A relative to some basis of V. The **trace** of the square matrix $\mathbf{A}$ is defined as the sum of its diagonal matrix elements:

$$\text{trace}[\mathbf{A}] := a_i^i. \tag{6.295}$$

Traces are used widely in physics, especially in quantum mechanics.

For example, the partition function of a finite quantum-mechanical system is

$$Z(\mu) = \text{trace}[e^{-\mu \mathbf{H}}] \tag{6.296}$$

in which $\mathbf{H}$ is the matrix of the Hamiltonian operator and $\mu = (kT)^{-1}$. All of the thermodynamic functions of a system can be calculated by differentiating Z with respect to μ.

Let $\mathbf{A}, \mathbf{B} \in \mathbb{F}^{n \times n}$. We show that

$$\text{trace}[\mathbf{AB}] = \text{trace}[\mathbf{BA}]. \tag{6.297}$$

Let $\mathbf{C} = \mathbf{BA}$. Using the formula $c_k^i = b_k^j a_j^i$ for the matrix elements of a product (Eq. [6.71]), one obtains

$$\text{trace}[\mathbf{C}] = c_i^i = b_i^j a_j^i = a_j^i b_i^j = \text{trace}[\mathbf{AB}] \tag{6.298}$$

because the order of the numerical factors a_j^i and b_i^j obviously is immaterial in the sum $a_j^i b_i^j$.

Next, we show that the trace of a product of three matrices is invariant under cyclic permutations of the factors:

$$\text{trace}[\mathbf{ABC}] = \text{trace}[\mathbf{CAB}] = \text{trace}[\mathbf{BCA}]. \tag{6.299}$$

By Eq. (6.71) one has

$$\text{trace}[\mathbf{ABC}] = a_i^j b_j^k c_k^i = c_k^i a_i^j b_j^k = \text{trace}[\mathbf{CAB}], \tag{6.300}$$

for example. Evidently the invariance of the trace under *cyclic* permutations of the factors in a matrix product follows from the fact that the row index of one matrix element is equal to the column index of the next (a_i^j is linked by j to b_j^k, for example) and from the the fact that the last matrix element is similarly linked to the first. One cannot apply a noncyclic permutation to the matrix factors in a trace any more than one can apply a noncyclic permutation to the links in a loop of chain.

It follows from Eq. (6.300) that the trace of the matrix $\mathbf{TAT}^{-1}$, which realizes the same linear mapping as $\mathbf{A}$ relative to a different basis, is

$$\text{trace}[\mathbf{TAT}^{-1}] = \text{trace}[\mathbf{T}^{-1}\mathbf{TA}] = \text{trace}[\mathbf{A}]. \tag{6.301}$$

This proves that *the trace of a square matrix is basis-independent.* Therefore the trace is a function not of a matrix but of the linear mapping of which the matrix is just the realization with respect to a specific basis. One can write trace[A] instead of trace[$\mathbf{A}$].

6.5.2 DETERMINANTS

Definition

Another important scalar-valued function of a linear mapping A is its **determinant**:

$$\det[\mathbf{A}] = \det[\mathbf{a}_1, \mathbf{a}_2, \ldots, \mathbf{a}_n] := \sum_{\pi \in S_n} \sigma(\pi) a_1^{\pi(1)} a_2^{\pi(2)} \cdots a_n^{\pi(n)} \tag{6.302}$$

in which $\dim[\mathcal{V}] = n$, π is any permutation of $(1:n)$, S_n is the symmetric group on n objects defined in Section 4.4, and $\pi(j)$ is the image of j under the permutation π. The function σ, which is called the **sign function** on permutations π, has the value $+1$ if π is an even permutation of $(1:n)$ and the value -1 if π is an odd permutation. The notation $\det[\mathbf{a}_1, \mathbf{a}_2, \ldots, \mathbf{a}_n]$ emphasizes that the determinant is a function of the column vectors of $\mathbf{A}$. Looking on a determinant in this possibly unfamiliar way is the key to streamlining several important derivations.

For example, the determinant of a 2×2 matrix,

$$\det[\mathbf{A}] = \det \begin{pmatrix} a_1^1 & a_2^1 \\ a_1^2 & a_2^2 \end{pmatrix} = \det \left[\begin{pmatrix} a_1^1 \\ a_1^2 \end{pmatrix}, \begin{pmatrix} a_2^1 \\ a_2^2 \end{pmatrix} \right] \tag{6.303}$$

is equal to

$$\det[\mathbf{A}] = a_1^1 a_2^2 - a_1^2 a_2^1. \tag{6.304}$$

The first term, $a_1^1 a_2^2$, corresponds to the identity permutation of $(1:2)$. The second term, $-a_1^2 a_2^1$, corresponds to the permutation (12). Because (12) is the transposition (exchange) of 1 and 2, it follows that $\sigma((12)) = -1$. Note that the right-hand side of Eq. (6.304) is equal to the contravariant component of the cross product $\mathbf{a}_1 \times \mathbf{a}_2$ along the vector $\mathbf{e}_1 \times \mathbf{e}_2$.

The definition Eq. (6.302) implies that

$$\det[\mathbf{A}] = \det[\mathbf{A}^T] \tag{6.305}$$

because one can write Eq. (6.302) as

$$\det[\mathbf{A}] = \sum_{\pi' \in S_n} \sigma(\pi') a_{\pi'(1)}^1 a_{\pi'(2)}^2 \cdots a_{\pi'(n)}^n \tag{6.306}$$

in which π' is the inverse of π. In this case it is natural to look upon det[$\mathbf{A}$] as a function of the row vectors of $\mathbf{A}$.

The determinant of a matrix $\mathbf{A}$ may or may not vanish, depending on the column vectors $\mathbf{a}_1, \mathbf{a}_2, \ldots, \mathbf{a}_n$. For example, the determinant of the unit matrix $\mathbf{1}$ (the i, j matrix element of which is equal to δ_i^j) is 1, but the determinant of a matrix in which one column or row is null obviously vanishes.

Basic Properties of Determinants

The determinant has three fundamentally important properties:

1. By Eq. (6.302) the determinant is **multilinear**: For every $i \in (1 : n)$,

$$\det[\mathbf{a}_1, \mathbf{a}_2, \ldots, \mathbf{a}_{i-1}, \lambda\mathbf{a}_i + \mu\mathbf{b}_i, \mathbf{a}_{i+1}, \ldots, \mathbf{a}_n]$$
$$= \lambda \det[\mathbf{a}_1, \mathbf{a}_2, \ldots, \mathbf{a}_{i-1}, \mathbf{a}_i, \mathbf{a}_{i+1}, \ldots, \mathbf{a}_n]$$
$$+ \mu \det[\mathbf{a}_1, \mathbf{a}_2, \ldots, \mathbf{a}_{i-1}, \mathbf{b}_i, \mathbf{a}_{i+1}, \ldots, \mathbf{a}_n]. \tag{6.307}$$

2. The determinant is also **antisymmetric** in all of its vector arguments:

$$\sum_{\pi \in S_n} \sigma(\pi) a_1^{\pi(1)} \cdots a_i^{\pi(j)} \cdots a_j^{\pi(i)} \cdots a_n^{\pi(n)}$$
$$= -\sum_{\pi' \in S_n} \sigma(\pi') a_1^{\pi'(1)} \cdots a_i^{\pi'(i)} \cdots a_j^{\pi'(j)} \cdots a_n^{\pi'(n)} \tag{6.308}$$

in which π and π' differ by the transposition of i and j:

$$\pi' = (ij)\pi \Rightarrow \sigma(\pi') = -\sigma(\pi). \tag{6.309}$$

3. Finally, the determinant of $\mathbf{A}$ is multiplied by $\sigma(\pi')$ if one permutes the columns according to a permutation π'.

To prove this, we use the group property of permutations:

$$\det[\mathbf{a}_{\pi'(1)}, \mathbf{a}_{\pi'(2)}, \ldots, \mathbf{a}_{\pi'(n)}] = \sum_{\pi \in S_n} \sigma(\pi) a_{\pi'(1)}^{\pi(1)} a_{\pi'(2)}^{\pi(2)} \cdots a_{\pi'(n)}^{\pi(n)}$$
$$= \sum_{\pi'' \in S_n} \sigma(\pi) a_1^{\pi''(1)} a_2^{\pi''(2)} \cdots a_n^{\pi''(n)} \tag{6.310}$$

in which $\pi'' = (\pi')^{-1}\pi$. Because σ is a representation of S_n, one has

$$\sigma(\pi) = \sigma(\pi')\sigma(\pi''). \tag{6.311}$$

Then

$$\det[\mathbf{a}_{\pi'(1)}, \mathbf{a}_{\pi'(2)}, \ldots, \mathbf{a}_{\pi'(n)}] = \sigma(\pi') \det[\mathbf{a}_1, \mathbf{a}_2, \ldots, \mathbf{a}_n], \tag{6.312}$$

which is what we set out to prove. Because σ is the sign representation of S_n, we say that Eq. (6.312) defines the **permutation symmetry** of the determinant, even though the determinant is antisymmetric under odd permutations.

Determinant of a Singular Matrix

It follows from the antisymmetry of the determinant under any transposition of columns, Eq. (6.308), that the determinant of a matrix in which two columns are equal is equal to its own negative and hence vanishes. For example, in

$$D = \det[\mathbf{a}_1, \ldots, \mathbf{a}_{i-1}, \mathbf{b}, \mathbf{a}_{i+1}, \ldots, \mathbf{a}_{j-1}, \mathbf{b}, \mathbf{a}_{j+1}, \ldots, \mathbf{a}_n] \tag{6.313}$$

the i-th and j-th columns are both equal to the vector $\mathbf{b}$. Then

$$D = -D \Rightarrow D = 0 \tag{6.314}$$

because D changes sign when the two equal columns are interchanged.

This observation permits us to evaluate the determinant of a singular matrix. If the matrix $\mathbf{A}$ is singular, then the column vectors $\mathbf{a}_i$ are linearly dependent. Hence there exist scalars β^i such that (for example)

$$\mathbf{a}_1 = \sum_{i=2}^{n} \beta^i \mathbf{a}_i, \tag{6.315}$$

which implies that

$$\det[\mathbf{A}] = \det\left[\sum_{i=2}^{n} \beta^i \mathbf{a}_i, \mathbf{a}_2, \ldots, \mathbf{a}_n\right] = \sum_{i=2}^{n} \beta^i \det[\mathbf{a}_i, \mathbf{a}_2, \ldots, \mathbf{a}_n] = 0 \tag{6.316}$$

because each of the determinants in the final sum has two identical columns. Thus *the determinant of a singular matrix vanishes*. On the other hand, suppose that $\mathbf{A}$ is nonsingular so that $\{\mathbf{a}_1, \ldots, \mathbf{a}_n\}$ is a basis of $\mathbb{F}^n$. Let $\{\mathbf{y}_1, \ldots, \mathbf{y}_n\}$ be any n elements of $\mathbb{F}^n$. Then there exist scalars y_i^j such that

$$\mathbf{y}_i = y_i^j \mathbf{a}_j. \tag{6.317}$$

Multilinearity implies that the determinant of the matrix whose columns are $\{\mathbf{y}_1, \ldots, \mathbf{y}_n\}$ is

$$\det[\mathbf{y}_1, \ldots, \mathbf{y}_n] = y_1^{j_1} \cdots y_n^{j_n} \det[\mathbf{a}_{j_1}, \ldots, \mathbf{a}_{j_n}]. \tag{6.318}$$

But, according to Eq. (6.312),

$$\det[\mathbf{a}_{j_1}, \ldots, \mathbf{a}_{j_n}] = \sigma(\pi) \det[\mathbf{a}_1, \ldots, \mathbf{a}_n] \tag{6.319}$$

in which π is the permutation such that $j_k = \pi(k)$. It follows from Eqs. (6.318) and (6.319) that

$$\det[\mathbf{a}_1, \ldots, \mathbf{a}_n] = 0 \Rightarrow \forall \mathbf{y}_1, \ldots, \mathbf{y}_n : \det[\mathbf{y}_1, \ldots, \mathbf{y}_n] = 0. \tag{6.320}$$

But we know that the determinant does not vanish identically for all possible vector arguments. This contradiction shows that the determinant of a nonsingular matrix is nonzero. Combining our results, we have

$$\det[\mathbf{A}] \neq 0 \Leftrightarrow \mathbf{A} \text{ is nonsingular} \tag{6.321}$$

or, equivalently,

$$\det[\mathbf{A}] = 0 \Leftrightarrow \mathbf{A} \text{ is singular.} \tag{6.322}$$

These tests for the singularity or nonsingularity of a matrix $\mathbf{A}$ are fundamental to our discussion of the eigenvalues and eigenvectors of a linear mapping in Chapter 9. Unfortunately the direct evaluation of a determinant using Eq. (6.302) is computationally feasible only if the dimensionality is very low (for example, 2).

Determinant of a Matrix Product

We now prove the important theorem that *the determinant of the product of two matrices is equal to the product of the determinants,*

$$\det[\mathbf{B}\mathbf{A}] = \det[\mathbf{B}]\det[\mathbf{A}]. \tag{6.323}$$

One's first step is to express the i-th column $\mathbf{c}_i$ of the product matrix $\mathbf{C} := \mathbf{B}\mathbf{A}$ in terms of the columns $\mathbf{b}_j$ of $\mathbf{B}$ and the elements of $\mathbf{A}$ with the formula $\mathbf{c}_i = a_i^j \mathbf{b}_j$ (see Eq. [6.66]). Then, by the property of multilinearity,

$$\begin{aligned}
\det[\mathbf{C}] &= \det\left[a_1^{j_1}\mathbf{b}_{j_1}, a_2^{j_2}\mathbf{b}_{j_2}, \ldots, a_n^{j_n}\mathbf{b}_{j_n}\right] \\
&= a_1^{j_1} a_2^{j_2} \cdots a_n^{j_n} \det[\mathbf{b}_{j_1}, \mathbf{b}_{j_2}, \ldots, \mathbf{b}_{j_n}].
\end{aligned} \tag{6.324}$$

The summation variables $j_1, j_2, \ldots, j_n$ take on the values $1, 2, \ldots, n$ independently. However, $\det[\mathbf{b}_{j_1}, \mathbf{b}_{j_2}, \ldots, \mathbf{b}_{j_n}]$ vanishes unless $j_1, j_2, \ldots, j_n$ are all different, that is, unless $\{j_1, j_2, \ldots, j_n\}$ is a permutation of $(1 : n)$. Let π be the permutation such that

$$\forall i \in (1 : n): \ \pi(i) := j_i. \tag{6.325}$$

Each permutation in S_n occurs once and only once in the multiple summation in Eq. (6.324). Then

$$\det[\mathbf{b}_{j_1}, \mathbf{b}_{j_2}, \ldots, \mathbf{b}_{j_n}] = \sigma(\pi)\det[\mathbf{b}_1, \mathbf{b}_2, \ldots, \mathbf{b}_n] = \sigma(\pi)\det[\mathbf{B}] \tag{6.326}$$

provided that $j_1, j_2, \ldots, j_n$ are all different. It follows that $\det[\mathbf{B}]$ is a common factor in every nonzero term of Eq. (6.324). Therefore

$$\det[\mathbf{C}] = \det[\mathbf{B}] \sum_{\pi \in S_n} \sigma(\pi) a_1^{\pi(1)} a_2^{\pi(2)} \cdots a_n^{\pi(n)} = \det[\mathbf{B}]\det[\mathbf{A}] \tag{6.327}$$

as we claimed in Eq. (6.323).

It follows from Eq. (6.323) that for every nonsingular $n \times n$ matrix $\mathbf{T}$,

$$1 = \det[\mathbf{1}] = \det[\mathbf{T}\mathbf{T}^{-1}] = \det[\mathbf{T}]\det[\mathbf{T}^{-1}] \tag{6.328}$$

and therefore that

$$\det[\mathbf{T}^{-1}] = \{\det[\mathbf{T}]\}^{-1}. \tag{6.329}$$

In words: The determinant of an inverse is the inverse of the determinant.

The principle of finite induction can be used to show that

$$\det[\mathbf{A}_1 \cdots \mathbf{A}_n] = \prod_{i=1}^{n} \det[\mathbf{A}_i]. \tag{6.330}$$

Then the determinant of the matrix $\mathbf{A}' = \mathbf{T}\mathbf{A}\mathbf{T}^{-1}$ of the same linear mapping with respect to a different basis is

$$\det[\mathbf{A}'] = \det[\mathbf{T}]\det[\mathbf{A}]\{\det[\mathbf{T}]\}^{-1} = \det[\mathbf{A}]. \tag{6.331}$$

Therefore *the value of the determinant of a matrix is the same in all bases.* Equivalently, *the determinant has the same value for all matrices that are similar to one another.* The determinant of a matrix $\mathbf{A}$ is a function only of the linear mapping A that the matrix $\mathbf{A}$ realizes. For this reason one can write $\det[A]$ rather than $\det[\mathbf{A}]$.

Determinants and the Handedness of Coordinate Systems

A linearly independent list of vectors

$$A := \{\mathbf{a}_1, \ldots, \mathbf{a}_n\} \tag{6.332}$$

in the vector space $\mathbb{R}^n$ is a basis and therefore defines a system of coordinate axes. A is said to define a **right-handed coordinate system** if and only if

$$\det[\mathbf{a}_1, \ldots, \mathbf{a}_n] > 0, \tag{6.333}$$

and a **left-handed coordinate system** if and only if

$$\det[\mathbf{a}_1, \ldots, \mathbf{a}_n] < 0. \tag{6.334}$$

For example, the list $\{\mathbf{i}, \mathbf{j}, \mathbf{k}\}$ in $\mathbb{R}^3$ defines a right-handed coordinate system because

$$\det[\mathbf{i}, \mathbf{j}, \mathbf{k}] = \det[\mathbf{1}] > 0. \tag{6.335}$$

One can say also that the coordinate system defined by $\{\mathbf{i}, \mathbf{j}, \mathbf{k}\}$ is right-handed because $\mathbf{i} \times \mathbf{j}$ points along $\mathbf{k}$.

A nonsingular matrix $\mathbf{A} \in \mathbb{R}^n$ is said to implement a **proper** (or **orientation-preserving**) transformation if and only if

$$\det[\mathbf{A}] > 0. \tag{6.336}$$

If

$$\det[\mathbf{A}] < 0, \tag{6.337}$$

then the transformation accomplished by $\mathbf{A}$ is called **improper**. For example,

$$\sigma_{XY} = \begin{pmatrix} 1 & 0 & 0 \\ 0 & 1 & 0 \\ 0 & 0 & -1 \end{pmatrix}, \tag{6.338}$$

which reflects every $\mathbf{r} \in \mathbb{R}^3$ in the XY plane, is an improper transformation of $\mathbb{R}^3$.

According to Eq. (6.366), an $n \times n$ permutation matrix $\mathbf{A}_\pi$ realizes a proper transformation if and only if π is an even permutation of $(1 : n)$. For every nonsingular matrix $\mathbf{A} \in \mathbb{R}^n$, the columns of $\mathbf{A}$ are the images of the canonical basis $\{\mathbf{e}_1, \ldots, \mathbf{e}_n\}$. One can write $\det[\mathbf{A}]$ as

$$\det[\mathbf{A}] = \det[\mathbf{Ae}_1, \ldots, \mathbf{Ae}_n] = \det[\mathbf{A}] \det[\mathbf{e}_1, \ldots, \mathbf{e}_n]. \tag{6.339}$$

Therefore $\{\mathbf{Ae}_1, \ldots, \mathbf{Ae}_n\}$ defines a coordinate system with the same handedness as the coordinate system defined by $\{\mathbf{e}_1, \ldots, \mathbf{e}_n\}$ if and only if $\mathbf{A}$ is a proper transformation of $\mathbb{R}^n$ (and therefore maps a right-handed system onto another right-handed system). For example, in $\mathbb{R}^3$ the linear mapping $A_{(12)(3)}$ that takes $\{\mathbf{i}, \mathbf{j}, \mathbf{k}\}$ to the permuted basis $\{\mathbf{j}, \mathbf{i}, \mathbf{k}\}$ is improper because $(12)(3)$ is an odd permutation of $(1 : 3)$.

Laplace Expansion of a Determinant

Every term of Eq. (6.302) contains exactly one factor from each row of the matrix $\mathbf{A}$. Pick a column of $\mathbf{A}$, say, column i. Let the coefficient of a_i^j in Eq. (6.302) be C_j^i, with C_j^i called the **cofactor** of column i and row j. Then we can write Eq. (6.302) as the **Laplace expansion** in terms of cofactors of column i:

$$\det[\mathbf{A}] = \sum_{j=1}^{n} a_i^j C_j^i. \tag{6.340}$$

(Note that we do not use the Einstein convention in this equation because the sum runs over j but over i.) For example, the cofactor of a_1^2 in the 2×2 determinant Eq. (6.303) is $-a_2^1$, according to Eq. (6.304).

One can easily develop an expression for the cofactor C_j^i with the help of the property of multilinearity, Eq. (6.307):

$$\det[\mathbf{a}_1, \mathbf{a}_2, \ldots, \mathbf{a}_i, \ldots, \mathbf{a}_n] = \det\left[\mathbf{a}_1, \mathbf{a}_2, \ldots, a_i^j \mathbf{e}_j, \ldots, \mathbf{a}_n\right]$$
$$= a_i^j \det[\mathbf{a}_1, \mathbf{a}_2, \ldots, \mathbf{e}_j, \ldots, \mathbf{a}_n] \tag{6.341}$$

in which $\mathbf{e}_j$ occupies column i. Then

$$
\begin{aligned}
C_j^i &= \det[\mathbf{a}_1, \mathbf{a}_2, \ldots, \mathbf{a}_{i-1}, \mathbf{e}_j, \mathbf{a}_{i+1}, \ldots, \mathbf{a}_n] \\
&= (-1)^{i-1} \det[\mathbf{e}_j, \mathbf{a}_1, \mathbf{a}_2, \ldots, \mathbf{a}_{i-1}, \mathbf{a}_{i+1}, \ldots, \mathbf{a}_n]
\end{aligned}
\tag{6.342}
$$

in which the second line follows because it takes $i - 1$ column interchanges to make $\mathbf{e}_j$ the first column. By the definition of the canonical basis $\{\mathbf{e}_i\}$, the first column of the matrix $[\mathbf{e}_j, \mathbf{a}_1, \mathbf{a}_2, \ldots, \mathbf{a}_{i-1}, \mathbf{a}_{i+1}, \ldots, \mathbf{a}_n]$ has only one nonzero element a 1 in row j. By Eq. (6.302),

$$
\begin{aligned}
&\det[\mathbf{e}_j, \mathbf{a}_1, \mathbf{a}_2, \ldots, \mathbf{a}_{i-1}, \mathbf{a}_{i+1}, \ldots, \mathbf{a}_n] \\
&\qquad = \sum_{\pi \in S_n} \sigma(\pi) \delta_j^{\pi(1)} a_1^{\pi(2)} \cdots a_{i-1}^{\pi(i)} a_{i+1}^{\pi(i+1)} \cdots a_n^{\pi(n)}.
\end{aligned}
\tag{6.343}
$$

The only nonvanishing terms in Eq. (6.343) have the form $\sigma(\pi) a_1^{\pi(2)} \cdots a_{i-1}^{\pi(i)} a_{i+1}^{\pi(i+1)} \cdots a_n^{\pi(n)}$ in which π is any permutation of $(1:n)$ that takes 1 onto j. If $\pi(1) = j$, then $\pi = \pi'(12 \ldots j)$ in which $(12 \ldots j)$ is the cyclic permutation that "rolls" the first $j - 1$ rows down by one row and sends row j to the top, and π' is a permutation of $(2:n)$. Then $\sigma(\pi) = (-1)^{j-1} \sigma(\pi')$, and

$$
\begin{aligned}
&\det[\mathbf{e}_j, \mathbf{a}_1, \mathbf{a}_2, \ldots, \mathbf{a}_{i-1}, \mathbf{a}_{i+1}, \ldots, \mathbf{a}_n] \\
&\qquad = (-1)^{j-1} \sum_{\pi' \in S_{n-1}} \sigma(\pi') a_1^{\pi'(2)} \cdots a_{i-1}^{\pi'(i)} a_{i+1}^{\pi'(i+1)} \cdots a_n^{\pi'(n)}.
\end{aligned}
\tag{6.344}
$$

The sum on the right-hand side of Eq. (6.344) is the determinant of the matrix obtained from $[\mathbf{a}_1, \mathbf{a}_2, \ldots, \mathbf{a}_n]$ by deleting row j and column i. Then the cofactor of row j and column i is

$$
C_j^i = (-1)^{i+j} M_j^i
\tag{6.345}
$$

in which the **minor** M_j^i of row j and column i is the determinant obtained by deleting row j and column i. In terms of minors of column i, the Laplace expansion reads

$$
\det[\mathbf{A}] = \sum_{j=1}^n (-1)^{i+j} a_i^j M_j^i.
\tag{6.346}
$$

Determinantal Formula for the Matrix Inverse

The cofactor expansion $\det[\mathbf{A}] = \sum_{j=1}^n a_i^j C_j^i$ (Eq. [6.340]) can be extended in a way that gives a formula for the inverse of a matrix:

$$
\sum_{j=1}^n a_k^j C_j^i = \delta_k^i \det[\mathbf{A}].
\tag{6.347}
$$

To prove this assertion, note that the sum $\sum_{j=1}^n a_k^j C_j^i$ is the determinant of a matrix in which the i-th and k-th columns are equal, and therefore it vanishes unless $k = i$.

If **A** is nonsingular, then det[**A**] ≠ 0. Therefore it is legal to divide both sides of Eq. (6.347) by det[**A**]:

$$\sum_{j=1}^{n} a_k^j \frac{C_j^i}{\det[\mathbf{A}]} = \delta_k^i. \tag{6.348}$$

It follows that

$$\mathbf{A}^{-1} = \left[\frac{C_j^i}{\det[\mathbf{A}]} \right] = \{\det[\mathbf{A}]\}^{-1} [C_j^i], \tag{6.349}$$

that is, *the inverse of* **A** *is the matrix whose element in row i and column j is the cofactor of row j and column i divided by* det[**A**].

It is unfortunate that the elegant equation (6.349) is computationally worthless for almost all matrices of dimension larger than approximately 3×3. The computational disadvantages of Eq. (6.349) are that the number of cofactors to compute and the number of terms in the determinant are both $n!$ and that the determinant is an alternating sum and therefore is subject to catastrophic cancellation (see Section 1.4.3). For example, the full expansion in minors of the determinant of a 50×50 matrix has $50! \approx 3 \times 10^{64}$ terms with alternating signs! There must be a better way; indeed there are several. Gaussian elimination (see Section 6.6.2) and the LU decomposition (see Section 6.6.4) are both superior to Eq. (6.349) for the computation of $\mathbf{A}^{-1}$. We show in Section 9.5.3 that for some matrices the solution of the system of linear equations $\mathbf{Ax} = \mathbf{b}$ (which is equivalent to the inversion of **A**) is ill-conditioned in much the same sense as floating-point subtraction (see Section 1.4.3). In these cases one can compute a well-conditioned approximation to $\mathbf{A}^{-1}$ using the singular-value decomposition derived in Section 9.4.1.

Vandermonde Determinants

A **Vandermonde determinant** is any determinant of the form

$$V_k(x_1, \ldots, x_k) := \det \begin{pmatrix} 1 & \cdots & 1 \\ x_1 & \cdots & x_k \\ \vdots & \ddots & \vdots \\ x_1^{k-1} & \cdots & x_k^{k-1} \end{pmatrix} \tag{6.350}$$

in which $x_1, \ldots, x_k \in \mathbb{R}$.

Systems of linear equations that involve the Vandermonde matrix shown in Eq. (6.350) (or its transpose) occur in many important problems. In the problem of moments, one must find a set of unknown positive constants

$$W := \left\{ w_i \;\middle|\; \sum_{i=0}^{k} w_i = 1; \; \forall i = 1, \ldots, k : w_i > 0 \right\} \tag{6.351}$$

subject to the constraints

$$\sum_{i=1}^{k} w_i = \mu_1$$
$$\sum_{i=1}^{k} w_i x_i = \mu_2 \tag{6.352}$$
$$\vdots$$
$$\sum_{i=1}^{k} w_i x_i^{k-1} = \mu_k$$

in which $\mu_j \in \mathbb{R}$ and $x_1, \ldots, x_k \in \mathbb{R}$ are all known. In probability theory, W is a discrete probability distribution, w_i is the probability of observing the value x_i of ξ, and $\{\mu_j\}$ are the moments of the distribution. The determinant of the coefficients of the unknowns $w_1, \ldots, w_k$ in Eq. (6.352) is the Vandermonde determinant defined in Eq. (6.350). Therefore the necessary and sufficient condition for the existence of a unique solution of Eq. (6.352) is that the Vandermonde determinant must be nonzero. The Vandermonde determinant also occurs in the problem of polynomial interpolation described in Section 7.6.

We show that the Vandermonde determinant is nonzero if and only if the scalars $x_1, \ldots, x_k$ are distinct. In the proof, $x_1, \ldots, x_k$ may be complex. We consider the Vandermonde determinant as a function of x_k, which we temporarily call x. Certainly $V_k(x_1, \ldots, x_{k-1}, x)$ vanishes if x is equal to any of the scalars $x_1, \ldots, x_{k-1}$, for then two of the columns of the determinant Eq. (6.350) are equal. Also, from the definition of a determinant, Eq. (6.302), we see that $V_k(x_1, \ldots, x_{k-1}, x)$ is a polynomial of degree $k - 1$ in x. Because a polynomial over the complex field $\mathbb{C}$ of degree $k - 1$ has exactly $k - 1$ roots, it follows that

$$V_k(x_1, \ldots, x_{k-1}, x) = \alpha(x - x_1) \cdots (x - x_{k-1}) \tag{6.353}$$

in which α is a constant. In the polynomial Eq. (6.353), the coefficient of x^{k-1} is α. But one sees directly from the Laplace expansion of Eq. (6.350) in terms of cofactors of the last column that the coefficient of x^{k-1} in $V_k(x_1, \ldots, x_{k-1}, x)$ is equal to $V_{k-1}(x_1, \ldots, x_{k-1})$. Therefore one has a two-term recurrence relation for the Vandermonde determinant:

$$V_k(x_1, \ldots, x_k) = V_{k-1}(x_1, \ldots, x_{k-1})(x_k - x_1) \cdots (x_k - x_{k-1}). \tag{6.354}$$

The value of $V_2(x_1, x_2)$ is

$$V_2(x_1, x_2) = \det \begin{pmatrix} 1 & 1 \\ x_1 & x_2 \end{pmatrix} = x_2 - x_1. \tag{6.355}$$

Then

$$V_k(x_1, \ldots, x_k) = \prod_{j=1}^{k-1} \prod_{i=j+1}^{k} (x_i - x_j). \tag{6.356}$$

This expression is nonzero if and only if the $x_1, \ldots, x_k$ are distinct.

6.5.3 EXERCISES FOR SECTION 6.5

6.5.1 A matrix $\mathbf{A} \in \mathbb{R}^{n \times n}$ is called **orthogonal** if and only if

$$\mathbf{A}^{-1} = \mathbf{A}^T. \tag{6.357}$$

Prove that if $\mathbf{A}$ is orthogonal, then

$$\det[\mathbf{A}] = \pm 1. \tag{6.358}$$

6.5.2 A matrix $\mathbf{A} \in \mathbb{C}^{n \times n}$ is called **unitary** if and only if

$$\mathbf{A}^{-1} = (\mathbf{A}^T)^* \tag{6.359}$$

in which the asterisk means the complex conjugate. Prove that if $\mathbf{A}$ is unitary, then

$$|\det[\mathbf{A}]| = 1 \text{ and therefore } \exists \theta \in \mathbb{R} :\ni: \det[\mathbf{A}] = e^{i\theta}. \tag{6.360}$$

6.5.3 Let

$$p(x) := a_0 + a_1 x + \cdots + a_{n-1} x^{n-1} \tag{6.361}$$

in which $a_{n-1} \neq 0$, be a polynomial of degree $n - 1$ in the unknown x. Let $\mathbf{C}_p$ be the **companion matrix**

$$\mathbf{C}_p := \begin{pmatrix} 0 & 1 & 0 & \cdots & 0 & 0 \\ 0 & 0 & 1 & \cdots & 0 & 0 \\ 0 & 0 & 0 & \cdots & 0 & 0 \\ \vdots & \vdots & \vdots & \ddots & \vdots & \vdots \\ 0 & 0 & 0 & \cdots & 0 & 1 \\ -a_0 & -a_1 & -a_2 & \cdots & -a_{n-2} & -a_{n-1} \end{pmatrix}. \tag{6.362}$$

Prove that

$$\det[\mathbf{C}_p - x\mathbf{1}] = (-1)^n p(x). \tag{6.363}$$

6.5.4 Let $\mathbf{i}$ and $\mathbf{j}$ be the usual Cartesian basis vectors of unit length along the X and Y axes, respectively, and let $\mathbf{k} = \mathbf{i} \times \mathbf{j}$. Show that

$$\forall \mathbf{a}, \mathbf{b} \in \mathbb{R}^2 : \mathbf{a} \times \mathbf{b} = (\det[\mathbf{a}, \mathbf{b}])\mathbf{k}. \tag{6.364}$$

6.5.5 Let $\mathbf{a}, \mathbf{b} \in \mathbb{R}^2$. Prove that $|\det[\mathbf{a}, \mathbf{b}]|$ is equal to the area of the parallelogram defined by $\mathbf{a}$ and $\mathbf{b}$.

6.5.6 Let $\mathbf{a}, \mathbf{b}, \mathbf{c} \in \mathbb{R}^3$. Show that

$$\mathbf{a} \cdot (\mathbf{b} \times \mathbf{c}) = \det[\mathbf{a}, \mathbf{b}, \mathbf{c}]. \tag{6.365}$$

6.5.7 Let $\mathbf{a}, \mathbf{b}, \mathbf{c} \in \mathbb{R}^3$. Prove that $|\det[\mathbf{a}, \mathbf{b}, \mathbf{c}]|$ is equal to the volume of the parallelepiped defined by $\mathbf{a}$, $\mathbf{b}$ and $\mathbf{c}$.

6.5.8   Let $\mathbf{A}_\pi$ be a permutation matrix (see Eq. [6.171] for the definition of the matrix elements). Prove that

$$\det[\mathbf{A}_\pi] = \sigma(\pi). \tag{6.366}$$

6.5.9   This problem concerns the determinant of a general 3×3 matrix,

$$\mathbf{A} = \begin{pmatrix} a_1^1 & a_2^1 & a_3^1 \\ a_1^2 & a_2^2 & a_3^2 \\ a_1^3 & a_2^3 & a_3^3 \end{pmatrix}. \tag{6.367}$$

(a) Write out Eq. (6.302) for this example. Identify the permutation of $(1, 2, 3)$ that is associated with each term of $\det[\mathbf{A}]$.

(b) Verify Eq. (6.306).

(c) Verify Eq. (6.308) for the special case in which $i = 1$ and $j = 2$.

(d) Verify Eq. (6.312) if π' is the cyclic permutation (123).

6.5.10   Prove that the vectors

$$\mathbf{v}_1 = \begin{pmatrix} 1 \\ 1 \\ 1 \\ \vdots \\ 1 \end{pmatrix}, \quad \mathbf{v}_2 = \begin{pmatrix} 1 \\ 2 \\ 3 \\ \vdots \\ n \end{pmatrix}, \ldots, \mathbf{v}_n = \begin{pmatrix} 1 \\ 2^{n-1} \\ 3^{n-1} \\ \vdots \\ n^{n-1} \end{pmatrix} \tag{6.368}$$

are linearly independent in $\mathbb{R}^n$.

6.6 SOLUTION OF LINEAR EQUATIONS

6.6.1 BASIC FACTS ABOUT LINEAR EQUATIONS

Over the past few decades, a theoretical and practical understanding of linear equations such as

$$\mathbf{A}\mathbf{x} = \begin{pmatrix} a_1^1 x^1 + a_2^1 x^2 + \cdots + a_m^1 x^m \\ \vdots \\ a_1^n x^1 + a_2^n x^2 + \cdots + a_m^n x^m \end{pmatrix} = \mathbf{b} = \begin{pmatrix} b^1 \\ \vdots \\ b^n \end{pmatrix}, \tag{6.369}$$

in which $\mathbf{A}$ is an $n \times m$ **coefficient matrix**, has become fundamental for physicists and electrical engineers because of the replacement of analytical methods by numerical methods in many areas.

Classification of Linear-Equation Problems

One of three mutually exclusive cases of Eq. (6.369) arises in most practical applications:

1. The direct problem: We know $\mathbf{A}$ and $\mathbf{x}$ and must find $\mathbf{b}$.
2. The inverse problem: We know $\mathbf{A}$ and $\mathbf{b}$ and must find the **solution vector** $\mathbf{x}$.
3. The fitting problem: We know $\mathbf{A}$ and $\mathbf{b}$ and must find a solution vector $\mathbf{x}$ that minimizes

the norm $\|\mathbf{r}\|$ of the **residual vector**

$$\mathbf{r} := \mathbf{b} - \mathbf{A}\mathbf{x}. \tag{6.370}$$

Straightforward matrix-vector multiplication solves the direct problem subject to rounding errors, which may be important in practice if some or all of the sums are alternating. Computational speed may also be a consideration in problems of very large dimensionality because the obvious method of multiplying an m-element real vector by an $n \times m$ real matrix requires mn multiplications and $(m - 1)n$ additions.

The fitting problem arises if one has more independent equations than unknowns ($n > m$), that is, if the system $\mathbf{A}\mathbf{x} = \mathbf{b}$ is **overdetermined**. In this case n may be the number of experimental observations, and m may be the number of independent parameters used to model the data. The most common method of solving the fitting problem is to minimize $\|\mathbf{r}\|_2$, which leads to the method of least squares, described in Section 9.6.1.

In the inverse problem, the most important questions concern the existence and uniqueness of a solution, and, if the solution exists and is unique, the accuracy that can be achieved in a practical computation. By Eq. (6.17) a solution of the inverse problem Eq. (6.369) for $\mathbf{x}$ exists only if

$$\mathbf{b} \in \text{range}[\mathbf{A}]. \tag{6.371}$$

If this necessary condition is not satisfied, then the system of linear equations given in Eq. (6.369) is called **inconsistent**. Clearly, $\mathbf{b}$ need not be the image under $\mathbf{A}$ of any $\mathbf{x}$ whenever rank[$\mathbf{A}$] is less than n because range[$\mathbf{A}$] is then a proper subspace of the space of n-dimensional column vectors into which $\mathbf{A}$ maps $\mathbf{x}$.

If the number of unknowns exceeds the number of equations ($n < m$), then the system is called **underdetermined**. In an underdetermined system with a nonzero right-hand side $\mathbf{b}$, if rank[$\mathbf{A}$] is less than n, then the system $\mathbf{A}\mathbf{x} = \mathbf{b}$ may have no solution even though there are more unknowns than equations because $\mathbf{b}$ still is not required to belong to range[$\mathbf{A}$]. One can always obtain a consistent (but still underdetermined) system by projecting $\mathbf{b}$ onto range[$\mathbf{A}$].

The uniqueness of the solution vector $\mathbf{x}$ in the inverse problem depends on whether $\mathbf{A}$ is singular or nonsingular. If $\mathbf{A}$ is singular and if we can somehow find a solution vector $\mathbf{x}$ such that $\mathbf{A}\mathbf{x} = \mathbf{b}$, then this solution is not unique because $\mathbf{x} + \mathbf{z}$ is also a solution for any vector $\mathbf{z}$ in the null space of $\mathbf{A}$.

An underdetermined system that is **homogeneous**, that is, for which $\mathbf{b} = \mathbf{0}$, always has solutions because $\mathbf{A}$ always has a nontrivial null space. The rank-nullity theorem, which we prove in Section 6.6.6 and again in Section 6.7.3, states that the sum of the rank and nullity of $\mathbf{A}$ is equal to m. If $\mathbf{A}$ carries an m-dimensional space into an n-dimensional space in which $n < m$, then some vectors must be carried onto the n-dimensional zero vector because rank [$\mathbf{A}$] cannot exceed n. Then the solution of a homogeneous overdetermined system $\mathbf{A}\mathbf{x} = \mathbf{0}$ is

$$\mathbf{x} = \mathbf{z} \tag{6.372}$$

in which $\mathbf{z}$ is any vector in null[$\mathbf{A}$].

If $\mathbf{A}$ is nonsingular, then it defines a vector isomorphism. Then $\mathbf{A}^{-1}$ exists, and the unique solution of the equation $\mathbf{A}\mathbf{x} = \mathbf{b}$ is

$$\mathbf{x} = \mathbf{A}^{-1}\mathbf{b}. \tag{6.373}$$

The direct computation of $\mathbf{A}^{-1}\mathbf{b}$ by Cramer's rule, which often is taught in high school, is ill-advised unless the dimensions of $\mathbf{A}$ are 2×2. Gaussian elimination, which we discuss in the next section, generalizes the method of "eliminating the unknowns" and is the most commonly used method of solving finite-dimensional inverse problems numerically.

One of the central practical difficulties in solving linear equations numerically is to obtain computationally useful criteria that guarantee that the coefficient matrix $\mathbf{A}$ is nonsingular. We summarize the criterion of diagonal dominance in Section 9.5.2. Another significant practical problem is to evaluate the sensitivity of the solution of a system of linear equations to changes in the right-hand side. We introduce the concept of condition number for this purpose in Section 9.5.3.

6.6.2 MATRIX FORMULATION OF GAUSSIAN ELIMINATION

Gaussian elimination is a practical method for solving a system of linear equations by successively modifying the coefficient matrix until it has become upper-triangular. Gaussian elimination is also useful for computing bases of the range and null space of square or rectangular matrices.

Echelon and Upper-Triangular Matrices

A square matrix $\mathbf{U}$ is called **upper-triangular** if and only if all of its elements below the main diagonal vanish:

$$\forall j > i : u_i^j = 0. \tag{6.374}$$

A **unit upper-triangular matrix** is an upper-triangular matrix in which every element on the main diagonal is equal to 1. Similarly, we define a **lower-triangular matrix** to be one in which all elements above the main diagonal are zero and a **unit lower-triangular matrix** to be a lower-triangular matrix whose diagonal elements are all equal to 1.

An upper-triangular matrix is a special case of an **echelon matrix**, which has an inverted-staircase form such as

$$\mathbf{E} = \begin{pmatrix} \underline{u_{j_1}^1} & \cdots & \cdots & \cdots \\ & \boxed{u_{j_2}^2} & \cdots & \cdots \\ & & \boxed{u_{j_3}^3} & \cdots \\ & \mathbf{0} & & \boxed{u_{j_4}^4} \end{pmatrix}. \tag{6.375}$$

The first nonzero element in a row of an echelon matrix is called a **pivot**. The distinguishing feature of an echelon matrix is that

$$j_1 < j_2 < \cdots < j_r \tag{6.376}$$

in which j_l is the number of the column in which the pivot of row l occurs and r is the number of the last nonzero row. For example,

$$\begin{pmatrix} 17 & 3 & 11 & 5 & 13 \\ 0 & 0 & 19 & 2 & 7 \\ 0 & 0 & 0 & 0 & 0 \end{pmatrix} \tag{6.377}$$

is an echelon matrix such that $r = 2$, $j_1 = 1$, and $j_2 = 3$.

Upper-triangular matrices have several useful properties. The product of two upper-triangular matrices is an upper-triangular matrix (see Exercise 6.6.1). The set of all non-singular $n \times n$ upper-triangular matrices is a group (see Exercise 6.6.2). We see in this section that every unit upper-triangular matrix is nonsingular and that the product of two unit upper-triangular matrices is another unit upper-triangular matrix; hence the set of all $n \times n$ unit upper-triangular matrices is a group. Similar statements hold for lower-triangular matrices.

The determinant of an upper-triangular matrix can be found easily from the Laplace expansion, Eq. (6.340), with $i = 1$:

$$\det[\mathbf{U}] = u_1^1 \ldots u_n^n \quad (\text{no } \Sigma). \tag{6.378}$$

Therefore an upper-triangular matrix is nonsingular if and only if its diagonal elements are all nonzero. In particular, every unit upper-triangular matrix is nonsingular.

If an upper-triangular matrix $\mathbf{U}$ is nonsingular, then the linear system

$$\mathbf{U}\mathbf{x} = \mathbf{b} \tag{6.379}$$

can be solved by the process of **back-substitution** starting with the last row, $u_n^n x^n = b^n$ (no Σ). The general result is

$$x^i = \frac{b^i - \sum_{k=i+1}^{n} u_k^i x^k}{u_i^i} \quad (\text{no } \Sigma) \tag{6.380}$$

for $i = n, n - 1, \ldots, 1$. Obviously the linear system

$$\mathbf{L}\mathbf{x} = \mathbf{b}, \tag{6.381}$$

in which $\mathbf{L}$ is a lower-triangular matrix, can be solved by forward-substitution starting with the first row, $l_1^1 x^1 = b^1$. Just as for an upper-triangular matrix, a lower-triangular matrix $\mathbf{L}$ is nonsingular if and only if all of its diagonal entries are nonzero because $\det[\mathbf{L}] = l_1^1 \ldots l_n^n$.

Gaussian Elimination

Let $\mathbf{A}$ be an $n \times n$ matrix over $\mathbb{R}$ or $\mathbb{C}$. We do not assume (unless we state otherwise) that $\mathbf{A}$ is nonsingular. Assume that $a_1^1 \neq 0$; if this is not true, then one must permute the rows or columns of $\mathbf{A}$ so that in the permuted matrix $a_1^1 \neq 0$. To apply a permutation π_r to the rows of $\mathbf{A}$, one forms the product $\mathbf{A}_{\pi_r^{-1}}\mathbf{A}$. (Permutation matrices are defined in Section 6.2.3.) To apply a permutation π_c to the columns of $\mathbf{A}$, one forms the product $\mathbf{A}\mathbf{A}_{\pi_c^{-1}}$.

For the sake of simplicity we assume momentarily that no row or column permutations are necessary. The first step of Gaussian elimination is to eliminate x^1 from the last $(n-1)$ equations of the system $\mathbf{A}\mathbf{x} = \mathbf{b}$ by subtracting l_1^i times the first equation from the i-th equation for $i = 2, \ldots, n$, in which

$$l_1^i = \frac{a_1^i}{a_1^1}. \tag{6.382}$$

One can accomplish the same result by premultiplying both $\mathbf{A}$ and $\mathbf{b}$ with a unit lower-triangular **multiplier matrix** $\mathbf{L}^{(1)}$:

$$\mathbf{A}^{(2)} := \mathbf{L}^{(1)}\mathbf{A} \quad \text{and} \quad \mathbf{b}^{(2)} := \mathbf{L}^{(1)}\mathbf{b} \Rightarrow \mathbf{A}^{(2)}\mathbf{x} = \mathbf{b}^{(2)}. \tag{6.383}$$

The multiplier matrix is

$$\mathbf{L}^{(1)} := \begin{pmatrix} 1 & 0 & 0 & \cdots & 0 \\ -l_1^2 & 1 & 0 & \cdots & 0 \\ -l_1^3 & 0 & 1 & \cdots & 0 \\ \vdots & \vdots & \vdots & \ddots & \vdots \\ -l_1^n & 0 & 0 & \cdots & 1 \end{pmatrix}, \tag{6.384}$$

and the transformed coefficient matrix is

$$\mathbf{A}^{(2)} = \begin{pmatrix} a_1^1 & a_2^1 & a_3^1 & \cdots & a_n^1 \\ 0 & a_2^{(2)2} & a_3^{(2)2} & \cdots & a_n^{(2)2} \\ 0 & a_2^{(2)3} & a_3^{(2)3} & \cdots & a_n^{(2)3} \\ \vdots & \vdots & \vdots & \ddots & \vdots \\ 0 & a_2^{(2)n} & a_3^{(2)n} & \cdots & a_n^{(2)n} \end{pmatrix} \tag{6.385}$$

in which

$$a_j^{(2)i} := a_j^i - l_1^i a_j^1 \tag{6.386}$$

Recall from Section 5.4.4 that the transformation from $\mathbf{A}$ to $\mathbf{A}^{(2)}$ is called a *Gauss transformation*. A Gauss transformation is *not* a similarity transformation because it is more than a basis change; one changes the vectors also.

Suppose now that $\mathbf{A}$ is nonsingular. The fastest way to see that the list of new columns, $\{\mathbf{a}_i^{(2)}\}$, is a basis of the column space of $\mathbf{A}$ is to show that $\det[\mathbf{A}^{(2)}]$ is nonzero because this implies at once that $\{\mathbf{a}_i^{(2)}\}$ is linearly independent. But

$$\det\left[\mathbf{A}^{(2)}\right] = \det[\mathbf{A}] \neq 0 \tag{6.387}$$

because the value of a determinant is unaffected by the operation of adding a multiple of one row to another row. Because the number of these linearly independent column vectors is equal to the dimension of the vector space, $\{\mathbf{a}_i^{(2)}\}$ is a basis.

If $\mathbf{A}$ is singular, then the list of new columns $\{\mathbf{a}_i^{(2)}\}$ spans the column space of $\mathbf{A}$. However, $\{\mathbf{a}_i^{(2)}\}$ is not a basis of $\mathbf{A}$ because the columns of $\mathbf{A}$ (and therefore of $\mathbf{A}^{(2)}$) are linearly dependent.

Supposing that $\mathbf{A}$ is nonsingular and continuing similarly to eliminate x^i from rows $i + 1, \ldots, n$ for $i = 2, \ldots, n$, one obtains

$$\begin{aligned} \mathbf{U} :&= \mathbf{A}^{(n)} \\ &= \mathbf{L}'\mathbf{A} \end{aligned} \tag{6.388}$$

and

$$\mathbf{b}^{(n)} := \mathbf{L}'\mathbf{b}, \tag{6.389}$$

in which

$$\mathbf{L}' := \mathbf{L}^{(n-1)} \cdots \mathbf{L}^{(1)}. \tag{6.390}$$

Therefore the original system of linear equations $\mathbf{A}\mathbf{x} = \mathbf{b}$ has become

$$\mathbf{U}\mathbf{x} = \mathbf{b}^{(n)}. \tag{6.391}$$

The only nonzero elements of the multiplier matrix $\mathbf{L}^{(i)}$ are in the i-th column below the main diagonal:

$$\mathbf{L}^{(i)} = \begin{pmatrix} 1 & \cdots & 0 & \cdots & 0 \\ \vdots & \ddots & \vdots & \ddots & \vdots \\ 0 & \cdots & 1 & \cdots & 0 \\ 0 & \cdots & -l_i^{i+1} & \cdots & 0 \\ \vdots & \ddots & \vdots & \ddots & \vdots \\ 0 & \cdots & -l_i^n & \cdots & 1 \end{pmatrix} \tag{6.392}$$

in which

$$l_i^k := \frac{a_k^{(k)i}}{a_k^{(k)k}} \quad (\text{no } \Sigma) \tag{6.393}$$

is the multiplier for row $i = k + 1, \ldots, n$. If the original coefficient matrix $\mathbf{A}$ is nonsingular, the pivots $a_k^{(k)k}$ occur along the main diagonal. After n steps of elimination, the original coefficient matrix has been transformed into

$$
\mathbf{U} = \mathbf{A}^{(n)} = \begin{pmatrix}
a_1^1 & a_2^1 & a_3^1 & \cdots & a_n^1 \\
0 & a_2^{(2)2} & a_3^{(2)2} & \cdots & a_n^{(2)2} \\
0 & 0 & a_3^{(3)3} & \cdots & a_n^{(3)3} \\
\vdots & \vdots & \vdots & \ddots & \vdots \\
0 & 0 & 0 & \cdots & a_n^{(n)n}
\end{pmatrix},
\tag{6.394}
$$

which is upper-triangular. Then the transformed system of linear equations $\mathbf{U}\mathbf{x} = \mathbf{b}^{(n)}$ can be solved by back-substitution as in Eq. (6.380).

If $\mathbf{A}$ is singular, then some elements $a_k^{(k)k}$ vanish. Obviously one skips to the first nonzero element in row k before calculating the multiplier. If rank$[\mathbf{A}] = r$, then r steps of Gaussian elimination lead to an echelon matrix $\mathbf{A}^{(r)}$.

Gaussian elimination gives one "for free" the value of the determinant of a nonsingular coefficient matrix $\mathbf{A}$. Because the determinant of a matrix is unchanged if one adds a multiple of any row to any other row,

$$
\det[\mathbf{A}] = \det[\mathbf{A}^{(2)}] = \cdots = \det[\mathbf{A}^{(n)}].
\tag{6.395}
$$

But $\mathbf{A}^{(n)}$ is upper-triangular. Therefore, by Eq. (6.378),

$$
\det[\mathbf{A}] = a_1^1 a_2^{(2)2} \cdots a_n^{(n)n},
\tag{6.396}
$$

which is just the product of the pivots.

In order to obtain the matrix $\mathbf{L}'$ and the transformed coefficient matrix $\mathbf{U}$ at the same time, one simply applies the same row operations to the identity matrix as one applies to $\mathbf{A}, \ldots, \mathbf{A}^{(n)}$. To understand why this method works, note that $\mathbf{L}'$ is equal to the product of the multiplier matrices. Therefore the matrix that results when $\mathbf{L}^{(1)}, \ldots, \mathbf{L}^{(n)}$ are successively applied to $\mathbf{1}$ is equal to $\mathbf{L}'$. If one is calculating by hand, it is convenient to augment the original matrix $\mathbf{A}$ by writing the n columns of the identity matrix $\mathbf{1}$ after the last column of $\mathbf{A}$.

This method is not limited to nonsingular matrices. For example, let

$$
\mathbf{A} = \begin{pmatrix}
1 & 2 & 1 \\
2 & 3 & 1 \\
1 & 3 & 2
\end{pmatrix}.
\tag{6.397}
$$

The augmented matrix is

$$
\mathbf{A}_{\text{aug}} = \begin{pmatrix}
1 & 2 & 1 & 1 & 0 & 0 \\
2 & 3 & 1 & 0 & 1 & 0 \\
1 & 3 & 2 & 0 & 0 & 1
\end{pmatrix}.
\tag{6.398}
$$

Gaussian elimination applied to $\mathbf{A}_{\text{aug}}$ results in the matrix

$$
\begin{pmatrix}
1 & 2 & 1 & 1 & 0 & 0 \\
0 & -1 & -1 & -2 & 1 & 0 \\
0 & 0 & 0 & -3 & 1 & 1
\end{pmatrix}.
\tag{6.399}
$$

Then

$$\mathbf{L}' = \begin{pmatrix} 1 & 0 & 0 \\ -2 & 1 & 0 \\ -3 & 1 & 1 \end{pmatrix}, \quad \mathbf{U} = \begin{pmatrix} 1 & 2 & 1 \\ 0 & -1 & -1 \\ 0 & 0 & 0 \end{pmatrix}. \tag{6.400}$$

One verifies by matrix multiplication that $\mathbf{L}'\mathbf{A} = \mathbf{U}$.

6.6.3 COMPUTATIONAL ASPECTS OF GAUSSIAN ELIMINATION

Exercises 6.6.3 and 6.6.4 derive the number of additions, multiplications, and divisions (which is usually called the **operation count**) required for Gaussian elimination. The operation count of an algorithm is an extremely important consideration in practical applications.

6.6.4 LU AND LDM^T DECOMPOSITIONS

LU Decomposition

Let $\mathbf{A}$ be a nonsingular $n \times n$ matrix. Then one can transform $\mathbf{A}$ by premultiplying with lower-triangular matrices to obtain an upper-triangular matrix as in Eq. (6.388). Each matrix multiplier $\mathbf{L}^{(i)}$ (Eq. [6.392]) is invertible. In fact,

$$\left[\mathbf{L}^{(i)}\right]^{-1} = \begin{pmatrix} 1 & \cdots & 0 & \cdots & 0 \\ \vdots & \ddots & \vdots & \ddots & \vdots \\ 0 & \cdots & 1 & \cdots & 0 \\ 0 & \cdots & +l_i^{i+1} & \cdots & 0 \\ \vdots & \ddots & \vdots & \ddots & \vdots \\ 0 & \cdots & +l_i^n & \cdots & 1 \end{pmatrix}, \tag{6.401}$$

which is self-evident because the above matrix simply undoes what $\mathbf{L}^{(i)}$ does. The matrix product

$$\mathbf{L} := \left[\mathbf{L}^{(1)}\right]^{-1} \cdots \left[\mathbf{L}^{(n-1)}\right]^{-1} \tag{6.402}$$

is unit lower-triangular because each $[\mathbf{L}^{(i)}]^{-1}$ is unit lower-triangular. Equations (6.388) and (6.402) imply the **LU decomposition**

$$\mathbf{A} = \mathbf{L}\mathbf{A}^{(n)} = \mathbf{L}\mathbf{U} \tag{6.403}$$

of an arbitrary square nonsingular matrix $\mathbf{A}$ into the product of a unit lower-triangular matrix and an upper-triangular matrix.

The LU decomposition has considerable practical usefulness if one must solve many systems of linear equations with the same coefficient matrix $\mathbf{A}$ and different right-hand sides $\mathbf{b}$. Exercises 6.6.3 and 6.6.4 show that the operation count for forward elimination is of the order of n times the operation count for back-substitution. The two triangular linear systems that result from $\mathbf{Ax} = \mathbf{b}$ and Eq. (6.403), namely,

$$\mathbf{Ly} = \mathbf{b} \tag{6.404}$$

and

$$\mathbf{Ux} = \mathbf{y}, \tag{6.405}$$

can be solved by forward-substitution and back-substitution, respectively. Therefore the LU decomposition makes it possible to avoid carrying out the computationally expensive process of forward elimination for every right-hand side.

An important property of the LU decomposition is that it is unique. Suppose that $\mathbf{A}$ has two LU decompositions,

$$\mathbf{A} = \mathbf{LU} = \mathbf{L'U'}. \tag{6.406}$$

Then

$$\mathbf{L}^{-1}\mathbf{L'} = \mathbf{U}[\mathbf{U'}]^{-1} \tag{6.407}$$

because all the required inverses exist. But $\mathbf{L}^{-1}\mathbf{L'}$ is a unit lower-triangular matrix, whereas $\mathbf{U}[\mathbf{U'}]^{-1}$ is an upper-triangular matrix. Therefore

$$\mathbf{L}^{-1}\mathbf{L'} = \mathbf{1}, \quad \mathbf{U}[\mathbf{U'}]^{-1} = \mathbf{1}, \tag{6.408}$$

whence $\mathbf{L'} = \mathbf{L}$ and $\mathbf{U'} = \mathbf{U}$.

$\mathbf{LDM}^T$ Decomposition

Assume that $\mathbf{A}$ is a square nonsingular matrix. Because $\det[\mathbf{A}]$ is nonzero, every pivot is nonzero after one applies Gaussian elimination to $\mathbf{A}$. Let the diagonal matrix of the pivots be

$$\mathbf{D} := \mathrm{diag}\big[a_1^1, a_2^{(2)2}, \dots, a_n^{(n)n}\big]. \tag{6.409}$$

Then

$$\mathbf{D}^{-1} = \mathrm{diag}\big[\big(a_1^1\big)^{-1}, \big(a_2^{(2)2}\big)^{-1}, \dots, \big(a_n^{(n)n}\big)^{-1}\big] \tag{6.410}$$

exists. The product

$$\mathbf{M}^T := \mathbf{D}^{-1}\mathbf{A}^{(n)} = \mathbf{D}^{-1}\mathbf{U} \tag{6.411}$$

is a unit upper-triangular matrix because each element of $\mathbf{D}^{-1}$ is the reciprocal of the pivot in the same diagonal position in $\mathbf{A}^{(n)}$. Therefore

$$\mathbf{U} = \mathbf{A}^{(n)} = \mathbf{L}^{(n-1)} \cdots \mathbf{L}^{(1)}\mathbf{A} = \mathbf{DM}^T. \tag{6.412}$$

From Eqs. (6.412) and (6.403) one sees that

$$\mathbf{A} = \mathbf{L}\mathbf{D}\mathbf{M}^T, \tag{6.413}$$

that is, every square, nonsingular matrix $\mathbf{A}$ can be expressed as the product of a unit upper-triangular matrix $\mathbf{M}^T$, a diagonal matrix with nonzero elements $\mathbf{D}$, and a unit lower-triangular matrix $\mathbf{L}$.

The $\mathbf{L}\mathbf{D}\mathbf{M}^T$ decomposition is not of much practical use but is significant theoretically because it is a kind of diagonalization of any nonsingular square matrix. In Section 9.5.4 we use the $\mathbf{L}\mathbf{D}\mathbf{M}^T$ decomposition to derive a diagonal form for any nonsingular Hermitian matrix.

6.6.5 BASES OF THE RANGE AND NULL SPACE

At the beginning of this chapter we defined the range and null space of a linear mapping in an abstract, basis-free manner. In this section we show how to use the LU decomposition to compute concrete bases of the range and null space of an arbitrary $n \times m$ matrix $\mathbf{A}$.

The method is to perform an LU decomposition of the *transposed* matrix $\mathbf{A}^T$, obtaining

$$\mathbf{A}^T = (\mathbf{U}^T)^{-1}\mathbf{V}^T \tag{6.414}$$

in which $(\mathbf{U}^T)^{-1}$ is an $m \times m$ unit lower-triangular matrix of the form given in Eq. (6.402), and $\mathbf{V}^T$ is an $m \times n$ echelon matrix. (Note that $(\mathbf{U}^T)^{-1}$ is invertible by Eq. [6.401].) Taking the transpose of both sides gives

$$\mathbf{A} = \mathbf{V}\mathbf{U}^{-1}. \tag{6.415}$$

Multiplying Eq. (6.415) on the right by $\mathbf{U}$, one obtains

$$\mathbf{A}\mathbf{U} = \mathbf{V}. \tag{6.416}$$

Here $\mathbf{U}$ is an $m \times m$ unit upper-triangular matrix; $\mathbf{V}$ is the $n \times m$ transpose of an echelon matrix. The nonzero columns of $\mathbf{U}$ are linearly independent because $\mathbf{U}$ is nonsingular. Therefore the columns of $\mathbf{U}$ are a basis of $\mathbb{F}^m$, and the columns of $\mathbf{V}$ span range$[\mathbf{A}] \subseteq \mathbb{F}^n$.

For example, let $n = 5$, $m = 3$ and

$$\mathbf{A} = \begin{pmatrix} 17 & -17 & 34 \\ 3 & -3 & 6 \\ 11 & 8 & 3 \\ 5 & -3 & 8 \\ 13 & -6 & 19 \end{pmatrix}. \tag{6.417}$$

Then

$$\mathbf{A}^T = \begin{pmatrix} 17 & 3 & 11 & 5 & 13 \\ -17 & -3 & 8 & -3 & -6 \\ 34 & 6 & 3 & 8 & 19 \end{pmatrix}, \tag{6.418}$$

from which one obtains via LU decomposition the echelon matrix

$$\mathbf{V}^T = \begin{pmatrix} 17 & 3 & 11 & 5 & 13 \\ 0 & 0 & 19 & 2 & 7 \\ 0 & 0 & 0 & 0 & 0 \end{pmatrix} \tag{6.419}$$

and the unit lower-triangular matrix

$$[\mathbf{U}^T]^{-1} = \begin{pmatrix} 1 & 0 & 0 \\ -1 & 1 & 0 \\ 2 & -1 & 1 \end{pmatrix}. \tag{6.420}$$

(Reader, please verify that $\mathbf{A}^T = [\mathbf{U}^T]^{-1}\mathbf{V}^T$.) Then

$$\mathbf{U} = \begin{pmatrix} 1 & 1 & -1 \\ 0 & 1 & 1 \\ 0 & 0 & 1 \end{pmatrix} \tag{6.421}$$

and

$$\mathbf{V} = \begin{pmatrix} 17 & 0 & 0 \\ 3 & 0 & 0 \\ 11 & 19 & 0 \\ 5 & 2 & 0 \\ 13 & 7 & 0 \end{pmatrix}. \tag{6.422}$$

We show subsequently that the nonzero columns of $\mathbf{V}$ are a basis of range[$\mathbf{A}$].

For a general matrix $\mathbf{A}$, the LU decomposition makes the last $m - r$ rows of the echelon matrix $\mathbf{V}^T$ null:

$$\mathbf{V}^T = \begin{bmatrix} (\mathbf{R}_{n \times r})^T \\ \mathbf{0}_{(m-r) \times n} \end{bmatrix}. \tag{6.423}$$

Then the column-partitioned form of $\mathbf{V}$ is

$$\mathbf{V} = \begin{bmatrix} \mathbf{R}_{n \times r}, \mathbf{0}_{n \times (m-r)} \end{bmatrix}. \tag{6.424}$$

For example, in Eq. (6.422) one has

$$\mathbf{R} = \begin{pmatrix} 17 & 0 \\ 3 & 0 \\ 11 & 19 \\ 5 & 2 \\ 13 & 7 \end{pmatrix}. \tag{6.425}$$

Equation (6.415) implies that for every m-component column vector $\mathbf{x}$, there exists a vector

$$\mathbf{y} = \mathbf{U}^{-1}\mathbf{x} \tag{6.426}$$

such that

$$\mathbf{Ax} = \mathbf{Vy}. \tag{6.427}$$

This is a matrix proof that the columns of $\mathbf{V}$ span range[$\mathbf{A}$].

Because the last $m - r$ columns of $\mathbf{V}$ are null vectors, the r columns of $\mathbf{R}$ span range[$\mathbf{A}$]. The columns of $\mathbf{R}$ are linearly independent because of the echelon form of $\mathbf{V}^T$, which ensures that no column of $\mathbf{V}$ can be linearly dependent on its predecessors. Therefore the columns of $\mathbf{R}$ are a basis of range[$\mathbf{A}$], and

$$r = \text{rank}[\mathbf{A}]. \tag{6.428}$$

It is clear from Eqs. (6.416) and (6.424) that the final $m - r$ columns of

$$\begin{aligned} \mathbf{U} &= [\mathbf{c}_1, \ldots, \mathbf{c}_r, \mathbf{n}_1, \ldots, \mathbf{n}_{m-r}] \\ &= \left[\mathbf{C}_{n \times r}, \mathbf{N}_{n \times (m-r)} \right] \end{aligned} \tag{6.429}$$

are vectors that $\mathbf{A}$ maps to zero. In the example used above,

$$\mathbf{C} = \begin{pmatrix} 1 & 1 \\ 0 & 1 \\ 0 & 0 \end{pmatrix} \quad \text{and} \quad \mathbf{N} = \begin{pmatrix} -1 \\ 1 \\ 1 \end{pmatrix}. \tag{6.430}$$

From a practical point of view, the construction derived here produces linearly independent, but generally nonorthogonal, basis vectors of range[$\mathbf{A}$] and null[$\mathbf{A}$]. If $\mathbf{A}$ is ill-conditioned in the sense defined in Section 9.5.3, then one should not attempt to use the LU decomposition to find bases of the range and null space. The singular-value decomposition derived in Section 9.4.1 should be used whenever $\mathbf{A}$ is ill conditioned or one needs orthogonal bases of the range and null space.

6.6.6 RANK-NULLITY THEOREM

Because $\mathbf{U}$ in Eq. (6.416) is nonsingular, its columns are linearly independent (recall Eq. [6.402]). Then $\{\mathbf{n}_1, \ldots, \mathbf{n}_{m-r}\}$ in Eq. (6.429) is a basis of null[$\mathbf{A}$] and $\{\mathbf{c}_1, \ldots, \mathbf{c}_r\}$ is a basis of a subspace that is complementary to null[$\mathbf{A}$]. Obviously

$$m - r = \text{nullity}[\mathbf{A}]. \tag{6.431}$$

Therefore

$$\text{rank}[\mathbf{A}] + \text{nullity}[\mathbf{A}] = \dim[\text{domain}[\mathbf{A}]], \tag{6.432}$$

which is the **rank-nullity theorem**.

For example, the null space of the 3×3 matrix $\mathbf{B}$ defined in Eq. (6.54) is one-dimensional (see Eq. [6.196]). Therefore nullity[$\mathbf{B}$] $= 1$. By Eq. (6.58), rank[$\mathbf{B}$] $= 2$. Therefore $\mathbf{B}$ satisfies the rank-nullity theorem.

The rank-nullity theorem can be used to find the nullity if the rank is known, or the rank if the nullity is known. For example, knowing that the rank of the 3×20 matrix $\mathbf{A}$ defined in Eq. (6.192) is 2, one concludes at once that nullity$[\mathbf{A}] = 18$.

6.6.7 EXERCISES FOR SECTION 6.6

6.6.1 Prove that the product of two $n \times n$ upper-triangular matrices is an upper-triangular matrix.

6.6.2 This problem concerns the set U of real, unit, upper-triangular 3×3 matrices:

$$U := \left\{ \begin{pmatrix} 1 & a & b \\ 0 & 1 & c \\ 0 & 0 & 1 \end{pmatrix} \middle| a, b, c \in \mathbb{R} \right\}. \tag{6.433}$$

(a) Calculate the inverse of the matrix

$$\begin{pmatrix} 1 & a & b \\ 0 & 1 & c \\ 0 & 0 & 1 \end{pmatrix}. \tag{6.434}$$

(b) Prove that U is a group.

(c) Prove that the set T of real 3×3 matrices of the form

$$\begin{pmatrix} 1 & 0 & b' \\ 0 & 1 & c' \\ 0 & 0 & 1 \end{pmatrix} \tag{6.435}$$

is a group.

(d) Prove that T is a *normal* subgroup of U.

6.6.3 (a) Show that the total number N_U of multiplications and divisions in the solution of an $n \times n$ upper-triangular system of linear equations

$$\mathbf{Ux} = \mathbf{b} \tag{6.436}$$

by back-substitution [Eq. (6.380)] is

$$N_U = \frac{n(n+1)}{2}. \tag{6.437}$$

Hint: You will need one of the formulas you derived in Exercise 2.4.1.

Comment: It is customary to count multiplications and divisions, but not additions and subtractions, because computers usually require more clock periods to multiply (or, especially, to divide) than they require to add or subtract.

(b) Show that the total number N_L of multiplications and divisions in the solution of an $n \times n$ unit lower-triangular system of linear equations

$$\mathbf{Ly} = \mathbf{b} \tag{6.438}$$

by forward-substitution is

$$N_L = \frac{n(n-1)}{2}.$$ (6.439)

(c) Combine the results of parts (a) and (b) to show that the number of operations required to solve a linear system if the LU decomposition of the coefficient matrix is already known is

$$N_{LU} = n^2.$$ (6.440)

6.6.4 (a) Show that the total number N_f of multiplications and divisions in the process of forward elimination (Eqs. [6.382] through [6.393]) resulting in the upper-triangular matrix $\mathbf{A}^{(n)}$, Eq. (6.394), is

$$N_f = \frac{n(n-1)(n+1)}{3}.$$ (6.441)

Hint: You will need the formulas you derived in Exercise 2.4.1.

(b) Combine the result of part (a) with the result of Exercise 6.6.3(c) to show that the total operation count for Gaussian elimination using LU decomposition is

$$N_G = \frac{n^3}{3} + n^2 - \frac{n}{3}.$$ (6.442)

6.6.5 Calculate by hand the LU decomposition of

$$\mathbf{A} = \begin{pmatrix} 7 & 2 & 23 \\ 3 & 11 & 13 \\ 17 & 19 & 5 \end{pmatrix}.$$ (6.443)

6.6.6 Calculate by hand a basis of the range and a basis of the null space of the matrix

$$\mathbf{A} = \begin{pmatrix} 31 & 93 & -31 \\ 7 & 21 & -7 \\ 2 & 6 & -2 \\ 5 & 44 & 53 \\ 23 & 72 & -17 \end{pmatrix}.$$ (6.444)

6.6.7 Let $\mathcal{W}$ be the subspace of $\mathbb{E}^4$ that is spanned by the vectors

$$\mathbf{f}_1 = \begin{pmatrix} 1 \\ 0 \\ 0 \\ 1 \end{pmatrix}, \quad \mathbf{f}_2 = \begin{pmatrix} 1 \\ -1 \\ -1 \\ 1 \end{pmatrix}.$$ (6.445)

(a) Construct a basis of $\mathcal{W}$ that is orthogonal with respect to the canonical inner product of vectors belonging to $\mathbb{E}^4$.

(b) Construct a basis of $\mathcal{W}^\perp$.

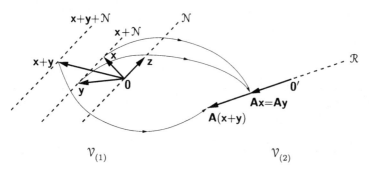

FIGURE 6.7

Left: $\mathcal{N}$ is the null space in $\mathcal{V}_{(1)}$ of the linear mapping **A**, the matrix of which is given in (6.446). The vectors **x** and **y** belong to the same coset of $\mathcal{N}$. *Right*: $\mathcal{R}$ is the range of **A** in $\mathcal{V}_{(2)}$. Both **x** and **y** have the same image, **Ax** = **Ay**.

6.7 COMPLEMENTS OF THE NULL SPACE

6.7.1 QUOTIENT SPACE $\mathcal{V}$/NULL[A]

Figure 6.7 illustrates the range and null space of the obviously singular linear mapping

$$\mathbf{A} = \begin{pmatrix} 1 & -1 \\ \frac{1}{3} & -\frac{1}{3} \end{pmatrix}. \tag{6.446}$$

Clearly the columns span a one-dimensional subspace:

$$\text{range}[\mathbf{A}] = \text{span}[\mathbf{r}'], \quad \text{where} \quad \mathbf{r}' = \begin{pmatrix} 1 \\ \frac{1}{3} \end{pmatrix}. \tag{6.447}$$

By inspection or computation, one sees that

$$\mathcal{N} := \text{null}[\mathbf{A}] = \text{span}[\mathbf{z}], \quad \text{where} \quad \mathbf{z} = \begin{pmatrix} 1 \\ 1 \end{pmatrix}. \tag{6.448}$$

In Fig. 6.7 one recognizes a pattern that is familiar from Section 5.5: The vectors **x** and **y** (which **A** maps onto the same image) belong to the same coset of a particular vector subspace, which in this case is the null space of **A**. This simple observation suggests that studying the cosets $\overline{x} + \text{null}[\mathbf{A}]$ and the quotient space $\mathcal{V}_{(1)}/\text{null}[\mathbf{A}]$ may be a good way to reveal the structure of a linear mapping A. From Eqs. (6.202) and (6.203) one sees that if $\overline{x}$ and $\overline{y}$ have the same image $\overline{x}'$ under A, then $\overline{x}$ and $\overline{y}$ belong to the same coset $\overline{x} + \text{null}[\mathbf{A}]$ of the null space:

$$A\overline{x} = A\overline{y} \Rightarrow \exists \overline{z} \in \text{null}[A] :\ni: \overline{y} = \overline{x} + \overline{z}. \tag{6.449}$$

The converse is also true, for if $\overline{y} = \overline{x} + \overline{z}$ where $\overline{z} \in \text{null}[A]$, then

$$A\overline{y} = A(\overline{x} + \overline{z}) = A\overline{x} + \overline{0}' = A\overline{x}. \tag{6.450}$$

Therefore

$$\forall \overline{x}, \overline{y} \in \mathcal{V}_{(1)} : A\overline{x} = A\overline{y} \Leftrightarrow \overline{y} \in \overline{x} + \text{null}[A] \tag{6.451}$$

(see Fig. 6.7). If A is singular, that is, if null[A] $\neq \mathcal{O}$, then null[A] is an infinite set and A therefore is infinitely many-to-one. For example, in Fig. 6.7 each of the infinitely many vectors that belong to $\mathbf{x} + \mathcal{N}$ is mapped onto the same image $\mathbf{x}' = \mathbf{Ax}$.

6.7.2 ISOMORPHISM OF THE RANGE TO A COMPLEMENT OF THE NULL SPACE

Let $A : \mathcal{V}_{(1)} \to \mathcal{V}_{(2)}$ be a linear mapping, and let

$$\mathcal{N} = \text{null}[\mathbf{A}] \quad \text{and} \quad \mathcal{R} = \text{range}[\mathbf{A}]. \tag{6.452}$$

We show that if $\mathcal{W}$ is any complement of $\mathcal{N}$, that is, if

$$\mathcal{V}_{(1)} = \mathcal{W} \oplus \mathcal{N}, \tag{6.453}$$

then

$$\mathcal{W} \cong \mathcal{R} \tag{6.454}$$

in which $\cong$ means *vector* isomorphism.

For example, in Fig. 6.7 every complement of the one-dimensional subspace $\mathcal{N}$ is one-dimensional and therefore is isomorphic to $\mathcal{R} = \text{range}[\mathbf{A}]$. If $\mathbf{u} \notin \mathcal{N}$, then

$$\mathcal{U} = \{\alpha\mathbf{u} \mid \alpha \in \mathbb{R}\} \tag{6.455}$$

is a complement of $\mathcal{N}$, and

$$\mathbb{R}^2/\mathcal{N} \cong \mathcal{U}. \tag{6.456}$$

The mapping $\mathbf{u} \mapsto \mathbf{r}'$ establishes a vector isomorphism of $\mathcal{U}$ and $\mathcal{R}$ because each space consists of the scalar multiples of a single vector and $\mathbf{u} \mapsto \mathbf{r}'$ implies that $\alpha\mathbf{u} \mapsto \alpha\mathbf{r}'$. Hence

$$\mathcal{R} \cong \mathcal{U}, \tag{6.457}$$

in which $\mathcal{U}$ is a complement of the null space $\mathcal{N}$.

To prove Eq. (6.454) in general, we choose a basis in $\mathcal{V}_{(1)}$ and carry out an LU decomposition of the transpose of the matrix of A relative to the chosen basis, as in Section 6.6.5. The result is

$$\mathbf{AU} = \mathbf{V} \tag{6.458}$$

in which

$$\mathbf{V} = \left[\mathbf{R}_{n \times r}, \mathbf{0}_{n \times (m-r)} \right] \tag{6.459}$$

and

$$\mathbf{U} = \left[\mathbf{C}_{n \times r}, \mathbf{N}_{n \times (m-r)} \right]. \tag{6.460}$$

The columns of $\mathbf{R}$ span $\mathcal{R}$ and the columns of $\mathbf{N}$ span $\mathcal{N}$. The columns of $\mathbf{C}$ span a subspace $\mathcal{C}$ of $\mathcal{V}_{(1)}$. By construction, $\mathcal{C}$ is complementary to $\mathcal{N}$. Then

$$\dim[\mathcal{C}] = \dim[\mathcal{V}] - \dim[\mathcal{N}] \tag{6.461}$$

by Eq. (5.254). The rank-nullity theorem implies that

$$\dim[\mathcal{R}] = \dim[\mathcal{V}] - \dim[\mathcal{N}]. \tag{6.462}$$

Therefore

$$\mathcal{C} \cong \mathcal{R} \tag{6.463}$$

because all vector spaces with the same dimension over the same field are isomorphic. If $\mathcal{W}$ is any vector complement of $\mathcal{N}$ in $\mathcal{V}$, then

$$\mathcal{W} \cong \mathcal{C} \tag{6.464}$$

because $\dim[\mathcal{W}] = \dim[\mathcal{V}] - \dim[\mathcal{N}] = \dim[\mathcal{C}]$. Equation (6.454) follows by the transitivity of the relation of vector isomorphism. Therefore *the range of a linear mapping is isomorphic to every complement of its null space.* This result is fundamental for solving the inverse problem $A\overline{x} = \overline{x}'$ (Eq. [6.101]) if A is a differential operator.

For an example, we refer once again to the matrix $\mathbf{B}$ defined in Eq. (6.54). Using the method of Section 6.6.5, one calculates the LU decomposition of $\mathbf{B}^T$ and finds that

$$\mathbf{BU} = \mathbf{V} \tag{6.465}$$

in which

$$\mathbf{V} = \begin{pmatrix} 1 & 0 & 0 \\ 2 & -1 & 0 \\ 1 & -1 & 0 \end{pmatrix} \quad \text{and} \quad \mathbf{U} = \begin{pmatrix} 1 & -2 & -3 \\ 0 & 1 & 1 \\ 0 & 0 & 1 \end{pmatrix}. \tag{6.466}$$

Then

$$\left\{ \mathbf{r}_1 = \begin{pmatrix} 1 \\ 2 \\ 1 \end{pmatrix}, \mathbf{r}_2 = \begin{pmatrix} 0 \\ -1 \\ -1 \end{pmatrix} \right\} \tag{6.467}$$

is a basis of $\mathcal{R}$, and

$$\left\{ \mathbf{c}_1 = \begin{pmatrix} 1 \\ 0 \\ 0 \end{pmatrix}, \mathbf{c}_2 = \begin{pmatrix} -2 \\ 1 \\ 0 \end{pmatrix} \right\} \tag{6.468}$$

is a basis of a complement $\mathcal{C}$ of the null space $\mathcal{N}$. The mapping M such that

$$\begin{pmatrix} 1 \\ 2 \\ 1 \end{pmatrix} \mapsto \begin{pmatrix} 1 \\ 0 \\ 0 \end{pmatrix}, \quad \begin{pmatrix} 0 \\ -1 \\ -1 \end{pmatrix} \mapsto \begin{pmatrix} -2 \\ 1 \\ 0 \end{pmatrix} \tag{6.469}$$

defines a vector isomorphism of $\mathcal{R}$ and $\mathcal{C}$ (why?).

6.7.3 RANK-NULLITY THEOREM (AGAIN)

If $\mathcal{V}_{(1)}$ is finite-dimensional, then the basic dimensionality relation for complementary subspaces, Eq. (5.255), takes the form

$$\dim[\mathcal{W}] + \text{nullity}[A] = \dim\big[\mathcal{V}_{(1)}\big] = n_1. \tag{6.470}$$

Because $\mathcal{W}$ is isomorphic to range[A],

$$\text{rank}[A] + \text{nullity}[A] = \dim[\text{domain}[A]] = \dim\big[\mathcal{V}_{(1)}\big]. \tag{6.471}$$

This is a second, more abstract proof of the rank-nullity theorem.

6.7.4 RIGHT INVERSES OF A LINEAR MAPPING

Another major consequence of the existence of an isomorphism between the range of a linear mapping A and any complement $\mathcal{W}$ of the null space of A is that, considered as a mapping from $\mathcal{W}$ to range[A], A is one-to-one, onto, and hence invertible. To show that $A : \mathcal{W} \to \text{range}[A]$ is one-to-one, let $A\bar{v} = A\bar{w}$, where both $\bar{v}$ and $\bar{w}$ belong to $\mathcal{W}$. Then the difference $\bar{v} - \bar{w}$ is equal to the zero vector $\bar{0}$ because $\bar{v} - \bar{w}$ belongs to both null[A] and $\mathcal{W}$. To show that A carries $\mathcal{W}$ *onto* range[A], let $\bar{x}'$ belong to range[A]. Then there exists a vector $\bar{x}$ in $\mathcal{V}_{(1)}$ such that $A\bar{x} = \bar{x}'$. By Eqs. (5.256) through (5.258), and because we have assumed that $\mathcal{W}$ is complementary to null[A],

$$\mathcal{V}_{(1)} = \mathcal{W} \oplus \text{null}[A], \tag{6.472}$$

there exist *unique* vectors $\bar{w} \in \mathcal{W}$ and $\bar{z} \in \text{null}[A]$ such that

$$\bar{x} = \bar{w} + \bar{z}. \tag{6.473}$$

Then

$$\bar{x}' = A\bar{w} + A\bar{z} = A\bar{w}, \tag{6.474}$$

showing that A is onto as well as one-to-one. Then, for every complement of the null space of a linear mapping A, there exists an inverse mapping $A^R : \text{range}[A] \to \mathcal{W}$ such that

$$\forall \bar{x}' \in \text{range}[A] : A^R \bar{x}' := \bar{w}. \tag{6.475}$$

Because Eqs. (6.474) and (6.475) imply that

$$\forall \overline{x}' \in \text{range}[A] : AA^R\overline{x}' = \overline{x}', \tag{6.476}$$

A^R is often called a **right inverse** of A. However, A^R does *not* satisfy Eq. (6.138) and is therefore not an inverse in the group-theoretical sense. In the context of ordinary and partial differential equations, the right inverse A^R is usually called the **Green operator**.

It follows from Eq. (6.476) that *if $\overline{x}' \in$ range[A]*, *then the linear equation $A\overline{x} = \overline{x}'$ has at least one solution*,

$$\overline{x} = A^R\overline{x}' = \overline{w} \in \mathcal{W}. \tag{6.477}$$

This solution is unique if and only if $\text{null}[A] = \mathcal{O}$. *If* $\text{null}[A] \neq \mathcal{O}$, *then* $A\overline{x} = \overline{x}'$ *has infinitely many solutions*

$$\overline{x} = \overline{w} + \overline{z}, \tag{6.478}$$

where $\overline{z} \in \text{null}[A]$.

6.7.5 EXAMPLES OF RIGHT INVERSES

Finite-Dimensional Vector Spaces

Let A be the matrix **B** defined in Eq. (6.54). The vector isomorphism M defined in Eq. (6.469) maps $\mathcal{R} = \text{range}[\mathbf{B}]$ to the subspace $\mathcal{C}$, a basis of which is given in Eq. (6.468). Because M maps a two-dimensional subspace of $\mathbb{R}^3$ onto another two-dimensional subspace, M cannot be extended uniquely to a mapping of $\mathbb{R}^3$ into $\mathbb{R}^3$.

We seek a 3×3 matrix $\mathbf{B}^R$ that solves the equations

$$\mathbf{B}^R\mathbf{r}_i = \mathbf{c}_i \tag{6.479}$$

for $i = 1, 2$, for which $\mathbf{c}_1$ and $\mathbf{c}_2$ are defined in Eq. (6.468). Because these equations do not define $\mathbf{B}^R$ uniquely, one must also define how $\mathbf{B}^R$ acts on a third vector that is linearly independent of $\mathbf{c}_1$ and $\mathbf{c}_2$. For this purpose one must choose a vector $\mathbf{k}$ that is linearly independent with respect to $\{\mathbf{r}_i\}$, where $\mathbf{r}_1$ and $\mathbf{r}_2$ are the basis of range[**B**] defined in Eq. (6.467). Let

$$\mathbf{B}^R\mathbf{k} = \mathbf{0} \quad \text{where} \quad \mathbf{k} = \begin{pmatrix} 1 \\ -1 \\ 1 \end{pmatrix}. \tag{6.480}$$

It is easy to verify that $\{\mathbf{r}_1, \mathbf{r}_2, \mathbf{k}\}$ is linearly independent. The equation to be solved therefore is

$$\mathbf{B}^R \begin{pmatrix} 1 & 0 & 1 \\ 2 & -1 & -1 \\ 1 & -1 & 1 \end{pmatrix} = \begin{pmatrix} 1 & -2 & 0 \\ 0 & 1 & 0 \\ 0 & 0 & 0 \end{pmatrix}. \tag{6.481}$$

Because the columns of the matrix that $\mathbf{B}^R$ multiplies are linearly independent, the matrix is nonsingular. Then

$$\mathbf{B}^R = \begin{pmatrix} 1 & -2 & 0 \\ 0 & 1 & 0 \\ 0 & 0 & 0 \end{pmatrix} \begin{pmatrix} 1 & 0 & 1 \\ 2 & -1 & -1 \\ 1 & -1 & 1 \end{pmatrix}^{-1}$$

$$= \frac{1}{3} \begin{pmatrix} -4 & 1 & 5 \\ 3 & 0 & -3 \\ 0 & 0 & 0 \end{pmatrix}.$$

(6.482)

Note that $\mathbf{B}^R$ is a singular matrix because it maps $\mathbf{k}$ to $\mathbf{0}$.

The general solution of the equation

$$\mathbf{B} \begin{pmatrix} x \\ y \\ z \end{pmatrix} = a\mathbf{r}_1 + b\mathbf{r}_2 = \begin{pmatrix} a \\ 2a - b \\ a - b \end{pmatrix}$$

(6.483)

therefore is

$$\begin{pmatrix} x \\ y \\ z \end{pmatrix} = \mathbf{B}^R(a\mathbf{r}_1 + b\mathbf{r}_2) + c\mathbf{n}$$

$$= a\mathbf{c}_1 + b\mathbf{c}_2 + c\mathbf{n}$$

(6.484)

$$= \begin{pmatrix} a - 3c \\ 2a - b + c \\ a - b + c \end{pmatrix}$$

in which

$$\mathbf{n} = \begin{pmatrix} -3 \\ 1 \\ 1 \end{pmatrix}$$

(6.485)

spans null[$\mathbf{B}$] and c is an arbitrary constant. Note that the right-hand side of Eq. (6.483) cannot be an arbitrary vector in $\mathbb{R}^3$. A right inverse of a linear mapping A can be applied only to a vector that belongs to the range of A.

The singular-value decomposition, which we shall introduce in Chapter 9, provides a numerically stable method for computing the right inverse.

Solutions of Differential Equations

The results in Eqs. (6.475) through (6.478) contain an important part of the theoretical foundation of physics and engineering. If A is a differential operator, one says that the general solution of the equation $A\bar{x} = \bar{x}'$ is the sum of a **particular solution** $\bar{w}$ and a solution $\bar{z}$ of the **homogeneous equation** $A\bar{z} = \bar{0}$. (Our proof is valid only if null[A] is finite-dimensional, but this case is general enough to cover many applications. Also, we have not proved that a particular solution exists if A is a differential operator.)

For example, let A be

$$A = \frac{d^2}{dx^2} \tag{6.486}$$

on the interval $(0, 1)$. According to Eq. (6.215), the null space of A is the span of the linearly independent functions 1 and x. Let $\mathcal{W}$ be any complement of null[A] in $\mathcal{V}$, that is,

$$\mathcal{V} = \text{null}[A] \oplus \mathcal{W}. \tag{6.487}$$

Then the formal solution of the differential equation

$$\frac{d^2 y}{dx^2}(x) = f(x) \tag{6.488}$$

in which $f \in \text{range}[A]$ is

$$y = A^R f + z \tag{6.489}$$

in which A^R, the right inverse of A that is determined by $\mathcal{W}$, maps range[A] onto $\mathcal{W}$, and $z \in \text{null}[A]$ is a solution of the homogeneous equation

$$Az = 0. \tag{6.490}$$

The only "details" that remain to be worked out are how to define $\mathcal{W}$ and how to construct the Green operator A^R.

6.7.6 EXERCISE FOR SECTION 6.7

6.7.1 Find a right inverse of each of the following matrices:

(a)

$$\begin{pmatrix} 1 & 2 & 1 \\ 2 & 3 & 3 \\ -1 & -1 & -2 \end{pmatrix} \tag{6.491}$$

(b)

$$\begin{pmatrix} 2 & 3 & 3 \\ 1 & 1 & 2 \\ 0 & 1 & -1 \end{pmatrix} \tag{6.492}$$

6.8 BIBLIOGRAPHY

Golub, Gene H., and Charles F. Van Loan. *Matrix Computations*. 2nd ed. Baltimore: Johns Hopkins University Press, 1989.

CHAPTER 7

LINEAR FUNCTIONALS

7.1 MOTIVATION FOR STUDYING FUNCTIONALS

A linear functional carries vectors (which may be column vectors, row vectors, or functions) onto scalars. For example, consider an electric dipole with a moment μ such that the magnitude of μ is fixed, but the direction may change. The energy of such a dipole in an electric field $\mathbf{E}$ is

$$W = -\mathbf{E} \cdot \mu. \tag{7.1}$$

One can look upon the electric field $\mathbf{E}$ as a linear mapping that maps the dipole moment μ to the scalar energy, W.

In physics and engineering one sometimes thinks of a functional as a "function on functions" that maps a function onto a single real or complex number. A familiar example of a scalar-valued linear "function of a function" is the **Riemann integral**,

$$I[f] = \int_a^b f(x)\,dx := \lim_{n \to \infty} \sum_{i=1}^n f(\xi_i)(x_i - x_{i-1}) \tag{7.2}$$

in which $x_0 := a < \xi_1 < x_1 < \cdots < \xi_n < x_n := b$. (A function f is called **Riemann-integrable** if and only if the limit in Eq. [7.2] is independent of how the x_i and ξ_i are chosen, as long as $\sup_{i \in (1:n)} |x_i - x_{i-1}| \to 0$ as $n \to \infty$.) Numerical approximations to the Riemann integral such as the trapezoid rule Eq. (6.237) are among the more important linear functionals encountered in practical applications.

The Dirac **delta functional** δ_ξ has the property that

$$\delta_\xi[f] := f(\xi). \tag{7.3}$$

In physics and engineering the delta functional is usually expressed in terms of a (purely formal) Riemann integral,

$$\delta_\xi[f] = \int_{-\infty}^{\infty} \delta(x - \xi)\, f(x)\,dx = f(\xi), \tag{7.4}$$

in which $\delta(\cdot)$ is called the Dirac delta "function." One can see at once that no Riemann-integrable function $\delta(\cdot)$ can have property Eq. (7.4); therefore the integration in Eq. (7.4) cannot be taken at face value. For example, if one chooses f such that $f(x) = 1$ for

$x \in [\xi - a, \xi + a]$ and $f(x) = 0$ for $x \notin [\xi - a, \xi + a]$, then

$$\int_{\xi-a}^{\xi+a} \delta(x - \xi)\, dx = 1 \tag{7.5}$$

for any $a > 0$, which implies that $\delta(\cdot)$ has unit area and that $\delta(\cdot)$ vanishes except at $x = \xi$. But Eq. (7.2) implies that the area under a function that vanishes everywhere except at a single point and is defined there (i.e., has a finite value at that point) is zero.

The operation of evaluating a function f at a point ξ in its domain, as Eq. (7.3) instructs one to do, is perfectly legitimate as long as one does not insist that $f(\xi)$ must be equal to the convolution of f with a universal Riemann-integrable function δ_ξ. One can see by inspection that the numerical integration rules defined in Section 3.4, and the functionals I and δ_ξ, all have the properties of additivity and homogeneity – that is, *linearity*. The area of mathematics called *functional analysis* was developed in order to study linear functionals such as I and δ_ξ, which are important in many applications. Functional analysis is an important tool for analyzing numerical methods as well as for a fundamental understanding of the function spaces used in mathematical physics.

7.2 DUAL SPACES

7.2.1 DEFINITIONS

A **linear functional** (or **linear form** or **1-form**) on a vector space $\mathcal{V}$ over a number field $\mathbb{F} = \mathbb{R}$ or $\mathbb{C}$ is a linear mapping $\underline{\phi}$ from $\mathcal{V}$ into $\mathbb{F}$. For example, the dot product with a fixed vector $\mathbf{a} \neq \mathbf{0}$ is a linear functional on $\mathbb{R}^2$ (see Fig. 7.1):

$$\forall \mathbf{r} \in \mathbb{R}^2 : \underline{\phi}_{\mathbf{a}}[\mathbf{r}] := \mathbf{a} \cdot \mathbf{r}. \tag{7.6}$$

More generally, a linear functional $\underline{\phi}$ on the vector space of n-dimensional column vectors $\mathbb{F}^n$ is a linear mapping from $\mathbb{F}^n$ to $\mathbb{F}$. Therefore the matrix that realizes $\underline{\phi}$ has dimensions

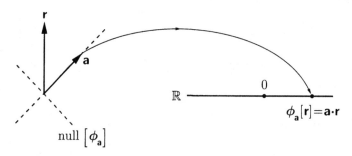

FIGURE 7.1

Illustration of the dot-product functional defined in Eq. (7.6).

$1 \times n$:

$$\underline{\phi} = (\phi_1, \ldots, \phi_n). \tag{7.7}$$

By the definition of matrix multiplication, $\underline{\phi}$ acts on $\mathbf{x} \in \mathbb{F}^n$ through the row-column product

$$\underline{\phi}[\mathbf{x}] = (\phi_1, \ldots, \phi_n) \begin{pmatrix} x^1 \\ \vdots \\ x^n \end{pmatrix} = \phi_1 x^1 + \cdots + \phi_n x^n, \tag{7.8}$$

which is a kind of scalar or dot product. We see in Section 8.1 that in the finite-dimensional inner-product spaces of interest in physics and engineering, every linear functional can be expressed as the inner product with a fixed vector.

Because a linear functional $\underline{\phi}$ is a linear mapping, $\underline{\phi}$ is additive,

$$\forall \overline{x}, \overline{x}' \in \mathcal{V} : \underline{\phi}[\overline{x} + \overline{x}'] = \underline{\phi}[\overline{x}] + \underline{\phi}[\overline{x}'], \tag{7.9}$$

and homogeneous,

$$\forall \overline{x} \in \mathcal{V} : \forall \alpha \in \mathbb{F} : \underline{\phi}[\alpha \overline{x}] = \alpha \underline{\phi}[\overline{x}]. \tag{7.10}$$

An additive functional $\underline{\phi} : \mathcal{V} \to \mathbb{C}$ that has the property that

$$\forall \overline{x} \in \mathcal{V} : \forall \alpha \in \mathbb{C} : \underline{\phi}[\alpha \overline{x}] = \alpha^* \underline{\phi}[\overline{x}] \tag{7.11}$$

is called **antilinear**. For example, if $\mathcal{V} = \mathbb{C}^n$, then the functional $\underline{\alpha}^i$ such that $\underline{\alpha}^i[\mathbf{x}] := x^{i*}$ is antilinear.

Let $\mathcal{V}^*$ denote the set of all linear functionals from $\mathcal{V}$ into $\mathbb{F}$. (Note that, if applied to a vector space, the asterisk means the dual, not complex conjugation!) The **null functional** $\underline{0} \in \mathcal{V}^*$ carries every vector onto the zero scalar:

$$\forall \overline{x} \in \mathcal{V} : \underline{0}[\overline{x}] := 0, \tag{7.12}$$

$\mathcal{V}^*$ is closed under addition,

$$\forall \underline{\phi}, \underline{\phi}' \in \mathcal{V}^* : (\underline{\phi} + \underline{\phi}')[\overline{x}] := \underline{\phi}[\overline{x}] + \underline{\phi}'[\overline{x}], \tag{7.13}$$

and under scalar multiplication,

$$\forall \underline{\phi} \in \mathcal{V}^* : \forall \alpha \in \mathbb{F} : (\alpha \underline{\phi})[\overline{x}] := \alpha \underline{\phi}[\overline{x}] \tag{7.14}$$

(see Exercise 7.2.2). It is easy to verify that $\mathcal{V}^*$ is a vector space over $\mathbb{F}$ with the above definitions (see Exercise 7.2.1). $\mathcal{V}^*$ is called the **algebraic dual** (or, if $\mathcal{V}$ is finite-dimensional, the **dual space**) of $\mathcal{V}$. In the older literature $\mathcal{V}^*$ is sometimes called the **conjugate** of $\mathcal{V}$.

If $\mathcal{V}$ is infinite-dimensional, as is often the case if $\mathcal{V}$ is a function space, then the set of *all* linear functionals on $\mathcal{V}$ turns out not to be very useful, because one almost always needs an axiom of *continuity* ($\overline{x}_n \to \overline{x} \Rightarrow \underline{\phi}[\overline{x}_n] \to \underline{\phi}[\overline{x}]$) in addition to the axioms of additivity and homogeneity. The space of all *continuous* linear functionals on $\mathcal{V}$ is usually written $\mathcal{V}'$ and is called the **topological dual** of $\mathcal{V}$. For finite-dimensional spaces, it turns out that $\mathcal{V}' = \mathcal{V}^*$.

Because $\mathcal{V}^*$ is a vector space, it has an algebraic dual $\mathcal{V}^{**}$, which is called the **second (algebraic) dual** of $\mathcal{V}$. We show in a subsequent section that if $\mathcal{V}$ is finite-dimensional, then the dual and second dual of $\mathcal{V}$ are isomorphic to $\mathcal{V}$.

7.2.2 RANGE AND NULL SPACE OF A LINEAR FUNCTIONAL

The range of a nonnull linear functional $\underline{\phi}$ is always $\mathbb{F}$. Because $\underline{\phi} \neq \underline{0}$, there exist a scalar $\alpha \in \mathbb{F}$ and a vector $\overline{x} \in \mathcal{V}$ such that

$$\underline{\phi}[\overline{x}] = \alpha \neq 0. \tag{7.15}$$

Let β be any element of $\mathbb{F}$ and let $\gamma = \beta/\alpha$. By the axiom of homogeneity,

$$\underline{\phi}[\gamma\overline{x}] = \gamma\underline{\phi}[\overline{x}] = \gamma\alpha = \beta. \tag{7.16}$$

Then $\beta \in \text{range}[\underline{\phi}]$. By varying γ, one gets all of $\mathbb{F}$. Obviously the range of the null functional $\underline{0}$ is just the set that consists of the zero scalar.

For example, every real number lies in the range of the linear functional $\underline{\phi}_a : \mathbb{R}^2 \to \mathbb{R}$ defined in Eq. (7.6), provided that $\mathbf{a} \neq \mathbf{0}$. To see this, let

$$\mathbf{v} := (\mathbf{a} \cdot \mathbf{a})^{-1}\mathbf{a}. \tag{7.17}$$

Then

$$\forall \alpha \in \mathbb{R} : \underline{\phi}_a[\alpha\mathbf{v}] = \alpha. \tag{7.18}$$

Therefore the image of $\mathbb{R}^2$ under $\underline{\phi}_a$ is the real line, $\mathbb{R}$ (see Fig. 7.1).

The null space, null$[\underline{\phi}]$, of a linear functional $\underline{\phi}$ on a vector space $\mathcal{V}$ is, of course, the set of vectors that $\underline{\phi}$ maps onto the zero scalar $0 \in \mathbb{F}$:

$$\text{null}[\underline{\phi}] := \{\overline{x} \in \mathcal{V} \mid \underline{\phi}[\overline{x}] = 0\}. \tag{7.19}$$

One knows already from Exercise 6.3.6 that null$[\underline{\phi}]$ is a vector subspace of $\mathcal{V}$.

For example, the null space of the functional

$$\underline{\phi} = (\phi_1, 0, 0) \tag{7.20}$$

is the set of vectors $\mathbf{x}$ in which the first component vanishes,

$$\mathbf{x} = \begin{pmatrix} 0 \\ x^2 \\ x^3 \end{pmatrix}, \tag{7.21}$$

because $\phi_i x^i = \phi_1 \cdot 0 + 0 \cdot x^2 + 0 \cdot x^3 = 0$. In other words, null$[\underline{\phi}]$ is the x^2-x^3 plane. In general, the equation of a plane in $\mathbb{R}^3$ that passes through the point $\mathbf{a}$ and is perpendicular to the unit vector $\mathbf{n}$ is

$$\mathbf{n} \cdot (\mathbf{x} - \mathbf{a}) = 0, \tag{7.22}$$

which is equivalent to requiring that $\mathbf{x} - \mathbf{a}$ must lie in the null space of the linear functional $\underline{\phi}_{\mathbf{n}}$.

If the dimension of a vector space $\mathcal{V}$ is finite and greater than 1, then every nonnull linear functional $\underline{\phi}$ on $\mathcal{V}$ has a nontrivial null space, that is, $\text{null}[\underline{\phi}] \neq \mathcal{O}$. For, if $\underline{\phi} \neq \underline{0}$, then there exists at least one nonzero component $\phi_j \neq 0$. Then $\underline{\phi}$ maps onto zero the infinitely many vectors $\overline{x}$ such that $x_j = (\phi_j)^{-1} \sum_{k \neq j} x_k \phi^k$. If $\underline{\phi} = \underline{0}$, then the null space of $\underline{\phi}$ is all of $\mathcal{V}$, which is certainly nontrivial.

The rank of a nonnull linear functional $\underline{\phi} : \mathcal{V} \to \mathbb{F}$ is 1, because $\mathbb{F}$ is a vector space of dimension 1 over $\mathbb{F}$. If $\mathcal{V}$ is of finite dimension n, then the rank-nullity theorem, Eq. (6.471), implies that

$$\forall \underline{\phi} \in \mathcal{V}^* :\ni: \underline{\phi} \neq 0 : \text{nullity}[\underline{\phi}] = n - 1. \tag{7.23}$$

For example, $\underline{\phi}_{\mathbf{a}}$ in Eq. (7.6) has nullity $2 - 1 = 1$, and $\underline{\phi}$ in Eq. (7.20) has nullity $3 - 1 = 2$. Unfortunately Eq. (7.23) is not useful if $\mathcal{V}$ is infinite-dimensional.

An equivalent approach that can be generalized to cases in which $\mathcal{V}$ is infinite-dimensional is to find the codimension of

$$\mathcal{N} := \text{null}[\underline{\phi}] \tag{7.24}$$

in $\mathcal{V}$. (Recall that the codimension of a vector subspace is the dimension of any one of its complementary subspaces; see Section 5.5.) Eq. (7.23) is equivalent to the statement that

$$\text{codim}_{\mathcal{V}}[\mathcal{N}] = 1, \tag{7.25}$$

which holds for both finite-dimensional and infinite-dimensional vector spaces.

7.2.3 EXERCISES FOR SECTION 7.2

7.2.1 Prove that Eqs. (7.13) and (7.14) imply that $\mathcal{V}^*$ satisfies the axioms for a vector space.

7.2.2 Prove that the functional $\underline{\phi}_{\mathbf{a}}$ on $\mathbb{R}^2$ defined in Eq. (7.6) has the following properties:

(a) $\underline{\phi}_{\mathbf{a}}$ satisfies Eq. (7.9) and (7.10).

(b) $\underline{\phi}_{\mathbf{a}}$ satisfies Eq. (7.14).

(c) $\underline{\phi}_{\mathbf{a}}$ and $\underline{\phi}_{\mathbf{b}}$ satisfy Eq. (7.13).

7.2.3 Prove that the null space, $\text{null}[\underline{\phi}]$, of a linear functional $\underline{\phi}$ on a vector space $\mathcal{V}$ satisfies the vector-space axioms in Eqs. (5.12) through (5.21) and is therefore a vector subspace of $\mathcal{V}$.

7.2.4 Let $\underline{\phi}$ be a linear functional on a vector space $\mathcal{V}$. Prove (by contradiction) that the set of vectors on which $\underline{\phi}$ is nonzero is not a vector subspace of $\mathcal{V}$.

7.3 COORDINATE FUNCTIONALS

7.3.1 DEFINITIONS

Assume for the moment that V is finite-dimensional and choose a basis $\{\bar{e}_i\}$. For any $\bar{x} \in V$, make the coordinate expansion $\bar{x} = x^i \bar{e}_i$. By the properties of additivity and homogeneity, Eqs. (7.9) and (7.10),

$$\underline{\phi}[\bar{x}] = \underline{\phi}[x^i \bar{e}_i] = x^i \underline{\phi}[\bar{e}_i] = x^i \phi_i \tag{7.26}$$

in which

$$\phi_i := \underline{\phi}[\bar{e}_i]. \tag{7.27}$$

In other words, $\underline{\phi}[\bar{x}]$ is a homogeneous linear expression in the contravariant components of $\bar{x}$. Conversely, ϕ_i can be defined as the coefficient of x^i in a homogeneous linear expression. Therefore the values ϕ_i of a linear functional on a basis $\{\bar{e}_i\}$ of V completely determine the functional.

In view of Eq. (7.27) it is natural to introduce the **coordinate functionals** $\underline{\epsilon}^i$ such that

$$\boxed{\underline{\epsilon}^i[\bar{e}_j] := \delta^i_j.} \tag{7.28}$$

The coordinate functionals deserve their name because the value of $\underline{\epsilon}^i$ on a vector $\bar{x}$ is equal to the i-th coordinate of $\bar{x}$ (rel$\{\bar{e}_j\}$):

$$x^i = x^j \delta^i_j = \underline{\epsilon}^i[x^j \bar{e}_j] = \underline{\epsilon}^i[\bar{x}]. \tag{7.29}$$

From Eqs. (7.26), (7.28), and (7.29) it follows that the coordinate functionals $\underline{\epsilon}^i$ span the dual space V^*:

$$\underline{\phi} = \underline{\phi}[\bar{e}_i]\underline{\epsilon}^i = \phi_i \underline{\epsilon}^i. \tag{7.30}$$

Because $\{\underline{\epsilon}^i\}$ is linearly independent (see Exercise 7.3.2), it is a basis of V^*. Therefore, for a finite-dimensional vector space V,

$$\dim[V^*] = n = \dim[V]. \tag{7.31}$$

In physics $\{\underline{\epsilon}^i\}$, which is often called the **dual basis**, is especially important in relativity theory and solid-state physics. Equations (7.27) and (7.30) give the coordinate expansion of an arbitrary linear functional $\underline{\phi} \in V^*$ relative to the dual basis.

From Eq. (7.26) one sees at once that the realizations of the coordinate functionals for the canonical basis in $\mathbb{F}^n$ (Eq. [5.122]) are the row vectors

$$\underline{\epsilon}^i = \mathbf{e}_i^T = (0, \quad \ldots, \quad \overset{1}{0}, \quad \ldots, \quad \overset{i-1}{0}, \quad \overset{i}{1}, \quad \overset{i+1}{0}, \quad \ldots, \quad \overset{n}{0}) \tag{7.32}$$

in which the exponent T indicates the transpose. For this reason $\{\underline{\epsilon}^i\}$ is called the **canonical basis of $\mathbb{F}^{n*}$**. It may be helpful to remember that the canonical basis vectors of $\mathbb{F}^n$ are the columns of the identity matrix $\mathbf{1}$, and the canonical basis functionals of $\mathbb{F}^{n*}$ are the rows of the identity matrix.

Let $\mathcal{V}$ be a finite-dimensional vector space of dimension n over $\mathbb{F}$, and let $\mathsf{N} : \mathcal{V}^* \to \mathbb{F}^{n*}$ be the mapping such that

$$\mathsf{N}\underline{\epsilon}^i := \underline{\epsilon}^i \tag{7.33}$$

It follows from the coordinate expansion Eq. (7.30) that N is homogeneous and additive and therefore is a linear mapping:

$$\mathsf{N}(\alpha\underline{\phi}) := \alpha\mathsf{N}\underline{\phi}, \quad \mathsf{N}(\underline{\phi} + \underline{\psi}) := \mathsf{N}\underline{\phi} + \mathsf{N}\underline{\psi} \tag{7.34}$$

Then, because $\dim[\mathcal{V}^*] = \dim[\mathbb{F}^{n*}]$, N is bijective and therefore is a vector isomorphism. Note that N is defined to map an arbitrary basis of coordinate functionals to the canonical basis of $\mathbb{F}^{n*}$, just as M is defined to map an arbitrary basis of $\mathcal{V}$ to the canonical basis of $\mathbb{F}^n$ in Eq. (5.202).

7.3.2 COORDINATE FUNCTIONALS ON $\mathbb{F}^n$

If $\{\mathbf{f}_i\}$ is a general basis of $\mathbb{F}^n$ (where $\mathbb{F} = \mathbb{R}$ or $\mathbb{C}$), then the coordinate functionals $\{\underline{\phi}^i\}$ that belong to (or, as one sometimes says, are induced by) the basis $\{\mathbf{f}_i\}$ are row vectors of the form seen in Eq. (7.7) such that

$$\underline{\phi}^i[\mathbf{f}_j] = (\phi_1^i, \ldots, \phi_n^i) \begin{pmatrix} f_j^1 \\ \vdots \\ f_j^n \end{pmatrix} = \delta_j^i. \tag{7.35}$$

Let $\mathbf{F}$ be the matrix whose j-th column is the j-th basis vector $\mathbf{f}_j$,

$$\mathbf{F} = [\mathbf{f}_1, \ldots, \mathbf{f}_n], \tag{7.36}$$

and let $\boldsymbol{\Phi}$ be the matrix whose i-th row is $\underline{\phi}^i$,

$$\boldsymbol{\Phi} = \begin{pmatrix} \underline{\phi}^1 \\ \vdots \\ \underline{\phi}^n \end{pmatrix} = \begin{pmatrix} \phi_1^1 & \cdots & \phi_n^1 \\ \vdots & \ddots & \vdots \\ \phi_1^n & \cdots & \phi_n^n \end{pmatrix}. \tag{7.37}$$

(The column matrix in this equation is a row partition of the square matrix.) Then Eq. (7.35) implies that

$$\boldsymbol{\Phi}\mathbf{F} = \mathbf{1}. \tag{7.38}$$

For example, let

$$\mathbf{f}_1 = \begin{pmatrix} 1 \\ 1 \\ 0 \end{pmatrix}, \quad \mathbf{f}_2 = \begin{pmatrix} 0 \\ 1 \\ 1 \end{pmatrix}, \quad \mathbf{f}_3 = \begin{pmatrix} 1 \\ 1 \\ 1 \end{pmatrix}; \tag{7.39}$$

then

$$\mathbf{F} = \begin{pmatrix} 1 & 0 & 1 \\ 1 & 1 & 1 \\ 0 & 1 & 1 \end{pmatrix}. \tag{7.40}$$

The inverse of $\mathbf{F}$ is

$$\mathbf{\Phi} = \begin{pmatrix} 0 & 1 & -1 \\ -1 & 1 & 0 \\ 1 & -1 & 1 \end{pmatrix}. \tag{7.41}$$

(One can verify Eq. [7.41] using the determinantal formula for the inverse, Eq. [6.349].) It follows from Eqs. (7.37), (7.38), and (7.41) that the coordinate functionals that belong to the basis Eq. (7.39) are

$$\underline{\phi}^1 = (0, 1, -1), \quad \underline{\phi}^2 = (-1, 1, 0), \quad \underline{\phi}^3 = (1, -1, 1). \tag{7.42}$$

Note that $\underline{\phi}^j[\mathbf{f}_i] = \delta_i^j$. We have established the following result: *The coordinate functionals (the elements of the dual basis) for an arbitrary basis in $\mathbb{F}^n$ are the rows of the matrix $\mathbf{\Phi}$ that is the inverse of the matrix $\mathbf{F}$ whose columns are the basis vectors.*

To compute $\mathbf{\Phi}$ one usually does not invert $\mathbf{F}$ directly except in the simplest cases. Instead, one solves the set of n^2 linear equations

$$f_j^i \phi_k^j = \delta_k^i \tag{7.43}$$

in the n^2 unknowns ϕ_k^j by a standard numerical method such as Gaussian elimination, which we discuss in Sections 5.4.4, 6.1.2, and 6.6.2.

7.3.3   ISOMORPHISM OF $\mathcal{V}^*$ TO $\mathcal{V}$

Linear functionals on $\mathbb{F}^n$ are row vectors, which one can map to column vectors by taking the transpose,

$$\mathsf{T}\underline{\phi} := \underline{\phi}^T \tag{7.44}$$

in which

$$\underline{\phi}^T = (\phi_1, \ldots, \phi_n)^T = \begin{pmatrix} \phi_1 \\ \vdots \\ \phi_n \end{pmatrix}. \tag{7.45}$$

The mapping T is additive,

$$\mathsf{T}(\underline{\phi} + \underline{\psi}) = (\underline{\phi} + \underline{\psi})^T = \underline{\phi}^T + \underline{\psi}^T = \mathsf{T}\underline{\phi} + \mathsf{T}\underline{\psi}, \tag{7.46}$$

and homogeneous,

$$T(\alpha\underline{\phi}) = (\alpha\underline{\phi})^T = \alpha\underline{\phi}^T = \alpha T\underline{\phi}. \tag{7.47}$$

Obviously T is bijective. It follows that T is a vector isomorphism and that

$$\mathbb{F}^{n*} \cong \mathbb{F}^n. \tag{7.48}$$

The diagram

$$
\begin{array}{ccc}
\mathcal{V} & \xleftarrow{\ M^{-1}TN\ } & \mathcal{V}^* \\[2pt]
\Big\downarrow M & & \Big\downarrow N \\[2pt]
\mathbb{F}^n & \xleftarrow{\ \ T\ \ } & \mathbb{F}^{n*}
\end{array}
\tag{7.49}
$$

shows that the mapping

$$M^{-1}TN : \underline{\epsilon}^i \mapsto \bar{e}_i \tag{7.50}$$

is a vector isomorphism, because it is the product of three vector isomorphisms. Therefore

$$\mathcal{V}^* \cong \mathcal{V} \tag{7.51}$$

for every finite-dimensional vector space $\mathcal{V}$.

T is one of many possible isomorphisms between $\mathcal{V}^*$ and $\mathcal{V}$. Unfortunately the identi-fication between $\mathcal{V}^*$ and $\mathcal{V}$ defined by Eq. (7.50) is basis-dependent. In Section 8.8.2 we encounter another isomorphism $G : \mathcal{V} \to \mathcal{V}^*$ that is more useful than Eq. (7.50) because G is defined in a basis-independent way by an inner product on $\mathcal{V}$.

If $\mathcal{V}$ is infinite-dimensional, it is *not* generally true that $\mathcal{V}^* \cong \mathcal{V}$ or that $\mathcal{V}' \cong \mathcal{V}$. For example, we show in Section 7.5 that the algebraic dual $(\mathbb{F}^{\mathbb{Z}^+})^*$ of the space of all sequences is isomorphic to $\phi(\mathbb{F})$, the space of all finite sequences. But $\phi(\mathbb{F})$ is not isomorphic to $\mathbb{F}^{\mathbb{Z}^+}$ (why?). Therefore, for the rest of this section and for most of the rest of this chapter we assume that $\mathcal{V}$ is finite-dimensional.

7.3.4 COORDINATE FUNCTIONALS ON TWO-DIMENSIONAL EUCLIDEAN SPACE

Additional examples of coordinate functionals on two- and three-dimensional space help to develop one's intuitive understanding of dual spaces. Let $\underline{\phi}^1$, $\underline{\phi}^2$ be the coordinate func-tionals induced by the basis $\{\mathbf{f}_1, \mathbf{f}_2\}$ of $\mathbb{R}^2$ depicted in Fig. 7.2. One can take advantage of the isomorphism $\underline{\phi} \mapsto \underline{\phi}^T$ defined in Eq. (7.44) to construct the coordinate functionals by first constructing the vectors in $\mathcal{V}$ that are the images of the coordinate functionals $\underline{\phi}^1$, $\underline{\phi}^2$ under T, as follows:

$$T\underline{\phi}^1 := \frac{\mathbf{f}_2 \times \mathbf{k}}{\|\mathbf{f}_1 \times \mathbf{f}_2\|}, \quad T\underline{\phi}^2 := \frac{\mathbf{k} \times \mathbf{f}_1}{\|\mathbf{f}_1 \times \mathbf{f}_2\|}, \tag{7.52}$$

in which

$$\|\mathbf{f}_1 \times \mathbf{f}_2\| := [(\mathbf{f}_1 \times \mathbf{f}_2) \cdot (\mathbf{f}_1 \times \mathbf{f}_2)]^{1/2} \tag{7.53}$$

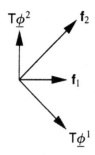

FIGURE 7.2

A basis $\{\mathbf{f}_1, \mathbf{f}_2\}$ of $\mathbb{R}^2$ and its reciprocal basis.

and $\mathbf{k}$ is the unit vector

$$\mathbf{k} := \frac{\mathbf{f}_1 \times \mathbf{f}_2}{\|\mathbf{f}_1 \times \mathbf{f}_2\|}, \tag{7.54}$$

computed in $\mathbb{R}^3$. Figure 7.2 shows $\mathsf{T}\underline{\phi}^1$ and $\mathsf{T}\underline{\phi}^2$ for a particular basis $\{\mathbf{f}_1, \mathbf{f}_2\}$.

Note that changing $\mathbf{f}_1$ or $\mathbf{f}_2$ (or both) changes both $\mathsf{T}\underline{\phi}^1$ and $\mathsf{T}\underline{\phi}^2$. However, if one holds the direction of $\mathbf{f}_1$ fixed then $\mathsf{T}\underline{\phi}^2$ always lies in a direction perpendicular to $\mathbf{f}_1$, even though the magnitude of $\mathsf{T}\underline{\phi}^2$ depends on $\mathbf{f}_2$ in general.

For example, let

$$\mathbf{f}_1 = \begin{pmatrix} 1 \\ 0 \end{pmatrix}, \quad \mathbf{f}_2 = \begin{pmatrix} \alpha \\ \beta \end{pmatrix} \tag{7.55}$$

in which we assume that $\beta > 0$ for the sake of convenience. Then one finds from Eq. (7.52) that

$$\mathsf{T}\underline{\phi}^1 = \frac{1}{\beta} \begin{pmatrix} \beta \\ -\alpha \end{pmatrix}, \quad \mathsf{T}\underline{\phi}^2 = \frac{1}{\beta} \begin{pmatrix} 0 \\ 1 \end{pmatrix}. \tag{7.56}$$

Clearly both coordinate functionals depend on both basis vectors. If $\mathbf{f}_1$ is known but $\mathbf{f}_2$ is unknown, then all that one can say with certainty is that the first component of $\mathsf{T}\underline{\phi}^1$ must be equal to 1 and that $\mathsf{T}\underline{\phi}^2$ must be equal to a scalar multiple of $\mathbf{e}_2$.

Clearly $\mathsf{T}\underline{\phi}^1$ and $\overline{\mathsf{T}}\underline{\phi}^2$ cannot be parallel; therefore they are linearly independent and are a basis of $\mathbb{R}^2$. By inspection, the construction of $\mathsf{T}\underline{\phi}^1$ and $\mathsf{T}\underline{\phi}^2$ in Eq. (7.52) satisfies the equation

$$(\mathsf{T}\underline{\phi}^j) \cdot \mathbf{f}_i = \delta_i^j. \tag{7.57}$$

Two bases $\{\mathbf{f}_i\}$, $\{\mathbf{g}^j\}$ of $\mathbb{R}^2$ or $\mathbb{R}^3$ such that

$$\mathbf{f}_i \cdot \mathbf{g}^j = \delta_i^j \tag{7.58}$$

are called **biorthogonal**. One says that $\{\mathbf{g}^j\}$ is the basis that is **reciprocal** to $\{\mathbf{f}_i\}$. (Note that the vectors $\mathbf{f}_i$ are not required to be mutually orthogonal.) Thus $\{\mathbf{f}_i\}$ and $\{\mathsf{T}\phi^j\}$ are biorthogonal bases of $\mathbb{R}^2$, and $\{\mathsf{T}\phi^j\}$ is the reciprocal basis (to $\{\mathbf{f}_i\}$). We generalize the concept of biorthogonality to a much wider class of vector spaces after we have defined inner products in Section 8.1 and the inner-product mapping in Section 8.8.

The definition of $\mathsf{T}\underline{\phi}^1$ and $\mathsf{T}\underline{\phi}^2$ and the properties of the cross product of vectors in $\mathbb{R}^3$ guarantee that $\underline{\phi}^1$ and $\underline{\phi}^2$ are the rows of the matrix that is the inverse of the matrix whose columns are $\mathbf{f}_1$ and $\mathbf{f}_2$.

7.3.5 COORDINATE FUNCTIONALS AND THE RECIPROCAL LATTICE

An understanding of coordinate functionals induced by a basis whose vectors are not necessarily mutually perpendicular in three-dimensional Euclidean space is important in the physics of periodic systems such as idealized crystalline solids. Let $\mathbf{f}_1, \mathbf{f}_2, \mathbf{f}_3$ be a basis of $\mathbb{R}^3$ (Fig. 7.3). Therefore the $\{\mathbf{f}_i\}$ are not coplanar and the scalar triple product $\mathbf{f}_1 \cdot (\mathbf{f}_2 \times \mathbf{f}_3)$ is nonzero (see Exercise 5.3.4). Let

$$\mathsf{T}\underline{\phi}^1 := \frac{\mathbf{f}_2 \times \mathbf{f}_3}{\mathbf{f}_1 \cdot (\mathbf{f}_2 \times \mathbf{f}_3)}, \quad \mathsf{T}\underline{\phi}^2 := \frac{\mathbf{f}_3 \times \mathbf{f}_1}{\mathbf{f}_1 \cdot (\mathbf{f}_2 \times \mathbf{f}_3)}, \quad \mathsf{T}\underline{\phi}^3 := \frac{\mathbf{f}_1 \times \mathbf{f}_2}{\mathbf{f}_1 \cdot (\mathbf{f}_2 \times \mathbf{f}_3)}. \tag{7.59}$$

By construction,

$$\forall i, j \in (1:3): (\mathsf{T}\underline{\phi}^j) \cdot \mathbf{f}_i = \delta_i^j. \tag{7.60}$$

Then the $\{\phi^j\}$ are the coordinate functionals generated by the basis $\{\mathbf{f}_i\}$ of $\mathbb{R}^3$. It follows that $\{\mathbf{f}_i\}$ and $\{\mathsf{T}\underline{\phi}^j\}$ are biorthogonal bases of $\mathbb{R}^3$, and that $\{\mathsf{T}\underline{\phi}^j\}$ is the basis that is reciprocal to the basis $\{\mathbf{f}_i\}$. Again, note that Eq. (7.60) implies that $\{\phi^j\}$ are the rows of the matrix that is the inverse of the matrix whose columns are $\mathbf{f}_1, \mathbf{f}_2$, and $\overline{\mathbf{f}}_3$.

Scattering by an Ideal Crystal Lattice

The scattering of waves such as X-rays, neutrons, or electrons by a periodic distribution of scatterers (such as the distribution of electrical charge in a perfect crystal) illustrates one route by which the dual space enters naturally into physics. An ideal finite **crystal lattice** is the set of points

$$\mathbf{r}_{m_1,m_2,m_3} := m_1 \mathbf{f}_1 + m_2 \mathbf{f}_2 + m_3 \mathbf{f}_3 \tag{7.61}$$

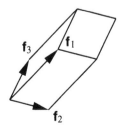

FIGURE 7.3

The unit cell defined by $\mathbf{f}_1, \mathbf{f}_2, \mathbf{f}_3$.

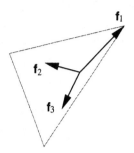

FIGURE 7.4

The (122) lattice plane, $r_{100} + W_{(122)}$.

in which every $m_i \in \mathbb{Z}$ and $m_i \leq M_i$ for some $M_i \in \mathbb{Z}^+$. The **unit cell** is the parallelepiped defined by f_1, f_2, and f_3 (see Fig. 7.3). In the simplest crystals atomic nuclei are located at the lattice points. In more complex crystals the nuclei are arranged within the unit cell in a pattern that is repeated throughout the crystal.

One can see either pictorially or algebraically that the points in an ideal crystal lattice fall in families of planes, which are affine subspaces of $\mathbb{R}^3$ according to Section 5.2.3. The affine subspace that passes through the lattice points r_{h00}, r_{0k0}, and r_{00l} is called the (**hkl**) **lattice plane**. Figure 7.4 indicates by dashed lines the intersections of the (122) plane with the planes defined by $\{f_1, f_2\}$, $\{f_1, f_3\}$ and $\{f_2, f_3\}$. By convention, if any of the integers h, k, or l is negative, then a bar is written over the integer instead of a minus sign. For example, the lattice plane $(1, -1, 1)$ is conventionally designated as the $(1\bar{1}1)$ plane.

We show now that the (*hkl*) lattice plane is the affine subspace $r_{h00} + W_{(hkl)}$, in which $W_{(hkl)}$ is the vector subspace that is spanned by the vector differences $r_{0k0} - r_{h00}$ and $r_{00l} - r_{h00}$:

$$W_{(hkl)} := \operatorname{span}\{r_{0k0} - r_{h00}, r_{00l} - r_{h00}\}. \tag{7.62}$$

In fact, $W_{(hkl)}$ contains all three of the vector differences among the points r_{h00}, r_{0k0}, and r_{00l}, for

$$r_{0k0} - r_{00l} = (r_{0k0} - r_{h00}) - (r_{00l} - r_{h00}). \tag{7.63}$$

Therefore the affine subspace $r_{h00} + W_{(hkl)}$ can also be written as $r_{0k0} + W_{(hkl)}$ or $r_{00l} + W_{(hkl)}$. Certainly r_{h00} belongs to the affine subspace $r_{h00} + W_{(hkl)}$: $r_{h00} = r_{h00} + \mathbf{0} \in r_{h00} + W_{(hkl)}$. Then

$$r_{0k0} = (r_{0k0} - r_{h00}) + r_{h00} \in r_{h00} + W_{(hkl)}. \tag{7.64}$$

Similarly, $r_{00l} \in r_{h00} + W_{(hkl)}$. Finally, because translation by any lattice vector leaves the lattice unchanged, it follows that every affine subspace of the form

$$r_{stu} + W_{(hkl)}, \tag{7.65}$$

in which s, t and u are integers, is also a lattice plane that is parallel to the (*hkl*) plane.

It can be shown that the amplitude of a wave (such as a beam of X-rays, neutrons, or electrons) scattered by a crystal is approximately proportional to

$$\tilde{\rho}(\mathbf{q}) := \int_C e^{i\mathbf{q}\cdot\mathbf{r}'} \rho(\mathbf{r}') \, d^3\mathbf{r}' \tag{7.66}$$

in which $\mathbf{q} = \mathbf{k}_i - \mathbf{k}_s$ is the incident minus the scattered wave vector, C is the region of space occupied by the crystal, and ρ is the density of scatterers. For example, for X-rays ρ is the charge density of the electrons in the crystal.

In an ideal, infinite crystal ρ has the periodicity of the crystal lattice:

$$\forall\, m_1, m_2, m_3 \in \mathbb{Z} : \rho(\mathbf{r} + \mathbf{r}_{m_1,m_2,m_3}) = \rho(\mathbf{r}). \tag{7.67}$$

In a finite crystal such that every $|m_i| \le M_i$, Eq. (7.67) holds as long as all of the vectors in the equation are inside the crystal. Then the integral in Eq. (7.66) can be expressed as the sum of the integrals over the unit cells in the crystal, with the result that

$$\tilde{\rho}(\mathbf{q}) = \sum_C e^{i(m_1\mathbf{q}\cdot\mathbf{f}_1 + m_2\mathbf{q}\cdot\mathbf{f}_2 + m_3\mathbf{q}\cdot\mathbf{f}_3)} \int_U e^{i\mathbf{q}\cdot\mathbf{r}'} \rho(\mathbf{r}') \, d^3\mathbf{r}' \tag{7.68}$$

in which U is a unit cell of the lattice and $\sum_C$ means that the sum runs over all values of m_1, m_2, and m_3 such that the lattice point $\mathbf{r}_{m_1,m_2,m_3}$ lies within the crystal.

If the origin is at the geometrical center of the crystal, then

$$\tilde{\rho}(\mathbf{q}) = g_{M_1}(\mathbf{q} \cdot \mathbf{f}_1)\, g_{M_2}(\mathbf{q} \cdot \mathbf{f}_2)\, g_{M_3}(\mathbf{q} \cdot \mathbf{f}_3) \int_U e^{i\mathbf{q}\cdot\mathbf{r}'} \rho(\mathbf{r}') \, d^3\mathbf{r}' \tag{7.69}$$

in which (by the formula for the sum of a geometric progression)

$$
\begin{aligned}
g_M(\mathbf{q} \cdot \mathbf{f}_j) &= \sum_{m_j=-M}^{M} e^{im_j\mathbf{q}\cdot\mathbf{f}_j} \\
&= \frac{\sin\left[\left(M + \frac{1}{2}\right)\mathbf{q} \cdot \mathbf{f}_j\right]}{\sin \frac{1}{2}\mathbf{q} \cdot \mathbf{f}_j}.
\end{aligned}
\tag{7.70}
$$

In deriving Eq. (7.70), we assume that the vertices of the crystal lie at the points $\mathbf{r}_{\pm M_1, \pm M_2, \pm M_3}$ for which every $M_i \in \mathbb{Z}^+$, $M_i > 1$.

Figure 7.5 illustrates the fact that the functions g_{M_i} are sharply peaked at the values of $\mathbf{q}$ that obey the **von Laue conditions**

$$\forall\, i = 1, 2, 3 : \mathbf{q} \cdot \mathbf{f}_i = 2\pi n_i, \tag{7.71}$$

in which each $n_i \in \mathbb{Z}$ (see Exercise 7.3.4). One recognizes that Eq. (7.71) is of the form

$$\underline{\phi}_{\mathbf{q}}[\mathbf{f}_i] = 2\pi n_i \tag{7.72}$$

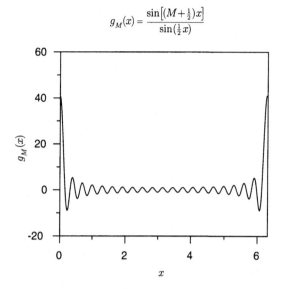

$$g_M(x) = \frac{\sin\left[(M+\tfrac{1}{2})x\right]}{\sin(\tfrac{1}{2}x)}$$

FIGURE 7.5

The function $\sin[(M+\tfrac{1}{2})x]/\sin(\tfrac{1}{2}x)$ plotted for $M = 20$.

(see Eq. [7.6]). It follows from the biorthogonality relation

$$(\mathsf{T}\underline{\phi}^j) \cdot \mathbf{f}_i = (\mathsf{T}(\mathsf{T}^{-1}\mathbf{f}_i)) \cdot (\mathsf{T}\underline{\phi}^j) = \delta_i^j \tag{7.73}$$

that $\mathsf{T}^{-1}\mathbf{f}_i = (f_i^1, f_i^2, f_i^3)$ is the i-th coordinate functional for the reciprocal basis $\{\mathsf{T}\underline{\phi}^j\}$. Then

$$\underline{\phi}_{\mathbf{q}} = 2\pi n_i \underline{\phi}^i. \tag{7.74}$$

Reciprocal Lattice

Any element κ of $\mathbb{R}^3$ with integer components relative to the reciprocal basis,

$$\kappa := n_1 \mathsf{T}\underline{\phi}^1 + n_2 \mathsf{T}\underline{\phi}^2 + n_3 \mathsf{T}\underline{\phi}^3, \quad \text{every } n_i \in \mathbb{Z}, \tag{7.75}$$

is called a **reciprocal-lattice vector**. The von Laue conditions, Eq. (7.71), state that the wave-vector change $\mathbf{q}$ due to scattering by a lattice-periodic charge distribution ρ must be equal to 2π times a reciprocal-lattice vector.

Certain reciprocal-lattice vectors have the physically important property of being perpendicular to a related set of lattice planes. For example, Fig. 7.6 shows that the reciprocal-lattice vector

$$\kappa_{(122)} := 4\mathsf{T}\underline{\phi}_1 + 2\mathsf{T}\underline{\phi}_2 + 2\mathsf{T}\underline{\phi}_3 \tag{7.76}$$

is perpendicular to the (122) lattice plane, which passes through the points $\mathbf{f}_1$, $2\mathbf{f}_2$, and $2\mathbf{f}_3$. To construct a reciprocal-lattice vector that is perpendicular to the (hkl) lattice plane, let

$$\kappa_{(hkl)} := kl\mathsf{T}\underline{\phi}_1 + hl\mathsf{T}\underline{\phi}_2 + hk\mathsf{T}\underline{\phi}_3. \tag{7.77}$$

Then $\kappa_{(hkl)}$ is perpendicular to $\mathcal{W}_{(hkl)}$ (and therefore to every lattice plane that is parallel to

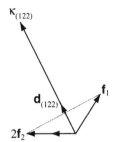

FIGURE 7.6

The reciprocal-lattice vector defined in Eq. (7.76) is perpendicular to the (122) lattice plane.

$\mathcal{W}_{(hkl)}$), because $\kappa_{(hkl)}$ is perpendicular to the vectors which span $\mathcal{W}_{(hkl)}$:

$$\kappa_{(hkl)} \cdot (\mathbf{r}_{0k0} - \mathbf{r}_{h00}) = hkl - hkl = 0. \tag{7.78}$$

Similarly, $\kappa_{(hkl)} \cdot (\mathbf{r}_{00l} - \mathbf{r}_{h00}) = 0$. In terms of $\kappa_{(hkl)}$, the unit vector $\mathbf{n}_{(hkl)}$ that is perpendicular to the (hkl) plane (and to every lattice plane which is parallel to the (hkl) plane) is

$$\mathbf{n}_{(hkl)} = \frac{\kappa_{(hkl)}}{\|\kappa_{(hkl)}\|}. \tag{7.79}$$

The vector $\kappa_{(hkl)}$ defined in Eq. (7.77) is inversely proportional to the distance between adjacent, parallel lattice planes that are parallel to the (hkl) plane. Translation of the (hkl) plane by any of the lattice vectors $\mathbf{r}_{h00}$, $\mathbf{r}_{0k0}$, or $\mathbf{r}_{00l}$ results in a lattice plane that is parallel to the (hkl) plane, such that there is no intervening lattice plane that is also parallel to the (hkl) plane. By elementary vector analysis, the distance $d_{(hkl)}$ between a plane and its translate by $\mathbf{r}_{h00}$ (for example) is equal to the dot product of the unit vector $\mathbf{n}_{(hkl)}$ by the translation vector $\mathbf{r}_{h00}$:

$$d_{(hkl)} = \mathbf{n}_{(hkl)} \cdot \mathbf{r}_{h00} = \frac{hkl}{\|\kappa_{(hkl)}\|}. \tag{7.80}$$

The relations we derive for lattice planes actually apply to all three-dimensional periodic structures.

7.3.6 ISOMORPHISM OF $\mathcal{V}$ TO $\mathcal{V}^{**}$

One can make essentially the same arguments for the second dual $\mathcal{V}^{**}$ of a finite-dimensional vector space $\mathcal{V}$ as have already been made for the dual $\mathcal{V}^*$. It turns out that for the second dual it is best to dispense with the roundabout procedure of first mapping an abstract vector space and its dual onto spaces of column or row vectors and then transposing the rows to get columns (or vice versa). Let us temporarily write an element of $\mathcal{V}^{**}$ as $\bar{\bar{x}}$. The coordinate functionals in $\mathcal{V}^{**}$ are the functionals $\bar{\bar{e}}_i$ on $\mathcal{V}^*$ such that

$$\bar{\bar{e}}_i[\underline{e}^j] := \delta_i^j. \tag{7.81}$$

Again, it is easy to show that $\{\bar{\bar{e}}_i\}$ is a basis of $\mathcal{V}^{**}$. Then the coordinate expansion of an element of $\mathcal{V}^{**}$ (rel$\{\bar{\bar{e}}_i\}$) is

$$\bar{\bar{x}} = x^i \bar{\bar{e}}_i, \tag{7.82}$$

in which

$$x^i := \overline{\overline{x}}[\underline{\epsilon}^i]. \tag{7.83}$$

From Eq. (7.81), the dimension of $\mathcal{V}^{**}$ is equal to the dimension of $\mathcal{V}^*$. Because all vector spaces with the same dimension over the same field $\mathbb{F}$ are isomorphic,

$$\mathcal{V}^{**} \cong \mathcal{V}^* \cong \mathcal{V}. \tag{7.84}$$

The isomorphism of $\mathcal{V}^{**}$ to $\mathcal{V}$ makes it possible for us to identify $\mathcal{V}^{**}$ with $\mathcal{V}$. The **natural embedding** $\mathsf{E} : \mathcal{V} \to \mathcal{V}^{**}$ such that

$$\mathsf{E}(\overline{e}_i) := \overline{\overline{e}}_i \tag{7.85}$$

makes the identification of $\mathcal{V}^{**}$ with $\mathcal{V}$ basis-free (in contrast to the basis-dependent relationship between $\mathcal{V}^*$ and $\mathcal{V}$ given in Eq. [7.50]), because Eqs. (7.29), (7.83), and (7.85) imply that E can be stated in the basis-free form $\mathsf{E}(\overline{x}) = \overline{\overline{x}}$:

$$\mathsf{E}(\overline{x}) = \mathsf{E}(x^i\overline{e}_i) = x^i\mathsf{E}(\overline{e}_i) = x^i\overline{\overline{e}}_i = \overline{\overline{x}}. \tag{7.86}$$

Note that $\overline{\overline{x}}$ is the vector in $\mathcal{V}^{**}$ whose contravariant components (rel$\{\overline{\overline{e}}_i\}$) are the same as the contravariant components of $\overline{x}$ (rel$\{\overline{e}_i\}$). We have merely made temporary use of the basis $\{\overline{e}_i\}$, and of the dual and second dual bases $\{\underline{\epsilon}^j\}$ and $\{\overline{\overline{e}}_i\}$ derived from $\{\overline{e}_i\}$, in order to make it clear exactly which element of $\mathcal{V}^{**}$ should be associated with an element of $\mathcal{V}$. One could equally well have used the basis-free equation

$$\forall \overline{x} \in \mathcal{V} : \forall \underline{\phi} \in \mathcal{V}^* : \overline{\overline{x}}[\underline{\phi}] := \underline{\phi}[\overline{x}] \tag{7.87}$$

to define the second-dual functional $\overline{\overline{x}} = \mathsf{E}(\overline{x})$ onto which the vector $\overline{x}$ is mapped.

If $\mathcal{V}$ is finite-dimensional, one can write

$$\overline{x}[\underline{\phi}] := \overline{\overline{x}}[\underline{\phi}] = \underline{\phi}[\overline{x}] \tag{7.88}$$

instead of Eq. (7.87), because one can identify $\mathcal{V}^{**}$ with $\mathcal{V}$. If $\mathcal{V}$ is infinite-dimensional, then one can say only that $\mathcal{V}^{**}$ is isomorphic to a vector subspace of $\mathcal{V}$.

An important consequence of Eq. (7.88) is that the basis vectors $\{\overline{e}_i\}$ of $\mathcal{V}$ are the coordinate functionals of the basis $\{\underline{\epsilon}^j\}$ of $\mathcal{V}^*$. We have already made use of this fact in deriving Eq. (7.74).

7.3.7 EXERCISES FOR SECTION 7.3

7.3.1   Show that $\underline{\phi}_{\mathbf{a}}$ satisfies Eq. (7.26) for the basis $\mathbf{e}_1 = \mathbf{i}$, $\mathbf{e}_2 = \mathbf{j}$ of unit vectors along the x and y axes, respectively.

7.3.2   Prove that the coordinate functionals defined in Eq. (7.28) are linearly independent.

7.3.3   Let $\mathbf{q}$ satisfy Eq. (7.71). Find the smallest values of $|(\mathbf{q}' - \mathbf{q}) \cdot \mathbf{e}_i|$ such that $f_i(\mathbf{q}') = 0$ for $i = 1, 2, 3$ (see Eq. [7.70]). This calculation establishes the width (in $\mathbf{q}$) of the peaks of f_i.

7.3.4 Find the coordinate functionals induced by the basis

$$\mathbf{f}_1 = \begin{pmatrix} 3 \\ 1 \end{pmatrix}, \quad \mathbf{f}_2 = \begin{pmatrix} 1 \\ 3 \end{pmatrix} \tag{7.89}$$

of $\mathbb{R}^2$.

7.3.5 Find the coordinate functionals induced by the basis

$$\mathbf{f}_1 = \begin{pmatrix} 1 \\ -2 \\ 1 \end{pmatrix}, \quad \mathbf{f}_2 = \begin{pmatrix} 1 \\ 1 \\ 1 \end{pmatrix}, \quad \mathbf{f}_3 = \begin{pmatrix} 1 \\ 0 \\ -1 \end{pmatrix} \tag{7.90}$$

of $\mathbb{R}^3$.

7.4 ANNIHILATOR OF A SUBSPACE

7.4.1 DEFINITIONS

The null space of the linear functional $\underline{\phi} = (1, 0, 0) \in \mathbb{R}^{3*}$ is the two-dimensional subspace $\mathcal{N}$ of $\mathbb{R}^3$ that is spanned by the basis vectors $\mathbf{e}_2$ and $\mathbf{e}_3$ defined in Eq. (5.122). Because $\dim[\mathcal{N}] = 2$, the codimension of $\mathcal{N}$ in $\mathbb{R}^3$ is $3 - 2 = 1$ (see Eq. [5.255]). Clearly ϕ is not the only linear functional whose null space is $\mathcal{N}$. In fact, every linear functional of the form $(\phi_1, 0, 0)$, in which $\phi_1 \in \mathbb{R}$, has the null space $\mathcal{N}$. This example suggests that for every vector subspace $\mathcal{W}$ of a vector space $\mathcal{V}$, there is a unique vector subspace of $\mathcal{V}^*$ that contains all of the linear functionals whose null space is $\mathcal{W}$.

One says that a linear functional $\underline{\phi}$ **annihilates** a vector subspace $\mathcal{W}$ of a vector space $\mathcal{V}$ if and only if

$$\forall \overline{w} \in \mathcal{W} : \underline{\phi}[\overline{w}] = 0. \tag{7.91}$$

The set of linear functionals that annihilate a subspace $\mathcal{W} \subseteq \mathcal{V}$ is called the **annihilator** of $\mathcal{W}$:

$$\text{ann}[\mathcal{W}] := \{\underline{\phi} \in \mathcal{V}^* \mid \forall \overline{x} \in \mathcal{W} : \underline{\phi}[\overline{x}] = 0\}. \tag{7.92}$$

For example, if $\mathcal{V} = \mathbb{R}^3$ and if $\mathcal{W}$ is, once again, the subspace of $\mathcal{V}$ that is spanned by $\mathbf{e}_2$ and $\mathbf{e}_3$ in Eq. (5.122), then $\text{ann}[\mathcal{W}]$ is the subspace of $\mathbb{R}^{3*}$ that consists of all linear functionals of the form $(\phi_1, 0, 0)$. This definition implies that $\mathcal{W}$ is included in the null space of every linear functional that belongs to $\text{ann}[\mathcal{W}]$. Moreover, $\text{ann}[\mathcal{W}]$ is automatically unique because it contains *all* of the linear functionals whose null space includes $\mathcal{W}$.

7.4.2 BASES OF THE ANNIHILATOR

Exercise 7.4.1 shows that $\text{ann}[\mathcal{W}]$ is a subspace of $\mathcal{V}^*$. In what follows, we assume that $\mathcal{V}$ is finite-dimensional. In order to eliminate the trivial cases $\mathcal{W} = \mathcal{O}$ and $\mathcal{W} = \mathcal{V}$ we assume also that $\mathcal{W}$ is a proper subspace of $\mathcal{V}$.

Let $\mathcal{V}$ be a vector space, let $\mathcal{W}$ be a vector subspace of $\mathcal{V}$, and let $B = \{\bar{e}_1, \ldots, \bar{e}_m\}$ be a basis of $\mathcal{W}$. Extend B to a basis $C = \{\bar{e}_1, \ldots, \bar{e}_m, \bar{e}_{m+1}, \ldots, \bar{e}_n\}$ of $\mathcal{V}$. Let $\{\underline{\epsilon}^1, \ldots, \underline{\epsilon}^m, \underline{\epsilon}^{m+1}, \ldots, \underline{\epsilon}^n\}$ be the coordinate functionals for C. It is almost trivial to see that

$$\forall i \in (m+1, n) : \underline{\epsilon}^i \in \text{ann}[\mathcal{W}]. \tag{7.93}$$

because the functionals $\underline{\epsilon}^{m+1}, \ldots, \underline{\epsilon}^n$ annihilate every basis element of $\mathcal{W}$:

$$\forall j \in (1, m) : \forall i \in (m+1, n) : \underline{\epsilon}^i[\bar{e}_j] = 0. \tag{7.94}$$

Therefore $\{\underline{\epsilon}^{m+1}, \ldots, \underline{\epsilon}^n\}$ annihilates $\mathcal{W}$.

Let $\underline{\psi} \in \mathcal{V}^*$ annihilate $\mathcal{W}$. Because $\{\underline{\epsilon}^k\}$ is a basis of $\mathcal{V}^*$, there exist scalars $\{\psi_k\}$ such that

$$\underline{\psi} = \psi_k \underline{\epsilon}^k. \tag{7.95}$$

Let $\bar{w}$ be any vector in $\mathcal{W}$. Then its coordinate expansion is

$$\bar{w} = \sum_{i=1}^{m} w^i \bar{e}_i . \tag{7.96}$$

Because $\underline{\psi} \in \text{ann}[\mathcal{W}]$,

$$\underline{\psi}[\bar{w}] = \sum_{i=1}^{m} \psi_i w^i = 0. \tag{7.97}$$

Then

$$\psi_1 = \cdots = \psi_m = 0, \tag{7.98}$$

implying that $\underline{\psi}$ belongs to the subspace of $\mathcal{V}^*$ that is spanned by $\{\underline{\epsilon}^{m+1}, \ldots, \underline{\epsilon}^n\}$. Therefore

$$\text{ann}[\mathcal{W}] = \text{span}\{\underline{\epsilon}^{m+1}, \ldots, \underline{\epsilon}^n\}. \tag{7.99}$$

Because $\dim[\mathcal{W}] = m$ and

$$\dim[\text{ann}[\mathcal{W}]] = n - m \tag{7.100}$$

by construction of the basis, it follows that

$$\dim[\mathcal{V}] = \dim[\mathcal{W}] + \dim[\text{ann}[\mathcal{W}]]. \tag{7.101}$$

From Eqs. (7.100) and (5.255) one sees that

$$\text{ann}[\mathcal{W}] \cong \mathcal{X} \tag{7.102}$$

in which $\mathcal{X}$ is any complement of $\mathcal{W}$ in $\mathcal{V}$.

To compute a basis for the annihilator of a subspace $\mathcal{W}$ one must, in effect, compute a basis of the orthogonal complement of $\mathcal{W}$ and then apply the inner-product mapping; see Eq. (8.527).

7.4.3 EXERCISES FOR SECTION 7.4

7.4.1   Let $\mathcal{W}$ be a subspace of a vector space $\mathcal{V}$. Show that ann[$\mathcal{W}$] is a subspace of $\mathcal{V}^*$.

7.5 OTHER REALIZATIONS OF DUAL SPACES

7.5.1 DUAL SPACE OF $\mathbb{C}$

According to Eq. (7.51), the dual $\mathbb{C}^*$ of the vector space $\mathbb{C}$ (over the real field $\mathbb{R}$) is isomorphic to $\mathbb{C}$, that is, $\mathbb{C}^*$ is a copy of the complex field $\mathbb{C}$. Let $z \in \mathbb{C}$ and $\phi \in \mathbb{C}^*$. Define

$$\phi[z] = \text{Re}[\phi z]. \tag{7.103}$$

We showed in Section 5.4.2 that $\{e_1 = 1, e_2 = i\}$ is a basis of $\mathbb{C}$. The coordinate functionals Eq. (7.28) generated by this basis are

$$\epsilon_1 = 1, \quad \epsilon_2 = -i. \tag{7.104}$$

The resulting coordinate expansions of $z \in \mathbb{C}$ and $\phi \in \mathbb{C}^*$ are

$$z = x + iy, \quad \phi = \xi - i\eta. \tag{7.105}$$

Then Eq. (7.26) takes the form

$$\phi[z] = \text{Re}[\phi z] = \xi x + \eta y. \tag{7.106}$$

7.5.2 DUAL OF $\mathbb{F}^{\mathbb{Z}^+}$

The coordinate functionals $\underline{\epsilon}^i$ (Eq. [7.28]) can be defined on any sequence space such as $\phi(\mathbb{F})$, $\mathbb{F}^{\mathbb{Z}^+}$, $m(\mathbb{F})$, $c(\mathbb{F})$, $c_0(\mathbb{F})$, and $l^2(\mathbb{F})$. For example, in $\mathbb{F}^{\mathbb{Z}^+}$,

$$\forall i \in \mathbb{Z}^+ : \forall \overline{x} \in \mathbb{F}^{\mathbb{Z}^+} : \underline{\epsilon}^i[\overline{x}] := x^i. \tag{7.107}$$

The *finite* linear combinations of the coordinate functionals on $\mathbb{F}^{\mathbb{Z}^+}$ span a vector space that is isomorphic to $\phi(\mathbb{F})$, the space of finite sequences, under the mapping Eq. (7.50). Therefore $(\mathbb{F}^{\mathbb{Z}^+})^* \cong \phi(\mathbb{F})$. This example shows that it is not true for a general sequence space that $\mathcal{V}^* \cong \mathcal{V}$.

7.5.3 BOUNDARY AND INITIAL CONDITIONS FOR DIFFERENTIAL EQUATIONS

Boundary and initial conditions on the solutions of ordinary and partial differential equations can be stated succinctly in terms of linear functionals. Let $\mathbb{F} = \mathbb{R}$ or $\mathbb{C}$, and consider any continuously differentiable function $f : [a, b] \to \mathbb{F}$. (In order for a differential operator of order k to be defined on f, f must belong to $\mathcal{C}^k([a, b]; \mathbb{F})$. See Section 5.2.2.) In terms of the delta functional defined in Eq. (7.4), homogeneous Dirichlet boundary conditions at $x = a$ and $x = b$ read

$$\delta_a[f] = 0, \quad \delta_b[f] = 0. \tag{7.108}$$

Let us define a new linear functional δ'_ξ that evaluates the first derivative at the point ξ,

$$\forall \xi \in [a, b] : \forall f \in \mathcal{C}^1([a, b]; \mathbb{F}) : \delta'_\xi[f] := -f'(\xi), \tag{7.109}$$

If δ_ξ were a continuously differentiable function (which it is not), and if f were to vanish at infinity (a behavior that is physically correct in many cases), then one could perform the following purely formal integration by parts:

$$\int_{-\infty}^{\infty} \delta_\xi' \, f(x) \, dx = \delta_\xi(x) \, f(x) \Big|_{-\infty}^{\infty} - \int_{-\infty}^{\infty} \delta_\xi \, f'(x) \, dx = -f'(\xi). \tag{7.110}$$

This formal calculation explains why the functional δ_ξ' is defined with a minus sign in Eq. (7.109). In terms of the functional δ_ξ', homogeneous Neumann conditions at a and b take the form

$$\delta_a'[f] = 0, \quad \delta_b'[f] = 0, \tag{7.111}$$

and homogeneous mixed conditions at a and b read

$$\delta_a[f] - \eta \delta_a'[f] = 0, \quad \delta_b[f] - \eta \delta_b'[f] = 0. \tag{7.112}$$

To give another example, the solution of a differential equation subject to the Dirichlet boundary conditions Eq. (7.108) must lie in the intersection of the null spaces of the functionals δ_a and δ_b. From Exercise 5.2.1 one knows that the intersection of two vector subspaces is a vector subspace.

7.6 POLYNOMIAL INTERPOLATION

7.6.1 LAGRANGIAN INTERPOLATION

Often one has a set of points through which one wishes to pass a smooth curve, for the purpose of obtaining the coordinates of intermediate points. Lagrange showed that there exists a unique polynomial of degree $n - 1$ that passes through each of a set of n points (x^1, y^1), $\ldots, (x^n, y^n)$. Although Lagrange's solution can be obtained by more elementary means than we use in this section, interpolation affords a good example of a practical use for coordinate functionals and other constructs discussed previously. As with many other problems, it is helpful to begin by restating the problem more abstractly. It is natural to consider $y^1, \ldots, y^n$ as the sampled values of a function (recall Eq. [6.104]). Because a polynomial of degree $n - 1$ has n coefficients, there is reason to hope that knowing the sampled values of a polynomial at n points will lead to a unique solution for the polynomial.

Now we reformulate the problem abstractly. Let $\{\underline{\phi}^1, \ldots, \underline{\phi}^m\}$ be linearly independent linear functionals on a vector space $\mathcal{V}$ of finite dimension n. The **interpolation problem** for $\{\underline{\phi}^1, \ldots, \underline{\phi}^m\}$ is to find a vector $\overline{p} \in \mathcal{V}$ such that

$$\forall \in (1, m) : \underline{\phi}^i[\overline{p}] = y^i \tag{7.113}$$

in which $y^1, \ldots, y^m \in \mathbb{F}$ are given. This equation defines a system of m linear equations in n unknowns. From Section 6.6 it follows that $m = n$ is a necessary condition for a unique solution.

For example, let V be the space of polynomials in t over $\mathbb{R}$ of degree not greater than $n - 1$. Let $\underline{\phi}^i$ be the functional $\underline{\delta}_{t_i}$ that evaluates a polynomial $p \in V$ at the point t_i,

$$\underline{\phi}^i[p] := p(t_i). \tag{7.114}$$

To solve the interpolation problem stated in Eq. (7.113) one must find a polynomial p of degree $n - 1$ or less such that

$$\forall i \in (1, m) : p(t_i) = y^i. \tag{7.115}$$

We return to this example after an abstract discussion of the solution of Eq. (7.113).

Assume that $m = n$. Then $\{\underline{\phi}^i\}$ is a basis of V^*, because $\dim[V^*] = \dim[V] = n$.

Let $\{\overline{f}_j\}$ be the basis (of $V^{**} \cong V$) that consists of the coordinate functionals on V^* induced by the basis $\{\underline{\phi}^i\}$. Then

$$\forall i \in (1, n) : \forall j \in (1, n) : \underline{\phi}^i[\overline{f}_j] = \delta^i_j. \tag{7.116}$$

Because $m = n$, $\{\overline{f}_j\}$ is uniquely determined by the basis $\{\underline{\phi}^i\}$ of V^*.

Let the coordinate expansion of the unknown vector $\overline{p}$ (rel$\{\overline{f}_j\}$) be

$$\overline{p} = p^j \overline{f}_j. \tag{7.117}$$

Then Eq. (7.113) implies that

$$\forall i \in (1, n) : p^j = y^j. \tag{7.118}$$

Therefore

$$\overline{p} = y^j \overline{f}_j \tag{7.119}$$

solves the interpolation problem Eq. (7.113) if $m = n$.

In the problem of polynomial interpolation a necessary condition for the linear independence of the functionals $\{\underline{\phi}^i\}$ defined in Eq. (7.114) is that the t_i must be distinct. Otherwise at least two of the functionals are equal and hence are linearly dependent, implying that $\{\underline{\phi}^i\}$ is not a basis of V^*. Therefore we assume that the t_i are distinct.

A general polynomial p over $\mathbb{F}$ of degree $n - 1$ has the form

$$p(t) = c_0 + c_1 t + \cdots + c_{n-1} t^{n-1} \tag{7.120}$$

in which t^i means t raised to the i-th power. For interpolation using this polynomial the system of equations defined in Eq. (7.113) is

$$\forall i \in (1, n) : p(t_i) = c_0 + c_1 t_i + \cdots + c_{n-1} t_i^{n-1} = y^i. \tag{7.121}$$

The unknowns in this system of n linear equations are $c_0, \ldots, c_n$. From Eq. (6.350) one recognizes the determinant of the coefficients, $\det[t_i^j]$, as the Vandermonde determinant $V_n(t_1, \ldots, t_n)$. Because we have assumed that the t_i are distinct, $V_n(t_1, \ldots, t_n) \neq 0$. Then the system Eq. (7.121) has a unique solution.

Because the coefficients $c_0, \ldots, c_n$ of the polynomial p that realizes $\overline{p}$ are uniquely determined, one has nothing to lose (and clarity to gain) by making a clever guess at the solution.

One already knows a basis $\{\underline{\phi}^i\} = \{\underline{\delta}_{t_i}\}$ of V^*. One must find the dual basis $\{\overline{f}_j\}$ of V, such that the abstract vector $\overline{f}_j$ is realized by a polynomial f_j. By Eq. (7.116), f_j must have the property that

$$f_j(t_i) = \delta_i^j. \tag{7.122}$$

This equation is solved by the polynomials

$$f_j(t) := \frac{\prod\limits_{k \neq j}(t - t_k)}{\prod\limits_{l \neq j}(t_j - t_l)}, \tag{7.123}$$

which are all of degree $n - 1$. In Eq. (7.123), the product on k (for example) runs from $k = 1$ to $k = n$, except that the factor $(t - t_j)$ corresponding to $k = j$ is omitted.

The polynomial p that solves Eq. (7.121),

$$p(t) = y^j f_j(t), \tag{7.124}$$

is called the **Lagrange polynomial**. Interpolation using the Lagrange polynomial is called **Lagrangian interpolation**.

7.6.2 EXERCISES FOR SECTION 7.6

7.6.1   Prove that a list $\{\underline{\phi}^1, \ldots, \underline{\phi}^n\}$ of linear functionals on an n-dimensional vector space V is linearly independent if and only if

$$\det[\underline{\phi}^j[\overline{x}_i]] \neq 0, \tag{7.125}$$

in which $\{\underline{x}_1, \ldots, \underline{x}_n\}$ is a basis of V. (The determinant $\det[\underline{\phi}^j[\overline{x}_i]]$ is sometimes called a **generalized Gram determinant**.)

7.6.2   Let V be the vector space $C^1([a, b]; \mathbb{R})$ of continuously differentiable, real-valued functions on the interval $[a, b]$. Show that the linear functionals $\underline{\delta}_c$, $\underline{\delta}'_c$ and $\underline{\iota}$ on V are linearly independent, with $\underline{\delta}_c$ and $\underline{\delta}'_c$ the function-evaluation and derivative-evaluation functionals defined in Eqs. (7.3) and (7.109), respectively, and

$$\forall f \in C^1([a, b]; \mathbb{R}) : \underline{\iota}[f] := \int_a^b f(t)\,dt. \tag{7.126}$$

7.7 TENSORS

Tensors are an important tool in several fields of physics and engineering, including relativity, elasticity, and electromagnetics. It is helpful to begin a discussion of tensors with a more fundamental concept, tensor products, which are useful in applications of quantum mechanics to a composite system such as a multielectron atom, or even a single electron with spin. In fact, tensor products occur wherever one has to work with products of functions that are defined on different spaces, or on different copies of the same space. For example, the electric energy density in a dielectric material is

$$U_E = \frac{\epsilon_0}{2} \mathbf{E} \cdot \mathbf{D}, \tag{7.127}$$

in which $\mathbf{E}$ is the electric field vector,

$$\mathbf{D} = \epsilon \mathbf{E} \tag{7.128}$$

is the electric displacement, and ϵ is the dielectric permittivity tensor.

7.7.1 DEFINITIONS AND BASIC PROPERTIES

Tensor Product of Vector Spaces

A **bilinear functional** on the Cartesian product $\mathcal{V}_{(1)} \times \mathcal{V}_{(2)}$ of two vector spaces over the same field $\mathbb{F}$ is a linear mapping $\mathsf{T} : \mathcal{V}_{(1)} \times \mathcal{V}_{(2)} \to \mathbb{F}$ such that for every $\overline{x}_{(1)} \in \mathcal{V}_{(1)}$ and for every $\overline{x}_{(2)} \in \mathcal{V}_{(2)}$,

$$\mathsf{T}\left[\overline{x}_{(1)} + \overline{y}_{(1)}, \overline{x}_{(2)}\right] = \mathsf{T}\left[\overline{x}_{(1)}, \overline{x}_{(2)}\right] + \mathsf{T}\left[\overline{y}_{(1)}, \overline{x}_{(2)}\right], \tag{7.129}$$

$$\mathsf{T}\left[\overline{x}_{(1)}, \overline{x}_{(2)} + \overline{y}_{(2)}\right] = \mathsf{T}\left[\overline{x}_{(1)}, \overline{x}_{(2)}\right] + \mathsf{T}\left[\overline{x}_{(1)}, \overline{y}_{(2)}\right], \tag{7.130}$$

and

$$\mathsf{T}\left[\alpha\overline{x}_{(1)}, \overline{x}_{(2)}\right] = \mathsf{T}\left[\overline{x}_{(1)}, \alpha\overline{x}_{(2)}\right] = \alpha\mathsf{T}\left[\overline{x}_{(1)}, \overline{x}_{(2)}\right]. \tag{7.131}$$

In other words, a bilinear functional is a functional that has two arguments and is additive and homogeneous in each argument.

In our first use of this definition, we use the duals $\mathcal{V}_{(1)}^*$ and $\mathcal{V}_{(2)}^*$ instead of $\mathcal{V}_{(1)}$ and $\mathcal{V}_{(2)}$ (after all, the duals *are* vector spaces!). The **tensor product**

$$\mathcal{V}_{(1)} \otimes \mathcal{V}_{(2)} \tag{7.132}$$

of two vector spaces $\mathcal{V}_{(1)}$ and $\mathcal{V}_{(2)}$ over the same field $\mathbb{F}$ is defined as the set of all bilinear functionals $\mathsf{T} : \mathcal{V}_{(1)}^* \times \mathcal{V}_{(2)}^* \to \mathbb{F}$ on the Cartesian product of the dual spaces.

We define a second-rank contravariant tensor as an element of $\mathcal{V}_{(1)} \otimes \mathcal{V}_{(2)}$. This approach is equivalent to the usual definition of a contravariant tensor in terms of the way in which the components transform under a change of basis. (The components of a tensor are defined in a subsequent section.) Each of these two ways of defining a tensor has significant advantages. The basis-free definition in terms of bilinear functionals permits one to see structure clearly; the component approach provides a powerful tool for calculation.

Defining tensors in terms of bilinear functionals makes the so-called contraction property true by definition. The usual approach eventually deduces the contraction property from the behavior of tensors under a change of basis. The approach that we use has the advantage of putting the contraction property in the center of the stage, where it belongs because it defines the structure of the objects that one calls tensors.

If $V_{(1)}$ and $V_{(2)}$ are the same vector space V, then one writes

$$V^{\otimes 2} := V \otimes V \tag{7.133}$$

for the tensor product of V with itself.

In quantum mechanics, the state space of a system that consists of two particles is the tensor product of the state spaces of the individual particles. Tensor products are also used to construct the state vector of a single particle with different kinds of degrees of freedom. For example, the state space of a particle with spin zero is the space $\mathcal{L}^2(\mathbb{R}^3; \mathbb{C})$ of square-integrable, complex-valued functions on $\mathbb{R}^3$. The state space of a particle with spin s (where s is equal to one half of a positive integer) is the tensor product $\mathcal{L}^2(\mathbb{R}^3; \mathbb{C}) \otimes \mathbb{C}^{2s+1}$ of the spin-zero state space, which describes the possible motions of the particle in three-dimensional space, with the space of spinors with $2s+1$ components, $\mathbb{C}^{2s+1}$, which describes the possible spin states.

Contravariant and Covariant Tensors

An element T of $V_{(1)} \otimes V_{(2)}$ is called a **second-rank contravariant tensor**. Similarly, an element C of the tensor product of the duals, $V_{(1)}^* \otimes V_{(2)}^*$, is called a **second-rank covariant tensor**. For a simple example of a rank-2 contravariant tensor, let $V_{(1)} = V_{(2)} = \mathbb{R}^2$, and let

$$\begin{aligned}
\underline{\phi} &= (\phi_1, \phi_2), \\
\underline{\psi} &= (\psi_1, \psi_2)
\end{aligned} \tag{7.134}$$

be functionals in $\mathbb{R}^{2*}$. Of course, the components ϕ_i, ψ_j are all real numbers. An example of a second-rank contravariant tensor T according to the definition given above is

$$\mathsf{T}[\underline{\phi}, \underline{\psi}] = t^{ij}\, \phi_i \psi_j, \tag{7.135}$$

in which $i, j \in (1:2)$ and every $t^{ij} \in \mathbb{R}$.

A second-rank tensor $\mathsf{S} \in V^{\otimes 2}$ is called **symmetric** if and only if

$$\forall \phi, \psi : \mathsf{S}[\underline{\phi}, \underline{\psi}] = \mathsf{S}[\underline{\psi}, \underline{\phi}]. \tag{7.136}$$

For example, if the dimensions of the vector spaces are $n_1 = n_2 = 2$, the tensor

$$\mathsf{S}[\underline{\phi}, \underline{\psi}] = \phi_1 \psi_2 + \phi_2 \psi_1 \tag{7.137}$$

is symmetric.

Similarly, a second-rank tensor $\mathsf{A} \in V^{\otimes 2}$ is called **antisymmetric** if and only if

$$\forall \phi, \psi : \mathsf{A}[\underline{\phi}, \underline{\psi}] = -\mathsf{A}[\underline{\psi}, \underline{\phi}]. \tag{7.138}$$

Important examples in which $n_1 = n_2 = 2$ include the general symmetric second-rank tensor

$$\mathsf{S}[\underline{\phi}, \underline{\psi}] = s^{11}\phi_1\psi_1 + s^{22}\phi_2\psi_2 + s^{12}(\phi_1\psi_2 + \phi_2\psi_1) \tag{7.139}$$

and the antisymmetric second-rank tensor

$$\mathsf{X}[\underline{\phi}, \underline{\psi}] = \phi_1 \psi_2 - \phi_2 \psi_1, \tag{7.140}$$

which is equal to the component of the cross product of $\underline{\phi}^T$ and $\underline{\psi}^T$ in the direction determined by the right-hand rule.

A vector field is another important example of an element of the tensor product of two vector spaces. The electric field vector $\mathbf{E}$, for instance, is an element of $\mathbb{R}^3$, each contravariant component of which is a real-valued function on $\mathbb{R}^3$. The value of $\mathbf{E}$ at a point $\mathbf{r} \in \mathbb{R}^3$ is

$$\mathbf{E}(\mathbf{r}) = \begin{pmatrix} E^1(\mathbf{r}) \\ E^2(\mathbf{r}) \\ E^3(\mathbf{r}) \end{pmatrix} \tag{7.141}$$

in which each E^i belongs to a vector space of functions, $\mathcal{F}$, which we do not need to specify precisely. The value of a functional $\underline{\phi} \in \mathbb{R}^{3*}$ on the vector $\mathbf{E}(\mathbf{r})$ is

$$\underline{\phi}[\mathbf{E}(\mathbf{r})] = \phi_1 E^1(\mathbf{r}) + \phi_2 E^2(\mathbf{r}) + \phi_3 E^3(\mathbf{r}). \tag{7.142}$$

One can consider $\underline{\phi}[\mathbf{E}(\mathbf{r})]$ as the value $g(\mathbf{r})$ of a function $g \in \mathcal{F}$, on which a linear functional (such as the integration functional, for example) can operate. It follows that if $\underline{\psi}$ is a functional belonging to the dual $\mathcal{F}'$ of the function space $\mathcal{F}$, then the equation

$$\mathbf{E}[\underline{\phi}, \underline{\psi}] := \phi_1 \underline{\psi}[E^1] + \phi_2 \underline{\psi}[E^2] + \phi_3 \underline{\psi}[E^3] \tag{7.143}$$

defines $\mathbf{E}$ as an element of the tensor product space $\mathbb{R}^3 \otimes \mathcal{F}$.

Mixed Tensors

An element of the tensor product of a vector space and the dual of another vector space, that is, an element of $\mathcal{V}_{(1)} \otimes \mathcal{V}_{(2)}^*$ or $\mathcal{V}_{(1)}^* \otimes \mathcal{V}_{(2)}$, is called a **mixed second-rank tensor**. For example, any linear mapping $\mathsf{A} : \mathcal{V}_{(1)} \to \mathcal{V}_{(2)}$ can be considered as a second-rank mixed tensor belonging to $\mathcal{V}_{(1)} \otimes \mathcal{V}_{(2)}^*$ if one defines

$$\mathsf{A}[\overline{x}_{(1)}, \underline{\phi}_{(2)}] := \underline{\phi}_{(2)}[\mathsf{A}\overline{x}_{(1)}]. \tag{7.144}$$

In this case, bilinearity and homogeneity follow from the additivity and homogeneity of linear mappings and linear functionals.

Sums and Scalar Multiples of Tensors

We define sums and scalar multiples of tensors as we did for ordinary linear functionals (see Eq. [7.13]). For a contravariant tensor, the definitions are

$$(\mathsf{T} + \mathsf{T}')[\underline{\phi}_{(1)}, \underline{\phi}_{(2)}] := \mathsf{T}[\underline{\phi}_{(1)}, \underline{\phi}_{(2)}] + \mathsf{T}'[\underline{\phi}_{(1)}, \underline{\phi}_{(2)}] \tag{7.145}$$

and

$$(\alpha \mathsf{T})[\underline{\phi}_{(1)}, \underline{\phi}_{(2)}] := \alpha \mathsf{T}[\underline{\phi}_{(1)}, \underline{\phi}_{(2)}]. \tag{7.146}$$

With these definitions, $\mathcal{V}_{(1)} \otimes \mathcal{V}_{(2)}$ satisfies the axioms for a vector space (Exercise 7.7.2).

7.7.2 COMPONENTS OF SECOND-RANK TENSORS

Components of Contravariant Tensors

Let us find the components of a second-rank contravariant tensor relative to the bases $\{\underline{\epsilon}^i_{(1)}\}$ and $\{\underline{\epsilon}^j_{(2)}\}$ of coordinate functionals in the dual spaces $V^*_{(1)}$ and $V^*_{(2)}$, respectively. Let T be any second-rank contravariant tensor. Then by Eqs. (7.145) and (7.146)

$$\mathsf{T}\big[\underline{\phi}_{(1)}, \underline{\phi}_{(2)}\big] = \mathsf{T}\big[\phi_{(1)i}\,\underline{\epsilon}^i_{(1)},\ \phi_{(2)j}\,\underline{\epsilon}^j_{(2)}\big] = t^{ij}\phi_{(1)i}\phi_{(2)j}, \tag{7.147}$$

in which

$$t^{ij} := \mathsf{T}\big[\underline{\epsilon}^i_{(1)}, \underline{\epsilon}^j_{(2)}\big]. \tag{7.148}$$

The numbers t^{ij} are called the **components** of the second-rank contravariant tensor T relative to the bases $\{\underline{\epsilon}^i_{(1)}\}$ and $\{\underline{\epsilon}^j_{(2)}\}$ of $V^*_{(1)}$ and $V^*_{(2)}$, respectively.

Because $(\underline{\phi}_{(1)}, \underline{\phi}_{(2)})$ can be any element of $V^*_{(1)} \times V^*_{(2)}$, the components t^{ij} specify the tensor T uniquely. Conversely, if one assembles any set of $n_1 n_2$ real numbers and labels them as t^{ij}, then they are the components of a unique second-rank contravariant tensor.

For example, in Eq. (7.140) one sees that relative to the canonical coordinate functionals $\underline{\epsilon}^1 = (1, 0)$ and $\underline{\epsilon}^2 = (0, 1)$, the definition implies that the components of the cross-product tensor X are

$$\begin{aligned} x^{12} &= \mathsf{X}[(1, 0), (0, 1)] = 1, \\ x^{21} &= \mathsf{X}[(0, 1), (1, 0)] = -1, \\ x^{11} &= x^{22} = 0 \end{aligned} \tag{7.149}$$

as one would expect. Similarly, the components of the totally symmetric tensor S are

$$s^{11} = s^{22} = s^{12} = s^{21} = 1. \tag{7.150}$$

The components t^{ij} and t^{ji} of symmetric or antisymmetric second-rank contravariant tensors satisfy important relations. If $\mathsf{S} \in V^{\otimes 2}$ is a symmetric second-rank contravariant tensor, then

$$\begin{aligned} \mathsf{S}[\underline{\phi}, \underline{\psi}] &= s^{ij}\phi_i\psi_j \\ \mathsf{S}[\underline{\psi}, \underline{\phi}] &= s^{ji}\psi_j\phi_i \end{aligned} \tag{7.151}$$

and therefore, for all values of i and j,

$$\boxed{s^{ji} = s^{ij}.} \tag{7.152}$$

Similarly, if $\mathsf{A} \in V^{\otimes 2}$ is an antisymmetric second-rank contravariant tensor, then

$$\boxed{a^{ji} = -a^{ij}.} \tag{7.153}$$

It follows that for an antisymmetric second-rank tensor one has $a^{ii} = 0$.

Conversely, any array of n^2 real numbers that satisfies Eq. (7.152) defines a unique symmetric second-rank tensor, and any array of n^2 real numbers that satisfies Eq. (7.153) defines a unique antisymmetric second-rank tensor.

For example, the components of the electromagnetic field tensor F are

$$f^{\mu\nu} = \frac{\partial A^\mu}{\partial x_\nu} - \frac{\partial A^\nu}{\partial x_\mu}, \tag{7.154}$$

in which $\{A^\mu \mid \mu \in (0:3)\}$ are the components of the vector potential. Clearly F is antisymmetric.

A general second-rank contravariant tensor $\mathsf{T} \in \mathcal{V}^{\otimes 2}$ can always be expressed as the sum of a symmetric second-rank contravariant tensor S and an antisymmetric second-rank contravariant tensor A. To see this, let

$$s^{ij} := \tfrac{1}{2}(t^{ij} + t^{ji})$$
$$\tag{7.155}$$
$$a^{ij} := \tfrac{1}{2}(t^{ij} - t^{ji}).$$

Define S as the tensor with components s^{ij}, and A as the tensor with components a^{ij}. Clearly

$$t^{ij} = s^{ij} + a^{ij}. \tag{7.156}$$

Therefore

$$\mathsf{T} = \mathsf{S} + \mathsf{A}. \tag{7.157}$$

By construction, S is symmetric and A is antisymmetric. This establishes the decomposition of T into symmetric and antisymmetric parts.

Components of Covariant or Mixed Tensors

To find the components of a second-rank covariant tensor, one interchanges the roles of the vector spaces $\mathcal{V}_{(1)}$, $\mathcal{V}_{(2)}$ and their duals. Thus the covariant versions of Eqs. (7.147) and (7.148) are

$$\mathsf{C}\big[\bar{x}_{(1)}, \bar{x}_{(2)}\big] = \mathsf{C}\big[x_{(1)i}\,\bar{e}_{(1)i}, x_{(2)j}\,\bar{e}_{(2)j}\big] = c_{ij}\,x_{(1)}^i x_{(2)}^j, \tag{7.158}$$

in which

$$c_{ij} := \mathsf{C}\big[\bar{e}_{(1)i}, \bar{e}_{(2)j}\big] \tag{7.159}$$

are the components of C.

For the mixed second-rank tensor A that corresponds to a linear mapping $A : \mathcal{V}_{(1)} \to \mathcal{V}_{(2)}$, Eq. (7.144), one has

$$
\begin{aligned}
\mathsf{A}\big[\bar{x}_{(1)}, \underline{\phi}_{(2)}\big] &= \underline{\phi}_{(2)}\big[A(x_{(1)}^i \bar{e}_{(1)i})\big] \\
&= x_{(1)}^i \underline{\phi}_{(2)}\big[A\bar{e}_{(1)i}\big] \\
&= x_{(1)}^i \underline{\phi}_{(2)}\big[a_i^j \bar{e}_{(2)j}\big] \\
&= a_i^j x_{(1)}^i \underline{\phi}_{(2)}\big[\bar{e}_{(2)j}\big] \\
&= a_i^j x_{(1)}^i \phi_{(2)j}.
\end{aligned}
\tag{7.160}
$$

It follows that the components of the mixed second-rank tensor associated with a linear mapping A through Eq. (7.144) are just the matrix elements of A. Taking $x^i_{(1)} = \delta^i_k$ (and therefore $\overline{x}_{(1)} = \overline{e}_{(1)k}$), $\phi_{(2)j} = \delta^l_j$ (and therefore $\underline{\phi}_{(2)} = \underline{\epsilon}^l_{(2)}$), one obtains the general formula

$$a^l_k = \underline{\epsilon}^l_{(2)}\left[A\overline{e}_{(1)k}\right] \tag{7.161}$$

for the matrix elements of A.

In order to be completely precise one should write the indices of components of second-rank mixed tensors in such a way that it is obvious whether the tensor belongs to $\mathcal{V}_{(1)} \otimes \mathcal{V}^*_{(2)}$ or to $\mathcal{V}^*_{(1)} \otimes \mathcal{V}_{(2)}$. For example, one should write $a_i{}^j$ instead of a^j_i in Eq. (7.160) in order to indicate that the components belong to an element of $\mathcal{V}_{(1)} \otimes \mathcal{V}^*_{(2)}$. In the same vein, the components of an element of $\mathcal{V}^*_{(1)} \otimes \mathcal{V}_{(2)}$ would be written as $a^j{}_i$. We circumvent the need to pay attention to the relative order of superscripts and subscripts by using different letters to denote elements of $\mathcal{V}_{(1)} \otimes \mathcal{V}^*_{(2)}$ and $\mathcal{V}^*_{(1)} \otimes \mathcal{V}_{(2)}$.

7.7.3 TENSOR PRODUCTS OF VECTORS

Definition of the Tensor Product of Two Vectors

The **tensor product** $\overline{x}_{(1)} \otimes \overline{x}_{(2)}$ of vectors $\overline{x}_{(1)} \in \mathcal{V}_{(1)}$ and $\overline{x}_{(2)} \in \mathcal{V}_{(2)}$ is the bilinear functional on $\mathcal{V}^*_{(1)} \times \mathcal{V}^*_{(2)}$ such that

$$\left(\overline{x}_{(1)} \otimes \overline{x}_{(2)}\right)\left[\underline{\phi}_{(1)}, \underline{\phi}_{(2)}\right] := \underline{\phi}_{(1)}\left[\overline{x}_{(1)}\right]\underline{\phi}_{(2)}\left[\overline{x}_{(2)}\right]. \tag{7.162}$$

For example, if $\mathcal{V}_{(1)} = \mathcal{V}_{(2)} = \mathbb{R}^2$, then

$$(\mathbf{e}_1 \otimes \mathbf{e}_2)[(\phi_1, \phi_2), (\psi_1, \psi_2)] = \phi_1\psi_2. \tag{7.163}$$

For general elements of $\mathbb{R}^2$,

$$\mathbf{x} = \begin{pmatrix} x^1 \\ x^2 \end{pmatrix}, \quad \mathbf{y} = \begin{pmatrix} y^1 \\ y^2 \end{pmatrix}, \tag{7.164}$$

one has

$$(\mathbf{x} \otimes \mathbf{y})[(\phi_1, \phi_2), (\psi_1, \psi_2)] = x^i y^j \phi_i \psi_j. \tag{7.165}$$

We show that one can think of the tensor product of two vectors of dimensions m and n, respectively, as a single vector with mn components $x^i y^j$. The structure of this single vector is $(x^1\mathbf{y}, x^2\mathbf{y}, \ldots, x^m\mathbf{y})^T$.

It is straightforward to show that the tensor product of two vectors $\overline{x}_{(1)} \in \mathcal{V}_{(1)}$ and $\overline{x}_{(2)} \in \mathcal{V}_{(2)}$ meets the definition of a second-rank tensor (see Exercise 7.7.4).

The reader should verify that the cross-product tensor defined in Eq. (7.140) is equal to

$$X = \mathbf{e}_1 \otimes \mathbf{e}_2 - \mathbf{e}_2 \otimes \mathbf{e}_1 \tag{7.166}$$

(see Exercise 7.7.3).

In general, the **outer tensor product** of two vectors is the second-rank contravariant tensor

$$\bar{x} \wedge \bar{y} := \bar{x} \otimes \bar{y} - \bar{y} \otimes \bar{x}. \tag{7.167}$$

This outer tensor product should not be confused with the dyadic product of two vectors defined in Section 9.1. One shows easily that the outer tensor product is antisymmetric:

$$\bar{x} \wedge \bar{y} = -\bar{y} \wedge \bar{x} \tag{7.168}$$

(see Exercise 7.7.8).

Basis of the Tensor Product Space

To find a basis of the tensor-product space $\mathcal{V}_{(1)} \otimes \mathcal{V}_{(2)}$, let $\{\bar{e}_{(1)i}\}$ and $\{\bar{e}_{(2)j}\}$ be bases of $\mathcal{V}_{(1)}$ and $\mathcal{V}_{(2)}$, respectively. By Eq. (7.162),

$$
\begin{aligned}
\left(\bar{x}_{(1)} \otimes \bar{x}_{(2)}\right)\left[\underline{\phi}_{(1)}, \underline{\phi}_{(2)}\right] &= \underline{\phi}_{(1)}\left[x_{(1)}^i \bar{e}_{(1)i}\right] \underline{\phi}_{(2)}\left[x_{(2)}^j \bar{e}_{(2)j}\right] \\
&= x_{(1)}^i x_{(2)}^j \underline{\phi}_{(1)}\left[\bar{e}_{(1)i}\right] \underline{\phi}_{(2)}\left[\bar{e}_{(2)j}\right] \tag{7.169} \\
&= x_{(1)}^i x_{(2)}^j \left(\bar{e}_{(1)i} \otimes \bar{e}_{(2)j}\right)\left[\underline{\phi}_{(1)}, \underline{\phi}_{(2)}\right].
\end{aligned}
$$

Therefore

$$\bar{x}_{(1)} \otimes \bar{x}_{(2)} = x_{(1)}^i x_{(2)}^j \bar{e}_{(1)i} \otimes \bar{e}_{(2)j}. \tag{7.170}$$

As one might expect, the tensor product of any two vectors can be expressed as a linear combination of the tensor products of the basis vectors.

It is important to understand that not every element of $\mathcal{V}_{(1)} \otimes \mathcal{V}_{(2)}$ is equal to the tensor product of an element of $\mathcal{V}_{(1)}$ and an element of $\mathcal{V}_{(2)}$. For example, if the cross product tensor X defined in Eq. (7.140) were equal to a tensor product $\mathbf{x}_{(1)} \otimes \mathbf{x}_{(2)}$, one would have

$$x^{11} = x_{(1)}^1 x_{(2)}^1 = 0 \quad \text{and} \quad x^{22} = x_{(1)}^2 x_{(2)}^2 = 0, \tag{7.171}$$

and therefore either

$$x_{(1)}^1 = 0 \quad \text{or} \quad x_{(2)}^1 = 0 \quad \text{or both.} \tag{7.172}$$

It follows that at least one of the components

$$x^{12} = x_{(1)}^1 x_{(2)}^2 = 1 \quad \text{and} \quad x^{21} = x_{(1)}^2 x_{(2)}^1 = -1 \tag{7.173}$$

must vanish. This contradiction establishes that X *cannot* be expressed as the tensor product of two vectors.

We show now that *every* bilinear functional T on $\mathcal{V}_{(1)}^* \times \mathcal{V}_{(2)}^*$ is equal to a linear combination of the tensor products of basis elements $\{\bar{e}_{(1)i} \otimes \bar{e}_{(2)j}\}$. By Eq. (7.162) and the definition of coordinate functionals,

$$\bar{e}_{(1)i} \otimes \bar{e}_{(2)j}\left[\underline{\epsilon}_{(1)}^k, \underline{\epsilon}_{(2)}^l\right] = \delta_i^k \delta_j^l. \tag{7.174}$$

Then, for every linear functional $\underline{\phi}_{(1)} \in V^*_{(1)}$ and for every linear functional $\underline{\phi}_{(2)} \in V^*_{(2)}$, one has

$$
\begin{aligned}
\mathsf{T}[\underline{\phi}_{(1)}, \underline{\phi}_{(2)}] &= t^{ij} \phi_{(1)i} \phi_{(2)j} \\
&= t^{ij} \delta^k_i \delta^l_j \phi_{(1)k} \phi_{(2)l} \\
&= t^{ij} \phi_{(1)k} \phi_{(2)l} \left(\overline{e}_{(1)i} \otimes \overline{e}_{(2)j} \right) \left[\underline{\epsilon}^k_{(1)}, \underline{\epsilon}^l_{(2)} \right] \\
&= t^{ij} \left(\overline{e}_{(1)i} \otimes \overline{e}_{(2)j} \right) \left[\underline{\phi}_{(1)}, \underline{\phi}_{(2)} \right].
\end{aligned}
\tag{7.175}
$$

Therefore

$$
\boxed{\mathsf{T} = t^{ij} \, \overline{e}_{(1)i} \otimes \overline{e}_{(2)j}.}
\tag{7.176}
$$

It follows that the set of tensor products of basis vectors, $\{\overline{e}_{(1)i} \otimes \overline{e}_{(2)j}\}$, spans the tensor-product space $V_{(1)} \otimes V_{(2)}$. One can show that the set $\{\overline{e}_{(1)i} \otimes \overline{e}_{(2)j}\}$ is linearly independent (see Exercise 7.7.5). Therefore $\{\overline{e}_{(1)i} \otimes \overline{e}_{(2)j}\}$ is a basis of $V_{(1)} \otimes V_{(2)}$.

To define an ordering of the tensor-product basis vectors $\{\overline{e}_{(1)i} \otimes \overline{e}_{(2)j}\}$, we choose the dictionary (lexicographic) order of the ordered pairs (i, j):

$$
(i, j) \prec (k, l) \text{ if } \begin{cases} i < k, \text{ or} \\ i = k \text{ and } j < l. \end{cases}
\tag{7.177}
$$

For example, if the dimensions of the vector spaces are $n_1 = 2$ and $n_2 = 3$, then the dictionary order of the pairs (i, j) is $(1, 1)$, $(1, 2)$, $(1, 3)$, $(2, 1)$, $(2, 2)$, $(2, 3)$.

One can show similarly that $\{\underline{\epsilon}^i_{(1)} \otimes \underline{\epsilon}^j_{(2)}\}$ is a basis of the space $V^*_{(1)} \otimes V^*_{(2)}$ of covariant second-rank tensors, and that $\{\overline{e}_{(1)i} \otimes \underline{\epsilon}^j_{(2)}\}$ is a basis of the space $V_{(1)} \otimes V^*_{(2)}$ of mixed second-rank tensors (see Exercises 7.7.6 and 7.7.7).

Because there are $n_1 n_2$ tensor-product basis vectors of the form $\{\overline{e}_{(1)i} \otimes \overline{e}_{(2)j}\}$, it follows that the dimension of the tensor-product space $V_{(1)} \otimes V_{(2)}$ is $n_1 n_2$. Therefore

$$
V_{(1)} \otimes V_{(2)} \cong \mathbb{F}^{n_1 n_2}.
\tag{7.178}
$$

This establishes that one can regard the tensor product of two vectors of dimensions n_1 and n_2 as a single vector of dimension $n_1 n_2$.

The set of antisymmetric second-rank contravariant tensors and the set of symmetric second-rank contravariant tensors are complementary vector spaces (see Exercise 7.7.9). For example, let $n = 2$ and consider $(\mathbb{C}^2)^{\otimes 2}$, which is the space of the spin-state vectors of two spin-$\frac{1}{2}$ particles in nonrelativistic quantum theory. The matrix

$$
\mathsf{S}_3 = \frac{1}{2}\sigma_3 = \frac{1}{2} \begin{pmatrix} 1 & 0 \\ 0 & -1 \end{pmatrix}
\tag{7.179}
$$

corresponds to the component along the z axis of the intrinsic angular momentum of a spin-$\frac{1}{2}$ particle. The eigenvectors of S_3 are the canonical basis vectors $\mathbf{e}_1 = (1, 0)^T$ (which represents a state in which the z-component of intrinsic angular momentum is equal to $\frac{1}{2}\hbar$), and $\mathbf{e}_2 = (0, 1)^T$ (which represents a state in which the z-component of intrinsic angular momentum is equal to $-\frac{1}{2}\hbar$). Because $(\mathbb{C}^2)^{\otimes 2}$ is equal to the direct sum of its symmetric and antisymmetric subspaces (see Exercise 7.7.9), it follows that every spin state of the two-particle system is equal to a linear combination of three symmetric and one antisymmetric spin states. The antisymmetric tensor that spans the subspace of antisymmetric tensors in $(\mathbb{C}^2)^{\otimes 2}$ is

$$T^{(0)} := \frac{1}{\sqrt{2}} \mathbf{e}_1 \wedge \mathbf{e}_2. \tag{7.180}$$

One can choose the three symmetric tensors that span the symmetric subspace of $(\mathbb{C}^2)^{\otimes 2}$ to be

$$
\begin{aligned}
T_1^{(1)} &:= \mathbf{e}_1 \otimes \mathbf{e}_1, \\
T_0^{(1)} &:= \tfrac{1}{2}[\mathbf{e}_1 \otimes \mathbf{e}_2 + \mathbf{e}_2 \otimes \mathbf{e}_1], \\
T_{-1}^{(1)} &:= \mathbf{e}_2 \otimes \mathbf{e}_2.
\end{aligned} \tag{7.181}
$$

The antisymmetric subspace of $(\mathbb{C}^2)^{\otimes 2}$ represents the spin-0 states of the two-particle system; the symmetric subspace represents the spin-1 states. The three symmetric tensors above represent, respectively, the spin-1 states in which the z-component of the total spin is equal to $+1, 0$, or -1.

7.7.4 TENSORS OF RANK m

One generalizes the definition of second-rank contravariant tensors by defining the tensor product

$$\bigotimes_{k=1}^{m} \mathcal{V}_{(k)} := \mathcal{V}_{(1)} \otimes \mathcal{V}_{(2)} \otimes \cdots \otimes \mathcal{V}_{(m)} \tag{7.182}$$

of m different vector spaces over the same number field, $\mathbb{F}$, as the set of all multilinear functionals T on $\mathcal{V}_{(1)}^* \times \mathcal{V}_{(2)}^* \times \cdots \times \mathcal{V}_{(m)}^*$. A **contravariant tensor of rank m** is any element of $\mathcal{V}_{(1)} \otimes \mathcal{V}_{(2)} \otimes \cdots \otimes \mathcal{V}_{(m)}$. One shows easily that $\mathcal{V}_{(1)} \otimes \mathcal{V}_{(2)} \otimes \cdots \otimes \mathcal{V}_{(m)}$ is a vector space. The vector space

$$\mathcal{V}^{\otimes m} := \underbrace{\mathcal{V} \otimes \mathcal{V} \otimes \ldots \otimes \mathcal{V}}_{m \text{ times}} \tag{7.183}$$

is called the **m-th tensor power** of the vector space $\mathcal{V}$.

The only step that is new with respect to second-rank tensors (the special case $m = 2$) is the observation that the iterated tensor product is associative, and therefore no parentheses are necessary (see the discussion on the associativity of Cartesian products in Section 2.2.4). For example, if $m = 3$ one has

$$\left(\mathcal{V}_{(1)} \otimes \mathcal{V}_{(2)}\right) \otimes \mathcal{V}_{(3)} \cong \mathcal{V}_{(1)} \otimes \left(\mathcal{V}_{(2)} \otimes \mathcal{V}_{(3)}\right) \tag{7.184}$$

in which $\cong$ stands for "is isomorphic to."

Generalizing Eq. (7.162), one defines the tensor product of m vectors drawn from vector spaces $\mathcal{V}_{(1)}, \ldots \mathcal{V}_{(m)}$ as

$$\left(\bigotimes_{k=1}^{m} \bar{x}_{(k)} \right) \left[\underline{\phi}_{(1)}, \ldots, \underline{\phi}_{(m)} \right] := \prod_{k=1}^{m} \underline{\phi}_{(k)} \left[\bar{x}_{(k)} \right]. \tag{7.185}$$

It is straightforward to show that

$$\left\{ \bigotimes_{k=1}^{m} \bar{e}_{(k)i_k} \right\} = \left\{ \bar{e}_{(1)i_1} \otimes \bar{e}_{(2)i_2} \cdots \otimes \bar{e}_{(m)i_m} \right\} \tag{7.186}$$

is a basis of the tensor-product space defined in Eq. (7.182) and that T is uniquely determined by its contravariant components,

$$t^{i_1 i_2 \ldots i_m} = \mathsf{T}\left[\underline{\epsilon}_{(1)}^{i_1}, \underline{\epsilon}_{(2)}^{i_2}, \ldots, \underline{\epsilon}_{(m)}^{i_m} \right]. \tag{7.187}$$

A tensor $\mathsf{S} \in \mathcal{V}^{\otimes m}$ is called **symmetric** if and only if, for all linear functionals in $\mathcal{V}^*$, and for all permutations π in the symmetric group of m objects,

$$\mathsf{S}[\underline{\phi}^{\pi 1}, \ldots, \underline{\phi}^{\pi m}] = \mathsf{S}[\underline{\phi}^1, \ldots, \underline{\phi}^m]. \tag{7.188}$$

A tensor $\mathsf{A} \in \mathcal{V}^{\otimes m}$ is called **antisymmetric** if and only if

$$\mathsf{A}[\underline{\phi}^{\pi 1}, \ldots, \underline{\phi}^{\pi m}] = \sigma(\pi) \mathsf{A}[\underline{\phi}^1, \ldots, [\underline{\phi}^m]. \tag{7.189}$$

For example, the determinant is an antisymmetric tensor D of rank $m = n$. The arguments of D are the rows of an $n \times n$ matrix.

The **outer tensor product** of m vectors belonging to $\mathcal{V}^{\otimes m}$,

$$\bar{x}_1 \wedge \bar{x}_2 \wedge \cdots \wedge \bar{x}_m := \sum \pi \in S_m \sigma(\pi) \bar{x}_1 \otimes \bar{x}_2 \otimes \cdots \otimes \bar{x}_m, \tag{7.190}$$

generalizes the definition of a rank-2 antisymmetric tensor given in Eq. (7.138).

7.7.5 LINEAR MAPPINGS OF TENSORS

Kronecker Product of Linear Mappings

Let $\mathsf{A}_{(1)} : \mathcal{V}_{(1)} \to \mathcal{W}_{(1)}, \ldots, \mathsf{A}_{(m)} : \mathcal{V}_{(m)} \to \mathcal{W}_{(m)}$ be linear mappings. The **Kronecker product** $\mathsf{A}_{(1)} \otimes \cdots \otimes \mathsf{A}_{(m)}$ is, by definition, the mapping such that

$$\forall \bar{x}_{(1)} \in \mathcal{V}_{(1)} : \cdots : \forall \bar{x}_{(m)} \in \mathcal{V}_{(m)} :$$
$$\left(\mathsf{A}_{(1)} \otimes \cdots \otimes \mathsf{A}_{(m)} \right) \left(\bar{x}_{(1)} \otimes \cdots \otimes \bar{x}_{(m)} \right) := \left(\mathsf{A}_{(1)} \bar{x}_{(1)} \right) \otimes \cdots \otimes \left(\mathsf{A}_{(m)} \bar{x}_{(m)} \right). \tag{7.191}$$

Because Eq. (7.186) gives a basis of $\mathcal{V}_{(1)} \otimes \mathcal{V}_{(2)} \otimes \cdots \otimes \mathcal{V}_{(m)}$, Eq. (7.191) defines the Kronecker product on every element of $\mathcal{V}_{(1)} \otimes \mathcal{V}_{(2)} \otimes \cdots \otimes \mathcal{V}_{(m)}$.

Let $\mathsf{T} \in \mathcal{V}_{(1)} \otimes \mathcal{V}_{(2)} \otimes \cdots \otimes \mathcal{V}_{(m)}$, and let $\{\bar{f}_{(k)j_k}\}$ be a basis of $\mathcal{W}_{(k)}$. Then there exist numbers $t^{i_1 i_2 \ldots i_m}$ such that

$$\mathsf{T} = t^{i_1 i_2 \ldots i_m} \bar{e}_{(1)i_1} \otimes \bar{e}_{(2)i_2} \cdots \otimes \bar{e}_{(m)i_m}. \tag{7.192}$$

Substituting this equation into Eq. (7.191) and using the definition of matrix elements of a linear mapping,

$$A_{(k)}\bar{e}_{(k)i_k} = a^{j_k}_{(k)i_k}\bar{f}_{(k)j_k},$$ (7.193)

one obtains

$$\left(A_{(1)} \otimes \cdots \otimes A_{(m)}\right)T = a^{j_1}_{(1)i_1} \cdots a^{j_m}_{(m)i_m} t^{i_1 i_2 \cdots i_m}\bar{f}_{(1)j_1} \otimes \cdots \otimes \bar{f}_{(m)j_m}.$$ (7.194)

In particular, if T is one of the basis tensors, one obtains

$$\left(A_{(1)} \otimes \cdots \otimes A_{(m)}\right)\bar{e}_{(1)i_1} \otimes \cdots \otimes \bar{e}_{(m)i_m} = a^{j_1}_{(1)i_1} \cdots a^{j_m}_{(m)i_m}\bar{f}_{(1)j_1} \otimes \cdots \otimes \bar{f}_{(m)j_m}.$$ (7.195)

Therefore the matrix elements of the Kronecker product linear mapping

$$A := A_{(1)} \otimes \cdots \otimes A_{(m)}$$ (7.196)

are

$$\boxed{a^{(j_1,\ldots,j_m)}_{(i_1,\ldots,i_m)} = a^{j_1}_{(1)i_1} \cdots a^{j_m}_{(m)i_m}.}$$ (7.197)

Note that the rows are indexed by the list $(j_1, \ldots, j_m)$; the columns are indexed by the list $(i_1, \ldots, i_m)$.

For example, let $m = 2$, and let A and B be $n \times n$ matrices. The elements of the Kronecker product matrix

$$C = A \otimes B$$ (7.198)

are

$$c^{(i,j)}_{(k,l)} = a^i_k b^j_l$$ (7.199)

according to Eq. (7.197). Because the heads of the row list (i, j) and the column list (k, l) both come from the matrix A, it follows that A controls the block structure of the Kronecker product matrix C.

Let us illustrate the block structure of the Kronecker product with 2×2 matrices. Taking

$$A = -i\sigma_2 = \begin{pmatrix} 0 & -1 \\ 1 & 0 \end{pmatrix}, \quad B = \sigma_1 = \begin{pmatrix} 0 & 1 \\ 1 & 0 \end{pmatrix},$$ (7.200)

one obtains

$$C = \begin{array}{c} \\ (11) \\ (12) \\ (21) \\ (22) \end{array} \begin{array}{cccc} (11) & (12) & (21) & (22) \\ \begin{pmatrix} 0 & 0 & 0 & -1 \\ 0 & 0 & -1 & 0 \\ 0 & 1 & 0 & 0 \\ 1 & 0 & 0 & 0 \end{pmatrix} \end{array} = \begin{pmatrix} 0 & -\sigma_1 \\ \sigma_1 & 0 \end{pmatrix}.$$ (7.201)

Similarly, one finds that

$$-i\sigma_2 \otimes \sigma_j = \begin{pmatrix} 0 & -\sigma_j \\ \sigma_j & 0 \end{pmatrix}. \tag{7.202}$$

for $j \in (1:3)$. The latter equation gives the three spacelike Dirac matrices in the Weyl representation.

We use the Kronecker product of linear mappings to express the components of a transformed tensor in terms of the components of the original tensor. Let the coordinate expansion of

$$\mathsf{T}' := \left(\mathsf{A}_{(1)} \otimes \cdots \otimes \mathsf{A}_{(m)}\right)\mathsf{T} \tag{7.203}$$

relative to the basis $\{\overline{f}_{(1)j_1} \otimes \cdots \otimes \overline{f}_{(m)j_m}\}$ of $\mathcal{W}_{(1)} \otimes \cdots \otimes \mathcal{W}_{(m)}$ be

$$\mathsf{T}' = t'^{j_1 j_2 \cdots j_m}\, \overline{f}_{(1)j_1} \otimes \cdots \otimes \overline{f}_{(m)j_m}. \tag{7.204}$$

Equating coefficients of the basis tensors $\overline{f}_{(1)j_1} \otimes \cdots \otimes \overline{f}_{(m)j_m}$ in Eqs. (7.194) and (7.204) yields

$$t'^{j_1 j_2 \cdots j_m} = a^{j_1}_{(1)i_1} \cdots a^{j_m}_{(m)i_m}\, t^{i_1 i_2 \cdots i_m}, \tag{7.205}$$

which is the law of transformation of tensor components under simultaneous linear mappings $\mathsf{A}_{(1)}, \ldots, \mathsf{A}_{(m)}$ of the spaces $\mathcal{V}_{(1)}, \ldots, \mathcal{V}_{(m)}$. Equation (7.205) is the usual definition of a contravariant rank-m tensor.

General Linear Mappings of $\mathcal{V}_{(1)} \otimes \cdots \otimes \mathcal{V}_{(m)}$

Not every linear mapping of the tensor-product space $\mathcal{V}_{(1)} \otimes \cdots \otimes \mathcal{V}_{(m)}$ can be expressed as a Kronecker product of m individual mappings carried out in $\mathcal{V}_{(1)}, \ldots, \mathcal{V}_{(m)}$. Let $\mathsf{A} : \mathcal{V}_{(1)} \otimes \cdots \otimes \mathcal{V}_{(m)} \to \mathcal{W}_{(1)} \otimes \cdots \otimes \mathcal{W}_{(m)}$ be a linear mapping. The images of the basis tensors $\overline{e}_{(1)i_1} \otimes \cdots \otimes \overline{e}_{(m)i_m}$ under A have coordinate expansions in $\mathcal{W}_{(1)} \otimes \cdots \otimes \mathcal{W}_{(m)}$,

$$\mathsf{A}\left(\overline{e}_{(1)i_1} \otimes \cdots \otimes \overline{e}_{(m)i_m}\right) = a^{j_1 \cdots j_m}_{i_1 \cdots i_m}\, \overline{f}_{(1)j_1} \otimes \cdots \otimes \overline{f}_{(m)j_m}. \tag{7.206}$$

The image

$$\mathsf{T}' = \mathsf{A}\mathsf{T}, \tag{7.207}$$

in which $\mathsf{T} \in \mathcal{V}_{(1)} \otimes \cdots \otimes \mathcal{V}_{(m)}$, also has a coordinate expansion relative to the basis $\{\overline{f}_{(1)j_1} \otimes \cdots \otimes \overline{f}_{(m)j_m}\}$ of $\mathcal{W}_{(1)} \otimes \cdots \otimes \mathcal{W}_{(m)}$. If the coordinates of T' are $t'^{j_1 j_2 \cdots j_m}$, then

$$\mathsf{T}' = \mathsf{A}\mathsf{T} = t'^{j_1 j_2 \cdots j_m}\, \overline{f}_{(1)j_1} \otimes \cdots \otimes \overline{f}_{(m)j_m}. \tag{7.208}$$

Therefore the components of a rank-m contravariant tensor transform according to the rule

$$t'^{j_1 j_2 \cdots j_m} = a^{j_1 \cdots j_m}_{i_1 \cdots i_m}\, t^{i_1 i_2 \cdots i_m} \tag{7.209}$$

under a general linear mapping A from $\mathcal{V}_{(1)} \otimes \cdots \otimes \mathcal{V}_{(m)}$ to $\mathcal{W}_{(1)} \otimes \cdots \otimes \mathcal{W}_{(m)}$.

The matrix elements $a_{i_1\ldots i_m}^{j_1\ldots j_m}$ of a general linear mapping of a tensor-product space defined in Eq. (7.206) can be arranged in a matrix, the rows of which are indexed by the lists $(j_1, \ldots, j_m)$, and the columns of which are indexed by the lists $(i_1, \ldots, i_m)$.

7.7.6 EXERCISES FOR SECTION 7.7

7.7.1   Show that if T is a second-rank tensor, then

$$T\big[\alpha \overline{x}_{(1)}, \alpha \overline{x}_{(2)}\big] = \alpha^2\, T\big[\overline{x}_{(1)}, \overline{x}_{(2)}\big]. \tag{7.210}$$

Conclude that a bilinear functional is also a homogeneous function of degree 2.

7.7.2   Let $\mathcal{V}_{(1)}$ and $\mathcal{V}_{(2)}$ be vector spaces over a field $\mathbb{F}$, in which $\mathbb{F} = \mathbb{R}$ or $\mathbb{F} = \mathbb{C}$. Demonstrate that $\mathcal{V}_{(1)} \otimes \mathcal{V}_{(2)}$ is a vector space over $\mathbb{F}$.

7.7.3   Verify that the cross-product tensor defined in Eq. (7.140) is equal to

$$X = e_1 \otimes e_2 - e_2 \otimes e_1. \tag{7.211}$$

7.7.4   Show that the tensor product of two vectors, $\overline{x}_{(1)} \otimes \overline{x}_{(2)}$, meets the definition of a second-rank tensor.

7.7.5   Use Eq. (7.148) to prove that the set of tensor products of basis vectors, $\{\overline{e}_{(1)i} \otimes \overline{e}_{(2)j}\}$, is linearly independent.

7.7.6   Show that $\{\underline{\epsilon}^i_{(1)} \otimes \underline{\epsilon}^j_{(2)}\}$ is a basis of the space $\mathcal{V}^*_{(1)} \otimes \mathcal{V}^*_{(2)}$ of covariant second-rank tensors.

7.7.7   Show that $\{\overline{e}_{(1)i} \otimes \underline{\epsilon}^j_{(2)}\}$ is a basis of the space $\mathcal{V}_{(1)} \otimes \mathcal{V}^*_{(2)}$ of mixed second-rank tensors.

7.7.8   Show that

$$\overline{x} \wedge \overline{y} = -\overline{y} \wedge \overline{x}. \tag{7.212}$$

7.7.9   This problem establishes the decomposition of the vector space of second-rank contravariant tensors into the direct sum of vector spaces of symmetric and antisymmetric contravariant second-rank tensors.

    (a) Show that the set of symmetric second-rank contravariant tensors is a vector space.

    (b) Show that the set of antisymmetric second-rank contravariant tensors is a vector space.

    (c) Show that the only second-rank contravariant tensor that is both symmetric and antisymmetric is equal to the null (zero) tensor.

    (d) Prove that the vector space of second-rank contravariant tensors is equal to the direct sum of the vector spaces of symmetric and antisymmetric second-rank tensors.

    (e) Show that the dimension of the antisymmetric subspace of $\mathcal{V}_{(1)} \otimes \mathcal{V}_{(2)}$ is equal to

$$\binom{n}{2} = \frac{n(n-1)}{2}.$$ (7.213)

(f) Show that the dimension of the symmetric subspace of $\mathcal{V}_{(1)} \otimes \mathcal{V}_{(2)}$ is equal to

$$n + \binom{n}{2} = \frac{n(n+1)}{2}.$$ (7.214)

(g) Using Eq. (7.176) and its analog for covariant tensors, prove that

$$\mathsf{T}[\mathsf{C}] = t^{ij} c_{ij}$$ (7.215)

in which T is a general second-rank contravariant tensor and C is a general second-rank covariant tensor.

7.7.10   Obtain the Kronecker product matrices

$$\boldsymbol{\Gamma}^{(E)}(g) \otimes \boldsymbol{\Gamma}^{(E)}(g)$$ (7.216)

for each element g of the group D_3. (See Eq. [4.185] for the matrices $\boldsymbol{\Gamma}^{(E)}(g)$.)

CHAPTER 8

INNER PRODUCTS AND NORMS

8.1 INNER-PRODUCT SPACES

8.1.1 DEFINITIONS

Some physically important quantities such as the energy density of an electromagnetic field, the total energy of a rotating solid body, or the total probability in quantum mechanics are scalar functions that are quadratic in the components of a vector and involve the dot product of that vector with another vector. For example, the electric-energy density in a linear dielectric is a real number that is proportional to

$$\mathbf{E} \cdot \mathbf{D} = \mathbf{E} \cdot (\epsilon \mathbf{E}) \tag{8.1}$$

in which ϵ is the dielectric-permittivity matrix. This equation defines a second-rank tensor according to the definition given in Section 7.7. A more elementary, and more useful, characterization of an inner product is as a generalization of the dot product in three-dimensional space. An inner product is linear in each of its two vector arguments in a real vector space or is linear in the second argument and antilinear in the first argument in a complex vector space. In this section we describe the most basic properties of inner products.

In Section 8.2 we show that inner products justify the use of familiar geometrical methods such as Pythagoras's theorem in quite general inner-product spaces. Section 8.3 uses the inner product to approximate a vector (or a function) as a linear combination of other vectors (or functions). Sections 8.4 and 8.5 use projection methods to obtain continuous and discrete least-squares approximations. Section 8.7 introduces the most important norms for both computation and theory in both finite-dimensional and infinite-dimensional inner-product spaces. In Section 8.8 we show that, in finite-dimensional vector spaces, every linear functional can be expressed as the inner product with a fixed vector.

Inner-Product Axioms

An **inner product** (or **scalar product** or **quadratic form**) is a functional $g : \mathcal{V} \times \mathcal{V} \to \mathbb{F}$ with two arguments, each drawn from the same vector space $\mathcal{V}$, and with values in the field $\mathbb{F}$, such that ($\forall \, \overline{x}, \overline{y}, \overline{z} \in \mathcal{V}$ and $\forall \alpha \in \mathbb{F}$)

$$g[\overline{x}, \overline{y} + \overline{z}] = g[\overline{x}, \overline{y}] + g[\overline{x}, \overline{z}], \tag{8.2}$$

$$g[\overline{x}, \alpha \overline{y}] = \alpha g[\overline{x}, \overline{y}], \tag{8.3}$$

and such that g is **symmetric**,

$$g[\bar{y}, \bar{x}] = g[\bar{x}, \bar{y}], \tag{8.4}$$

in a real vector space, or **Hermitian**,

$$g[\bar{y}, \bar{x}] = g[\bar{x}, \bar{y}]^*, \tag{8.5}$$

in a complex vector space. (Recall that when applied to a complex scalar, an asterisk indicates the complex conjugate, as is customary in physics and engineering.) Note that Eqs. (8.4) and (8.5) describe how the value of g changes (or does not change) if its arguments are interchanged.

For example, the ordinary dot product in real three-dimensional Euclidean space is an inner product, because $\mathbf{a} \cdot \mathbf{b} \in \mathbb{R}$, $\mathbf{a} \cdot (\mathbf{b} + \mathbf{c}) = \mathbf{a} \cdot \mathbf{b} + \mathbf{a} \cdot \mathbf{c}$, $\mathbf{a} \cdot (\beta \mathbf{b}) = \beta(\mathbf{a} \cdot \mathbf{b})$, and $\mathbf{a} \cdot \mathbf{b} = \mathbf{b} \cdot \mathbf{a}$.

It follows from Eq. (8.2) and either Eq. (8.4) or Eq. (8.5) that

$$g[\bar{x} + \bar{y}, \bar{z}] = g[\bar{x}, \bar{z}] + g[\bar{y}, \bar{z}]. \tag{8.6}$$

One deduces from Eq. (8.4) that if g is symmetric, then g is linear in each argument (with the other argument held fixed):

$$g[\alpha\bar{x}, \bar{y}] = \alpha g[\bar{x}, \bar{y}] = g[\bar{x}, \alpha\bar{y}]. \tag{8.7}$$

However, it follows from Eqs. (8.3) and (8.5) that if g is Hermitian, then g is linear in the second argument (with the first argument held fixed),

$$g[\bar{x}, \alpha\bar{y}] = \alpha g[\bar{x}, \bar{y}], \tag{8.8}$$

and antilinear in the first argument (with the second argument held fixed),

$$g[\alpha\bar{x}, \bar{y}] = \alpha^* g[\bar{x}, \bar{y}]. \tag{8.9}$$

A vector space that is equipped with an inner product that satisfies Eqs. (8.2) and (8.3) and either Eq. (8.4) or Eq. (8.5) is called an **inner-product space**. An inner product that obeys Eqs. (8.3), (8.4), and (8.7) is called **bilinear** because it is linear in both arguments. A real vector space that is equipped with a symmetric inner product is called a **real inner-product space**. Clearly a bilinear inner product is a bilinear functional on $\mathcal{V} \times \mathcal{V}$ and is therefore a covariant second-rank tensor (see Section 7.7). We have more to say in subsequent sections about the tensor properties of inner products.

An inner product that obeys Eqs. (8.3), (8.5), and (8.9) is called **sesquilinear** (meaning *half-linear*). A complex vector space that is equipped with a Hermitian inner product is called a **complex inner-product space**. For many purposes it is convenient to consider a sesquilinear inner product as a covariant second-rank tensor, even though it does not strictly satisfy the definition.

Notation for Inner Products

It is customary to write the inner product of $\bar{x}$ and $\bar{y}$ as an ordered pair enclosed between delimiters of some sort. In other words, for this particular kind of functional one usually omits the name of the functional (such as g in the previous section) and writes only its arguments. Popular delimiter choices include parentheses $(\bar{x}, \bar{y})$ and square brackets $[\bar{x}, \bar{y}]$. In physics square brackets usually denote the commutator of two operators, $[A, B] = AB - BA$, and therefore would be a poor choice for the inner product. Using parentheses for the inner product may cause confusion with the standard notation for an open interval, $(a, b) = \{x \in \mathbb{R} \mid a < x < b\}$. This book uses the notation

$$\langle \bar{x}, \bar{y} \rangle := g[\bar{x}, \bar{y}], \tag{8.10}$$

because it is common in engineering and because the angle brackets suggest Dirac's inner-product notation $\langle \psi \mid \phi \rangle$, which is widely used in physics. See the notes at the end of the chapter for further comments on Dirac's notation.

Comments on the Inner-Product Axioms

The axioms given in Eqs. (8.4) and (8.5) are not at all arbitrary, although they may seem so at first sight. The reason why only inner products that satisfy Eqs. (8.4) and (8.5) are likely to have physical significance is that often one wants to interpret $g[\bar{x}, \bar{y}]$ geometrically, as we do in Section 8.2. If the number field is $\mathbb{R}$, then one normally wants to interpret $g[\bar{x}, \bar{y}]$ as being proportional to the cosine of the angle between $\bar{x}$ and $\bar{y}$, in analogy with the familiar relation $\mathbf{a} \cdot \mathbf{b} = ab \cos \theta$. Because the angle between $\bar{x}$ and $\bar{y}$ must be the same as the angle between $\bar{y}$ and $\bar{x}$ (otherwise "angle" would make no sense), g must be symmetric in a real inner-product space. Even if the scalars are complex numbers one can use $g[\bar{x}, \bar{x}]$ to represent an inherently real quantity such as energy, because axiom Eq. (8.5) guarantees that $g[\bar{x}, \bar{x}]$ is equal to $g[\bar{x}, \bar{x}]^*$ and is therefore real.

8.1.2 CANONICAL INNER PRODUCTS

The standard or simplest form of a mathematical object is often called *canonical*. For example, the canonical form of the equation of a circle (in $\mathbb{R}^2$) is

$$x^2 + y^2 = 1. \tag{8.11}$$

In this section we define the canonical inner products on several spaces of fundamental importance in physics and engineering.

Space of Complex Numbers

In the vector space $\mathbb{C}$ over the real field $\mathbb{R}$, the **canonical inner product** of $z = x + iy$ and $z' = x' + iy'$ is

$$\begin{aligned}\langle z', z \rangle &:= \mathrm{Re}[z'^* z] \\ &= x'x + y'y.\end{aligned} \tag{8.12}$$

Then $\langle z, z \rangle = x^2 + y^2 = |z|^2$. Note that $\langle z, z' \rangle = \langle z', z \rangle$; therefore the canonical inner product on $\mathbb{C}$ is symmetric. Because $|z|^2 \geq 0$ and $|z|^2 = 0 \Leftrightarrow z = 0$, it follows that the canonical inner product in $\mathbb{C}$ is positive-definite (see Section 8.1.3).

Spaces of Column Vectors

The **canonical inner product in** $\mathbb{R}^n$ is

$$\mathbf{x} \cdot \mathbf{y} := \mathbf{x}^T \mathbf{y}$$

$$= \sum_{j=1}^{n} x^j y^j,$$

(8.13)

which generalizes the ordinary dot product in the simplest possible way. For example, the canonical inner product in $\mathbb{R}^4$ of the vectors $\mathbf{x} = (1, 2, 3, 4)^T$ and $\mathbf{y} = (1, -1, -1, 1)^T$ is $\mathbf{x}^T \mathbf{y} = 1 \cdot 1 + 2(-1) + 3(-1) + 4 \cdot 1 = 0$.

The canonical inner product in $\mathbb{R}^n$ is symmetric,

$$\mathbf{y} \cdot \mathbf{x} = \mathbf{x} \cdot \mathbf{y},$$

(8.14)

and is also positive-definite (see Section 8.1.3). The real vector space $\mathbb{R}^n$ equipped with this inner product is written $\mathbb{E}^n$ and is called **Euclidean n-space**. We reserve the dot-product notation for the canonical inner products in Euclidean n-space $\mathbb{E}^n$ and Minkowski space $\mathcal{M}$.

The **canonical inner product in** $\mathbb{C}^n$ is

$$\langle \mathbf{x}, \mathbf{y} \rangle := \mathbf{x}^\dagger \mathbf{y}$$

$$= \mathbf{x}^* \cdot \mathbf{y}$$

$$= \sum_{j=1}^{n} x^{j*} y^j.$$

(8.15)

Note that this inner product is Hermitian:

$$\langle \mathbf{y}, \mathbf{x} \rangle = \langle \mathbf{x}, \mathbf{y} \rangle^*.$$

(8.16)

The complex vector space $\mathbb{C}^n$ under its canonical inner product is usually denoted $\mathbb{U}^n$ and is called **unitary n-space**.

We define other, noncanonical inner products on $\mathbb{R}^n$ and $\mathbb{C}^n$ in Section 8.1.3.

Application to finite, discrete quantum systems $\mathbb{U}^n$ is the state space of a finite, discrete quantum system (see Eqs. [5.23] and [5.26]). For example, let $\overline{x}$ be a state vector. According to Eq. (8.15), the normalization condition Eq. (5.26) is equivalent to the self–inner product

$$\langle \overline{x}, \overline{x} \rangle = 1$$

(8.17)

in $\mathbb{U}^n$. If $\overline{x}$ and $\overline{y}$ are two normalized state vectors, then the probability of finding the system in the state $\overline{x}$ if it has been prepared in the state $\overline{y}$ is (according to one of the postulates of quantum mechanics)

$$P(\bar{x} \mid \bar{y}) := |\langle \bar{x}, \bar{y} \rangle|^2. \tag{8.18}$$

The inner product $\langle \bar{x}, \bar{y} \rangle$ of any two normalized state vectors is called a **probability amplitude**. Equation (8.18) postulates a fundamental physical significance for inner products. We see in a subsequent section that the polarization identity, Eq. (8.145), implies the rather surprising result that all transition probabilities are determined by the set of all (unnormalized) self–inner products.

Matrix Spaces

Let **A** and **B** be any two matrices in the vector space $\mathbb{R}^{n \times m}$. The **canonical inner product in** $\mathbb{R}^{n \times m}$ is

$$\langle \mathbf{A}, \mathbf{B} \rangle := \text{trace}[\mathbf{A}^T \mathbf{B}] = \sum_{i=1}^{m} \sum_{j=1}^{n} a_i^j b_i^j \tag{8.19}$$

in which the trace of a matrix is defined in Eq. (6.295). Evidently this is the inner product that one obtains by first mapping **A** and **B** to vectors as in Eq. (5.190) or Eq. (5.192) and then using the canonical inner product in $\mathbb{E}^{mn}$. Note that the inner product Eq. (8.19) is symmetric,

$$\langle \mathbf{B}, \mathbf{A} \rangle = \langle \mathbf{A}, \mathbf{B} \rangle, \tag{8.20}$$

because the trace of the transpose of a matrix is equal to the trace of the matrix: $\text{trace}[\mathbf{A}^T \mathbf{B}] = \text{trace}[(\mathbf{A}^T \mathbf{B})^T] = \text{trace}[\mathbf{B}^T \mathbf{A}]$.

The **canonical inner product in** $\mathbb{C}^{n \times m}$ is

$$\langle \mathbf{A}, \mathbf{B} \rangle := \text{trace}[\mathbf{A}^\dagger \mathbf{B}] = \sum_{i=1}^{m} \sum_{j=1}^{n} a_i^{j*} b_i^j. \tag{8.21}$$

This inner product is Hermitian,

$$\langle \mathbf{B}, \mathbf{A} \rangle = \langle \mathbf{A}, \mathbf{B} \rangle^*, \tag{8.22}$$

by virtue of the properties of the trace and the definition

$$\mathbf{A}^\dagger = \mathbf{A}^{T*}. \tag{8.23}$$

For example, it is easy to verify that the Pauli matrices $\sigma_1, \sigma_2, \sigma_3 \in \mathbb{C}^{2 \times 2}$ defined in Eq. (5.149) have the properties

$$\sigma_i^\dagger = \sigma_i,$$

$$\text{trace}[\sigma_i] = 0, \tag{8.24}$$

$$\sigma_i \sigma_j = \begin{cases} i\sigma(\pi)\sigma_k = -\sigma_j \sigma_i, & \text{if } j \neq i; \\ 1, & \text{if } j = i. \end{cases}$$

in which π is the permutation that maps the list $(1, 2, 3)$ to the list (i, j, k). Then the canonical inner product of two Pauli matrices is

$$\begin{aligned}
\langle \sigma_i, \sigma_j \rangle &= \mathrm{trace}[\sigma_i^\dagger \sigma_j] \\
&= \mathrm{trace}[\sigma_i \sigma_j] \\
&= \begin{cases} \sigma(\pi)\,\mathrm{trace}[\sigma_k] = 0, & \text{if } j \neq i; \\ \mathrm{trace}[\mathbf{1}] = 2, & \text{if } j = i. \end{cases}
\end{aligned} \tag{8.25}$$

8.1.3 METRIC TENSOR

Definitions

The n^2 inner products, corresponding to all possible ordered pairs of basis vectors,

$$g_{ij} := g[\overline{e}_i, \overline{e}_j] = \langle \overline{e}_i, \overline{e}_j \rangle, \tag{8.26}$$

are the elements of the **Gram matrix**

$$\mathbf{G} := [g_{ij}] \tag{8.27}$$

of the basis $\{\overline{e}_i\}$.[1] We explain elsewhere why $\mathbf{G}$ often is called the *metric tensor*.

For example, if $\mathcal{V} = \mathbb{R}^3$ and if the inner product is the ordinary dot product, then the Gram matrix of the Cartesian basis $\mathbf{i}, \mathbf{j}, \mathbf{k}$ is

$$\mathbf{G} = [g_{ij}] = \begin{pmatrix} 1 & 0 & 0 \\ 0 & 1 & 0 \\ 0 & 0 & 1 \end{pmatrix}. \tag{8.28}$$

For a less trivial example, let $\mathcal{V} = \mathbb{U}^n$ and let $\mathbf{F}$ be the matrix with columns $\mathbf{f}_i$,

$$\mathbf{F} = (\mathbf{f}_1, \dots, \mathbf{f}_n), \tag{8.29}$$

in which $\{\mathbf{f}_1, \dots, \mathbf{f}_n\}$ is a basis of $\mathbb{U}^n$. Then

$$\mathbf{G} = \mathbf{F}^\dagger \mathbf{F} \tag{8.30}$$

is the Gram matrix of the basis $\{\mathbf{f}_1, \dots, \mathbf{f}_n\}$. This equation expresses the elements of $\mathbf{G}$ in terms of the contravariant components of the basis vectors. Obviously, in $\mathbb{E}^n$ the Gram matrix is equal to

$$\mathbf{G} = \mathbf{F}^T \mathbf{F} \tag{8.31}$$

in which $\mathbf{F}$ is the matrix of basis vectors.

---

[1] The Gram matrix is named after the Danish mathematician Jörgen Pedersen Gram.

To see an example from differential geometry, let r, θ, ϕ be the spherical polar coordinates of a point in $\mathbb{R}^3$. Then an infinitesimal displacement $d\mathbf{s}$ can be expressed as

$$d\mathbf{s} = dr(\hat{\mathbf{r}}) + d\theta(r\hat{\boldsymbol{\theta}}) + d\phi(r \sin\theta\,\hat{\boldsymbol{\phi}}). \tag{8.32}$$

If one regards dr, $d\theta$ and $d\phi$ as the contravariant components of a vector relative to the basis $\{\hat{\mathbf{r}}, r\hat{\boldsymbol{\theta}}, r \sin\theta\hat{\boldsymbol{\phi}}\}$ of the tangent space, then the metric tensor is

$$\mathbf{G} = \begin{pmatrix} 1 & 0 & 0 \\ 0 & r^2 & 0 \\ 0 & 0 & r^2 \sin^2\theta \end{pmatrix}. \tag{8.33}$$

For example,

$$g_{r,\theta} = \hat{\mathbf{r}} \cdot (r\hat{\boldsymbol{\theta}}) = 0 \tag{8.34}$$

and

$$g_{\theta,\theta} = (r\hat{\boldsymbol{\theta}}) \cdot (r\hat{\boldsymbol{\theta}}) = r^2. \tag{8.35}$$

Tensor Properties

The properties of bilinearity or sesquilinearity, Eqs. (8.4) and (8.5), and the definition of the elements of the Gram matrix imply that the inner product $\langle \bar{x}, \bar{y} \rangle$ in $\mathcal{V}$ can be expressed in terms of the $\{g_{ij}\}$ and the contravariant components of both $\bar{x}$ and $\bar{y}$:

$$\langle \bar{x}, \bar{y} \rangle = \langle x^i \bar{e}_i, y^j \bar{e}_j \rangle = \begin{cases} g_{ij}\, x^i y^j & \text{if } \mathbb{F} = \mathbb{R}; \\ g_{ij}\, x^{i*} y^j & \text{if } \mathbb{F} = \mathbb{C}. \end{cases} \tag{8.36}$$

It is obvious from this equation that $\mathbf{G}$ is a second-rank covariant tensor (see Section 7.7). Because the inner product of a vector $\bar{x}$ with itself often is defined as the square of the length of $\bar{x}$, one calls $\mathbf{G}$ the **metric tensor**.

Equation (8.36) is equivalent to the matrix equations

$$\langle \bar{x}, \bar{y} \rangle = \langle \mathbf{x}, \mathbf{y} \rangle = \begin{cases} x^i g_{ij} y^j = \mathbf{x}^T \mathbf{G}\mathbf{y} & \text{if } \mathbb{F} = \mathbb{R}, \text{ or} \\ x^{i*} g_{ij} y^j = \mathbf{x}^\dagger \mathbf{G}\mathbf{y} & \text{if } \mathbb{F} = \mathbb{C}, \end{cases} \tag{8.37}$$

in which

$$\mathbf{x} := \begin{pmatrix} x^1 \\ \vdots \\ x^n \end{pmatrix} = M\bar{x} \tag{8.38}$$

is the column vector onto which $\bar{x}$ is mapped under the vector isomorphism $M : \bar{e}_i \mapsto \mathbf{e}_i$ defined in Eq. (5.202).

According to Eq. (8.26), an inner product on a vector space $\mathcal{V}$ and a basis $\{\bar{e}_i\}$ of $\mathcal{V}$ determine a unique Gram matrix. Conversely, a basis $\{\bar{e}_i\}$ of a vector space $\mathcal{V}$ and an $n \times n$ matrix $\mathbf{G} = [g_{ij}]$ determine a unique inner product on $\mathcal{V}$, provided that $\mathbf{G}$ is symmetric if $\mathbb{F} = \mathbb{R}$ or Hermitian if $\mathbb{F} = \mathbb{C}$:

$$g_{ij} = \langle \bar{e}_i, \bar{e}_j \rangle = \begin{cases} \langle \bar{e}_j, \bar{e}_i \rangle = g_{ji} \Rightarrow \mathbf{G}^T = \mathbf{G} & \text{if } \mathbb{F} = \mathbb{R}; \\ \langle \bar{e}_j, \bar{e}_i \rangle^* = g_{ji}^* \Rightarrow \mathbf{G}^\dagger = \mathbf{G} & \text{if } \mathbb{F} = \mathbb{C}. \end{cases} \tag{8.39}$$

It follows that if a square matrix (or a second-rank covariant tensor) $\mathbf{G}$ satisfies Eq. (8.39), then the bilinear or sesquilinear functional defined in Eq. (8.36) satisfies Eq. (8.4) or Eq. (8.5) and is therefore an inner product.

For example, the relationship between the angular-momentum vector $\mathbf{L}$ and the angular-velocity vector $\boldsymbol{\omega}$ of a rigid body is

$$\mathbf{L} = \mathbf{I}\boldsymbol{\omega} \tag{8.40}$$

where $\mathbf{I}$ is the moment-of-inertia tensor. One knows from the physical definition of the moments and products of inertia that $\mathbf{I}$ is a real, symmetric matrix. Therefore $\mathbf{I}$ can be used as a Gram matrix to define an inner product. The kinetic energy of a rigid body is then

$$T = \tfrac{1}{2}\boldsymbol{\omega} \cdot \mathbf{L} = \tfrac{1}{2}\boldsymbol{\omega} \cdot (\mathbf{I}\boldsymbol{\omega}) = \tfrac{1}{2}\langle \boldsymbol{\omega}, \boldsymbol{\omega} \rangle \tag{8.41}$$

in which we have introduced the inner product $\langle \boldsymbol{\omega}', \boldsymbol{\omega} \rangle := \boldsymbol{\omega}' \cdot (\mathbf{I}\boldsymbol{\omega})$.

Positive-Definite Inner Products

If an inner-product space $\mathcal{V}$ is such that

$$\forall \bar{x} \in \mathcal{V} : \langle \bar{x}, \bar{x} \rangle \geq 0, \tag{8.42}$$

then the inner product is called **positive-semidefinite**. If, in addition, the inner product is such that

$$\langle \bar{x}, \bar{x} \rangle = 0 \Leftrightarrow \bar{x} = \bar{0}, \tag{8.43}$$

then the inner product is called **positive-definite**. If an inner product is positive-definite, then

$$\forall \bar{x} \neq \bar{0} : \langle \bar{x}, \bar{x} \rangle > 0, \tag{8.44}$$

for if the inner product $\langle \bar{x}, \bar{x} \rangle$ vanishes, then $\bar{x} = \bar{0}$. An inner-product space in which Eqs. (8.42) and (8.43) hold is called a **Euclidean space** if $\mathbb{F} = \mathbb{R}$, or a **unitary space** if $\mathbb{F} = \mathbb{C}$. For example, $\mathbb{E}^n$ and $\mathbb{R}^{n \times m}$ (under the canonical inner products) are Euclidean spaces. $\mathbb{U}^n$ and $\mathbb{C}^{n \times m}$ (under the canonical inner products) are unitary spaces.

If $\langle \bar{x}, \bar{x} \rangle$ can be zero or negative for some nonzero vectors $\bar{x}$, as in relativity theory, then the inner product is called **indefinite**.

If one draws an analogy with three-dimensional Euclidean space, in which the square of the length of a vector $\mathbf{a}$ is $\mathbf{a} \cdot \mathbf{a}$, one sees that the natural definition of the length of a vector in a general Euclidean or unitary space is

$$\|\mathbf{x}\|_{ip} := \langle \mathbf{x}, \mathbf{x} \rangle^{1/2}, \tag{8.45}$$

which is called the **inner-product norm** of $\overline{x}$. We use the notation $\|\overline{x}\|_{ip}$, which should be read "inner-product norm of $\overline{x}$," for the sake of consistency with more general definitions of vector length in Section 8.7.

In quantum physics, one usually calculates an inner-product norm only for the purpose of normalizing a state vector (see Eq. [8.17]). The interpretation of a norm as a distance is a much more far-reaching concept, because one needs to compute distances between vectors in fields ranging from functional analysis to computation to communications theory.

We have already shown that every $n \times n$ matrix $\mathbf{G}$ that satisfies Eq. (8.39) is the Gram matrix of some inner product. A necessary condition for the inner product defined by $\mathbf{G}$ to satisfy Eq. (8.43) is that $\mathbf{G}$ must be nonsingular, for if $\mathbf{G}\mathbf{x} = \mathbf{0}$ for some nonzero vector $\mathbf{x}$, then $\mathbf{x}^T \mathbf{G}\mathbf{x} = 0$, implying that $\mathbf{G}$ is not positive-definite. Therefore a necessary condition for positive-definiteness is that the **Gramian** (or **Gram determinant**) $\det[\mathbf{G}]$ must not vanish:

$$\det[\mathbf{G}] \neq 0. \tag{8.46}$$

However, the example of the inner product of special relativity given in Section 8.1.4 shows that a nonsingular Gram matrix can define an indefinite inner product. A sufficient condition for the positive-definiteness of $\mathbf{G}$ is that the eigenvalues of $\mathbf{G}$ all must be positive (see Section 9.3).

Computation of Finite-Dimensional Inner Products

The computation of inner products is a key step in many algorithms of numerical linear algebra. Because computations of the canonical inner products on $\mathbb{E}^n$, $\mathbb{C}^n$, or $\mathbb{F}^{n \times n}$ may require subtractions, they are potentially subject to loss of accuracy through the catastrophic cancellation of significant digits. For this reason it is desirable to accumulate the sum indicated in Eq. (8.13) (for example) with a larger number of significant digits than in the working floating-point representation, if this is possible in hardware. The accumulation of inner products is one of the applications that can benefit from the 80-bit-fraction extended-precision format provided in hardware implementations of the IEEE standards for floating-point arithmetic.

For example, consider the computation of the canonical inner product of the $\mathbb{R}^3$ vectors

$$\mathbf{x} = \begin{pmatrix} 1.00000_2 \times 2^3 \\ 1.00000_2 \times 2^{-2} \\ 1.00000_2 \times 2^3 \end{pmatrix}, \quad \mathbf{y} = \begin{pmatrix} 1.00000_2 \times 2^3 \\ 1.00000_2 \times 2^{-2} \\ -1.00000_2 \times 2^3 \end{pmatrix} \tag{8.47}$$

in a hypothetical floating-point representation for which $\beta = 2$ and $p = 6$. The inner product is

$$
\begin{aligned}
\text{round}(\mathbf{x}^T \mathbf{y}) &= \text{round}(\text{round}(\text{round}(1.00000_2 \times 2^6 + 1.00000_2 \times 2^{-4}) \\
&\quad - 1.00000_2 \times 2^6)) \\
&= \text{round}(1.00000_2 \times 2^6 - 1.00000_2 \times 2^6) \\
&= 0.00000_2,
\end{aligned}
\tag{8.48}
$$

assuming that two extra digits are kept in computations of sums. The correct inner product is $1.00000_2 \times 2^{-4}$; therefore the relative error in the computation is 1. A computation with $p = 12$ significant bits produces the exact result.

It can be shown that an error bound for the computation of an inner product is

$$
|\text{round}(\mathbf{x}^T \mathbf{y}) - \mathbf{x}^T \mathbf{y}| \lesssim 2 \|\mathbf{x}\|_{\text{ip}} \|\mathbf{y}\|_{\text{ip}} \epsilon_{\text{mach}}
\tag{8.49}
$$

in which $\|\mathbf{x}\|_{\text{ip}} := (\mathbf{x}^T \mathbf{x})^{1/2}$ is the inner-product norm defined in Eq. (8.45) and $\epsilon_{\text{mach}} = \beta^{1-p}$. In the example above, $\epsilon_{\text{mach}} = 2^{-5}$ and $\|\mathbf{x}\|_{\text{ip}} = \|\mathbf{y}\|_{\text{ip}} \approx \sqrt{2}\, 2^3$. Therefore $\|\mathbf{x}\|_{\text{ip}} \|\mathbf{y}\|_{\text{ip}} \approx 2^7$, implying that the actual error found in this example is smaller by a factor of 2^7 than the error bound given in Eq. (8.49). One finds in practice that Eq. (8.49) usually overestimates the error.

Sequence Spaces

The natural generalization of the canonical inner products on $\mathbb{C}^n$ or $\mathbb{R}^n$ to a general sequence space fails because the components of a general element of $\mathbb{F}^{\mathbb{Z}^+}$ do not decrease in magnitude as the sequence index n becomes large without bound; hence the sequence of partial sums of a series such as $\sum_1^\infty a_i^* b_i$ may fail to converge. However, it is possible to generalize the canonical inner product on $\mathbb{C}^n$ such that the sequence space $\mathcal{L}^2(\mathbb{C})$ formally resembles $\mathbb{C}^n$. The **canonical inner product on $l^2(\mathbb{C})$** is

$$
\langle \bar{a}, \bar{b} \rangle := \sum_{i=1}^\infty a_i^* b_i,
\tag{8.50}
$$

which is Hermitian by inspection. The convergence of infinite sums such as Eq. (8.21) is discussed in Chapter 10.

$\mathbb{C}^n$ and $\mathcal{L}^2$ Spaces

The **canonical inner product on $\mathbb{C}^0$** ([a, b]; $\mathbb{C}$),

$$
\langle f, g \rangle := \int_a^b f(x)^* g(x)\, dx,
\tag{8.51}
$$

makes $\mathbb{C}^0([a, b]; \mathbb{C})$ into a unitary space. $\mathbb{C}^0([a, b]; \mathbb{C})$ is also a unitary space under the more general inner product

$$\langle f, g \rangle := \int_a^b f(x)^* \, g(x) \, w(x) \, dx, \tag{8.52}$$

provided that the weight function w is real and strictly positive on the interval (a, b). Exercise 8.1.4 demonstrates that the inner product defined by Eq. (8.52) is positive-definite.

The **canonical inner product on** $\mathcal{L}_w^2([a, b]; \mathbb{C})$ is also defined by Eq. (8.52), except that to be perfectly correct one should replace the weight function w with a strictly positive measure and the Riemann integral with a Lebesgue integral.

Nearly all of the special functions of mathematical physics belong to unitary spaces with inner products of the form Eq. (8.52). The weight function w and the interval (a, b) are generally different for each family of special functions.

The formula

$$\langle f, g \rangle := \sum_{m=0}^n \int_a^b f^{(m)}(x) \, g^{(m)}(x) \, w_m(x) \, dx, \tag{8.53}$$

in which $f^{(m)} := d^m f/dx^m$ is the m-th derivative of f and all of the functions w_m are assumed to be real and strictly positive on the interval (a, b), defines a Euclidean inner product on $C^n([a, b]; \mathbb{R})$. Inner products of this type are useful for studying solutions of differential equations, because the inner product $\langle f - g, f - g \rangle$ is small (and therefore $f \approx g$) only if the difference $f - g$ is a fairly smooth function in the sense that its derivatives through order n must be small in magnitude.

Spaces of Sampled Functions

In Section 6.1.5 we introduced the concept of a sampling mapping, which maps an infinite-dimensional function space into $\mathbb{F}^n$ by mapping a function f onto an n-component vector of sampled values. The **uniform-sampling mapping** $S_N : \mathcal{C}^0([0, 1); \mathbb{C}) \to \mathbb{C}^N$ is the mapping such that

$$f \mapsto S_N f := \begin{pmatrix} f(x_0) \\ \vdots \\ f(x_{N-1}) \end{pmatrix}, \tag{8.54}$$

in which the sample points are equally spaced:

$$\forall k \in (0 : N - 1) : x_k := \frac{k}{N}. \tag{8.55}$$

An inner product on $\mathbb{C}^N$ that approximates the canonical inner product on $\mathcal{C}^0([0, 1]; \mathbb{C})$ defined in Eq. (8.51) is

$$\langle S_N f, S_N g \rangle := \frac{1}{N} \sum_{k=0}^{N-1} f(x_k)^* \, g(x_k). \tag{8.56}$$

We call this the **canonical inner product of uniformly sampled functions**.

Equation (8.56) defines a kind of average. In particular, $\langle S_N f, S_N f \rangle$ is equal to the mean of $|f|^2$, calculated on the sample points x_k:

$$\langle S_N f, S_N f \rangle = \frac{1}{N} \sum_{k=0}^{N-1} |f(x_k)|^2. \tag{8.57}$$

A more general inner product that is useful if the sampling points are not uniformly distributed as in Eq. (8.55) is

$$\langle S_N f, S_N g \rangle := \frac{1}{\sum_{l=0}^{N-1} w_l} \sum_{k=0}^{N-1} w_k \, f(x_k)^* \, g(x_k). \tag{8.58}$$

8.1.4 INDEFINITE INNER PRODUCTS

If the condition

$$\forall \overline{y} \in \mathcal{V} : \langle \overline{x}, \overline{y} \rangle = 0 \Leftrightarrow \overline{x} = \overline{0} \tag{8.59}$$

holds in an inner-product space $\mathcal{V}$, then $\mathcal{V}$ is called **pseudo-Euclidean** (if $\mathbb{F} = \mathbb{R}$), or **pseudo-unitary** (if $\mathbb{F} = \mathbb{C}$). Evidently Euclidean or unitary spaces are special cases of pseudo-Euclidean or pseudo-unitary spaces. Equation (8.59) defines a much weaker condition than definiteness, Eq. (8.43), because it is possible for an indefinite inner product to satisfy Eq. (8.59).

We show in Section 8.8.2 that requiring that Eq. (8.59) must hold is equivalent to requiring that **G** must be the matrix of a nonsingular linear mapping.

Mathematicians usually define all inner-product spaces to have properties (8.42) and (8.43). For the purposes of physics such a definition is not broad enough to be useful, because vectors that violate Eqs. (8.42) and (8.43) are of fundamental importance in relativity theory. Therefore we consider as inner products functionals that satisfy the weaker condition Eq. (8.59).

Minkowski Space

The space-time vectors of special relativity are vectors $x \in \mathbb{R}^4$ with the special identifications $x^0 := ct$, $x^1 := x$, $x^2 := y$, and $x^3 := z$, in which c is the speed of light, t is the time, and x, y, z are Cartesian coordinates in three-dimensional space. We use bold italic lowercase letters to represent vectors that we interpret in terms of special relativity.

The inner product in special relativity is

$$x \cdot y := x^0 y^0 - (x^1 y^1 + x^2 y^2 + x^3 y^3). \tag{8.60}$$

The metric tensor (relative to the canonical basis in $\mathbb{R}^4$) induced by the inner product

Eq. (8.60) is

$$G = [g_{\mu\nu}] = \begin{pmatrix} 1 & 0 & 0 & 0 \\ 0 & -1 & 0 & 0 \\ 0 & 0 & -1 & 0 \\ 0 & 0 & 0 & -1 \end{pmatrix}. \tag{8.61}$$

Because a displacement in time has a positive self–inner product, this metric is sometimes called "the metric of the future."

The self–inner product

$$x \cdot x = (x^0)^2 - (x^1)^2 - (x^2)^2 - (x^3)^2 \tag{8.62}$$

is manifestly not positive-definite. In fact, $x \cdot x$ vanishes whenever x is "light-like", that is, equal to the difference of the four-vectors of two space-time points that can be connected by a light signal. Also, $x \cdot x$ is negative for "space-like" vectors such that $(x^0)^2 < (x^1)^2 + (x^2)^2 + (x^3)^2 = r \cdot r$. The real vector space $\mathbb{R}^4$ with the inner product Eq. (8.60) is sometimes called **Minkowski space**, which we denote with the symbol $\mathcal{M}$. Evidently $\mathcal{M}$ is pseudo-Euclidean.

8.1.5 ORTHOGONALITY

Definition
Two nonzero vectors $\overline{x}$ and $\overline{y}$ in an inner-product space $\mathcal{V}$ are called **orthogonal** (written $\overline{x} \perp \overline{y}$) when

$$\langle \overline{x}, \overline{y} \rangle = 0. \tag{8.63}$$

If the inner product happens to be the usual dot product $\mathbf{x} \cdot \mathbf{y}$ in three-dimensional space, the definition Eq. (8.63) coincides with the usual analytic-geometry condition for two vectors $\mathbf{x}$ and $\mathbf{y}$ to be perpendicular.

In a pseudo-unitary or pseudo-Euclidean space such as Minkowski space $\mathcal{M}$, a nonzero vector x can be orthogonal to itself. In $\mathcal{M}$, a light-like vector is orthogonal to itself by definition, because $x \cdot x = 0$ defines x as light-like. In a Euclidean or unitary space, on the other hand, Eq. (8.43) guarantees that the only vector that is orthogonal to itself is the zero vector.

Mutually Orthogonal Sets
Orthogonality is not merely a relation between two vectors. If $X = \{\overline{x}_i \mid i \in I\}$ is a subset of $\mathcal{V}$, then a vector $\overline{y} \in \mathcal{V}$ is called **orthogonal to the set** X (written $\overline{y} \perp X$) if and only if

$$\forall i \in I : \langle \overline{y}, \overline{x}_i \rangle = 0, \tag{8.64}$$

that is, if and only if $\overline{y}$ is orthogonal to every element of X. For example, in $\mathbb{E}^3$ the canonical basis vector $\mathbf{e}_3$ is orthogonal to the set comprised of the other two canonical basis vectors, $\{\mathbf{e}_1, \mathbf{e}_2\}$.

Orthogonality is a powerful theoretical and computational tool for ensuring the linear independence of a set of vectors. We claim: *If a vector $\overline{y}$ such that $\langle \overline{y}, \overline{y} \rangle \neq 0$ is orthogonal to a linearly independent set X, then the set $S := \{\overline{y}\} \cup X$ is linearly independent.*

We begin the proof by supposing that S is linearly dependent. Then there exist scalars β, α^i such that

$$\beta \overline{y} + \sum_i \alpha^i \overline{x}_i = \overline{0}. \tag{8.65}$$

In this equation $\beta \neq 0$ (for, if $\beta = 0$, then X is linearly dependent), and the set of α^i such that $\alpha^i \neq 0$ is finite. Therefore only finitely many terms contribute to the sum. Take the inner product of the sum with $\overline{y}$, obtaining

$$\beta \langle \overline{y}, \overline{y} \rangle + \sum_i \alpha^i \langle \overline{y}, \overline{x}_i \rangle = \beta \langle \overline{y}, \overline{y} \rangle = 0. \tag{8.66}$$

Because β is nonzero, one has $\langle \overline{y}, \overline{y} \rangle = 0$, contradicting the assumption that $\langle \overline{y}, \overline{y} \rangle \neq 0$.

For example, in Minkowski space $\mathcal{M}$ the vector e_3, for which $e_3 \cdot e_3 = -1$, is orthogonal to the set

$$X = \{e_0, e_1, e_2\}. \tag{8.67}$$

The set

$$\{e_3\} \cup X = \{e_0, e_1, e_2, e_3\} \tag{8.68}$$

is linearly independent.

The vectors $\overline{x}_i$ that belong to a set $X = \{\overline{x}_i \mid i \in I\}$ are called **mutually orthogonal** if and only if

$$\forall i, j \in I :\ni: i \neq j : \overline{x}_i \perp \overline{x}_j. \tag{8.69}$$

Let $p(n)$ be the following assertion: *Every set $\{\overline{x}_1, \ldots, \overline{x}_n\}$ of n mutually orthogonal vectors for which every self–inner product $\langle \overline{x}_i, \overline{x}_i \rangle \neq 0$ is linearly independent.*

We prove $p(n)$ using the principle of finite induction. Clearly $p(1)$ is true. Suppose that $p(n)$ is true. Let the vectors $\overline{x}_1, \ldots, \overline{x}_n, \overline{x}_{n+1}$ be mutually orthogonal, and let every self–inner product $\langle \overline{x}_i, \overline{x}_i \rangle$ be nonzero for $i \in (1 : n + 1)$. By the inductive hypothesis, $X_n := \{\overline{x}_1, \ldots, \overline{x}_n\}$ is linearly independent. Because $\overline{x}_{n+1}$ is orthogonal to the linearly independent set X_n, the set $X_{n+1} := X_n \cup \{\overline{x}_{n+1}\}$ is linearly independent. Therefore $p(n + 1)$ holds.

Orthogonal Bases

The following theorem is an immediate consequence of our proof of $p(n)$ and of the fact that in Euclidean and unitary spaces $\langle \overline{x}, \overline{x} \rangle = 0$ if and only if $\overline{x} = 0$: *If $\mathcal{V}$ is a Euclidean or unitary space of finite dimension n, then every set of n mutually orthogonal vectors in $\mathcal{V}$ is a basis.*

A basis of mutually orthogonal vectors, such that

$$\forall i, j \in (1:n) :\ni: i \neq j : \langle \bar{e}_i, \bar{e}_j \rangle = 0 \tag{8.70}$$

is called an **orthogonal basis**. Orthogonal bases are the subject of Section 8.2.2. An obvious advantage of an orthogonal basis is that the Gram matrix of a mutually orthogonal set of basis vectors $\{\bar{b}_1, \ldots, \bar{b}_n\}$ is diagonal:

$$\mathbf{G} = \begin{pmatrix} \langle \bar{b}_1, \bar{b}_1 \rangle & & \mathbf{0} \\ & \ddots & \\ \mathbf{0} & & \langle \bar{b}_n, \bar{b}_n \rangle \end{pmatrix}. \tag{8.71}$$

Another, computational, advantage of orthogonal bases is that orthogonality guarantees linear independence. For example, least-squares fits with nonorthogonal bases can be quite ill conditioned because of numerical cancellation of significant digits in a floating-point representation with a fixed precision p; see Exercise 8.4.3.

Orthogonality replaces the process of computing the contravariant components of a vector by Gaussian elimination with the simpler operation of taking an inner product. To compute the contravariant component x^i of a vector $\bar{x}$ in a Euclidean or unitary space $\mathcal{V}$ relative to an orthogonal basis $\{\bar{b}_1, \ldots, \bar{b}_n\}$ of $\mathcal{V}$, one forms the inner product of $\bar{b}_i$ with $\bar{x} = x^j \bar{b}_j$:

$$\langle \bar{b}_i, \bar{x} \rangle = \sum_{j=1}^{n} x^j \langle \bar{b}_i, \bar{b}_j \rangle. \tag{8.72}$$

Then the contravariant components of a vector relative to an orthogonal basis $\{\bar{b}_j\}$ are

$$x^i = \frac{\langle \bar{b}_i, \bar{x} \rangle}{\langle \bar{b}_i, \bar{b}_i \rangle}. \tag{8.73}$$

It follows that, relative to an orthogonal basis, the coordinate expansion of a vector $\bar{x}$ in a Euclidean or unitary space of dimension n is

$$\bar{x} = \sum_{i=1}^{n} \frac{\langle \bar{b}_i, \bar{x} \rangle}{\langle \bar{b}_i, \bar{b}_i \rangle} \bar{b}_i. \tag{8.74}$$

This expression is called the **orthogonal expansion** of the vector $\bar{x}(\mathrm{rel}\{\bar{b}_i\})$.

For example, let

$$B := \left\{ \mathbf{b}_1 = \begin{pmatrix} 2 \\ 1 \end{pmatrix}, \quad \mathbf{b}_2 = \begin{pmatrix} -1 \\ 2 \end{pmatrix} \right\} \tag{8.75}$$

in $\mathbb{E}^2$. The vectors $\mathbf{b}_1$ and $\mathbf{b}_2$ are orthogonal under the canonical inner product Eq. (8.13).

Therefore B is an orthogonal basis of $\mathbb{E}^2$, and the component x^2 of

$$\mathbf{x} = \begin{pmatrix} 7 \\ 2 \end{pmatrix} \tag{8.76}$$

with respect to B is

$$x^2 = \frac{-1 \cdot 7 + 2 \cdot 2}{(-1) \cdot (-1) + 2 \cdot 2} = -\frac{3}{5}. \tag{8.77}$$

Similarly, one finds that

$$x^1 = \frac{16}{5}. \tag{8.78}$$

Checking the calculation, one gets

$$x^1 \mathbf{b}_1 + x^2 \mathbf{b}_2 = \frac{16}{5} \begin{pmatrix} 2 \\ 1 \end{pmatrix} - \frac{3}{5} \begin{pmatrix} -1 \\ 2 \end{pmatrix} = \frac{1}{5} \begin{pmatrix} 35 \\ 10 \end{pmatrix} = \begin{pmatrix} 7 \\ 2 \end{pmatrix}. \tag{8.79}$$

The Pauli matrices are an important special case of an orthogonal basis. The set

$$\Sigma = \{\mathbf{1}, \sigma_1, \sigma_2, \sigma_3\} \tag{8.80}$$

is a basis of $\mathbb{C}^{2\times 2}$ because the number of linearly independent matrices in the set Σ is equal to the dimension (four) of $\mathbb{C}^{2\times 2}$. By Eq. (8.25), Σ is an orthogonal basis. Let the coordinate expansion of a 2×2 matrix $\mathbf{A} \in \mathbb{C}^{2\times 2}$ relative to the basis Σ be

$$\mathbf{A} = a^0 \mathbf{1} + a^1 \sigma_1 + a^2 \sigma_2 + a^3 \sigma_3. \tag{8.81}$$

It is possible to calculate the contravariant components a^μ by taking the inner product of $\mathbf{A}$ with each of the Pauli matrices, exactly as if one were dealing with vectors. One finds from Eq. (8.74) and from the equation that defines the inner products of the Pauli matrices, Eq. (8.25), that

$$\forall \mu \in (0:3) : a^\mu = \begin{cases} \frac{1}{2}\text{trace}[\mathbf{A}], & \text{if } \mu = 0; \\ \frac{1}{2}\text{trace}[\sigma_\mu \mathbf{A}], & \text{if } \mu \in (1:3). \end{cases} \tag{8.82}$$

For example, if

$$\mathbf{A} = \begin{pmatrix} e^{-i\psi/2}\cos\beta/2 & -e^{i\chi/2}\sin\beta/2 \\ e^{-i\chi/2}\sin\beta/2 & e^{i\psi/2}\cos\beta/2 \end{pmatrix}, \tag{8.83}$$

then one finds from Eq. (8.74) that

$$a^3 = \frac{\langle \sigma_3, \mathbf{A} \rangle}{\langle \sigma_3, \sigma_3 \rangle} = -i\sin(\psi/2)\cos(\beta/2) \tag{8.84}$$

under the canonical inner product in $\mathbb{C}^{2\times 2}$. The full coordinate expansion of $\mathbf{A}$ relative to Σ is

$$\mathbf{A} = \cos(\psi/2)\cos(\beta/2)\mathbf{1} - i\sin(\chi/2)\sin(\beta/2)\boldsymbol{\sigma}_1$$
$$- i\cos(\chi/2)\sin(\beta/2)\boldsymbol{\sigma}_2 - i\sin(\psi/2)\cos(\beta/2)\boldsymbol{\sigma}_3. \tag{8.85}$$

$\mathbf{A}$ is the most general 2×2 matrix with determinant equal to 1. The orthogonal expansion given in Eq. (8.85) is useful in deriving the general form of the irreducible representations of the matrix group $SU(2)$.

8.1.6 EXERCISES FOR SECTION 8.1

8.1.1 Prove that $\mathbb{R}^n$ is a Euclidean space under the canonical inner product defined in Eq. (8.13).

Hint: Recall that to prove that a vector space $\mathcal{V}$ is a Euclidean or unitary space, one must show that the proposed inner product satisfies Eqs. (8.2) and (8.3) and either Eq. (8.4) or Eq. (8.5) and is positive-definite.

8.1.2 Prove that $\mathbb{C}^n$ is a unitary space under the canonical inner product defined in Eq. (8.15).

8.1.3 Prove that under the canonical inner product defined in Eq. (8.21), $\mathbb{C}^{n \times n}$ is a unitary space.

8.1.4 Show that if the weight function $w(x)$ in Eq. (8.52) is real, continuous and strictly positive in (a, b), and if one interprets the integral as a Riemann integral, then the inner product $\langle f, g \rangle$ is positive-definite.

8.1.5 Show that if, in $\mathbb{E}^2$, $\mathbf{x}' = \mathsf{R}\mathbf{x}$ and $\mathbf{y}' = \mathsf{R}\mathbf{y}$, in which R is the rotation through an angle θ (see Eq. [6.125]), then

$$\langle \mathbf{x}', \mathbf{y}' \rangle = \langle \mathbf{x}, \mathbf{y} \rangle \tag{8.86}$$

under the canonical inner product defined in Eq. (8.13).

8.1.6 Show that if, in the two-dimensional version of Minkowski space, $x' = \Lambda x$ and $y' = \Lambda y$, in which Λ is the two-dimensional Lorentz transformation defined in Eqs. (6.126) and (6.127), then

$$\langle x', y' \rangle = \langle x, y \rangle \tag{8.87}$$

under the inner product

$$x \cdot y := x^0 y^0 - x^1 y^1. \tag{8.88}$$

8.1.7 Let $\mathcal{V}$ be a Euclidean or unitary space, let $\mathcal{W}$ be a vector subspace of $\mathcal{V}$, and let $\{\bar{e}_1, \ldots, \bar{e}_m\}$ be a basis of $\mathcal{W}$. Prove that a vector $\bar{x} \in \mathcal{V}$ is orthogonal to $\mathcal{W}$ if and only if

$$\forall i \in (1 : m) : \langle \bar{e}_i, \bar{x} \rangle = 0. \tag{8.89}$$

8.2 GEOMETRY OF INNER-PRODUCT SPACES

8.2.1 PYTHAGORAS'S THEOREM

Length and *angle* are fundamental concepts of geometry. In Eq. (8.45) we generalized the concept of length to Euclidean or unitary n-space, using the analogy between the inner products in $\mathbb{E}^n$ and $\mathbb{U}^n$ and the dot product in three-dimensional space. In this section we generalize Pythagoras's theorem and the concept of angle to arbitrary finite-dimensional Euclidean and unitary spaces.

Generalized Pythagorean Theorem

From one's experience with real three-dimensional Euclidean space, in which $\mathbf{a} \cdot \mathbf{b} = ab \cos \theta$, one expects that if the angle between two abstract vectors $\bar{x}$ and $\bar{y}$ can be defined, then it ought to be

$$\theta = \cos^{-1} \left[\frac{\langle \bar{x}, \bar{y} \rangle}{\|\bar{x}\|_{\mathrm{ip}} \|\bar{y}\|_{\mathrm{ip}}} \right]. \tag{8.90}$$

For example, if

$$\mathbf{x} = \begin{pmatrix} 1 \\ 2 \\ 3 \\ 4 \end{pmatrix} \quad \text{and} \quad \mathbf{y} = \begin{pmatrix} 1 \\ -1 \\ -1 \\ 1 \end{pmatrix} \tag{8.91}$$

in $\mathbb{E}^4$, then

$$\theta = \cos^{-1} 0 = \pm \frac{\pi}{2}. \tag{8.92}$$

Immediately one encounters two difficulties in applying this simple idea: $\langle \bar{x}, \bar{y} \rangle$ is real by definition only if $\mathbb{F} = \mathbb{R}$, and even if $\langle \bar{x}, \bar{y} \rangle$ is real its magnitude might perhaps exceed $\|\bar{x}\|_{\mathrm{ip}} \|\bar{y}\|_{\mathrm{ip}}$ and thereby give rise to a cosine greater than unity. The Cauchy-Schwarz-Bunyakovsky inequality, which we derive in the next subsection, asserts that if the inner product is positive-definite, then the latter possibility cannot arise.

Pythagoras's theorem $c^2 = b^2 + a^2$ for a right triangle can be generalized easily to apply to any inner-product space, although it can be interpreted in terms of distances only if the inner product is positive-definite (Eqs. [8.42] and [8.43]). Let $\bar{h}_1, \ldots, \bar{h}_n$ be mutually orthogonal:

$$\langle \bar{h}_i, \bar{h}_j \rangle = \begin{cases} 0 & \text{if } i \neq j, \\ \|\bar{h}_i\|_{\mathrm{ip}}^2 & \text{if } i = j, \end{cases} \tag{8.93}$$

and let

$$\bar{x} = \sum_{i=1}^{n} \bar{h}_i. \tag{8.94}$$

Then the norm squared of $\overline{x}$ is

$$\|\overline{x}\|_{ip}^2 = \langle \overline{x}, \overline{x} \rangle = \sum_{i=1}^n \sum_{j=1}^n \langle \overline{h}_i, \overline{h}_j \rangle = \sum_{i=1}^n \|\overline{h}_i\|_{ip}^2. \tag{8.95}$$

This result, which holds for every finite set of mutually orthogonal vectors, is called the **generalized Pythagorean theorem**.

Cauchy-Schwarz-Bunyakovsky Inequality

In real three-dimensional Euclidean space, the magnitude of the inner product of two vectors is not greater than the product of the lengths of the vectors:

$$|\mathbf{a} \cdot \mathbf{b}| = ab|\cos \theta| \le ab. \tag{8.96}$$

We know that this is true because we know from

$$\cos^2 \theta + \sin^2 \theta = 1, \tag{8.97}$$

which is a special case of the Pythagorean theorem, that

$$|\cos \theta| \le 1. \tag{8.98}$$

Therefore it makes sense to use the generalized Pythagorean theorem to obtain an upper bound for $|\langle \overline{x}, \overline{y} \rangle|$.

Given elements $\overline{x} \ne \overline{0}$ and $\overline{y} \ne \overline{0}$ of a Euclidean or unitary space $\mathcal{V}$, we claim that one can always find a scalar λ and a vector $\overline{z}$ such that

$$\overline{y} = \lambda\overline{x} + \overline{z} \text{ and } \langle \lambda\overline{x}, \overline{z} \rangle = 0 \tag{8.99}$$

(see Fig. 8.1). The requirement that

$$\langle \overline{x}, \overline{z} \rangle = 0 \tag{8.100}$$

implies that

$$\langle \overline{x}, \overline{z} \rangle = \langle \overline{x}, \overline{y} - \lambda\overline{x} \rangle = \langle \overline{x}, \overline{y} \rangle - \lambda\langle \overline{x}, \overline{x} \rangle = 0, \tag{8.101}$$

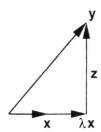

FIGURE 8.1

Illustration of the projection $\lambda\overline{x}$ of $\overline{y}$ along $\overline{x}$.

which is true if and only if

$$\lambda = \frac{\langle \overline{x}, \overline{y} \rangle}{\|\overline{x}\|_{\text{ip}}^2}.$$

(8.102)

The denominator in Eq. (8.102) cannot vanish, because $\overline{x}$ is nonzero and the inner product is positive-definite.

Let

$$\overline{u} := \frac{1}{\|\overline{x}\|_{\text{ip}}} \overline{x}.$$

(8.103)

Because

$$\langle \overline{u}, \overline{u} \rangle = 1,$$

(8.104)

one calls $\overline{u}$ the **unit vector in the direction of** $\overline{x}$. The vector

$$P_{\overline{u}} \overline{y} := \langle \overline{u}, \overline{y} \rangle \overline{u} = \lambda \overline{x}$$

(8.105)

is called the **orthogonal projection of** $\overline{y}$ **along the direction of** $\overline{x}$, and $P_{\overline{u}}$ is called the **orthogonal projector on** $\overline{u}$.

We now apply the generalized Pythagorean theorem to calculate the norm squared of $\overline{y}$ in terms of the squared norms of the orthogonal vectors $\overline{x}$ and $\overline{z}$:

$$\|\overline{y}\|_{\text{ip}}^2 = \|\lambda \overline{x}\|_{\text{ip}}^2 + \|\overline{z}\|_{\text{ip}}^2 = |\lambda|^2 \|\overline{x}\|_{\text{ip}}^2 + \|\overline{y} - \lambda \overline{x}\|_{\text{ip}}^2$$

(8.106)

$$= \frac{|\langle \overline{x}, \overline{y} \rangle|^2}{\|\overline{x}\|_{\text{ip}}^2} + \|\overline{y} - P_{\overline{u}} \overline{y}\|_{\text{ip}}^2.$$

(8.107)

Therefore

$$|\langle \overline{x}, \overline{y} \rangle|^2 = \|\overline{x}\|_{\text{ip}}^2 \|\overline{y}\|_{\text{ip}}^2 - \|\overline{x}\|_{\text{ip}}^2 \|\overline{y} - P_{\overline{u}} \overline{y}\|_{\text{ip}}^2.$$

(8.108)

Then one has the **Cauchy-Schwarz-Bunyakovsky inequality**

$$|\langle \overline{x}, \overline{y} \rangle| \leq \|\overline{x}\|_{\text{ip}} \|\overline{y}\|_{\text{ip}},$$

(8.109)

which holds in all Euclidean or unitary spaces. According to Eq. (8.108), equality holds in Eq. (8.109) if and only if $\|\overline{y} - \lambda \overline{x}\|_{\text{ip}} = 0$, that is,

$$|\langle \overline{x}, \overline{y} \rangle| = \|\overline{x}\|_{\text{ip}} \|\overline{y}\|_{\text{ip}} \Leftrightarrow \overline{y} = P_{\overline{u}} \overline{y}.$$

(8.110)

The geometrical interpretation of Eq. (8.110) is that the Cauchy-Schwarz-Bunyakovsky inequality is an equality if and only if $\overline{y}$ is parallel to $\overline{x}$.

In $\mathbb{C}$, an elementary derivation of the Cauchy-Schwarz-Bunyakovsky inequality goes as follows:

$$|\langle z', z \rangle| = |\text{Re}[z'^* z]| = |\text{Re}[\rho' e^{-i\phi'} \rho e^{i\phi}]| = \rho \rho' |\cos \theta| \leq \rho \rho',$$

(8.111)

in which $z = \rho e^{i\phi}$ is the polar decomposition of z and $\theta = \phi - \phi'$ is the angle between z and z' in the complex plane. An important corollary is the triangle inequality for complex numbers,

$$|z + z'| \le |z| + |z'|,$$ (8.112)

which is a special case of a more general triangle inequality derived in a subsequent section.
Cauchy's inequality for $\mathbb{U}^n$ is

$$\left| \sum_i^n y_i^* x_i \right| \le \left[\sum_i^n |x_i|^2 \right]^{1/2} \left[\sum_i^n |y_i|^2 \right]^{1/2}.$$ (8.113)

Schwarz's inequality for $\mathcal{L}_w^2([a, b]; \mathbb{C})$ is

$$\left| \int_a^b f(x)^* g(x) w(x) \, dx \right| \le \left[\int_a^b |f(x)|^2 w(x) \, dx \right]^{1/2} \left[\int_a^b |g(y)|^2 w(y) \, dy \right]^{1/2}.$$ (8.114)

Bunyakovsky apparently was the first to use Eq. (8.114) for the Riemann integral.

Reflection in a Line

Let $\bar{x}$ and $\bar{y}$ be vectors in a Euclidean or unitary space. The vector

$$R_{\bar{u}} \bar{y} := \bar{y} - 2(\bar{y} - P_{\bar{u}} \bar{y})$$ (8.115)

is called the **reflection of $\bar{y}$ in the line** defined by the unit vector $\bar{u} = \bar{x}/||\bar{x}||_{ip}$.

This definition agrees with the usual definition of reflection in two- or three-dimensional space, for reflection of a vector $\bar{y}$ in the line defined by a vector $\bar{x}$ leaves the component of $\bar{y}$ that is parallel to $\bar{x}$ ($P_{\bar{u}} \bar{y} = \lambda \bar{x}$ in Eq. [8.99]) unchanged, but changes the sign of the component of $\bar{y}$ that is orthogonal to $\bar{x}$ ($\bar{z} = \bar{y} - \lambda \bar{x}$ in Eq. [8.99]). The original vector is $\bar{y} = P_{\bar{u}} \bar{y} + \bar{z}$. Therefore the reflected vector $R_{\bar{u}} \bar{y}$ is equal to

$$P_{\bar{u}} \bar{y} - \bar{z} = \bar{y} - 2\bar{z}.$$ (8.116)

For example, consider a complex number $z = x + iy$ (which is an element of the real vector space $\mathbb{R}^2$ with coordinates x and y relative to the basis $\{1, i\}$). The complex conjugate $z^* = x - iy$ is the reflection of z in the real axis (the line that consists of the scalar multiples of 1), for $z^* = z - 2iy$.

Covariant Components of a Vector

The inner product

$$x_i := \langle \bar{e}_i, \bar{x} \rangle$$ (8.117)

of a basis element $\bar{e}_i$ with a vector $\bar{x}$ is called the **covariant component** of $\bar{x}$ along $\bar{e}_i$. If $\bar{e}_i$

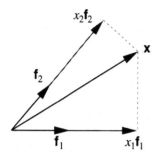

FIGURE 8.2

The covariant components x_1, x_2 of a vector $x \in \mathbb{E}^2$. The basis vectors f_i have unit length.

is a unit vector, $\langle \bar{e}_i, \bar{e}_i \rangle = 1$, then $x_i \bar{e}_i$ (no Σ) is equal to the orthogonal projection

$$P_{\bar{e}_i} \bar{x} = \langle \bar{e}_i, \bar{x} \rangle \bar{e}_i \tag{8.118}$$

of $\bar{x}$ on $\bar{e}_i$. Figure 8.2 gives an example of the covariant components of a vector in $\mathbb{E}^2$.
In terms of the metric tensor one has

$$\begin{aligned}
x_i &= \langle \bar{e}_i, \bar{x} \rangle \\
&= \langle \bar{e}_i, x^j \bar{e}_j \rangle \\
&= g_{ij} x^j.
\end{aligned} \tag{8.119}$$

Therefore the matrix that transforms a vector of contravariant components to a vector of covariant components is the Gram matrix:

$$\mathbf{x}_{\text{covariant}} = \mathbf{G} \mathbf{x}_{\text{contravariant}}. \tag{8.120}$$

This equation defines a set of n linear equations that relate the covariant components x_i to the contravariant components x^j.
For example, in Minkowski space $\mathcal{M}$ the covariant components of a four-vector x are

$$x_0 = g_{00} x^0 = x^0 \tag{8.121}$$

and

$$x_i = g_{ii} x^i = -x^i \quad \text{for} \quad i = 1, 2, 3 \quad (\text{no } \Sigma). \tag{8.122}$$

In $\mathbb{E}^n$ (and in $\mathbb{E}^n$ alone!) the covariant components are equal to the contravariant components,

$$x_i = g_{ii} x^i = x^i \quad (\text{no } \Sigma). \tag{8.123}$$

Equations (8.119) and (8.36) imply that the inner product $\langle \bar{x}, \bar{y} \rangle$ in $\mathcal{V}$ can be expressed in terms of the contravariant components of $\bar{x}$ and the covariant components of $\bar{y}$ (or vice versa):

$$\langle \bar{x}, \bar{y} \rangle = \begin{cases} \langle x^i \bar{e}_i, y^j \bar{e}_j \rangle = x^i y_i = x_i y^i & \text{if } \mathbb{F} = \mathbb{R}; \\ \langle x^i \bar{e}_i, y^j \bar{e}_j \rangle = x^{i*} y_i = x_i^* y^i & \text{if } \mathbb{F} = \mathbb{C}. \end{cases} \tag{8.124}$$

In $\mathbb{E}^n$, in which the contravariant and covariant components along each orthogonal coordinate axis are equal, the inner product generalizes the dot product in terms of Cartesian components in $\mathbb{E}^3$:

$$(\bar{x}, \bar{y}) = \sum_i^n x^i y^i. \tag{8.125}$$

The terms *covariant* and *contravariant* refer to behavior under a change of basis

$$\bar{e}'_j = b^i_j \bar{e}_i. \tag{8.126}$$

The *covariant* components x'_j transform *with* (i.e, in the same way) as the basis vectors:

$$x'_j = \langle \bar{e}'_j, \bar{x} \rangle = \begin{cases} b^i_j \langle \bar{e}_i, \bar{x} \rangle = b^i_j x_i & \text{if } \mathbb{F} = \mathbb{R}; \\ b^i_j{}^* \langle \bar{e}_i, \bar{x} \rangle = b^i_j{}^* x_i & \text{if } \mathbb{F} = \mathbb{C}. \end{cases} \tag{8.127}$$

However, the *contra*variant components x^j transform by the matrix $\mathbf{T}$, the inverse of the matrix $\mathbf{B} = [b^i_j]$:

$$x'^j = t^j_i x^i \tag{8.128}$$

(see Section 6.2.2). In this sense the contravariant components of $\bar{x}$ transform *contrarily* to the basis vectors.

For example, let

$$[b^i_j] = \begin{pmatrix} \cos\theta & -\sin\theta \\ \sin\theta & \cos\theta \end{pmatrix} \tag{8.129}$$

in $\mathbb{E}^2$. Then

$$\begin{aligned} \mathbf{e}'_1 &= \cos\theta\,\mathbf{e}_1 + \sin\theta\,\mathbf{e}_2 \\ \mathbf{e}'_2 &= -\sin\theta\,\mathbf{e}_1 + \cos\theta\,\mathbf{e}_2, \end{aligned} \tag{8.130}$$

and

$$\begin{aligned} x'^1 &= \cos\theta\,x^1 - \sin\theta\,x^2 \\ x'^2 &= \sin\theta\,x^1 + \cos\theta\,x^2. \end{aligned} \tag{8.131}$$

The plane rotation according to which the contravariant components transform in this example is the inverse of the rotation by which the basis vectors transform, as one should expect because we have insisted that vectors must remain fixed while the basis vectors transform.

Other Geometrical Properties of Euclidean and Unitary Spaces

Several other important geometrical properties of Euclidean and unitary spaces can be expressed naturally in terms of the inner-product norm Eq. (8.45). The **law of cosines**,

$$\|\bar{x} - \bar{y}\|^2_{\text{ip}} = \|\bar{x}\|^2_{\text{ip}} + \|\bar{y}\|^2_{\text{ip}} - 2\|\bar{x}\|_{\text{ip}}\|\bar{y}\|_{\text{ip}} \cos\theta_{\bar{x},\bar{y}}, \tag{8.132}$$

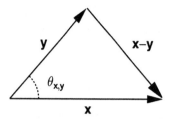

FIGURE 8.3

Illustration of the law of cosines.

follows directly from axioms in Eqs. (8.2) through (8.6) and the definition

$$2\cos\theta_{\bar{x},\bar{y}} := \frac{\langle \bar{x}, \bar{y}\rangle + \langle \bar{y}, \bar{x}\rangle}{\|\bar{x}\|_{ip}\|\bar{y}\|_{ip}} \tag{8.133}$$

(see Fig. 8.3). (The Cauchy-Schwarz-Bunyakovsky inequality, Eq. (8.109), implies that $|\cos\theta_{\bar{x},\bar{y}}| \leq 1$; see subsequent text.) For example, in Fig. 8.4, by Pythagoras's theorem, $\|\bar{x} \pm \bar{y}\|_{ip}^2 = (\|\bar{x}\|_{ip} \pm p)^2 + h^2$, in which $p = \|\bar{y}\|_{ip}\cos\theta_{\bar{x},\bar{y}}$.

The **triangle inequality**

$$\|\bar{x} + \bar{y}\|_{ip} \leq \|\bar{x}\|_{ip} + \|\bar{y}\|_{ip} \tag{8.134}$$

can be derived as follows: The counterpart of Eq. (8.132) for $\bar{x} + \bar{y}$ is

$$\|\bar{x} + \bar{y}\|_{ip}^2 = \|\bar{x}\|_{ip}^2 + \|\bar{y}\|_{ip}^2 + 2\|\bar{x}\|_{ip}\|\bar{y}\|_{ip}\cos\theta_{\bar{x},\bar{y}}. \tag{8.135}$$

The triangle inequality for real or complex numbers, $|\alpha + \beta| \leq |\alpha| + |\beta|$, and the Cauchy-Schwarz-Bunyakovsky inequality, Eq. (8.109), imply that

$$\begin{aligned} 2\|\bar{x}\|_{ip}\|\bar{y}\|_{ip}|\cos\theta_{\bar{x},\bar{y}}| &= |\langle\bar{x},\bar{y}\rangle + \langle\bar{y},\bar{x}\rangle| \\ &\leq |\langle\bar{x},\bar{y}\rangle| + |\langle\bar{y},\bar{x}\rangle| \\ &\leq 2\|\bar{x}\|_{ip}\|\bar{y}\|_{ip}. \end{aligned} \tag{8.136}$$

[If we apply this derivation to $\mathbb{C}$, we use only the triangle inequality for *real* numbers!] From Eqs. (8.135) and (8.136),

$$\|\bar{x} + \bar{y}\|_{ip}^2 \leq \|\bar{x}\|_{ip}^2 + \|\bar{y}\|_{ip}^2 + 2\|\bar{x}\|_{ip}\|\bar{y}\|_{ip} = (\|\bar{x}\|_{ip} + \|\bar{y}\|_{ip})^2. \tag{8.137}$$

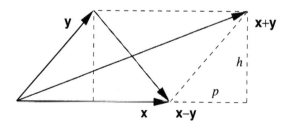

FIGURE 8.4

Illustration of the parallelogram law.

Taking the positive square root of Eq. (8.137) gives Eq. (8.134). One shows similarly that

$$\|\bar{x} + \bar{y}\|_{ip} \geq |\|\bar{x}\|_{ip} - \|\bar{y}\|_{ip}|. \tag{8.138}$$

It is important to know that although the Cauchy-Schwarz-Bunyakovsky inequality Eq. (8.109) implies the triangle inequality, the converse is not true. In Chapter 10 we encounter norms that satisfy the triangle inequality but do not obey Eq. (8.45) for any inner product.

The **parallelogram law**,

$$\|\bar{x} + \bar{y}\|_{ip}^2 + \|\bar{x} - \bar{y}\|_{ip}^2 = 2(\|\bar{x}\|_{ip}^2 + \|\bar{y}\|_{ip}^2), \tag{8.139}$$

follows from adding Eqs. (8.132) and (8.135). In Fig. 8.4, Pythagoras's theorem implies that $\|\bar{y}\|_{ip}^2 = h^2 + p^2$, $\|\bar{x} + \bar{y}\|_{ip}^2 = h^2 + (\|\bar{x}\|_{ip} + p)^2$ and $\|\bar{x} - \bar{y}\|_{ip}^2 = h^2 + (\|\bar{x}\| - p)^2$. Adding these two expressions gives Eq. (8.139) for this example.

Subtracting Eq. (8.132) from Eq. (8.135) gives the cosine of $\theta_{\bar{x},\bar{y}}$ directly in terms of norms:

$$\|\bar{x}\|_{ip}\|\bar{y}\|_{ip} \cos\theta_{\bar{x},\bar{y}} = \tfrac{1}{4}(\|\bar{x} + \bar{y}\|_{ip}^2 - \|\bar{x} - \bar{y}\|_{ip}^2). \tag{8.140}$$

If $\mathcal{V}$ is Euclidean, then the inner product is symmetric and is completely determined by $\|\bar{x}\|_{ip}\|\bar{y}\|_{ip} \cos\theta_{\bar{x},\bar{y}}$:

$$\langle \bar{x}, \bar{y} \rangle = \|\bar{x}\|_{ip}\|\bar{y}\|_{ip} \cos\theta_{\bar{x},\bar{y}} = \tfrac{1}{4}(\|\bar{x} + \bar{y}\|_{ip}^2 - \|\bar{x} - \bar{y}\|_{ip}^2). \tag{8.141}$$

In a Euclidean space, all scalar multiples of $\bar{x}$ and $\bar{y}$ give the same value of $|\cos\theta_{\bar{x},\bar{y}}|$:

$$\cos\theta_{\alpha\bar{x},\beta\bar{y}} = \frac{\langle \alpha\bar{x}, \beta\bar{y} \rangle}{\|\alpha\bar{x}\|_{ip}\|\beta\bar{y}\|_{ip}}$$

$$= \operatorname{sgn}(\alpha)\operatorname{sgn}(\beta)\frac{\langle \bar{x}, \bar{y} \rangle}{\|\bar{x}\|_{ip}\|\bar{y}\|_{ip}} \tag{8.142}$$

$$= \operatorname{sgn}(\alpha)\operatorname{sgn}(\beta)\cos\theta_{\bar{x},\bar{y}}.$$

Then $\theta_{\alpha\bar{x},\beta\bar{y}} = \theta_{\bar{x},\bar{y}}$ or $\pi - \theta_{\bar{x},\bar{y}}$, depending on the algebraic signs of α and β (as indicated by the sign function sgn). However, if $\mathcal{V}$ is unitary, then $\|\bar{x}\|_{ip}\|\bar{y}\|_{ip} \cos\theta_{\bar{x},\bar{y}}$ determines only the real part of the inner product $\langle \bar{x}, \bar{y} \rangle$:

$$\operatorname{Re}[\langle \bar{x}, \bar{y} \rangle] = \tfrac{1}{2}[\langle \bar{x}, \bar{y} \rangle + \langle \bar{y}, \bar{x} \rangle] = \|\bar{x}\|_{ip}\|\bar{y}\|_{ip} \cos\theta_{\bar{x},\bar{y}}. \tag{8.143}$$

Unlike a Euclidean space, a unitary space has the property that not all scalar multiples of $\bar{x}$ and $\bar{y}$ give the same value of $|\cos\theta_{\bar{x},\bar{y}}|$. In fact,

$$\cos\theta_{\bar{x},i\bar{y}} = \frac{\langle \bar{x}, i\bar{y} \rangle + \langle i\bar{y}, \bar{x} \rangle}{2\|\bar{x}\|_{ip}\|\bar{y}\|_{ip}} = i\frac{\langle \bar{x}, \bar{y} \rangle - \langle \bar{y}, \bar{x} \rangle}{2\|\bar{x}\|_{ip}\|\bar{y}\|_{ip}} = -\operatorname{Im}\frac{\langle \bar{x}, \bar{y} \rangle}{\|\bar{x}\|_{ip}\|\bar{y}\|_{ip}}. \tag{8.144}$$

Therefore, in a unitary space one can express the full inner product in terms of norms as follows:

$$\langle \overline{x}, \overline{y} \rangle = \|\overline{x}\|_{\text{ip}} \|\overline{y}\|_{\text{ip}} (\cos \theta_{\overline{x}, \overline{y}} - i \cos \theta_{\overline{x}, i\overline{y}})$$

$$= \tfrac{1}{4} \left\{ \|\overline{x} + \overline{y}\|_{\text{ip}}^2 - \|\overline{x} - \overline{y}\|_{\text{ip}}^2 + i \left(\|\overline{x} + i\overline{y}\|_{\text{ip}}^2 - \|\overline{x} - i\overline{y}\|_{\text{ip}}^2 \right) \right\}. \tag{8.145}$$

Equation (8.145) is called the **polarization identity**.

8.2.2 ORTHONORMAL BASES

Definitions and Examples

In a Euclidean or unitary space $\mathcal{V}$ of finite dimension n, a basis $\{\overline{e}_1, \ldots, \overline{e}_n\}$ is called **orthonormal** if and only if

$$g_{ij} = \langle \overline{e}_i, \overline{e}_j \rangle = \begin{cases} 1, & \text{if } j = i; \\ 0, & \text{if } j \neq i. \end{cases} \tag{8.146}$$

For example, the Cartesian basis $\{\mathbf{i}, \mathbf{j}, \mathbf{k}\}$ of $\mathbb{E}^3$ is orthonormal.

A useful orthonormal basis of $\mathbb{E}^n$ or $\mathbb{U}^n$ that is closely related to the Chebyshev polynomials is the **sampled cosine basis**

$$C = \{\mathbf{c}_1, \ldots, \mathbf{c}_n\} \tag{8.147}$$

in which the contravariant components of $\mathbf{c}_i$ are proportional to the sampled values of $\cos mx$ on a uniform grid,

$$c_i^j := \begin{cases} \sqrt{\dfrac{2}{n}} \cos[(i-1)x_j], & \text{if } i > 1, \\[2mm] \dfrac{1}{\sqrt{n}}, & \text{if } i = 1, \end{cases} \tag{8.148}$$

and

$$x_j := \frac{2j-1}{n} \frac{\pi}{2} \tag{8.149}$$

(see Exercise 8.2.15 for a proof that this basis is orthonormal). Figure 8.5 shows some of the vectors $\mathbf{c}_i$ for $n = 20$ in graphical form.

Contravariant Components Relative to an Orthonormal or Pseudoorthonormal Basis

A basis $\{\overline{e}_1, \ldots, \overline{e}_n\}$ of a pseudo-Euclidean or pseudo-unitary space $\mathcal{V}$ is called **pseudoorthonormal** if and only if

$$g_{ij} = \langle \overline{e}_i, \overline{e}_j \rangle = \begin{cases} \sigma_i, & \text{if } j = i; \\ 0, & \text{if } j \neq i, \end{cases} \tag{8.150}$$

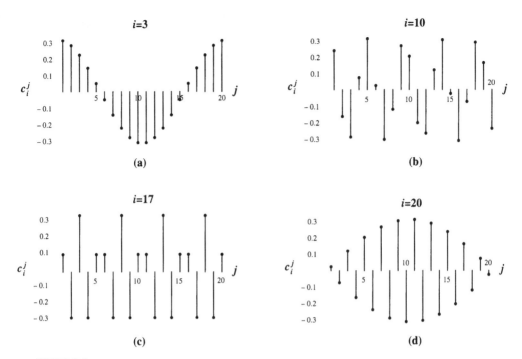

FIGURE 8.5

Illustration of the components c_i^j of several members of the cosine basis for $n = 20$ defined in Eq. 8.148. (a) $i = 3$; (b) $i = 10$; (c) $i = 17$; (d) $i = 20$.

in which

$$\sigma_i = \pm 1. \tag{8.151}$$

For example, the basis of Minkowski space $\mathcal{M}$ that consists of unit vectors along the x, y, z and ct axes is pseudoorthonormal; see Eq. (8.61).

Orthonormal or pseudoorthonormal bases are most convenient for many purposes. The contravariant component x^i of a vector $\overline{x} = x^j\,\overline{e}_j$ relative to an orthonormal or pseudoorthonormal basis of $\mathcal{V}$ is equal to the orthogonal projection of $\overline{x}$ along the basis vector $\overline{e}_i$:

$$x^i = \sigma_i\,\langle \overline{e}_i, \overline{x} \rangle. \tag{8.152}$$

This statement is *not* true for a basis that is not orthonormal or pseudoorthonormal; see Section 8.2.1. One sees also from Eqs. (8.70) and (8.71) that if $\{\overline{e}_i\}$ is an orthonormal or pseudoorthonormal basis, then the generalized Pythagorean theorem applies to the contravariant components of a vector $\overline{x}(\mathrm{rel}\{\overline{e}_i\})$:

$$\langle \overline{x}, \overline{x} \rangle = \sum_{i=1}^{n} \sigma_i |x^i|^2. \tag{8.153}$$

An orthonormal or pseudoorthonormal basis does not necessarily exist in a general inner-product space with an indefinite metric. Consider the example of an inner product that vanishes identically: Every basis is trivially orthogonal, but there are no unit vectors because all norms are zero. However, we show in this section and in Appendix C that for the class of pseudo-Euclidean or pseudo-unitary spaces, which includes the spaces that are essential for relativity theory and for relativistic quantum field theory as well as all Euclidean and unitary spaces of finite or countable dimension, it is possible to construct an orthonormal or pseudoorthonormal basis.

If $\{\bar{e}_i\}$ is an orthonormal or pseudoorthonormal basis of an inner-product space and $A : \mathcal{V} \to \mathcal{V}$ is a linear mapping, then it follows from the definition Eq. (6.16) of the matrix elements a^i_j of A as the contravariant components of the basis-vector images,

$$A\bar{e}_j = a^i_j\, \bar{e}_i, \tag{8.154}$$

that

$$a^i_j = \sigma_i \langle \bar{e}_i, A\bar{e}_j \rangle. \tag{8.155}$$

For a Euclidean or unitary space, for which $\sigma_i = 1$, this equation is often used as the definition of the matrix elements of a linear mapping or a linear operator.

Gram-Schmidt Construction

Let $\mathcal{V}$ be an n-dimensional Euclidean or unitary space, and let $A := \{\bar{a}_1, \bar{a}_2, \ldots, \bar{a}_n\}$ be a basis of $\mathcal{V}$. With the **Gram-Schmidt construction** one calculates an orthonormal basis by successively orthogonalizing the elements of A. The method is to pick a first vector (arbitrarily) from A and then subtract the projection of the first vector from each of the other vectors. In the resulting basis, every vector (other than the first vector itself) is orthogonal to the first vector, by construction (see Fig. 8.6). One then proceeds similarly with a second vector chosen from the new basis, and so on until one has a basis of n mutually orthogonal vectors.

At the first step of the Gram-Schmidt construction, the vectors are

$$\forall i \in (1:n) : \bar{a}_i^{(1)} := \bar{a}_i. \tag{8.156}$$

In order to obtain a matrix formulation of the Gram-Schmidt construction, it is useful to construct an auxiliary family of vectors, the first of which is

$$\bar{q}_1 := \bar{a}_1^{(1)} = \bar{a}_1. \tag{8.157}$$

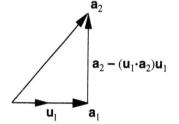

FIGURE 8.6

Illustration of the Gram-Schmidt construction in two-dimensional Euclidean space.

The inner-product norm of $\bar{q}_1$ is

$$
\begin{aligned}
d_1 &:= \|\bar{q}_1\|_{\text{ip}} \\
&= \langle \bar{q}_1, \bar{q}_1 \rangle^{1/2}.
\end{aligned}
$$
(8.158)

The (row 1, column 1) element of an auxiliary matrix that we construct is

$$
r_{11} := 1.
$$
(8.159)

Then

$$
\bar{u}_1 := \frac{1}{d_1} \bar{q}_1
$$
(8.160)

is a unit vector,

$$
\langle \bar{u}_1, \bar{u}_1 \rangle = 1,
$$
(8.161)

and $\{\bar{u}_1, \bar{a}_2^{(1)}, \ldots, \bar{a}_n^{(1)}\}$ is a basis of $\mathcal{V}$.

To complete the first step, we calculate the projections of $\{\bar{a}_2^{(1)}, \ldots, \bar{a}_n^{(1)}\}$ on the unit vector $\bar{u}_1$, obtaining

$$
\begin{aligned}
\forall j \in (2:n) : r_{1j} &:= \frac{1}{d_1} \langle \bar{q}_1, \bar{a}_j^{(1)} \rangle \\
&= \langle \bar{u}_1, \bar{a}_j^{(1)} \rangle.
\end{aligned}
$$
(8.162)

As in the construction of Fig. 8.6, one subtracts the projection of $\bar{a}_j^{(1)}$ along $\bar{u}_1$ from $\bar{a}_j^{(1)}$, obtaining a vector that is orthogonal to $\bar{u}_1$:

$$
\begin{aligned}
\forall j \in (2:n) : \bar{a}_j^{(2)} &:= \bar{a}_j^{(1)} - r_{1j} \bar{u}_1 \\
&= \bar{a}_j^{(1)} - \langle \bar{u}_1, \bar{a}_j^{(1)} \rangle \bar{u}_1 \\
\Rightarrow \langle \bar{u}_1, \bar{a}_j^{(2)} \rangle &= 0.
\end{aligned}
$$
(8.163)

Clearly

$$
A^{(2)} := \{ \bar{u}_1, \bar{a}_2^{(2)}, \ldots, \bar{a}_n^{(2)} \}
$$
(8.164)

is linearly independent and is a basis of $\mathcal{V}$.

At the beginning of the k-th step, one has constructed a basis

$$
A_k := \{ \bar{u}_1, \ldots, \bar{u}_{k-1}, \bar{a}_k^{(k)}, \ldots, \bar{a}_n^{(k)} \}
$$
(8.165)

such that each of the vectors $\bar{u}_1, \ldots, \bar{u}_{k-1}$ is orthogonal to all of its predecessors and successors in the list A_k, and each $\bar{u}_i$ is a unit vector. Applying the Gram-Schmidt construction to the next vector in the list, one obtains

$$\bar{q}_k := \bar{a}_k^{(k)}, \tag{8.166}$$

$$d_k := \langle \bar{q}_k, \bar{q}_k \rangle^{1/2} = \|\bar{q}_k\|_{\text{ip}}, \tag{8.167}$$

$$r_{kk} := 1, \tag{8.168}$$

$$\bar{u}_k := \frac{1}{d_k} \bar{q}_k, \tag{8.169}$$

and

$$\forall j \in (k+1 : n) : r_{kj} := \frac{1}{d_k} \langle \bar{q}_k, \bar{a}_j^{(k)} \rangle = \langle \bar{u}_k, \bar{a}_j^{(k)} \rangle. \tag{8.170}$$

Clearly $\bar{u}_k$ is a unit vector. The starting vectors for the next step are

$$\begin{aligned} \forall j \in (k+1 : n) : \bar{a}_j^{(k+1)} :&= \bar{a}_j^{(k)} - r_{kj} \bar{u}_k \\ &= \bar{a}_j^{(k)} - \langle \bar{u}_k, \bar{a}_j^{(k)} \rangle \bar{u}_k. \end{aligned} \tag{8.171}$$

By construction, each of the unit vectors $\bar{u}_i$ is orthogonal to every vector $\bar{a}_j^{(k+1)}$ (for $k+1 \le j \le n$):

$$\forall i \in (1 : k) : \forall j \in (k+1 : n) : \langle \bar{u}_i, \bar{a}_j^{(k+1)} \rangle = 0. \tag{8.172}$$

Also by construction, the vectors $\{\bar{u}_1, \ldots, \bar{u}_k\}$ are mutually orthogonal.

After the n-th step, one has constructed an orthonormal basis

$$A_n := \{\bar{u}_1, \ldots, \bar{u}_n\}, \tag{8.173}$$

as well as a unit upper-triangular matrix

$$\mathbf{R} := [r_{kj}] \tag{8.174}$$

in which all of the matrix elements that one has not calculated are defined as zero,

$$\forall j < k : r_{kj} := 0. \tag{8.175}$$

If one maps $\mathcal{V}$ onto $\mathbb{F}^n$ by the correspondence $\bar{e}_i \mapsto \mathbf{e}_i$ as usual, then one has a list of column vectors $\{\mathbf{q}_1, \ldots, \mathbf{q}_n\}$. The vectors $\mathbf{q}_i$ are mutually orthogonal by construction, for $\mathbf{q}_i = d_i \mathbf{u}_i$ in which where $\bar{u}_i \mapsto \mathbf{u}_i$.

The definition

$$\bar{a}_j^{(k)} = \bar{a}_j^{(k+1)} + r_{kj} \bar{q}_k \tag{8.176}$$

defines a downward recurrence relation for the vector $\bar{a}_j$, the solution of which is

$$\bar{a}_j = \sum_{k=j}^{n} r_{kj}\, \bar{q}_k. \tag{8.177}$$

If one lets **A** be the matrix whose columns are the vectors $\mathbf{a}_i$ where $\bar{a}_i \mapsto \mathbf{a}_i$,

$$\mathbf{A} := [\mathbf{a}_1, \ldots, \mathbf{a}_n], \tag{8.178}$$

then the entire Gram-Schmidt construction can be summarized by the single matrix equation

$$\mathbf{A} = \mathbf{QR}. \tag{8.179}$$

Equation (8.179) is called the **QR decomposition** of a nonsingular $n \times n$ matrix **A** into the product of a unit upper-triangular matrix **R** and a column-orthogonal matrix **Q**.

A great theoretical advantage of the Gram-Schmidt construction is that each vector $\bar{u}_i$ is constructed once and for all. One is spared from reorthogonalizing vectors that one has already constructed because basis vectors $\bar{u}_j$, for which $j > i$, are calculated as linear combinations of vectors $\bar{a}_l^{(j)}$, which are orthogonal to $\bar{u}_i$. For this reason it is possible to apply the Gram-Schmidt construction to any countable basis in an $\mathcal{L}^2$ or ℓ^2 space under the canonical inner product.

From a computational point of view the theoretical advantage of the Gram-Schmidt construction is its Achilles heel. Because of cancellation of significant digits in the subtractions used at every step of the Gram-Schmidt construction, the vectors that one constructs late in the process may not be numerically orthogonal to those constructed in the early steps, even though one uses linear combinations of vectors that are orthogonal to $\mathbf{u}_1$ to compute $\mathbf{u}_n$ (for example).

8.2.3 ORTHOGONAL POLYNOMIALS

Definition

A **family of orthogonal polynomials** is a set $\{p_n \mid n = 0, 1, \ldots\}$ of real- or complex-valued polynomials p_n, such that the degree of p_n is n and $\{p_n\}$ is mutually orthogonal under one of the inner products defined in Eq. (8.52) or (8.56). Many of the special functions of mathematical physics, such as the Legendre polynomials, are orthogonal polynomials.

In this section we derive the Legendre polynomials by applying the Gram-Schmidt procedure, show that the members of any family of orthogonal polynomials must satisfy a three-term recurrence relation, and then discuss the specific example of Chebyshev polynomials.

Legendre Polynomials via the Gram-Schmidt Procedure

Many applications require an orthogonal set of functions on the surface of the unit sphere. The simplest inner product, in which all area elements on the unit sphere are weighted equally, is

$$\langle\!\langle f, g \rangle\!\rangle = \int_0^{2\pi} \int_0^{\pi} f(\theta, \phi)^* \, g(\theta, \phi) \, \sin\theta \, d\theta \, d\phi$$

$$= \int_0^{2\pi} \int_{-1}^{1} f(\theta, \phi)^* \, g(\theta, \phi) \, d(\cos\theta) \, d\phi. \tag{8.180}$$

In simple cases f and g are products of functions of θ and ϕ alone. Thus one is led to consider an inner product of the form

$$\langle f, g \rangle = \int_{-1}^{1} f(\theta)^* \, g(\theta) \, d(\cos\theta), \tag{8.181}$$

which is an inner product on the interval $[-1, 1]$ of the form of Eq. (8.52) with a unit weight function, $w(x) = 1$.

One obtains an orthonormal basis $\{p_l(x)\}$ of the vector space of polynomials on $[-1, 1]$ with degree $\leq n$ by applying the Gram-Schmidt procedure to the nonorthogonal basis $\{1, x, \ldots, x^n\}$. The result is (see Exercise 8.2.1)

$$\forall l \in (0 : n) : p_l(x) = \sqrt{\frac{2l + 1}{2}} \, P_l(x), \tag{8.182}$$

in which P_l is the Legendre polynomial of order l. No matter how great n is, one can always compute P_n (although not by the Gram-Schmidt method, for purely practical reasons). Therefore one can orthonormalize the basis $\{1, x, \ldots, x^n, \ldots\}$ of the infinite-dimensional space $\mathcal{L}^2([0, 1]; \mathbb{R})$. Practical computations of $P_n(x)$ and other orthogonal polynomials use the recurrence relations derived in the next subsection.

Three-Term Recurrence Relations

Orthogonal polynomials always obey three-term recurrence relations, from which one may conveniently (and, usually, accurately) compute numerical values. Recurrence relations are also an essential theoretical tool for deriving many of the abstract properties of a family of polynomials.

For example, a recurrence relation for the values of the Legendre polynomials is

$$(l + 1)P_{l+1} = (2l + 1)x \, P_l(x) - l P_{l-1}(x). \tag{8.183}$$

We show that the existence of three-term recurrence relations follows directly from the property of orthogonality if applied to a family of polynomials such that the degree of $p(n)$ is n. Let $\mathcal{P}_n$ be the vector space over $\mathbb{F} = \mathbb{C}$ or $\mathbb{R}$ spanned by the monomials $1, x, \ldots, x^n$. The monomials are linearly independent on every finite interval of $\mathbb{R}$ (see Exercise 5.3.9). Therefore $\{1, x, \ldots, x^n\}$ is a basis of $\mathcal{P}_n$, and

$$\dim[\mathcal{P}_n] = n + 1. \tag{8.184}$$

Let $\{p_j \mid j \in (0 : n)\}$ be a set of orthonormal polynomials in $\mathcal{P}_n$ under one of the inner products Eq. (8.52) or Eq. (8.56), such that the degree of p_j is j. No p_j is linearly dependent on its predecessors (which are of lower degree than j); therefore $\{p_0, \ldots, p_n\}$ is linearly independent. Every set of $n + 1$ linearly independent vectors in a vector space $\mathcal{V}$ of dimension $n + 1$ is a basis of $\mathcal{V}$ (see Section 5.4.1). Hence $\{p_0, \ldots, p_n\}$ is a basis of $\mathcal{P}_n$.

The polynomial $x p_n$ (with values $x p_n(x)$) is of degree $n + 1$ and therefore is linearly independent of all polynomials of lower degree, including $p_0, \ldots, p_n$. Therefore the list $\{p_0, \ldots, p_n, x p_n\}$ is a basis of $\mathcal{P}_{n+1}$. It follows that there exist scalars $\alpha_n, c_{n0}, \ldots, c_{nn}$ such that

$$p_{n+1} = \alpha_n x p_n - \sum_{i=0}^{n} c_{ni} p_i. \tag{8.185}$$

Taking the inner product of each term of Eq. (8.185) with p_j (where $j = 0, \ldots, n$), and then using orthonormality, one obtains

$$c_{nj} = \alpha_n \langle p_j, x p_n \rangle. \tag{8.186}$$

For the inner products defined in Eqs. (8.52) and (8.58),

$$\langle p_j, x p_n \rangle = \langle x p_j, p_n \rangle. \tag{8.187}$$

The degree of $x p_j$ is $j + 1$ by the definition of p_j. If $j + 1 < n$, then $x p_j$ can be expressed as a linear combination of $p_0, \ldots, p_{j+1}$, each of which is orthogonal to p_n. Then

$$\forall j \in (0 : n - 2) : c_{nj} = 0. \tag{8.188}$$

It follows from Eqs. (8.185) and (8.188) that the orthogonal polynomials p_j obey the three-term recurrence relation

$$p_{n+1} = \alpha_n x p_n - c_{nn} p_n - c_{n,n-1} p_{n-1}. \tag{8.189}$$

One can cast Eq. (8.189) into a form which uses the information in Eq. (8.186). One has

$$c_{nn} = \alpha_n \langle p_n, x p_n \rangle,$$
$$c_{n,n-1} = \alpha_n \langle p_{n-1}, x p_n \rangle = \alpha_n \langle x p_{n-1}, p_n \rangle. \tag{8.190}$$

Then

$$\boxed{p_{n+1} = \alpha_n (x - \beta_n) p_n - \gamma_n p_{n-1}} \tag{8.191}$$

where

$$\beta_n := \frac{c_{nn}}{\alpha_n} = \langle p_n, x p_n \rangle,$$
$$\gamma_n := \alpha_n \langle x p_{n-1}, p_n \rangle. \tag{8.192}$$

The recurrence relation Eq. (8.191) and the scalars $\{\alpha_n, \beta_n, \gamma_n\}$ define the family of polynomials $\{p_n\}$ uniquely up to an overall multiplicative constant.

For example, for the Legendre polynomials $\beta_n = 0$, $\alpha_n = (2n+1)/(n+1)$ and $\gamma_n = n/(n+1)$. For the Chebyshev polynomials, which are orthogonal on the interval $[-1, 1]$, one has $\beta_n = 0$, $\alpha_n = 2$, and $\gamma_n = 1$. The Chebyshev polynomials are discussed in greater detail in the next subsection.

In several important families such as the Legendre, Chebyshev, and Hermite polynomials, the interval on which the polynomials are defined is $[-a, a]$ and the weight function w in Eq. (8.52) is even (that is, $w(-x) = w(x)$). Exercise 8.2.9 is to prove that in this case, p_n contains only terms of even degree if n is even, or only terms of odd degree if n is odd. It follows that p_n has parity $(-1)^n$:

$$p_n(-x) = (-1)^n p_n(x). \tag{8.193}$$

The same theorem holds for polynomials defined on a finite set if both the sampling points x_k and the weights w_i in Eq. (8.58) are symmetrically distributed about $x = 0$. Then

$$\beta_n = \langle p_n, x p_n \rangle = 0 \tag{8.194}$$

because the integral or sum in the inner product involves only odd powers of x, which have odd parity and whose integral or sum therefore vanishes.

Chebyshev Polynomials

The easy route to the Chebyshev polynomials goes through the family of trigonometric functions $\{\cos n\theta\}$, which obey the recurrence relation

$$\cos(n+1)\theta + \cos(n-1)\theta = 2\cos\theta\cos n\theta. \tag{8.195}$$

Iterating this relation, one obtains an expression for $\cos n\theta$ as a polynomial in powers of $\cos\theta$. Therefore it is natural to define a new variable $x = \cos\theta$, making $\cos n\theta$ a polynomial in x:

$$x := \cos\theta \Rightarrow x \in [-1, 1],$$
$$T_n(x) := \cos[n\cos^{-1} x] \Rightarrow T_n(\cos\theta) = \cos n\theta. \tag{8.196}$$

Then the **Chebyshev polynomials** $\{T_n\}$ obey the recurrence relation

$$T_{n+1}(x) = 2x T_n(x) - T_{n-1}(x). \tag{8.197}$$

It follows directly from the definition Eq. (8.196) that T_n has parity $(-1)^n$ (see Exercise 8.2.11). An explicit expression for the Chebyshev polynomials follows from the recurrence relation and the principle of finite induction:

$$T_n(x) = \frac{n}{2} \sum_{m=0}^{\lfloor n/2 \rfloor} (-1)^m \frac{(n-m-1)!}{m!(n-2m)!} (2x)^{n-2m} \tag{8.198}$$

(see Exercise 8.2.16). (Recall that the notation $\lfloor n/2 \rfloor$ means the integer part of $n/2$.) Equation (8.198) is not of much use in numerical computations because the alternation of the signs of the terms leads to a troublesome loss of significant digits (see Section 1.4.3). The inverse expression for the monomial x^n in terms of $T_n, T_{n-2}, \ldots, T_1$ or T_0,

$$x^n = 2^{1-n} \left[T_n(x) + \binom{n}{1} T_{n-2}(x) + \cdots + \binom{n}{k} \begin{cases} T_1(x) & \text{if } n = 2k+1 \\ \frac{1}{2} T_0(x) & \text{if } n = 2k \end{cases} \right], \qquad (8.199)$$

follows from the relation $T_n(\cos \theta) = \cos(n\theta)$ and the Euler formula $\cos \theta = \frac{1}{2}(e^{i\theta} + e^{-i\theta})$ (see Exercise 8.2.17). Equation (8.199) can be quite useful in numerical applications, especially in improving the numerical properties of partial sums of power series.

Because we have obtained the Chebyshev polynomials without making use of orthogonality, we must now find a weight function that makes $\langle T_m, T_n \rangle = 0$ if $m \neq n$ (or a set of discrete weights w_i that lead to the same result). The trigonometric orthogonality relations satisfied by $\cos n\theta$ for $\theta \in [0, \pi]$ are the natural starting points. From

$$\int_0^\pi \cos n\theta \cos m\theta \, d\theta = 0 \text{ if } m \neq n \qquad (8.200)$$

one finds by changing variables as in Eq. (8.196) that a weight function w that makes T_m orthogonal to T_n if $m \neq n$ is

$$w(x) = \frac{1}{[1 - x^2]^{1/2}}. \qquad (8.201)$$

With this weight function, one finds straightforwardly that

$$\langle T_m, T_n \rangle = \int_{-1}^1 T_m(x) \, T_n(x) \frac{1}{[1 - x^2]^{1/2}} \, dx = \begin{cases} 0, & \text{if } m \neq n; \\ \dfrac{\pi}{2}, & \text{if } m = n \neq 0; \\ \pi, & \text{if } m = n = 0 \end{cases} \qquad (8.202)$$

(see Exercise 8.2.18).

We show in Section 8.4.5 that the Chebyshev polynomials are also useful for approximating functions the values of which are known only at a certain finite set of sampling points. The *Chebyshev-sampling mapping*

$$\mathsf{C}_N : \mathcal{C}^0([-1, 1]; \mathbb{R}) \to \mathbb{R}^N \qquad (8.203)$$

is defined as a sampling mapping (see Eq. [6.104]) in which the sampling points $x_{N,j}$ are chosen as the zeros of T_N. Because $T_N(x) = \cos(Nx)$, it follows that

$$x_{N,j} := \cos \left(\frac{(2j+1)\pi}{2N} \right) \quad (j \in (0, N-1)). \qquad (8.204)$$

The numbers $\{x_{N,j}\}$ are called the **Chebyshev abscissae**.

The vector of sampled values produced by Chebyshev sampling of a function $f \in \mathcal{C}^0$ $([-1, 1]; \mathbb{R})$ is

$$C_N f := \begin{pmatrix} f(x_{N,1}) \\ \vdots \\ f(x_{N,N}) \end{pmatrix} \tag{8.205}$$

in which C_N is the **Chebyshev-sampling mapping** on N points. In particular, the vector $\mathbf{c}_j$ defined in Eq. (8.148) is proportional to the vector obtained through Chebyshev sampling of the Chebyshev polynomial T_{j-1}:

$$\mathbf{c}_j = \begin{cases} \sqrt{\dfrac{2}{N}} C_N T_{j-1}, & \text{if } j > 0; \\[2ex] \sqrt{\dfrac{1}{N}} C_N T_{j-1}, & \text{if } j = 0. \end{cases} \tag{8.206}$$

(See Exercise 8.2.15.)

The discrete inner product of Chebyshev-sampled functions,

$$\langle\!\langle f, g \rangle\!\rangle := \langle C_N f, C_N g \rangle_{\mathbb{E}^N}$$
$$= \sum_{i=1}^{N} f(x_i) g(x_i), \tag{8.207}$$

approximates the inner product Eq. (8.202) by a sum of sampled values in which the density of the sampling points is equal to the weight function, $[1 - x^2]^{-1/2}$. Evidently Eq. (8.207) is a special case of the inner product Eq. (8.56) for nonuniformly sampled functions.

Under the inner product Eq. (8.207), the Chebyshev polynomials obey a discrete orthogonality relation that is quite important in computing a numerical Chebyshev approximation of a function $f : [-1, 1] \to \mathbb{R}$. To obtain this orthogonality relation, recall that $T_n(x) = \cos(n \cos^{-1} x)$. The same calculation (Exercise 8.2.15) that shows that the vectors $\mathbf{c}_i$ defined in Eq. (8.148) are orthonormal shows that under the inner product Eq. (8.207) the Chebyshev polynomials are mutually orthogonal,

$$\langle\!\langle T_i, T_j \rangle\!\rangle = 0 \text{ if } j \neq i, \tag{8.208}$$

and that the square of the inner-product norm of T_j in $\mathbb{E}^n$ is

$$\langle\!\langle T_j, T_j \rangle\!\rangle = \| C_N T_j \|_{\text{ip}}^2 = \begin{cases} N/2, & \text{if } j \neq 0; \\ N, & \text{if } j = 0. \end{cases} \tag{8.209}$$

We use these relations in Section 8.4 to construct an approximation to a function that is known only at the points x_j.

8.2.4 EXERCISES FOR SECTION 8.2

8.2.1 Let V be $\mathcal{L}^2([-1, 1]; \mathbb{R})$ and let $w(x) = 1$ for all $x \in [-1, 1]$. The set of functions $S = \{1, x, x^2, x^3, x^4\}$ spans a subspace of V. Orthonormalize S using the Gram-Schmidt procedure, and compare your results with the standard definition of the Legendre polynomials,

$$
\begin{aligned}
P_0(x) &= 1 \\
P_1(x) &= x \\
P_2(x) &= \tfrac{1}{2}(3x^2 - 1) \\
P_3(x) &= \tfrac{1}{2}(5x^3 - 3x) \\
P_4(x) &= \tfrac{1}{8}(35x^4 - 30x^2 + 3).
\end{aligned}
\tag{8.210}
$$

8.2.2 Calculate $\cos\theta_{\bar{x},\lambda\bar{x}}$ for a unitary space. Conclude that $\cos\theta_{\bar{x},\lambda\bar{x}} = 1$ if and only if λ is real.

8.2.3 Prove that the canonical basis $\{e_i\}$ defined in Eq. (5.122) is orthonormal in $\mathbb{R}^n$ under the canonical inner product.

8.2.4 Prove that the canonical basis $\{e_i\}$ defined in Eq. (5.122) is orthonormal in $\mathbb{U}^n$ under the canonical inner product.

8.2.5 Prove that the canonical basis $\{E_i^j\}$ defined in Eq. (5.168) is orthonormal in $\mathbb{C}^{n\times n}$ under the canonical inner product Eq. (8.21).

8.2.6 Verify that the sequences $\bar{e}_i$ defined in Eq. (5.169) are orthonormal under the inner product Eq. (8.50) on $\mathcal{L}^2(\mathbb{C})$.

8.2.7 Prove that the **Fourier functions**

$$
\forall m \in \mathbb{Z} : e_m(x) := \frac{1}{\sqrt{2\pi}} e^{imx}
\tag{8.211}
$$

are orthonormal in $\mathcal{L}^2([-\pi, \pi]; \mathbb{C})$ under the inner product Eq. (8.52) if the weight function is defined as

$$
w(x) := 1.
\tag{8.212}
$$

8.2.8 Prove that the **discrete Fourier functions**

$$
\forall m \in (0, N-1) : f_m(x) := e^{2\pi imx},
\tag{8.213}
$$

if uniformly sampled on the interval $[0, 1]$ as in Eq. (8.54),

$$
\begin{aligned}
\mathbf{f}_m &:= \mathsf{S}_N f_m \\
&= \begin{pmatrix} 1 \\ e^{2\pi im/N} \\ \vdots \\ e^{2\pi im(N-1)/N} \end{pmatrix},
\end{aligned}
\tag{8.214}
$$

are an orthonormal basis of $\mathbb{C}^N$ under the inner product of uniformly sampled functions defined in Eq. (8.56).

Hint: Establish that

$$\langle S_N f_m, S_N f_n \rangle = \frac{1}{N} \mathbf{f}_m^\dagger \mathbf{f}_n = \frac{1}{N} \begin{cases} N, & \text{if } m \equiv n \pmod{N}. \\ \dfrac{1 - e^{2\pi i (n-m)}}{1 - e^{2\pi i (n-m)/N}}, & \text{if } m \neq n. \end{cases} \tag{8.215}$$

8.2.9   Let $\{p_n\}$ be a family of polynomials defined on $[-a, a]$ that are orthonormal under the inner product Eq. (8.52), and let the weight function w be even,

$$w(-x) = w(x). \tag{8.216}$$

Prove that p_n contains only terms of even degree if n is even, or only terms of odd degree if n is odd.

Hint: The easy way is to prove Eq. (8.194) using the principle of finite induction and the recurrence relation Eq. (8.191), starting at $n = 0$. Next use the recurrence relation again to show inductively that if n is even, then p_n has only terms of even degree, and so forth.

8.2.10   Using the recurrence relation Eq. (8.197) and assuming that

$$T_0(x) = 1, \tag{8.217}$$

show that

$$\begin{aligned} T_0(x) &= 1 \\ T_1(x) &= x \\ T_2(x) &= 2x^2 - 1 \\ T_3(x) &= 4x^3 - 3x \\ T_4(x) &= 8x^4 - 8x^2 + 1. \end{aligned} \tag{8.218}$$

8.2.11   Prove that

$$\cos[n \cos^{-1}(-x)] = (-1)^n \cos[n \cos^{-1}(x)]. \tag{8.219}$$

Deduce the parity of the Chebyshev polynomials,

$$T_n(-x) = (-1)^n T_n(x). \tag{8.220}$$

8.2.12   Prove that the coefficient of the leading term x^n in $T_n(x)$ is 2^{n-1} for all $n \geq 1$.

8.2.13   Prove that for every $x \in [-1, 1]$,

$$|T_n(x)| \leq 1. \tag{8.221}$$

8.2.14   Let $\mathcal{V}$ be a Euclidean space. Let a function $f : \mathcal{V} \to \mathbb{R}$ (which we intend to interpret as a vector length) have the following properties for all vectors $\overline{x}, \overline{y} \in \mathcal{V}$

and all scalars $\alpha \in \mathbb{R}$:

$$f(\bar{x}) \geq 0, \tag{8.222}$$

$$f(\bar{x}) = 0 \Leftrightarrow \bar{x} = \bar{0}, \tag{8.223}$$

$$f(\alpha \bar{x}) = |\alpha| \, f(\bar{x}), \tag{8.224}$$

and

$$f(\bar{x} + \bar{y}) \leq f(\bar{x}) + f(\bar{y}). \tag{8.225}$$

(These are reasonable properties if f is a vector length, and in fact are the axioms for a vector norm; see Section 8.7.) Prove that the function $g : \mathcal{V} \times \mathcal{V} \to \mathbb{R}$ such that

$$g(\bar{x}, \bar{y}) := \tfrac{1}{4}([f(\bar{x} + \bar{y})]^2 - [f(\bar{x} - \bar{y})]^2) \tag{8.226}$$

is an inner product on $\mathcal{V}$ if and only if the function f satisfies the parallelogram law,

$$[f(\bar{x} + \bar{y})]^2 + [f(\bar{x} - \bar{y})]^2 = 2([f(\bar{x})]^2 + [f(\bar{y})]^2). \tag{8.227}$$

8.2.15 Prove that the vectors $c_1, \ldots, c_n$ defined in Eq. (8.148) are orthonormal and are therefore a basis of $\mathbb{E}^n$ or $\mathbb{U}^n$.

Hint: Express the cosines which occur in the inner product $\langle c_i, c_j \rangle$ in terms of exponentials and use the formula for the sum of a geometric progression.

8.2.16 Use the principle of finite induction and the recurrence relation for the Chebyshev polynomials, Eq. (8.197), to prove Eq. (8.198).

8.2.17 Use the relation $T_n(\cos \theta) = \cos(n\theta)$ and the Euler expression $\cos \theta = \tfrac{1}{2}(e^{i\theta} + e^{-i\theta})$ to prove Eq. (8.199).

8.2.18 Prove Eq. (8.202).

8.2.19 The goal of this problem is to compare the results of a numerical and an analytical Gram-Schmidt procedure. By a numerical Gram-Schmidt procedure, we mean that the procedure is to be applied to a set of n-dimensional vectors that are obtained by sampling certain continuous functions. Let $\mathcal{V} = \mathcal{C}^0([-1, 1]; \mathbb{R})$, and let S_n be the uniform sampling mapping defined in Eq. (8.54).

(a) Create a computer program that carries out the Gram-Schmidt procedure (or, equivalently, the QR decomposition). Design your program to accept as input both a value of n and an initial set of vectors. Assume that the inner product is the canonical inner product in $\mathbb{E}^n$. (If your instructor permits, you may make use of published programs.)

(b) Let

$$x_j := S_n x^j \tag{8.228}$$

be the vector obtained by uniform sampling of the function $f(x) = x^j$ on the interval $[-1, 1]$. Using the computer program created in part (a), apply the Gram-Schmidt procedure to the set $\{x_0, \ldots, x_{10}\}$.

(c) Using an appropriate numerical method, express the orthonormal vectors that you computed in part (b) as linear combinations of the uniformly sampled monomials, $\{x_0, \ldots, x_{10}\}$.

(d) Compare the results you obtained in part (c) with the analytic formulas for the Legendre polynomials of orders 0 through 10. (Don't forget the normalization constant!)

8.3 PROJECTION METHODS

8.3.1 PROJECTION OF A VECTOR ONTO A SUBSPACE: DEFINITION

In real three-dimensional Euclidean space, one projects a vector on a plane or a line by "dropping a perpendicular" from the tip of the vector. For example, computer graphics software that renders three-dimensional objects must project the polygons that make up a three-dimensional object onto a two-dimensional rendering plane. An orthogonal projection on a horizontal plane in three-dimensional space is called a **plan**, and an orthogonal projection on a vertical plane is called an **elevation**.

To generalize the concept of projection onto a two-dimensional plane to abstract vector spaces, suppose that $\mathcal{W}$ is an m-dimensional subspace of an n-dimensional Euclidean or unitary space $\mathcal{V}$. By definition, the vector $\overline{w} \in \mathcal{W}$ is the **orthogonal projection** of the vector $\overline{x} \in \mathcal{V}$ onto a vector subspace $\mathcal{W}$ if and only if $\overline{x} - \overline{w}$ is orthogonal to every element of $\mathcal{W}$. We show that the requirement that

$$\overline{x} - \overline{w} \perp \mathcal{W} \tag{8.229}$$

defines the orthogonal projection $\overline{w}$ uniquely. In so doing, we obtain a useful formula for $\overline{w}$.

The inner product $\langle \overline{w}, \overline{w}' \rangle$ of any two vectors in $\mathcal{W}$ obeys Eqs. (8.42) and (8.43) because $\mathcal{V}$ is Euclidean or unitary. Then $\mathcal{W}$ is itself a Euclidean or unitary space and therefore possesses an orthonormal basis $\{\overline{e}_1, \ldots, \overline{e}_m\}$, which can be constructed (at least in principle) by the Gram-Schmidt method. Using $\{\overline{e}_i\}$ and referring to Eq. (8.64) and Exercise 8.1.7, one sees that

$$\overline{x} - \overline{w} \perp \mathcal{W} \Leftrightarrow [\forall i \in (1 : m) : \langle \overline{e}_i, \overline{x} - \overline{w} \rangle = 0] \tag{8.230}$$

$$\Leftrightarrow [\forall i \in (1 : m) : \langle \overline{e}_i, \overline{w} \rangle = \langle \overline{e}_i, \overline{x} \rangle]. \tag{8.231}$$

Eq. (8.231) determines the projection of $\overline{x}$ onto $\mathcal{W}$:

$$\overline{w} = \sum_{i=1}^{m} \langle \overline{e}_i, \overline{x} \rangle \overline{e}_i. \tag{8.232}$$

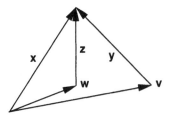

FIGURE 8.7

Orthogonal projection of a vector **x** on a vector subspace $\mathcal{W}$, represented here by a plane. Note that the orthogonal projection **w** is the best approximation to **x** by an element of $\mathcal{W}$.

Equation (8.232) establishes the existence of a vector $\overline{w} \in \mathcal{W}$ such that $\overline{x} - \overline{w} \perp \mathcal{W}$. It is also true that the projection $\overline{w}$ given in Eq. (8.232) is the *unique* element of $\mathcal{W}$ that satisfies the requirement Eq. (8.229), because Eq. (8.231) determines the contravariant components of $\overline{w}$ uniquely.

Figure 8.7 illustrates the concept of projection on a subspace if $\mathcal{V} = \mathbb{E}^3$ and $\mathcal{W} = \mathbb{E}^2$.

8.3.2 ORTHOGONAL PROJECTORS

The mapping $\mathsf{P}_{\mathcal{W}} : \overline{x} \mapsto \overline{w}$ is called the **orthogonal projector** on the subspace $\mathcal{W}$. Equivalently,

$$\mathsf{P}_{\mathcal{W}}\overline{x} := \sum_{i=1}^{m} \langle \overline{e}_i, \overline{x} \rangle \overline{e}_i. \tag{8.233}$$

Because $\langle \overline{e}_i, \overline{x} + \overline{y} \rangle = \langle \overline{e}_i, \overline{x} \rangle + \langle \overline{e}_i, \overline{y} \rangle$ and $\langle \overline{e}_i, \alpha \overline{x} \rangle = \alpha \langle \overline{e}_i, \overline{x} \rangle$, an orthogonal projector $\mathsf{P}_{\mathcal{W}}$ is a linear mapping.

For example, the projector $\mathsf{P}_{\overline{u}}$ defined in Eq. (8.105) projects on the one-dimensional subspace spanned by $\overline{u}$. From Eqs. (8.103) and (8.105),

$$\mathsf{P}_{\overline{x}}\overline{y} := \frac{\langle \overline{x}, \overline{y} \rangle}{\langle \overline{x}, \overline{x} \rangle} \overline{x} \tag{8.234}$$

is the projector on the one-dimensional subspace spanned by an arbitrary vector $\overline{x}$ such that $\langle \overline{x}, \overline{x} \rangle \neq 0$.

In order to find the matrix $\mathbf{P}_{\mathbf{u}}$ of a projector on $\mathbf{u} \in \mathbb{U}^n$, in which $\mathbf{u}$ is a unit vector ($\langle \mathbf{u}, \mathbf{u} \rangle = 1$), let us make a coordinate expansion of $\mathbf{P}_{\mathbf{u}}\mathbf{x}$ relative to the canonical basis:

$$\mathbf{P}_{\mathbf{u}}\mathbf{x} = \langle \mathbf{u}, \mathbf{x} \rangle \mathbf{u} = \sum_{i,j=1}^{n} u^{i*} x^i u^j \mathbf{e}_j. \tag{8.235}$$

(Results for $\mathbb{E}^n$ follow if one suppresses complex conjugation.) By inspection one sees that the (row i, column j) matrix element of $\mathbf{P_u}$ is

$$p^i_j = u^{i*}u^j. \tag{8.236}$$

It follows that the matrix elements of the projector $\mathbf{P_x}$ on an arbitrary vector $\mathbf{x} \in \mathbb{E}^n$ or $\mathbf{x} \in \mathbb{U}^n$ are

$$p^i_j = \frac{x^{i*}x^j}{\|\mathbf{x}\|^2}. \tag{8.237}$$

For example, the matrix of the projector along the direction of the vector

$$\mathbf{x} = \begin{pmatrix} 1 \\ -2 \\ 1 \end{pmatrix} \in \mathbb{E}^3 \tag{8.238}$$

is

$$\mathbf{P_x} = \frac{1}{3}\begin{pmatrix} 1 & -2 & 1 \\ -2 & 4 & -2 \\ 1 & -2 & 1 \end{pmatrix}. \tag{8.239}$$

Exercise 9.3.3 derives the eigenvalues and eigenvectors of any $n \times n$ matrix $\mathbf{P_x}$ that is an orthogonal projector on the direction of a vector $\mathbf{x}$.

From Eq. (8.233) it follows that

$$\mathsf{P_W} = \sum_{i=1}^{m} \mathsf{P}_i, \tag{8.240}$$

in which

$$\mathsf{P}_i := \mathsf{P}_{\bar{e}_i} \tag{8.241}$$

is the orthogonal projector on the one-dimensional subspace spanned by $\bar{e}_i$.

In many practical situations $\mathcal{V}$ is either $\mathbb{E}^n$ or $\mathbb{U}^n$. To compute the matrix of $\mathsf{P_W}$ given an *orthogonal* basis $\{\mathbf{f}_1, \ldots, \mathbf{f}_n\}$ of $\mathcal{W}$, one uses Eq. (8.240) with P_i replaced by $\mathbf{P}_{\mathbf{f}_i}$. Each projection matrix $\mathbf{P}_{\mathbf{f}_i}$ can be computed from Eq. (8.237). For example, let $\mathcal{W}$ be the subspace of $\mathbb{E}^3$ spanned by the orthogonal vectors

$$\mathbf{f}_1 := \begin{pmatrix} 1 \\ 1 \\ 1 \end{pmatrix}, \quad \mathbf{f}_2 := \begin{pmatrix} 1 \\ -2 \\ 1 \end{pmatrix}. \tag{8.242}$$

Then

$$\mathbf{P_W} = \mathbf{P_{f_1}} + \mathbf{P_{f_2}}$$

$$= \frac{1}{3}\begin{pmatrix} 1 & 1 & 1 \\ 1 & 1 & 1 \\ 1 & 1 & 1 \end{pmatrix} + \frac{1}{6}\begin{pmatrix} 1 & -2 & 1 \\ -2 & 4 & -2 \\ 1 & -2 & 1 \end{pmatrix} \tag{8.243}$$

$$= \frac{1}{2}\begin{pmatrix} 1 & 0 & 1 \\ 0 & 2 & 0 \\ 1 & 0 & 1 \end{pmatrix}.$$

Projection matrices such as $\mathbf{P_W}$ in the preceding equation are useful because they automate the process of projection on a subspace. If one knows $\mathbf{P_W}$, then one can replace the steps of computing the inner product of $\mathbf{x}$ with each of the basis vectors of W and adding the resulting vectors with the single step of matrix multiplication.

Equation (8.152) implies that for an orthonormal basis $\{\bar{e}_i\}$,

$$\bar{x} = \sum_{i=1}^{n} \langle \bar{e}_i, \bar{x} \rangle \bar{e}_i = \sum_{i=1}^{n} P_i \bar{x}. \tag{8.244}$$

Then

$$\sum_{i=1}^{n} P_i = 1. \tag{8.245}$$

Clearly $\mathbf{1}$ is the projector on $\mathcal{V}$. This equation is often called the **completeness relation** for the projectors P_i.

One shows easily using the orthonormality of the basis $\{\bar{e}_i\}$ that every projector P_W is **idempotent**:

$$P_W^2 := P_W P_W = P_W \tag{8.246}$$

(see Exercise 8.3.1). The reader should take the time to check that each of the projection matrices above is idempotent.

8.3.3 ORTHOGONAL COMPLEMENT

Complementary subspaces, as defined in Section 5.5, have only the zero vector in common and are such that every vector in the vector space can be expressed as the sum of two vectors, one from each of the complementary subspaces. Among the infinitely many subspaces that are complementary to any given proper subspace, there is one that is especially useful, namely, the *orthogonal* complement, every vector of which is orthogonal to the other subspace.

Orthogonal Decomposition

Let W be a vector subspace of a Euclidean or unitary space V. The set $W^\perp$, which consists of all vectors that are orthogonal to W,

$$W^\perp := \{\bar{x} \in V \mid \bar{x} \perp W\}, \tag{8.247}$$

is called the **orthogonal complement** of W. The orthogonal complement $W^\perp$ is unique because, by definition, it includes *all* vectors in V that are orthogonal to W. Although a nonnull proper vector subspace W of a Euclidean or unitary space V has infinitely many complements, W has only one *orthogonal* complement, $W^\perp$.

For example, if $V = E^3$ and if W is the vector subspace spanned by the Cartesian unit vector **k**, then $W^\perp$ is the XY plane (because a vector **z** is orthogonal to a subspace W if and only if **z** is orthogonal to every vector in a basis of W).

By Eqs. (8.2) through (8.6), all linear combinations of vectors in $W^\perp$ belong to $W^\perp$:

$$\bar{x} \perp W, \bar{y} \perp W \Rightarrow \alpha\bar{x} + \beta\bar{y} \perp W. \tag{8.248}$$

The zero vector of V belongs to $W^\perp$. Because $W^\perp \subseteq V$, $W^\perp$ satisfies all of the other axioms for a vector space. Hence $W^\perp$ is a vector subspace of V.

Complementarity

We show that V is the direct sum of W and $W^\perp$, and therefore that W and $W^\perp$ are complementary subspaces as defined in Section 5.5.1. We assume for now that V is of finite dimension n and that dim$[W] = m$. (The infinite-dimensional case is treated in Chapter 10.) To prove that

$$V = W \oplus W^\perp \tag{8.249}$$

(see Eqs. [5.256] through [5.258]), one must show that for every vector $\bar{x} \in V$ there exist *unique* vectors $\bar{w} \in W$ and $\bar{z} \in W^\perp$ such that

$$\bar{x} = \bar{w} + \bar{z}. \tag{8.250}$$

Certainly $\bar{w}$ and $\bar{z}$ exist, for one can take

$$\bar{w} = P_W\bar{x}, \quad \bar{z} = \bar{x} - \bar{w}. \tag{8.251}$$

The definition of P_W implies that $P_W\bar{x} \in W$ and that $\bar{x} - \bar{w} \perp W$ (see Eq. [8.229]). Then $\bar{z} = \bar{x} - \bar{w} \in W^\perp$. To verify in another way that $\bar{z} = \bar{x} - \bar{w}$ belongs to $W^\perp$, let $\{\bar{e}_1, \ldots, \bar{e}_m\}$ be an orthonormal basis of W, and let $\bar{w}'$ be any vector in W. Then

$$\bar{w}' = \sum_{i=1}^{m} \langle \bar{e}_i, \bar{w}' \rangle \bar{e}_i = \sum_{i=1}^{m} w'^i \bar{e}_i, \tag{8.252}$$

and therefore

$$\forall \overline{w}' \in W : \langle \overline{z}, \overline{w}' \rangle = \sum_{i=1}^{m} w'^i \langle \overline{z}, \overline{e}_i \rangle = 0 \tag{8.253}$$

because $\langle \overline{z}, \overline{e}_i \rangle = 0$ for every $i \in (1 : m)$. Then $\overline{z} = \overline{x} - P_W \overline{x} \in W^\perp$. One calls Eqs. (8.250) and (8.251) the **orthogonal decomposition** of a vector $\overline{x}$ with respect to a subspace W.

Uniqueness

The uniqueness of the orthogonal decomposition given in Eqs. (8.250) and (8.251) follows from the definition of orthogonality and the assumed positive-definiteness of the inner product defined on V. Suppose that, for some vector $\overline{x} \in V$, one has

$$\overline{x} = \overline{w} + \overline{z} \quad \text{and} \quad \overline{x} = \overline{w}' + \overline{z}' \tag{8.254}$$

in which $\overline{w}, \overline{w}' \in W$ and $\overline{z}, \overline{z}' \in W^\perp$. Then

$$(\overline{w} - \overline{w}') + (\overline{z} - \overline{z}') = \overline{0}. \tag{8.255}$$

Now take the inner product of this equation with $\overline{w} - \overline{w}'$, recalling that $\overline{z} - \overline{z}'$ belongs to $W^\perp$ and therefore is orthogonal to $\overline{w} - \overline{w}'$:

$$\langle \overline{w} - \overline{w}', \overline{w} - \overline{w}' \rangle + \langle \overline{w} - \overline{w}', \overline{z} - \overline{z}' \rangle = \langle \overline{w} - \overline{w}', \overline{w} - \overline{w}' \rangle = 0. \tag{8.256}$$

It follows from this equation and Eqs. (8.43) and (8.255) that

$$\overline{w} = \overline{w}' \text{ and therefore that } \overline{z}' = \overline{z}. \tag{8.257}$$

Therefore the decomposition given in Eq. (8.250) is unique, implying that W and $W^\perp$ are complementary subspaces.

In three dimensions, the only case in which a subspace W and its orthogonal complement $W^\perp$ are neither the zero subspace nor all of three-dimensional space occurs if W is a line $\mathcal{L}$ (through the origin) and $W^\perp$ is the plane perpendicular to $\mathcal{L}$ through the origin (or the same case with W and $W^\perp$ interchanged).

Note that two planes in three-dimensional space that are perpendicular according to the usual solid-geometry definition are not orthogonal vector subspaces if the inner product is the usual dot product. First, a plane $\mathcal{P}$ in $\mathbb{R}^3$ is an affine subspace but is not a vector subspace unless $\mathcal{P}$ contains the origin. Second, two perpendicular planes $\mathcal{P}, \mathcal{P}'$ in $\mathbb{R}^3$ that both contain the origin, and which are therefore vector subspaces, intersect in a line $\mathcal{L}$ that passes through the origin and is therefore a vector subspace. Every nonzero vector $\mathbf{x}$ along $\mathcal{L}$ belongs to both of the subspaces $\mathcal{P}$ and $\mathcal{P}'$, but $\mathbf{x} \cdot \mathbf{x} \neq 0$. Therefore $\mathcal{P}$ and $\mathcal{P}'$ are not orthogonal as vector subspaces.

Basis of the Orthogonal Complement

One can compute a basis of the orthogonal complement of an m-dimensional subspace W of $\mathbb{E}^n$ or $\mathbb{U}^n$ by the method of LU decomposition, which we used in Section 6.6.5 to find a basis of the null space of a linear mapping $\mathbf{A}$. To keep the notation general we explain the method for $\mathbb{U}^n$; the results can be specialized for $\mathbb{E}^n$ by suppressing complex conjugation. The notation used here differs from that used in our discussion of the LU decomposition in Section 6.6.4, but the concepts are the same.

The canonical inner product in $\mathbb{U}^n$ is $\langle \mathbf{x}, \mathbf{y} \rangle = \mathbf{x}^\dagger \mathbf{y}$. The statement that $\mathbf{y}$ is orthogonal to $\mathbf{x}$ is equivalent to the statement that $\mathbf{y}$ belongs to the null space of a linear mapping with the matrix $\mathbf{x}^\dagger$. It follows that if $\{\mathbf{n}_1, \ldots, \mathbf{n}_{n-m}\}$ is a basis of $W^\perp$ under the canonical inner product, then $\{\mathbf{n}_1, \ldots, \mathbf{n}_{n-m}\}$ is a basis of the null space of $\mathbf{A}_m^\dagger$, in which $\mathbf{A}_m = [\mathbf{a}_1, \ldots, \mathbf{a}_m]$ is a basis of W. Symbolically,

$$\mathbf{y} \in W^\perp \Leftrightarrow \mathbf{A}_m^\dagger \mathbf{y} = \mathbf{0}. \tag{8.258}$$

A method of finding a basis of $W^\perp$ now suggests itself. To find a basis of null $[\mathbf{A}_m^\dagger]$ one can perform an LU decomposition of the $n \times m$ matrix $\mathbf{A}_m$, obtaining

$$\mathbf{A}_m = (\mathbf{U}^\dagger)^{-1} \mathbf{V}^\dagger \tag{8.259}$$

in which $(\mathbf{U}^\dagger)^{-1}$ is an $n \times n$ unit lower-triangular matrix such as $\mathbf{L}$ in Eq. (6.403), and $\mathbf{V}^\dagger$ is an $n \times m$ upper echelon matrix. Then

$$\mathbf{A}_m^\dagger \mathbf{U} = \mathbf{V} \tag{8.260}$$

in which $\mathbf{V}$ is the $m \times n$ transpose of an upper echelon matrix. The column-partitioned forms of $\mathbf{V}$ and $\mathbf{U}$ are

$$\mathbf{V} = \left[\mathbf{R}_{m \times m}, \mathbf{0}_{m \times (n-m)} \right], \quad \mathbf{U} = \left[\mathbf{C}_{n \times m}, \mathbf{N}_{n \times (n-m)} \right] \tag{8.261}$$

in which $\mathbf{N}_{n \times (n-m)}$ is a basis of $W^\perp$.

For example, to find the orthogonal complement of the subspace W of $\mathbb{E}^3$ spanned by the vectors

$$\mathbf{a}_1 = \begin{pmatrix} 1 \\ 0 \\ 1 \end{pmatrix}, \quad \mathbf{a}_2 = \begin{pmatrix} 0 \\ 1 \\ 0 \end{pmatrix} \tag{8.262}$$

one performs an LU decomposition of the matrix $\mathbf{A}_2 = [\mathbf{a}_1, \mathbf{a}_2]$, obtaining

$$\mathbf{A}_2 = \begin{pmatrix} 1 & 0 \\ 0 & 1 \\ 1 & 0 \end{pmatrix} = \begin{pmatrix} 1 & 0 & 0 \\ 0 & 1 & 0 \\ 1 & 0 & 1 \end{pmatrix} \begin{pmatrix} 1 & 0 \\ 0 & 1 \\ 0 & 0 \end{pmatrix} \tag{8.263}$$

for the equation $\mathbf{A}_m = (\mathbf{U}^\dagger)^{-1}\mathbf{V}^\dagger$. Then the equation $\mathbf{A}_m^\dagger \mathbf{U} = \mathbf{V}$ reads

$$\begin{pmatrix} 1 & 0 & 1 \\ 0 & 1 & 0 \end{pmatrix} \begin{pmatrix} 1 & 0 & -1 \\ 0 & 1 & 0 \\ 0 & 0 & 1 \end{pmatrix} = \begin{pmatrix} 1 & 0 & 0 \\ 0 & 1 & 0 \end{pmatrix}. \tag{8.264}$$

It follows that the final column of $\mathbf{U}$, namely,

$$\mathbf{a}_3 = \begin{pmatrix} -1 \\ 0 \\ 1 \end{pmatrix}, \tag{8.265}$$

is a basis of the orthogonal complement of $\mathcal{W}$. This is obviously correct because $\mathbf{a}_3$ is orthogonal to $\mathbf{a}_1$ and $\mathbf{a}_2$; therefore $\mathbf{a}_3$ is both linearly independent of the first two and an element of $\mathcal{W}^\perp$.

The singular-value decomposition derived in Section 9.4.1, which constructs *orthonormal* bases, provides a method of computing a basis of $\mathcal{W}^\perp$ preferable to the simple LU decomposition discussed here if the matrix $\mathbf{A}_m$ is ill conditioned, that is, if the vectors $\mathbf{a}_1, \ldots, \mathbf{a}_m$ are nearly linearly dependent.

Orthogonal Complement of the Orthogonal Complement

Let $\mathcal{W}$ be a vector subspace of a Euclidean or unitary space $\mathcal{V}$, and let $\mathcal{W}^\perp$ be the orthogonal complement of $\mathcal{W}$. Because

$$\bar{z} \perp \bar{w} \in \mathcal{W} \Rightarrow \bar{w} \perp \bar{z} \in \mathcal{W}^\perp, \tag{8.266}$$

one sees that

$$\mathcal{W} \subseteq (\mathcal{W}^\perp)^\perp. \tag{8.267}$$

This result remains true in infinite-dimensional Euclidean and unitary spaces. Assume now that $\mathcal{V}$ is finite-dimensional. To see whether $\mathcal{W} = (\mathcal{W}^\perp)^\perp$, let $\bar{x} \in (\mathcal{W}^\perp)^\perp$. The orthogonal decomposition of $\bar{x}$ with respect to $\mathcal{W}$ is

$$\bar{x} = \bar{w} + \bar{z} \tag{8.268}$$

in which

$$\bar{w} := \mathsf{P}_\mathcal{W} \bar{x} \in \mathcal{W}, \quad \bar{z} := \bar{x} - \mathsf{P}_\mathcal{W} \bar{x} \in \mathcal{W}^\perp. \tag{8.269}$$

Then

$$\langle \bar{x}, \bar{z} \rangle = 0 = \langle \bar{w} + \bar{z}, \bar{z} \rangle = \|\bar{z}\|^2 \Rightarrow \bar{z} = \bar{0}, \tag{8.270}$$

which implies that $\bar{x} \in \mathcal{W}$. Then

$$(\mathcal{W}^\perp)^\perp \subseteq \mathcal{W}. \tag{8.271}$$

Similarly, $(W^\perp)^\perp \subseteq W$ if W is finite-dimensional; therefore

$$(W^\perp)^\perp = W. \tag{8.272}$$

If V is an infinite-dimensional Hilbert space, then Eqs. (8.271) and (8.272) hold if and only if W is complete.

Definition of a Subspace as an Orthogonal Complement

One of the more useful ways to define a subspace W of $\mathbb{E}^n$ or $\mathbb{U}^n$ is to define W as the orthogonal complement of a subspace that is spanned by $m < n$ linearly independent vectors $\mathbf{f}_1, \ldots, \mathbf{f}_m$. (The $\{\mathbf{f}_i\}$ are not required to be orthogonal.) If a vector $\mathbf{x} \in W$ is orthogonal to the subspace spanned by the $\{\mathbf{f}_i\}$, then one obtains m linear relations,

$$\forall i \in (1 : m) : \mathbf{f}_i \cdot \mathbf{x} = 0, \tag{8.273}$$

in the n components x^j of $\mathbf{x}$. For example, the easy way to characterize the plane in $\mathbb{E}^3$ that passes through a point $\mathbf{x}_0$ and is perpendicular to a unit vector $\mathbf{n}$ is to observe that a vector $\mathbf{x}$ lies in the plane if and only if

$$(\mathbf{x} - \mathbf{x}_0) \cdot \mathbf{n} = 0, \tag{8.274}$$

that is, if and only if $\mathbf{x} - \mathbf{x}_0$ belongs to the orthogonal complement of the one-dimensional subspace that is spanned by $\mathbf{n}$. This equation gives the linear relation

$$n^1\left(x^1 - x_0^1\right) + n^2\left(x^2 - x_0^2\right) + n^3\left(x^3 - x_0^3\right) = 0, \tag{8.275}$$

which (one hopes) is familiar from analytic geometry.

Projectors on Orthogonal Complements

Let W be a subspace of a Euclidean or unitary space V. (If W is infinite-dimensional, we assume that Eq. [8.272] holds.) One writes the projector on the orthogonal complement $W^\perp$ as $P_W^\perp$. Let $\bar{x}$ be any vector in V. We have shown that there exist unique vectors $\bar{w}$ and $\bar{z}$ such that $\bar{x} = \bar{w} + \bar{z}$, and that $\bar{z} \in W^\perp$. By the definition of an orthogonal projector,

$$\bar{w} = P_W \bar{x} \tag{8.276}$$

and

$$\bar{z} = P_W^\perp \bar{x}. \tag{8.277}$$

Then

$$\forall \bar{x} \in V : \bar{x} = \bar{w} + \bar{z}$$
$$= \left(P_W + P_W^\perp\right)\bar{x}, \tag{8.278}$$

from which it follows immediately that

$$\boxed{P_w + P_w^\perp = 1.}$$ (8.279)

This relation and Eq. (8.284) (below) summarize the complementarity of $\mathcal{W}$ and $\mathcal{W}^\perp$.

For example, if $\mathbf{P}_w$ is the projection matrix defined in Eq. (8.243), then the matrix of the projector on the orthogonal complement is

$$
\begin{aligned}
\mathbf{P}_w^\perp &= \mathbf{1} - \mathbf{P}_w \\
&= \begin{pmatrix} 1 & 0 & 0 \\ 0 & 1 & 0 \\ 0 & 0 & 1 \end{pmatrix} - \frac{1}{2} \begin{pmatrix} 1 & 0 & 1 \\ 0 & 2 & 0 \\ 1 & 0 & 1 \end{pmatrix} \\
&= \frac{1}{2} \begin{pmatrix} 1 & 0 & -1 \\ 0 & 0 & 0 \\ -1 & 0 & 1 \end{pmatrix}.
\end{aligned}
$$ (8.280)

The technique of finding a projector on the orthogonal complement of a subspace by subtracting the projector on the subspace from the unit matrix is useful if one does not need to compute a basis of the orthogonal complement. For example, suppose that one wishes to project a three-dimensional object (represented by the position vectors of the vertexes of a set of polygons) onto a plane that is parallel to a two-dimensional subspace $\mathcal{W}^\perp$, in which the plane is defined in terms of a unit normal $\mathbf{n}$ as in Eq. (8.274). Instead of computing a basis of $\mathcal{W}^\perp$, one can simply compute the projector onto the one-dimensional subspace $\mathcal{W}$ spanned by the unit normal $\mathbf{n}$,

$$P_w = \mathbf{n}\mathbf{n}^T.$$ (8.281)

Column vector–row vector products such as $\mathbf{n}\mathbf{n}^T$ are studied more thoroughly in Section 9.1. For now, we simply note that such a matrix product is legitimate according to Eq. (6.76). Then

$$
\begin{aligned}
\mathbf{P}_w^\perp &= \mathbf{1} - \mathbf{P}_w \\
&= \mathbf{1} - \mathbf{n}\mathbf{n}^T.
\end{aligned}
$$ (8.282)

Another important relation between projectors follows from the orthogonality of $\mathcal{W}$ and $\mathcal{W}^\perp$. One has

$$P_w^\perp P_w \bar{x} = P_w^\perp P_w (\bar{w} + \bar{z}) = P_w^\perp \bar{w} = \bar{0} \Rightarrow P_w^\perp P_w = \mathbf{0},$$ (8.283)

in which $\mathbf{0}$ is the zero mapping, $\mathbf{0}\bar{x} := \bar{0}$. Similarly, $P_W P_W^\perp = \mathbf{0}$. Then

$$P_W^\perp P_W = P_W P_W^\perp = \mathbf{0}. \qquad (8.284)$$

For example, the projection matrices P_W and $P_W^\perp$ in Eq. (8.280) satisfy this relation.

Orthogonality and complementarity are independent concepts. Projectors P and Q such that $PQ = QP = \mathbf{0}$ are projectors on orthogonal, but not necessarily complementary, subspaces. For example, the projectors P_i, P_j on the subspaces spanned by different vectors of an orthonormal basis are orthogonal:

$$P_i P_j = P_j P_i = \begin{cases} \mathbf{0} & \text{if } i \neq j; \\ P_i & \text{if } i = j. \end{cases} \qquad (8.285)$$

However, P_i and P_j are not such that $P_i + P_j = \mathbf{1}$ in general, because the subspaces on which they project are not necessarily complementary.

8.3.4 EXERCISES FOR SECTION 8.3

8.3.1 Let P_W project on a subspace W of a Euclidean or unitary space V. Prove that

$$P_W^2 := P_W P_W = P_W. \qquad (8.286)$$

8.3.2 Let $V = \mathbb{E}^3$, and let

$$\mathbf{a}_1 := \begin{pmatrix} 1 \\ -1 \\ 1 \end{pmatrix}. \qquad (8.287)$$

  (a) Find the orthogonal projection of $\mathbf{a}_1$ on the subspace X spanned by the canonical basis vectors $\mathbf{e}_1$, $\mathbf{e}_2$.

  (b) Find the orthogonal projection of $\mathbf{a}_1$ on the subspace Y spanned by

$$\mathbf{a}_2 := \begin{pmatrix} 1 \\ 1 \\ 1 \end{pmatrix}, \quad \mathbf{a}_3 := \begin{pmatrix} 1 \\ 1 \\ -1 \end{pmatrix}. \qquad (8.288)$$

  (c) Let W be the subspace of V that is spanned by the vector $\mathbf{a}_1$. Find a basis of $W^\perp$. Is $W^\perp$ equal to either X or Y? Explain.

8.3.3 Let V be the finite-dimensional subspace of $\mathcal{L}^2([0, 1]; \mathbb{R})$ spanned by the Legendre polynomials P_0, P_1, P_2, and let W be the subspace of V spanned by the polynomial $p = P_0 - P_1 + P_2$. Find a basis of $W^\perp$.

8.3.4 Find the matrix of the orthogonal projector along the direction of the vector

$$x = \begin{pmatrix} 1 \\ -2 \\ 1 \end{pmatrix} \in \mathbb{E}^3 \tag{8.289}$$

relative to the canonical basis.

8.3.5 Verify that the matrices given in Eqs. (8.239), (8.243), and (8.280) are idempotent (see Eq. [8.246]).

8.3.6 Find an orthonormal basis of the orthogonal complement of the one-dimensional subspace of $\mathbb{E}^3$ spanned by the vector a_1 defined in Eq. (8.287). (Note that the basis that is requested is not unique.)

8.4 LEAST-SQUARES APPROXIMATIONS

8.4.1 MOTIVATION

In the eighteenth century two approaches to approximating functions were developed that are still moderately useful: Taylor series and Lagrange's interpolation formula. Both of these approaches had the goal of exactly matching the values of a function, at specific points (in the case of Lagrange's formula) or in a specific interval of convergence (in the case of Taylor series). A modern approach, spline interpolation, has the goal of improving on the accuracy of Lagrange's method for the values of a function at points between those at which the function's values are known exactly.

In the twentieth century global approaches have tended to supplant pointwise approaches to the approximation of functions for many applications. Suppose that the function to be approximated is a vector $\overline{x}$ in an Euclidean or unitary space $\mathcal{V}$ such as $\mathcal{L}^2([-\pi, \pi]; \mathbb{C})$, and let $\mathcal{W}$ be a subspace of $\mathcal{V}$ such as the subspace that is spanned by the Fourier functions

$$f_m(x) := \frac{e^{imx}}{\sqrt{2\pi}}, \tag{8.290}$$

in which $-M \le m \le M$. In $\mathcal{V}$, it makes sense to take the geometrical point of view that the "distance" between $\overline{x}$ and another vector $\overline{y}$ is $\|\overline{x} - \overline{y}\|_{ip}$, and that therefore the **best approximation to $\overline{x}$ by an element $\overline{w}$ of $\mathcal{W}$** has the property that for all $\overline{y} \in \mathcal{W}$,

$$\|\overline{x} - \overline{y}\|_{ip} \ge \|\overline{x} - \overline{w}\|_{ip}. \tag{8.291}$$

For example, to obtain the best approximation to a function $y \in \mathcal{L}^2([-\pi, \pi]; \mathbb{C})$ with a linear combination $\alpha^i f_i$, one minimizes

$$\left\| x - \sum_{i=1}^m \alpha^i f_i \right\|_{ip}^2 = \int_{-\pi}^{\pi} |x(t) - \alpha^m f_m(t)|^2 dt. \tag{8.292}$$

This approach imposes no upper bound on the maximum deviation, $\sup_t |x(t) - \alpha^m f_m(t)|$.

The idea of minimizing a functional in order to solve a physical problem is not new; the calculus of variations was invented in the eighteenth century. The (relatively) new features of the now widely used approach described in this section are the choice of the functional to minimize, the method of minimization, and the geometrical interpretation of the results. It is also possible to minimize $\|x - \sum_1^m \alpha^i f_i\|^2$ by differentiation (Exercise 8.4.1), but a more geometrically motivated approach arrives at the same result in a more intuitively appealing and more generalizable way.

The technique of expressing a vector as the sum of two orthogonal vectors (as in Eq. [8.99]) and then applying the generalized Pythagorean theorem, which we used in the course of deriving the Cauchy-Schwarz-Bunyakovsky inequality, is a powerful tool for deriving other inequalities. In this section we use that technique to solve the problem of finding the best approximation (in the inner-product norm) to a vector by an element of a given vector subspace. The vector to be approximated may, of course, be a function from $\mathbb{R}^m$ or $\mathbb{C}^m$ to $\mathbb{R}^n$ or $\mathbb{C}^n$. For technical reasons we assume here that the inner-product space $\mathcal{V}$ is finite-dimensional; see Chapter 10 for the infinite-dimensional case.

8.4.2 ABSTRACT FORMULATION

Suppose that one wants to find the best approximation of a known vector $\bar{x} \in \mathcal{V}$ by an element $\bar{y}$ of $\mathcal{W}$, in the sense that the approximation minimizes $\|\bar{x} - \bar{y}\|_{\text{ip}}$. Resolve $\bar{x} - \bar{y}$ into the sum of a vector that lies in $\mathcal{W}$ and a vector that lies in $\mathcal{W}^{\perp}$:

$$\bar{x} - \bar{y} = (\bar{x} - P_{\mathcal{W}}\bar{x}) + (P_{\mathcal{W}}\bar{x} - \bar{y}). \tag{8.293}$$

By construction, $\bar{x} - P_{\mathcal{W}}\bar{x}$ belongs to $\mathcal{W}^{\perp}$ and $P_{\mathcal{W}}\bar{x} - \bar{y}$ belongs to $\mathcal{W}$. From Eq. (8.293) and the generalized Pythagorean theorem Eq. (8.95) one obtains **Bessel's identity**

$$\|\bar{x} - \bar{y}\|_{\text{ip}}^2 = \|\bar{x} - P_{\mathcal{W}}\bar{x}\|_{\text{ip}}^2 + \|P_{\mathcal{W}}\bar{x} - \bar{y}\|_{\text{ip}}^2, \tag{8.294}$$

which implies that

$$\|\bar{x} - \bar{y}\|_{\text{ip}}^2 \geq \|\bar{x} - P_{\mathcal{W}}\bar{x}\|_{\text{ip}}^2. \tag{8.295}$$

Therefore the minimum value of $\|\bar{x} - \bar{y}\|_{\text{ip}}$ is $\|\bar{x} - P_{\mathcal{W}}\bar{x}\|_{\text{ip}}$. By Eq. (8.294), this minimum is achieved when

$$\|\bar{x} - \bar{y}\|_{\text{ip}}^2 = \|\bar{x} - P_{\mathcal{W}}\bar{x}\|^2 \Leftrightarrow \bar{y} = P_{\mathcal{W}}\bar{x}. \tag{8.296}$$

This implies the following major result: *The minimum distance between a vector $\bar{x} \in \mathcal{V}$ and a vector $\bar{y} \in \mathcal{W} \subset \mathcal{V}$ is achieved if $\bar{y}$ is the projection of $\bar{x}$ onto $\mathcal{W}$. Equivalently, one obtains the* **least-squares approximation** *to $\bar{x}$ by projecting $\bar{x}$ onto $\mathcal{W}$.* This statement is intuitively clear in three-dimensional space; see Fig. 8.7.

In Chapter 10 we show that even if $\mathcal{V}$ is an infinite-dimensional Hilbert space it is possible to prove, without introducing a basis, that there exists a vector $\bar{z} \in \mathcal{W}$ that minimizes

$\|\bar{x} - \bar{y}\|_{ip}$. It remains true in the infinite-dimensional case that the minimum is achieved when $\bar{y} = P_W \bar{x}$.

8.4.3 INEQUALITIES FOR LEAST-SQUARES APPROXIMATIONS

One obtains an upper bound on the norm of the approximating vector $P_W \bar{x}$ known as **Bessel's inequality** by applying Bessel's identity, Eq. (8.294), with $\bar{y}$ set equal to zero:

$$\|\bar{x}\|_{ip}^2 = \|\bar{x} - P_W \bar{x}\|_{ip}^2 + \|P_W \bar{x}\|_{ip}^2 \geq \|P_W \bar{x}\|_{ip}^2 = \sum_i^n |\langle \bar{e}_i, \bar{x} \rangle|^2. \tag{8.297}$$

In this equation and inequality, $\|\bar{x} - P_W \bar{x}\|_{ip}^2$ is a measure of the error in the approximation $\bar{x} \approx \bar{y} = P_W \bar{x}$. Evidently $\|\bar{x}\|_{ip}$ is an upper bound of the norm of the least-squares approximating vector (or function).

If V is finite-dimensional and W is a proper subspace of V, then Bessel's identity implies that $\|\bar{x}\|_{ip}$ is not a least upper bound unless $\bar{x}$ happens to belong to W. We show in Chapter 10 that if V is an infinite-dimensional Hilbert space, then $\|P_W \bar{x}\|_{ip}$ can attain the bound $\|\bar{x}\|_{ip}$ if and only if the family of vectors $\{\bar{e}_i\}$ used to construct orthogonal projections on finite-dimensional subspaces has the property of *completeness* in V, and that if $\{\bar{e}_i\}$ is complete, then $\|\bar{x} - P_W \bar{x}\|_{ip}^2$ approaches 0 as $n = \dim[W] \to \infty$.

An important inequality for the inner product of two vectors $\bar{x}$ and $\bar{x}'$ in terms of their projections on W follows from the fact that $\bar{x} - P_W \bar{x}$ is orthogonal to any vector in W, including $P_W \bar{x}'$. Then

$$\begin{aligned} \langle \bar{x}, \bar{x}' \rangle &= \langle P_W \bar{x} + (\bar{x} - P_W \bar{x}), \, P_W \bar{x}' + (\bar{x}' - P_W \bar{x}') \rangle \\ &= \langle P_W \bar{x}, P_W \bar{x}' \rangle + \langle \bar{x} - P_W \bar{x}, \bar{x}' - P_W \bar{x}' \rangle. \end{aligned} \tag{8.298}$$

This equation, the triangle inequality for complex numbers and the Cauchy-Schwarz-Bunyatovsky inequality imply that

$$\begin{aligned} |\langle \bar{x}, \bar{x}' \rangle| &\leq |\langle P_W \bar{x}, P_W \bar{x}' \rangle| + |\langle \bar{x} - P_W \bar{x}, \bar{x}' - P_W \bar{x}' \rangle| \\ &\leq |\langle P_W \bar{x}, P_W \bar{x}' \rangle| + \|\bar{x} - P_W \bar{x}\|_{ip} \|\bar{x}' - P_W \bar{x}'\|_{ip} \end{aligned} \tag{8.299}$$

and that

$$|\langle P_W \bar{x}, P_W \bar{x}' \rangle| - \|\bar{x} - P_W \bar{x}\|_{ip} \|\bar{x}' - P_W \bar{x}'\|_{ip} \leq |\langle \bar{x}, \bar{x}' \rangle|. \tag{8.300}$$

In words, the error that the approximations $\bar{x} \approx \bar{y} = P_W \bar{x}$ and $\bar{x}' \approx \bar{y}' = P_W \bar{x}'$ contribute to an evaluation of the inner product $\langle \bar{x}, \bar{x}' \rangle$ is not greater than the product of the norms of the error vectors. Within this margin of error, then, one can evaluate an inner product $\langle \bar{x}, \bar{x}' \rangle$ as the sum of the products of the coordinates of $\bar{x}$ and $\bar{x}'$ with respect to the orthonormal basis of W (with complex conjugation, if necessary):

$$\langle \bar{x}, \bar{x}' \rangle = \sum_{i=1}^n \langle \bar{e}_i, \bar{x} \rangle^* \langle \bar{e}_i, \bar{x}' \rangle + O(\|\bar{x} - P_W \bar{x}\|_{ip} \|\bar{x}' - P_W \bar{x}'\|_{ip}). \tag{8.301}$$

8.4.4 APPROXIMATION BY FINITE FOURIER SUMS

An example of projection and least-squares approximation that is of fundamental impor-
tance in physics and engineering arises if one takes V to be $\mathcal{L}^2([-\pi, \pi]; \mathbb{C})$, W to be the
subspace of V spanned by the Fourier functions $\{e_m\}$ defined in Eq. (8.211), and the inner
product to be the canonical inner product on $\mathcal{L}^2([-\pi, \pi]; \mathbb{C})$ with unit weight function defined in Eq. (8.52). According to Eq. (8.233), the projection of a function $f \in \mathcal{L}^2([-\pi, \pi]; \mathbb{C})$
on W is

$$
\begin{aligned}
(P_W f)(u) &= \sum_{m=-M}^{M} \langle e_m, f \rangle e_m(u) \\
&= \frac{1}{2\pi} \sum_{m=-M}^{M} e^{imu} \int_{-\pi}^{\pi} e^{-imu'} f(u') \, du',
\end{aligned}
\tag{8.302}
$$

which is a partial sum of the Fourier series of f. For this reason the coefficients $\langle \bar{e}_i, \bar{x} \rangle$ in
Eq. (8.232) are sometimes called the **Fourier coefficients** of $\bar{x}$, even if the Fourier basis is
not in use. It follows from the inequality Eq. (8.295) that: *The Fourier partial sum shown in
Eq. (8.302) provides the best approximation to f (in the sense of least squares) that can be
obtained using the Fourier functions Eq. (8.211). In general, a finite linear combination of any
functions that are orthogonal under the scalar product Eq. (8.52) provides the least-squares
approximation to a square-integrable function if and only if one calculates the coefficients
by projection as in Eq. (8.232) or Eq. (8.302).*

We show in Chapter 10 that if f is square-integrable on $[-\pi, \pi]$, then the sequence of
partial sums Eq. (8.302) converges to a function that has the same values as f almost everywhere. In this case one writes the limit symbolically as

$$
f(x) = \sum_{m \in \mathbb{Z}} c_m e^{imx},
\tag{8.303}
$$

which is the **Fourier series** of f.

If one applies Bessel's inequality, Eq. (8.297), to Eq. (8.302), one obtains

$$
\sum_{-M}^{M} |c_m|^2 \le \int_{-\pi}^{\pi} |f(x)|^2 \, dx
\tag{8.304}
$$

for any finite positive integer M. If one lets $M \to \infty$, one can see, without knowing anything
about the definition of convergence, that the sums $\sum_{-M}^{M} |c_m|^2$ are nondecreasing and are
bounded from above. We show in Chapter 10 that this implies that the sequence $\sum_{-M}^{M} |c_m|^2$
converges to a finite limit. In other words, the sequence of Fourier coefficients belongs to
$\ell^2(\mathbb{Z}; \mathbb{C})$.

8.4.5 CHEBYSHEV APPROXIMATIONS

The method of least-squares approximation can be used, not only with the Fourier basis employed in Eq. (8.302), but also with any finite basis of orthogonal functions. Here we give a brief discussion of approximations using the Chebyshev polynomials defined in Section 8.2.3.

Let $\mathcal{W}$ be the N-dimensional vector subspace of $\mathcal{C}^0([-1, 1]; \mathbb{R})$ that is spanned by the Chebyshev polynomials $T_0, \ldots, T_{N-1}$. The least-squares approximation by a vector in $\mathcal{W}$ to a function $f \in \mathcal{C}^0([-1, 1]; \mathbb{R})$ is

$$P_{\mathcal{W}} f = \sum_{m=0}^{N-1} \frac{1}{\|T_m\|_{\text{ip}}^2} \langle T_m, f \rangle T_m. \tag{8.305}$$

The inner product used in this equation is defined in Eq. (8.202).

Chebyshev approximations based on Eq. (8.305) are useful in numerical computations in which a frequently evaluated function must be approximated by a polynomial on a given interval with an error that is less than some specified maximum. We illustrate this application of Chebyshev polynomials in Section 10.6.

In practical computations one is often limited to a finite set of function values, as well as to a finite basis. Chebyshev sampling (Eq. [8.205]) is an example of a discretization mapping, as defined in Section 6.1.5. A useful approximation mapping A_n that maps a vector $\mathbf{a}$ in $\mathbb{E}^n$ to a polynomial in $\mathcal{C}^0([-1, 1]; \mathbb{R})$ follows from the least-squares method if one interprets $\mathbf{a}$ as the vector of Chebyshev-sampled values of some continuous function a.

Let $\mathcal{W}$ be the n-dimensional subspace spanned by the sampled Chebyshev polynomials $C_n T_i$, in which $i \in (0, n-1)$. Let

$$\mathbf{f}_n = C_n f$$

$$= \begin{pmatrix} f(x_{n,1}) \\ \vdots \\ f(x_{n,n}) \end{pmatrix} \tag{8.306}$$

be the vector whose components are the values of a continuous function f at the first n zeros of T_n (see Eq. [8.204]). The least-squares approximation to $\mathbf{f}_n$ with elements of $\mathcal{W}$ is

$$\tilde{\mathbf{f}}_n = \sum_{i=1}^{n} \frac{1}{\|C_n T_i\|_{\mathbb{E}^n}^2} \langle\!\langle T_i, f \rangle\!\rangle C_n T_i, \tag{8.307}$$

in which $\langle\!\langle T_i, f \rangle\!\rangle = \langle C_n T_i, C_n f \rangle_{\mathbb{E}^n}$. Then the function

$$(A_n \mathbf{f}_n)(x) := \sum_{i=1}^{n} \frac{1}{\|C_n T_i\|_{\mathbb{E}^n}^2} \langle\!\langle T_i, f \rangle\!\rangle T_i(x) \tag{8.308}$$

is a polynomial approximation to the function f. We call Eq. (8.308) the **discrete Chebyshev approximation** to f.

8.4.6 MAPPING A FUNCTION TO ITS FOURIER COEFFICIENTS

Although the subject of this section is approximation by finite linear combinations of orthogonal functions, we need to quote a result from Chapter 10 on the convergence of Fourier series in order to motivate our discussion of some of the properties of sampled finite linear combinations of Fourier functions. For the sake of greater generality we shift and scale the interval on which the function g is defined from $[-\pi, \pi]$ to $[a, b]$. Let $\overline{g}$ be a sequence of complex numbers,

$$\overline{g} = (\dots, g_{-1}, g_0, g_1, \dots). \tag{8.309}$$

If $\sum_{-M}^{M} |g_m|^2$ converges to a limit as $M \to \infty$ (in other words, if the Fourier series $\sum_{-\infty}^{\infty} g_m e^{2\pi imx/(b-a)}$ converges in the 2-norm), then a corollary of the Riesz-Fischer theorem (Kolmogorov and Fomin 1975, p. 153) states that there exists a function g such that $g \in \mathcal{L}^2([a, b]; \mathbb{C})$ and g_m is the m-th Fourier coefficient of g,

$$g_m = \frac{1}{b-a} \int_a^b e^{-2\pi imx/(b-a)} g(x) \, dx, \tag{8.310}$$

in which $m \in \mathbb{Z}$. In other words, the terms of every square-summable sequence are the Fourier coefficients of some square-integrable function. The converse, that the sequence of Fourier coefficients of a square-integrable function belongs to $\ell^2(\mathbb{Z}; \mathbb{C})$, follows from Eq. (8.297).

Therefore, for every sequence $\overline{g} = (\dots, g_{-1}, g_0, g_1, \dots) \in \ell^2(\mathbb{Z}; \mathbb{C})$ there exists a mapping

$$\mathsf{F}: \overline{g} \mapsto g \in \mathcal{L}^2([a, b]; \mathbb{C}) \tag{8.311}$$

in which

$$g(x) = \sum_{m \in \mathbb{Z}} g_m e^{2\pi imx/(b-a)} \tag{8.312}$$

in the sense of convergence in the 2-norm (as discussed in Chapter 10). We call F the **Fourier mapping**. Figure 8.8 illustrates the Fourier mapping for a square pulse defined on the interval $[-1, 1]$.

Equation (8.312) implies that the Fourier mapping F is homogeneous,

$$\alpha \overline{g} = (\dots, \alpha g_{-1}, \alpha g_0, \alpha g_1, \dots) \mapsto \alpha g, \tag{8.313}$$

and additive,

$$\overline{f} + \overline{g} = (\dots, f_{-1}, f_0, f_1, \dots) + (\dots, g_{-1}, g_0, g_1, \dots) \mapsto f + g. \tag{8.314}$$

Therefore F is linear.

Equation (8.310), which gives the Fourier coefficients of a square-integrable function g, defines a linear mapping $\mathsf{F}^\dagger : \mathcal{L}^2([a, b]; \mathbb{C}) \to \ell^2(\mathbb{Z}; \mathbb{C})$ from a function g to the sequence $\overline{g}$

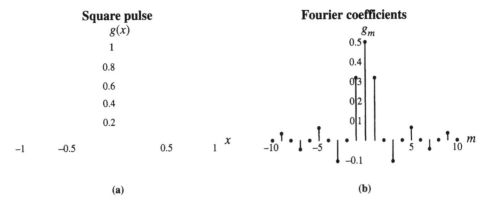

FIGURE 8.8

Illustration of the Fourier mapping. (*a*) The function g is a square pulse defined on the interval $[-1, 1]$. (*b*) A few of the terms of the sequence $\bar{g}$ of Fourier coefficients of (*a*).

of Fourier coefficients of g,

$$\mathsf{F}^\dagger : g \mapsto \bar{g}. \tag{8.315}$$

It can be shown that $\mathsf{F}^\dagger$ is the Hermitian adjoint of F.

We show in Chapter 10 that Fourier series are unique in the sense that if two functions have the same Fourier coefficients, $\mathsf{F}^\dagger g = \mathsf{F}^\dagger h$, then $g = h$ except (possibly) on a set of measure zero.

8.4.7 EXERCISES FOR SECTION 8.4

8.4.1   The traditional method for obtaining the values of the arguments that minimize the value of a differentiable function is to set the partial derivatives equal to zero and solve the resulting equations. The function to be minimized in the method of least-squares approximation is

$$F(\alpha^1, \ldots, \alpha^m) = \left\| \bar{x} - \sum_{i=1}^m \alpha^i \bar{e}_i \right\|^2 = \left\langle \bar{x} - \sum_{i=1}^m \alpha^i \bar{e}_i, \bar{x} - \sum_{i=1}^m \alpha^i \bar{e}_i \right\rangle. \tag{8.316}$$

Minimize F by differentiating with respect to α^j, and show that the minimum is attained if

$$\alpha^j = \langle \bar{e}_j, \bar{x} \rangle, \tag{8.317}$$

in agreement with Eq. (8.232).

8.4.2   Let $\mathcal{V} = \mathbb{E}^3$ and let **x** be the vector defined in Eq. (8.287). Let $\mathcal{Y}$ be the subspace spanned by the vectors defined in Eq. (8.288). Find the vector **y** in the subspace $\mathcal{Y}$ that minimizes $\|\mathbf{x} - \mathbf{y}\|^2$.

8.4.3   In many models of physical and engineering systems, an output f should depend on an input t through an N-term polynomial:

$$f(t) = \sum_{m=0}^{N-1} \alpha^m t^m. \tag{8.318}$$

If $x : [a, b] \to \mathbb{R}$ is a known, integrable function, but is not necessarily a polynomial, the problem of determining the coefficients α^m that give a best fit to x on the interval $[a, b]$ reduces to a least-squares approximation problem if one defines the best fit as in Eq. (8.292), taking

$$f_m(t) = t^m. \tag{8.319}$$

Prove that the coefficients $\{\alpha_m \mid m \in (0 : N - 1)\}$ that minimize the integral $\int_a^b |x(t) - \alpha^m f_m(t)|^2 \, dt$ solve the system of linear equations

$$\sum_{m=0}^{N-1} h_{nm} \alpha^m = \langle f_n, f \rangle \tag{8.320}$$

in which

$$h_{nm} = \langle f_m, f_n \rangle = \frac{1}{m + n + 1}. \tag{8.321}$$

Comment: The matrix whose elements are h_{nm}, in which $m, n \in (0 : N - 1)$, is called the **Hilbert matrix** of order N. It is very difficult (and perhaps impossible) to obtain an accurate numerical solution of Eq. (8.320) if N is larger than approximately 5 (in single precision) or 8 (in double precision).

8.5   DISCRETE FOURIER TRANSFORM

Despite the great importance of Fourier series in the classical methods of mathematical physics and engineering, the discrete Fourier transform of a function is faster to compute numerically, and therefore is much more useful for practical purposes, than the Fourier series. To mention only two of many application areas, discrete Fourier transforms are embedded in many numerical methods for solving ordinary and partial differential equations and are an essential ingredient of many digital signal processing methods. In this section, we develop the discrete Fourier transform using the concepts of orthogonal projections and orthonormal bases.

8.5.1   APPROXIMATION OF FOURIER COEFFICIENTS

For simplicity we work at first with an interval $[a, b]$ of unit length. Suppose that a function $g \in \mathcal{C}^0([0, 1]; \mathbb{C})$ is sampled uniformly at N points in the interval $[0, 1]$. The sampling points are

$$\forall k \in (0 : N - 1) : x_k = \frac{k}{N}. \tag{8.322}$$

The result of the sampling operation is to map the function g to a vector $\mathbf{g} \in \mathbb{C}^N$,

$$g \mapsto \mathbf{g} = S_N g = \begin{pmatrix} g(x_0) \\ g(x_1) \\ \vdots \\ g(x_{N-1}) \end{pmatrix}, \tag{8.323}$$

in which S_N is the uniform sampling mapping defined in Eq. (6.104).

If one uses the rectangle rule to evaluate the integral that occurs in the definition of the m-th Fourier coefficient of g, Eq. (8.310), one finds that

$$g_m \approx \frac{1}{N} \sum_{k=0}^{N-1} e^{-2\pi i m x_k} g(x_k). \tag{8.324}$$

Rather than expand g in a Taylor series to estimate the error in this approximation, we use the orthogonality properties of the discrete Fourier basis to obtain an exact relation between the Fourier coefficient given by Eq. (8.310) and the discrete sum that appears in Eq. (8.324).

8.5.2 DISCRETE FOURIER BASIS

The Fourier basis on the interval $[0, 1]$ is $\{f_m(x) = e^{2\pi i m x}\}$. If the functions $\{f_m \mid m \in (0 : N - 1)\}$ are sampled at the N points $\{x_k = k/N \mid k \in (0 : N - 1)\}$, one obtains the **discrete Fourier basis** vectors,

$$\mathbf{f}_m := S_N f_m$$

$$= \begin{pmatrix} 1 \\ e^{2\pi i m/N} \\ \vdots \\ e^{2\pi i m(N-1)/N} \end{pmatrix}. \tag{8.325}$$

Exercise 8.2.8 shows that the N vectors $\mathbf{f}_m$ are mutually orthogonal in $\mathbb{U}^N$,

$$\forall m, n \in (0 : N - 1) :\ni: m \neq n : \mathbf{f}_m^\dagger \mathbf{f}_n = 0, \tag{8.326}$$

and therefore are a basis of $\mathbb{U}^N$. The normalization of the discrete Fourier basis is

$$\mathbf{f}_m^\dagger \mathbf{f}_m = N. \tag{8.327}$$

We choose basis vectors that are *not* of unit length, in order to facilitate a discrete approximation to the Fourier coefficients. (For comparison with Eq. [8.233], note that the sampled discrete Fourier basis is orthogonal but is not normalized to unity.) The orthonormal basis that one can substitute directly into Eq. [8.233] is $\{N^{-1/2}\mathbf{f}_m \mid m \in (0 : N - 1)\}$. The normalization chosen here is the one that is most commonly used in electrical engineering.

On a general interval $[a, b]$, the definition of the Fourier functions f_m is

$$f_m(x) := e^{2\pi m(x-a)/(b-a)}. \tag{8.328}$$

If one samples these functions (for $m \in (0 : N - 1)$) at the N points

$$x_k = a + k\frac{b - a}{N}, \tag{8.329}$$

then one obtains exactly the same discrete Fourier basis vectors as on the unit interval $[0, 1]$.

Because the discrete Fourier basis vectors span $\mathbb{U}^N$, one can expand the vector of sampled values, $\mathbf{g}$, in terms of the $\{\mathbf{f}_m\}$. Using Eq. (8.327), one finds easily that

$$\mathbf{g} = \frac{1}{N} \sum_{m=0}^{N-1} \gamma_m \mathbf{f}_m \tag{8.330}$$

in which

$$\gamma_m = \mathbf{f}_m^\dagger \mathbf{g}. \tag{8.331}$$

Thus the operation by which one evaluates γ_m is just the canonical inner product in unitary N-space ($\mathbb{U}^N$).

An explicit formula for γ_m, valid for any interval $[a, b]$, is

$$\gamma_m = \sum_{k=0}^{N-1} e^{-2\pi i m(x_k - a)/(b-a)} g(x_k) \tag{8.332}$$

in which $x_k = a + k(b - a)/N$. For the special case of the unit interval, one has

$$\gamma_m = \sum_{k=0}^{N-1} e^{-2\pi i mk/N} g(k/N). \tag{8.333}$$

These equations give γ_m as N times the approximation Eq. (8.324) to g_m that one obtains by using the rectangle rule to evaluate the integral Eq. (8.310).

In communications theory and digital signal processing, one usually calls Eq. (8.332) the **discrete Fourier transform** and Eq. (8.330) the **inverse discrete Fourier transform**. The space $\mathbb{C}^N$, to which all vectors of sampled function values belong, is called the **signal space**, because the components of the vector $\mathbf{g}$ (relative to the canonical basis) often are the values of a time-varying signal g at the sample times $x_0, \ldots, x_{N-1}$.

The N complex numbers γ_m produced by the discrete Fourier transform are the components of a vector that also belongs to $\mathbb{C}^N$,

$$\tilde{\mathbf{g}} := \begin{pmatrix} \gamma_0 \\ \gamma_1 \\ \vdots \\ \gamma_{N-1} \end{pmatrix}. \tag{8.334}$$

One calls the space to which $\tilde{\mathbf{g}}$ belongs the **transform space**. Evidently the transform space is isomorphic to the signal space; one distinguishes between them for the sake of clarity.

The mapping

$$\mathbf{g} \mapsto \tilde{\mathbf{g}} \tag{8.335}$$

is linear, by inspection of Eq. (8.331). The matrix of this mapping is the adjoint of the matrix whose columns are the discrete Fourier basis vectors,

$$\mathbf{F} = (\mathbf{f}_0, \ldots, \mathbf{f}_{N-1}). \tag{8.336}$$

The matrix-vector equation

$$\boxed{\tilde{\mathbf{g}} = \mathbf{F}^\dagger \mathbf{g}} \tag{8.337}$$

summarizes the discrete Fourier transform.

The orthogonality and normalization relations for the discrete Fourier basis, Eqs. (8.326) and (8.327), imply that

$$\mathbf{F}^\dagger \mathbf{F} = N\mathbf{1}. \tag{8.338}$$

Then $\mathbf{F}^{-1} = N^{-1}\mathbf{F}^\dagger$, which implies that

$$\mathbf{F}\mathbf{F}^\dagger = N\mathbf{1}. \tag{8.339}$$

The matrix version of the inverse discrete Fourier transform,

$$g = \frac{1}{N} F\tilde{g},$$

$$(8.340)$$

follows by applying $F^\dagger$ to both sides of Eq. (8.337) and using Eq. (8.339). In other words, the inverse discrete Fourier transform of the discrete Fourier transform returns the original vector of sampled values, provided that rounding and truncation errors can be neglected.

If one were to evaluate the discrete Fourier transform as an inner product, then for each of the N complex numbers $\{\gamma_m\}$ one would need N complex floating-point multiplications (which would require $4N$ real floating-point multiplications) and $N - 1$ complex floating-point additions (which would require $2(N - 1)$ real floating-point additions). The total operation count to compute the discrete Fourier transform (all of the $\{\gamma_m\}$) using an inner-product algorithm would be $4N^2$ real floating-point multiplications and $2N(N - 1)$ real floating-point additions. The sum would be $O(N^2)$.

Much of the usefulness of the discrete Fourier transform derives from the fact that it can be computed with only $O(N \ln N)$ floating-point operations using the *fast Fourier transform*. Most engineers and physicists regard the fast Fourier transform as a "black box" that produces the complex numbers γ_m. One should be aware, however, that if one's problem requires the computation of many thousands of discrete Fourier transform–inverse transform pairs, it is important to pay attention to certain computational details. For example, one must compute the sines and cosines of the sampling angles to the full available precision in the first quadrant, $[0, \pi/4]$, and then use trigononometric identities to obtain $e^{2\pi i m k/N}$ in the remaining three octants. It is also useful for long computations to create a lookup table of the values of $e^{2\pi i m k/N}$.

8.5.3 PERIODIC EXTENSION

Often one wants to consider the sampled values of a function that has a definite symmetry (even or odd) about the point $x = 0$. In such a case the use of an asymmetric interval such as $[0, 1]$ is a nuisance.

A function g that is defined only on a finite interval $[a, b]$ can be defined on the entire real line through the process of **periodic extension**, in which one defines the values of f outside of $[a, b]$ by the equation

$$\forall x \in \mathbb{R} : g(x) := g(x')$$

$$(8.341)$$

in which

$$x' = x - (b - a) \left\lfloor \frac{x - a}{b - a} \right\rfloor.$$

$$(8.342)$$

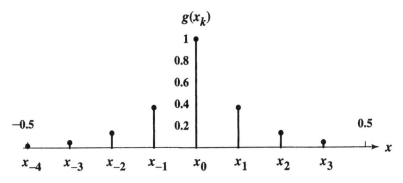

FIGURE 8.9

The function $g(x) = e^{-8|x|}$, sampled uniformly at $N = 8$ points on the interval $[-\frac{1}{2}, \frac{1}{2}]$.

In this equation, the second term subtracts the correct integral multiple of $b - a$ to make x' lie in the interval $[a, b]$. In other words, one extends the domain on which g is defined by translating the interval $[a, b]$ through integral multiples of the length $b - a$ and "stenciling" copies of g on the translated intervals.

For example, in Fig. 8.9, the function $g(x) = e^{-8|x|}$ has been sampled on the interval $[-\frac{1}{2}, \frac{1}{2}]$ at $N = 8$ points

$$x_k = \frac{k(b - a)}{N} \tag{8.343}$$

in which $k \in (-4 : 3)$ and $b - a = 1$. The vector of sampled values is

$$\mathbf{g}' = \begin{pmatrix} x_{-4} \\ x_{-3} \\ x_{-2} \\ x_{-1} \\ x_0 \\ x_1 \\ x_2 \\ x_3 \end{pmatrix}. \tag{8.344}$$

To define g on the interval $[\frac{1}{2}, 1]$, for comparison with numerically computed inverse Fourier transforms, one makes use of periodic extension,

$$g(x) = g(x - 1). \tag{8.345}$$

The result is

$$g(x_k) = g(x_{k-8}) \tag{8.346}$$

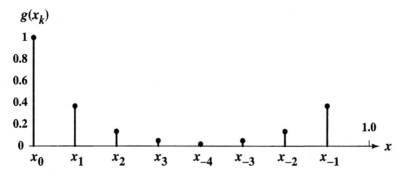

FIGURE 8.10

The function $g(x) = e^{-8|x|}$, sampled uniformly at $N = 8$ points on the interval $[0, 1]$. The values of g on the interval $[\frac{1}{2}, 1)$ have been obtained by periodicity from the values of g on the interval $[-\frac{1}{2}, 0)$.

for $k = 4, 5, 6, 7$. The vector of sampled values of g on the interval $[0, 1]$ therefore is

$$
\mathbf{g} = \begin{pmatrix} x_0 \\ x_1 \\ x_2 \\ x_3 \\ x_4 \\ x_5 \\ x_6 \\ x_7 \end{pmatrix} = \begin{pmatrix} x_0 \\ x_1 \\ x_2 \\ x_3 \\ x_{-4} \\ x_{-3} \\ x_{-2} \\ x_{-1} \end{pmatrix} \tag{8.347}
$$

(see Fig. 8.10). The right-hand side of this equation is what the inverse fast Fourier transform computes.

8.5.4 ALIASING

The interpretation of the discrete Fourier transform in terms of frequencies has profound practical implications for computation, data analysis, and sampling of continuous signals.

If one samples a function at uniformly spaced points

$$
x_k = a + \frac{k(b - a)}{N}, \tag{8.348}
$$

then the frequency of sampling (defined as the number of sampling points per unit interval) is $(\Delta x)^{-1}$, in which the distance between adjacent points is

$$
\Delta x = x_{k+1} - x_k = \frac{b - a}{N}. \tag{8.349}
$$

The corresponding angular frequency is

$$\omega_S = \frac{2\pi}{\Delta x} = \frac{2\pi N}{b - a}. \tag{8.350}$$

Either $(\Delta x)^{-1}$ or ω_S can be called the **sampling frequency**. Usually it is clear from the context whether a circular frequency or an angular frequency is meant.

The angular frequency of the Fourier-series term $g_l e^{2\pi i l x/(b-a)}$ for a function g defined on an interval $[a, b]$ is

$$\omega_l = \frac{2\pi l}{b - a}. \tag{8.351}$$

If the Fourier function with frequency ω_l on the interval $[a, b]$,

$$f_l(x) = e^{i\omega_l x}, \tag{8.352}$$

is sampled at the points x_k, one obtains

$$f_l(x_k) = e^{i\omega_l x_k} = w^{kl} \tag{8.353}$$

in which

$$w = e^{2\pi i/N}. \tag{8.354}$$

Clearly w is an N-th root of unity,

$$w^N = 1. \tag{8.355}$$

If a Fourier function f_{l+nN}, the frequency of which differs from ω_l by an integral multiple of the sampling frequency, is sampled at the points x_k, one obtains

$$\begin{aligned} f_{l+nN}(x_k) &= w^{k(l+nN)} \\ &= w^{kl}(w^N)^{kn} \\ &= w^{kl} \\ &= f_l(x_k). \end{aligned} \tag{8.356}$$

Restricted to the points x_k, then, f_l and f_{l+nN} are the same function, because they take the same values at each point of the domain $\{x_k = a + k(b - a)/N\}$. One can regard f_l as an "alias" for f_{l+nN}. The property that

$$\mathbf{f}_{l+nN} = S_N f_{l+nN} = S_N f_l = \mathbf{f}_l \tag{8.357}$$

is called **aliasing**. Because of aliasing, the vectors $\mathbf{f}_l$ of the sampled Fourier basis are periodic (with period N) in the *index* l. Therefore l can be any integer, not just one of the integers in the list $(0 : N - 1)$. Equivalently, the indexes $l = 0, \ldots, N - 1$ that label unique sampled Fourier basis vectors can be considered as elements of $\mathbb{Z}_N$, the set of integers modulo N.

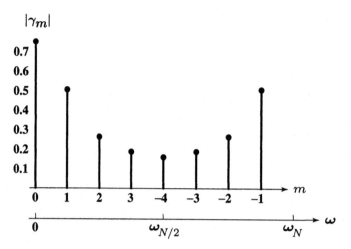

FIGURE 8.11

The modulus of the discrete Fourier transform of the function $g(x) = e^{-8|x|}$, as computed by the fast Fourier transform.

Equation (8.357) implies that, because of aliasing, the components γ_m of the discrete Fourier transform are periodic in m:

$$\gamma_{m+nN} = \gamma_m. \tag{8.358}$$

Hence a plot of $|\gamma_m|$ versus m can be arranged (in transform space) as either Fig. 8.9 or Fig. 8.10 is arranged in signal space. For example, Fig. 8.11 shows $|\gamma_m|$ versus m on the interval $[0, \omega_N)$ for the function plotted in Fig. 8.10, and Fig. 8.12 shows $|\gamma_m|$ versus m on the interval $[-\omega_N/2, \omega_N/2)$ for the same function.

To understand aliasing in terms of frequency, one must look at the frequencies associated with different Fourier basis functions that, if sampled, are aliased to the same vector of sampled values. Equation (8.357) implies that the difference of the frequencies associated with the Fourier functions f_l and f_{l+nN}, which give the same vector $\mathbf{f}_l$ of sampled values, is

$$\frac{2\pi(l + nN)}{b - a} - \frac{2\pi l}{b - a} = n\omega_N \tag{8.359}$$

in which

$$\boxed{\omega_N := \frac{2\pi N}{b - a}} \tag{8.360}$$

is called the **Nyquist frequency**. (Evidently ω_N is equal to the sampling frequency ω_S.) Another way to describe aliasing, then, is to say that frequencies ω_{l+nN} and ω_l that differ by

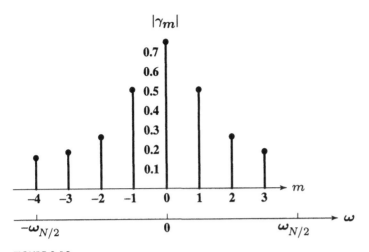

FIGURE 8.12

The modulus of the discrete Fourier transform of the function $g(x) = e^{-8|x|}$. The values of $|\gamma_m|$ on the interval $[-\frac{1}{2}\omega_N, 0)$ have been obtained by periodicity from the values of $|\gamma_m|$ on the interval $[\frac{1}{2}\omega_N, \omega_N)$.

an integral multiple of the Nyquist frequency, ω_N, cannot be distinguished by sampling at the points $\{x_k = a + k(b - a)/N\}$.

We call the interval $[-\omega_N/2, \omega_N/2)$ the **Nyquist interval**. Figures 8.11 and 8.12 illustrate the fact that one can always consider the discrete Fourier transform as a function on the discrete frequencies that lie within the Nyquist interval.

8.5.5 SAMPLING THEOREM AND ALIAS MAPPING

Aliasing induces a mapping from the space of sequences of Fourier coefficients $\overline{g}$ to transform space (the elements of which are vectors $\tilde{\mathbf{g}}$ with components γ_m). From a practical point of view, the most important consequence of the existence of such a mapping is a quantitative value for the minimum frequency at which a signal with a finite bandwidth must be sampled in order to reconstruct the signal exactly from its discrete Fourier transform.

We return to the interval $[0, 1]$ and heuristically derive the formula for the mapping from Fourier sequence space to transform space by substituting the Fourier series Eq. (8.312) into Eq. (8.332) and setting $b - a = 1$:

$$
\begin{aligned}
\gamma_m &= \sum_{k=0}^{N-1} e^{-2\pi i km/N} \sum_{l=-\infty}^{\infty} g_l e^{2\pi i kl/N} \\
&= \sum_{l=-\infty}^{\infty} g_l \sum_{k=0}^{N-1} e^{2\pi i k(l-m)/N} \\
&= \sum_{l=-\infty}^{\infty} g_l \mathbf{f}_m^{\dagger} \mathbf{f}_l .
\end{aligned}
\tag{8.361}
$$

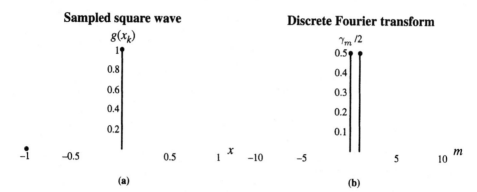

(a)

(b)

FIGURE 8.13

Illustration of the discrete Fourier transform of a square wave sampled at $N = 2$ points. (a) The values of $g(x_k)$. (b) The values of the discrete Fourier transform, $\gamma_m/2$, for the sampled values shown in (a).

Equation (8.357) implies that the inner product of two discrete Fourier basis functions $\mathbf{f}_l$ and $\mathbf{f}_m$, in which l is not restricted to lie between 0 and $N - 1$, is

$$\mathbf{f}_m^\dagger \mathbf{f}_l = N \sum_{n=-\infty}^{\infty} \delta_{l,m+nN}. \tag{8.362}$$

Hence

$$\gamma_m = N \sum_{n=-\infty}^{\infty} g_{m+nN}. \tag{8.363}$$

For example, if $N = 2$ then the discrete Fourier transform of the sampled square wave shown in Fig. 8.13 has only $\gamma_0 = 1$ and $\gamma_1 = 1$. All of the Fourier coefficients g_m for odd m add together to give $\gamma_1/2 = \frac{1}{2}$. (Note if m is even, the only nonzero Fourier coefficient g_m is g_0 in this example.)

The complex amplitude γ_m of the discrete Fourier transform is equal to the sum of the complex Fourier coefficients at all of the frequencies $\omega_l = 2\pi l/(b - a)$ that differ from ω_m by an integral multiple of the Nyquist frequency ω_N. If a signal has a nonzero Fourier amplitude at a frequency ω_l that is higher than ω_N, the amplitude g_l is *not* thrown away but is *added* to the amplitude g_m such that m is congruent to l (modulo N) and $0 \leq m < N$. This consequence of aliasing has the practical effect that the components γ_m of the discrete Fourier transform are not equal to N times the corresponding Fourier coefficients g_m if the function g has nonzero Fourier coefficients at frequencies that are higher than ω_N. Great care must be taken in interpreting discrete Fourier transforms of signals that have not been filtered to remove frequencies above ω_N.

Equation (8.363), which gives the components of the discrete Fourier transform $\tilde{\mathbf{g}}$ in terms of the sequence of Fourier coefficients $\overline{g}$, defines a mapping A such that

$$\tilde{\mathbf{g}} = A\overline{g}. \tag{8.364}$$

Clearly A is linear. We refer to A as the **alias mapping**.

In practical applications of the discrete Fourier transform, it is important to know if a signal can, or cannot, be reconstructed exactly after the procedures of time-sampling and discrete Fourier transformation have been applied. If exact reconstruction is not possible, then some information (in both colloquial and technical senses) has been lost.

Not all signals can be reconstructed uniquely, because it is possible for a nonzero signal to vanish at every sample point x_k. For example, if $s(x) = \sin(\omega_N x)$, then $s(x_k) = \sin(2\pi k) = 0$. In other words, A is singular. Therefore the uniform sampling mapping S_N maps two periodic signals g and g' that differ by $\sin(\omega_N x)$ to the same vector of sampled values, $\mathbf{g}$. Therefore g and g' have the same discrete Fourier transforms. It follows that g cannot be reconstructed uniquely from its discrete Fourier transform.

By definition, the null space of the alias mapping,

$$\mathcal{N} := \text{null}[A], \tag{8.365}$$

consists of Fourier-coefficient sequences $\overline{g}_0$ such that

$$\tilde{\mathbf{g}}_0 = A\overline{g}_0 = \mathbf{0}. \tag{8.366}$$

From the equation for the inverse discrete Fourier transform, Eq. (8.340), it follows that the discrete Fourier transform $\tilde{\mathbf{g}}_0 = \mathbf{0}$ if and only if the corresponding vector of sampled values $\mathbf{g}_0 = \mathbf{0}$. Therefore the null space $\mathcal{N}$ of the alias mapping consists of the Fourier-coefficient sequences of signals that vanish at every sample point.

Equation (8.363) implies that an equivalent characterization of the null space $\mathcal{N}$ is

$$\mathcal{N} = \left\{ \overline{g} \mid \forall m \in (0 : N-1) : \sum_{n \in \mathbb{Z}} g_{m+nN} = 0 \right\}. \tag{8.367}$$

Because the linear mapping A from sequences of Fourier coefficients to discrete Fourier transforms is singular, and because every singular linear mapping defines a nonsingular linear mapping from a complement of its null space to its range (see Section 6.7), it follows that the possibility of reconstructing a function from its discrete Fourier transform rests on finding a physically useful complement of $\mathcal{N}$. We show that a subspace of the space $\ell^2(\mathbb{Z}; \mathbb{C})$ of Fourier-coefficient sequences that is complementary to $\mathcal{N}$ is the space of Fourier coefficients of **band-limited signals**,

$$\mathcal{B} := \{ \overline{g} \mid \forall m :\ni: \omega_m \notin [-\omega_N/2, \omega_N/2) : g_m = 0 \}. \tag{8.368}$$

A sequence $\overline{g}$ belongs to $\mathcal{B}$ if and only if every Fourier coefficient that corresponds to a frequency outside of the Nyquist interval vanishes.

To show that $\mathcal{B}$ and $\mathcal{N}$ are complementary subspaces of $\ell^2(\mathbb{Z}; \mathbb{C})$, one must show that

$$\mathcal{B} \cap \mathcal{N} = 0 \tag{8.369}$$

and that

$$\mathcal{B} \oplus \mathcal{N} = \ell^2(\mathbb{Z}; \mathbb{C}). \tag{8.370}$$

If $\overline{g} \in \mathcal{B}$, then within the Nyquist interval the infinite sum in Eq. (8.363) reduces to a single term. Equations (8.367) and (8.368) would imply that $g_m = 0$ for every integer m; hence $\mathcal{B} \cap \mathcal{N} = 0$. To show that $\mathcal{B} \oplus \mathcal{N} = \ell^2(\mathbb{Z}; \mathbb{C})$, let $\overline{g} \in \ell^2(\mathbb{Z}; \mathbb{C})$, let $\overline{g}_0$ be the sequence such that

$$g_{0,m} = \begin{cases} -\gamma_m/N + g_m, & \text{if } \omega_m \in [-\omega_N/2, \omega_N/2), \\ g_m, & \text{if } \omega_m \notin [-\omega_N/2, \omega_N/2), \end{cases} \tag{8.371}$$

and let $\overline{g}'$ be the sequence such that

$$g'_m = \begin{cases} \gamma_m/N, & \text{if } \omega_m \in [-\omega_N/2, \omega_N/2), \\ 0, & \text{if } \omega_m \notin [-\omega_N/2, \omega_N/2). \end{cases} \tag{8.372}$$

Then

$$\overline{g} = \overline{g}_0 + \overline{g}' \tag{8.373}$$

in which $\overline{g}_0 \in \mathcal{N}$ and $\overline{g}' \in \mathcal{B}$. Therefore $\mathcal{B}$ is complementary to $\mathcal{N}$, and the linear mapping A, restricted to $\mathcal{B}$,

$$\mathsf{A} : \mathcal{B} \to \mathbb{C}^N, \tag{8.374}$$

is nonsingular.

It follows that a *band-limited* signal can be reconstructed exactly from its discrete Fourier transform $\tilde{\mathbf{g}}$ by computing

$$g(x) = \sum_{\omega_m \in [-\omega_N/2, \omega_N/2)} g_m e^{2\pi i m x/(b-a)} \tag{8.375}$$

using

$$g_m = \gamma_m/N \tag{8.376}$$

for all m such that ω_m lies in the Nyquist interval. This important result is known as the **sampling theorem**. An equivalent, more common form of the theorem states that a signal can be reconstructed exactly from its discrete Fourier transform if and only if the sampling rate $(\omega_S = \omega_N)$ is at least twice as large as the absolute value of the highest frequency in the signal $(\omega_N/2$ in our case).

In nonlinear systems, an initial signal that lies within the Nyquist interval can be altered in such a way that signals are generated at new frequencies. Some or all of the new frequencies

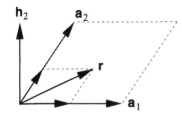

FIGURE 8.14

A 2-parallelepiped. The vector **r** belonging to the parallelepiped is equal to $t^1 \mathbf{a}_1 + t^2 \mathbf{a}_2$. The scalars t^1 and t^2 may independently take any values between 0 and 1.

may lie outside the Nyquist interval. As a result of aliasing, the aliased nonlinearly generated signals may add to valid signals at frequencies within the Nyquist interval and cause numerical instability. One application of the sampling theorem is to determine what frequencies must be removed by digital filtering in numerical simulations of nonlinear phenomena.

A function g, the vector **g** of its sampled values, the sequence $\overline{g}$ of its Fourier coefficients, and its discrete Fourier transform $\tilde{\mathbf{g}}$ are related by the following commutative diagram:

$$
\begin{array}{ccc}
g & \xrightarrow{\ \mathsf{F}^\dagger\ } & \overline{g} \\
{\scriptstyle\mathsf{S}_N}\Big\downarrow & & \Big\downarrow{\scriptstyle\mathsf{A}} \\
\mathbf{g} & \xrightarrow[\ \mathsf{F}^\dagger\ ]{} & \tilde{\mathbf{g}}
\end{array}
\tag{8.377}
$$

This diagram shows in an abstract way how the sampling mapping S_N induces the alias mapping A. Note that both the sampling mapping and the alias mapping are many-to-one.

8.5.6 EXERCISE FOR SECTION 8.5

8.5.1   Create a computer program to sample the Fourier functions f_l and f_{l+nN} uniformly at $N = 8$ or $N = 16$ points. Plot both the continuous graph and the sampled values of f_l and f_{l+nN} (for each l), and verify that they take the same values at the sampling points.

8.6 VOLUME OF AN m-PARALLELEPIPED

8.6.1 PARALLELEPIPEDS

The **m-parallelepiped** $P_{\mathbf{a}_1,\dots,\mathbf{a}_m}$ **defined by the vectors** $\mathbf{a}_1, \dots, \mathbf{a}_m$ in n-dimensional Euclidean space $\mathbb{E}^n$ is defined as the set

$$
P_{\mathbf{a}_1,\dots,\mathbf{a}_m} := \left\{ \sum_{i=1}^m t^i \mathbf{a}_i \mid \forall i \in (1:m) : t^i \in [0,1] \right\}.
\tag{8.378}
$$

Figure 8.14, for example, shows a 2-parallelepiped and a vector $\mathbf{r} = t^1 \mathbf{a}_1 + t^2 \mathbf{a}_2$ lying in the parallelepiped. Figure 7.3 shows the 3-parallelepiped $P_{\mathbf{f}_1,\mathbf{f}_2,\mathbf{f}_3}$ in $\mathbb{E}^3$. A **rectangular m-parallelepiped** is defined by mutually orthogonal vectors.

8.6.2 RECURSIVE DEFINITION OF VOLUME

We define the *m*-dimensional volume

$$V_{a_1,\ldots,a_m} := V_{a_1,\ldots,a_{m-1}} \, \|h_m\| \tag{8.379}$$

of the parallelepiped $P_{a_1,\ldots,a_m}$ recursively as the $(m-1)$-dimensional volume $V_{a_1,\ldots,a_{m-1}}$, multiplied by the length

$$\|h_m\| := \left(h_m^T h_m\right)^{1/2} \tag{8.380}$$

of the vector

$$h_m := a_m - P_{W_{m-1}} a_m \tag{8.381}$$

that is orthogonal to the vector subspace W_{m-1} spanned by the vectors $a_1, \ldots, a_{m-1}$,

$$W_{m-1} := \text{span}\{a_1, \ldots, a_{m-1}\}. \tag{8.382}$$

If $m = 1$, the volume V_{a_1} is defined as $\|a_1\|$, which is the length of a_1.

For example, in Fig. 8.14 $V_{a_1,a_2} = \|h_2\| \, V_{a_1} = \|h_2\| \, \|a_1\|$, in agreement with the plane-geometry formula for the area of a parallelogram.

The recursive equation (8.379) defining the volume can be solved immediately for an *m*-parallelepiped $P_{h_1,\ldots,h_m}$ defined by *m* mutually orthogonal vectors $h_1, \ldots, h_m \in \mathbb{E}^n$:

$$V_{h_1,\ldots,h_m} := \|h_1\| \cdots \|h_m\|. \tag{8.383}$$

This result for a rectangular *m*-parallelepiped is an obvious generalization of the formula "volume = height × width × depth" for a rectangular solid in $\mathbb{E}^3$.

The definition given in Eq. (8.379) generalizes the solid-geometry theorem that the volume of a 3-parallelepiped is equal to the product of its base (here, $V_{a_1,\ldots,a_{m-1}}$) by its height (here, $\|h_m\|$). Note that if the list $\{a_1, \ldots, a_m\}$ is linearly dependent, then the parallelepiped $P_{a_1,\ldots,a_m}$ collapses to a figure of height zero, because a_m is equal to a linear combination of its predecessors and therefore lies in W_{m-1}. Then the component of a_m orthogonal to W_{m-1} vanishes: $h_m = 0$. Therefore a parallelepiped's volume vanishes, $V_{a_1,\ldots,a_m} = 0$, if the defining vectors are linearly dependent.

8.6.3 VOLUME AS A DETERMINANT

We prove by induction on *m* that the volume of an *m*-parallelepiped obeys the equation

$$(V_{a_1,\ldots,a_m})^2 = \det\left[A_m^T A_m\right] \tag{8.384}$$

in which A_m is an $n \times m$ real matrix, the columns of which are the vectors that define the *m*-parallelepiped:

$$A_m := [a_1, \ldots, a_m]. \tag{8.385}$$

If the list $\{a_1, \ldots, a_m\}$ is linearly dependent, then $\mathbf{A}_m$ is singular. It follows that $\mathbf{A}_m^T \mathbf{A}_m$ is singular, and therefore that both the determinant and the volume must vanish. We assume from now on that $\{a_1, \ldots, a_m\}$ is linearly independent.

If $m = 1$, then $\mathbf{A}_1 = \mathbf{a}_1$ is an $n \times 1$ column vector and $\mathbf{A}_m^T \mathbf{A}_m = \mathbf{a}_1^T \mathbf{a}_1 = \|\mathbf{a}_1\|^2$. Therefore the theorem is true for $m = 1$. Assume then that Eq. (8.384) holds for $m - 1$. The determinant in Eq. (8.384) is the determinant of the canonical inner products of the vectors $\mathbf{a}_i$ in $\mathbb{E}^m$:

$$\det\left[\mathbf{A}_m^T \mathbf{A}_m\right] = \det\left[\mathbf{a}_i^T \mathbf{a}_j\right] = \det\left[\mathbf{a}_i \cdot \mathbf{a}_j\right]. \tag{8.386}$$

The k-th column of $\mathbf{A}_m^T \mathbf{A}_m$ is equal to

$$\mathbf{b}_k := \begin{pmatrix} \mathbf{a}_1 \cdot \mathbf{a}_k \\ \vdots \\ \mathbf{a}_{m-1} \cdot \mathbf{a}_k \\ \mathbf{a}_m \cdot \mathbf{a}_k \end{pmatrix}. \tag{8.387}$$

Define the vector $\mathbf{h}_m$ as in Eq. (8.381), and take $\{a_1, \ldots, a_{m-1}\}$ as a basis of $\mathcal{W}_{m-1}$. Then

$$\mathbf{a}_m = \mathbf{h}_m + P_{\mathcal{W}_{m-1}} \mathbf{a}_m$$
$$= \mathbf{h}_m + \sum_{l=1}^{n-1} a_m^l \, \mathbf{a}_l. \tag{8.388}$$

The scalars a_m^l are the contravariant components of $P_{\mathcal{W}_{m-1}} \mathbf{a}_m$ relative to the basis we have chosen in $\mathcal{W}_{m-1}$. Note that although Eq. (8.388) expresses the orthogonal decomposition of $\mathbf{a}_m$ with respect to the subspace $\mathcal{W}_{m-1}$, one is not required to use an orthonormal basis in $\mathcal{W}_{m-1}$.

The m-th column of $\mathbf{A}_m^T \mathbf{A}_m$ is equal to

$$\mathbf{b}_m = \begin{pmatrix} \mathbf{a}_1 \cdot \mathbf{a}_m \\ \vdots \\ \mathbf{a}_{m-1} \cdot \mathbf{a}_m \\ \mathbf{a}_m \cdot \mathbf{a}_m \end{pmatrix} = \|\mathbf{h}_m\|^2 \mathbf{e}_m + \sum_{l=1}^{m-1} a_m^l \, \mathbf{b}_l \tag{8.389}$$

in which we have used Eq. (8.388) to substitute for $\mathbf{a}_m$ in each of the inner products $\mathbf{a}_i \cdot \mathbf{a}_m$ occurring in $\mathbf{b}_m$. The term $\|\mathbf{h}_m\|^2 \mathbf{e}_m$ occurs because

$$\mathbf{a}_i \cdot \mathbf{h}_m = \delta_m^i \, \mathbf{a}_m \cdot \mathbf{h}_m = \delta_m^i \, \|\mathbf{h}_m\|^2 \tag{8.390}$$

in view of Eq. (8.388) and of the orthogonality of $\mathbf{h}_m$ to $\mathcal{W}_{m-1}$. By Eq. (8.389), the determinant of $\mathbf{A}_m^T \mathbf{A}_m$ is equal to

$$\det\left[\mathbf{A}_m^T \mathbf{A}_m\right] = \det[\mathbf{b}_1, \ldots, \mathbf{b}_m]$$
$$= \det\left[\mathbf{b}_1, \ldots, \mathbf{b}_{m-1}, \|\mathbf{h}_m\|^2 \mathbf{e}_m + \sum_{l=1}^{m-1} a_m^l \, \mathbf{b}_l\right]. \tag{8.391}$$

As in the derivation of Eq. (6.316), none of the terms in the summation contributes to the determinant because each column $\mathbf{b}_l$ (in which $l \in (1 : m - 1)$] occurs twice. Therefore

$$\det\left[\mathbf{A}_m^T \mathbf{A}_m\right] = \|\mathbf{h}_m\|^2 \det[\mathbf{b}_1, \ldots, \mathbf{b}_{m-1}, \mathbf{e}_m]. \tag{8.392}$$

Expanding the determinant in minors of the last column, which consists of zeros except for a 1 in the last row, one sees that

$$\det\left[\mathbf{A}_m^T \mathbf{A}_m\right] = \|\mathbf{h}_m\|^2 \det\left[\mathbf{A}_{m-1}^T \mathbf{A}_{m-1}\right] \tag{8.393}$$

in which $\mathbf{A}_{m-1}$ is the matrix of the vectors which span $\mathcal{W}_{m-1}$,

$$\mathbf{A}_{m-1} := [\mathbf{a}_1, \ldots, \mathbf{a}_{m-1}]. \tag{8.394}$$

Then

$$\det\left[\mathbf{A}_{m-1}^T \mathbf{A}_{m-1}\right] = (V_{\mathbf{a}_1,\ldots,\mathbf{a}_{m-1}})^2. \tag{8.395}$$

Equation (8.384) follows now from the recursive definition of the volume of an m-parallelepiped, Eq. (8.379).

If $m = n$, then $\mathbf{A}_n = [\mathbf{a}_1, \ldots, \mathbf{a}_n]$ is square. Then

$$\det\left[\mathbf{A}_n^T \mathbf{A}_n\right] = (\det[\mathbf{A}_n])^2. \tag{8.396}$$

It follows from this equation and Eq. (8.384) that the volume of an n-parallelepiped is

$$V_{\mathbf{a}_1,\ldots,\mathbf{a}_n} = |\det[\mathbf{A}_n]| \tag{8.397}$$

in n-dimensional Euclidean space. This result generalizes the formula

$$V_{\mathbf{a}_1,\mathbf{a}_2,\mathbf{a}_3} = |\mathbf{a}_1 \cdot (\mathbf{a}_2 \times \mathbf{a}_3)| \tag{8.398}$$

for the volume of a 3-parallelepiped in $\mathbb{E}^3$.

8.6.4 DETERMINANT AS A VOLUME RATIO

A nonsingular linear mapping $A : \mathbb{E}^n \to \mathbb{E}^n$ transforms an m-parallelepiped $P_{\mathbf{a}_1,\ldots,\mathbf{a}_m}$ into another m-parallelepiped $P_{A\mathbf{a}_1,\ldots,A\mathbf{a}_m}$, in which $\mathbf{A}$ is the $n \times n$ matrix that realizes A. If $m = n$, there is a simple formula for the transformation of the volume $V_{\mathbf{a}_1,\ldots,\mathbf{a}_n}$ under $\mathbf{A}$:

$$V_{A\mathbf{a}_1,\ldots,A\mathbf{a}_n} = |\det[\mathbf{A}\mathbf{A}_n]| = |\det[\mathbf{A}]| \, |\det[\mathbf{A}_n]|. \tag{8.399}$$

It follows that the absolute value of the determinant of a matrix $\mathbf{A} \in \mathbb{R}^{n \times n}$ is equal to the ratio of the volume of the image of any n-parallelepiped to its original volume:

$$|\det[\mathbf{A}]| = \frac{V_{A\mathbf{a}_1,\ldots,A\mathbf{a}_n}}{V_{\mathbf{a}_1,\ldots,\mathbf{a}_n}}. \tag{8.400}$$

In Eqs. (6.333) and (6.334) we related the handedness of the coordinate system defined by n linearly independent vectors $\mathbf{a}_1, \ldots, \mathbf{a}_n \in \mathbb{E}^n$ to the sign of $\det[\mathbf{A}_n] = \det[\mathbf{a}_1, \ldots, \mathbf{a}_n]$. The sign of $\det[\mathbf{A}_n]$ also determines the **orientation** of the n-parallelepiped $P_{\mathbf{a}_1, \ldots, \mathbf{a}_n}$ according to the following convention: If $\det[\mathbf{A}_n] > 0$, then $P_{\mathbf{a}_1, \ldots, \mathbf{a}_n}$ is **positively oriented**; if $\det[\mathbf{A}_n] < 0$, then $P_{\mathbf{a}_1, \ldots, \mathbf{a}_n}$ is **negatively oriented**. Because $\det[\mathbf{A}\mathbf{A}_n] = \det[\mathbf{A}] \det[\mathbf{A}_n]$, it follows that the sign of $\det[\mathbf{A}]$ determines whether a linear mapping A changes or preserves the orientation of the coordinate system.

Summary: *The absolute value of the determinant of a linear mapping,* $|\det[\mathbf{A}]|$, *is the* **volume expansion factor**. *If* $\det[\mathbf{A}] > 0$, *then* **A** *preserves the orientation of the coordinate system in* $\mathbb{E}^n$. *If* $\det[\mathbf{A}] < 0$, *then* **A** *changes the orientation of the coordinate system.*

8.6.5 JACOBIAN DETERMINANT

The fact that the determinant is equal to the ratio of volumes before and after a linear mapping, Eq. (8.400), underlies a useful formula for changing variables in a multidimensional integral over a region R, such as

$$\int_R f(\mathbf{r}) \, d^n r. \tag{8.401}$$

In interesting cases, the coordinates of points $\mathbf{r} \in R$ are curvilinear, as in Eqs. (8.32) and (8.33).

A heuristic derivation of the Jacobian formula starts by dividing the region of integration into parallelepipeds bounded by infinitesimal displacements $d\mathbf{r}_1, \ldots, d\mathbf{r}_n$ along the directions of the coordinates $x^1, \ldots, x^k$. Because there are n coordinates, the location $\mathbf{r}_\iota$ of each parallelepiped can be indexed with a list of n integers,

$$\iota := (i_1, i_2, \ldots, i_n). \tag{8.402}$$

The volume of the parallelepiped at $\mathbf{r}_\iota$ is

$$V_{d\mathbf{s}_1, \ldots, d\mathbf{s}_n}(\mathbf{r}_\iota) = |\det[d\mathbf{s}_1, \ldots, d\mathbf{s}_n]|. \tag{8.403}$$

A Riemann sum that approximates the integral is

$$\int_R f(\mathbf{r}) \, d^n r \approx \sum_\iota f(\mathbf{r}_\iota) \, V_{d\mathbf{s}_1, \ldots, d\mathbf{s}_n}(\mathbf{r}_\iota). \tag{8.404}$$

Let us change to a set of new coordinates $x'^1, \ldots, x'^n$ by expressing the old coordinates $x^1, \ldots, x^n$ in terms of the new ones,

$$x^k = \phi^k(x'^1, \ldots, x'^n), \tag{8.405}$$

and use the same points $\mathbf{r}_\iota$ (expressed in terms of the new coordinates) to evaluate the function f. Although the mapping of coordinates may be nonlinear, the mapping of infinitesimal displacements

$$ds = \begin{pmatrix} dx^1 \\ \vdots \\ dx^n \end{pmatrix} \tag{8.406}$$

is linear. If one uses the chain rule of partial differentiation in the form

$$dx^k = \sum_{l=1}^{n} \frac{\partial \phi^k}{\partial x^{\prime l}} dx^{\prime l}, \tag{8.407}$$

then one recognizes a linear mapping

$$ds' = Jds, \tag{8.408}$$

the matrix of which is

$$J = \begin{pmatrix} \dfrac{\partial \phi^1}{\partial x^{\prime 1}} & \cdots & \dfrac{\partial \phi^1}{\partial x^{\prime n}} \\ \vdots & \ddots & \vdots \\ \dfrac{\partial \phi^n}{\partial x^{\prime 1}} & \cdots & \dfrac{\partial \phi^n}{\partial x^{\prime n}} \end{pmatrix}. \tag{8.409}$$

J is called the Jacobian matrix.

The volume of an infinitesimal parallelepiped in the new coordinates,

$$\begin{aligned} V_{ds_1',\ldots,ds_n'} &= |\det[ds_1', \ldots, ds_n']| \\ &= |\det[Jds_1, \ldots, Jds_n]| \\ &= |\det[J]| V_{ds_1,\ldots,ds_n}, \end{aligned} \tag{8.410}$$

differs from the volume of an elementary parellepiped in the old coordinates, $V_{ds_1,\ldots,ds_n}$, by the factor $|\det[J]|$. In terms of the new coordinates, the Riemann sum is

$$\begin{aligned} \int_R f(\mathbf{r}) \, d^n r &\approx \sum_\iota f(\phi(\mathbf{r}_\iota')) \frac{V_{ds_1,\ldots,ds_n}}{V_{ds_1',\ldots,ds_n'}} V_{ds_1',\ldots,ds_n'} \\ &\approx \sum_\iota f(\phi(\mathbf{r}_\iota)) |\det[J]| V_{ds_1',\ldots,ds_n'}. \end{aligned} \tag{8.411}$$

In the limit as one refines the Riemann sum, the expression in the second line converges to

$$\boxed{\int_R f(\mathbf{r}) \, d^n r = \int_{R'} f(\phi(\mathbf{r}')) \, |\det[J]| \, d^n r'.} \tag{8.412}$$

8.6.6 EXERCISE FOR SECTION 8.6

8.6.1 Find the volume of the parallelepiped P_{a_1, a_2, a_3}, for which the vectors a_1, a_2, a_3 $\in \mathbb{E}^3$ are defined in Eqs. (8.287) and (8.288).

8.7 VECTOR AND MATRIX NORMS

8.7.1 VECTOR NORMS

The norm of a vector is a nonnegative real number that one can interpret as the length of the vector. Vector norms are especially important for estimating the accuracy of numerical computations and for establishing abstract properties such as continuity.

The inner-product norm

$$\|\overline{x}\|_{ip} = \langle \overline{x}, \overline{x} \rangle^{1/2} \tag{8.413}$$

plays a major role in the geometry of Euclidean and unitary spaces. In these spaces the inner-product norm is positive-definite, is homogeneous in the sense that $\|\alpha \overline{x}\|_{ip} = |\alpha| \, \|\overline{x}\|_{ip}$, and obeys the triangle inequality (Eq. [8.134]). However, the inner-product norm has other properties that the more general vector norms that we are about to define do not have. For example, we prove in Chapter 9 that the inner-product norm has the property of invariance under unitary transformations U:

$$\|U\overline{x}\|_{ip} = \langle U\overline{x}, U\overline{x} \rangle^{1/2} = \langle \overline{x}, \overline{x} \rangle^{1/2} = \|\overline{x}\|_{ip}. \tag{8.414}$$

This property and the parallelogram law, Eq. (8.139), are unique to the inner-product norm.

Unfortunately the evaluation of the inner-product norm is too expensive from a computational point of view to permit its use in the innermost loop of a computationally intensive program. Pythagoras's theorem implies that the inner-product norm of a vector $\overline{x} = x^i \, \overline{e}_i$ is equal to

$$\|\overline{x}\|_{ip} = \left[\sum_{i=1}^{n} |x^i|^2 \right]^{1/2} \tag{8.415}$$

in terms of coordinates x^i relative to an orthonormal basis $\{\overline{e}_i\}$. To evaluate this expression numerically requires n floating-point multiplications, $n - 1$ floating-point additions, and a square root, the argument of which must be checked for validity (real and nonnegative) before the square root can be computed. Because a floating-point multiplication requires many more central processing unit clock periods than a floating-point addition, the computational effort required to evaluate an inner-product norm is dominated by the floating-point multiplications and the square root. For example, the notoriously slow FORTRAN function cabs evaluates Eq. (8.415) for $n = 2$ in order to compute $|z|$ for $z \in \mathbb{C}$.

In order to define a single nonnegative real number to represent the length of a vector without sacrificing computational speed or theoretical generality, one must give up the geometrical properties of the inner-product norm. Let $\mathcal{V}$ be a vector space over the field $\mathbb{F} = \mathbb{R}$ or $\mathbb{C}$. A **norm** is a functional $\| \cdot \| : \mathcal{V} \to \mathbb{R}$ that satisfies the following axioms:

1. *Positive-definiteness:*

$$\forall \bar{x} \in \mathcal{V} : \begin{cases} \|\bar{x}\| \geq 0; \\ \|\bar{x}\| = 0 \Leftrightarrow \bar{x} = \bar{0}. \end{cases} \tag{8.416}$$

2. *Homogeneity:*

$$\forall \alpha \in \mathbb{F} : \forall \bar{x} \in \mathcal{V} : \|\alpha \bar{x}\| = |\alpha| \, \|\bar{x}\|. \tag{8.417}$$

3. *The triangle inequality:*

$$\forall \bar{x}, \bar{y} \in \mathcal{V} : \|\bar{x} + \bar{y}\| \leq \|\bar{x}\| + \|\bar{y}\|. \tag{8.418}$$

These axioms comprise some, but not all, of the properties of vector length that are familiar from one's everyday experience. In particular, vector norms other than the inner-product norm do not obey the parallelogram law and do not provide for a concept of orthogonality.

A vector space $\mathcal{V}$ on which a norm is defined is called a **normed vector space**. Familiar examples of normed vector spaces include the real numbers $\mathbb{R}$ under the **absolute-value norm**

$$\forall x \in \mathbb{R} : \|x\| := |x| \tag{8.419}$$

and the complex numbers $\mathbb{C}$ under the **modulus norm**

$$\forall z \in \mathbb{C} : \|z\| := |z| \tag{8.420}$$

in which

$$|z| := [(\mathrm{Re}[z])^2 + (\mathrm{Im}[z])^2]^{1/2}. \tag{8.421}$$

Finite-Dimensional p-Norms

Computationally useful norms on finite-dimensional vector spaces include the **maximum norm** or $\infty-$ norm,

$$\|\bar{x}\|_\infty := \max_i |x^i|, \tag{8.422}$$

and the **1-norm**,

$$\|\bar{x}\|_1 := \sum_{i=1}^{n} |x^i|. \tag{8.423}$$

It is clear from the properties of the absolute value of a real number or the modulus of a complex number that these norms satisfy the axioms stated in Eqs. (8.416) through (8.418).

TABLE 8.1 The Values of $\|\mathbf{e}_1\|_p$ for $p = 1, 2$, and ∞ in the Bases C and F

$\|\mathbf{e}_1\|_p$	Basis	
	C	F
$p = 1$	1	$2^{1/2}$
$p = 2$	1	1
$p = \infty$	1	$2^{-1/2}$

Although the algorithms for evaluating the 1 norm and the inner-product norm are both $O(n)$, the 1 norm is significantly cheaper to evaluate than the inner-product norm. Each evaluation of an inner-product norm requires $O(n)$ floating-point multiplications; each evaluation of a 1-norm requires $O(n)$ floating-point additions, which are much faster than floating-point multiplications.

The 1-norm and the ∞-norm may be evaluated in any basis. However, $\|\bar{x}\|_1$ and $\|\bar{x}\|_\infty$ depend on the basis to which $\bar{x}$ is referred. For example, Table 8.1 summarizes the results of evaluating $\|\mathbf{e}_1\|_1$ and $\|\mathbf{e}_1\|_\infty$ for the canonical-basis vector

$$\mathbf{e}_1 := \begin{pmatrix} 1 \\ 0 \end{pmatrix} \tag{8.424}$$

relative to the canonical basis $C := \{\mathbf{e}_1, \mathbf{e}_2\}$ and the basis

$$F := \left\{ \mathbf{f}_1 = \frac{1}{\sqrt{2}} \begin{pmatrix} 1 \\ 1 \end{pmatrix}, \ \mathbf{f}_2 = \frac{1}{\sqrt{2}} \begin{pmatrix} 1 \\ -1 \end{pmatrix} \right\}. \tag{8.425}$$

F is the result of applying a unitary transformation to the vectors of C (rotation through the angle $\pi/4$); hence the 2-norm $\|\mathbf{e}_1\|_2$ is the same in both bases. Note that in order to evaluate norms relative to F one must use the coordinate expansion of $\mathbf{e}_1$ (rel F),

$$\mathbf{e}_1 = \frac{1}{\sqrt{2}} \mathbf{f}_1 + \frac{1}{\sqrt{2}} \mathbf{f}_2. \tag{8.426}$$

The 1-norm and the ∞-norm are special cases of the **finite-dimensional p-norm**

$$\|\bar{x}\|_p := \left[\sum_{i=1}^{n} |x^i|^p \right]^{1/p}. \tag{8.427}$$

The number field may be either $\mathbb{R}$ or $\mathbb{C}$. Although it is obvious that the p-norm is homogeneous and positive-definite, the proof of the triangle inequality for $\| \cdot \|_p$ is nontrivial. As in the special cases $p = 1$ and $p = \infty$, any p-norm may be evaluated in terms of contravariant

TABLE 8.2 Frequently Used Vector Norms

Space	Vector norm	Notation	Definition
$\mathbb{F}^n$	Inner-product norm	$\|\bar{x}\|_{ip}$	$\langle \bar{x}, \bar{x} \rangle^{1/2}$
$\mathbb{F}^n$	p-Norm	$\|\bar{x}\|_p$	$\left[\sum_{i=1}^{n} |x^i|^p \right]^{1/p}$
$\mathbb{F}^n$	1-Norm	$\|\bar{x}\|_1$	$\sum_{i=1}^{n} |x^i|$
$\mathbb{F}^n$	∞-Norm	$\|\bar{x}\|_\infty$	$\max_i |x^i|$
$\mathcal{C}^0((a, b); \mathbb{F})$	Supremum norm	$\|f\|_C$	$\sup_{x \in (a,b)} |f(x)|$

Note. $\mathbb{F}$ may be either $\mathbb{R}$ or $\mathbb{C}$. The notation $|x|$ means the absolute value if $\mathbb{F} = \mathbb{R}$ or the modulus if $\mathbb{F} = \mathbb{C}$.

components referred to an arbitrary basis, but $\|\bar{x}\|_p$ is basis-dependent except if p is equal to 2 and the basis in which the norm is evaluated is orthonormal.

The p-norm algorithm Eq. (8.427) is of $O(n)$, like the algorithm for the 2-norm, and is significantly more expensive to compute than the ∞-norm.

One can easily show that for a given vector $\bar{x}$, the value of the ∞-norm is not greater than the value of any of the p-norms of $\bar{x}$ (if p is finite),

$$\|\bar{x}\|_\infty \le \|\bar{x}\|_p, \tag{8.428}$$

and that the value of the p-norm is bounded by $n^{1/p}\|\bar{x}\|_\infty$.

$$\|\bar{x}\|_p \le n^{1/p}\|\bar{x}\|_\infty \tag{8.429}$$

(see Exercises 8.7.7 and 8.7.8). However, because of the basis-dependence of $\|\bar{x}\|$, one is usually interested in the order of magnitude of $\|\bar{x}\|$ rather than in its exact value. For example, the entries in Table 8.7.1 are all of order 1.

Supremum Norm

An important example of an infinite-dimensional normed vector space is $\mathcal{V} = \mathcal{C}^0((a, b); \mathbb{R})$ under the **supremum** (or **maximum** or **Chebyshev**) **norm**,

$$\forall f \in \mathcal{C}^0((a, b); \mathbb{R}) : \|f\|_C := \sup_{x \in (a,b)} |f(x)|. \tag{8.430}$$

The supremum norm is useful for estimating the error in numerical approximations, as well as for theoretical purposes. We return to this particular normed vector space often.

Table 8.2 summarizes the vector norms discussed in this section.

8.7.2 NORM OF A LINEAR MAPPING

In this section we extend the concept of a norm from vectors to linear mappings. The norm of a linear mapping A is a nonnegative real number $\|A\|$ that tells quantitatively how "large" A is, just as the norm of a vector is a quantitative measure of how "long" the vector

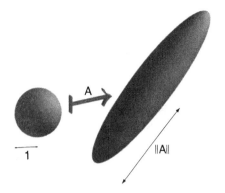

FIGURE 8.15

Illustration of the norm of a linear mapping in $\mathbb{R}^3$.

is. For example, the inner-product norm is just the Euclidean length of a vector. Figure 8.15 illustrates a linear mapping $A : \mathbb{E}^3 \mapsto \mathbb{E}^3$ that transforms the three-dimensional unit sphere into an ellipsoid by stretching along the z-axis, contracting along the x- and y-axes, and finally rotating about the x-axis. Clearly the images of some vectors are shorter than the lengths of the original vectors, and the images of other vectors (for example, a vector along the z-axis) are longer. Obviously a single number cannot describe the complicated linear mapping illustrated in Fig. 8.15. If one must give one number that says how "large" A is, it makes sense to use for $\|A\|$ the maximum possible dilatation of a vector. It will soon become clear that this is equivalent to the assertion that $\|A\|$ is the upper bound of values of $\|A\bar{x}\|$, assuming that $\|\bar{x}\| = 1$.

We show that for each vector norm, there exists a linear-mapping norm that is *compatible* with the vector norm in the sense that the vector and linear-mapping norms obey the inequality

$$\|A\bar{x}\| \leq \|A\| \, \|\bar{x}\|. \tag{8.431}$$

This inequality turns out to be indispensable for estimating the sensitivity of the solution vector of a system of linear equations to the input data as well as for many other purposes, both practical and theoretical.

Let $\mathcal{V}$ and $\mathcal{X}$ be normed vector spaces, and let $A : \mathcal{V} \to \mathcal{X}$ be a linear mapping. The **norm** of A is defined as the least upper bound of the quotient of $\|A\bar{x}\|$ (calculated in $\mathcal{X}$) by $\|\bar{x}\|$ (calculated in $\mathcal{V}$), in which $\bar{x}$ is any nonzero vector in $\mathcal{V}$:

$$\|A\| := \sup_{\bar{x} \neq \bar{0}} \frac{\|A\bar{x}\|_x}{\|\bar{x}\|_v}. \tag{8.432}$$

The value of $\|A\bar{x}\|/\|\bar{x}\|$ obviously is independent of the magnitude of $\|\bar{x}\|$. Defined as above, $\|A\|$ is automatically independent of any particular vector $\bar{x}$, because $\|A\|$ is the supremum over all nonzero vectors in $\mathcal{V}$. Note that in order to prove that one has indeed found the norm of a linear mapping, one must show that the proposed norm is the *least* upper bound; it is not enough merely to find *an* upper bound. In practice this means that one must show (or it must be obvious) that no smaller upper bound can be found.

Because of the linearity of A, the norm of A is equal to the least upper bound of the norm $\|A\bar{u}\|$ (calculated in $\mathcal{X}$), in which $\bar{u}$ is any vector of norm 1 in $\mathcal{V}$:

$$\|A = \sup_{\|\bar{u}\|_v=1} \|A\bar{u}\|_x. \tag{8.433}$$

One says that the linear-mapping norm is **induced** by, or is **consistent with**, the vector norm. The induced linear-mapping norm depends on the vector norms, which in turn depend on the bases in which they are evaluated. Usually we omit the subscripts $\mathcal{X}$ and $\mathcal{V}$, which indicate in which spaces the norms of $A\bar{x}$ and $\bar{x}$ are to be evaluated.

The definition Eq. (8.432) and the axioms for the vector norm imply certain fundamental properties:

1. The norm of $A\bar{x}$ is not greater than the product of the norm of A by the norm of $\bar{x}$:

$$\|A\bar{x}\| \le \|A\| \, \|\bar{x}\|. \tag{8.434}$$

This property follows directly from the definition of the supremum in Section 2.5.2.

2. The norm of a linear mapping is positive-definite:

$$\forall A \in \hom[\mathcal{V}, \mathcal{X}] : \begin{cases} \|A\| \ge 0; \\ \|A\| = 0 \Leftrightarrow A = \mathbf{0}. \end{cases} \tag{8.435}$$

That $\|A\| \ge 0$ follows from the definition of a vector norm. That $\|A\| = 0 \Leftrightarrow A = \mathbf{0}$ follows from the fact that if $\|A\| = 0$, then $\|A\bar{x}\| = 0$ for all vectors $\bar{x} \in \mathcal{V}$. Then Eq. (8.416) implies that $A\bar{x} = \bar{0}$ for all $\bar{x}$; but this is the definition of the null mapping $A = \mathbf{0}$.

3. The norm of a linear mapping is homogeneous:

$$\|\alpha A\| = |\alpha| \, \|A\| \tag{8.436}$$

This property follows straightforwardly from the homogeneity property of the vector norm, Eq. (8.417).

4. The linear-mapping norm obeys the triangle inequality:

$$\|A + B\| \le \|A\| + \|B\| \tag{8.437}$$

This property follows from the fact that the vector norm obeys the triangle inequality

$$\|(A + B)\bar{x}\| \le \|A\bar{x}\| + \|B\bar{x}\| \le \|A\| \, \|\bar{x}\| + \|B\| \, \|\bar{x}\|, \tag{8.438}$$

which implies that $\|A\| + \|B\|$ is an upper bound of $\|(A + B)\bar{x}\|/\|\bar{x}\|$.

Properties 2 through 4 imply that $\hom[\mathcal{V}, \mathcal{X}]$ is itself a normed vector space.

5. The norm of the product of two linear mappings $A : \mathcal{V} \to \mathcal{V}$ and $B : \mathcal{V} \to \mathcal{V}$ obeys the inequality

$$\|AB\| \le \|A\| \, \|B\| \tag{8.439}$$

because

$$\|AB\bar{x}\| \leq \|A\| \|B\bar{x}\| \leq \|A\| \|B\| \|\bar{x}\|. \tag{8.440}$$

In an important special case X is $\mathbb{F}$, the field over which V is a vector space. Then the linear mapping A is simply a linear functional $\underline{\phi}$. Then

$$\|\underline{\phi}\| = \sup_{\|\bar{u}\|=1} |\underline{\phi}[\bar{u}]| \tag{8.441}$$

in which $|\cdot|$ denotes the absolute-value norm if $\mathbb{F} = \mathbb{R}$ or the modulus norm if $\mathbb{F} = \mathbb{C}$.

For example, let $V = \mathbb{R}^4$ under the ∞ norm, and let $\underline{\phi} : \mathbb{R}^4 \to \mathbb{R}$ be the linear functional such that

$$\underline{\phi}[\mathbf{x}] = 10x^1 - x^2 + 3x^3 + 7x^4 \tag{8.442}$$

for any vector $\mathbf{x} \in \mathbb{R}^4$, relative to the canonical basis. The norm in X is the absolute-value norm because $X = \mathbb{R}$. By Eq. (8.432), the norm $\|\underline{\phi}\|_\infty$ of $\underline{\phi}$ induced by the ∞ norm on $\mathbb{R}^4$ is the smallest nonnegative real number such that

$$|\underline{\phi}[\mathbf{x}]| = |10x^1 - x^2 + 3x^3 + 7x^4| \\ \leq \|\underline{\phi}\|_\infty \max\{|x^1|, |x^2|, |x^3|, |x^4|\}. \tag{8.443}$$

But

$$|10x^1 - x^2 + 3x^3 + 7x^4| \leq 10|x^1| + |x^2| + 3|x^3| + 7|x^4| \\ \leq (10 + 1 + 3 + 7) \max\{|x^1|, |x^2|, |x^3|, |x^4|\}. \tag{8.444}$$

This inequality establishes that 21 is an upper bound of $|\underline{\phi}[\mathbf{x}]|$ if $\|\mathbf{x}\|_\infty = 1$. Before one can assert that $\|\underline{\phi}\|_\infty = 21$, one must show that 21 is the *least* upper bound. One can do so by exhibiting a vector for which the inequalities Eq. (8.444) become equalities. For example, let

$$\mathbf{x} = \begin{pmatrix} 1 \\ -1 \\ 1 \\ 1 \end{pmatrix}. \tag{8.445}$$

Then $\|\mathbf{x}\|_\infty = 1$. One sees at once that for this particular vector, the inequalities in Eq. (8.444) become equalities. Then

$$\|\underline{\phi}\|_\infty = \sum_{j=1}^{4} |\phi_j| = 21. \tag{8.446}$$

This example illustrates a general formula derived below for the ∞ norm of a matrix.

8.7.3 MATRIX NORMS

Let $\mathcal{V}$ be a normed vector space of finite dimension n. Unless the vector norm is a p-norm with $p = 2$, and is computed relative to an orthonormal basis, the linear-mapping norm induced by a given vector norm is basis-dependent. Therefore, to compute the norm of a linear mapping $A : \mathcal{V} \rightarrow \mathcal{V}$ one must compute a norm of the $n \times n$ matrix $\mathbf{A}$ that realizes A relative to a specific basis of $\mathcal{V}$. We define

$$\|\mathbf{A}\| := \|A\| \tag{8.447}$$

in which the linear-mapping norm is the norm induced by the vector norm on $\mathcal{V}$, computed in a fixed basis $\{\bar{e}_i\}$ of $\mathcal{V}$.

Matrix p-Norms

Apart from the special cases $p = 1$ and $p = \infty$, there is no neat, closed formula for the p-norm of a matrix. The best method for computing the p-norm of a matrix $\mathbf{A}$ is to compute the singular value decomposition of $\mathbf{A}$, as described in Section 9.4.2. The computation of the SVD of an $n \times n$ matrix requires $O(n^2)$ floating-point multiplications.

The most easily computed, and therefore the most commonly used, matrix p-norms are the 1 norm and the ∞ norm, $\|\cdot\|_1$ and $\|\cdot\|_\infty$. In the following, $\mathbf{A}$ is an $n \times n$ matrix over $\mathbb{F} = \mathbb{R}$ or $\mathbb{C}$.

We begin with the ∞-norm:

$$
\begin{aligned}
\|\mathbf{A}\mathbf{x}\|_\infty &= \max_{i \in (1:n)} \left| \sum_{j=1}^{n} a_j^i x^j \right| \\
&\leq \max_{i \in (1:n)} \sum_{j=1}^{n} |a_j^i| \, |x^j| \\
&\leq \left[\max_{i \in (1:n)} \sum_{j=1}^{n} |a_j^i| \right] \left[\max_{k \in (1:n)} |x^k| \right].
\end{aligned}
\tag{8.448}
$$

For every matrix $\mathbf{A}$ there exists a vector $\mathbf{x}$ such that the upper bound in the last line of Eq. (8.449) is achieved. We begin the proof of this statement with the observation that the sum $\sum_{j=1}^{n} |a_j^i|$ runs over the columns of row i of the matrix $\mathbf{A}$. Let i_0 be the integer such that

$$\sum_{j=1}^{n} |a_j^{i_0}| = \max_{i \in (1:n)} \sum_{j=1}^{n} |a_j^i|. \tag{8.449}$$

In other words, let the sum of absolute values or moduli of the matrix elements be greatest in row i_0. If $\mathbb{F} = \mathbb{R}$ the inequalities in Eq. (8.449) become equalities if one chooses the vector x such that

$$
\forall j \in (1:n) : x^j = \begin{cases} 1, & \text{if } a_j^{i_0} > 0; \\ -1, & \text{if } a_j^{i_0} < 0; \\ 0, & \text{if } a_j^{i_0} = 0. \end{cases} \tag{8.450}
$$

The ∞-norm of $\bar{x}$ is

$$
\|\bar{x}\|_\infty = \max_{k \in (1:n)} |x^k| = 1. \tag{8.451}
$$

Then

$$
\|\mathbf{A}\|_\infty = \max_{i \in (1:n)} \sum_{j=1}^{n} |a_j^i|. \tag{8.452}
$$

The proof for $\mathbb{F} = \mathbb{C}$ is similar, except that the components of x are complex numbers of modulus 1, the phases of which are chosen to cancel the phases of the matrix elements $\{a_j^{i_0}\}$.

Next we consider the matrix 1-norm:

$$
\begin{aligned}
\|\mathbf{Ax}\|_1 &= \sum_{i=1}^{n} \left| \sum_{j=1}^{n} a_j^i x^j \right| \\
&\le \sum_{i=1}^{n} \sum_{j=1}^{n} |a_j^i| \, |x^j| \\
&\le \left[\max_j \sum_{i=1}^{n} |a_j^i| \right] \left[\sum_{k=1}^{n} |x^k| \right].
\end{aligned} \tag{8.453}
$$

To show that the upper bound in this equation is achieved for some vector x, let j_0 be the integer such that

$$
\sum_{i=1}^{n} |a_{j_0}^i| = \max_j \sum_{i=1}^{n} |a_j^i|. \tag{8.454}
$$

(Here the sums run over the entries in a particular column of **A**.) For both $\mathbb{F} = \mathbb{R}$ and $\mathbb{F} = \mathbb{C}$ the inequalities in Eq. (8.454) become equalities for the vector $x = e_{j_0}$ (that is, if $x^j = \delta_{j_0}^j$). Then

$$
\|\mathbf{A}\|_1 = \max_j \sum_{i=1}^{n} |a_j^i|. \tag{8.455}
$$

The norms $\|\mathbf{A}\|_1$ and $\|\mathbf{A}\|_\infty$ are quite clearly basis-dependent. For practice in evaluating matrix norms, see Exercise 8.7.4.

A numerical evaluation of either Eq. (8.452) or Eq. (8.455) involves $n - 1$ floating-point additions and a linear search involving $O(n)$ operations. The algorithms for both the matrix ∞ norm and the matrix 1 norm therefore are $O(n^2)$, as for the computation of a general p-norm. However, the 1-norm and the ∞-norm require only floating-point additions and comparisons, which are computationally much cheaper than the floating-point multiplications required in the computation of a general p-norm.

Frobenius Norm

Equation (8.155), which gives the matrix elements of a linear mapping relative to an orthonormal basis, permits one to obtain an upper bound for the 2-norm $\|A\|_2$ of a linear mapping, or, equivalently, an upper bound for the 2-norm of a matrix. Because $\|A\|_2$ is defined only for linear mappings whose domain is a Euclidean or unitary inner-product space, one can introduce an orthonormal basis $\{\bar{e}_i\}$ relative to which the 2-norm of a vector is given by Eq. (8.415). From Eqs. (6.42) and (8.415) one has

$$\|\mathbf{A}\mathbf{x}\|_2 = \left[\sum_{i=1}^{n} \left| \sum_{j=1}^{n} a^i_j x^j \right|^2 \right]^{1/2}. \tag{8.456}$$

By the triangle inequality for complex numbers,

$$\|\mathbf{A}\mathbf{x}\|_2 \le \left[\sum_{i=1}^{n} \left(\sum_{j=1}^{n} |a^i_j|^2 |x^j|^2 \right) \right]^{1/2} = \left[\sum_{j=1}^{n} \left(\sum_{i=1}^{n} |a^i_j|^2 |x^j|^2 \right) \right]^{1/2}. \tag{8.457}$$

Then Cauchy's inequality Eq. (8.113) implies that

$$\|\mathbf{A}\mathbf{x}\|_2 \le \|\mathbf{A}\|_F \|\mathbf{x}\|_2, \tag{8.458}$$

in which

$$\|\mathbf{A}\|_F := \left[\sum_{i,j=1}^{n} |a^i_j|^2 \right]^{1/2} \tag{8.459}$$

is called the **Frobenius norm**.

The Frobenius norm may overestimate $\|A\|_2$, for it follows from Eq. (8.458) and from the fact that $\|A\|_2$ is the *least* upper bound of $\|Ax\|_2 m / \|x\|_2$ that

$$\|\mathbf{A}\|_2 \le \|\mathbf{A}\|_F. \tag{8.460}$$

For example, the Frobenius norm of any unitary matrix is $\|U\| = \sqrt{n}$, but the 2-norm is $\|U\|_2 = 1$ according to Eqs. (6.159) and (8.432).

For an $n \times n$ matrix, numerical evaluation of the Frobenius norm requires n^2 floating-point multiplications, $n^2 - 1$ floating-point additions and a square root. Although the algorithms given in Eqs. (8.452) and (8.455) for the ∞-norm and the 1-norm are also $O(n^2)$, the ∞- and

1-norm involve only floating-point additions and therefore are significantly faster to compute than the Frobenius norm.

The matrix norm that is induced by the canonical matrix inner product

$$\langle \mathbf{A}, \mathbf{B} \rangle := \text{trace}[\mathbf{A}^\dagger \mathbf{B}], \tag{8.461}$$

which we originally defined in Eq. (8.21), is the Frobenius norm. The cyclic invariance of the trace implies that the Frobenius norm is invariant under unitary similarity transformations (Eq. [6.157] with $\mathbf{T}^{-1} = \mathbf{T}^\dagger$).

8.7.4 NORM OF AN INTEGRAL

In Section 7.2 we show that the definite integral of a function $f : (a, b) \to \mathbb{R}$ is a linear functional on the infinite-dimensional space $\mathcal{C}^0((a, b); \mathbb{R})$. Let us calculate the norm of the linear functional

$$\underline{\iota}[f] := \int_a^b f(x)\, w(x)\, dx \tag{8.462}$$

in which $w(x) > 0$ for all x in (a, b), using the supremum norm on $\mathcal{C}^0((a, b); \mathbb{R})$ and the absolute-value norm on $\mathbb{R}$. One has

$$|\underline{\iota}[f]| \leq \int_a^b |f(x)|\, w(x)\, dx \leq \sup_{x \in (a,b)} |f(x)| \int_a^b w(x)\, dx. \tag{8.463}$$

This inequality establishes an upper bound of $|\underline{\iota}[f]|$. In the special case that $\forall x \in (a, b)$: $f(x) = c > 0$ in which c is a constant, each of the inequalities becomes an equality. Then there can be no smaller upper bound, and

$$\|\underline{\iota}\|_C = \int_a^b w(x)\, dx. \tag{8.464}$$

Table 8.3 summarizes the linear-mapping norms discussed in this chapter.

8.7.5 EXERCISES FOR SECTION 8.7

8.7.1 Let the normed vector space $\mathcal{C}$ be $\mathcal{C}^0((a, b); \mathbb{R})$ under the supremum norm. Let $\underline{\delta}_\xi$ be the Dirac delta functional defined in Eq. (7.3), in which $a < \xi < b$. Find the norm of $\underline{\delta}_\xi$ that is consistent with the supremum norm on $\mathcal{C}$ (see Eq. [8.432] and the subsequent discussion). Exhibit a function f for which the upper bound you have found is achieved.

8.7.2 Let the normed vector space $\mathcal{C}$ be $\mathcal{C}^0((a, b); \mathbb{R})$ under the supremum norm. Let $\underline{\tau}$ be the linear functional on $\mathcal{C}$ such that

$$\forall f \in \mathcal{C} : \underline{\tau}[f] := f(a) + f(b) - 2f((a + b)/2). \tag{8.465}$$

TABLE 8.3 Frequently Used Norms for Matrices and More General
Linear Mappings

Mapping	Linear-mapping norm	Notation	Definition
$A : \mathcal{V} \to \mathcal{X}$	Consistent with norms on $\mathcal{V}$, $\mathcal{X}$	$\|A\|$	$\sup_{\bar{x} \neq \bar{0}} \dfrac{\|A\bar{x}\|_{\mathcal{X}}}{\|\bar{x}\|_{\mathcal{V}}}$
$A : \mathbb{F}^m \to \mathbb{F}^n$	Inner-product norm	$\|A\|_{ip}$	$\sup_{x \neq 0} \dfrac{\|Ax\|_{ip,\mathbb{F}^n}}{\|x\|_{ip,\mathbb{F}^m}}$
$A : \mathbb{F}^m \to \mathbb{F}^n$	p-Norm	$\|A\|_p$	$\sup_{x \neq 0} \dfrac{\|Ax\|_p}{\|x\|_p}$
$A : \mathbb{F}^m \to \mathbb{F}^n$	1-Norm	$\|A\|_1$	$\max_j \sum_{i=1}^{m} \|a_j^i\|$
$A : \mathbb{F}^m \to \mathbb{F}^n$	∞-Norm	$\|A\|_\infty$	$\max_i \sum_{j=1}^{n} \|a_j^i\|$
$A : \mathbb{F}^m \to \mathbb{F}^n$	Frobenius norm	$\|A\|_F$	$\left[\sum_{i=1}^{m} \sum_{j=1}^{n} \|a_j^i\|^2\right]^{1/2}$
$\phi : \mathcal{C} \to \mathbb{R}$	Supremum norm	$\|\phi\|_C$	$\sup_{f \neq 0} \dfrac{\|\phi[f]\|}{\|f\|_C}$

Note: $\mathbb{F}$ may be either $\mathbb{R}$ or $\mathbb{C}$. The notation $|x|$ means the absolute value if $\mathbb{F} = \mathbb{R}$ or the modulus if $\mathbb{F} = \mathbb{C}$. $\mathcal{C}$ is the vector space $\mathcal{C}^0((a, b); \mathbb{R})$ under the supremum norm.

Find the norm of $\underline{\tau}$ that is consistent with the supremum norm on $\mathcal{C}$ (see Eq. [8.432] and the subsequent discussion). Exhibit a function f for which the upper bound you have found is achieved.

8.7.3   Let the normed vector space $\mathcal{C}$ be $\mathcal{C}^0((0, 1); \mathbb{R})$ under the supremum norm. Let $\underline{\xi}$ be the linear functional on $\mathcal{C}$ such that

$$\forall f \in \mathcal{C} : \underline{\xi}[f] := \int_0^1 f(x)\, e^{-\alpha x}\, dx \tag{8.466}$$

in which $\alpha > 0$. Find the norm of $\underline{\xi}$ that is consistent with the supremum norm on $\mathcal{C}$ (see Eq. [8.432] and the subsequent discussion). Exhibit a function $f \in \mathcal{C}$ for which the upper bound you have found is achieved.

8.7.4   Find the 1-norm, the ∞-norm, and the Frobenius norm of the matrix

$$A = \begin{pmatrix} 17 & 31 & 19 & 7 \\ 23 & 47 & 43 & 57 \\ 11 & 2 & 13 & 53 \\ 5 & 41 & 3 & 29 \end{pmatrix}. \tag{8.467}$$

8.7.5   Let $\mathcal{V}$ be an inner-product space. Let $\underline{x}$ be the linear functional such that

$$\forall \bar{y} \in \mathcal{V} : \underline{x}[\bar{y}] := \langle \bar{x}, \bar{y} \rangle \tag{8.468}$$

in which $\bar{x}$ is a fixed element of $\mathcal{V}$. Find the norm of $\underline{x}$ that is consistent with the inner-product norm on $\mathcal{V}$ (see Eq. [8.432] and the subsequent discussion).

8.7.6   Let

$$\mathbf{x} = \begin{pmatrix} 1 \\ -2 \\ 3 \end{pmatrix} \in \mathbb{R}^3. \tag{8.469}$$

(a) Find the inner-product norm of $\mathbf{x}$, using the canonical inner-product norm on $\mathbb{R}^3$.

(b) Find the 1-norm of $\mathbf{x}$.

(c) Find the ∞-norm of $\mathbf{x}$.

8.7.7   Show that

$$\|\bar{x}\|_\infty \leq \|\bar{x}\|_p, \tag{8.470}$$

and that the inequality becomes an equality for a vector in which all contravariant components have the same absolute value or modulus.

8.7.8   Show that

$$\|\bar{x}\|_p \leq n^{1/p} \|\bar{x}\|_\infty \tag{8.471}$$

and that the inequality becomes an equality for a vector in which all contravariant components have the same absolute value or modulus.

8.8   INNER PRODUCTS AND LINEAR FUNCTIONALS

8.8.1   INTRODUCTION

Every inner product–in fact, every bilinear or sesquilinear functional–defines a linear functional if the first argument is held fixed. For example, let

$$\mathbf{r}' = \begin{pmatrix} 1 \\ -2 \\ 3 \end{pmatrix} \in \mathbb{E}^3. \tag{8.472}$$

The inner product $\mathbf{r}' \cdot \mathbf{r}$ defines a linear functional

$$\underline{\phi} = (1, -2, 3) \tag{8.473}$$

on $\mathbb{E}^3$ such that

$$\begin{aligned} \underline{\phi}[\mathbf{r}] &:= \mathbf{r}' \cdot \mathbf{r} \\ &= x^1 - 2x^2 + 3x^3. \end{aligned} \tag{8.474}$$

We show in this section that in a finite-dimensional Euclidean or unitary space, it is also true that every linear functional can be expressed as the inner product with a certain vector.

In Section 8.8.2 we generalize the idea behind Eq. (8.474) by showing that in a Euclidean or unitary space, the linear functional defined by taking the inner product with a fixed vector $\bar{x}'$ is the image of $\bar{x}'$ under a certain linear mapping $G : \mathcal{V} \to \mathcal{V}^*$, which we call the *inner-product mapping*. The matrix of G turns out to be the Gram matrix **G**.

One can understand why, if a Euclidean or unitary space $\mathcal{V}$ is finite-dimensional, the inner product with a fixed vector defines a linear functional, and why a linear functional is just the inner product with a certain vector, as follows: The dual space $\mathcal{V}^*$ has the same dimension as $\mathcal{V}$, and therefore is isomorphic to $\mathcal{V}$. If the Gram matrix **G** (the matrix of inner products of the basis vectors) is nonsingular (as it must be for a useful inner product), then it defines a nonsingular linear mapping G from $\mathcal{V}$ to $\mathcal{V}^*$ of which **G** is the matrix. The image under G of $\mathcal{V}$ (which is a subspace of $\mathcal{V}^*$) has the same dimension as $\mathcal{V}^*$, and therefore is equal to $\mathcal{V}^*$. Then $G : \mathcal{V} \to \mathcal{V}^*$ is bijective. It follows that every linear functional on $\mathcal{V}$ can be represented as the inner product with some vector in $\mathcal{V}$. The infinite-dimensional version of this theorem is known as the *Riesz representation theorem*.

8.8.2 INNER-PRODUCT MAPPING

Definition and Examples

Generalizing Eq. (8.474), one sees that if $\bar{x}$ is a fixed vector belonging to an inner-product space $\mathcal{V}$, then the inner product $\langle \bar{x}, \bar{y} \rangle$ is the value of a linear functional for the argument $\bar{y} \in \mathcal{V}$. This observation leads one to define a mapping $\underline{x} : \mathcal{V} \to \mathbb{F}$ such that

$$\forall \bar{y} \in \mathcal{V} : \underline{x}[\bar{y}] := \langle \bar{x}, \bar{y} \rangle. \tag{8.475}$$

One sees from the inner-product axioms (8.2) and (8.3) that $\underline{x}$ satisfies the axioms for a linear functional, Eqs. (7.9) through (7.14):

$$\underline{x}[\bar{y} + \bar{y}'] = \langle \bar{x}, \bar{y} + \bar{y}' \rangle = \langle \bar{x}, \bar{y} \rangle + \langle \bar{x}, \bar{y}' \rangle = \underline{x}[\bar{y}] + \underline{x}[\bar{y}'], \tag{8.476}$$

$$(\underline{x} + \underline{x}')[\bar{y}] = \langle \bar{x} + \bar{x}', \bar{y} \rangle = \langle \bar{x}, \bar{y} \rangle + \langle \bar{x}', \bar{y} \rangle = \underline{x}[\bar{y}] + \underline{x}'[\bar{y}], \tag{8.477}$$

$$\underline{x}[\alpha \bar{y}] = \langle \bar{x}, \alpha \bar{y} \rangle = \alpha \langle \bar{x}, \bar{y} \rangle = \alpha \, \underline{x}[\bar{y}] = (\alpha \, \underline{x})[\bar{y}]. \tag{8.478}$$

Then $\underline{x}$ is a linear functional, as advertised. The coordinate expansion of $\underline{x}$ (relative to the basis of coordinate functionals) is

$$\underline{x}[\bar{y}] = \begin{cases} x_j y^j = x_j \epsilon^j [\bar{y}] \Rightarrow \underline{x} = x_j \epsilon^j, & \text{if } \mathbb{F} = \mathbb{R}; \\ x_j^* y^j = x_j^* \epsilon^j [\bar{y}] \Rightarrow \underline{x} = x_j^* \epsilon^j, & \text{if } \mathbb{F} = \mathbb{C}. \end{cases} \tag{8.479}$$

It follows that for every vector $\bar{x} \in \mathcal{V}$, the linear functional $\underline{x} \in \mathcal{V}^*$ is unique. (For a check of the consistency of Eq. [8.475] with Eqs. [7.26] and [7.88], see Exercise 8.8.1.)

For example, evaluating a linear functional $\phi \in \mathbb{R}^{n*}$ on $\mathbf{x} \in \mathbb{R}^n$ by Eq. (7.8) produces the same result as taking the inner product of $\underline{\phi}^T$ and $\overline{x}$ using the canonical inner product, Eq. (8.13).

We call the mapping $G : V \to V^*$ such that

$$G\overline{x} := \underline{x} \in V^* \tag{8.480}$$

the **inner-product mapping**. The definition implies that for all $\overline{y}$,

$$\boxed{(G\overline{x})[\overline{y}] = \langle \overline{x}, \overline{y} \rangle.} \tag{8.481}$$

The additivity of G, $G(\overline{x} + \overline{x}') = G\overline{x} + G\overline{x}'$, then follows from the additivity of the inner product, Eq. (8.6). To investigate the homogeneity of G, we apply the definition:

$$(G(\alpha\overline{x}))[\overline{y}] = \langle \alpha\overline{x}, \overline{y} \rangle = \begin{cases} \alpha \langle \overline{x}, \overline{y} \rangle = \alpha(G\overline{x})[\overline{y}] & \text{if } \mathbb{F} = \mathbb{R}; \\ \alpha^* \langle \overline{x}, \overline{y} \rangle = \alpha^*(G\overline{x})[\overline{y}] & \text{if } \mathbb{F} = \mathbb{C}. \end{cases} \tag{8.482}$$

Therefore the inner-product mapping G is linear if $\mathbb{F} = \mathbb{R}$ but is antilinear if $\mathbb{F} = \mathbb{C}$. One should bear in mind, however, that the functional $\underline{x} = G\overline{x}$ is always *linear*:

$$\underline{x}[\overline{y} + \overline{z}] = \langle \overline{x}, \overline{y} \rangle + \langle \overline{x}, \overline{z} \rangle \tag{8.483}$$

and

$$\underline{x}[\alpha\overline{y}] = \langle \overline{x}, \alpha\overline{y} \rangle = \alpha \langle \overline{x}, \overline{y} \rangle. \tag{8.484}$$

For example, under the canonical inner product $\langle z', z \rangle := \text{Re}[z'^*z]$ in $\mathbb{C}$, the inner-product mapping is

$$Gz' = z'^*, \tag{8.485}$$

which is simply complex conjugation. In this case, G is its own inverse.

Equation (7.8) implies that the inner-product mapping induced on $\mathbb{E}^n$ by the canonical inner product Eq. (8.13) is linear,

$$G\mathbf{x} = \mathbf{x}^T; \tag{8.486}$$

the inner-product mapping induced on $\mathbb{U}^n$ by the canonical inner product Eq. (8.15) is antilinear,

$$G\mathbf{x} = \mathbf{x}^\dagger. \tag{8.487}$$

However, one should note carefully that the right-hand sides of Eqs. (8.486) and (8.487) are *linear* functionals. For example, $G\mathbf{x}[\mathbf{y}] = \mathbf{x}^\dagger\mathbf{y}$ is linear in $\mathbf{y}$.

The matrix realization of the inner-product mapping is a little more complicated if the inner product is not the canonical inner product on $\mathbb{R}^n$ or $\mathbb{C}^n$. For example, the inner-product mapping on $\mathbb{C}^n$ induced by the inner product $\langle \mathbf{x}, \mathbf{y} \rangle = \mathbf{x}^\dagger \mathbf{G} \mathbf{y}$ (Eq. [8.37]) is

$$\underline{\mathbf{x}} = \mathbf{G} \mathbf{x} = (\mathbf{G} \mathbf{x})^\dagger = \mathbf{x}^\dagger \mathbf{G}^\dagger = \mathbf{x}^\dagger \mathbf{G}. \tag{8.488}$$

To check that this result is correct, note that $\underline{\mathbf{x}}[\mathbf{y}] = \mathbf{x}^\dagger \mathbf{G} \mathbf{y}$, in agreement with Eq. (8.37).

Mapping from "Kets" to "Bras" in Quantum Mechanics

Another important example of the inner-product mapping is the mapping of a "ket vector" $|\psi\rangle$ onto a "bra vector" $\langle\psi|$ in Dirac's formulation of nonrelativistic quantum mechanics. In Dirac's notation, the coordinate functionals are written $\langle e_i|$, and the inner-product mapping carries the "ket vector"

$$|\psi\rangle = \sum_i x^i |e_i\rangle \tag{8.489}$$

onto the "bra vector"

$$\langle\psi| = \sum_i x^{i*} \langle e_i|. \tag{8.490}$$

Because the state space of a quantum-mechanical system is unitary, the inner-product mapping is antilinear and therefore takes the complex conjugate of every probability amplitude.

Image of a Basis Under G

The images of a basis $\{\bar{e}_i\}$ of $\mathcal{V}$ under the inner-product mapping G are the linear functionals

$$\underline{e}_i := G\bar{e}_i. \tag{8.491}$$

The linear functionals $\{\underline{e}_i\}$ are useful in both computations and theoretical derivations because $\underline{e}_i$ picks out the i-th covariant component of any vector on which it acts:

$$\underline{e}_i[\bar{x}] = \langle \bar{e}_i, \bar{x} \rangle = x_i. \tag{8.492}$$

To find the matrix of G, we make a coordinate expansion of $\underline{e}_i$ relative to the basis of coordinate functionals in $\mathcal{V}^*$, using Eq. (8.26):

$$\underline{e}_i[\bar{x}] = \underline{e}_i[x^j \bar{e}_j] = x^j \langle \bar{e}_i, \bar{e}_j \rangle = g_{ij} \underline{\epsilon}^j[\bar{x}]. \tag{8.493}$$

(Because the basis $\{\underline{\epsilon}^j\}$ is not necessarily orthonormal, one is not entitled to make use of orthogonality is calculating matrix elements.) Therefore the matrix elements of G relative to the bases $\{\bar{e}_i\}$ of $\mathcal{V}$ and $\{\underline{\epsilon}^j\}$ of $\mathcal{V}^*$ are the elements of the metric tensor:

$$G\bar{e}_i = \underline{e}_i = g_{ij}\underline{\epsilon}^j. \tag{8.494}$$

It follows that the matrix of G is the Gram matrix $\mathbf{G}$. The elements g_{ij} of $\mathbf{G}$ carry two subscripts instead of one subscript and one superscript because $\mathbf{G}$ is the matrix of a linear mapping from $\mathcal{V}$ to a vector space of linear functionals on $\mathcal{V}$.

We have assumed that the finite-dimensional inner-product space $\mathcal{V}$ is Euclidean or unitary. Then the list $\{\underline{e}_i\}$ is linearly independent, for the Gram matrix $\mathbf{G}$ is nonsingular in a Euclidean or unitary space, implying that the inner-product mapping G is nonsingular. Then $\{\underline{e}_i\}$ is a basis of $\mathcal{V}^*$, because $\{\underline{e}_i\}$ has $n = \dim[\mathcal{V}^*]$ linearly independent elements. It follows that the inner-product mapping G is bijective.

It is also true that $\{\underline{e}_i\}$ is linearly independent, and the inner-product mapping is bijective, in Minkowski space and other finite-dimensional inner-product spaces in which the condition Eq. (8.59) holds. The reader is invited to consult Appendix C for further details.

8.8.3 INVERSE INNER-PRODUCT MAPPING

Definition
If the inner-product mapping

$$G : \overline{x} \mapsto \underline{x} \qquad (8.495)$$

is bijective, as we have just shown is the case in every Euclidean or unitary space, then there exists a bijective **inverse inner-product mapping**

$$G^{-1}\underline{x} := \overline{x}. \qquad (8.496)$$

If G^{-1} acts on a linear functional $\underline{\phi}$, the image is a vector $\overline{x}$ such that for every vector $\overline{y} \in \mathcal{V}$,

$$\underline{\phi}[\overline{y}] = \langle \overline{x}, \overline{y} \rangle. \qquad (8.497)$$

With the help of the inverse inner-product mapping, one can represent every linear functional on $\mathcal{V}$ as the inner product with a certain vector.

The inverse of a linear (antilinear) mapping is also linear (antilinear); hence G^{-1} is linear if $\mathcal{V} = \mathbb{R}^n$ and is antilinear if $\mathcal{V} = \mathbb{C}^n$.

For example, Eq. (8.494) implies that in $\mathbb{R}^n$, the inverse inner-product mapping induced by the inner product Eq. (8.13) is just the operation of taking the transpose of a row vector:

$$G^{-1}\mathbf{x}^T = \mathbf{x}. \qquad (8.498)$$

We use this special case of the inverse inner-product mapping in our discussion of the reciprocal lattice in $\mathbb{E}^3$ (see Eqs. [7.44] and [7.46]).

Equation (8.487) implies that if $\mathcal{V} = \mathbb{U}^n$, then G^{-1} is the operation of taking the adjoint of a row vector:

$$G^{-1}\mathbf{x}^\dagger = \mathbf{x}. \qquad (8.499)$$

If $\mathcal{V}$ is $\mathbb{C}^n$ and the inner product is $\langle \mathbf{x}, \mathbf{y} \rangle = \mathbf{x}^\dagger \mathbf{G} \mathbf{y}$ (see Eq. [8.37]), then the inverse inner-product mapping is

$$\mathbf{x} = G^{-1}\underline{\mathbf{x}} = \underline{\mathbf{x}}^{\dagger}G^{-1} \tag{8.500}$$

(see Exercise 8.8.5).

Reciprocal Basis and Biorthogonality

We showed previously that the linear functionals $\{\underline{e}_i\}$, which are the images of the basis $\{\overline{e}_i\}$ under the inner-product mapping, produce the contravariant components of any vector to which they are applied. The *vectors* $\{\overline{e}'^j\}$ that result if the *inverse* inner-product mapping is applied to the basis of coordinate functionals $\{\underline{\epsilon}^j\}$ play an equally important, though somewhat underappreciated, role as the elements of the *reciprocal basis*, such that the inner product of a reciprocal basis vector and a basis vector is either 1 or 0. In an oblique coordinate system, a basis and its reciprocal basis are not the same; see, for example, the discussion of reciprocal lattices in Sections 7.3.4 and 7.3.5.

Our goal is to find vectors $\overline{e}'^j$ such that the inner product $\langle \overline{e}'^j, \overline{x} \rangle$ has the same value as $\underline{\epsilon}^j[\overline{x}]$, the value of the linear functional $\underline{\epsilon}^j$ on the vector $\overline{x}$. Because the inverse inner-product mapping G^{-1} is nonsingular, the list $\{G^{-1}\underline{\epsilon}^j\}$, which contains the images of n linearly independent elements of $\mathcal{V}^*$, is a basis of $\mathcal{V}$. The basis

$$\overline{e}'^j := G^{-1}\underline{\epsilon}^j \tag{8.501}$$

is called the **reciprocal basis** of $\mathcal{V}$ induced by the basis $\{\overline{e}_i\}$. Calculating the inner product of an element of the basis with an element of the reciprocal basis, one obtains

$$\langle G^{-1}\underline{\epsilon}^j, \overline{e}_i \rangle = GG^{-1}\underline{\epsilon}^j[\overline{e}_i] = \delta_i^j. \tag{8.502}$$

Thus taking the inner product with the vector $G^{-1}\epsilon^j$ gives the same result as evaluating the linear functional $\underline{\epsilon}^j$.

For example, if $\{\mathbf{f}_1, \mathbf{f}_2, \mathbf{f}_3\}$ is a basis of $\mathbb{E}^3$, then the reciprocal basis is

$$\left\{ \mathbf{e}'^1 = \frac{\mathbf{f}_2 \times \mathbf{f}_3}{\mathbf{f}_1 \cdot (\mathbf{f}_2 \times \mathbf{f}_3)}, \quad \mathbf{e}'^2 = \frac{\mathbf{f}_3 \times \mathbf{f}_1}{\mathbf{f}_1 \cdot (\mathbf{f}_2 \times \mathbf{f}_3)}, \quad \mathbf{e}'^3 = \frac{\mathbf{f}_1 \times \mathbf{f}_2}{\mathbf{f}_1 \cdot (\mathbf{f}_2 \times \mathbf{f}_3)} \right\}. \tag{8.503}$$

(recall Eq. [7.59]). One can see at once that $\mathbf{e}'^j \cdot \mathbf{e}_i = \delta_i^j$.

Generalizing Eq. (7.58), one calls two bases $\{\overline{e}_i\}$ and $\{\overline{f}^j\}$ of an inner-product space $\mathcal{V}$ **biorthogonal** if and only if

$$\langle \overline{f}^j, \overline{e}_i \rangle = \delta_i^j. \tag{8.504}$$

The bases $\{\overline{e}'^j = G^{-1}\underline{\epsilon}^j\}$ and $\{\overline{e}_i\}$ are biorthogonal.

It follows from Eq. (8.504) that the condition for two bases $\{\mathbf{f}'^j\}$ and $\{\mathbf{f}_i\}$ to be biorthogonal in $\mathbb{C}^n$ under the inner product defined in Eq. (8.37), is

$$F'^{\dagger}GF = 1 \tag{8.505}$$

in which

$$\mathbf{F}' := [\mathbf{f}'^1, \ldots, \mathbf{f}'^n], \quad \mathbf{F} := [\mathbf{f}_1, \ldots, \mathbf{f}_n]. \tag{8.506}$$

If $\{\mathbf{f}_j\}$ is an arbitrary basis of $\mathbb{C}^n$, then the coordinate functionals are the rows of the matrix $\Phi = \mathbf{F}^{-1}$ according to Eqs. (7.37) and (7.38). Then Eq. (8.500) implies that the reciprocal basis vectors are the columns of the matrix

$$\mathbf{F}' = \mathbf{G}^{-1}\Phi^\dagger = \mathbf{G}^{-1}\mathbf{F}^{-1\dagger} \tag{8.507}$$

that realizes Eq. (8.501). In order to verify that Eq. (8.507) satisfies Eq. (8.505) one must make use of the fact that $\mathbf{G}^{-1}$ is symmetric (if $\mathcal{V}$ is Euclidean) or Hermitian (if $\mathcal{V}$ is unitary):

$$\mathbf{G}^\dagger = \mathbf{G} \Rightarrow (\mathbf{G}^{-1})^\dagger = \mathbf{G}^{-1}. \tag{8.508}$$

Mapping from Covariant to Contravariant Components

The matrix of the inverse inner-product mapping $\mathbf{G}^{-1}$ turns out to map the column vector of contravariant components of a vector

$$\mathbf{x} = \begin{pmatrix} x^1 \\ x^2 \\ \vdots \\ x^n \end{pmatrix} \tag{8.509}$$

to the column vector of covariant components of the same vector,

$$\mathbf{x}' = \begin{pmatrix} x_1 \\ x_2 \\ \vdots \\ x_n \end{pmatrix} \tag{8.510}$$

If expressed in terms of components, the inverse inner-product mapping reads

$$\bar{x} = \mathbf{G}^{-1}\underline{x} = \mathbf{G}^{-1}\mathbf{G}\bar{x} = \mathbf{G}^{-1}(g_{ij}\,x^i\underline{\epsilon}^j) = g_{ij}\,x^i\mathbf{G}^{-1}\underline{\epsilon}^j = g_{ji}\,x^i\mathbf{G}^{-1}\underline{\epsilon}^j \tag{8.511}$$

if $\mathcal{V}$ is Euclidean, or

$$\bar{x} = \mathbf{G}^{-1}\underline{x} = \mathbf{G}^{-1}\mathbf{G}\bar{x} = \mathbf{G}^{-1}(g_{ij}\,x^{i*}\,\underline{\epsilon}^j) = g^*_{ij}\,x^i\mathbf{G}^{-1}\underline{\epsilon}^j = g_{ji}\,x^i\mathbf{G}^{-1}\underline{\epsilon}^j \tag{8.512}$$

if $\mathcal{V}$ is unitary. By definition, $\mathbf{G}^{-1}\underline{\epsilon}^j$ is an element of $\mathcal{V}$ and can therefore be expressed as a linear combination of the basis vectors of $\mathcal{V}$. Then there exist scalars $g^{jl} \in \mathbb{F}$ such that

$$\mathbf{G}^{-1}\underline{\epsilon}^j = g^{lj}\bar{e}_l\,. \tag{8.513}$$

It follows from Eq. (8.508) that

$$g^{lj} = \begin{cases} g^{jl}, & \text{if } \mathcal{V} \text{ is Euclidean;} \\ g^{jl*}, & \text{if } \mathcal{V} \text{ is unitary.} \end{cases} \tag{8.514}$$

From Eqs. (8.511) or Eq. (8.512) and (8.513) and (8.514) one finds the componentwise expression of the fact that $\mathbf{G}^{-1}$ is the inverse of $\mathbf{G}$:

$$\mathbf{G}^{-1}\mathbf{G}\bar{e}_k = \bar{e}_k = \mathbf{G}^{-1}(g_{kj}\underline{\epsilon}^j) = g_{kj}\mathbf{G}^{-1}\underline{\epsilon}^j = g_{kj}g^{jl}\bar{e}_l = \delta_k^l\bar{e}_l$$

$$\Rightarrow g_{kj}g^{jl} = \delta_k^l = g^{lj}g_{jk} \tag{8.515}$$

if $\mathcal{V}$ is Euclidean, or

$$\mathbf{G}^{-1}\mathbf{G}\bar{e}_k = \bar{e}_k = \mathbf{G}^{-1}(g_{kj}\underline{\epsilon}^j) = g_{kj}^*\mathbf{G}^{-1}\underline{\epsilon}^j = g_{kj}^* g^{jl*}\bar{e}_l = \delta_k^l\bar{e}_l \tag{8.516}$$

$$\Rightarrow g_{kj}^* g^{jl*} = \delta_k^l = g^{lj}g_{jk} \tag{8.517}$$

if $\mathcal{V}$ is unitary.

In terms of the basis of $\mathcal{V}$ rather than the basis of the dual space $\mathcal{V}^*$, the statement that $\bar{x} = \mathbf{G}^{-1}\underline{x}$ reads

$$x^l\bar{e}_l = \bar{x} = \mathbf{G}^{-1}\underline{x} = \mathbf{G}^{-1}(x_j\underline{\epsilon}^j) = x_j\mathbf{G}^{-1}\underline{\epsilon}^j = x_j g^{lj}\bar{e}_l \tag{8.518}$$

(if $\mathcal{V}$ is Euclidean), or

$$x^l\bar{e}_l = \bar{x} = \mathbf{G}^{-1}\underline{x} = \mathbf{G}^{-1}\left(x_j^*\underline{\epsilon}^j\right) = x_j\mathbf{G}^{-1}\underline{\epsilon}^j = x_j g^{lj}\bar{e}_l \tag{8.519}$$

(if $\mathcal{V}$ is unitary). Therefore

$$x^i = g^{ij}x_j. \tag{8.520}$$

Just as g_{ij} transforms contravariant into covariant components, g^{ij} transforms covariant into contravariant components.

Inner Product in the Dual Space

The mapping $\mathbf{G}^{-1}$ induces an **inner product in the dual space** $\mathcal{V}^*$:

$$\langle \underline{x}, \underline{y} \rangle := \bar{x}[\underline{y}] = \underline{y}[\bar{x}] = \langle \bar{y}, \bar{x} \rangle. \tag{8.521}$$

(Exercise 8.8.3 is to verify that Eq. [8.521] is consistent with the inner-product axioms.) The component form of the dual-space inner product is

$$\langle \underline{x}, \underline{y} \rangle = \underline{y}[\bar{x}] = \begin{cases} x^i y_i = g^{ij}x_i y_j, & \text{if } \mathbb{F} = \mathbb{R} : \\ x^i y_i^* = g^{ij}x_i y_j^*, & \text{if } \mathbb{F} = \mathbb{C}. \end{cases} \tag{8.522}$$

The antilinearity of the mapping from $\bar{x}$ to $\underline{x}$ if $\mathcal{V}$ is unitary creates some potential confusion, because complex conjugates appear in possibly unexpected places in Eq. (8.522). Therefore we give the component form of the dual-space inner product directly in terms of the coordinate expansions Eq. (7.30) of two linear functionals $\underline{\phi}$, $\underline{\psi}$ on a unitary space $\mathcal{V}$:

$$\langle \underline{\phi}, \underline{\psi} \rangle = g^{ij} \phi_i^* \psi_j. \tag{8.523}$$

It follows directly from Eq. (8.522) that the dual-space inner product is positive-definite. It follows immediately that if $\mathcal{V}$ is a Euclidean space, then so is $\mathcal{V}^*$, and that if $\mathcal{V}$ is a unitary space, then so is $\mathcal{V}^*$.

From Eq. (8.523), the inner product of two coordinate functionals is

$$\langle \underline{\epsilon}^i, \underline{\epsilon}^j \rangle = g^{kl} \delta_k^i \delta_l^j = g^{ij}, \tag{8.524}$$

in full correspondence with Eq. (8.26). Therefore g^{ij} is the metric tensor of the dual space $(\mathrm{rel}\{\underline{\epsilon}^i\})$.

Relation of the Orthogonal Complement to the Annihilator

It should already be clear that the statement that a linear functional ϕ annihilates a subspace $\mathcal{W}$ of a Euclidean or unitary space $\mathcal{V}$ expresses a kind of orthogonality of ϕ to $\mathcal{W}$. The definitions of the inner-product mapping and its inverse make this statement precise.

Because $\mathcal{W}$ is a subspace of $\mathcal{V}$, the images $G\mathcal{W}$ and $G\mathcal{W}^\perp$ under the inner-product mapping G are subspaces of $\mathcal{V}^*$. In fact,

$$\forall \overline{w} \in \mathcal{W}, \overline{z} \in \mathcal{W}^\perp : \langle \overline{z}, \overline{w} \rangle = \underline{z}[\overline{w}] = 0 \Rightarrow \underline{z} \in \mathrm{ann}[\mathcal{W}], \tag{8.525}$$

implying that $G\mathcal{W}^\perp$ is a subspace of the annihilator of $\mathcal{W}$. Similarly, $G\mathcal{W}$ is a subspace of $\mathrm{ann}[\mathcal{W}^\perp]$. One can see almost trivially that $G\mathcal{W}^\perp$ is identical with $\mathrm{ann}[\mathcal{W}]$. Because G is onto, every element $\underline{z}$ of $\mathrm{ann}[\mathcal{W}]$ is the image $G\overline{z}$ of an element of $\mathcal{V}$, implying that

$$\forall \underline{z} \in \mathrm{ann}[\mathcal{W}] : \forall \overline{w} \in \mathcal{W} : \underline{z}[\overline{w}] = \langle \overline{z}, \overline{w} \rangle = 0 \Rightarrow \overline{z} \in \mathcal{W}^\perp. \tag{8.526}$$

Therefore every element of $\mathrm{ann}[\mathcal{W}]$ is the image of some element of $\mathcal{W}^\perp$ under G. One can write

$$G\mathcal{W}^\perp = \mathrm{ann}[\mathcal{W}]. \tag{8.527}$$

Similarly,

$$G\mathcal{W} = \mathrm{ann}[\mathcal{W}^\perp]. \tag{8.528}$$

One can use the dual-space inner product to summarize the relationship between a subspace $\mathcal{W}$ and its annihilator $\mathrm{ann}[\mathcal{W}]$. From Eq. (8.521), two vectors $\overline{w} \in \mathcal{W}$ and $\overline{z} \in \mathcal{W}^\perp$ are orthogonal (in $\mathcal{V}$) if and only if $\underline{w}$ and $\underline{z}$ are orthogonal in $\mathcal{V}^*$. Then

$$
\begin{aligned}
\underline{z} \in (G\mathcal{W})^{\perp} &\Leftrightarrow \forall \underline{w} \in G\mathcal{W} : (\underline{z}, \underline{w}) = 0 \\
&\Leftrightarrow \forall \overline{w} \in \mathcal{W} : (\overline{z}, \overline{w}) = 0 \\
&\Leftrightarrow \overline{z} \in \mathcal{W}^{\perp} \\
&\Leftrightarrow \underline{z} \in G\mathcal{W}^{\perp},
\end{aligned}
\tag{8.529}
$$

which implies that

$$
G\mathcal{W}^{\perp} = (G\mathcal{W})^{\perp}.
\tag{8.530}
$$

Then a linear functional belongs to ann[$\mathcal{W}$] if and only if it belongs to the orthogonal complement (in $\mathcal{V}^*$) of $G\mathcal{W}$, and the converse is true also.

8.8.4 EXERCISES FOR SECTION 8.8

8.8.1   Show that Eq. (8.475) is consistent with Eqs. (7.26) and (7.88).

8.8.2   Prove that Eq. (8.59) implies that the linear functionals $\{\underline{e}_i\}$ defined in Eq. (8.492) are linearly independent.

8.8.3   Verify that the dual-space inner product (8.521) is consistent with the inner-product axioms, Eqs. (8.2) through (8.9).

8.8.4   Show that the inner product induced on $\mathbb{C}^{n*}$ by the inner product Eq. (8.37) is

$$
\langle \underline{x}, \underline{y} \rangle = \underline{y} G^{-1} \underline{x}^{\dagger}.
\tag{8.531}
$$

Then obtain the inner product on $\mathbb{R}^{n*}$ by removing the complex conjugates from Eq. (8.531).

8.8.5   Verify that the inner-product mapping on $\mathbb{C}^n$ induced by the inner product Eq. (8.37) is

$$
\underline{x} = Gx = (Gx)^{\dagger},
\tag{8.532}
$$

and that the inverse inner-product mapping is

$$
x = G^{-1}\underline{x} = \underline{x}^{\dagger} G^{-1}.
\tag{8.533}
$$

8.9 BIBLIOGRAPHY AND ENDNOTES

8.9.1 BIBLIOGRAPHY

Birkhoff, Garrett, and Saunders Mac Lane. *A Survey of Modern Algebra*. 3 ed. New York: Macmillan, 1965, Chapter VII.

Golub, Gene H., and Charles F. Van Loan. *Matrix Computations*. 2 ed. Baltimore: Johns Hopkins University Press, 1989.

Halmos, Paul R. *Finite-Dimensional Vector Spaces*. 2 ed. Princeton: Van Nostrand, 1958, Chapter III.

Kolmogorov, Andrei Nikolaevich, and Sergei Vasilevich Fomin. *Introductory Real Analysis*. New York: Dover, 1975, Chapters 4, 5.

Loomis, Lynn H., and Shlomo Sternberg. *Advanced Calculus*. Reading: Addison-Wesley, 1968, Chapters 1 and 2.

8.9.2 ENDNOTES

Section 8.1

Mathematicians usually write $g[\alpha\bar{x}, \bar{y}] = \alpha g[\bar{x}, \bar{y}]$ and $g[\bar{x}, \alpha\bar{y}] = \alpha^* g[\bar{x}, \bar{y}]$ instead of Eq. (8.3). A Hermitian inner product in mathematicians' notation is equal to the complex conjugate of a formally identical Hermitian inner product in physicists' notation.

A reason for not using Dirac's notation $\langle \psi \mid \phi \rangle$ for the inner product is that in Dirac's notation the expression for the inner product of a vector with the image of a second vector under a linear mapping, $\langle \psi \mid A \mid \phi \rangle$, is potentially ambiguous. Because $\mid \phi \rangle$ denotes a vector in Dirac's system, it is unclear which vector A acts upon in the expression $\langle \psi \mid A \mid \phi \rangle$. To show precisely on which vector A acts, one should either modify Dirac's notation to read $\langle \psi \mid A\phi \rangle$ or write $\langle \psi, A\phi \rangle$.

Rounding errors in the computation of inner products are discussed in Golub and Van Loan (1989, pp. 63–65).

Other widespread conventions for the metric tensor $g_{\mu\nu}$ of special relativity include using $+1$ for the spatial components ($\mu = 1, 2, 3$) and -1 for the ct component, which is often indexed with $\mu = 4$ instead of $\mu = 0$. Starting with Minkowski, some authors have tried to avoid the subject of the metric tensor altogether by using a four-dimensional Euclidean metric ($\mathbf{G} = \mathbf{1}$) along with a purely imaginary time coordinate $x^4 = x_4 = ict$. The Minkowski approach has the drawback that it does not provide a useful foundation for the study of general relativity, in which the concept of the metric tensor is essential because space-time is curved, not flat. However, Minkowski's approach is very useful in deriving relations between the irreducible representations of the Lorentz group and the real orthogonal group in four dimensions.

Section 8.2

There seem to be almost as many ways to spell "Chebyshev" as there are written human languages. Several different spellings are used commonly in English. The Cyrillic spelling is very simple: Чебышев. The transliteration used here conforms to the conventions used in Russian mathematical and scientific journals.

CHAPTER 9

LINEAR MAPPINGS II

9.1 DYADS

9.1.1 MOTIVATION

The product of an $n \times 1$ matrix (a column vector) and a $1 \times m$ matrix (a row vector) is an $n \times m$ matrix; recall Eq. (6.76). The product of a column vector and a row vector is called a *dyad* or an *outer product*, to distinguish it from the inner product formed by multiplying a row vector into a column vector. For example, in Eq. (8.281) we expressed the orthogonal projector onto the subspace spanned by a unit vector $\mathbf{n}$ as the outer product $\mathbf{n}\mathbf{n}^T$.

It turns out that any matrix can be expressed as a sum of dyads, and that this gives one both theoretical insights and useful computational tools. The starting point for identifying the abstract concept that underlies an outer product is the observation that an $n \times 1$ matrix maps from $\mathbb{F}^n$ to $\mathbb{F}$, and a $1 \times m$ matrix maps from $\mathbb{F}$ to $\mathbb{F}^m$.

9.1.2 DEFINITION OF A DYAD

Let $\mathcal{V}_{(1)}$ and $\mathcal{V}_{(2)}$ be finite-dimensional vector spaces, and let $\overline{y}'$ be any vector in $\mathcal{V}_{(2)}$. With $\overline{y}'$ we associate the mapping $\overline{y}' : \mathbb{F} \to \mathcal{V}_{(2)}$ such that

$$\forall \alpha \in \mathbb{F} : \overline{y}'(\alpha) := \alpha \overline{y}' \in \mathcal{V}_{(2)}. \tag{9.1}$$

This mapping is linear, for $\overline{y}'(\alpha + \beta) = \overline{y}'(\alpha) + \overline{y}'(\beta)$, and so forth. The **dyad** $\overline{y}' \circ \underline{\phi} : \mathcal{V}_{(1)} \to \mathcal{V}_{(2)}$ is the linear mapping that results if a linear functional $\underline{\phi} : \mathcal{V}_{(1)} \to \mathbb{F}$ is evaluated and subsequently $\overline{y}'$ is applied to the image under $\underline{\phi}$. By definition, then, a **dyad** is a composite mapping (or a composite function) such that

$$\forall \overline{x} \in \mathcal{V}_{(1)} : (\overline{y}' \circ \underline{\phi})[\overline{x}] := \overline{y}'(\underline{\phi}[\overline{x}])$$
$$= (\underline{\phi}[\overline{x}])\overline{y}' \in \mathcal{V}_{(2)} \tag{9.2}$$

in which $\underline{\phi}[\overline{x}] \in \mathbb{F}$. (A dyad such as $\overline{y}' \circ \underline{\phi}$ should not be confused with the value of a linear functional on $\mathcal{V}_{(1)}^*$.)

If $\mathcal{V}_{(1)} = \mathbb{F}^{n_1}$ and $\mathcal{V}_{(2)} = \mathbb{F}^{n_2}$, the symbol $\circ$ is unnecessary, because the operation involved in constructing such a dyad is ordinary matrix multiplication. For example, if

$$\underline{\phi} = \left(\phi_1, \ldots, \phi_{n_1}\right) \in \mathbb{R}^{n_1 *} \tag{9.3}$$

and

$$\mathbf{y}' = \begin{pmatrix} y'^1 \\ \vdots \\ y'^{n_2} \end{pmatrix} \in \mathbb{R}^{n_2}, \tag{9.4}$$

then the dyad $\mathbf{y}'\underline{\phi}$ acts on $\mathbf{x} \in \mathbb{R}^{n_1}$ as follows:

$$(\mathbf{y}'\underline{\phi})[\mathbf{x}] = \mathbf{y}'(\underline{\phi}[\mathbf{x}]) = (\underline{\phi}[\mathbf{x}])\mathbf{y}' = (\phi_i x^i)y'^j \mathbf{e}'_j. \tag{9.5}$$

It follows that $\mathbf{y}'\underline{\phi}$ is the $n_2 \times n_1$ matrix product of the $n_2 \times 1$ matrix $\mathbf{y}'$ and the $1 \times n_1$ matrix $\underline{\phi}$:

$$\mathbf{y}'\underline{\phi} = \begin{pmatrix} y'^1 \phi_1 & \cdots & y'^1 \phi_{n_1} \\ \vdots & \ddots & \vdots \\ y'^{n_2} \phi_1 & \cdots & y'^{n_2} \phi_{n_1} \end{pmatrix}$$

$$= \begin{pmatrix} y'^1 \underline{\phi} \\ \vdots \\ y'^{n_2} \underline{\phi} \end{pmatrix} \tag{9.6}$$

$$= (\phi_1 \mathbf{y}', \dots, \phi_{n_1} \mathbf{y}').$$

Note the structure: Each column of $\mathbf{y}'\underline{\phi}$ is a scalar multiple of the column vector $\mathbf{y}'$, and each row is a scalar multiple of the row vector $\underline{\phi}$. To give a numerical example, let

$$\underline{\phi} = (4, 5, 6) \in \mathbb{E}^{3*}, \quad \mathbf{y}' = \begin{pmatrix} 1 \\ -2 \\ 3 \end{pmatrix} \in \mathbb{E}^3. \tag{9.7}$$

Then

$$\mathbf{y}'\underline{\phi} = \begin{pmatrix} 1 \\ -2 \\ 3 \end{pmatrix} (4, 5, 6) = \begin{pmatrix} 4 & 5 & 6 \\ -8 & -10 & -12 \\ 12 & 15 & 18 \end{pmatrix}. \tag{9.8}$$

In Dirac's notation a dyad is always a "ket-bra" such as $|\psi\rangle\langle\phi|$, which is of the same form as $\mathbf{y}'\underline{\phi}$ in Eq. (9.2).

The rank of a dyad is unity,

$$\mathrm{rank}[\mathbf{y}'\underline{\phi}] = 1, \tag{9.9}$$

because the range has unit dimension:

$$\mathrm{range}[\mathbf{y}'\underline{\phi}] = \mathrm{span}[\mathbf{y}']. \tag{9.10}$$

Conversely, if an $n_2 \times n_1$ matrix has unit rank, there exists some nonzero vector $\mathbf{y}'$ such that

$$\mathrm{range}[\mathbf{A}] = \mathrm{span}[\mathbf{y}']. \tag{9.11}$$

Then the columns $\mathbf{a}_1, \ldots, \mathbf{a}_{n_1}$ of $\mathbf{A}$, which span range[$\mathbf{A}$], are scalar multiples of $\mathbf{y}'$. Let

$$\forall j \in (1 : n_1) : \quad \mathbf{a}_j = \phi_j \mathbf{y}', \tag{9.12}$$

and let $\underline{\phi}$ be the row vector with components $\phi_1, \ldots, \phi_{n_1}$. Then

$$\mathbf{A} = \mathbf{y}'\underline{\phi}. \tag{9.13}$$

The expression of a rank-1 matrix as a dyad is not unique, for

$$\mathbf{A} = (\alpha \mathbf{y}')(\alpha^{-1}\underline{\phi}) \tag{9.14}$$

for every nonzero scalar α.

Because one may safely assume that any inner-product mapping $G_{(1)}$ defined on $\mathcal{V}_{(1)} = \mathbb{F}^{n_1}$ is nonsingular, it follows that the linear functional $\underline{\phi}$ that enters into a dyad $\mathbf{y}'\underline{\phi}$ can be expressed as the operation of taking the inner product with a certain vector $\mathbf{x}$:

$$\exists \mathbf{x} \in \mathbb{F}^{n_1} :\ni: \forall \mathbf{y} \in \mathbb{F}^{n_2} : \underline{\phi}[\mathbf{y}] = \langle \mathbf{x}, \mathbf{y} \rangle = \underline{\mathbf{x}}[\mathbf{y}] \Rightarrow \underline{\phi} = \underline{\mathbf{x}} = G_{(1)}\mathbf{x} = \mathbf{x}^\dagger G \tag{9.15}$$

(see Eqs. [8.475] and [8.488]). Then

$$\mathbf{y}'\underline{\phi} = \mathbf{y}'\underline{\mathbf{x}}. \tag{9.16}$$

If $\mathcal{V}_{(1)} = \mathbb{E}^{n_1}$, then

$$\mathbf{y}'\underline{\mathbf{x}} = \begin{pmatrix} y'^1 x^1 & \cdots & y'^1 x^{n_1} \\ \vdots & \ddots & \vdots \\ y'^{n_2} x^1 & \cdots & y'^{n_2} x^{n_1} \end{pmatrix}. \tag{9.17}$$

The projector matrix $\mathbf{P_u}$ defined in Eq. (8.236) is of the same dyadic form as Eq. (9.17). Therefore

$$\mathbf{P_u} = \mathbf{u}\underline{\mathbf{u}} = [u^i u^j] = \begin{pmatrix} (u^1)^2 & \cdots & u^1 u^n \\ \vdots & \ddots & \vdots \\ u^n u^1 & \cdots & (u^n)^2 \end{pmatrix} \tag{9.18}$$

in $\mathbb{E}^n$. For example, if

$$\mathbf{u} = \frac{1}{\sqrt{3}} \begin{pmatrix} 1 \\ 1 \\ 1 \end{pmatrix}, \tag{9.19}$$

then

$$\mathbf{P_u} = \frac{1}{3} \begin{pmatrix} 1 & 1 & 1 \\ 1 & 1 & 1 \\ 1 & 1 & 1 \end{pmatrix}. \tag{9.20}$$

Note that in the dyadic form of the projector $\mathbf{P_u}$, the operations that define the orthogonal projector along the vector $\mathbf{u}$, namely, orthogonal projection on $\mathbf{u}$ followed by multiplication of $\mathbf{u}$ by the resulting scalar, neatly correspond to the column vector and row vector that make up the dyad.

A dyadic expression such as Eq. (9.18) offers no numerical economies if one must compute every matrix element of $\mathbf{P_u}$. We show in the next section that one can achieve significant savings of computational effort if a matrix is representable as a dyad or as a sum of a few dyads. Also, it is much more convenient in a symbolic calculation to manipulate a dyad (or a few dyads) than to work directly with the elements of a matrix.

9.1.3 DYADIC EXPANSIONS

Because $\mathrm{hom}\,[\mathcal{V}_{(1)}, \mathcal{V}_{(2)}]$ is a vector space and dyads are linear mappings, any linear combination of dyads is also a linear mapping. For some purposes it is most convenient to write a matrix or a linear operator as a dyad or as a linear combination of dyads.

For example, one can see at once that

$$\mathbf{e}'_i \boldsymbol{\epsilon}^j = \mathbf{E}^j_i \tag{9.21}$$

in which $\mathbf{E}^j_i$ is the $n_2 \times n_1$ matrix in which the only nonzero element is a 1 located at the intersection of row i and column j (see Eq. [5.125]). Let $\mathbf{A} \in \mathbb{F}^{n_2 \times n_1}$ be any $n_2 \times n_1$ matrix over the field $\mathbb{F} = \mathbb{R}$ or $\mathbb{C}$. Then, because $\{\mathbf{e}'_i \boldsymbol{\epsilon}^j\}$ is a basis of $\mathbb{F}^{n_2 \times n_1}$ according to Eq. (5.167),

$$\mathbf{A} = a^i_j \mathbf{E}^j_i = a^i_j \mathbf{e}'_i \boldsymbol{\epsilon}^j. \tag{9.22}$$

Evidently basis-coordinate-functional dyads such as $\mathbf{e}'_i \boldsymbol{\epsilon}^j$ are the most fundamental building blocks from which matrices can be assembled. For example,

$$\mathbf{D} = \begin{pmatrix} 0 & 1 & 1 \\ 0 & 0 & 1 \\ 0 & 0 & 0 \end{pmatrix} = \mathbf{e}_1 \boldsymbol{\epsilon}^2 + \mathbf{e}_1 \boldsymbol{\epsilon}^3 + \mathbf{e}_2 \boldsymbol{\epsilon}^3$$

$$= \begin{pmatrix} 1 \\ 0 \\ 0 \end{pmatrix} (0, 1, 0) + \begin{pmatrix} 1 \\ 0 \\ 0 \end{pmatrix} (0, 0, 1) + \begin{pmatrix} 0 \\ 1 \\ 0 \end{pmatrix} (0, 0, 1). \tag{9.23}$$

Equation (9.22) implies that every matrix $\mathbf{A} \in \mathbb{F}^{n_2 \times n_1}$ can be expressed as a linear combination of dyads in not merely one, but at least three different ways, depending on whether one performs no partial sum (as in Eq. [9.22]), a partial sum on j, or a partial sum on i.

If one performs a partial sum on j in Eq. (9.22), then one obtains the expansion

$$\mathbf{A} = \mathbf{a}_i \boldsymbol{\epsilon}^i, \tag{9.24}$$

in which the vectors that enter into the dyads are the columns of **A**. For example,

$$\mathbf{D} = \begin{pmatrix} 0 & 1 & 1 \\ 0 & 0 & 1 \\ 0 & 0 & 0 \end{pmatrix} = \mathbf{d}_2\underline{\epsilon}^2 + \mathbf{d}_3\underline{\epsilon}^3$$

$$= \begin{pmatrix} 1 \\ 0 \\ 0 \end{pmatrix} (0, 1, 0) + \begin{pmatrix} 1 \\ 1 \\ 0 \end{pmatrix} (0, 0, 1) \tag{9.25}$$

in which

$$\mathbf{d}_2 = \begin{pmatrix} 1 \\ 0 \\ 0 \end{pmatrix}, \quad \mathbf{d}_3 = \begin{pmatrix} 1 \\ 1 \\ 0 \end{pmatrix}. \tag{9.26}$$

(To verify this equation, apply **D** to an arbitrary vector in $\mathbb{R}^3$.)

If, instead, one performs a partial sum on i in Eq. (9.22), one obtains the expansion

$$\mathbf{A} = \mathbf{e}'_j\underline{a}^j \tag{9.27}$$

in which

$$\underline{a}^j := a^j_i\,\underline{\epsilon}^i \tag{9.28}$$

is the j-th row of the matrix **A**. For example,

$$\mathbf{D} = \begin{pmatrix} 0 & 1 & 1 \\ 0 & 0 & 1 \\ 0 & 0 & 0 \end{pmatrix} = \mathbf{e}_1\underline{d}^1 + \mathbf{e}_2\underline{d}^2$$

$$= \begin{pmatrix} 1 \\ 0 \\ 0 \end{pmatrix} (0, 1, 1) + \begin{pmatrix} 0 \\ 1 \\ 0 \end{pmatrix} (0, 0, 1) \tag{9.29}$$

in which

$$\underline{d}^1 = (0, 1, 1), \quad \underline{d}^2 = (0, 0, 1). \tag{9.30}$$

Equations (9.22), (9.24), and (9.27) enable one to disassemble a matrix into convenient pieces such as individual elements, columns, or rows.

The dyadic expansion of a general finite-dimensional linear mapping is

$$A = a^j_i\,\bar{e}'_j \circ \underline{\epsilon}^i = \bar{a}_i \circ \underline{\epsilon}^i = \bar{e}'_j \circ \underline{a}^j. \tag{9.31}$$

One calculates the product of two dyads $\bar{y}' \circ \underline{\phi} : V_{(1)} \to V_{(2)}$ and $\bar{y}'' \circ \underline{\psi}' : V_{(2)} \to V_{(3)}$ through successive applications of the definition given in Eq. (9.2):

$$\forall \bar{x} \in V_{(1)} : (\bar{y}'' \circ \underline{\psi}')(\bar{y}' \circ \underline{\phi}[\bar{x}]) = (\underline{\phi}[\bar{x}])(\underline{\psi}'[\bar{y}'])\,\bar{y}''. \tag{9.32}$$

Note that, in effect, one merely has to rearrange a few parentheses.

If two matrices $\mathbf{A}$ and $\mathbf{B}$ are expressible as dyads,

$$\mathbf{A} = \mathbf{y}'\underline{\phi}, \quad \mathbf{B} = \mathbf{y}''\underline{\psi}', \tag{9.33}$$

then the computation of a resultant vector

$$\mathbf{x}'' = \mathbf{BAx} = (\underline{\phi}[\mathbf{x}])(\underline{\psi}'[\mathbf{y}'])\,\mathbf{y}'' \tag{9.34}$$

using the dyadic expressions in Eq. (9.33) requires far fewer steps than would be required to form the matrix product $\mathbf{BA}$ and then the resultant $\mathbf{BAx}$ by the usual procedure of matrix multiplication. If the dimensions of $\mathbf{A}$ and $\mathbf{B}$ are $n_2 \times n_1$ and $n_3 \times n_2$, respectively, then an element-by-element computation of the matrix product $\mathbf{BA}$ requires $n_1 n_3 (2n_2 - 1)$ operations. (Note that in order to compute each one of the $n_1 n_3$ matrix elements of $\mathbf{BA}$, one must compute the product of a column of $\mathbf{A}$ by a row of $\mathbf{B}$. Each row-column product requires n_2 multiplications and $n_2 - 1$ additions, hence $2n_2 - 1$ operations.)

On the other hand, to compute $\mathbf{BAx}$ using Eq. (9.34) requires only $2(n_1 + n_2) + n_3 - 1$ operations: $2n_1 - 1$ operations are required for $\underline{\phi}[\mathbf{x}]$ and $2n_2 - 1$ operations for $\underline{\psi}'[\mathbf{y}']$, one operation is required to form the product $\alpha = (\underline{\phi}[\mathbf{x}])(\underline{\psi}'[\mathbf{y}'])$, and n_3 operations are required to multiply each element of $\mathbf{y}''$ by α. If the matrices $\mathbf{A}$ and $\mathbf{B}$ are square, then $n_1 = n_2 = n_3 = n$ and the operation counts are of order $2n^3$ for element-by-element computation of $\mathbf{BAx}$ and of order $5n$ for computation of $\mathbf{BAx}$ using the dyadic forms of $\mathbf{A}$ and $\mathbf{B}$. The ratio of the operation count for an element-by-element computation to the operation count for a dyadic product is $2n^2/5$, which works out to 10^3 if $n = 50$. Clearly one should use dyadic products whenever matrices can be expressed as dyads or as linear combinations of m dyads, in which $m \ll n_1 n_2$. The singular-value decomposition derived in Section 9.4.1 provides a dyadic expansion of any matrix in which the number of dyads is equal to the smaller of n_1 and n_2.

9.1.4 RESOLUTIONS OF THE IDENTITY MAPPING

The coordinate expansion of any vector $\bar{x}$ in a vector space $\mathcal{V}$, relative to a basis $\{\bar{e}_i\}$, is

$$\forall \bar{x} \in \mathcal{V}: \ \bar{x} = x^i \bar{e}_i = x^j \bar{e}_i \delta^i_j = \bar{e}_i x^j \underline{\epsilon}^i[\bar{e}_j] = \bar{e}_i \underline{\epsilon}^i[x^j \bar{e}_j] = \bar{e}_i \circ \underline{\epsilon}^i[\bar{x}], \tag{9.35}$$

in which $\bar{e}_i \circ \underline{\epsilon}^i$ stands for the dyadic sum

$$\bar{e}_i \circ \underline{\epsilon}^i := \bar{e}_1 \circ \underline{\epsilon}^1 + \cdots + \bar{e}_n \circ \underline{\epsilon}^n. \tag{9.36}$$

It follows from Eqs. (9.35) and (9.36) that $\bar{e}_i \circ \underline{\epsilon}^i$ is equal to the identity mapping:

$$\bar{e}_i \circ \underline{\epsilon}^i = \mathbf{1}. \tag{9.37}$$

Because we use $\mathbf{1}$ to denote the identity mapping on any vector space, it follows by duality that

$$\underline{\epsilon}^i \circ \bar{e}_i = \mathbf{1} \tag{9.38}$$

in which $\bar{e}_i$ is considered as a linear functional on $\mathcal{V}^*$. Equations (9.37) and (9.38) are called **resolutions of the identity**. For example,

$$\begin{pmatrix} 1 \\ 0 \\ 0 \end{pmatrix}(1,0,0) + \begin{pmatrix} 0 \\ 1 \\ 0 \end{pmatrix}(0,1,0) + \begin{pmatrix} 0 \\ 0 \\ 1 \end{pmatrix}(0,0,1) = \begin{pmatrix} 1 & 0 & 0 \\ 0 & 1 & 0 \\ 0 & 0 & 1 \end{pmatrix} \tag{9.39}$$

is a resolution of the identity in $\mathbb{R}^3$.

Another widely used class of resolutions of the identity makes use of biorthogonal bases $\{\bar{u}_i\}$ and $\{\bar{v}^j\}$ such that $\langle \bar{v}^j, \bar{u}_i \rangle = \delta_i^j$ (see Eq. [8.504]). One has

$$\forall \bar{x} \in \mathcal{V}: \bar{x} = x^i \bar{u}_i = \langle \bar{v}^j, x^i \bar{u}_i \rangle \bar{u}_j = \bar{u}_j \circ \underline{v}^j [\bar{x}]. \tag{9.40}$$

Then

$$\bar{e}_i \circ \underline{f}^i = \mathbf{1}. \tag{9.41}$$

Note that it is *not* necessary for $\{\bar{u}_i\}$ to be orthonormal. For example, if

$$\mathbf{u}_1 = \begin{pmatrix} 1 \\ 0 \end{pmatrix}, \quad \mathbf{u}_2 = \begin{pmatrix} \alpha \\ \beta \end{pmatrix} \in \mathbb{E}^2 \tag{9.42}$$

then the basis of $\mathbb{E}^2$ that is biorthogonal to $\{\mathbf{u}_i\}$ is

$$\mathbf{v}^1 = \frac{1}{\beta} \begin{pmatrix} \beta \\ -\alpha \end{pmatrix}, \quad \mathbf{v}^2 = \frac{1}{\beta} \begin{pmatrix} 0 \\ 1 \end{pmatrix}. \tag{9.43}$$

One verifies by matrix multiplication that Eq. (9.41) holds in this example:

$$\mathbf{u}_1 \underline{v}^1 + \mathbf{u}_2 \underline{v}^2 = \frac{1}{\beta} \begin{pmatrix} 1 \\ 0 \end{pmatrix} (\beta, -\alpha) + \frac{1}{\beta} \begin{pmatrix} \alpha \\ \beta \end{pmatrix} (0,1) = \mathbf{1}. \tag{9.44}$$

In the special case in which the basis $\{\bar{u}_i\}$ is orthonormal, one has $\bar{v}^i = \bar{u}_i$. Then Eqs. (8.245) and (9.18) imply that

$$\sum_{i=1}^n \mathsf{P}_i = \mathbf{1} \Rightarrow \bar{u}_i \circ \underline{u}_i = \mathbf{1} \tag{9.45}$$

in which P_i projects on the subspace spanned by $\bar{u}_i$.

Resolutions of the identity such as are given in Eqs. (9.41) or Eq. (9.45) can be very useful in shortening formal calculations. For example, the usual equation for the matrix elements of a product AB becomes

$$\langle \bar{u}_k, \mathsf{AB}\bar{u}_l \rangle = \langle \bar{u}_k, \mathsf{A}\bar{u}_i \circ \underline{u}^i \mathsf{B}\bar{u}_l \rangle = \sum_{i=1}^n \langle \bar{u}_k, \mathsf{A}\bar{u}_i \rangle \langle \bar{u}^i, \mathsf{B}\bar{u}_l \rangle \tag{9.46}$$

if one uses a resolution of the identity in terms of an orthonormal basis $\{\bar{u}_i\}$. The technique of inserting a resolution of the identity at convenient places is used frequently in quantum mechanics, in which the orthonormal basis consists of normalized state vectors.

9.1.5 EXERCISE FOR SECTION 9.1

9.1.1   Prove that the inner-product norm of a dyad, relative to an orthonormal basis, is given by the formula

$$\|\overline{y}' \circ \underline{x}\|_{\text{ip}} = \|\overline{y}'\|_{\text{ip}} \|\underline{x}\|_{\text{ip}} . \tag{9.47}$$

9.2 TRANSPOSE AND ADJOINT

In Chapters 5 and 6, we allude to the fact that writing vectors as rows instead of columns, and implementing linear mappings using the row matrix–rectangular matrix products $\phi'\mathbf{A}$, is just as valid as writing vectors as columns and implementing linear mappings with rectangular matrix-column vector products such as $\mathbf{Ax}$. An $n \times m$ rectangular matrix $\mathbf{A}$ maps an m-element column vector $\mathbf{x} \in \mathbb{F}^m$ to an n-element column vector $\mathbf{x}' = \mathbf{Ax} \in \mathbb{F}^n$. The same matrix, $\mathbf{A}$, *also* implements a linear mapping from an n-element row vector $\phi' \in \mathbb{F}^{n*}$ to an m-element row vector $\phi = \phi'\mathbf{A} \in \mathbb{F}^{m*}$. Thus one can regard a matrix $\mathbf{A}$ either as the matrix of a linear mapping from $\mathbb{F}^m$ to $\mathbb{F}^n$ or as the matrix of a linear mapping from $\mathbb{F}^{n*}$ to $\mathbb{F}^{m*}$.

9.2.1 TRANSPOSE

Any linear mapping $A : \mathcal{V}_{(1)} \to \mathcal{V}_{(2)}$ induces another linear mapping A^T, called the *transpose* of A, between the dual spaces $\mathcal{V}_{(1)}^*$ and $\mathcal{V}_{(2)}^*$. To understand the possibly counterintuitive feature that A^T maps $\mathcal{V}_{(2)}^*$ into $\mathcal{V}_{(1)}^*$, consider the image $A\overline{x} \in \mathcal{V}_{(2)}$ of a vector $\overline{x} \in \mathcal{V}_{(1)}$ under a linear mapping A. Every linear functional $\underline{\phi}' \in \mathcal{V}_{(2)}^*$ is defined on $A\overline{x}$. Moreover, $\underline{\phi}'[A(\cdot)]$ is additive and homogeneous in its argument, for

$$\underline{\phi}'[A(\alpha\overline{x} + \beta\overline{y})] = \underline{\phi}'[\alpha A\overline{x} + \beta A\overline{y}] = \alpha\underline{\phi}'[A\overline{x}] + \beta\underline{\phi}'[A\overline{y}]. \tag{9.48}$$

This establishes that for every $\underline{\phi}' \in \mathcal{V}_{(2)}^*$, $\underline{\phi}'[A(\cdot)]$ is a linear functional on domain $[A] = \mathcal{V}_{(1)}$ and therefore belongs to $\mathcal{V}_{(1)}^*$. The **transpose** (or **dual** or **conjugate**) of a linear mapping $A : \mathcal{V}_{(1)} \to \mathcal{V}_{(2)}$ is the mapping $A^T : \mathcal{V}_{(2)}^* \to \mathcal{V}_{(1)}^*$ such that

$$\forall \underline{\phi}' \in \mathcal{V}_{(2)}^* : \forall \overline{x} \in \mathcal{V}_{(1)} : (A^T\underline{\phi}')[\overline{x}] := \underline{\phi}'[A\overline{x}]. \tag{9.49}$$

(The functional $A^T\underline{\phi}'$ is uniquely defined because Eq. [9.49] specifies the value of $A^T\underline{\phi}'$ on every element of $\mathcal{V}_{(1)}$.) Defined in this way, A^T is both additive and homogeneous by virtue of the definitions of the addition and scalar multiplication of linear functionals in Eqs. (7.13) and (7.14) (see Exercise 9.2.1). Therefore $A^T : \mathcal{V}_{(2)}^* \to \mathcal{V}_{(1)}^*$ is a linear mapping.

The commutative diagram

$$
\begin{array}{ccc}
\mathcal{V}_{(1)} & \xrightarrow{\ A\ } & \mathcal{V}_{(2)} \\
\downarrow{\scriptstyle G_{(1)}} & & \downarrow{\scriptstyle G_{(2)}} \\
\mathcal{V}_{(1)}^* & \xleftarrow{\ A^T\ } & \mathcal{V}_{(2)}^*
\end{array}
\tag{9.50}
$$

illustrates the relationship between a linear mapping A and its transpose A^T.

It follows from Eq. (9.49) that the transpose of the sum of two linear mappings is

$$(A + B)^T = A^T + B^T \tag{9.51}$$

(see Exercise 9.2.2). By induction, Eq. (9.51) implies that the transpose of any finite sum of linear mappings is equal to the sum of the transposed linear mappings. It is straightforward to show that the transpose of the product of two linear mappings A and B is

$$(BA)^T = A^T B^T \tag{9.52}$$

(see Exercise 9.2.3).

By the definition Eq. (9.49), the image under A^T of a coordinate functional $\underline{\epsilon}'^i \in V^*_{(2)}$ is the linear functional in $V^*_{(1)}$ such that

$$\forall \bar{x} \in V_{(1)} : (A^T \underline{\epsilon}'^i)[\bar{x}] = \underline{\epsilon}'^i[A\bar{x}] = \underline{\epsilon}'^i[Ax^j \bar{e}_j] = x^j a^i_j = a^i_j \underline{\epsilon}^j[\bar{x}]. \tag{9.53}$$

Therefore, by Eq. (9.28),

$$A^T \underline{\epsilon}'^i = \underline{a}^i \tag{9.54}$$

in which

$$\underline{a}^i := a^i_j \underline{\epsilon}^j. \tag{9.55}$$

(Compare Eq. [9.54] with Eq. [6.15].) It follows from Eqs. (7.30) and (9.54) that the analog of Eq. (6.17) is

$$\forall \underline{\phi}' \in V^*_{(2)} : A^T \underline{\phi}' = \phi'_i \underline{a}^i. \tag{9.56}$$

One sees from Eqs. (7.88) and (9.56) that $A^T \underline{\phi}' = a^i_j \underline{\epsilon}^j \bar{e}'_i[\underline{\phi}']$. Therefore

$$A^T = a^i_j \underline{\epsilon}^j \bar{e}'_i. \tag{9.57}$$

One sees also that

$$A^T = \underline{a}^i \bar{e}'_i \tag{9.58}$$

and

$$A^T = \underline{\epsilon}^j \bar{a}_j \tag{9.59}$$

respectively.

One finds the matrix realization of the transpose of a linear mapping by calculating the matrix elements of A^T. The matrix-element formula

$$a^j_i = \underline{\epsilon}^j[A\bar{e}_i] \tag{9.60}$$

expresses the (row j, column i) matrix element of a linear mapping as the value of the j-th coordinate functional (in the dual of the new space) on the image of the i-th basis vector (of the old space). The (row j, column i) matrix element of $A^T : V_{(2)}^* \to V_{(1)}^*$ is therefore

$$\bar{e}_j[A^T \underline{\epsilon}^{\prime i}] = (A\bar{e}_j)[\underline{\epsilon}^{\prime i}] = \underline{\epsilon}^{\prime i}[A\bar{e}_j] = a_j^i, \tag{9.61}$$

which is the (row i, column j) matrix element of A. (Eq. [7.88] implies the second equality.)

To summarize: The matrix $\mathbf{A}^T$ of the transposed linear mapping A^T (relative to the bases of coordinate functionals $\{\underline{\epsilon}^{\prime i}\}$ and $\{\underline{\epsilon}^j\}$) is obtained from $\mathbf{A}$ by interchanging rows and columns. Because the columns of $\mathbf{A}^T$ are the rows of $\mathbf{A}$, range $[\mathbf{A}^T]$ is the row space of $\mathbf{A}$. If the dimensions of $\mathbf{A}$ are $n_2 \times n_1$, then the dimensions of $\mathbf{A}^T$ are $n_1 \times n_2$. In particular, the transpose of a column vector is a row vector, and vice versa.

If one takes $a_i^j = \delta_i^j$ in Eqs. (9.31) and (9.59), one finds that

$$(\bar{e}_j^\prime \circ \underline{\epsilon}^i)^T = \underline{\epsilon}^i \circ \bar{e}_j^\prime, \tag{9.62}$$

in which $\bar{e}_j^\prime$ is a functional on $V_{(2)}^*$ according to Eq. (7.88). The coordinate expansion of a dyad $\bar{v}^\prime \underline{\phi}$ relative to the dyad basis $\{\bar{e}_j^\prime \circ \underline{\epsilon}^i\}$ is

$$\bar{v}^\prime \underline{\phi} = v^{\prime j} \phi_i \, \bar{e}_j^\prime \circ \underline{\epsilon}^i. \tag{9.63}$$

From Eqs. (9.51), (9.62) and (9.63) one sees that the transpose of the dyad $\bar{v}^\prime \circ \underline{\phi}$ is the dyad

$$(\bar{v}^\prime \circ \underline{\phi})^T = v^{\prime j} \phi_i \, \underline{\epsilon}^i \circ \bar{e}_j^\prime = \underline{\phi} \circ \bar{v}^\prime. \tag{9.64}$$

In this equation, $\bar{v}^\prime$ is to be regarded as a linear functional on $V_{(2)}^*$.

The matrix realization of a linear functional $\underline{\phi}^\prime \in V_{(2)}^*$ is the row vector $\underline{\phi}^\prime := (\phi_1^\prime, \ldots, \phi_{n_2}^\prime)$. From Eq. (9.56), the matrix realization of $A^T \underline{\phi}^\prime$ is the linear combination $\phi_i^\prime \mathbf{a}^i$ of the rows of the matrix $\mathbf{A}$, which is equal to $\underline{\phi}^\prime \mathbf{A}$ by the rule for matrix-matrix multiplication. Then the matrix form of the equation

$$\underline{\phi} = A^T \underline{\phi}^\prime \tag{9.65}$$

is

$$\underline{\phi} = \underline{\phi}^\prime \mathbf{A}. \tag{9.66}$$

The reason why $\mathbf{A}$ appears instead of $\mathbf{A}^T$ and why the usual order of matrix and vector is reversed is that we choose to realize linear functionals by row vectors. If one takes the transpose of Eq. (9.66) and use Eq. (9.52), then one obtains

$$\underline{\phi}^T = \mathbf{A}^T \underline{\phi}^{\prime T}, \tag{9.67}$$

in which the transposed matrix $\mathbf{A}^T$ maps the column vector $\underline{\phi}^{\prime T}$ to the column vector $\underline{\phi}^T$ and in which the usual matrix-vector order is restored. Authors such as Birkhoff and Mac Lane, who realize abstract vectors as row vectors instead of column vectors, normally write abstract linear mappings to the right of the vectors on which they act in order to have the same order as in the matrix realization, Eq. (9.66).

From Eq. (9.59), the range of A^T is spanned by the row functionals $\{\underline{a}^j\}$. Therefore the range of A^T is realized by the row space of A,

$$\text{range}[A^T] = \text{span}\{\underline{a}^j\} = \text{ann}[\text{null}[A]], \tag{9.68}$$

which implies that the row rank of A is equal to the (column) rank of A^T. It follows that the (column) rank of A is equal to the (column) rank of A^T:

$$\text{rank}[A] = \text{rank}[A^T]. \tag{9.69}$$

In a similar way one finds that the annihilator of the null space of A^T is the range of A (which is realized by the column space of **A**):

$$\text{ann}[\text{null}[A^T]] = \text{span}\{\overline{a}_i\} = \text{range}[A]. \tag{9.70}$$

(To establish this equation, one goes through the derivation of the counterpart of Eq. [9.67] in the dual space instead of in $\mathcal{V}_{(1)}$.) The diagram

$$
\begin{array}{ccc}
\mathcal{V}_{(1)} & \xrightarrow{\;A\;} & \mathcal{V}_{(2)} \\
\Big\downarrow{M_{(1)}} & & \Big\downarrow{M_{(2)}} \\
\mathbb{F}^{n_1} & \xrightarrow{\;A\;} & \mathbb{F}^{n_2} \\
\Big\downarrow{T_{(1)}} & & \Big\downarrow{T_{(2)}} \\
\mathbb{F}^{n_1*} & \xleftarrow{\;A^T\;} & \mathbb{F}^{n_2*} \\
\Big\downarrow{M^T_{(1)}} & & \Big\downarrow{M^T_{(2)}} \\
\mathcal{V}^*_{(1)} & \xleftarrow{\;A^T\;} & \mathcal{V}^*_{(2)}
\end{array}
\tag{9.71}
$$

summarizes the relationship among A, A^T and their matrices.

If a linear mapping A is invertible, then so is A^T. One sees easily that

$$(A^T)^{-1} = (A^{-1})^T, \tag{9.72}$$

that is, the inverse of the transpose is equal to the transpose of the inverse.

9.2.2 ADJOINT

In this section we show that evaluating the inner product of a vector in an image space, $\overline{x}'\mathcal{V}_{(2)}$, with an image vector $A\overline{y}$ under a linear mapping A defines a new vector $A^H\overline{x}'$ in the *original* vector space, $\mathcal{V}_{(1)}$. We show that the mapping A^H is linear and is realized by the adjoint matrix if the underlying vector spaces are unitary.

Definitions Write the inner product in $\mathcal{V}_{(i)}$ as $\langle \bar{x}, \bar{y} \rangle_{(i)}$, and assume that both $\mathcal{V}_{(1)}$ and $\mathcal{V}_{(2)}$ are finite-dimensional, complex vector spaces equipped with positive-definite inner products unless we specify different choices. We indicate vectors that belong to $\mathcal{V}_{(2)}$ with a prime. By the axioms for an inner product and a linear mapping, $\langle \bar{x}', A\bar{y} \rangle_{(2)}$ is an additive, homogeneous, scalar-valued function of $\bar{y}$ and therefore defines an element of $\mathcal{V}_{(1)}^*$ that we write temporarily as

$$\underline{z} := \langle \bar{x}', A(\cdot) \rangle_{(2)}. \tag{9.73}$$

Because every linear functional on a finite-dimensional vector space can be represented as the inner product with a fixed vector $\bar{z}$ (or, equivalently, because the inner-product mapping $G_{(1)} : \mathcal{V}_{(1)} \to \mathcal{V}_{(1)}^*$ is one-to-one and onto), it follows that for every $\bar{x}' \in \mathcal{V}_{(2)}$ there exists a vector $\bar{z} \in \mathcal{V}_{(1)}$ such that

$$\underline{z} = G_{(1)}\bar{z}. \tag{9.74}$$

The *adjoint* of A is the mapping $A^H : \mathcal{V}_{(2)} \to \mathcal{V}_{(1)}$ such that

$$A^H \bar{x}' = \bar{z}. \tag{9.75}$$

We need a definition of the adjoint that is valid in both finite-dimensional and infinite-dimensional spaces. Therefore we define a vector $\bar{z}$ (which is a function of $\bar{x}$) such that

$$\forall \bar{x}' \in \mathcal{V}_{(2)} : \forall \bar{y} \in \text{domain}[A] : \langle \bar{z}, \bar{y} \rangle_{(1)} := \langle \bar{x}', A\bar{y} \rangle_{(2)}. \tag{9.76}$$

To check whether $\bar{z}$ is defined uniquely, suppose that for some $\bar{x}' \in \mathcal{V}_{(2)}$ there exists another vector $\bar{z}'$ such that

$$\forall \bar{y} \in \text{domain}[A] : \langle \bar{z}', \bar{y} \rangle_{(1)} = \langle \bar{x}', A\bar{y} \rangle_{(2)}. \tag{9.77}$$

Then

$$\forall \bar{y} \in \text{domain}[A] : \langle \bar{z} - \bar{z}', \bar{y} \rangle_{(1)} = 0, \tag{9.78}$$

which is true if and only if

$$\bar{z} - \bar{z}' \in \{\text{domain}[A]\}^{\perp}. \tag{9.79}$$

Then Eq. (9.76) defines $\bar{z}$ uniquely, and it is possible to define a single-valued mapping A^H if and only if the orthogonal complement of domain [A] consists of the zero vector alone:

$$\{\text{domain}[A]\}^{\perp} = \mathcal{O}_{(1)}. \tag{9.80}$$

If $\mathcal{V}_{(1)}$ is finite-dimensional, then one can ensure that Eq. (9.80) is satisfied by making sure that A is defined on a set of vectors that spans $\mathcal{V}_{(1)}$. If $\mathcal{V}_{(1)}$ is infinite-dimensional, then Eq. (9.80) holds if and only if domain[A] is dense in $\mathcal{V}_{(1)}$ (see Chapter 10 for the concept of denseness). Assume for now that Eq. (9.80) holds. The **adjoint** of A is the mapping A^H such that

$$A^H \bar{x}' := \bar{z}. \tag{9.81}$$

This definition implies that for every $\bar{y} \in$ domain[A] and for every $\bar{x}' \in$ domain[A^H],

$$\langle A^H \bar{x}', \bar{y} \rangle_{(1)} = \langle \bar{x}', A\bar{y} \rangle_{(2)}. \tag{9.82}$$

Equation (9.82) applies to both finite-dimensional and infinite-dimensional vector spaces. It follows immediately from Eq. (9.82) that

$$\langle A\bar{x}, \bar{y}' \rangle_{(2)} = \langle \bar{x}, A^H \bar{y}' \rangle_{(1)} \tag{9.83}$$

if one uses the fact that the inner product is either symmetric or Hermitian.

Examples   The adjoint of the linear mapping $\mathbf{x} : \mathbb{F} \to \mathbb{F}^n$ such that

$$\forall \alpha \in \mathbb{F} : \mathbf{x}[\alpha] := \alpha \mathbf{x} \tag{9.84}$$

must be such that

$$\mathbf{x}^H[\mathbf{y}] = \langle \mathbf{x}, \mathbf{y} \rangle. \tag{9.85}$$

(The left-hand side can be considered as the inner product of the scalar 1 with the scalar $\mathbf{x}^H[\mathbf{y}]$.) Then one obtains the adjoint of a vector (considered as a linear mapping!) by finding the image of the vector under the inner-product mapping:

$$\mathbf{x}^H = \underline{\mathbf{x}}. \tag{9.86}$$

To give another example, consider the linear mapping $A : \mathbb{C} \to \mathbb{C}$ such that

$$Az = \alpha z \tag{9.87}$$

in which α is a fixed complex number. The adjoint of this mapping must obey Eq. (9.82), which reads

$$\langle A^H z', z \rangle = \langle z', Az \rangle \tag{9.88}$$

in which the inner product is $\langle z', z \rangle = \mathrm{Re}(z'^* z)$. Therefore

$$\forall z, z' \in \mathbb{C} : \mathrm{Re}((A^H z')^* z) = \mathrm{Re}(z'^* \alpha z), \tag{9.89}$$

which implies that

$$A^H z' = \alpha^* z' \tag{9.90}$$

for all $z' \in \mathbb{C}$. According to Eq. (6.94) the matrix of A is

$$A = \begin{pmatrix} \alpha' & -\alpha'' \\ \alpha'' & \alpha' \end{pmatrix} \tag{9.91}$$

in which $\alpha' = \text{Re}(\alpha)$ and $\alpha'' = \text{Im}(\alpha)$. Because complex conjugation changes the sign of α'', it follows that the matrix of A^H is

$$\mathbf{A}^H = \begin{pmatrix} \alpha' & \alpha'' \\ -\alpha'' & \alpha' \end{pmatrix}, \tag{9.92}$$

which is just the transpose of $\mathbf{A}$.

Another important special case is the Hermitian adjoint of a matrix in $\mathbb{R}^{n_2 \times n_1}$ or $\mathbb{C}^{n_2 \times n_1}$ under the inner products $\langle \mathbf{x}, \mathbf{y} \rangle = \mathbf{x}^T \mathbf{G} \mathbf{y}$ or $\langle \mathbf{x}, \mathbf{y} \rangle = \mathbf{x}^\dagger \mathbf{G} \mathbf{y}$. For the sake of simplicity we consider only the latter. Let us find the adjoint of a linear mapping $\mathbf{A} : \mathbb{C}^{n_1} \to \mathbb{C}^{n_2}$ assuming that the inner products are

$$\langle \mathbf{x}, \mathbf{y} \rangle_{(1)} = \mathbf{x}^\dagger \mathbf{G}_{(1)} \mathbf{y} \tag{9.93}$$

in $\mathbb{C}^{n_1}$ and

$$\langle \mathbf{x}', \mathbf{y}' \rangle_{(2)} = \mathbf{x}'^\dagger \mathbf{G}_{(2)} \mathbf{y}' \tag{9.94}$$

in $\mathbb{C}^{n_2}$. The definition Eq. (9.82) implies that for all $\mathbf{x}' \in \mathbb{C}^{n_2}$ and for all $\mathbf{y} \in \mathbb{C}^{n_1}$,

$$\begin{aligned} \langle \mathbf{x}', \mathbf{A}\mathbf{y} \rangle_{(2)} &= \langle \mathbf{A}^H \mathbf{x}', \mathbf{y} \rangle_{(1)} \\ &= (\mathbf{A}^H \mathbf{x}')^\dagger \mathbf{G}_{(1)} \mathbf{y} \\ &= \mathbf{x}'^\dagger \mathbf{A}^{H\dagger} \mathbf{G}_{(1)} \mathbf{y} \\ &= \mathbf{x}'^\dagger \mathbf{G}_{(2)} \mathbf{A}\mathbf{y}. \end{aligned} \tag{9.95}$$

Then

$$\mathbf{A}^{H\dagger} \mathbf{G}_{(1)} = \mathbf{G}_{(2)} \mathbf{A}. \tag{9.96}$$

Taking the transpose and complex conjugate of both sides and recalling that $\mathbf{G}_{(1)}^\dagger = \mathbf{G}_{(1)}$ and that $\mathbf{G}_{(2)}^\dagger = \mathbf{G}_{(2)}$, one obtains

$$\mathbf{G}_{(1)} \mathbf{A}^H = \mathbf{A}^\dagger \mathbf{G}_{(2)}. \tag{9.97}$$

Therefore

$$\mathbf{A}^H = \mathbf{G}_{(1)}^{-1} \mathbf{A}^\dagger \mathbf{G}_{(2)}. \tag{9.98}$$

In $\mathbb{R}^{n_1}$ and $\mathbb{R}^{n_2}$, one drops the complex conjugates.

Even if $\mathbf{G}_{(1)}$ and $\mathbf{G}_{(2)}$ are diagonal, one cannot necessarily simplify Eq. (9.98). If $\mathcal{V}_{(1)}$ and $\mathcal{V}_{(2)}$ are pseudo-Euclidean or pseudo-unitary, then factors of σ_i appear on the diagonals of the Gram matrices $\mathbf{G}_{(1)}$ and $\mathbf{G}_{(2)}$ and affect the signs of elements of $\mathbf{A}^H$. If, and *only* if, $\mathcal{V}_{(1)}$ and $\mathcal{V}_{(2)}$ are both Euclidean or unitary *and* if one introduces orthonormal bases $\{\bar{e}_i\}$ and $\{\bar{e}'_j\}$, then the matrix $\mathbf{A}^H$ (rel$\{\overline{e}_i\}$ and $\{\overline{e}'_j\}$) is the transpose, complex conjugate of $\mathbf{A}$:

$$\mathbf{A}^H = \mathbf{A}^\dagger. \tag{9.99}$$

From now on, we shall write $\mathbf{A}^\dagger$ for $\mathbf{A}^H$ when $\mathcal{V}_{(1)}$ and $\mathcal{V}_{(2)}$ are $\mathbb{E}^n$ or $\mathbb{U}^n$.

9.2.3 OTHER REALIZATIONS OF THE ADJOINT

l^2 Spaces

The adjoint of the right-shift operator S (Eq. [6.205]), relative to the inner product defined in Eq. (8.50), is $S^\dagger$:

$$\langle \overline{a}, S\overline{b} \rangle = a_2^* b_1 + a_3^* b_2 + a_4^* b_3 + \cdots = \langle S^\dagger \overline{a}, \overline{b} \rangle. \tag{9.100}$$

It is equally straightforward to show that the creation operator $a^\dagger$ is the adjoint of the annihilation operator A.

Spaces of Differentiable Functions

Some of the most important practical applications of the definition of the adjoint of a linear mapping, Eq. (9.82), involve infinite-dimensional vector spaces. For example, let $\mathcal{V}_{(1)} = \mathcal{V}_{(2)} = \mathcal{V} = \mathcal{C}^0([a, b]; \mathbb{R})$ and let q be the linear mapping such that

$$\forall f \in \mathcal{C}^0([a, b]; \mathbb{R}) : \forall x \in [a, b] : qf(x) := xf(x). \tag{9.101}$$

In quantum mechanics, q is called the **position operator**. The adjoint of q, relative to the inner product Eq. (8.52), must satisfy the equation

$$\begin{aligned}
\langle q^\dagger f, g \rangle &= \int_a^b [q^\dagger f(x)]^* g(x)\, dx \\
&= \int_a^b f(x)^* x g(x)\, dx \\
&= \langle f, qg \rangle
\end{aligned} \tag{9.102}$$

for all functions f and g in $\mathcal{C}^0([a, b]; \mathbb{R})$. Because one can interpret the integrals in Eq. (9.101) as Riemann integrals in this case, it is straightforward to show that

$$q^\dagger = q \tag{9.103}$$

as one would expect.

For a less trivial example, let $\mathcal{V} = \mathcal{C}^1([a, b]; \mathbb{C})$ and let p be the linear mapping in $\mathcal{V}$ such that

$$\forall f \in \mathcal{C}^1([a, b]; \mathbb{C}) : \forall x \in [a, b] : pf(x) := -i\frac{df}{dx}(x), \tag{9.104}$$

and let the inner product be Eq. (8.52) with a weight function $w = 1$. Apart from a factor of $\hbar$, p is the **momentum operator** of quantum mechanics. In deference to the standard notation of quantum mechanics, we write the adjoint of p as $p^\dagger$.

Many students believe that $p^\dagger = p$ always. However, the adjoint $p^\dagger$ must satisfy the equation

$$\int_a^b [p^\dagger f(x)]^* g(x)\, dx = \int_a^b f(x)^* \left[-i \frac{dg}{dx}(x) \right] dx \tag{9.105}$$

for all functions f and g in $\mathcal{C}^1([a, b]; \mathbb{C})$. Integrating by parts, one finds that

$$\langle p^\dagger f, g \rangle = \underline{\beta}[f, g] + \int_a^b \left[-i \frac{df}{dx}(x) \right]^* g(x)\, dx \tag{9.106}$$

in which

$$\underline{\beta}[f, g] := -i f(x)^* g(x) |_a^b \tag{9.107}$$

is called the **boundary functional** or the **bilinear concomitant** of p. In other words, $p^\dagger$ is not equal to p unless the boundary functional vanishes on all functions in the space that we want to consider.

The subspace of $\mathcal{V}$ on which the boundary functionals vanish is generally a *proper* subspace. For example, the set of functions f that satisfy the homogeneous Dirichlet boundary conditions

$$f(a) = f(b) = 0 \tag{9.108}$$

is a vector space $\mathcal{D}$. The intersection $\mathcal{X} := \mathcal{D} \cap \mathcal{C}^1([a, b]; \mathbb{C})$ is also a vector space according to Exercise 5.2.1. On $\mathcal{X}$, therefore, $p^\dagger = p$.

Adjoint of a Sturm-Liouville Differential Operator

Let $\mathcal{V} = \mathcal{C}^\infty([a, b]; \mathbb{R})$ and let $L : \mathcal{V} \to \mathcal{V}$ be the linear n-th-order differential operator such that for all $g \in \mathcal{V}$

$$(Lg)(x) := \frac{1}{w(x)} \sum_{k=0}^n a_k(x) \frac{d^k g}{dx^k}(x) \tag{9.109}$$

in which $a_k \in \mathcal{C}^\infty([a, b]; \mathbb{R})$. Integrating by parts in the inner product

$$\langle f, Lg \rangle = \int_a^b f(x)(Lg(x))\, w(x)\, dx \tag{9.110}$$

gives

$$\langle L^H f, g \rangle = \langle f, Lg \rangle = \langle L^F f, g \rangle + \underline{\beta}_L[f, g] \tag{9.111}$$

in which

$$(L^F f)(x) := \sum_{k=0}^n (-1)^k \frac{d^k(a_k f)}{dx^k}(x) \tag{9.112}$$

is called the **formal adjoint** (or **Lagrange adjoint**) of L, and

$$\underline{\beta}_L[f, g] := \sum_{k=1}^{n} \sum_{j=0}^{k-1} (-1)^j \frac{d^{k-j-1} f}{dx^{k-j-1}}(x) \frac{d^j(a_k g)}{dx^j}(x) \Big|_a^b \tag{9.113}$$

is a boundary functional.

Many of the differential equations that one encounters in mathematical physics involve Sturm-Liouville operators, which are of the form

$$\forall f \in \mathcal{C}^\infty([a, b]; \mathbb{R}) : \forall x \in [a, b] :$$

$$Lf(x) := \frac{1}{w(x)} \frac{d}{dx} \left[p(x) \frac{df}{dx}(x) \right] + \frac{q(x)}{w(x)} f(x) \tag{9.114}$$

in which w is the weight function in the inner product (Eq. [8.52]) and p, q are real-valued functions. (Note that for a Sturm-Liouville operator the only nonzero coefficient functions in Eq. [9.109] are $a_2 = p$, $a_1 = p'$ and $a_0 = q$.)

We show that for a Sturm-Liouville differential operator, $L^F = L$. Calculating the adjoint L^H now involves two integrations by parts. In a condensed notation in which a prime denotes differentiation, one has

$$\langle f, Lg \rangle = \int_a^b f[(pg')' + qg] \, dx$$

$$= fpg'|_a^b + \int_a^b (-pf'g' + fqg) \, dx \tag{9.115}$$

$$= (fpg' - pf'g)|_a^b + \int_a^b [(pf')' + qf]g \, dx$$

$$= \underline{\beta}_L[f, g] + \langle Lf, g \rangle$$

in which the boundary functional is

$$\underline{\beta}_L[f, g] := p(x) \left[f(x) \frac{dg}{dx}(x) - \frac{df}{dx}(x) g(x) \right] \Big|_a^b. \tag{9.116}$$

(It is worth one's while to verify that Eq. [9.116] is consistent with the $n = 2$ case of Eq. [9.113].) Therefore the adjoint of a Sturm-Liouville operator acts as follows:

$$\langle L^H f, g \rangle = \underline{\beta}_L[f, g] + \langle Lf, g \rangle. \tag{9.117}$$

Note that L^H is not equal to L unless the boundary functional vanishes for all functions under consideration. What this means in practice is that one must deal with a "smaller" space than $\mathcal{C}^2([a, b]; \mathbb{C})$ if one wants to ensure that $L^H = L$.

9.2.4 PROPERTIES OF THE ADJOINT

Linearity of A^H

We have not yet established in general that A^H is a linear mapping. The additivity of the inner product implies that for all $\overline{x}', \overline{x}'' \in \text{domain}[A^H]$,

$$
\begin{aligned}
\langle A^H(\overline{x}' + \overline{y}'), \overline{y} \rangle_{(1)} &= \langle \overline{x}' + \overline{y}', A\overline{y} \rangle_{(2)} \\
&= \langle \overline{x}', A\overline{y} \rangle_{(2)} + \langle \overline{y}', A\overline{y} \rangle_{(2)} \\
&= \langle A^H \overline{x}', \overline{y} \rangle_{(1)} + \langle A^H \overline{y}', \overline{y} \rangle_{(1)} \\
&\Rightarrow A^H(\overline{x}' + \overline{y}') - (A^H \overline{x}' + A^H \overline{y}') \in \{\text{domain}[A]\}^\perp.
\end{aligned}
\tag{9.118}
$$

Because we have assumed that Eq. (9.80) holds in order to define A^H, it follows that

$$
A^H(\overline{x}' + \overline{y}') = A^H \overline{x}' + A^H \overline{y}'.
\tag{9.119}
$$

To show that A^H is homogeneous, we carry out the following short calculation, which holds for every $\overline{x}' \in \text{domain}[A^H]$ and every $\overline{y} \in \text{domain}[A]$:

$$
\begin{aligned}
\langle A^H(\alpha \overline{x}'), \overline{y} \rangle_{(1)} &= \langle \alpha \overline{x}', A\overline{y} \rangle_{(2)} = \alpha^* \langle \overline{x}', A\overline{y} \rangle_{(2)} \\
&= \alpha^* \langle A^H \overline{x}', \overline{y} \rangle_{(1)} \\
&= \langle \alpha A^H \overline{x}', \overline{y} \rangle_{(1)} \\
&\Rightarrow A^H(\alpha \overline{x}') - \alpha A^H \overline{x}' \in \{\text{domain}[A]\}^\perp.
\end{aligned}
\tag{9.120}
$$

Then, assuming again that Eq. (9.80) holds, one has

$$
A^H(\alpha \overline{x}') = \alpha A^H \overline{x}'.
\tag{9.121}
$$

Therefore A^H is a linear mapping, and $\text{domain}[A^H]$ is a vector subspace of $\mathcal{V}_{(2)}$.

Relation of A^T to A^H

To find a formula that relates A^T to A^H if $\mathcal{V}_{(1)}$ and $\mathcal{V}_{(2)}$ are finite-dimensional vector spaces, we use the properties of the inner-product mappings and the definition of A^T:

$$
\begin{aligned}
\langle \overline{x}', A\overline{y} \rangle_{(2)} &= (G_{(2)} \overline{x}')[A\overline{y}] \\
&= (A^T G_{(2)} \overline{x}')[\overline{y}] \\
&= (G_{(1)} G_{(1)}^{-1} A^T G_{(2)} \overline{x}')[\overline{y}] \\
&= \langle G_{(1)}^{-1} A^T G_{(2)} \overline{x}', \overline{y} \rangle_{(1)}.
\end{aligned}
\tag{9.122}
$$

Therefore

$$
A^H = G_{(1)}^{-1} A^T G_{(2)}.
\tag{9.123}
$$

The diagram

$$
\begin{array}{ccc}
\mathcal{V}^*_{(1)} & \xleftarrow{\quad A^T \quad} & \mathcal{V}^*_{(2)} \\
\Big\downarrow G^{-1}_{(1)} & & \Big\uparrow G_{(2)} \\
\mathcal{V}_{(1)} & \xleftarrow{\quad A^H \quad} & \mathcal{V}_{(2)}
\end{array}
\tag{9.124}
$$

shows the sequence of the mappings in Eq. (9.123).

Dyadic Expansion of A^H

The dyadic expansion

$$
A^H = g^{lj}_{(1)} g_{(2)ik} a^{i\,*}_{j}\, \overline{e}_l \circ \underline{\epsilon}'^k
\tag{9.125}
$$

is equivalent to, and is sometimes more useful than, Eq. (9.123). To establish Eq. (9.125) one substitutes the dyadic expansion of A^T, Eq. (9.57), into Eq. (9.123), with the following results:

$$
A^H \overline{x}' = G^{-1}_{(1)}\left(a^i_j \underline{\epsilon}^j \overline{e}_i\big[G_{(2)}(x'^k \overline{e}'_k)\big]\right) = G^{-1}_{(1)}\left(a^i_j \underline{\epsilon}^j \overline{e}_i [x'^{k\,*} \underline{e}k']\right)
\tag{9.126}
$$

$$
= a^{i\,*}_j x'^k G^{-1}_{(1)}\left(\underline{\epsilon}^j \overline{e}_i [g_{(2)km} \underline{\epsilon}'^m]\right) = a^{i\,*}_j x'^k G^{-1}_{(1)}\left(\underline{\epsilon}^j g_{(2)ki}\right)
\tag{9.127}
$$

$$
= a^{i\,*}_j x'^k g^*_{(2)ki} G^{-1}_{(1)}(\underline{\epsilon}^j) = a^{i\,*}_j x'^k g_{(2)ik} g^{lj}_{(1)} \overline{e}_l
\tag{9.128}
$$

$$
= g^{lj}_{(1)} g_{(2)ik} a^{i\,*}_j\, (\overline{e}_l \circ \underline{\epsilon}'^k)[\overline{x}'].
\tag{9.129}
$$

In Eq. (9.126) we use the definition of the inner-product mapping, Eq. (8.480); in Eq. (9.127) we use the coordinate expansion of $\overline{e}'_k$, Eq. (8.494); and in Eq. (9.129) we use Eq. (8.513) for the image of $\underline{\epsilon}^j$ under the inverse inner-product mapping.

Range and Null Space of A^H

The range and null space of A^H can be deduced directly from Eq. (9.125). However, the shortest way is to use what we already know, specifically, Eqs. (9.123) and (9.68) through (9.70). It is straightforward to show that if $\mathcal{V}_{(2)}$ is a finite-dimensional Euclidean or unitary space, then

$$
\mathrm{null}[A^H] = \{\mathrm{range}[A]\}^{\perp}.
\tag{9.130}
$$

A vector $\overline{z}' \in \mathcal{V}_{(2)}$ belongs to the null space of A^H if and only if

$$
\forall \overline{y} \in \mathcal{V}_{(1)}: \ \langle A^H \overline{z}', \overline{y}\rangle_{(1)} = \langle \overline{z}', A\overline{y}\rangle_{(2)} = 0.
\tag{9.131}
$$

That is, $\overline{z}' \in \mathrm{null}[A^H]$ if and only if $\overline{z}' \perp \mathrm{range}[A]$, which is the content of Eq. (9.130).

If $\mathrm{domain}[A] = \mathcal{V}_{(1)}$, then Eqs. (9.59), (9.68), and (8.527) imply that

$$
\mathrm{range}[A^H] = \mathrm{null}[A]^{\perp}.
\tag{9.132}
$$

Equations (9.130) and (9.132) completely characterize A^H in terms of A.

If a linear mapping A is invertible, then so is A^H. One shows easily that

$$(A^H)^{-1} = (A^{-1})^H. \tag{9.133}$$

From Eqs. (9.52) and (9.82), one finds that the adjoint of the product of two linear mappings $A : V_{(1)} \to V_{(2)}$ and $B : V_{(2)} \to V_{(3)}$ is the product of the adjoints, taken in the reversed order:

$$\begin{aligned}(BA)^H &= G_{(1)}^{-1}(BA)^T G_{(3)} = G_{(1)}^{-1} A^T B^T G_{(3)} \\ &= G_{(1)}^{-1} A^T G_{(2)} G_{(2)}^{-1} B^T G_{(3)} = A^H B^H.\end{aligned} \tag{9.134}$$

9.2.5 HERMITIAN AND SELF-ADJOINT MAPPINGS

Definition of a Hermitian Mapping

In many important cases a linear mapping A maps a vector space V into itself:

$$V = V_{(1)} = V_{(2)}. \tag{9.135}$$

A linear mapping $A : V \to V$ is called **Hermitian** if and only if

$$\forall \overline{x}, \overline{y} \in \text{domain}[A] : \ \langle \overline{x}, A\overline{y} \rangle = \langle A\overline{x}, \overline{y} \rangle. \tag{9.136}$$

For example, the matrix realization of Eq. (9.136) in $\mathbb{C}^n$ is

$$\forall x, y \in \mathbb{C}^n : (Ax)^\dagger Gy = x^\dagger GAy, \tag{9.137}$$

which is equivalent to the condition

$$A^\dagger G = GA. \tag{9.138}$$

It follows that, relative to an orthonormal basis of $\mathbb{C}^n$ (for which the Gram matrix **G** is equal to the identity matrix **1**), **A** is the matrix of a Hermitian linear mapping if and only if

$$A^\dagger = A. \tag{9.139}$$

A matrix that satisfies this equation is called a **Hermitian matrix**.

The Pauli spin matrices $\sigma_1, \sigma_2, \sigma_3$ are Hermitian. For example,

$$\sigma_2^\dagger = \begin{pmatrix} 0 & -i \\ i & 0 \end{pmatrix}. \tag{9.140}$$

If the elements of a Hermitian matrix **A** are all real, then **A** is a Hermitian linear mapping on V if and only if

$$A^T = A. \tag{9.141}$$

In this case **A** is called a **real symmetric matrix**. We see in Section 9.3 that Hermitian and real symmetric matrices have real eigenvalues, a property of great importance in many applications in physics and engineering, in which physical quantities are inherently real.

Symmetric and Self-adjoint Operators

If V is an infinite-dimensional vector space, then there is no guarantee that A and A^H have the same domain. In that case the strongest conclusion that one can draw from the definition Eq. (9.136) without further assumptions is that

$$\forall \, \bar{y} \in \text{domain}[A] : \forall \bar{x} \in \text{domain}[A] \cap \text{domain}[A^H] : \langle (A^H - A)\bar{x}, \bar{y} \rangle = 0. \qquad (9.142)$$

Then

$$\forall \bar{x} \in \text{domain}[A] \cap \text{domain}[A^H] : (A^H - A)\bar{x} \in \{\text{domain}[A]\}^{\perp}. \qquad (9.143)$$

If Eq. (9.80) holds, then

$$\forall \bar{x} \in \text{domain}[A] \cap \text{domain}[A^H] : (A^H - A)\bar{x} = \bar{0}, \qquad (9.144)$$

which says that A and A^H agree on the intersection of their domains. (In an infinite-dimensional space, it is mathematically possible for this intersection to consist of the zero vector alone.) If V is finite-dimensional, then domain$[A]$ = domain$[A^H]$ = V, and Eq. (9.144) implies that

$$\forall \bar{x}, \bar{y} \in V : \langle (A^H - A)\bar{x}, \bar{y} \rangle = 0. \qquad (9.145)$$

Then Eq. (8.59) implies that

$$\forall \bar{x} \in V : (A^H - A)\bar{x} = \bar{0}. \qquad (9.146)$$

Therefore $A^H = A$. Regardless of the dimension of the underlying vector space V, a linear mapping A such that

$$\text{domain}[A^H] = \text{domain}[A] \text{ and } A^H = A \qquad (9.147)$$

is called **self-adjoint**. The statement "A is self-adjoint" always implies the statement "A is Hermitian," but the converse is true only in a finite-dimensional vector space. In an infinite-dimensional vector space, a Hermitian linear mapping is required to be identical with its adjoint only on the intersection of their domains.

If L is a linear differential operator, then its formal adjoint L^F satisfies the definition given in Eq. (9.111). If

$$L^F = L \qquad (9.148)$$

then L is called **formally self-adjoint**. A formally self-adjoint differential operator L is self-adjoint if and only if the boundary functional vanishes on all elements of the space on which L is defined.

In a finite-dimensional vector space,

$$(A^H)^H = A. \tag{9.149}$$

This relation also holds for *bounded* operators on an infinite-dimensional Hilbert space.

The importance of self-adjoint linear mappings in mathematics comes from the fact that if a self-adjoint mapping has eigenvalues, then they are real.

Projection Operators

We show that every projection operator P on a finite-dimensional subspace $\mathcal{W}$ of a Euclidean or unitary space $\mathcal{V}$ is self-adjoint by showing that

$$\forall \bar{x}, \bar{x}' \in V : \langle \bar{x}', P_W \bar{x} \rangle = \langle P_W \bar{x}', P_W \bar{x} \rangle. \tag{9.150}$$

Because $\bar{x}' - P_W \bar{x}' \perp \mathcal{W}$, one has

$$\forall \bar{x}, \bar{x}' \in V : \langle \bar{x}', P_W \bar{x} \rangle = \langle P_W \bar{x}' + (\bar{x}' - P_W \bar{x}'), P_W \bar{x} \rangle = \langle P_W \bar{x}', P_W \bar{x} \rangle. \tag{9.151}$$

But

$$\langle P_W \bar{x}', \bar{x} \rangle = \langle P_W \bar{x}', P_W \bar{x} + (\bar{x} - P_W \bar{x}) \rangle = \langle P_W \bar{x}', P_W \bar{x} \rangle. \tag{9.152}$$

Therefore

$$P_W^H = P_W. \tag{9.153}$$

From Eqs. (9.123) and (9.136) one finds a formula for the transpose of a projector:

$$P_W^T = G P_W G^{-1}. \tag{9.154}$$

From Eqs. (9.68) and (9.154) it follows that P_W^T projects on ann $[\mathcal{W}^\perp]$.

9.2.6 ISOMETRIC AND UNITARY MAPPINGS

In this subsection we shall define, and derive some of the properties of, a class of linear mappings that preserve length.

A linear mapping U that is defined on all of a vector space $\mathcal{V}$ and that preserves lengths,

$$\forall \bar{x} \in \mathcal{V} : \|U\bar{x}\| = \|\bar{x}\|, \tag{9.155}$$

is called **isometric**. There exist norms that are not defined (and are not definable) in terms of an inner product. However, if the norms in Eq. (9.155) are inner-product norms, and if $\mathcal{V}$ is either Euclidean or unitary, then Eqs. (8.141), (8.145), and (9.155) imply that

$$\forall \bar{x}, \bar{y} \in \mathcal{V} : \langle U\bar{x}, U\bar{y} \rangle = \langle \bar{x}, \bar{y} \rangle \tag{9.156}$$

for every isometric mapping U. We adopt Eq. (9.156) as the definition of an isometric mapping in a pseudo-Euclidean or pseudo-unitary space such as Minkowski space. By the definition of the adjoint, Eq. (9.82), one has

$$\forall \bar{x}, \bar{y} \in \mathcal{V}: \langle U^H U \bar{x}, \bar{y} \rangle = \langle \bar{x}, \bar{y} \rangle. \tag{9.157}$$

Then Eq. (8.59) implies that

$$\forall \bar{x} \in V: (U^H U - 1)\bar{x} = \bar{0}, \tag{9.158}$$

which implies that

$$U^H U = 1. \tag{9.159}$$

In other words, if U is isometric then U^H is a left inverse of U.

An isometric mapping U on a finite- or infinite-dimensional vector space $\mathcal{V}$ is called **unitary** (if $\mathcal{V}$ is unitary) or **orthogonal** (if $\mathcal{V}$ is Euclidean) if and only if

$$UU^H = 1. \tag{9.160}$$

Because this equation implies that U^H is also a right inverse, it follows that for a unitary mapping,

$$U^{-1} = U^H. \tag{9.161}$$

It follows that the adjoint U^H of a unitary mapping U is also unitary.

Isometric Operators in Infinite-Dimensional Spaces

We use the example of the right-shift operator S on $l^2(\mathbb{F})$ defined in Eq. (6.205) to establish some important facts about isometric operators in infinite-dimensional spaces. For example, according to Eq. (6.209),

$$S^\dagger S = 1. \tag{9.162}$$

Then S is isometric according to Eq. (9.159).

In an infinite-dimensional space not all isometric mappings are unitary. For example, in $l^2(\mathbb{F})$, the right-shift operator is isometric but $SS^\dagger$ is not equal to the identity mapping:

$$SS^\dagger(x^1, x^2, x^3, \ldots) = (0, x^2, x^3, \ldots). \tag{9.163}$$

Therefore S is not unitary. We see in a subsequent section that dimensionality arguments preclude such a situation in a finite-dimensional inner-product space of the sort that is useful in physics or engineering.

Isometric Mappings in Finite-Dimensional Spaces

We show that if $\mathcal{V}$ is a finite-dimensional pseudo-Euclidean or pseudo-unitary space, then every isometric mapping in $\mathcal{V}$ is unitary. Let $\{\bar{e}_i\}$ be an orthonormal or pseudoorthonormal basis of $\mathcal{V}$. Then $\{U\bar{e}_i\}$ is a list of mutually orthogonal vectors with nonzero norms. Hence $\{U\bar{e}_i\}$ is linearly independent. Because the number of elements of $\{U\bar{e}_i\}$ is equal to the number of elements of $\{\bar{e}_i\}$, which is equal to dim$[\mathcal{V}]$, it follows that $\{U\bar{e}_i\}$ is a basis of $\mathcal{V}$. Then rank$[U] = $ dim$[\mathcal{V}]$. By the rank-nullity theorem, nullity$[U] = 0$. Therefore U is nonsingular and has an inverse, U^{-1}. Then

$$UU^{-1} = UU^H. \tag{9.164}$$

Applying U^{-1} to both sides of this equation, one obtains

$$U^{-1} = U^H, \tag{9.165}$$

which implies that U is unitary.

Unitary Mappings in Finite-Dimensional Spaces

We have just shown that if U is unitary and if $\{\bar{e}_i\}$ is a basis of $\mathcal{V}$, then $\{U\bar{e}_i\}$ is also a basis. It follows that the (row i, column j) metric-tensor element, relative to $\{U\bar{e}_i\}$, is

$$\langle U\bar{e}_i, U\bar{e}_j \rangle = \langle \bar{e}_i, \bar{e}_j \rangle = g_{ij}. \tag{9.166}$$

In this sense a unitary mapping preserves the metric tensor.

From Eqs. (9.159) and (9.160) one sees that the inverse of the matrix U of a unitary mapping U is the matrix U^H of the linear mapping that is adjoint to U:

$$U^H U = UU^H = 1. \tag{9.167}$$

In terms of the Gram matrix G, Eqs. (9.98) (for U^H) and (9.167) imply that

$$G^{-1}U^\dagger GU = UG^{-1}U^\dagger G = 1. \tag{9.168}$$

Then the matrix expression of the property of preserving the metric tensor, Eq. (9.166), is

$$U^\dagger GU = G. \tag{9.169}$$

For example, a **Lorentz transformation** is a unitary mapping in Minkowski space $\mathcal{M}$ that preserves the inner product of special relativity, Eq. (8.60) (recall Exercise 6.1.14). Because the matrix Λ of a Lorentz transformation is real, the version of Eq. (9.169) that applies to Λ is

$$\Lambda^T G\Lambda = G \tag{9.170}$$

in which G is the matrix of the metric tensor of special relativity,

$$G = \begin{pmatrix} 1 & 0 & 0 & 0 \\ 0 & -1 & 0 & 0 \\ 0 & 0 & -1 & 0 \\ 0 & 0 & 0 & -1 \end{pmatrix}. \tag{9.171}$$

See Exercise 9.2.4 for an important example.

If V is Euclidean or unitary, and if the basis $\{\bar{e}_i\}$ with respect to which $\mathbf{U}$ is computed is orthonormal, then $\mathbf{G} = \mathbf{1}$, making Eq. (9.168) simplify to

$$\mathbf{U}^\dagger\mathbf{U} = \mathbf{U}\mathbf{U}^\dagger = \mathbf{1}. \tag{9.172}$$

A matrix that satisfies this equation is called a **unitary matrix**.

The Pauli spin matrices σ_1, σ_2, σ_3 are unitary, because $\sigma_i^\dagger = \sigma_i$ and $\sigma_i^2 = \mathbf{1}$. For example,

$$\sigma_2^\dagger\sigma_2 = \begin{pmatrix} 0 & -i \\ i & 0 \end{pmatrix}\begin{pmatrix} 0 & -i \\ i & 0 \end{pmatrix} = \begin{pmatrix} 1 & 0 \\ 0 & 1 \end{pmatrix}. \tag{9.173}$$

If the elements of a unitary matrix $\mathbf{U}$ are all real, then $\mathbf{U}$ satisfies the equation

$$\mathbf{U}^T\mathbf{U} = \mathbf{U}\mathbf{U}^T = \mathbf{1} \tag{9.174}$$

and is called a **real orthogonal matrix**. For example, the matrix of a finite rotation in the plane is real orthogonal:

$$\begin{pmatrix} \cos\theta & -\sin\theta \\ \sin\theta & \cos\theta \end{pmatrix}^T \begin{pmatrix} \cos\theta & -\sin\theta \\ \sin\theta & \cos\theta \end{pmatrix} = \begin{pmatrix} 1 & 0 \\ 0 & 1 \end{pmatrix}. \tag{9.175}$$

The (row i, column j) matrix element of the product $\mathbf{U}^\dagger\mathbf{U}$ is equal to the product of the i-th row of $\mathbf{U}^\dagger$ with the j-th column of $\mathbf{U}$, which in turn is equal to the product of the complex conjugate of the transpose of the i-th column of $\mathbf{U}$ with the j-th column of $\mathbf{U}$. Then the unitarity condition $\mathbf{U}^\dagger\mathbf{U} = \mathbf{1}$ (relative to an orthonormal basis) implies that the column vectors of a unitary matrix $\mathbf{U}$ are orthonormal. Of course, the column vectors of $\mathbf{U}$ are the images $\mathbf{U}e_i$ of the basis vectors e_i. Similarly, the equation $\mathbf{U}\mathbf{U}^\dagger = \mathbf{1}$ implies that relative to an orthonormal basis, the row vectors of a unitary matrix $\mathbf{U}$ are orthonormal.

Unitary Groups

From the facts that every unitary mapping has an inverse, that the composition of linear mappings is associative, that there exists an identity mapping, and that the product of two unitary mappings U and V is unitary,

$$(\mathrm{VU})^H = \mathrm{U}^H\mathrm{V}^H = \mathrm{U}^{-1}\mathrm{V}^{-1} = (\mathrm{VU})^{-1}, \tag{9.176}$$

one sees that the set of all unitary mappings in a pseudo-Euclidean or pseudo-unitary space is a group. This group, which is called the **unitary group** of V, is a proper subgroup of the general linear group of V. If $V_{(1)}$ and $V_{(2)}$ are n-dimensional and are isomorphic as inner-product spaces, then their unitary groups are isomorphic.

It is equally straightforward to show that the set of all $n \times n$ unitary matrices with elements in $\mathbb{F}$ is a group that is isomorphic to the unitary group of any n-dimensional Euclidean space (if $\mathbb{F} = \mathbb{R}$) or of any n-dimensional unitary space (if $\mathbb{F} = \mathbb{C}$). The standard notation for the group of $n \times n$ complex unitary matrices is $U(n)$. The notation for the group of real orthogonal $n \times n$ matrices is $O(n)$.

9.2.7 EXERCISES FOR SECTION 9.2

9.2.1 Show that the mapping A^T defined in Eq. (9.49) is additive and homogeneous.

9.2.2 Prove that the transpose of the sum of two linear mappings is the sum of the transposes:

$$(A + B)^T = A^T + B^T. \tag{9.177}$$

9.2.3 Show that

$$(BA)^T = A^T B^T \tag{9.178}$$

in which $A : \mathcal{V}_{(1)} \to \mathcal{V}_{(2)}$ and $B : \mathcal{V}_{(2)} \to \mathcal{V}_{(3)}$ are linear mappings.

9.2.4 Evaluate the matrix of the linear mapping that is adjoint to the linear mapping whose matrix is

$$\mathbf{F} = \begin{pmatrix} 0 & E_x & E_y & E_z \\ -E_x & 0 & B_z & -B_y \\ -E_y & -B_z & 0 & B_x \\ -E_z & B_y & -B_x & 0 \end{pmatrix} \tag{9.179}$$

if

$$\mathbf{G}_{(1)} = \mathbf{G}_{(2)} = \begin{pmatrix} 1 & 0 & 0 & 0 \\ 0 & -1 & 0 & 0 \\ 0 & 0 & -1 & 0 \\ 0 & 0 & 0 & -1 \end{pmatrix}. \tag{9.180}$$

The matrix in Eq. (9.179) is a realization of the electromagnetic-field tensor in a relativistically covariant formulation of electrodynamics.

9.2.5 Prove that on the interval $[-1, 1]$, the Chebyshev operator

$$L := (1 - x^2)^{1/2} \frac{d}{dx} \left[(1 - x^2)^{1/2} \frac{d}{dx} \right] \tag{9.181}$$

is formally self-adjoint with respect to the inner product

$$\langle f, g \rangle = \int_a^b f(x)^* g(x) w(x) \, dx \tag{9.182}$$

with the weight function

$$w(x) = \frac{1}{(1 - x^2)^{1/2}}. \tag{9.183}$$

9.2.6 (a) Prove that the n-th–order Bessel operator

$$L := \frac{1}{x} \left[\frac{d}{dx} \left(x \frac{d}{dx} \right) - \frac{n^2}{x} \right] \tag{9.184}$$

is formally self-adjoint with respect to the inner product

$$\langle f, g \rangle = \int_a^b f(x)^* g(x) w(x) \, dx \tag{9.185}$$

in which $a = 0$, b is finite, and the weight function is

$$\forall x \in [a, b] : w(x) = x. \tag{9.186}$$

(b) State all of the boundary conditions you can think of that ensure the self-adjointness of the differential operator in (a).

9.2.7 What conditions must the elements of the matrix **D** of the deformation calculated in Exercise 6.1.11 satisfy in order for **D** to be unitary with respect to the inner product Eq. (8.13)?

9.2.8 What conditions must the elements of the matrix **S** of the shear calculated in Exercise 6.1.12 satisfy in order for **S** to be unitary with respect to the inner product Eq. (8.13)?

9.2.9 Prove that the rotation matrix **R** calculated in Exercise 6.1.13 is the matrix of a unitary mapping in $\mathbb{E}^2$.

9.2.10 Show that the Lorentz-transformation matrix Λ calculated in Exercise 6.1.14 satisfies Eq. (9.166), in which

$$\mathbf{G} := \begin{pmatrix} -1 & 0 \\ 0 & 1 \end{pmatrix}. \tag{9.187}$$

(The order of the 1s and -1s is reversed because we choose $x^2 = ct$.)

9.2.11 A linear operator C on $l^2(\mathbb{C})$ is defined as follows:

$$C(x_1, x_2, \ldots, x_n, \ldots) := (x_2, x_1 + x_3, \ldots, x_{n-1} + x_{n+1}, \ldots). \tag{9.188}$$

(a) Find $C\bar{e}_n$.

(b) Show that C is Hermitian.

9.2.12 Derive Eqs. (9.111) through (9.113).

9.2.13 Demonstrate that the Pauli spin matrices

$$\sigma_1 := \begin{pmatrix} 0 & 1 \\ 1 & 0 \end{pmatrix}, \quad \sigma_2 := \begin{pmatrix} 0 & -i \\ i & 0 \end{pmatrix}, \quad \sigma_3 := \begin{pmatrix} 1 & 0 \\ 0 & -1 \end{pmatrix} \tag{9.189}$$

are both unitary and Hermitian with respect to the canonical inner product of vectors belonging to $\mathbb{U}^2$.

9.3 EIGENVALUES AND EIGENVECTORS

9.3.1 SECULAR EQUATION

Diagonal matrices are the easiest ones to use, either in matrix multiplication or in a system of linear equations. In this section we derive conditions that, if satisfied by some linear mapping $A : V \rightarrow V$, imply that one can choose a basis relative to which the matrix realization $\mathbf{A}$ is diagonal. In Section 9.4 we show that it is possible to find bases for the domain and range of a general linear mapping A relative to which the matrix of A is diagonal. We assume throughout the remaining sections of this chapter that the underlying vector space is finite-dimensional.

If there exists a basis $\{\bar{e}_i\}$ relative to which the matrix of a linear mapping $A : V \rightarrow V$ is diagonal,

$$\mathbf{A} = \text{diag}[\lambda_1, \ldots, \lambda_n], \tag{9.190}$$

then the image of each $\bar{e}_i$ under A is a multiple of $\bar{e}_i$: $A\bar{e}_i = \lambda_i\bar{e}_i$. A *nonzero* vector $\bar{x}$ in a finite-dimensional vector space V over a number field $\mathbb{F}$ is called an **eigenvector** of a linear mapping $A : V \rightarrow V$ if and only if

$$\exists \lambda \in \mathbb{F} :\ni: A\bar{x} = \lambda\bar{x}. \tag{9.191}$$

The scalar λ is called the **eigenvalue** of A corresponding to the eigenvector $\bar{x}$. Clearly

$$\bar{x} \neq \bar{0} \text{ and } A\bar{x} = \lambda\bar{x} \Leftrightarrow \bar{x} \in \text{null}[A - \lambda\mathbf{1}] \Leftrightarrow A - \lambda\mathbf{1} \text{ is singular}, \tag{9.192}$$

because $A - \lambda\mathbf{1}$ carries $\bar{x} \neq \bar{0}$ onto $\bar{0}$ if and only if Eq. (9.191) is true. A (nonzero) eigenvector corresponds to every eigenvalue λ of A, for $A - \lambda\mathbf{1}$ is singular if and only if $A - \lambda\mathbf{1}$ carries a nonzero vector $\bar{x}$ onto $\bar{0}$.

It follows from Eq. (9.192) that $\bar{x}$ is an eigenvector of A with eigenvalue λ if and only if

$$\det[A - \lambda\mathbf{1}] = 0. \tag{9.193}$$

Equation (9.193) is called the **secular equation** or the **characteristic equation** of the linear mapping A. The secular equation is a polynomial equation of degree $n = \dim[V]$ in λ.

Therefore it is convenient to choose a number field $\mathbb{F}$ that is **algebraically closed**, that is, such that every polynomial equation with coefficients in $\mathbb{F}$ has at least one root in $\mathbb{F}$. (The field $\mathbb{R}$ is not algebraically closed; for example, $\lambda^2 + 1 = 0$ has no real solutions.) If $\mathbb{F}$ is algebraically closed, then a polynomial equation of degree n over $\mathbb{F}$ has n roots in $\mathbb{F}$. Because the complex field $\mathbb{C}$ is algebraically closed (according to the "fundamental theorem of algebra"), we assume from now on that $\mathbb{F} = \mathbb{C}$.

If two or more linearly independent eigenvectors have the same eigenvalue, then that eigenvalue is called **degenerate**. The maximum number of linearly independent eigenvectors that have a given eigenvalue is called the **degeneracy** (or the **geometrical multiplicity**) of the eigenvalue. This definition is equivalent to saying that the geometrical multiplicity of an eigenvalue λ_i is equal to the dimension of the null space of $A - \lambda_i\mathbf{1}$.

The set of all eigenvectors that belong to a particular eigenvalue λ,

$$\mathcal{W}_\lambda := \{\bar{x} \mid A\bar{x} = \lambda\bar{x}\}, \tag{9.194}$$

is a vector subspace of $\mathcal{V}$, called the **eigensubspace** of λ. Linearity implies that $\mathcal{W}_\lambda$ is invariant under A in the sense defined in Section 6.1.4. The **degeneracy** of the eigenvalue λ is defined as the dimension of $\mathcal{W}_\lambda$.

Eigenvalue is a mongrel word. It is a half-translation of the German noun *Eigenwert*, meaning "proper value," that is, a value that belongs (to a particular linear mapping, etc.). There have been sporadic efforts to replace "eigenvalue" with the clearer terms "proper value" or "characteristic value," but "eigenvalue" is so strongly rooted in everyday usage in engineering and physics that all attempts to eradicate it have failed.

The set of eigenvalues of a linear mapping A is called its **spectrum**. If **A** is the matrix of A relative to some basis of $\mathcal{V}$ and if **T** is any nonsingular matrix, then

$$\mathbf{A} - \lambda\mathbf{1} \text{ is singular} \iff \mathbf{T}(A - \lambda\mathbf{1})\mathbf{T}^{-1} \text{ is singular.} \tag{9.195}$$

Therefore the spectrum of A (or of its matrix realization **A**) does not depend on a choice of basis of the underlying space $\mathcal{V}$.

The number of times a root λ_i of the secular equation is repeated is called the **algebraic multiplicity** of λ_i. Every linear mapping of $\mathcal{V}$ into $\mathcal{V}$ has exactly n eigenvalues if one counts each degenerate eigenvalue a number of times equal to its algebraic multiplicity. For some linear mappings, the geometrical multiplicities and algebraic multiplicities of some eigenvalues are not equal.

The algebraic multiplicity of an eigenvalue λ_i must be at least as great as the geometrical multiplicity. To see this, let $\mathcal{W}_{\lambda_i}$ be the eigensubspace of λ_i, and let $\bar{v}_1, \ldots, \bar{v}_m$ be a basis of $\mathcal{W}_{\lambda_i}$. Then $\mathcal{W}_{\lambda_i}$ is an invariant subspace of A. Let $\{\bar{w}_1, \ldots, \bar{w}_{n-m}\}$ be a basis of $\mathcal{W}_{\lambda_i}^\perp$. Then $\mathcal{V} = \mathcal{W}_{\lambda_i} \oplus \mathcal{W}_{\lambda_i}^\perp$, and

$$B := \{\bar{v}_1, \ldots, \bar{v}_m, \bar{w}_1, \ldots, \bar{w}_{n-m}\} \tag{9.196}$$

is a basis of $\mathcal{V}$. Relative to B, the matrix of $A - \lambda\mathbf{1}$ is

$$
\mathbf{A} - \lambda\mathbf{1} = \begin{array}{c} \\ 1 \\ \vdots \\ m \\ m+1 \\ \vdots \\ n \end{array}
\begin{array}{c}
\begin{array}{cccccc} 1 & \cdots & m & m+1 & \cdots & n \end{array} \\
\left(\begin{array}{cccccc}
\lambda_i - \lambda & \cdots & 0 & a_{m+1}^1 & \cdots & a_n^1 \\
\vdots & \ddots & \vdots & \vdots & \ddots & \vdots \\
0 & \cdots & \lambda_i - \lambda & a_{m+1}^m & \cdots & a_n^m \\
0 & \cdots & 0 & a_{m+1}^{m+1} - \lambda & \cdots & a_n^{m+1} \\
\vdots & \ddots & \vdots & \vdots & \ddots & \vdots \\
0 & \cdots & 0 & a_{m+1}^n & \cdots & a_n^n - \lambda
\end{array} \right)
\end{array}. \tag{9.197}
$$

according to Eq. (6.84). Therefore

$$\det[\mathbf{A} - \lambda\mathbf{1}] = (\lambda - \lambda_i)^m p(\lambda) \tag{9.198}$$

in which p is a polynomial of degree $n - m$. In particular, if the algebraic multiplicity of λ_i is equal to 1, then the geometrical multiplicity of λ_i is equal to 1.

If the geometrical multiplicity of some eigenvalue of a matrix is less than the algebraic multiplicity, then the matrix is called **defective**. To see that the algebraic multiplicity may exceed the geometrical multiplicity, consider the defective matrix

$$\mathbf{A} = \begin{pmatrix} 0 & 1 & 1 \\ 0 & 0 & 1 \\ 0 & 0 & 0 \end{pmatrix}. \tag{9.199}$$

The secular equation is $\det[\mathbf{A} - \lambda\mathbf{1}] = \lambda^3 = 0$, implying that the eigenvalue $\lambda_1 = 0$ has an algebraic multiplicity of 3. However, all of the eigenvectors that have this eigenvalue are scalar multiples of the one vector

$$\mathbf{v}_1 = \begin{pmatrix} 1 \\ 0 \\ 0 \end{pmatrix}, \tag{9.200}$$

implying that the geometrical multiplicity of λ_1 is 1.

Outside of textbooks one usually does not compute the spectrum of a real symmetric or Hermitian matrix $\mathbf{A}$ by solving the secular equation for $\mathbf{A}$, because the numerical computation of repeated roots of a polynomial is ill conditioned. (The exception to this rule is that the computation of the roots of the secular equation of a symmetric tridiagonal matrix is well conditioned.) Instead, one computes the eigenvalues of $\mathbf{A}$ by first tridiagonalizing $\mathbf{A}$ by a method such as the Householder transformation. The interested reader should consult Golub and Van Loan (1989).

9.3.2 DIAGONALIZATION OF HERMITIAN MATRICES

Hermitian linear mappings on unitary spaces play a fundamental role in quantum mechanics and in the ordinary and partial differential equations of engineering and physics. If A is Hermitian and $\mathcal{W}_\lambda$ is the eigensubspace corresponding to one of the eigenvalues of A, then it is easy to see that the orthogonal complement $\mathcal{W}_\lambda^\perp$ is also invariant under A:

$$\bar{x} \in \mathcal{W}_\lambda \text{ and } \bar{z} \in \mathcal{W}_\lambda^\perp \Rightarrow \langle A\bar{z}, \bar{x} \rangle = \langle \bar{z}, A\bar{x} \rangle = 0$$
$$\Rightarrow A\bar{z} \in \mathcal{W}_\lambda^\perp. \tag{9.201}$$

Although this result holds whether or not the inner product is positive-definite, the usefulness of Eq. (9.201) is almost nil unless one can be sure that the only vector $\bar{x}$ that belongs simultaneously to $\mathcal{W}_\lambda$ and $\mathcal{W}_\lambda^\perp$ is the null vector, $\bar{x} = \bar{0}$. In the following derivation, therefore, either one must require the inner-product space to be unitary or, if one cannot avoid a pseudo-unitary space (as in special relativity), one must restrict oneself to Hermitian linear mappings that have no eigenvectors of zero norm.

Decomposition into Orthogonal Eigensubspaces

For the time being we take the simplest course and assume that $\mathcal{V}$ is a unitary space. The problem of calculating the components of the eigenvectors of A relative to an initial basis falls naturally into two parts: finding the eigenvectors that belong to unequal eigenvalues, and finding orthonormal bases of the degenerate eigensubspaces W_λ such that $\dim[W_\lambda] > 1$. If W_λ is the eigensubspace belonging to any eigenvalue λ of a Hermitian linear mapping A, then (by Eq. [9.201]) both W_λ and $W_\lambda^\perp$ are invariant subspaces of A. Equation (6.84) implies then that, relative to a basis of V that is a union of bases of W_λ and $W_\lambda^\perp$, the matrix of A is block-diagonal,

$$\mathbf{A} = \begin{pmatrix} \mathbf{A}_{W_\lambda} & \mathbf{0} \\ \mathbf{0} & \mathbf{A}_{W_\lambda^\perp} \end{pmatrix}, \tag{9.202}$$

in which $\mathbf{A}_{W_\lambda}$ is the matrix of the restriction of A to W_λ (as explained in the comments following Eq. [6.84]). Because W_λ is an eigensubspace of A, the matrix $\mathbf{A}_{W_\lambda}$ is a multiple of the identity matrix,

$$\mathbf{A}_{W_\lambda} = \lambda \mathbf{1}, \tag{9.203}$$

relative to a basis of eigenvectors of A.

If $\dim[W] > 1$, then one has to calculate an orthonormal basis of W by the Gram-Schmidt procedure. This is always possible, because we have assumed that V is a unitary space. We can assume, then, that we have found an orthonormal basis of W and have extended it to an orthonormal basis of V. Relative to this basis, the matrix of the restriction of A to the orthogonal complement of W, $A \upharpoonright W^\perp$, is Hermitian:

$$[\mathbf{A} \upharpoonright W^\perp]^H = \mathbf{A} \upharpoonright W^\perp. \tag{9.204}$$

Now that one has a Hermitian matrix that maps $W^\perp$ into itself, one can repeat for $\mathbf{A} \upharpoonright W^\perp$ the process just carried out for $\mathbf{A}$. Once again, if the dimension of the chosen eigensubspace is greater than 1, one uses the Gram-Schmidt procedure to find an orthonormal basis. Because V is finite-dimensional, this recursive procedure finally ends when $\mathbf{A}$ has been fully diagonalized. In the process, one finds an orthonormal basis of eigenvectors $\bar{e}_i$, such that

$$\begin{aligned} A\bar{e}_i &= \lambda_i \bar{e}_i, \\ \langle \bar{e}_i, \bar{e}_j \rangle &= \delta_{ij}, \\ V &= \mathrm{span}[\bar{e}_1, \ldots, \bar{e}_n]. \end{aligned} \tag{9.205}$$

Still supposing that A is a Hermitian linear mapping, let

$$A\bar{x} = \lambda \bar{x}, \quad \bar{x} \neq \bar{0}, \quad \lambda \in \mathbb{C}. \tag{9.206}$$

Then

$$\langle \bar{x}, A\bar{x} \rangle = \lambda \langle \bar{x}, \bar{x} \rangle = \langle A\bar{x}, \bar{x} \rangle = \lambda^* \langle \bar{x}, \bar{x} \rangle. \tag{9.207}$$

Cancelling the nonzero factor $\langle \bar{x}, \bar{x} \rangle$, one sees that every eigenvalue of a Hermitian linear mapping on a unitary space is real:

$$\lambda^* = \lambda. \tag{9.208}$$

Because of Eq. (9.208) and its analog for Hilbert spaces, one postulates in quantum mechanics that every physically observable quantity must correspond to a Hermitian linear operator defined on a Hilbert space. The eigenvalues of the Hamiltonian operator of a discrete quantum system are the energies of the system's stationary states.

It follows from Eq. (9.207) that eigenvectors $\bar{x}$ and $\bar{y}$ that belong to different eigenvalues λ, μ of a Hermitian linear mapping A are orthogonal:

$$
\begin{aligned}
\langle \bar{x}, A\bar{y} \rangle &= \mu \langle \bar{x}, \bar{y} \rangle \\
&= \langle A\bar{x}, \bar{y} \rangle \\
&= \lambda^* \langle \bar{x}, \bar{y} \rangle \\
&= \lambda \langle \bar{x}, \bar{y} \rangle,
\end{aligned}
\tag{9.209}
$$

implying that

$$
(\lambda - \mu)\langle \bar{x}, \bar{y} \rangle = 0.
\tag{9.210}
$$

Therefore either $\lambda = \mu$ or $\bar{x} \perp \bar{y}$.

Regardless of the degeneracy of an eigenvalue λ of a matrix $\mathbf{A}$, the eigenvector-eigenvalue equation reads

$$
(\mathbf{A} - \lambda \mathbf{1})\mathbf{x} = \mathbf{0}.
\tag{9.211}
$$

The equation states that every eigenvector $\mathbf{x}$ belongs to the null space of $\mathbf{A} - \lambda\mathbf{1}$. To find a basis of null$[\mathbf{A} - \lambda\mathbf{1}]$ one can use either the LU decomposition (see Section 6.6.5) or the singular-value decomposition (see Section 9.4.1). If one uses the LU decomposition to find a basis of a degenerate eigensubspace $\mathcal{W}_\lambda$, then one must do additional work to find an orthonormal basis of $\mathcal{W}_\lambda$. The textbook Gram-Schmidt method of orthogonalization is a mathematically correct but computationally unsound approach to the construction of an orthonormal basis of a degenerate eigensubspace $\mathcal{W}_\lambda$. The singular-value decomposition derived in Section 9.4.1 is preferable if $\dim[\mathcal{W}_\lambda]$ is greater than, say, two.

On a pseudo-unitary space the derivation leading up to Eq. (9.205) applies to any Hermitian linear mapping that has no eigenvectors of zero norm. For example, in special relativity only light-like vectors have zero norm. Therefore, in Minkowski space there exists a pseudoorthonormal basis of eigenvectors for every Hermitian linear mapping that has no light-like eigenvectors.

Matrix Version of the Orthogonal Decomposition

The matrix elements of a Hermitian linear mapping A relative to the orthonormal basis of eigenvectors $\{\bar{e}_i\}$ defined in Eq. (9.205) are

$$
a^i_j = a_{ij} = \langle \bar{e}_i, A\bar{e}_j \rangle = \lambda_j \langle \bar{e}_i, \bar{e}_j \rangle = \lambda_i \delta_{ij}
$$

$$
\Rightarrow \mathbf{A} = \Lambda := \begin{pmatrix} \lambda_1 & & \mathbf{0} \\ & \ddots & \\ \mathbf{0} & & \lambda_n \end{pmatrix} \quad (\mathrm{rel}\{\bar{e}_i\}).
\tag{9.212}
$$

If $\{\overline{f}_j\}$ is any other orthonormal basis of $\mathcal{V}$, then one can expand the eigenvector $\overline{e}_i$ as

$$\overline{e}_i = \overline{v}_i := v_i^j \overline{f}_j. \tag{9.213}$$

The column vector $\mathbf{v}_i$ is the realization $(\mathrm{rel}\{\overline{f}_j\})$ of the unit eigenvector $\overline{e}_j$. The matrix whose i-th column is $\mathbf{v}_i$,

$$\mathbf{V} = [\mathbf{v}_1, \ldots, \mathbf{v}_n], \tag{9.214}$$

is unitary, for the orthonormality of the basis implies that $\mathbf{V}^H = \mathbf{V}^\dagger$, and

$$\mathbf{V}^\dagger \mathbf{V} = [\mathbf{v}_i^\dagger \mathbf{v}_j] = [\delta_{ij}] = \mathbf{1} \Rightarrow \mathbf{V}^\dagger = \mathbf{V}^{-1}. \tag{9.215}$$

The eigenvalue equation

$$\mathbf{A}\mathbf{v}_i = \lambda_i \mathbf{v}_i \quad (i = 1, \ldots, n) \tag{9.216}$$

can be written compactly in matrix form as

$$\mathbf{A}\mathbf{V} = \mathbf{V}\Lambda. \tag{9.217}$$

By the unitarity of the eigenvector matrix $\mathbf{V}$, we have

$$\mathbf{V}^\dagger \mathbf{A}\mathbf{V} = \mathbf{V}^{-1}\mathbf{A}\mathbf{V} = \Lambda \tag{9.218}$$

and

$$\mathbf{A} = \mathbf{V}\Lambda\mathbf{V}^\dagger. \tag{9.219}$$

Therefore the matrix of eigenvectors, $\mathbf{V}$, is the matrix of a unitary similarity transformation that diagonalizes the Hermitian matrix $\mathbf{A}$.

The columns of the matrix of eigenvectors, $\mathbf{V}$, define a basis relative to which the matrix of $\mathbf{A}$ is diagonal. In this basis one can describe an intuitively appealing picture of the image of the unit sphere

$$\mathbf{x}^T \mathbf{x} = 1 \tag{9.220}$$

under a real symmetric matrix $\mathbf{A}$, all eigenvalues of which we assume are nonzero. Let

$$\mathbf{y} := \mathbf{A}\mathbf{x} \tag{9.221}$$

be the image of $\mathbf{x}$ under $\mathbf{A}$. Because all of the eigenvalues of $\mathbf{A}$ are real, the equation of the unit sphere becomes

$$(\mathbf{A}^{-1}\mathbf{y})^T \mathbf{A}^{-1}\mathbf{y} = 1 = \sum_{i=1}^{n} \left(\frac{y^i}{\lambda_i}\right)^2 \tag{9.222}$$

relative to a basis of eigenvectors. This equation describes a hyperellipsoid (that is, an n-dimensional ellipsoid) with semiaxes equal to $|\lambda_1|, \ldots, |\lambda_n|$. Therefore the magnitudes of the eigenvalues of a real symmetric matrix are equal to the semiaxes of the hyperellipsoid into which $\mathbf{A}$ carries the unit sphere (see Fig. 8.15).

Hermitian Matrices and Inner Products

The Gram matrix

$$\begin{aligned} \mathbf{G} &= [g_{ij}] \\ &= \mathbf{F}^\dagger \mathbf{F}, \end{aligned} \tag{9.223}$$

in which $\mathbf{F} = [\mathbf{f}_1, \ldots, \mathbf{f}_n]$ is the matrix of basis vectors, is Hermitian:

$$\mathbf{G}^\dagger = \mathbf{G}. \tag{9.224}$$

It follows that $\mathbf{G}$ is self-adjoint, for

$$\mathbf{G}^H = \mathbf{G}^{-1}\mathbf{G}^\dagger\mathbf{G} = \mathbf{G}. \tag{9.225}$$

Then the eigenvalues $\gamma_1, \ldots, \gamma_n$ of $\mathbf{G}$ are real. Let $\mathbf{U}$ be the matrix of orthonormal eigenvectors of $\mathbf{G}$. Then

$$\mathbf{G} = \mathbf{U}\mathbf{\Gamma}\mathbf{U}^\dagger \tag{9.226}$$

in which $\mathbf{\Gamma} = \mathrm{diag}[\gamma_1, \ldots, \gamma_n]$. In terms of $\mathbf{\Gamma}$, the inner product is

$$\langle \bar{y}, \bar{x} \rangle = \mathbf{y}^\dagger \mathbf{G} \mathbf{x} = (\mathbf{U}^\dagger \mathbf{y})^\dagger \mathbf{\Gamma} \mathbf{U}^\dagger \mathbf{x}. \tag{9.227}$$

Because $\mathbf{x}$ and $\mathbf{y}$ are arbitrary vectors, it follows that: *The inner product on $\mathbb{C}^n$ defined by a Hermitian matrix $\mathbf{G}$ is positive-definite if and only if every eigenvalue of $\mathbf{G}$ is positive. The inner product is semidefinite if and only if every eigenvalue of $\mathbf{G}$ is nonnegative. The inner product is indefinite if and only if some eigenvalues of $\mathbf{G}$ are positive and the rest are negative.* The number of positive eigenvalues of $\mathbf{G}$ minus the number of negative eigenvalues is called the **signature** of the inner product.

Principal-Axis Theorem

A **homogeneous real quadratic form** in n variables is a polynomial in $x^1, \ldots, x^n$ in which each term is quadratic and the coefficient of each term is real, such as

$$q(x^1, \ldots, x^n) = x^i b_{ij} x^j = \mathbf{x}^T \mathbf{B} \mathbf{x}. \tag{9.228}$$

Homogeneous real quadratic forms that are important in physics and engineering include the electromagnetic energy density,

$$\begin{aligned} u(\mathbf{r}, t) &= \frac{1}{2}(\mathbf{E} \cdot \mathbf{D} + \mathbf{H} \cdot \mathbf{B}) \\ &= \frac{1}{2} \sum_{i=1}^{3} \sum_{j=1}^{3} (E^i \epsilon_{ij} E^j + H^i \mu_{ij} H^j), \end{aligned} \tag{9.229}$$

in which ϵ_{ij} and μ_{ij} are elements of the electric-permittivity and magnetic-permeability tensors, and the kinetic energy of a rigid body,

$$T = \frac{1}{2} \sum_{i=1}^{3} \omega^i I_{ij} \omega^j, \tag{9.230}$$

in which I_{ij} is an element of the moment-of-inertia tensor. The word *tensor*, which is traditional in the context of these and other quadratic forms, can be replaced by *matrix* for our purposes.

The matrix of coefficients $\mathbf{B} = [b_{ij}]$ in any homogeneous real quadratic form can be written trivially as the sum of two matrices $\mathbf{A}$ and $\mathbf{S}$, one of which is antisymmetric and the other of which is symmetric:

$$\mathbf{B} = \mathbf{A} + \mathbf{S}, \tag{9.231}$$

$$\mathbf{A} := \tfrac{1}{2}(\mathbf{B} - \mathbf{B}^T), \tag{9.232}$$

$$\mathbf{S} := \tfrac{1}{2}(\mathbf{B} + \mathbf{B}^T), \tag{9.233}$$

$$\mathbf{A}^T = -\mathbf{A}, \tag{9.234}$$

$$\mathbf{S}^T = \mathbf{S}. \tag{9.235}$$

Clearly only the symmetric combinations

$$s_{ij} = \tfrac{1}{2}(b_{ij} + b_{ji}) \tag{9.236}$$

of the original coefficients contribute to the quadratic form $q = x^i b_{ij} x^j$, because the contribution from the antisymmetric coefficients $\mathbf{A}$ changes sign if the dummy summation indices i and j are interchanged. Therefore one can assume without loss of generality that the matrix of coefficients of a homogeneous real quadratic form is symmetric.

Every real symmetric matrix $\mathbf{B}$ defines both a real, homogeneous quadratic form and an inner product. It is trivial to check that

$$\langle\!\langle \mathbf{y}, \mathbf{x} \rangle\!\rangle := \mathbf{y}^T \mathbf{B} \mathbf{x} \tag{9.237}$$

satisfies all of the axioms of an inner product. Therefore one calls a quadratic form **positive-definite** or **semidefinite** if and only if the corresponding inner product is positive definite or semidefinite. The **signature** of a quadratic form is the signature of the corresponding inner product.

The theorem that every self-adjoint matrix can be diagonalized by a unitary similarity transformation can be restated as the **principal-axis theorem**: *For every real quadratic form q there exists an orthonormal basis in which q is diagonal,*

$$q = x'^T \mathbf{B}' x',$$

$$x' := \mathbf{V}^T x,$$

$$\mathbf{B}' := \mathbf{V}^T \mathbf{B} \mathbf{V} \tag{9.238}$$

$$= \text{diag}[\lambda_1, \ldots, \lambda_n],$$

$$\mathbf{V}\mathbf{V}^T = \mathbf{V}^T \mathbf{V} = \mathbf{1}.$$

The columns of the matrix $\mathbf{V}$, which define an orthonormal basis relative to which the matrix of the linear mapping defined by $\mathbf{B}$ is diagonal, are called the **principal axes** of $\mathbf{B}$. The intuitive meaning of the principal axes if the eigenvalues λ_i of $\mathbf{B}$ are distinct is that

$$\mathbf{y} = \mathbf{Bx} \tag{9.239}$$

is parallel to $\mathbf{x}$ if and only if $\mathbf{x}$ is parallel to one of the principal axes, that is, if and only if $\mathbf{x} = \mathbf{v}_i$ for some i. (The equation $\mathbf{y} = \mathbf{Bx} = \beta\mathbf{x}$ implies that $\mathbf{x}$ is an eigenvector of $\mathbf{B}$.)

For example, in principal axes the rotational kinetic energy of a rigid body is

$$T = \tfrac{1}{2}\left(I_1\omega_1^2 + I_2\omega_2^2 + I_3\omega_3^2\right) \tag{9.240}$$

and the components of the angular momentum $\mathbf{L} = \mathbf{I}\omega$ are

$$L_1 = I_1\omega_1, \quad L_2 = I_2\omega_2, \quad L_3 = I_3\omega_3. \tag{9.241}$$

In principal axes, the expression for the rotational kinetic energy in terms of the angular momentum is also simple:

$$T = \frac{1}{2}\left(\frac{L_1^2}{I_1} + \frac{L_2^2}{I_2} + \frac{L_3^2}{I_3}\right). \tag{9.242}$$

If the principal moments of inertia I_1, I_2, I_3 are distinct, then $\mathbf{L}$ is parallel to ω if and only if two of the angular-velocity components $\omega_1, \omega_2, \omega_3$ relative to the principal axes vanish and the third angular-velocity component is nonzero.

If the coefficient matrix $\mathbf{B}$ of a quadratic form q is nonsingular, then one can associate n-dimensional conic sections with $\mathbf{B}$ and q. For example, the equation

$$T = \text{constant} \tag{9.243}$$

defines an ellipsoid in ω space with semiaxes equal to

$$a = \sqrt{\frac{2T}{I_1}}, \quad b = \sqrt{\frac{2T}{I_2}}, \quad c = \sqrt{\frac{2T}{I_3}}. \tag{9.244}$$

The same equation $T = \text{constant}$ defines another ellipsoid in $\mathbf{L}$ space with semiaxes equal to

$$a = \sqrt{2T I_1}, \quad b = \sqrt{2T I_2}, \quad c = \sqrt{2T I_3}. \tag{9.245}$$

In general, if $\mathbf{B}$ is positive definite then the equation $q = 1$ becomes, relative to principal axes and in terms of the components of $\mathbf{x}$,

$$q = \mathbf{x}^T\mathbf{Bx} = \sum_{i=1}^{n}\lambda_i(x^i)^2 = 1. \tag{9.246}$$

This equation defines the surface of a hyperellipsoid with semiaxes equal to $(\lambda_1)^{-1/2}, \ldots,$ $(\lambda_n)^{-1/2}$. In terms of the components of $\mathbf{y} = \mathbf{Bx}$ relative to principal axes, the equation $q = 1$

reads

$$q = (\mathbf{B}^{-1}\mathbf{y})^T \mathbf{BB}^{-1}\mathbf{y} = \sum_{i=1}^{n} \frac{(y^i)^2}{\lambda_i},$$

(9.247)

which defines a hyperellipsoid with semiaxes equal to $(\lambda_1)^{1/2}, \ldots, (\lambda_n)^{1/2}$. If some eigenvalues of $\mathbf{B}$ are positive and some are negative, then $\mathbf{B}$ defines an n-dimensional hyperboloid.

Spectral Decomposition of a Self-adjoint Linear Mapping

It is useful to have a projector formulation of the diagonalization of a self-adjoint linear mapping on a Hermitian space. Let P_i project on the subspace $\mathcal{W}_i$ spanned by $\bar{e}_i$ (Eq. [9.205]). The orthogonality of the basis vectors $\{\bar{e}_i\}$ implies the orthogonality of the projectors P_i. Also,

$$\sum_{i=1}^{n} P_i = 1.$$

(9.248)

Equation (9.212) implies the **spectral decomposition** of A:

$$A = \sum_{i=1}^{n} \lambda_i P_i.$$

(9.249)

Equation (9.249) follows from the observations that

$$\bar{x} \in \mathcal{V} \Rightarrow \bar{x} = x^i \bar{e}_i = \sum_i P_i \bar{x}$$

(9.250)

and that

$$A\bar{x} = A(x^i \bar{e}_i) = x^i A\bar{e}_i = \sum_i x^i \lambda_i \bar{e}_i = \sum_i \lambda_i P_i \bar{x}.$$

(9.251)

9.3.3 NORMAL LINEAR MAPPINGS

The derivation for a self-adjoint linear mapping leads us to ask, "What are the necessary and sufficient conditions for the diagonalizability of a general linear mapping A on a Hermitian space?" Many answers exist, depending on the properties one requires of the basis of eigenvectors that results from diagonalizing A. Suppose for the moment that we limit ourselves to *orthonormal* bases of eigenvectors. A linear mapping is called **normal** if and only if it commutes with its Hermitian adjoint:

$$AA^H = A^H A.$$

(9.252)

We generalize the invariance of the orthogonal complement of an eigensubspace, $\mathcal{W}_\lambda^\perp$, to normal linear mappings.

If a linear mapping A has an orthonormal basis $\{\bar{e}_i\}$ of eigenvectors, then Eq. (9.250) holds and it is easy to show that A is normal. To find the Hermitian adjoint of A, one needs the Hermitian adjoint of $\lambda_i P_i$. It follows from Eqs. (9.153) and (9.154) that

$$(\lambda_i P_i)^H = G^{-1}(\lambda_i P_i)^T G = \lambda_i^* G^{-1} P_i^T G = \lambda_i^* G^{-1} G P_i G^{-1} G = \lambda_i^* P_i. \tag{9.253}$$

Therefore

$$A^H = \sum_{i=1}^{n} \lambda_i^* P_i, \tag{9.254}$$

and

$$A A^H = \sum_{i,j=1}^{n} \lambda_i \lambda_j^* P_i P_j = \sum_{i,j=1}^{n} \lambda_j^* \lambda_i P_j P_i = A^H A, \tag{9.255}$$

in which the last equality follows from the orthogonality of the $\{P_i\}$, Eq. (8.285).

Whether or not A is normal, every eigensubspace $\mathcal{W}_\lambda$ is invariant under A. We turn to the slightly more complicated task of showing that if A is normal, then $\mathcal{W}_\lambda^\perp$ is also invariant under A. We begin by establishing the nature of the spectrum of the adjoint of a normal linear mapping. If one replaces P_i by $\mathbf{1}$ in Eq. (9.253), one sees that

$$(A - \lambda \mathbf{1})^H = A^H - \lambda^* \mathbf{1}. \tag{9.256}$$

Then $B := A - \lambda \mathbf{1}$ is normal if and only if A is normal. Because B is normal, one has, for every $\bar{x} \in \mathcal{V}$,

$$\begin{aligned}
\|B\bar{x}\|^2 &= \langle B\bar{x}, B\bar{x} \rangle \\
&= \langle B^H B\bar{x}, \bar{x} \rangle \\
&= \langle B B^H \bar{x}, \bar{x} \rangle \\
&= \langle B^H \bar{x}, B^H \bar{x} \rangle \\
&= \|B^H \bar{x}\|^2.
\end{aligned} \tag{9.257}$$

Then

$$\begin{aligned}
(A - \lambda \mathbf{1})\bar{x} = \bar{0} &\Leftrightarrow \|(A - \lambda \mathbf{1})\bar{x}\|^2 = 0 \\
&\Leftrightarrow \|(A^H - \lambda^* \mathbf{1})\bar{x}\|^2 = 0 \\
&\Leftrightarrow (A^H - \lambda^* \mathbf{1})\bar{x} = \bar{0}.
\end{aligned} \tag{9.258}$$

Therefore an eigenvector $\bar{x}$ corresponding to the eigenvalue λ of a normal linear mapping A is also an eigenvector of A^H corresponding to λ^*.

Now let $\bar{x}$ belong to the eigensubspace $\mathcal{W}_\lambda$ and let $\bar{z}$ belong to $\mathcal{W}_\lambda^\perp$. Recalling that λ need not be real, one has

$$\langle \bar{x}, A\bar{z} \rangle = \langle A^H \bar{x}, \bar{z} \rangle = \lambda \langle \bar{x}, \bar{z} \rangle = 0 \Rightarrow A\bar{z} \in \mathcal{W}_\lambda^\perp. \tag{9.259}$$

Then $\mathcal{W}_\lambda^\perp$ is invariant under A, as claimed, and therefore one can construct an orthonormal basis of eigenvectors by exactly the same procedure we used for a self-adjoint linear mapping. It follows that: *A linear mapping* A *on a unitary space has a spectral decomposition Eq. (9.249) in terms of orthogonal projectors if and only if* A *is normal. Equivalently, a matrix* **A** *can be diagonalized by a unitary similarity transformation if and only if* **A** *is normal.*

From the definition in Eqs. (9.159) and (9.160) it follows that a unitary mapping U is normal. Therefore one can choose an orthonormal basis of eigenvectors of U (or, equivalently, one can diagonalize the matrix realization **U** by a unitary similarity transformation). From Eq. (9.249) (for normal linear mappings) and Eq. (9.254) it follows that an eigenvalue λ of a unitary transformation must obey the equation $|\lambda|^2 = 1$. Therefore *the eigenvalues of a unitary or orthogonal linear mapping or its matrix are complex numbers of modulus one,* $\lambda = e^{i\theta}$.

Nonnormal Linear Mappings

Even if a linear mapping A on a unitary space is not normal, it may be possible to find a nonorthogonal basis of eigenvectors of A. If so, then the matrix of A is diagonal relative to that nonorthogonal basis. Conversely, if **A** $= \text{diag}[\lambda_1, \ldots, \lambda_n]$ in some basis $\{\overline{v}_i\}$ of $\mathcal{V}$, then each of the basis vectors is an eigenvector. Let $\overline{u}$ be any eigenvector of A, and let λ be its eigenvalue. Because $\{\overline{v}_i\}$ is a basis of eigenvectors,

$$\overline{u} = \sum_{i=1}^{n} u^i \overline{v}_i \Rightarrow A\overline{u} = \sum_{i=1}^{n} u^i A\overline{v}_i = \sum_{i=1}^{n} \lambda_i u^i \overline{v}_i. \tag{9.260}$$

But $\overline{u}$ is an eigenvector of A:

$$A\overline{u} = \lambda \overline{u} \Rightarrow \sum_{i=1}^{n}(\lambda_i - \lambda)u^i \overline{v}_i = \overline{0} \Rightarrow \forall i = 1, \ldots, n : (\lambda_i - \lambda)u^i = 0. \tag{9.261}$$

Because $\overline{u}$ must be nonzero to be an eigenvector, u^i cannot vanish for all i. Then $\lambda_i = \lambda$ for some i and $u^j = 0$ for $\lambda_j \neq \lambda$. We have shown that: *A square matrix* **A** *over* $\mathbb{C}$ *is similar to a diagonal matrix if and only if the eigenvectors of the linear mapping* A *span* $\mathcal{V}$. *If there exists a basis in which* **A** *is diagonal, then every eigenvalue of* A *is one of the diagonal elements of* **A**.

Because we have made no use of inner products in deriving this result, it holds also if the inner-product space is pseudo-unitary.

9.3.4 EXERCISES FOR SECTION 9.3

9.3.1 Let $\mathcal{V} = l^2(\mathbb{C})$, let a be the harmonic-oscillator annihilation operator defined in Eq. (6.96), and let $\overline{x}$ be the sequence in $l^2(\mathbb{C})$ whose n-th element is

$$x^n = e^{-|\alpha|^2/2} \frac{\alpha^{n-1}}{\sqrt{(n-1)!}}, \tag{9.262}$$

in which $\alpha \in \mathbb{C}$.

(a) Show that

$$\|\bar{x}\|_2 = 1 \tag{9.263}$$

in which $\|\cdot\|_2$ is the 2-norm in $l^2(\mathbb{C})$,

$$\|\bar{x}\|_2^2 := \sum_{n=1}^{\infty} |x^n|^2. \tag{9.264}$$

(b) Show that

$$\mathbf{a}\bar{x} = \alpha\bar{x}. \tag{9.265}$$

(A unit eigenvector of the annihilation operator $\mathbf{a}$ is called a **coherent state**.)

*(c) Prove that $\mathbf{a}^\dagger$ has no eigenvectors in $l^2(\mathbb{C})$.

9.3.2 Consider the linear operator $C : l^2(\mathbb{C}) \to l^2(\mathbb{C})$ such that

$$C(x^1, x^2, \ldots, x^n, \ldots) := (x^2, x^1 + x^3, \ldots, x^{n-1} + x^{n+1}, \ldots), \tag{9.266}$$

in which x^j is the j-th coordinate of $\bar{x} \in l^2(\mathbb{C})$. Show that if $\bar{x} \neq \bar{0}$ and if

$$C\bar{x} = \lambda\bar{x} \tag{9.267}$$

in which $\lambda \neq 0$, then $\|\bar{x}\|_2^2$ is infinite.

Hint: Calculate the first six components of $\bar{x}$ in terms of x^1, and then generalize. Conclude that C has no normalizable eigenvectors in $l^2(\mathbb{C})$.

9.3.3 An $n \times n$ matrix $\mathbf{A}$ over $\mathbb{R}$ has the following matrix elements, relative to an *orthonormal* basis:

$$a_{ij} = x_i x_j. \tag{9.268}$$

Show that the eigenvalues of $\mathbf{A}$ are

$$\lambda = 0 \quad (n - 1 \text{ times}) \tag{9.269}$$

and

$$\lambda = \|\bar{x}\|^2 = (x_1)^2 + \cdots + (x_n)^2 \quad (\text{once}). \tag{9.270}$$

Hint: Express $\mathbf{A}$ as a dyad.

9.3.4 Obtain the eigenvalues, and a set of orthonormal eigenvectors, for each of the nondiagonal Pauli spin matrices,

$$\sigma_x = \begin{pmatrix} 0 & 1 \\ 1 & 0 \end{pmatrix}, \quad \sigma_y = \begin{pmatrix} 0 & -i \\ i & 0 \end{pmatrix}. \tag{9.271}$$

9.3.5   Find the eigenvalues, and a set of orthonormal eigenvectors, of the matrix

$$
\mathbf{A} = \begin{pmatrix} 1 & -1 & -1 \\ -1 & 1 & -1 \\ -1 & -1 & 1 \end{pmatrix}. \tag{9.272}
$$

Hint: If you can find two orthonormal eigenvectors $\mathbf{u}$ and $\mathbf{v}$, then the third can be taken as $\mathbf{u} \times \mathbf{v}$.

9.3.6   Find the eigenvalues, and three column vectors that constitute an orthonormal set of eigenvectors, of the matrix

$$
\begin{pmatrix} 1 & 1 & 1 \\ 1 & 1 & 1 \\ 1 & 1 & 1 \end{pmatrix}. \tag{9.273}
$$

9.3.7   For each of the following defective matrices, find all of the eigenvalues, the algebraic multiplicity of each eigenvalue, a linearly independent set of eigenvectors, and the geometrical multiplicity of each eigenvalue:

$$
\begin{pmatrix} 0 & 1 \\ 0 & 0 \end{pmatrix}, \quad \begin{pmatrix} 1 & 1 \\ 0 & 1 \end{pmatrix}, \quad \begin{pmatrix} 0 & 0 & 1 \\ 0 & 0 & 0 \\ 0 & 0 & 0 \end{pmatrix}, \quad \begin{pmatrix} 0 & 1 & 0 \\ 0 & 0 & 1 \\ 0 & 0 & 0 \end{pmatrix}. \tag{9.274}
$$

9.3.8   Let a Gram matrix $\mathbf{G} = \mathbf{F}^{\dagger}\mathbf{F}$ be positive-definite, let the eigenvalues of $\mathbf{G}$ be γ_i (in which $i \in (1 : n)$), and let

$$
\boldsymbol{\Gamma} = \mathrm{diag}[\gamma_1, \ldots, \gamma_n]. \tag{9.275}
$$

(a) Show that if $\mathbf{U}$ is the matrix the columns of which are the orthonormal eigenvectors of $\mathbf{G}$, then $\mathbf{U}$ is unitary and

$$
\boldsymbol{\Gamma} = \mathbf{U}^{\dagger}\mathbf{G}\mathbf{U}. \tag{9.276}
$$

(b) Show that it is possible to define a diagonal matrix with positive elements, $\boldsymbol{\Gamma}^{1/2}$, such that

$$
\left(\boldsymbol{\Gamma}^{1/2}\right)^2 = \boldsymbol{\Gamma}. \tag{9.277}
$$

(c) Show that the columns of the matrix $\mathbf{F}\mathbf{U}\boldsymbol{\Gamma}^{-1/2}$ are an orthonormal basis of vectors, relative to which the Gram matrix is the identity matrix.

Hint: Prove that the matrix $\mathbf{F}\mathbf{U}\boldsymbol{\Gamma}^{-1/2}$ is unitary.

9.3.9   Let

$$
\mathbf{u} = \begin{pmatrix} 1 \\ 0 \\ -1 \end{pmatrix}, \quad \mathbf{v} = \begin{pmatrix} 1 \\ 2 \\ 3 \end{pmatrix}. \tag{9.278}
$$

(a) Calculate

$$A = vu^T. \tag{9.279}$$

(b) Without writing down or solving a secular equation, determine the eigen-vectors v_1, v_2, v_3 and eigenvalues $\lambda_1, \lambda_2, \lambda_3$ of A. Ensure that eigenvectors that belong to degenerate eigenvalues (if there are any) are orthogonal.

(c) Find a matrix Z such that

$$Z^{-1}AZ = \begin{pmatrix} \lambda_1 & 0 & 0 \\ 0 & \lambda_2 & 0 \\ 0 & 0 & \lambda_3 \end{pmatrix}. \tag{9.280}$$

(d) Show that A is not a normal matrix. Explain why it is possible to diagonalize A despite the fact that A is not normal.

9.3.10 Let a matrix $A \in \mathbb{R}^{n \times n}$ be defined by Eq. (9.279), in which u and v are general vectors in $\mathbb{R}^n$ such that

$$u^T v \neq 0. \tag{9.281}$$

(a) Without writing down or solving a secular equation, determine the eigen-vectors z_i and eigenvalues λ_i of A, in which $i \in (1 : n)$.

(b) Show that there exists a matrix Z such that

$$Z^{-1}AZ = \text{diag}[\lambda_1, \ldots, \lambda_n]. \tag{9.282}$$

Is Z unitary? Why, or why not?

(c) Show that A is not a normal matrix. Explain why it is possible to diagonalize A despite the fact that A is not normal.

9.4 SINGULAR-VALUE DECOMPOSITION

9.4.1 DERIVATION OF THE SINGULAR-VALUE DECOMPOSITION

Our results on the diagonalizability of some linear mappings that map V into itself can be generalized, in a certain sense, to any linear mapping $A : V_{(1)} \rightarrow V_{(2)}$. The matrix A of such a linear mapping is rectangular, with dimensions $n_2 \times n_1$. For convenience we assume in the following that $n_1 \leq n_2$. If this is not the case, then we simply work with A^H instead of A. We also assume from now on that $V_{(1)}$ and $V_{(2)}$ are both unitary spaces, because we need to be able to assume that every nonzero vector has a nonzero norm in order to be sure that A is bounded.

The linear mapping $A^H A$ is Hermitian $((A^H A)^H = A^H A)$. Therefore there exists an or-thonormal basis $\{\bar{v}_i\}$ of $V_{(1)}$ such that every basis vector $\bar{v}_i$ is an eigenvector of $A^H A$:

$$A^H A \bar{v}_i = \lambda_i \bar{v}_i. \tag{9.283}$$

The vectors $\{\overline{v}_i\}$ are called the **right singular vectors** of A. One can assume that the eigenvalues and their corresponding eigenvectors have been ordered so that

$$\lambda_1 \geq \lambda_2 \geq \cdots \geq \lambda_{n_1}. \tag{9.284}$$

The mapping $A^H A$ is positive-semidefinite, for

$$\forall \overline{x} \in \mathcal{V}_{(1)} : \langle \overline{x}, A^H A \overline{x} \rangle = \langle A \overline{x}, A \overline{x} \rangle \geq 0. \tag{9.285}$$

Equality holds if and only if $\overline{x}$ belongs to null[A]. It follows that every eigenvalue λ_i is either positive or zero.

Let us see what happens if we take apart the product mapping $A^H A$ by performing first A, then A^H. Let

$$\overline{w}_i := A \overline{v}_i \tag{9.286}$$

for $i = 1, 2, \ldots, n_1 = \dim[\mathcal{V}_{(1)}]$. Clearly the list $\{\overline{w}_i\}$ spans range[A]. By the definition of the eigenvalues λ_i,

$$A^H \overline{w}_i = \lambda_i \overline{v}_i. \tag{9.287}$$

Every $\overline{w}_i$ is an eigenvector of AA^H with the eigenvalue λ_i:

$$AA^H \overline{w}_i = \lambda_i A \overline{v}_i = \lambda_i \overline{w}_i. \tag{9.288}$$

We may not have found all of the eigenvectors of AA^H yet, for $n_2 \geq n_1$. However, it turns out that we have found all of the eigenvectors which correspond to *nonzero* eigenvalues of AA^H.

From the orthogonality of the eigenvectors $\overline{v}_i$ of $A^H A$, one deduces that the eigenvectors $\{\overline{w}_i\}$ of AA^H that correspond to nonzero eigenvalues λ_i are mutually orthogonal:

$$\langle \overline{w}_i, \overline{w}_j \rangle_{(2)} = \langle A \overline{v}_i, A \overline{v}_j \rangle_{(2)} = \langle \overline{v}_i, A^H A \overline{v}_i \rangle_{(1)} = \lambda_i \delta_{ij}. \tag{9.289}$$

Clearly the set $\{\overline{w}_i \mid \lambda_i > 0\}$ is linearly independent, for we have assumed that both $\mathcal{V}_{(1)}$ and $\mathcal{V}_{(2)}$ are unitary spaces, in which mutual orthogonality implies linear independence.

From the linearly independent set $\{\overline{w}_i \mid \lambda_i > 0\}$ one can select a basis of range[A]. Let r be the greatest value of i such that $\lambda_i > 0$. Then one can define

$$\forall i \in (1 : r) : \sigma_i := \sqrt{\lambda_i}, \tag{9.290}$$

$$\forall i \in (1 : r) : \overline{u}_i := \frac{1}{\sigma_i} \overline{w}_i = \frac{1}{\sigma_i} A \overline{v}_i, \tag{9.291}$$

and

$$\forall i \in (r + 1 : n_1) : \sigma_i := 0. \tag{9.292}$$

By Eqs. (9.286) and (9.291), $\{\overline{u}_i \mid i \in (1 : r)\}$ spans range[A]. Also, the set $\{\overline{u}_i\}$ is orthonormal and is therefore linearly independent:

$$\langle \overline{u}_i, \overline{u}_j \rangle_{(2)} = \frac{1}{\sigma_i \sigma_j} \langle \overline{w}_i, \overline{w}_j \rangle_{(2)} = \delta_{ij}. \tag{9.293}$$

It follows that $\{\overline{u}_i \mid i \in (1:r)\}$ is a basis of range[A] and that

$$r = \text{rank}[A]. \tag{9.294}$$

The nonnegative numbers σ_i are called the **singular values** of A. The name is appropriate because the mapping $A - \sigma_i \mathbf{1}$ is of lower rank than A for every nonzero σ_i.

Let us extend the basis $\{\overline{u}_i \mid i \in (1:r)\}$ of the range of A to an orthonormal basis $\{\overline{u}_i \mid i \in (1:n_2)\}$ of $\mathcal{V}_2$. Then $\{\overline{u}_i \mid i \in (r+1:n_2)\}$ is a basis of $\{\text{range}[A]\}^{\perp}$, and

$$\forall i \in (1:n_1) : A\overline{v}_i = \sigma_i \overline{u}_i \tag{9.295}$$

because $\sigma_i = 0$ for every $i > r$. Therefore $\{\overline{v}_i \mid i \in (r+1:n_1)\}$ is a basis of null[A] and $\{\overline{v}_i \mid i \in (1:r)\}$ is a basis of $\{\text{null}[A]\}^{\perp}$.

One sees at once that

$$\forall i \in (1:r) : A^H \overline{u}_i = \frac{1}{\sigma_i} A^H \overline{w}_i = \sigma_i \overline{v}_i. \tag{9.296}$$

One can see now that for every $j \in (1:n_1)$ and for every $i \in (r+1:n_2)$,

$$\langle \overline{v}_j, A^H \overline{u}_i \rangle = \langle A\overline{v}_j, \overline{u}_i \rangle = \begin{cases} \langle \overline{w}_j, \overline{u}_i \rangle = \sigma_j \langle \overline{u}_j, \overline{u}_i \rangle = 0, & \text{if } j \leq r; \\ \langle \overline{0}, \overline{u}_i \rangle = 0, & \text{if } j > r. \end{cases} \tag{9.297}$$

Then

$$\forall i \in (r+1:n_2) : A^H \overline{u}_i = \overline{0}_{(1)}. \tag{9.298}$$

Because $\sigma_i = 0$ for all $i \in (r+1:n_1)$, we can restate our results in the equations

$$\forall i \in (1:n_1) : A^H \overline{u}_i = \sigma_i \overline{v}_i \tag{9.299}$$

$$\forall i \in (n_1+1:n_2) : A^H \overline{u}_i = \overline{0}_{(1)}. \tag{9.300}$$

The vectors $\{\overline{u}_i\}$ are called the **left singular vectors** of A.

9.4.2 MATRIX VERSION OF THE SINGULAR-VALUE DECOMPOSITION

To obtain a matrix realization of these results, we choose orthonormal bases $\{\overline{e}_i \mid i \in (1:n_1)\}$ in $\mathcal{V}_{(1)}$ and $\{\overline{e}'_k \mid k \in (1:n_2)\}$ in $\mathcal{V}_{(2)}$. Let the coordinate expansions of $\overline{v}_i$ and $\overline{u}_k$ be

$$\overline{v}_i = v_i^j \overline{e}_j, \quad \overline{u}_k = u_k^l \overline{e}'_l. \tag{9.301}$$

The matrices

$$
\mathbf{V} := \begin{array}{c} \\ 1 \\ \vdots \\ n_1 \end{array} \begin{pmatrix} v_1^1 & \cdots & v_{n_1}^1 \\ \vdots & \ddots & \vdots \\ v_1^{n_1} & \cdots & v_{n_1}^{n_1} \end{pmatrix}, \qquad \mathbf{U} := \begin{array}{c} \\ 1 \\ \vdots \\ n_2 \end{array} \begin{pmatrix} u_1^1 & \cdots & u_{n_2}^1 \\ \vdots & \ddots & \vdots \\ u_1^{n_2} & \cdots & u_{n_2}^{n_2} \end{pmatrix} \tag{9.302}
$$

are unitary because of the orthonormality of the sets $\{\bar{v}_i\}$ and $\{\bar{u}_k\}$. The n_1 equations $A\bar{v}_i = \sigma_i \bar{u}_i$ are equivalent to the single matrix equation

$$
\mathbf{AV} = \mathbf{U\Sigma} \tag{9.303}
$$

in which

$$
\Sigma = \begin{array}{c} \\ 1 \\ \vdots \\ n_1 \\ n_1 + 1 \\ \vdots \\ n_2 \end{array} \begin{pmatrix} \sigma_1 & \cdots & 0 \\ \vdots & \ddots & \vdots \\ 0 & \cdots & \sigma_{n_1} \\ 0 & \cdots & 0 \\ \vdots & \ddots & \vdots \\ 0 & \cdots & 0 \end{pmatrix}. \tag{9.304}
$$

It follows at once that

$$
\boxed{\mathbf{A} = \mathbf{U\Sigma V}^H.} \tag{9.305}
$$

This factorization of an arbitrary matrix $\mathbf{A} \in \mathbb{C}^{n_2 \times n_1}$ is called the **singular-value decomposition (SVD)**. The SVD for $\mathbf{A}^H$,

$$
\mathbf{A}^H = \mathbf{V\Sigma}^H \mathbf{U}^H, \tag{9.306}
$$

follows if one interchanges $\mathcal{V}_{(1)}$ and $\mathcal{V}_{(2)}$, which means that one must replace $\mathbf{A}$, $\mathbf{U}$, and $\mathbf{V}$ with $\mathbf{A}^H$, $\mathbf{V}$, and $\mathbf{U}$, respectively.

If the matrix elements of $\mathbf{A}$, $\mathbf{U}$, and $\mathbf{V}$ are computed in orthonormal bases, then one can replace $\mathbf{A}^H$ with $\mathbf{A}^\dagger$, and so forth, in the SVD. For example,

$$
\mathbf{A} = \begin{pmatrix} 0 & 2 \\ 0 & 0 \\ 3 & 0 \end{pmatrix} \Rightarrow \mathbf{A}^H = \begin{pmatrix} 0 & 0 & 3 \\ 2 & 0 & 0 \end{pmatrix} = \mathbf{A}^\dagger \tag{9.307}
$$

in $\mathbb{E}^2$ and $\mathbb{E}^3$. Then

$$
\mathbf{A}^H \mathbf{A} = \begin{pmatrix} 9 & 0 \\ 0 & 4 \end{pmatrix}. \tag{9.308}
$$

The eigenvectors that belong to the eigenvalues $\lambda_1 = 9$ and $\lambda_2 = 4$ are

$$
\mathbf{v}_1 = \begin{pmatrix} 1 \\ 0 \end{pmatrix}, \quad \mathbf{v}_2 = \begin{pmatrix} 0 \\ 1 \end{pmatrix}. \tag{9.309}
$$

The singular values are $\sigma_1 = 3$ and $\sigma_2 = 2$; hence the rank of $\mathbf{A}$ is $r = 2$. One easily finds that for $i = 1, 2$, $\mathbf{A}\mathbf{v}_i = \sigma_i \mathbf{u}_i$ in which

$$\mathbf{u}_1 = \begin{pmatrix} 0 \\ 0 \\ 1 \end{pmatrix}, \quad \mathbf{u}_2 = \begin{pmatrix} 1 \\ 0 \\ 0 \end{pmatrix}. \tag{9.310}$$

We choose

$$\mathbf{u}_3 = \begin{pmatrix} 0 \\ 1 \\ 0 \end{pmatrix}, \tag{9.311}$$

which is orthogonal to both $\mathbf{u}_1$ and $\mathbf{u}_2$. Then $\mathbf{A}^\dagger \mathbf{u}_i = \sigma_i \mathbf{v}_i$ for $i = 1, 2$, and $\mathbf{A}^\dagger \mathbf{u}_3 = \mathbf{0}$.

9.4.3 THE FUNDAMENTAL SUBSPACES OF A LINEAR MAPPING

One of the remarkable aspects of the SVD is that it gives us the ranges and null spaces of A and A^H as the spaces that are spanned by the column partitions

$$\mathbf{U} = \left(\underbrace{\mathbf{u}_1, \ldots, \mathbf{u}_r}_{\text{basis of range}[A]}, \underbrace{\mathbf{u}_{r+1}, \ldots, \mathbf{u}_{n_2}}_{\text{basis of null}[A^H]} \right) \tag{9.312}$$

and

$$\mathbf{V} = \left(\underbrace{\mathbf{v}_1, \ldots, \mathbf{v}_r}_{\text{basis of range}[A^H]}, \underbrace{\mathbf{v}_{r+1}, \ldots, \mathbf{v}_{n_1}}_{\text{basis of null}[A]} \right) \tag{9.313}$$

(see the comments between Eqs. [9.293] and [9.294] and after Eq. [9.295]). From Eqs. (9.312) and (9.313), one has

$$\begin{aligned}
\text{range}[A] &= \text{span}[\bar{u}_1, \ldots, \bar{u}_r], \\
\text{null}[A] &= \text{span}[\bar{v}_{r+1}, \ldots, \bar{v}_{n_1}], \\
\text{range}[A^H] &= \text{span}[\bar{v}_1, \ldots, \bar{v}_r], \\
\text{null}[A^H] &= \text{span}[\bar{u}_{r+1}, \ldots, \bar{u}_{n_2}].
\end{aligned} \tag{9.314}$$

In the example of Section 9.4.2, range$[\mathbf{A}] = \text{span}[\mathbf{u}_1, \mathbf{u}_2]$. Because the bases in the above equations are orthonormal,

$$\begin{aligned}
\dim[\text{range}[A]] &= r, \\
\dim[\text{null}[A]] &= n_1 - r, \\
\dim[\text{range}[A^H]] &= r, \\
\dim[\text{null}[A^H]] &= n_2 - r.
\end{aligned} \tag{9.315}$$

Because **A** may be *any* matrix, this constitutes yet another proof of the **rank-nullity theorem**:

$$\text{rank} + \text{nullity} = \text{dimension of domain.} \tag{9.316}$$

A quick inspection gives the more detailed information that

$$\begin{aligned}
\text{null}[A] &= \{\text{range}[A^H]\}^\perp, \\
\text{null}[A^H] &= \{\text{range}[A]\}^\perp, \\
\text{range}[A^H] &= \{\text{null}[A]\}^\perp, \\
\text{range}[A] &= \{\text{null}[A^H]\}^\perp.
\end{aligned} \tag{9.317}$$

These relations, which are valid for all linear mappings on finite-dimensional, unitary inner-product spaces, constitute the **fundamental theorem of linear algebra**.

9.4.4 INVERSE AND PSEUDO-INVERSE IN THE SVD

If a matrix **A** is square and nonsingular, then one sees immediately from the SVD that the inverse of **A** is

$$A^{-1} = \sum_{i=1}^{n} v_i \frac{1}{\sigma_i} u_i{}^H, \tag{9.318}$$

from which one gets the matrix formula

$$\boxed{A^{-1} = V\Sigma^{-1}U^H.} \tag{9.319}$$

In the more general case in which A is singular, we know from Section 6.3.3 that the range of A is isomorphic to any complement of the null space of A. Even if A is singular it is possible to compute a generalized inverse that maps an optimally chosen (i.e., orthogonal) complement of null[A] onto range[A] in a one-to-one manner.

We show that the matrix of the linear mapping that carries range[A] onto the *orthogonal* complement of null[A] is

$$A^+ := V\Sigma^+U^H \tag{9.320}$$

in which

$$\Sigma^+ := \begin{array}{c} \\ 1 \\ \vdots \\ r \\ r+1 \\ \vdots \\ n_1 \end{array} \begin{pmatrix} 1 & \cdots & r & r+1 & \cdots & n_2 \\ 1/\sigma_1 & \cdots & 0 & 0 & \cdots & 0 \\ \vdots & \ddots & \vdots & \vdots & \ddots & \vdots \\ 0 & \cdots & 1/\sigma_r & 0 & \cdots & 0 \\ 0 & \cdots & 0 & 0 & \cdots & 0 \\ \vdots & \ddots & \vdots & \vdots & \ddots & \vdots \\ 0 & \cdots & 0 & 0 & \cdots & 0 \end{pmatrix}. \tag{9.321}$$

The matrix A^+ is called the **pseudo-inverse** (or the **Moore-Penrose inverse**) of **A**.

Because $\mathbf{U}$ is the matrix of orthonormal basis vectors of $\mathcal{V}_{(2)}$, and because $\mathbf{U}$ can be column-partitioned into a basis of null$[\mathbf{A}]$ and a basis of $\{$null$[\mathbf{A}]\}^{\perp}$ as in Eq. (9.312), one can determine the image of $\mathcal{V}_{(2)}$ under $\mathbf{A}^{+}$ by calculating

$$
\begin{aligned}
\mathbf{A}^{+}\mathbf{U} &= \mathbf{V}\mathbf{\Sigma}^{+}\mathbf{U}^{H}\mathbf{U} \\
&= \mathbf{V}\mathbf{\Sigma}^{+} \\
&= (\mathbf{v}_{1}, \ldots, \mathbf{v}_{r}, \mathbf{0}, \ldots, \mathbf{0})\mathbf{\Sigma}^{+}.
\end{aligned}
\tag{9.322}
$$

It follows that $\mathbf{A}^{+}$ maps the range of $\mathbf{A}$ (which is spanned by $\mathbf{u}_{1}, \ldots, \mathbf{u}_{r}$) onto the orthogonal complement of null$[\mathbf{A}]$ (which is spanned by $\mathbf{v}_{1}, \ldots, \mathbf{v}_{r}$).

To demonstrate the generalized inverse property of $\mathbf{A}^{+}$, we calculate $\mathbf{A}^{+}\mathbf{A}$ in orthonormal bases (so that $\mathbf{U}^{H} = \mathbf{U}^{\dagger}$, etc.):

$$
\begin{aligned}
\mathbf{A}^{+}\mathbf{A} &= \mathbf{V}\mathbf{\Sigma}^{+}\mathbf{U}^{\dagger}\mathbf{U}\mathbf{\Sigma}\mathbf{V}^{\dagger} \\
&= \mathbf{V}\mathbf{\Sigma}^{+}\mathbf{\Sigma}\mathbf{V}^{\dagger}.
\end{aligned}
\tag{9.323}
$$

It is easy to see from the definitions of $\mathbf{\Sigma}^{+}$ and $\mathbf{\Sigma}$ that

$$
\mathbf{\Sigma}^{+}\mathbf{\Sigma} =
\begin{array}{c}
\\
1 \\
\vdots \\
r \\
r+1 \\
\vdots \\
n_{1}
\end{array}
\begin{array}{ccccccc}
1 & \cdots & r & r+1 & \cdots & n_{1} \\
\left(\begin{array}{ccccc}
1 & \cdots & 0 & 0 & \cdots & 0 \\
\vdots & \ddots & \vdots & \vdots & \ddots & \vdots \\
0 & \cdots & 1 & 0 & \cdots & 0 \\
0 & \cdots & 0 & 0 & \cdots & 0 \\
\vdots & \ddots & \vdots & \vdots & \ddots & \vdots \\
0 & \cdots & 0 & 0 & \cdots & 0
\end{array}\right).
\end{array}
\tag{9.324}
$$

Then

$$
\begin{aligned}
\mathbf{A}^{+}\mathbf{A} &= (\mathbf{v}_{1}, \ldots, \mathbf{v}_{r}, \mathbf{0}, \ldots, \mathbf{0})
\begin{pmatrix}
\mathbf{v}_{1}^{\dagger} \\
\vdots \\
\mathbf{v}_{r}^{\dagger} \\
\mathbf{0}^{\dagger} \\
\vdots \\
\mathbf{0}^{\dagger}
\end{pmatrix} \\
&= \mathbf{v}_{1}\mathbf{v}_{1}^{\dagger} + \cdots + \mathbf{v}_{r}\mathbf{v}_{r}^{\dagger} \\
&= \mathbf{P}_{\{\text{null}[\mathbf{A}]\}^{\perp}} \\
&= \mathbf{P}_{\text{range}[\mathbf{A}^{H}]}
\end{aligned}
\tag{9.325}
$$

in which $\mathbf{P}_{\{\text{null}[\mathbf{A}]\}^{\perp}}$ and $\mathbf{P}_{\text{range}[\mathbf{A}^{H}]}$ are the projectors on $\{$null$[\mathbf{A}]\}^{\perp}$ and range$[\mathbf{A}^{H}]$, respectively.

One can show similarly that

$$
\Sigma\Sigma^+ = \begin{array}{c} \\ 1 \\ \vdots \\ r \\ r+1 \\ \vdots \\ n_2 \end{array} \begin{array}{cccccc} 1 & \cdots & r & r+1 & \cdots & n_2 \\ \begin{pmatrix} 1 & \cdots & 0 & 0 & \cdots & 0 \\ \vdots & \ddots & \vdots & \vdots & \ddots & \vdots \\ 0 & \cdots & 1 & 0 & \cdots & 0 \\ 0 & \cdots & 0 & 0 & \cdots & 0 \\ \vdots & \ddots & \vdots & \vdots & \ddots & \vdots \\ 0 & \cdots & 0 & 0 & \cdots & 0 \end{pmatrix} \end{array}
\tag{9.326}
$$

and that

$$
\mathbf{AA}^+ = \mathbf{P}_{\text{range}[\mathbf{A}]},
\tag{9.327}
$$

in which $\mathbf{P}_{\text{range}[\mathbf{A}]}$ is the projector on the range of $\mathbf{A}$.

9.4.5 DATA COMPRESSION USING THE SVD

Equation (1.13) is an $M \times N$ matrix of 8-bit integers that represent shades of gray on a scale from 0 to 255_{10}. To represent an $M \times N$ screenful of data without any compression requires MN bytes. One can turn to the SVD to compress the graphical information that is present in the matrix $\mathbf{P}$ with reasonable visual fidelity.

One SVD technique for the compression of graphical information is to obtain the SVD of $\mathbf{P}$ and to select only the largest singular values and the corresponding right and left singular vectors for transmission. If the choice is well made, the remaining singular values are so small that the human eye cannot easily see a difference between the original image and the image obtained from the pixel matrix $\mathbf{P}$ that has been reconstituted from only some of the singular values and singular vectors.

9.4.6 EXERCISES FOR SECTION 9.4

9.4.1   Find by hand calculation the SVD of the matrix

$$
\mathbf{A} = \begin{pmatrix} 1 & 0 \\ 1 & 1 \\ 0 & 1 \end{pmatrix}.
\tag{9.328}
$$

9.4.2   A matrix $\mathbf{A}$ has the singular values

$$
\sigma_1 = 2, \quad \sigma_2 = 1,
\tag{9.329}
$$

the right singular vectors

$$
\mathbf{v}_1 = \begin{pmatrix} 1 \\ 0 \end{pmatrix}, \quad \mathbf{v}_2 = \begin{pmatrix} 0 \\ 1 \end{pmatrix},
\tag{9.330}
$$

and the left singular vectors

$$\mathbf{u}_1 = \begin{pmatrix} \dfrac{1}{\sqrt{2}} \\ \dfrac{1}{\sqrt{2}} \end{pmatrix}, \quad \mathbf{u}_2 = \begin{pmatrix} \dfrac{1}{\sqrt{2}} \\ -\dfrac{1}{\sqrt{2}} \end{pmatrix}. \tag{9.331}$$

Find **A**.

9.5 LINEAR EQUATIONS II

9.5.1 NUMERICAL VERSUS ANALYTICAL METHODS

A full discussion of the reasons why numerical methods have supplanted analytical methods in many areas of physics and engineering would require a historical summary of a large number of developments in the physical sciences since 1950. We must settle for an outline of the most essential points. In many problems in physics and engineering one must obtain an analytical or numerical result for $\bar{x}$ from the equation

$$A\bar{x} = \bar{b}, \tag{9.332}$$

in which A is a linear differential or integral operator that acts on functions $\bar{x}$ in an infinite-dimensional vector space $\mathcal{V}$. The most general method of solving such equations that is taught in traditional courses is the purely analytical **method of eigenfunction expansions**, which goes as follows: If the eigenvectors or eigenfunctions $\{\bar{v}_i\}$ of the linear mapping A span $\mathcal{V}$, then one solves the equation $A\bar{x} = \bar{b}$ by expanding $\bar{b}$ and $\bar{x}$ relative to $\{\bar{v}_i\}$:

$$\bar{b} = b^i \bar{v}_i, \quad \bar{x} = x^i \bar{v}_i. \tag{9.333}$$

Then

$$x^i = \frac{b^i}{\lambda_i} \quad (\text{no } \Sigma), \tag{9.334}$$

in which λ_i is the eigenvalue that belongs to $\bar{v}_i$. In the days when the only tools of numerical computation were paper and pencil, and when tables of values of the so-called special functions were a precious resource, mathematicians and physicists spared no effort in discovering bases of functions with respect to which important linear operators such as the Laplacian and the Helmholtz operator are diagonal. The method of eigenfunction expansions is still useful in some cases, with the considerable improvement that one's computer can calculate numerical values of special functions on demand. Other methods that were used widely in the precomputer era, such as conformal mapping and contour integration, have all but disappeared from day-to-day practice in physics and engineering.

The principal limitations of the method of eigenfunction expansions are: First, well-studied eigenfunctions are available only for highly symmetrical geometries. Realistic models usually do not give rise to known eigenfunctions. Second, the numerical evaluation of the innocent-looking expression $\bar{x} = x^i \bar{v}_i$ can pose major computational difficulties if (as often happens) the eigenfunction expansion has an infinite set of nonzero terms with alternating + and − signs. In the few cases in which the symmetry of a realistic problem permits the use of known eigenfunctions, the eigenfunction-expansion method can be quite powerful even if the final result requires numerical evaluation, provided only that the eigenfunction series has only a few terms for source functions $\bar{b}$ of interest.

Today, after the revolution wrought by the ready availability of powerful computers, one usually maps the function-space problem that one wants to solve into a finite-dimensional vector space and performs a numerical solution of an equation of the form of Eq. (9.191). Apart from possible problems associated with sampling continuous functions at a finite set of points, the numerical approach has the considerable advantage that one can incorporate realistic geometries and boundary conditions almost as easily as highly symmetrical geometries.

9.5.2 DIAGONAL DOMINANCE

In principle, one can determine whether a square matrix $\mathbf{A}$ is nonsingular by checking whether det[A] is nonzero. In practice the computationally least expensive way to compute det[A] turns out to be to solve the linear system $\mathbf{Ax} = \mathbf{b}$; see Section 6.6.2. Therefore one needs criteria for the nonsingularity of $\mathbf{A}$ that are more easily checked that the nonvanishing of det[A]. Strict diagonal dominance is such a criterion. An $n \times n$ matrix $\mathbf{A}$ is called **row diagonally dominant** if and only if

$$\forall i = 1, \ldots, n : |a_i^i| \geq \sum_{j \neq i} |a_j^i| \tag{9.335}$$

or **column diagonally dominant** if and only if

$$\forall i = 1, \ldots, n : |a_i^i| \geq \sum_{j \neq i} |a_i^j|. \tag{9.336}$$

(There is no summation on i in these equations.) If $\mathbf{A}$ is row diagonally dominant, then $\mathbf{A}^T$ is column diagonally dominant, and conversely. $\mathbf{A}$ is called **strictly row diagonally dominant** if and only if strict inequality (> instead of ≥) holds in (9.335). For the rest of our discussion of diagonal dominance, we assume strict row diagonal dominance.

A strictly row diagonally dominant square matrix is nonsingular. To see why, let us suppose the contrary, that is, let us suppose that there exist a singular, strictly row diagonally dominant square matrix $\mathbf{A}$ and a nonzero vector $\mathbf{x}$ such that $\mathbf{Ax} = \mathbf{0}$. Then

$$\forall i = 1, \ldots, n : \sum_{j=1}^{n} a_j^i x^j = 0, \tag{9.337}$$

which implies that

$$a_i^i x^i = -\sum_{j \neq i} a_j^i x^j. \tag{9.338}$$

From the properties of inequalities it follows that

$$|a_i^i x^i| = |a_i^i| \, |x^i| = \left| \sum_{j \neq i} a_j^i x^j \right|$$

$$\leq \sum_{j \neq i} |a_j^i| \, |x^j| \leq \left\{ \max_k |x^k| \right\} \sum_{j \neq i} |a_j^i|. \tag{9.339}$$

Now choose i such that

$$|x^i| = \max_k |x^k|. \tag{9.340}$$

Then $|x^i| \neq 0$ because $\mathbf{x} \neq \mathbf{0}$. Cancelling $|x^i|$ from Eq. (9.339), one obtains the inequality

$$|a_i^i| \leq \sum_{j \neq i} |a_j^i|, \tag{9.341}$$

which contradicts the assumption that $\mathbf{A}$ is strictly row diagonally dominant. Therefore $\mathbf{A}$ is nonsingular.

9.5.3 CONDITION NUMBER OF THE LINEAR-EQUATION PROBLEM

Consider a system of linear equations stated in matrix form as $\mathbf{Ax} = \mathbf{b}$. Mathematically, a unique solution to this linear system is guaranteed if the coefficient matrix $\mathbf{A}$ is nonsingular. In practice, a physicist or an engineer usually needs a *well-conditioned solution*, in which "small" changes in the right-hand side, $\mathbf{b}$, do not produce "large" changes in the solution vector, $\mathbf{x}$. In order to quantify the concept of "small" and "large" changes, let $\mathbf{x}$ be the solution if $\mathbf{b}$ is the right-hand side and let $\mathbf{x}'$ be the solution if $\mathbf{b}'$ is the right-hand side. Let

$$\Delta\mathbf{x} := \mathbf{x}' - \mathbf{x} \tag{9.342}$$

$$\Delta\mathbf{b} := \mathbf{b}' - \mathbf{b}. \tag{9.343}$$

Because $\|\mathbf{x}\|$ is measures the magnitude of a vector, it is reasonable to regard

$$\frac{\|\Delta\mathbf{x}\|}{\|\mathbf{x}\|} \quad \text{and} \quad \frac{\|\Delta\mathbf{b}\|}{\|\mathbf{b}\|} \tag{9.344}$$

as measures of the relative changes in $\mathbf{x}$ and $\mathbf{b}$ if $\mathbf{b}$ is replaced by $\mathbf{b}'$ in Eq. (6.369). A "small" change in $\mathbf{b}$, then, is one that produces a relative change that is small in the sense that $\epsilon = \|\Delta\mathbf{b}\|/\|\mathbf{b}\| \ll 1$. For example, if $\mathbf{b}$ is a computed vector, then one expects ϵ to be of the order of the relative error in rounding, $\epsilon = \epsilon_{\text{mach}}/2$. If $\mathbf{b}$ is derived from experimental measurements, then ϵ is of the order of the experimental uncertainty and is ordinarily much larger than ϵ_{mach}.

A criterion for the solution $\mathbf{x}$ to be well conditioned under small changes in the right-hand side $\mathbf{b}$ is that the relative change in the solution, $\|\Delta\mathbf{x}\|/\|\mathbf{x}\|$, must not be larger than some chosen multiple of the relative change in the right-hand side, $\|\Delta\mathbf{b}\|/\|\mathbf{b}\|$:

$$\sup_{\Delta\mathbf{b}}\left\{\frac{\|\Delta\mathbf{x}\|/\|\mathbf{x}\|}{\|\Delta\mathbf{b}\|/\|\mathbf{b}\|}\right\} < C \text{ if } \frac{\|\Delta\mathbf{b}\|}{\|\mathbf{b}\|} = \epsilon. \tag{9.345}$$

For a given right-hand side $\mathbf{b}$ and a given "error" ϵ, the norms $\|\mathbf{b}\|$ and $\|\Delta\mathbf{b}\| = \epsilon\|\mathbf{b}\|$ are constant in Eq. (9.345). Then the least upper bound must be calculated over all directions of $\Delta\mathbf{b}$. The value of the least upper bound so obtained depends on ϵ.

The values of ϵ and C cannot be specified precisely unless additional information is available, such as the choice of vector and matrix norms, the choice of basis, the experimental uncertainties in the right-hand side $\mathbf{b}$, the number of significant digits kept in one's numerical computations, and what relative error in the solution vector one is willing to accept.

The criterion (9.345) for the solution to be well conditioned can be strengthened and put into a form that does not involve specific vectors by making use of the matrix norm that is consistent with the chosen vector norm. The basic relation

$$\|\mathbf{b}\| = \|\mathbf{A}\mathbf{x}\| \le \|\mathbf{A}\| \, \|\mathbf{x}\| \tag{9.346}$$

implies that a condition that guarantees that the criterion Eq. (9.345) will hold is that

$$\|\mathbf{A}\| \sup_{\Delta\mathbf{b}}\left\{\frac{\|\Delta\mathbf{x}\|}{\|\Delta\mathbf{b}\|}\right\} < C. \tag{9.347}$$

Because $\Delta\mathbf{b} = \mathbf{A}\Delta\mathbf{x}$, and therefore $\Delta\mathbf{x} = \mathbf{A}^{-1}\Delta\mathbf{b}$, one has the inequality

$$\|\Delta\mathbf{x}\| \le \|\mathbf{A}^{-1}\| \, \|\Delta\mathbf{b}\|. \tag{9.348}$$

It follows that a sufficient condition for Eq. (9.347) to hold is that

$$\text{cond}[\mathbf{A}] := \|\mathbf{A}\| \, \|\mathbf{A}^{-1}\| < C, \tag{9.349}$$

in which cond[$\mathbf{A}$] is called the **condition number** of the coefficient matrix $\mathbf{A}$. If $\mathbf{A}$ is singular, then cond[$\mathbf{A}$] is defined as ∞. This definition is reasonable, for if $\mathbf{A}$ is singular, Eq. (9.348) and the fact that $\Delta\mathbf{b} = \mathbf{A}\Delta\mathbf{x} = \mathbf{0}$ imply that $\|\mathbf{A}^{-1}\| \times \|\Delta\mathbf{b}\|$ (which equals zero) gives a finite result, $\|\Delta\mathbf{x}\|$, which is impossible if $\|\mathbf{A}^{-1}\|$ is any finite real number.

It follows from Eqs. (9.346) through (9.348) that the condition number gives a worst-case estimate of the ratio of relative changes in $\mathbf{x}$ and $\mathbf{b}$ for a given coefficient matrix $\mathbf{A}$, a given basis, and a given p-norm:

$$\boxed{\frac{\|\Delta\mathbf{x}\|}{\|\mathbf{x}\|} \le \text{cond}[\mathbf{A}]\frac{\|\Delta\mathbf{b}\|}{\|\mathbf{b}\|}.} \tag{9.350}$$

Equality holds for *some* right-hand side $\mathbf{b}$ and *some* change $\Delta\mathbf{b}$. To see what may happen if the criterion Eq. (9.349) is violated, suppose that $\|\Delta\mathbf{b}\| = \epsilon\|\mathbf{b}\|$ and that one has chosen

$C < \epsilon^{-1}$. Then

$$\|\Delta \mathbf{b}\| = \epsilon \|\mathbf{b}\| \text{ and } \mathrm{cond}[\mathbf{A}] = \frac{1}{\epsilon} > C \Rightarrow \frac{\|\Delta \mathbf{x}\|}{\|\mathbf{x}\|} \leq 1. \tag{9.351}$$

In the $p = \infty$ norm, for example, Eq. (9.351) implies that the change $\|\Delta \mathbf{x}\|$ may be as large as the maximum of the magnitudes of the components of $\mathbf{x}$.

It follows that in the worst case, some or all of the components of the solution vector may have no significant digits in common with the correct answer! Exercise 9.6.3 involves a 2×2 coefficient matrix $\mathbf{A}$ that is exceptionally ill conditioned; one does not have to solve a 1000×1000 system to see a bad condition number.

What is particularly insidious is that the norm of the **residual vector**

$$\mathbf{r} := \mathbf{b} - \mathbf{A}\mathbf{x}, \tag{9.352}$$

may be of order $\epsilon_{\mathrm{mach}} \|\mathbf{b}\|$, but at the same time $\|\Delta \mathbf{x}\|/\|\mathbf{x}\|$ may be so large that the computed result is worthless. Exercise 9.6.3 provides an informative example.

One can regard $\{\mathrm{cond}[\mathbf{A}]\}^{-1}$ as a quantitative measure of the closeness of $\mathbf{A}$ to singularity, because $\{\mathrm{cond}[\mathbf{A}]\}^{-1} = 0$ if $\mathbf{A}$ is singular. Then Eq. (9.351) shows how nearly singular $\mathbf{A}$ must be for the relative error of the solution vector to be of order unity for some choice of $\mathbf{b}$ and $\Delta \mathbf{b}$.

In Eqs. (9.344) through (9.351) we have not specified which norm is to be used. If one uses a p-norm, then the value of $\mathrm{cond}[\mathbf{A}]$ depends on the value of p and the basis in which the norms are evaluated. In practice the order of magnitude of $\mathrm{cond}[\mathbf{A}]$, which does not depend on the choice of norm or basis, is much more important than the exact value of the condition number, as Eq. (9.350) suggests. Therefore we are free to choose whichever matrix norm makes it easiest to evaluate

$$\mathrm{cond}_p[\mathbf{A}] = \|\mathbf{A}\|_p \|\mathbf{A}^{-1}\|_p. \tag{9.353}$$

The trick is to evaluate $\|\mathbf{A}^{-1}\|$ efficiently, because Eqs. (8.452) and (8.455) give readily evaluated formulas for $\|\mathbf{A}\|_1$ and $\|\mathbf{A}\|_\infty$. A good estimate of the condition number can be obtained by skillfully exploiting Eq. (9.348) for $p = \infty$.

If the coefficient matrix $\mathbf{A}$ is a normal matrix, then it can be diagonalized by a unitary similarity transformation. In that case one obtains

$$\mathrm{cond}_2[\mathbf{A}] = \left| \frac{\lambda_{\max}}{\lambda_{\min}} \right|. \tag{9.354}$$

The obvious limitations of this approach are that $\mathbf{A}$ may not be normal, and that computation of a full set of eigenvalues requires significantly more time than Gaussian elimination.

The SVD given in Eqs. (9.305) and (9.319) gives a simple formula for $\mathrm{cond}_2[\mathbf{A}]$. From Eqs. (9.305), (9.319), and (9.349), one sees that

$$\mathrm{cond}_2[\mathbf{A}] = \frac{\sigma_1}{\sigma_n}. \tag{9.355}$$

Unfortunately, for a square matrix $\mathbf{A}$ the computation of the singular-value matrix $\mathbf{\Sigma}$, which is required in order to evaluate Eq. (9.355), requires significantly more numerical operations than solving the linear system $\mathbf{A}\mathbf{x} = \mathbf{b}$ by the method of Gaussian elimination.

9.5.4 THE LDL† AND CHOLESKY DECOMPOSITIONS

The LDL† Decomposition

In Section 6.6.4 we derived the decomposition $\mathbf{A} = \mathbf{L}\mathbf{D}\mathbf{M}^T$, in which $\mathbf{D}$ is the diagonal matrix of pivots obtained in Gaussian elimination. This decomposition is not often used in practice, but it is important theoretically because it is a kind of diagonalization of the nonsingular matrix $\mathbf{A}$. For example, suppose that $\mathbf{A}$ is a nonsingular Hermitian matrix. Then there exist unit lower-triangular matrices $\mathbf{L}$, $\mathbf{M}$ and a nonsingular diagonal matrix $\mathbf{D}$ such that $\mathbf{A} = \mathbf{L}\mathbf{D}\mathbf{M}^T$. Assuming that the underlying basis is orthonormal (which implies that $\mathbf{A}^H = \mathbf{A}^\dagger$), and because $\mathbf{A}$ is Hermitian, one has

$$\mathbf{A}^\dagger = \mathbf{M}^{\dagger T}\mathbf{D}^\dagger\mathbf{L}^\dagger = \mathbf{A} = \mathbf{L}\mathbf{U}. \tag{9.356}$$

The matrix $\mathbf{M}^{\dagger T}$ is unit lower-triangular, and the matrix $\mathbf{D}^\dagger\mathbf{L}^\dagger$ is upper-triangular. From the uniqueness of the LU decomposition it follows that

$$\mathbf{M}^{\dagger T} = \mathbf{L} \Rightarrow \mathbf{M}^T = \mathbf{L}^\dagger. \tag{9.357}$$

Therefore

$$\mathbf{A} = \mathbf{L}\mathbf{D}\mathbf{L}^\dagger \tag{9.358}$$

for every nonsingular Hermitian matrix $\mathbf{A}$.

The Cholesky Decomposition

Now suppose that $\mathbf{A}$ is a positive-definite Hermitian matrix. Although this case sounds specialized, it occurs rather often, especially in the linear least-squares problem discussed in Section 9.6.1. From the assumption of positive-definiteness it follows that

$$\forall \mathbf{x} \in \mathbb{C}^n :\ni: \mathbf{x} \neq \mathbf{0} : \mathbf{x}^\dagger\mathbf{A}\mathbf{x} > 0 \Rightarrow (\mathbf{L}^\dagger\mathbf{x})^\dagger\mathbf{D}\mathbf{L}^\dagger\mathbf{x} > 0. \tag{9.359}$$

Because $\mathbf{L}^\dagger$ is invertible, it follows that one can find a vector $\mathbf{x}$ such that any one chosen component of $\mathbf{L}^\dagger\mathbf{x}$ is 1, and all the others vanish. It follows from this observation and from Eq. (9.359) that each of the diagonal elements $a_1^1, \ldots, a_n^{(n)n}$ of $\mathbf{D}$ is positive. Therefore one can define the square root of $\mathbf{D}$,

$$\mathbf{D}^{1/2} := \text{diag}\left[\sqrt{a_1^1}, \ldots, \sqrt{a_n^{(n)n}}\right], \tag{9.360}$$

and can rewrite Eq. (9.358) as follows:

$$\mathbf{A} = \mathbf{L}\mathbf{D}^{1/2}\mathbf{D}^{1/2}\mathbf{L}^\dagger. \tag{9.361}$$

If we define the lower-triangular matrix

$$\mathbf{C} := \mathbf{L}\mathbf{D}^{1/2}, \tag{9.362}$$

then Eq. (9.361) implies that

$$\mathbf{A} = \mathbf{C}\mathbf{C}^\dagger \tag{9.363}$$

for every positive-definite Hermitian matrix $\mathbf{A}$. Equation (9.363) is called the **Cholesky decomposition** of $\mathbf{A}$. The computation of the Cholesky decomposition turns out to take about half as many operations as Gaussian elimination (Golub and Van Loan 1989).

One sees either from Eq. (9.362) or from the fact that $\mathbf{C}$ is lower-triangular that the columns of $\mathbf{C}^\dagger$ are linearly independent, and therefore that $\mathbf{C}$ is nonsingular. Comparing Eqs. (8.30) and (9.363), one then recognizes that the columns of $\mathbf{C}^\dagger$ are the basis vectors relative to which $\mathbf{A}$ is the Gram matrix.

9.6 SELECTED APPLICATIONS OF LINEAR EQUATIONS

9.6.1 THE LINEAR LEAST-SQUARES PROBLEM

Every working physicist or engineer will encounter the linear least-squares problem at some point in his or her career. Suppose that according to a certain model of a physical process or an engineering system, a dependent variable b, which depends on an independent variable t, is a linear combination of m linearly independent (but not necessarily orthogonal) functions $f_i(t)$:

$$b(t) = \sum_{i=1}^{m} x^i f_i(t). \tag{9.364}$$

A common example is a **polynomial fit**, in which $f_i(t) = t^{i-1}$. Suppose also that we have measured $b(t)$ for n different values of t. Then the model Eq. (9.364) implies that

$$\forall j \in (1:n) \; : \; b(t_j) = \sum_{i=1}^{m} a_i^j x^i \tag{9.365}$$

in which $\mathbf{b} := \mathrm{col}[b^1, \ldots, b^n]$ is called the **observation vector**,

$$a_i^j := f_i(t_j), \tag{9.366}$$

and in which x^i, $i \in (1:m)$, are **parameters** to be determined. The $n \times m$ **design matrix** $\mathbf{A} = \mathrm{matrix}[a_i^j]$ is a quantitative expression of the model Eq. (9.364) as applied to a particular discrete set of values of the independent variable t. Obviously the linear equations Eq. (9.365) can be summarized in a matrix equation of the form $\mathbf{A}\mathbf{x} = \mathbf{b}$ in which the coefficient matrix is rectangular instead of square.

In order to determine the parameters, we must have more observations (equations) than unknowns:

$$m < n. \tag{9.367}$$

Then

$$\text{rank}[A] \leq \dim[\text{span}[\bar{e}_1, \ldots, \bar{e}_m]] = m \Rightarrow \text{rank}[A] < \dim[\mathcal{V}] = n \tag{9.368}$$

in which $\mathcal{V}$ is the vector space ($\mathbb{R}^n$ or $\mathbb{C}^n$) to which the observation vector $\mathbf{b}$ belongs. For an accurate determination of $\mathbf{x}$ one usually wants to have $n \gg m$, which implies that rank$[\mathbf{A}] \ll \dim[\mathcal{V}]$. Equations (9.368) and (6.371) imply that the linear equations (9.365) have no solution in general, because $\mathbf{b}$ generally lies outside of range$[\mathbf{A}]$. In more classical language, the linear system Eq. (9.365) is overdetermined.

However, it is possible to determine a vector $\mathbf{x}$ of **best-fit parameters** by requiring that the residual vector $\mathbf{r}$ defined in Eq. (9.352) have the smallest possible p-norm $\|\mathbf{r}\|_p$. The 2-norm is popular for this purpose because $\|\mathbf{r}\|_2$ is differentiable and because minimizing $\|\mathbf{r}\|_2$ leads to a set of linear equations for the parameter vector $\mathbf{x}$. The statement of the **linear least-squares problem** is therefore

$$\|\mathbf{b} - \mathbf{Ax}\|_2 = \text{minimum}. \tag{9.369}$$

Almost always one can assume that the n components of the observation vector are relative to an orthogonal basis of $\mathcal{V}$. (This does *not* mean that we assume anything about the orthogonality of the m functions f_i!) The subspace $\mathcal{W}$ in our discussion of least-squares approximation in Section 8.4 is the range of the coefficient matrix $\mathbf{A}$ in the problem defined by Eq. (9.369). If $\mathbf{A}$ were square and nonsingular, and if its column vectors were orthonormal (which never happens in practice), then we could obtain the vector $\mathbf{x}$ that minimizes $\|\mathbf{b} - \mathbf{Ax}\|_2$ by projecting $\mathbf{b}$ onto $\mathcal{W}$, as in Eq. (8.232), and then applying $\mathbf{A}^{-1}$. Because $\mathbf{A}$ is generally rectangular in a data-fitting problem, the most that one can say here is that

$$\mathbf{b} - \mathbf{Ax} \perp \mathcal{W} := \text{range}[\mathbf{A}]. \tag{9.370}$$

Then, according to Eq. (9.130), $\mathbf{b}$ lies in the null space of $\mathbf{A}^\dagger$:

$$\mathbf{b} - \mathbf{Ax} \in \text{null}[\mathbf{A}^\dagger]. \tag{9.371}$$

This means that

$$\mathbf{A}^\dagger(\mathbf{b} - \mathbf{Ax}) = \mathbf{0}. \tag{9.372}$$

This equation holds if and only if

$$\boxed{\mathbf{A}^\dagger\mathbf{Ax} = \mathbf{A}^\dagger\mathbf{b}.} \tag{9.373}$$

The m linear equations Eq. (9.373) are called the **normal equations** of the linear least-squares problem.

A unique solution of the normal equations exists if and only if the $m \times m$ matrix $A^\dagger A$ is nonsingular, which is true if and only if

$$\text{rank}[A^\dagger A] = m. \tag{9.374}$$

Clearly $\text{rank}[A^\dagger] = \text{rank}[A^T]$. According to Eq. (9.69), $\text{rank}[A^\dagger] = \text{rank}[A]$. Therefore Eq. (9.374) holds if and only if $\text{rank}[A] = m$, that is, if and only if A is nonsingular. If $A^\dagger A$ is nonsingular, then the unique solution of the normal equations Eq. (9.373) is

$$x = (A^\dagger A)^{-1} A^\dagger b. \tag{9.375}$$

Of course, constructing the inverse of the coefficient matrix is not a computationally sound approach to solving a system of linear equations. Gaussian elimination is an adequate method for solving the normal equations if the matrix $A^\dagger A$ is well conditioned. However, Gaussian elimination takes no advantage of the fact that the coefficient matrix $A^\dagger A$ in Eq. (9.375) is Hermitian and positive-definite. Cholesky decomposition is about a factor of two faster than Gaussian elimination for solving the normal equations.

Much more often than one might expect, it turns out that the matrix $A^\dagger A$ is nearly singular. For practical purposes, "nearly singular" means that

$$\frac{\sigma_n}{\sigma_1} < \frac{1}{C}, \tag{9.376}$$

in which σ_1 and σ_n are, respectively, the maximum and minimum singular values of $A^\dagger A$, and C is the maximum factor by which we can tolerate the relative error in the solution to exceed the relative error in b. If

$$\frac{\sigma_n}{\sigma_1} < \epsilon_{\text{mach}} \tag{9.377}$$

in which ϵ_{mach} is machine epsilon in the floating-point representation on one's computer, then the situation is completely hopeless; there may be no correct significant digits in some components of the solution, for some right-hand sides b.

Neither Gaussian elimination nor Cholesky decomposition is well adapted to least-squares problems in which the coefficient matrix is nearly singular. If one suspects that $A^\dagger A$ is nearly singular, then one should use the SVD to test this hypothesis. If the hypothesis is confirmed, then one should use fewer parameters. Sometimes it may be necessary to reexamine one's entire approach to the physics or engineering model to which one is trying to fit experimental data in order to come up with a new statement of the model that will reliably yield a well-conditioned matrix $A^\dagger A$.

9.6.2 LINEAR DIFFERENCE EQUATIONS

A **linear difference equation of order k with constant coefficients** is a relation of the form

$$\forall n \in \mathbb{Z}^+ :\ni: n > k \ : \ \alpha_0 y_n + \alpha_1 y_{n-1} + \cdots + \alpha_k y_{n-k} = \gamma_n \tag{9.378}$$

among the elements of a sequence $\overline{y} \in \mathbb{F}^\infty$, such that $\alpha_0 \neq 0$ and $\alpha_k \neq 0$. The number field $\mathbb{F}$ is either the real field $\mathbb{R}$ or the complex field $\mathbb{C}$. The phrase "constant coefficients" means that the coefficients $\alpha_i \in \mathbb{F}$ are independent of n.

Finite-difference approximations to linear ordinary differential equations result in difference equations of the form of Eq. (9.378), although not necessarily with constant coefficients. In a finite-difference approximation, y_n is the approximate value computed for the sampled value $y(nh)$, in which y is the function that exactly solves the differential equation of which Eq. (9.378) is an approximation. The step size h is equal to the distance between the points at which y is sampled. The coefficients α_i are determined by the coefficients in the differential equation and by the method used to approximate the derivatives.

For the purpose of analyzing the stability of finite-difference methods for solving ordinary differential equations, one studies the solutions of homogeneous linear difference equations of order k with constant coefficients, which are of the form of Eq. (9.378) with every γ_n equal to zero. If one defines

$$\mathbf{x}_n := \begin{pmatrix} y_{n-k+1} \\ \vdots \\ y_n \end{pmatrix},$$

(9.379)

then every homogeneous linear difference equation of order k with constant coefficients is equivalent to the two-term matrix recurrence relation

$$\mathbf{x}_n = \mathbf{A}\mathbf{x}_{n-1}$$

(9.380)

in which

$$\mathbf{A} = \begin{pmatrix} 0 & 1 & 0 & \cdots & 0 & 0 \\ 0 & 0 & 1 & \cdots & 0 & 0 \\ 0 & 0 & 0 & \cdots & 0 & 0 \\ \vdots & \vdots & \vdots & \ddots & \vdots & \vdots \\ 0 & 0 & 0 & \cdots & 0 & 1 \\ -\dfrac{\alpha_k}{\alpha_0} & -\dfrac{\alpha_{k-1}}{\alpha_0} & -\dfrac{\alpha_{k-2}}{\alpha_0} & \cdots & -\dfrac{\alpha_2}{\alpha_0} & -\dfrac{\alpha_1}{\alpha_0} \end{pmatrix}.$$

(9.381)

Note that the diagonal immediately above the main diagonal contains 1s, and that the only other nonzero elements of $\mathbf{A}$ are in the last row. (The reader should verify this recurrence relation before proceeding further.)

The solution of the recurrence relation Eq. (9.380) is

$$\forall n \geq k : \mathbf{x}_n = \mathbf{A}^{n-k}\mathbf{x}_k.$$

(9.382)

Let ξ_i be the i-th eigenvalue of $\mathbf{A}$, and let $\mathbf{v}_i$ be the eigenvector that belongs to ξ_i:

$$\forall i \in (1:k) : \mathbf{A}\mathbf{v}_i = \xi_i \mathbf{v}_i.$$

(9.383)

If the eigenvectors of $\mathbf{A}$ span $\mathbb{C}^k$, then the matrix of eigenvectors,

$$\mathbf{V} = [\mathbf{v}_1, \ldots, \mathbf{v}_k],$$

(9.384)

diagonalizes both $\mathbf{A}$ and $\mathbf{A}^n$ by the similarity transformation

$$(\mathbf{V}^{-1}\mathbf{A}\mathbf{V})^n = \mathbf{V}^{-1}\mathbf{A}^n\mathbf{V} = \begin{pmatrix} \xi_1^n & 0 & \cdots & 0 \\ 0 & \xi_2^n & \cdots & 0 \\ \vdots & \vdots & \ddots & \vdots \\ 0 & 0 & \cdots & \xi_k^n \end{pmatrix}. \tag{9.385}$$

It follows that if $\mathbf{x}_k$ is a scalar multiple of an eigenvector $\mathbf{v}_l$,

$$\mathbf{x}_k = \kappa_l \mathbf{v}_l, \tag{9.386}$$

then Eq. (9.382) implies that

$$\mathbf{x}_n = \xi_l^{n-k} \mathbf{x}_k. \tag{9.387}$$

The componentwise form of this equation is

$$\forall n > k : \forall r \in (0 : k - 1) : y_{n-r} = \xi_l^{n-k} y_{k-r}. \tag{9.388}$$

Therefore, if $\mathbf{x}_k$ is a scalar multiple of an eigenvector $\mathbf{v}_l$, the sequence $\bar{y}$ is a scalar multiple of the geometric sequence

$$\bar{y}^{(l)} := (1, \xi_l, \ldots, \xi_l^n, \ldots). \tag{9.389}$$

Relative to a basis of eigenvectors, then, the solution of the difference equation Eq. (9.378) (with $\gamma_n = 0$) is essentially trivial.

The secular equation of $\mathbf{A}$ is

$$\det[\mathbf{A} - \xi\mathbf{1}] = 0 = \det \begin{pmatrix} -\xi & 1 & 0 & \cdots & 0 & 0 \\ 0 & -\xi & 1 & \cdots & 0 & 0 \\ 0 & 0 & -\xi & \cdots & 0 & 0 \\ \vdots & \vdots & \vdots & \ddots & \vdots & \vdots \\ 0 & 0 & 0 & \cdots & -\xi & 1 \\ -\dfrac{\alpha_k}{\alpha_0} & -\dfrac{\alpha_{k-1}}{\alpha_0} & -\dfrac{\alpha_{k-2}}{\alpha_0} & \cdots & -\dfrac{\alpha_2}{\alpha_0} & -\dfrac{\alpha_1}{\alpha_0} - \xi \end{pmatrix}. \tag{9.390}$$

Expanding the determinant in cofactors of the last row, one finds that the secular equation is

$$\left(\frac{\alpha_1}{\alpha_0} + \xi\right)(-\xi)^k + \frac{\alpha_2}{\alpha_0}(-\xi)^{k-2} - \cdots + (-1)^k \frac{\alpha_k}{\alpha_0} = 0. \tag{9.391}$$

Then ξ is an eigenvalue of $\mathbf{A}$ if and only if ξ is one of the roots of the **characteristic equation**

$$\rho(\xi) = 0 \tag{9.392}$$

in which

$$\rho(\xi) := \sum_{i=0}^{k} \alpha_i \xi^{k-i}. \tag{9.393}$$

The roots of the characteristic equation are called **characteristic roots**.

By the fundamental theorem of algebra, the characteristic equation has k roots $\xi_1, \ldots, \xi_k$, some or all of which may be complex. Equal characteristic roots are called **degenerate**. Characteristic roots which are distinct are called **nondegenerate**. We assume for the time being that the characteristic roots $\xi_1, \ldots, \xi_k$ are nondegenerate. We discuss the case of degenerate roots in the next subsection.

The equations that determine the components of the eigenvector $\mathbf{v}_l$ that belongs to ξ_l are

$$
\begin{aligned}
\xi_l v_l^1 &= v_l^2, \\
\xi_l v_l^2 &= v_l^3, \\
&\vdots \\
\xi_l v_l^{k-1} &= v_l^k, \\
\xi_l v_l^k &= -\frac{\alpha_k}{\alpha_0} v_l^2 - \frac{\alpha_{k-1}}{\alpha_0} v_l^3 - \cdots - \frac{\alpha_1}{\alpha_0} v_l^k.
\end{aligned}
\tag{9.394}
$$

Because ξ_l is a root of the characteristic equation, the solution of these equations is

$$
v_l^s = \xi_l^{s-1} v_l^1.
\tag{9.395}
$$

One can take $v_l^1 = 1$ without loss of generality. Then

$$
\mathbf{v}_l = \begin{pmatrix} 1 \\ \xi_l \\ \vdots \\ \xi_l^{k-1} \end{pmatrix}.
\tag{9.396}
$$

We show that the eigenvectors $\mathbf{v}_1, \ldots, \mathbf{v}_k$ are linearly independent if and only if the characteristic roots are nondegenerate.

The columns $\mathbf{v}_1, \ldots, \mathbf{v}_k$ of the matrix $\mathbf{V}$ are linearly independent (and therefore span $\mathbb{C}^k$) if and only if $\mathbf{V}$ is nonsingular. We show in Section 6.5.2 that a matrix is nonsingular if and only if its determinant is nonzero. It follows from Eq. (9.396) that $\det[\mathbf{V}]$ is a Vandermonde determinant and therefore is nonzero if and only if $\xi_1, \ldots, \xi_k$ are nondegenerate.

If $\xi_1, \ldots, \xi_k$ are nondegenerate, then $\mathbf{v}_1, \ldots, \mathbf{v}_k$ span $\mathbb{C}^k$, and there exist scalars κ_l such that

$$
\mathbf{x}_k = \sum_{l=1}^{k} \kappa_l \mathbf{v}_l.
\tag{9.397}
$$

It follows from this equation and Eqs. (9.386) and (9.387) that

$$
\forall n > k : \mathbf{x}_n = \sum_{l=1}^{k} \kappa_l \xi_l^{n-k} \mathbf{v}_l.
\tag{9.398}
$$

Then every sequence $\overline{y}$ that solves Eq. (9.378) is a linear combination

$$
\overline{y} = \sum_{l=1}^{k} \kappa_l \overline{y}^{(l)}
\tag{9.399}
$$

of the geometric sequences defined in Eq. (9.389). The coefficients κ_l must be determined from initial conditions.

Let the initial conditions be

$$y_1 = \mu_1, \ldots, y_k = \mu_k. \tag{9.400}$$

From Eq. (9.399) it follows that

$$\forall n \in \mathbb{Z}^+ : y_n = \sum_{l=1}^{k} \kappa_l \xi_l^n. \tag{9.401}$$

Then the coefficients κ_l must obey the k linear equations

$$\sum_{l=1}^{k} \kappa_l = \mu_1$$
$$\sum_{l=1}^{k} \kappa_l \xi_l = \mu_2 \tag{9.402}$$
$$\vdots$$
$$\sum_{l=1}^{k} \kappa_l \xi_l^{k-1} = \mu_k$$

The determinant of the coefficients is a Vandermonde determinant. Therefore the linear equations Eq. (9.402) have a unique solution for $\kappa_1, \ldots, \kappa_k$ if and only if the characteristic roots are nondegenerate.

Degenerate Characteristic Roots

If some roots of the characteristic equation $\rho(\xi) = 0$ are repeated, then there are not enough distinct characteristic roots to make k linearly independent geometric sequences of the form of Eq. (9.389). Assume for simplicity that one characteristic root ξ_0 is repeated r times. Then there are $k - r + 1$ linearly independent geometric sequences, one for each of the $k - r$ distinct characteristic roots that are different from ξ_0 and one for ξ_0. We show that, for every integer l such that

$$l \in (1 : r - 1), \tag{9.403}$$

the sequence $\overline{y}^{(0,l)}$ whose n-th term is

$$y_n^{(0,l)} = n(n-1)\cdots(n-l+1)\xi_0^n \tag{9.404}$$

satisfies the homogeneous case of the linear difference equation (9.378). To do so, we make use of the identity

$$n(n-1)\cdots(n-l+1)\xi^n = \xi^l \frac{d^l}{d\xi^l} \xi^n, \tag{9.405}$$

which permits us to replace the n-dependent coefficient of ξ in Eq. (9.404) with the simpler operator $\xi^l (d/d\xi)^l$. Then

$$\alpha_0 y_n^{(0, l)} + \alpha_1 y_{n-1}^{(0, l)} + \cdots + \alpha_k y_{n-k}^{(0, l)}$$

$$= \xi^l \frac{d^l}{d\xi^l} [\alpha_0 \xi^n + \alpha_1 \xi^{n-1} + \cdots + \alpha_k \xi^{n-k}] \Big|_{\xi=\xi_0}$$

$$= \xi^l \frac{d^l}{d\xi^l} [\xi^{n-k} \rho(\xi)] \Big|_{\xi=\xi_0} \tag{9.406}$$

$$= \xi^l \left\{ \left[\frac{d^l}{d\xi^l} \xi^{n-k} \right] \rho(\xi) + \binom{l}{1} \left[\frac{d^{l-1}}{d\xi^{l-1}} \xi^{n-k} \right] \frac{d\rho}{d\xi} + \cdots + \binom{l}{l} \xi^{n-k} \frac{d^l \rho}{d\xi^l} \right\} \Big|_{\xi=\xi_0}$$

(The last equality follows from Leibniz's rule for differentiating a product.) By assumption, ρ has a zero of order r at $\xi = \xi_0$. Therefore, for every value of l in $(1 : r - 1)$,

$$\rho(\xi_0) = 0, \quad \frac{d\rho}{d\xi}(\xi_0) = 0, \ldots, \frac{d^l \rho}{d\xi^l}(\xi_0) = 0 \tag{9.407}$$

and, for every $n > k$,

$$\alpha_0 y_n^{(0, l)} + \alpha_1 y_{n-1}^{(0, l)} + \cdots + \alpha_k y_{n-k}^{(0, l)} = 0. \tag{9.408}$$

Because there are $r - 1$ sequences $\overline{y}^{(0, l)}$ and $k - r + 1$ linearly independent power sequences (9.389), it follows that the dimension of the subspace that consists of the sequences that obey the homogeneous version of Eq. (9.378) is not greater than k and is equal to k if and only if the sequences $\overline{y}^{(0, l)}$ are linearly independent.

The list $\{\overline{y}^{(0, 1)}, \ldots, \overline{y}^{(0, r-1)}\}$ is linearly independent if and only if the polynomials $\{x(x - 1) \ldots (x - r + l) \mid l = 1, \ldots, r - 1\}$ are linearly independent. It is easy to show inductively that

$$\begin{pmatrix} 1 \\ x \\ x^2 \\ \vdots \\ x^{r-1} \end{pmatrix} = \mathbf{L} \begin{pmatrix} 1 \\ x \\ x(x - 1) \\ \vdots \\ x(x - 1) \cdots (x - r + 1) \end{pmatrix} \tag{9.409}$$

in which $\mathbf{L}$ is a unit lower-triangular matrix. The matrix elements

$$\mathfrak{S}_{k-1}^{(m)} := l_m^k \tag{9.410}$$

of $\mathbf{L}$, which have the property that

$$x^k = \sum_{j=0}^{k} \mathfrak{S}_k^{(j)} x(x - 1) \cdots (x - j + 1), \tag{9.411}$$

are called **Stirling numbers of the second kind**. The $\{\mathfrak{S}_n^{(m)}\}$ can be calculated from the recurrence relation

$$\mathfrak{S}_{n+1}^{(m)} = m\,\mathfrak{S}_n^{(m)} + \mathfrak{S}_n^{(m-1)}. \tag{9.412}$$

By Eq. (6.378) $\mathbf{L}$ is nonsingular. Then the linear independence of $\{1, x, x^2, \ldots, x^{r-1}\}$ implies the linear independence of $\{1, x, x(x-1), \ldots, x(x-1)\cdots(x-r+1)\}$. It follows that $\{1, x, x(x-1), \ldots, x(x-1)\cdots(x-r+1)\}$ is a basis of the vector space spanned by $\{1, x, x^2, \ldots, x^{r-1}\}$. Therefore the sequence whose n-th element is $p(n)\xi_0^n$, in which p is any polynomial of degree not greater than $r-1$, is a solution of the homogeneous case of the linear difference equation (9.378) if the characteristic root ξ_0 is repeated r times.

9.6.3 SOLUTION OF TRIDIAGONAL SYSTEMS

Tridiagonal systems of linear equations occur often in physics and engineering, especially in the numerical solution of ordinary and partial differential equations. It is a waste of computer time and memory to apply Gaussian elimination to a tridiagonal system, because Gaussian elimination does not conserve the tridiagonal structure (thereby wasting memory) and because Gaussian elimination of a matrix that is already tridiagonal requires many more floating-point operations than a direct LU decomposition. We summarize the Thomas algorithm for the LU decomposition of a tridiagonal matrix. Let us write a tridiagonal system of linear equations as

$$\mathbf{Tx} = \mathbf{b}, \tag{9.413}$$

in which

$$\mathbf{T} = \begin{pmatrix} b_1 & c_1 & 0 & 0 & \cdots \\ a_2 & b_2 & c_2 & 0 & \\ 0 & a_3 & b_3 & c_3 & \\ \vdots & & & & \ddots \end{pmatrix} \tag{9.414}$$

The LU decomposition of $\mathbf{T}$ is

$$\mathbf{T} = \mathbf{T}_L \mathbf{T}_U, \tag{9.415}$$

in which

$$\mathbf{T}_L = \begin{pmatrix} \alpha_1 & 0 & 0 & 0 & \cdots \\ \beta_2 & \alpha_2 & 0 & 0 & \\ 0 & \beta_3 & \alpha_3 & 0 & \\ \vdots & & & & \ddots \end{pmatrix}, \quad \mathbf{T}_U = \begin{pmatrix} 1 & \gamma_1 & 0 & 0 & \cdots \\ 0 & 1 & \gamma_2 & 0 & \\ 0 & 0 & 1 & \gamma_3 & \\ \vdots & & & & \ddots \end{pmatrix} \tag{9.416}$$

It is easy to verify that the equations that define the elements of $\mathbf{T}_L$ and $\mathbf{T}_U$ are as follows:

For row 1, column 1 of $\mathbf{T}$:

$$\alpha_1 = b_1. \tag{9.417}$$

For row 1, column 2 of $\mathbf{T}$:

$$\alpha_1 \gamma_1 = c_1 \Rightarrow \gamma_1 = c_1/b_1. \tag{9.418}$$

For row $n + 1$, column n of $\mathbf{T}$:

$$\beta_n = a_n. \tag{9.419}$$

For row n, column n of $\mathbf{T}$:

$$\beta_n \gamma_{n-1} + \alpha_n = b_n \Rightarrow \alpha_n = b_n - a_n \gamma_{n-1}. \tag{9.420}$$

For row n, column $n + 1$ of $\mathbf{T}$:

$$\alpha_n \gamma_n = c_n \Rightarrow \gamma_n = \frac{c_n}{\alpha_n}. \tag{9.421}$$

The first step in solving Eq. (9.413) is to solve the system $\mathbf{T}_L \mathbf{f} = \mathbf{b}$ by forward elimination. The resulting components of $\mathbf{f}$ are

$$f_1 = \frac{d_1}{\alpha_1}, \quad f_i = \frac{d_i - \beta_i f_{i-1}}{\alpha_i} \tag{9.422}$$

for rows $i = 2, \ldots, n - 1$. We then solve $\mathbf{T}_U \mathbf{u} = \mathbf{f}$ by back-substitution:

$$u_n = f_n, \quad u_j = f_j - \gamma_j u_{j+1} \tag{9.423}$$

for rows $j = n - 1, n - 2, \ldots, 1$.

The advantages of the Thomas algorithm are that the operation count is $\sim O(n^2)$, instead of the $\sim O(n^3/3)$ operations required for Gaussian elimination of a full matrix, and that the memory required for the Thomas algorithm is $\sim O(3n)$, instead of n^2 as for a full $n \times n$ matrix.

9.6.4 EXERCISES FOR SECTION 9.6

9.6.1   Using the principle of finite induction, prove the recurrence relation for the Stirling numbers of the second kind, Eq. (9.412).

9.6.2   Write and test a computer program that implements the Thomas algorithm.

9.6.3   (Nievergelt [1991].) Consider the linear system

$$\mathbf{A}\mathbf{x} = \begin{pmatrix} 888,445 & 887,112 \\ 887,112 & 885,781 \end{pmatrix} \begin{pmatrix} x \\ y \end{pmatrix} = \begin{pmatrix} 1 \\ 0 \end{pmatrix}. \tag{9.424}$$

(a) Solve this equation *by hand* using Cramer's rule and the information that $\det[\mathbf{A}] = 1$.

(b) Solve this equation using a calculator, and compare your answer with the result you obtained by hand.

(c) Verify that **A** is of the form

$$\mathbf{A} = \begin{pmatrix} \frac{1}{2}\alpha^2 + \alpha + 1 & \frac{1}{2}\alpha^2 \\ \frac{1}{2}\alpha^2 & \frac{1}{2}\alpha^2 - \alpha + 1 \end{pmatrix}. \tag{9.425}$$

Also use this equation for **A** to verify algebraically that $\det[\mathbf{A}] = 1$.

(d) Calculate $\mathbf{A}^{-1}$ *algebraically*, and then substitute to obtain a numerical answer.

(e) Show algebraically that the eigenvalues of **A** are

$$\lambda_1 \approx \alpha^2 + 2$$
$$\lambda_2 \approx -\frac{1}{\alpha^2 + 2},$$

to $O(\alpha^{-6})$. Use this result and Eq. (9.354) to estimate the condition number of **A**.

(f) Estimate how many significant digits would have to be kept in a brute-force solution of Eq. (9.424) in order to obtain three-digit accuracy in the solution.

9.7 BIBLIOGRAPHY

Birkhoff, Garrett, and Saunders Mac Lane. *A Survey of Modern Algebra*. 3 ed. New York: Macmillan, 1965, chapter VII.

Golub, Gene H., and Charles F. Van Loan. *Matrix Computations*. 2 ed. Baltimore: Johns Hopkins University Press, 1989.

Halmos, Paul R. *Finite-Dimensional Vector Spaces*. 2 ed. Princeton: Van Nostrand, 1958, chapter III.

Kolmogorov, Andrei Nikolaevich, and Sergei Vasilevich Fomin. *Introductory Real Analysis*. New York: Dover, 1975, chapters 4, 5.

Nievergelt, Y. *American Mathematical Monthly* 98 (1991):539–543.

CHAPTER 10

CONVERGENCE IN NORMED VECTOR SPACES

10.1 METRICS AND NORMS

The purpose of this chapter is to generalize the concepts of limit, convergence, and continuity from real-valued functions of a single real variable to vector-valued functions of vector variables. We go far beyond ordinary multivariable calculus in some respects, because the "vectors" may belong to infinite-dimensional spaces and the "functions" may be operators such as the Laplacian, the Helmholtz operator, and the wave operator.

We are not pursuing generality for its own sake. Our goal for this chapter is to lay a foundation for reasoning correctly about vectors and operators in separable Hilbert spaces such as one encounters in quantum mechanics, as well as in separable Banach spaces, which are important for estimating the accuracy of finite-dimensional numerical approximations to infinite-dimensional operators.

10.1.1 METRIC SPACES

In linear algebra and linear functional analysis one generalizes many concepts of three-dimensional Euclidean geometry to spaces of arbitrary finite or countable dimensionality. Along the way one generalizes the Euclidean definition of distance by generalizing the concept of vector norm. If one studies the consequences of the existence of a distance between points *without* assuming that the points represent elements of a vector space, then the structure that one studies is called a *metric space*. Metric spaces were first defined and studied by Maurice Fréchet in his doctoral dissertation (1906).

Axioms for a Metric Space

Let S be a set; in the present context we call the elements of S **points**. A function $\rho : S \times S \to \mathbb{R}$ is called a **metric** (or **distance**) if and only if the following axioms are obeyed for all points $x, y \in S$:

1. *Positive-definiteness:* The distance between two different points is always positive,

$$x \neq y \Rightarrow \rho(x, y) > 0, \tag{10.1}$$

and the distance from a point to itself is always zero,

$$\rho(x, x) = 0. \tag{10.2}$$

2. *Symmetry:* The distance from y to x is equal to the distance from x to y,

$$\rho(x, y) = \rho(y, x). \tag{10.3}$$

3. *Triangle inequality:*

$$\forall z \in S : \rho(x, y) \le \rho(x, z) + \rho(z, y), \tag{10.4}$$

that is, the distance from x to y cannot be shortened by going through any other point z.

A **metric space** is an ordered pair (S, ρ) in which S is a set and ρ is a metric defined on S. We refer to S itself as a metric space if it is clear from the context that only one metric is meant.

Examples of Metric Spaces

The set of real numbers $\mathbb{R}$ is a metric space under the **absolute-value metric**

$$\rho(x, y) = |x - y|. \tag{10.5}$$

For another example, let S be any set and let ρ be the **trivial metric**

$$\forall x, y \in S : \rho(x, y) := \begin{cases} 0, & \text{if } y = x; \\ 1, & \text{if } y \neq x. \end{cases} \tag{10.6}$$

It is straightforward to show that S is a metric space under ρ.

An example of some practical importance concerns ordered n-tuples. Let $X = (x_1, x_2, \ldots, x_n)$ and $Y = (y_1, y_2, \ldots, y_n)$, in which the components belong to the same set A, that is, $\forall i \in (1 : n) : x_i, y_i \in A$. (One sometimes calls A the **alphabet** to which the components belong.) It can be shown that the **Hamming distance** between X and Y, which is defined as the number of values of i such that $y_i \neq x_i$, is a metric on the set of n-tuples with components in A. For instance, let $A = \{0, 1\}$ and $n = 8$; then ordered 8-tuples with components in A are strings of 8 bits, that is, bytes. The Hamming distance between the bit strings 11010001 and 11111111 is equal to 4. The Hamming distance can be used to construct error-correcting codes and to decode fixed-length codewords in the presence of multiple errors.

From the definition of a metric space (S, ρ) one sees that *every* subset $T \subseteq S$ is also a metric space under the metric ρ. For example, the interval $[a \ b] \subset \mathbb{R}$ is a metric space under the metric Eq. (10.5). A less trivial example is the Cantor set, which is a proper subset of $\mathbb{R}$ and is a metric space under the metric Eq. (10.5).

Distance Between Sets

Let S be a metric space and let X be a subset of S. If $y \in S$, then the **distance** $\rho(y, X)$ **from y to X** is defined as the greatest lower bound of the set of distances from y to the points of X:

$$\rho(y, X) := \inf_{x \in X} \rho(y, x). \tag{10.7}$$

For example, let $S = \mathbb{R}$, $y = 2$ and let X be the open interval $(0, 1)$, $X = \{x \in \mathbb{R} \mid 0 < x < 1\}$. Then $\rho(y, X) = 1$.

If X and Y are non-empty subsets of a metric space S, then the **distance between X and Y** is defined as the greatest lower bound of the distances from the points of Y to the set X:

$$\rho(Y, X) := \inf_{y \in Y} \inf_{x \in X} \rho(y, x). \tag{10.8}$$

For example, let $X = (0, 1) = \{x \in \mathbb{R} \mid 0 < x < 1\}$, $Y = (1, 2) = \{y \in \mathbb{R} \mid 1 < y < 2\}$ and $Z = (2, 3) = \{z \in \mathbb{R} \mid 2 < z < 3\}$. Then $\rho(Y, X) = 0 = \rho(Z, Y)$ and $\rho(Z, X) = 1$.

Metric Vector Spaces

A metric space $(\mathcal{V}, \rho)$, in which $\mathcal{V}$ is a vector space, is called a **metric vector space**. The concept of a metric vector space is useful because many important properties such as the convergence of sequences of vectors, the continuity of functions of vectors and the theory of distributions depend on the metric-space axioms.

A metric ρ on a vector space $\mathcal{V}$ is called **translation-invariant** if and only if

$$\forall \bar{x}, \bar{y}, \bar{z} \in \mathcal{V} : \ \rho(\bar{x}, \bar{y}) = \rho(\bar{x} + \bar{z}, \bar{y} + \bar{z}). \tag{10.9}$$

A translation-invariant distance function depends only on the difference of its arguments. Take $\bar{z} = -\bar{y}$; then the previous equation becomes

$$\rho(\bar{x}, \bar{y}) = \rho(\bar{x} - \bar{y}, \bar{0}). \tag{10.10}$$

Section 10.1.2 defines a large class of translation-invariant metrics.

10.1.2 NORMED VECTOR SPACES

We recall from Section 8.7 that a vector norm $\| \cdot \|$ is a generalization of vector length that obeys the following axioms:

1. *Positive-definiteness:*

$$\bar{x} \neq \bar{0} \Rightarrow \|\bar{x}\| > 0 \tag{10.11}$$

and

$$\|\bar{0}\| = 0. \tag{10.12}$$

2. *Homogeneity:*

$$\|\alpha \bar{x}\| = |\alpha| \|\bar{x}\|. \tag{10.13}$$

3. *The triangle inequality:*

$$\|\bar{x} + \bar{y}\| \leq \|\bar{x}\| + \|\bar{y}\|. \tag{10.14}$$

A vector space on which a norm that obeys these axioms is defined is called a **normed vector space** (or a **normed linear space**). A normed vector space is a metric space under the metric that is defined by the norm,

$$\rho(\overline{x}, \overline{y}) := \|\overline{x} - \overline{y}\|. \tag{10.15}$$

If no real-valued function that obeys axioms Eq. (10.11) through (10.14) can be defined on a vector space $\mathcal{V}$, then one says that $\mathcal{V}$ is **not normable**.

For example, the set of complex numbers $\mathbb{C}$ is a normed vector space under the modulus norm

$$\rho(z_1, z_2) = |z_1 - z_2| = [(x_1 - x_2)^2 + (y_1 - y_2)^2]^{1/2}, \tag{10.16}$$

in which $z_j = x_j + iy_j$ and x_j, y_j belong to $\mathbb{R}$.

We show in Section 8.7 that in Euclidean and unitary spaces the inner-product norm

$$\|\overline{x}\|_{ip} := \langle \overline{x}, \overline{x} \rangle^{1/2} \tag{10.17}$$

fulfills axioms (10.11) through (10.14). Each inner product (that is, each different metric tensor or Gram matrix) defines a different norm. By the way, the expression for the square of the inner-product metric,

$$\rho(\overline{x}, \overline{0})^2 = \langle \overline{x}, \overline{x} \rangle = g_{ij} x^i x^j, \tag{10.18}$$

justifies the name of the metric tensor.

Every normed vector space is also a metric vector space under the definition

$$\rho(\overline{x}, \overline{y}) := \|\overline{x} - \overline{y}\| \Rightarrow \rho(\overline{x} + \overline{y}, \overline{y}) = \rho(\overline{x}, \overline{0}) = \|\overline{x}\|. \tag{10.19}$$

Clearly the metric $\|\overline{x} - \overline{y}\|$ is translation-invariant.

For example, one knows from Section 8.7 that every finite-dimensional Euclidean or unitary space is a normed vector space under any of the p-norms

$$\|\overline{x}\|_p := \left[\sum_{i=1}^{n} |x^i|^p \right]^{1/p}. \tag{10.20}$$

In the special case $p = \infty$ then one has the maximum norm,

$$\|\overline{x}\|_\infty := \max_i |x^i|, \tag{10.21}$$

which is particularly convenient for computational purposes.

Exercise 8.2.14 shows that if $p \neq 2$ there is no inner product such that $\langle \overline{x}, \overline{x} \rangle^{1/2} = \|\overline{x}\|_p$, because the p-norm violates the parallelogram law $\|\overline{x} + \overline{y}\|^2 + \|\overline{x} - \overline{y}\|^2 = 2(\|\overline{x}\|^2 + \|\overline{y}\|^2)$ unless $p = 2$.

Although a normed vector space is a metric space, the converse is not necessarily true because there are metric vector spaces that are not normable. The metric-space axioms do not require translation invariance or the property of homogeneity, Eq. (10.13). We show in a subsequent section that the so-called Fréchet metric on $\mathbb{F}^\infty$ is translation-invariant but is not homogeneous. In fact, it can be shown that $\mathbb{F}^\infty$ is not normable. The space $\mathcal{D}([a, b]; \mathbb{R})$ of infinitely differentiable functions of compact support in $[a, b]$ and the space $\mathcal{S}$ of rapidly decreasing functions on $\mathbb{R}$, both of which we introduce in this chapter, are examples of vector spaces that are not normable. Both $\mathcal{D}([a, b]; \mathbb{R})$ and $\mathcal{S}$ are fundamental in the study of distributions such as the Dirac delta "function".

Other Forms of the Triangle Inequality

Quite often one needs to use the triangle inequality in forms that do not look the same as Eq. (10.14). One sees at once that each of the relations

$$\|\bar{x}\| - \|\bar{y}\| \leq \|\bar{x} \pm \bar{y}\| \tag{10.22}$$

and

$$\|\bar{y}\| - \|\bar{x}\| \leq \|\bar{x} \pm \bar{y}\| \tag{10.23}$$

is equivalent to Eq. (10.14). Then

$$|\|\bar{x}\| - \|\bar{y}\|| \leq \|\bar{x} - \bar{y}\|. \tag{10.24}$$

This form of the triangle inequality is useful if one wants to obtain an upper bound rather than a lower bound.

10.1.3 EXAMPLES OF METRIC AND NORMED VECTOR SPACES

Sequence Spaces

The vector space $m(\mathbb{C})$ of bounded complex sequences $\bar{z} = (z_1, z_2, \ldots)$ (in which every $|z_i| \leq M_{\bar{z}}$ for some real number $M_{\bar{z}}$) is a normed vector space under the **discrete supremum norm**

$$\rho(\bar{z}, \bar{z}') = \sup_{i \in \mathbb{Z}^+} |z_i - z_i'|. \tag{10.25}$$

The supremum norm is not useful for the larger vector space $\mathbb{F}^\infty$ of *all* sequences with elements in $\mathbb{F}$, because $|z_i - z_i'|$ has no upper bound if $\bar{z}$ and $\bar{z}'$ are arbitrary sequences. However, under the **Fréchet metric**

$$\forall \bar{x}, \bar{y} \in \mathbb{F}^\infty : \quad \rho(\bar{x}, \bar{y}) := \sum_{i=1}^{\infty} \frac{1}{2^i} \frac{|x_i - y_i|}{1 + |x_i - y_i|} \tag{10.26}$$

$\mathbb{F}^\infty$ is a metric vector space. By comparing the series Eq. (10.26) with the geometric series $\sum_{1}^{\infty} 2^{-i} = 1$ one sees that

$$\rho(\bar{x}, \bar{y}) \leq 1 \tag{10.27}$$

for *every* pair of sequences. The Fréchet metric satisfies the axioms of positive-definiteness and symmetry and is translation-invariant. It takes a little work to establish the triangle inequality (see Exercise 10.1.4). However, the metric defined in Eq. (10.26) does not have the property of homogeneity, because

$$\rho(\alpha\overline{x}, \alpha\overline{y}) = \sum_{i=1}^{\infty} \frac{1}{2^i} \frac{|\alpha|(|x_i - y_i|)}{1 + |\alpha|(|x_i - y_i|)} \neq |\alpha|\rho(\overline{x}, \overline{y}). \tag{10.28}$$

Therefore the Fréchet metric cannot be derived from a norm.

ℓ^p Spaces

The vector space $\ell^p(\mathbb{Z}^+, \mathbb{F})$ such that

$$\sum_{i=1}^{\infty} |x_i|^p < \infty \tag{10.29}$$

(in which $1 \leq p \leq \infty$ and every $x^i \in \mathbb{F}$) is a normed vector space. The norm of the sequence $\overline{x} = (x_1, x_2, \ldots)$ is

$$\|\overline{x}\|_p := \left[\sum_{i=1}^{\infty} |x_i|^p \right]^{1/p}. \tag{10.30}$$

The properties of positive-definiteness and homogeneity are immediate, but the proof that the norm defined in Eq. (10.30) satisfies the triangle inequality is difficult. As is true for the finite-dimensional p-norms Eq. (10.20), the infinite-dimensional p-norm can be derived from an inner product via Eq. (10.17) if and only if $p = 2$.

Space of Continuous Real-Valued Functions on [a, b]

The vector space $\mathcal{C}^0([a, b]; \mathbb{R})$ of continuous real-valued functions on the interval $[a, b]$ is a normed vector space under the **supremum** (or **uniform** or **Chebyshev**) **norm**

$$\|f\|_C = \sup_{x \in [a,b]} |f(x)|. \tag{10.31}$$

P. L. Chebyshev used this norm in his studies of the best polynomial approximation to a continuous function on an interval $[a, b]$.

Spaces of Continuously Differentiable Functions

The space $\mathcal{C}^m([a, b]; \mathbb{R})$ of m-times continuously differentiable real-valued functions on $[a, b]$ is a vector subspace of $\mathcal{C}^0([a, b]; \mathbb{R})$ and therefore is a normed vector space under the supremum norm. A different norm on $\mathcal{C}^m([a, b]; \mathbb{R})$,

$$\|f\| := \sum_{m=0}^{n} q_m[f] \tag{10.32}$$

in which

$$q_m[f] := \sup_{x \in [a,b]} |f^{(m)}(x)|, \tag{10.33}$$

incorporates a requirement of *smoothness*, for two functions f and g are "close" under such a norm if and only if the difference $f - g$ and its first n derivatives are small. The greater n is, the smoother $f - g$ must be for a given value of $\|f - g\|$.

If one lets n become infinite, then the sum in Eq. (10.32) does not converge for many functions in $C^\infty([a, b]; \mathbb{R})$. However, one can incorporate the requirement of smoothness by defining the metric

$$\rho(f, g) := \sum_{m=0}^\infty \frac{1}{2^{m+1}} \frac{q_m[f - g]}{1 + q_m[f - g]}, \tag{10.34}$$

which is clearly translation-invariant but not homogeneous and hence is not a norm. A subspace of $C^\infty([a, b]; \mathbb{R})$ under this metric is important in distribution theory.

Another subspace of $C^\infty([a, b]; \mathbb{R})$ that is important for distribution theory is that of the real-valued *rapidly decreasing functions* on $\mathbb{R}$. Let

$$q_{l,m}[f] := \sup_{x \in \mathbb{R}} \left| |x|^l f^{(m)}(x) \right|. \tag{10.35}$$

A function $f : \mathbb{R} \to \mathbb{R}$ and its derivatives are called **rapidly decreasing** if and only if

$$\forall \epsilon > 0 : \forall l, m = 0, 1, \ldots : \exists B_{l,m} > 0 :\ni: \forall |x| > B_{l,m} : \sup_{|x| < B_{l,m}} \left| |x|^l f^{(m)}(x) \right| < \epsilon. \tag{10.36}$$

For example, $f(x) = e^{-x^2}$ is a rapidly decreasing function. Let S be the space of rapidly decreasing functions. Clearly S is a metric vector space if one defines the distance between two functions f, g as

$$\rho^S(f, g) := \sum_{l,m=0}^\infty \frac{1}{2^{l+m+2}} \frac{q_{l,m}[f - g]}{1 + q_{l,m}[f - g]}. \tag{10.37}$$

The metric ρ^S is translation-invariant but is not homogeneous.

It can be shown that the Fourier transform of a rapidly decreasing function is also rapidly decreasing. For example, the Fourier transform $\tilde{f}$ of the Gaussian function $f(x) = e^{-x^2/\sigma^2}$ is a Gaussian, for $\tilde{f}(k) = \sqrt{2\pi}\sigma e^{-\sigma^2 k^2/4}$. An important approach to defining distributions such as the Dirac delta "function" uses sequences of rapidly decreasing functions.

L^p Spaces

The vector space $L^p([a, b]; \mathbb{R})$ such that

$$\int_a^b |f(x)|^p \, dx < \infty \tag{10.38}$$

is a normed vector space under the norm

$$\|f\|_p := \left[\int_a^b |f(x)|^p \, dx \right]^{1/p} \tag{10.39}$$

in which $f : [a, b] \to \mathbb{R}$ is any function such that $|f|$ is Riemann-integrable. (The integrals in Eqs. [10.38] and [10.39] are really Lebesgue integrals, but it does little harm for most practical purposes to regard them as Riemann integrals.)

The greater the value of p, the greater the contribution to the integral of any subinterval of $[a, b]$ in which $|f|$ achieves its maximum value. If one lets $p \to \infty$, then $\|f\|_p$ approaches the supremum norm $\sup_{x \in [a,b]} |f(x)|$.

10.1.4 OPEN SETS

Neighborhoods

Certain kinds of subsets of a metric space play especially important roles in topology and analysis, as well as in many applications in physics and engineering. If $x \in S$ and $r > 0$, the **open sphere** (or **open ball**) $S_r(x)$ is

$$S_r(x) := \{z \in S \mid \rho(x, z) < r\}. \tag{10.40}$$

Note carefully that the definition requires *strict* inequality. One also calls $S_r(x)$ a **neighborhood** of the point x.

For example, on the real line $\mathbb{R}$ under the absolute-value norm, every neighborhood is an open interval:

$$S_r(x) = \{y \in \mathbb{R} \mid |y - x| < r\} = (x - r, x + r). \tag{10.41}$$

In the complex plane $\mathbb{C}$ with distance defined by the modulus norm, a neighborhood is

$$S_r(z) = \{z' \in \mathbb{C} \mid |z - z'| < r\}, \tag{10.42}$$

which is an **open disk** of radius r, centered at z. Exercises 10.1.1 and 10.1.2 give additional examples.

If S is a normed vector space $\mathcal{V}$, then the axiom of homogeneity, Eq. (10.13), implies that the norm of a vector is unchanged by inversion in the origin:

$$\|-\overline{x}\| = \|\overline{x}\|. \tag{10.43}$$

Therefore every open sphere $S_r(\overline{x})$ in $\mathcal{V}$ is invariant under the operation of inversion in the point $\overline{x}$,

$$\forall \overline{y} \in S_r(\overline{x}) : \overline{y} \mapsto \overline{y}' = 2\overline{x} - \overline{y}, \tag{10.44}$$

which carries $\overline{y} - \overline{x}$ onto $\overline{x} - \overline{y}$. To use physicists' terminology, in a normed vector space every open sphere centered at $\overline{0}$ is invariant under the parity operation ($\overline{x} \mapsto -\overline{x}$).

A subset O of a metric space S is called **open** if and only if it includes at least one neighborhood of every one of its points:

$$\forall x \in O : \exists r > 0 :\ni: \ S_r(x) \subseteq O. \tag{10.45}$$

From an intuitive point of view, an open set is like a loaf of bread without the crust. However, one should not carry this analogy too far. Exercise 10.1.8 gives an example in which every point is an open set.

An **interior point** of a subset $X \subseteq S$ is any point $x \in X$ such that, for some $r > 0$, $S_r(x) \subseteq X$. The set of interior points of X is called the **interior** of X and is written $X°$. Therefore one could rephrase the definition of an open set by saying that O is open if and only if $O° = O$, that is, if and only if O consists entirely of interior points. The whole metric space S is an open set, because all of its points are interior points.

To show that an open sphere $S_r(x)$ is an open set, one must show that

$$\forall y \in S_r(x) : \exists s > 0 :\ni: \ S_s(y) \subseteq S_r(x). \tag{10.46}$$

Let

$$s = r - \rho(x, y). \tag{10.47}$$

The strict inequality in Eq. (10.40) implies that $s > 0$. From the triangle inequality one has

$$\forall z \in S_s(y) : \ \rho(z, x) \le \rho(z, y) + \rho(y, x) < s + \rho(x, y) = r, \tag{10.48}$$

which establishes that $S_s(y) \subseteq S_r(x)$. Clearly any value of s such that $0 < s \le r - \rho(x, y)$ will do; see Fig. 10.1.

Unions and Intersections of Open Sets

It follows from the definition of an open set that *every union of open sets is open,* because any point that belongs to the union

$$\bigcup_{i \in I} O_i \tag{10.49}$$

belongs to at least one of the open sets O_i. For example, the union of all of the open sets O_i such that $O_i \subseteq X$ is the set that consists of all of the interior points of X. Therefore

$$X° = \bigcup_{O \subseteq X} O. \tag{10.50}$$

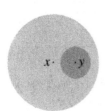

FIGURE 10.1

$S_r(x)$ and $S_s(y)$ if $0 < s < r - \rho(x, y)$.

It follows also from the definition of an open set that *the intersection of a finite collection of open sets O_i is open.* Let

$$O := \bigcap_{i=1}^{n} O_i. \tag{10.51}$$

Then $x \in O$ if and only if $x \in O_i$ for every $i \in (1 : n)$. For every such i, there exists a real number $r_i > 0$ such that $S_{r_i}(x) \subseteq O_i$. Let

$$r := \min_{1 \le i \le n} r_i. \tag{10.52}$$

Then

$$\forall i \in (1 : n) : \quad S_r(x) \subseteq O_i \tag{10.53}$$

and therefore

$$S_r(x) \subseteq O, \tag{10.54}$$

which shows that O is open. The theorem is true even if O is empty, because $\emptyset$, having no points, is trivially open.

A simple example shows that the intersection of an *infinite* set of open sets need not be open. Let $S = \mathbb{R}$, $\rho(x, y) = |x - y|$, and

$$O_i := S_{r_i}(x), \quad r_i = \frac{1}{i}. \tag{10.55}$$

Then the intersection of the sets O_i is

$$\bigcap_{i=1}^{\infty} O_i = \{x\}, \tag{10.56}$$

which is not an open subset of $\mathbb{R}$ under the absolute-value norm.

Other, more general definitions of an open set and a neighborhood are normally used in courses on general topology. The definitions given above suffice for understanding separable Hilbert spaces at a level that is useful for most purposes in physics and engineering.

10.1.5 EXERCISES FOR SECTION 10.1

10.1.1 The following questions concern neighborhoods in $\mathbb{R}^2$ and the vectors

$$\mathbf{x} = \begin{pmatrix} x^1 \\ x^2 \end{pmatrix}, \quad \mathbf{y} = \begin{pmatrix} y^1 \\ y^2 \end{pmatrix}. \tag{10.57}$$

When you are asked to make a sketch of a set, use shading to indicate the points that belong to the set. Outline the area that represents the set with a dotted line. Do not draw a solid outline unless the points represented by the outline actually belong to the set.

(a) Sketch $S_1(\mathbf{0})$, in which $\mathbf{0}$ is the zero vector, under the norm

$$\|\mathbf{x} - \mathbf{y}\|_1 = |x^1 - y^1| + |x^2 - y^2|. \tag{10.58}$$

(b) Sketch $S_1(\mathbf{0})$ under the norm

$$\|\mathbf{x} - \mathbf{y}\|_2 = [|x^1 - y^1|^2 + |x^2 - y^2|^2]^{1/2}. \tag{10.59}$$

(c) Sketch $S_1(\mathbf{0})$ under the norm

$$\|\mathbf{x} - \mathbf{y}\|_{10} = [|x^1 - y^1|^{10} + |x^2 - y^2|^{10}]^{1/10}. \tag{10.60}$$

(d) Sketch $S_1(\mathbf{0})$ under the norm

$$\|\mathbf{x} - \mathbf{y}\|_\infty = \max[|x^1 - y^1|, |x^2 - y^2|]. \tag{10.61}$$

(e) Explain the progression of shapes from (a) through (d).

10.1.2 This problem concerns neighborhoods in $\mathcal{C}^0([0, 1]; \mathbb{R})$ under the supremum norm, Eq. (10.31). Let $x \in [0, 1]$ and let $y = f(x)$, in which $f \in \mathcal{C}^0([0, 1]; \mathbb{R})$. (Follow the directions on sketching neighborhoods in Exercise 10.1.1.)

(a) Let $f \in S_1(\mathbf{0})$, in which $\mathbf{0}$ is the zero function. Sketch the region of the x-y plane to which the graph of f is confined.

(b) Let $y = f(x) = \sin(\pi x)$, and suppose that $g \in S_1(f)$. Sketch the region of the x-y plane to which the graph of g is confined.

10.1.3 Prove that if y is a common point of two neighborhoods,

$$y \in S_r(x), \quad y \in S_{r'}(x'), \tag{10.62}$$

then the distance $\rho(x, x')$ is less than the sum of r and r':

$$\rho(x, x') \leq r + r'. \tag{10.63}$$

10.1.4 The goal of this problem is to establish the triangle inequality for the Fréchet metric, Eq. (10.26).

(a) Show (without calculating a derivative) that for all $t > 0$,

$$f(t) := 1 - \frac{1}{1+t} = \frac{t}{1+t} \tag{10.64}$$

is a strictly increasing function of t.

(b) Show that for all real numbers ξ, η,

$$\frac{|\xi + \eta|}{1 + |\xi + \eta|} \leq \frac{|\xi|}{1 + |\xi|} + \frac{|\eta|}{1 + |\eta|}. \tag{10.65}$$

Hint: First establish the result for positive ξ and η. Prove that the result holds also when ξ and η are both negative. For the remaining case, in which ξ and η have opposite signs, assume that $|\xi| \geq |\eta|$; therefore $|\xi + \eta| \leq |\xi|$. Then use (a).

(c) Using (b), show that the Fréchet metric defined in Eq. (10.26) obeys the triangle inequality.

10.1.5 Say whether each of the following functionals in $\mathbb{R}^2$ is a norm, and explain your answers:

(a)

$$f_1(x^1, x^2) := |x^1| \tag{10.66}$$

(b)

$$f_2(x^1, x^2) := \frac{|x^1|}{1 + |x^1|} + |x^2| \tag{10.67}$$

(c)

$$f_3(x^1, x^2) := \left[|x^1|^{1/2} + |x^2|^{1/2}\right]^{1/2} \tag{10.68}$$

10.1.6 Let $\mathcal{V}$ be a real or complex vector space of finite dimension n. For any $\bar{x} \in \mathcal{V}$, let

$$f(\bar{x}) := \left[\sum_{i=1}^{n} |x^i|^p\right]^{1/p} \tag{10.69}$$

in which p is any real number such that

$$0 < p < 1. \tag{10.70}$$

Is f a norm, or not? Justify your answer with a proof.

10.1.7 Show that the Hamming distance between codewords of a fixed length l satisfies the axioms for a metric, Eqs. (10.1) through (10.4).

10.1.8 Let S be an arbitrary set. Define a metric on S as follows:

$$\rho(x, y) := \begin{cases} 0, & \text{if } y \neq x; \\ 1, & \text{if } y = x. \end{cases} \tag{10.71}$$

(a) Demonstrate that S satisfies the axioms for a metric space.

(b) Let $x \in S$. Prove that $\{x\}$ is open.

10.2 LIMIT POINTS

10.2.1 LIMIT POINTS AND CLOSED SETS

Among the reasons why metric spaces are important in physics and engineering (as well as in pure mathematics) is that the concept of a metric space helps one to generalize properties such as limit, convergence and continuity from the real line to vector spaces of arbitrary finite

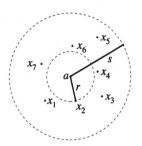

FIGURE 10.2

The minimum distance from a to an element of $L = \{x_1, \ldots, x_7\}$ is r.

dimension or of countable dimensionality. We begin this process of generalization now by laying the foundation for the concept of a limit.

Definition of a Limit Point

A point a of a metric space S is called a **limit point** (or an **accumulation point** or a **condensation point**) of a subset $L \subseteq S$ if and only if every neighborhood of a contains a point $x \neq a$ such that $x \in L$. Note that a limit point of L may or may not belong to L. For example, 0 is the only limit point of the sequence $(1, 2^{-1}, \ldots, 2^{-n}, \ldots)$, but 0 is not a term of the sequence. The sequence

$$\left(\tfrac{1}{2}, -\tfrac{1}{2}, \tfrac{3}{4}, -\tfrac{3}{4}, \ldots, 1 - 2^{-n}, -1 + 2^{-n}, \ldots \right), \tag{10.72}$$

has two limit points, $+1$ and -1.

Every neighborhood of a limit point a of $L \subseteq S$ includes an infinite subset of L. To see why, suppose (for the sake of establishing a contradiction) that a is a limit point of L and that there exists a neighborhood $S_s(a)$ such that the intersection $S_s(a) \cap L$ is a finite set. Let $\{x_1, x_2, \ldots, x_n\}$ be the points of L that belong to $S_s(a)$, such that $x_i \neq a$ for every $i \in (1 : n)$, as shown in Fig. 10.2. Let r be the distance of the set $\{x_1, x_2, \ldots, x_n\}$ from a:

$$r := \min_{1 \leq i \leq n} \rho(a, x_i). \tag{10.73}$$

Certainly $r > 0$, because every $x_i \neq a$. But the neighborhood $S_r(a)$ contains no point of L that is different from a, contradicting the assumption that a is a limit point of L. The converse is obvious (if every neighborhood of a contains an infinite subset of L, then by definition a is a limit point of L). Therefore *a is a limit point of $L \subseteq S$ if and only if every neighborhood of a contains an infinite subset of L.*

Closed Sets

A subset C of a metric space S is called **closed** if and only if C contains *every* limit point of C. For example, the interval $[0, 1]$ is a closed subset of $\mathbb{R}$ under the absolute-value norm. The half-open interval $(0, 1]$ is not closed because it does not contain its limit point 0. A metric space S is closed, because every limit point must belong to S by definition.

A finite subset of S is closed because it has no limit points, and therefore it satisfies the definition trivially. For example, let $S = \mathbb{R}$ under the absolute-value norm. Then the set $\{0\}$ is closed. By the same argument, the empty set is closed.

Let S be a metric space. We prove that *a set $C \subseteq S$ is closed if and only if its complement $S \setminus C$ is open, and a set $O \subseteq S$ is open if and only if its complement is closed.* Let us show first that the complement of a closed set is open. Let $C \subseteq S$ be closed. If $S \setminus C = \emptyset$, then we are done, because $\emptyset$ is open. If $S \setminus C$ is non-empty, let $x \in S \setminus C$. By the definitions of a complement and of a closed set, x cannot be a limit point of C. Then there exists at least one real number $s > 0$ such that $S_s(x)$ contains (at most) a finite subset $\{z_1, z_2, \ldots, z_n\} \subset C$ such that $\forall i \in (1 : n) : z_i \neq x$. It may happen that $S_s(x) \cap C = \emptyset$. In that case $S_s(x) \subseteq S \setminus C$, and there is nothing left to prove. If $n > 0$, let

$$t := \min_{1 \le i \le n} \rho(z_i, x) \tag{10.74}$$

and choose r such that $0 < r < t$. Then $S_r(x)$ contains no points of C, which implies that $S_r(x) \subseteq S \setminus C$. Therefore $S \setminus C$ is open. To establish that the complement of every open set is closed, let $O \subseteq S$ be open and let l be a limit point of $S \setminus O$. If $l \in O$, then there exists some $s > 0$ such that $S_s(l) \subset O$. But l is a limit point of $S \setminus O$, which implies that $S_s(l)$ contains at least one point of $S \setminus O$ (a contradiction). Then $l \in S \setminus O$, showing that $S \setminus O$ is closed. Note that S and the empty set $\emptyset$ are both open and closed.

The Closure of a Set

It follows from DeMorgan's Laws, $S \setminus (\cap_i A_i) = \cup_i (S \setminus A_i)$ and $S \setminus (\cup_i A_i) = \cap_i (S \setminus A_i)$, (Eqs. [2.98] through [2.106]) and from the discussion associated with Eq. (10.51) that *every intersection of closed sets is closed* and that *the union of a finite set of closed sets is closed.* However, the union of an infinite collection of closed sets need not be closed. For example, every open set is the union of its points, but each point is a closed set.

The **closure** $\overline{M}$ of a subset $M \subseteq S$ is defined as the intersection of all closed sets that include M:

$$\overline{M} := \bigcap_{Q \in K} Q \tag{10.75}$$

in which

$$K := \{Q \subseteq S \mid Q \text{ is closed and } M \subseteq Q\}. \tag{10.76}$$

Certainly $\overline{M}$ is closed, because $\overline{M}$ is an intersection of closed sets. Moreover,

$$M \subseteq \overline{M}, \tag{10.77}$$

because M is included in every set in K and is therefore included in the intersection Eq. (10.75). If C is closed and $M \subseteq C$, then $C \in K$ and therefore $\overline{M} \subseteq C$. Defining $\overline{M}$ as the intersection of a certain collection of sets ensures that $\overline{M}$ is unique and that $\overline{M}$ is the *smallest* closed set that includes M. Finally,

$$M \text{ closed } \Leftrightarrow M = \overline{M}, \tag{10.78}$$

for if M is closed, then $M \in K$.

Because the closure $\overline{M}$ of a set M is closed, it follows that $\overline{M}$ contains every limit point of M. Let L be the set of limit points of M. L is a subset of $\overline{M}$. Then the union $L \cup M$ is included in $\overline{M}$. Moreover, if y is a limit point of L, then y is a limit point of M, for every neighborhood S_ϵ of y contains an infinite subset N of L. Every neighborhood $S_\eta(z)$ of any point $z \in N$ includes an infinite subset of M. Choose η such that $0 < \eta < \epsilon - \rho(y, z)$; then $S_\eta(z)$ is a subset of $S_\epsilon(y)$. It follows that $S_\epsilon(y)$ includes an infinite subset of M and therefore y is a limit point of M. Then the union $L \cup M$ is a closed set of which M is a subset. From the definition of $\overline{M}$ it follows that $\overline{M} \subseteq L \cup M$. Because we have already established the opposite inclusion, $L \cup M \subseteq \overline{M}$, it follows that

$$\overline{M} = L \cup M, \tag{10.79}$$

that is, that the closure of a set is equal to the union of the set and the collection of all of its limit points.

Let X be a nonempty set of real numbers that is bounded from above, that is, such that X has an upper bound. The closure $\overline{X}$ of X has the important property that $\overline{X}$ contains the least upper bound $\sup X$ of X:

$$X \subset \mathbb{R} \quad \text{and} \quad \exists M \in \mathbb{R} :\ni: \forall x \in X : x \leq M \Rightarrow \sup X \in \overline{X}. \tag{10.80}$$

To see this, let $y = \sup X$. If $y \in X$, then $y \in \overline{X}$ and we are through. Assume then that $y \notin X$. For every $\epsilon > 0$, a point of X belongs to the open interval $(y - \epsilon, y + \epsilon)$ (for otherwise $y - \epsilon$ would be an upper bound of X that is less than y). Then y is a limit point of X and therefore belongs to the closure $\overline{X}$.

The Support of a Function

The **support** supp[f] of a function $f : S \to \mathbb{F}$ (in which $\mathbb{F} = \mathbb{R}$ or $\mathbb{C}$) is defined as the closure of the subset of S on which f is nonzero:

$$\text{supp}[f] := \overline{\{x \in S \mid f(x) \neq 0\}}. \tag{10.81}$$

For example, if $S = \mathbb{R}$ then supp $[\sin(\cdot)] = \overline{\mathbb{R} \setminus \pi\mathbb{Z}} = \mathbb{R}$, in which the elements of the set $\pi\mathbb{Z}$ are the integral multiples of π.

Functions of bounded support on $\mathbb{R}^n$, that is, functions that vanish for all arguments that fall outside of a bounded, closed subset of $\mathbb{R}^n$, can be used to represent signals and waves, that are localized in space. For example, a pure sine wave $f(t) = \sin \omega t$ is not localized, because its support is all of $\mathbb{R}$. The support of the function f such that

$$f(t) = \begin{cases} \exp(-(t - t_0)^{-2}(t - t_1)^{-2}) \sin \omega t, & \text{if } t \in (t_0, t_1), \text{ or} \\ 0, & \text{if } t \notin (t_0, t_1), \end{cases} \tag{10.82}$$

is the bounded, closed set $[t_0, t_1]$. Clearly f represents a localized wave; see Fig. 10.3.

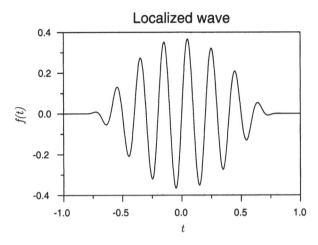

FIGURE 10.3

Plot of the localized wave defined by the function $f(t) = \exp(-(t+1)^{-2}(t-1)^{-2})\sin \omega t$ if $-1 < t < 1$ and $f(t) = 0$ for all other values of t. The support of f is the closed set $[-1, 1]$.

Closed Spheres

If S is a metric space, $x \in S$, and $r > 0$, the **closed sphere** $\overline{S}_r(x)$ is

$$\overline{S}_r(x) := \{z \in S \mid \rho(x, z) \leq r\}. \tag{10.83}$$

For example, in $\mathbb{R}$ under the absolute-value norm a closed sphere is a closed interval:

$$\overline{S}_r(x) = [x - r, x + r] = \{y \in \mathbb{R} \mid x - r \leq y \leq x + r\}. \tag{10.84}$$

Note the difference with respect to an open sphere: The closed sphere $\overline{S}_r(x)$ includes the points z such that $\rho(x, z) = r$ (if there are any), and the open sphere $S_r(x)$ includes only those points z such that $\rho(x, z) < r$.

We show that a closed sphere is a closed set by showing that the complement $S \setminus \overline{S}_r(x)$ is open. If $S \setminus \overline{S}_r(x)$ is the empty set $\emptyset$, then $\overline{S}_r(x) = S$, which is both closed and open. If $S \setminus \overline{S}_r(x)$ is non-empty, then, by the definition of $\overline{S}_r(x)$,

$$S \setminus \overline{S}_r(x) = \{y \in S \mid \rho(x, y) > r\}. \tag{10.85}$$

Let $y \in S \setminus \overline{S}_r(x)$, and choose s such that

$$0 < s < \rho(x, y) - r. \tag{10.86}$$

Let $z \in S_s(y)$. By the triangle inequality,

$$\rho(y, x) \leq \rho(y, z) + \rho(z, x). \tag{10.87}$$

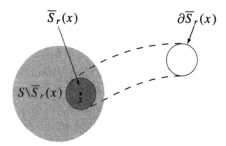

FIGURE 10.4

$\overline{S}_r(x)$, $S\backslash\overline{S}_r(x)$ and $\partial\overline{S}_r(x)$ in $S \subset \mathbb{R}^2$ under the 2-norm. In this example, but not in all metric spaces, $\overline{S_r(x)} = \overline{S}_r(x)$.

Then

$$\rho(x, z) \geq \rho(x, y) - \rho(y, z) > \rho(x, y) - s. \tag{10.88}$$

By Eq. (10.86),

$$\rho(x, y) - s > \rho(x, y) - [\rho(x, y) - r] = r. \tag{10.89}$$

From Eqs. (10.88) and (10.89),

$$\rho(x, z) > r, \tag{10.90}$$

which implies that $z \in S \setminus \overline{S}_r(x)$. Then $S \setminus \overline{S}_r(x)$ is open, and therefore $\overline{S}_r(x)$ is closed. Figure 10.4 illustrates $S_r(x)$ and $\overline{S}_r(x)$ for the 2-norm for $S \subset \mathbb{R}^2$.

It can be shown that, if S is a vector space over $\mathbb{R}$ or $\mathbb{C}$, and if the metric ρ is a p-norm, then $\overline{S_r(x)} = \overline{S}_r(x)$.

In the previous section we define the interior X° of a subset X of a metric space S as the collection of all interior points of X. The complement of the closure $\overline{X}$ relative to S, $S \setminus \overline{X}$, is the set of points that are not in X and are not limit points of X. For this reason $S \setminus \overline{X}$ is often called the **exterior** of X. The **boundary** ∂X of X is the set of points that belong to both the closure of X and the closure of the complement $S \setminus X$:

$$\partial X = \overline{X} \cap \overline{S \setminus X}. \tag{10.91}$$

Figure 10.4 illustrates $\partial\overline{S}_r(x)$ under the 2-norm for $S \subset \mathbb{R}^2$.

10.2.2 DENSE SETS AND SEPARABLE SPACES

Density

A subset A of a metric space S is called (**everywhere-**) **dense** in S if and only if $\overline{A} = S$. For example, the set $\mathbb{Q}$ of rational numbers is dense in the metric space $\mathbb{R}$ of real numbers under the absolute-value norm, because every real number x is the limit of at least one sequence $\{\rho_i\}$ of rational numbers that converges to x (see Chapter 2). The set of "rational" complex numbers $\xi + i\eta$ such that ξ and η are rational is dense in $\mathbb{C}$ under the modulus norm. Under any p-norm, the set of points with rational coordinates, $\mathbb{Q}^n$, is dense in $\mathbb{R}^n$ (see Exercise 10.2.2), the set of points with rational complex coordinates is dense in $\mathbb{C}^n$, the set of $m \times m$ matrices with rational matrix elements is dense in $\mathbb{R}^{m \times m}$, and the set of $m \times m$ matrices with rational complex elements is dense in $\mathbb{C}^{n \times n}$.

Separability

A metric space S is called **separable** if and only if S contains a finite or countable everywhere-dense subset. A metric space that is not separable is called **inseparable**. We show when we discuss Hilbert and Banach spaces that separable normed vector spaces that are complete (see Section 10.7) have many properties in common with finite-dimensional vector spaces. This fact is what makes many of the theoretical methods of physics and engineering work.

We give a few examples of separable normed vector spaces. The first two are (by now) obvious: The space $\mathbb{R}^n$ is separable because the set $\mathbb{Q}^n$ is countable (see Chapter 2), and the space $\mathbb{C}^n$ is separable because the set of rational complex n-tuples is countable.

We show next that the space $l^p(\mathbb{R})$ is separable. The proof depends on the fact that the space $\phi(\mathbb{Q})$ of finite rational sequences is countable. Certainly $\phi(\mathbb{Q})$ is equivalent to the union over n of the sets $\mathbb{Q}^n$:

$$
\begin{aligned}
\phi(\mathbb{Q}) &= \bigcup_{n=1}^{\infty} \{(\rho_1, \rho_2, \ldots, \rho_n, 0, \ldots) \mid \forall i \in (1:n): \ \rho_i \in \mathbb{Q}\} \\
&\sim \bigcup_{n=1}^{\infty} \{(\rho_1, \ldots, \rho_n)^T \mid \forall i \in (1:n): \ \rho_i \in \mathbb{Q}\} \\
&\sim \bigcup_{n=1}^{\infty} \mathbb{Q}^n.
\end{aligned}
\tag{10.92}
$$

The right-hand side of Eq. (10.92) is a countable union of countable sets and is therefore countable (see Chapter 2). To show that $\phi(\mathbb{Q})$ is dense in $l^p(\mathbb{R})$, let $\epsilon > 0$ and let $\bar{x} = (x_1, x_2, \ldots)$ belong to $l^p(\mathbb{R})$. Because the elements of $\bar{x}$ must obey Eq. (10.30), it follows from the properties of convergent numerical series that there exists an integer n such that

$$
\sum_{i=n+1}^{\infty} |x_i|^p < \frac{\epsilon^p}{2}.
\tag{10.93}
$$

Because $\mathbb{Q}$ is dense in $\mathbb{R}$, there exist rational numbers $\rho_1, \ldots, \rho_n$ such that

$$
\forall i \in (1:n): \ |x_i - \rho_i|^p < \frac{\epsilon^p}{2n}.
\tag{10.94}
$$

The sequence

$$
\bar{r} := (\rho_1, \ldots, \rho_n, 0, \ldots)
\tag{10.95}
$$

belongs to $\phi(\mathbb{Q})$. From Eq. (10.94) it follows that

$$
\sum_{i=1}^{n} |x_i - \rho_i|^p < \frac{\epsilon^p}{2}.
\tag{10.96}
$$

From this equation and Eq. (10.93) one has

$$\|\bar{x} - \bar{r}\|_p^p = \sum_{i=1}^{n} |x_i - \rho_i|^p + \sum_{i=n+1}^{\infty} |x_i|^p < \frac{\epsilon^p}{2} + \frac{\epsilon^p}{2} = \epsilon^p \tag{10.97}$$

and therefore that $\|\bar{x} - \bar{r}\|_p < \epsilon$, as we claim. One can show similarly that $l^p(\mathbb{C})$ is separable.

Uncountable Dense Sets

A subset of a metric space S need not be countable in order to be everywhere dense in S. For example, the **Weierstraß approximation theorem** (which we prove in Appendix F) asserts that *the set $\mathcal{P}(\mathbb{R})$ of all polynomial functions with real coefficients is dense (under the supremum norm) in the space $\mathcal{C}^0([a, b]; \mathbb{R})$ of all continuous functions from the bounded, closed interval $[a, b]$ into $\mathbb{R}$.*

The Weierstraß approximation theorem is the foundation for all polynomial-approximation methods, because it implies that there exists a polynomial that differs from a given continuous function on $[a, b]$ by as little as one pleases. Of course, the theorem guarantees nothing about the accuracy of the polynomial approximation outside the interval $[a, b]$. In other words, the Weierstraß approximation theorem tells us that polynomials are good for interpolation but is silent on the subject of extrapolation.

Dense Subsets of Dense Sets

Sometimes it happens that it is awkward to prove directly that a *countable* subset U of a metric space S is dense in S, but there exists a subset T such that

$$U \subseteq T \subseteq S \tag{10.98}$$

and such that one can prove relatively easily that U is dense in T and that T is dense in S. Then it follows immediately that U is dense in S and that S is separable. Let $s \in S$ and choose $\epsilon > 0$. Because T is dense in S and U is dense in T, there exist an element $t \in T$ such that $\rho(s, t) < \epsilon/2$ and an element $u \in U$ such that $\rho(t, u) < \epsilon/2$. Then $\rho(s, u) \leq \rho(s, t) + \rho(t, u) < \epsilon/2 + \epsilon/2 = \epsilon$, which shows that U is dense in S.

For example, on a bounded, closed interval $[a, b]$ the set $\mathcal{P}(\mathbb{Q})$ of all polynomial functions with rational coefficients is dense (under the supremum norm) in the set $\mathcal{P}(\mathbb{R})$ of all polynomial functions with real coefficients (see Exercise 10.2.3). The coefficients of a polynomial in $\mathcal{P}(\mathbb{Q})$ define a unique finite sequence in $\phi(\mathbb{Q})$ and conversely. Therefore $\mathcal{P}(\mathbb{Q})$ is countable. But $\mathcal{P}(\mathbb{R})$ is dense in $\mathcal{C}^0([a, b]; \mathbb{R})$. Hence $\mathcal{P}(\mathbb{Q})$ is dense in $\mathcal{C}^0([a, b]; \mathbb{R})$. Therefore $\mathcal{C}^0([a, b]; \mathbb{R})$ is separable.

It can be shown that $\mathcal{C}^0([a, b]; \mathbb{R})$ is dense in $L^p([a, b]; \mathbb{R})$. Therefore $L^p([a, b]; \mathbb{R})$ is separable.

Nonsingular Linear Mappings

By way of giving an important example of an uncountable dense set, we show that the set of nonsingular linear mappings is dense (under the 2-norm) in the finite-dimensional vector space hom[$\mathcal{V}$, $\mathcal{V}$] of all linear mappings of a finite-dimensional Euclidean or unitary space $\mathcal{V}$ into itself. It can be significant in least-squares fitting and in other numerical applications that every singular square matrix is arbitrarily close to (an infinite set of) nonsingular matrices.

It is enough to show that for every singular linear mapping $A : \mathcal{V} \to \mathcal{V}$ and for every $\epsilon > 0$, there exists a nonsingular linear mapping B such that

$$\|A - B\|_2 < \epsilon. \tag{10.99}$$

First we construct a linear mapping B that meets this requirement, and then we show by contradiction that B is nonsingular.

Let $\mathcal{N} := \text{null}[A]$, and let $\{\overline{e}_1, \ldots, \overline{e}_n\}$ be an orthonormal basis of $\mathcal{N}$. Let $\mathcal{R} := \text{range}[A]$. The rank-nullity theorem $\{\dim[\mathcal{R}] + \dim[\mathcal{N}] = \dim[\mathcal{V}]\}$ and the fact that $\mathcal{V}$ is the direct sum of $\mathcal{R}$ and $\mathcal{R}^\perp$ (whence $\dim[\mathcal{R}^\perp] + \dim[\mathcal{R}] = \dim[\mathcal{V}]$) imply that

$$\dim[\mathcal{R}^\perp] = \dim[\mathcal{N}] = n. \tag{10.100}$$

Let $\{\overline{f}_1, \ldots, \overline{f}_n\}$ be an orthonormal basis of $\mathcal{R}^\perp$. Then the linear mapping

$$B = A + \frac{1}{2}\epsilon \sum_{i=1}^{n} \overline{f}_i \underline{e}_i \tag{10.101}$$

satisfies Eq. (10.99), because one can show using the definition of the norm of a linear mapping and the Pythagorean theorem that the 2-norm of the sum of dyads in Eq. (10.101) is equal to 1. The idea behind this definition of B is that A is an invertible mapping from $\mathcal{N}^\perp$ to $\mathcal{R}$, and $\sum_{i=1}^n \overline{f}_i \underline{e}_i$ is an invertible mapping from $\mathcal{N}$ to $\mathcal{R}^\perp$; therefore one expects the sum Eq. (10.101) to be invertible.

To prove that B is nonsingular, we begin with the observation that

$$B\overline{x} = \overline{0} \Leftrightarrow A\overline{x} = -\frac{1}{2}\epsilon \sum_{i=1}^{n} \langle \overline{e}_i, \overline{x} \rangle \overline{f}_i. \tag{10.102}$$

We assume that $\overline{x} \neq \overline{0}$ and derive a contradiction. Equation (10.102) implies that an element of $\mathcal{R}$, $A\overline{x}$, is equal to an element of $\mathcal{R}^\perp$. In a Euclidean or unitary space this is possible if and only if $A\overline{x} = \overline{0}$. Then

$$\sum_{i=1}^{n} \langle \overline{e}_i, \overline{x} \rangle \overline{f}_i = \overline{0}. \tag{10.103}$$

But $A\overline{x} = \overline{0}$ implies that $\overline{x} \in \mathcal{N}$, which implies in turn that

$$\sum_{i=1}^{n} \langle \overline{e}_i, \overline{x} \rangle \overline{f}_i \neq \overline{0} \tag{10.104}$$

because $\{\overline{e}_i\}$ is a basis of $\mathcal{N}$. The contradiction between Eqs. (10.103) and (10.104) establishes that $\overline{x} = \overline{0}$. Therefore B is nonsingular.

10.2.3   EXERCISES FOR SECTION 10.2

10.2.1   Prove that a is a limit point of a metric space S if and only if

$$a \in \overline{S \setminus \{a\}}. \tag{10.105}$$

10.2.2   Prove that $\mathbb{Q}^n$ is dense in $\mathbb{R}^n$ under any p-norm.

10.2.3   Prove that, on a finite interval $[a, b]$ and under the supremum norm, the set $\mathcal{P}(\mathbb{Q})$ of all polynomial functions with rational coefficients is dense in the set $\mathcal{P}(\mathbb{R})$ of all polynomial functions with real coefficients.

10.2.4   Let $\mathbf{A}$ be the singular matrix

$$\mathbf{A} := \frac{1}{3} \begin{pmatrix} 1 & 1 & 1 \\ 1 & 1 & 1 \\ 1 & 1 & 1 \end{pmatrix}. \tag{10.106}$$

Construct a nonsingular matrix $\mathbf{B}$ such that $\|\mathbf{A} - \mathbf{B}\|_2 < \epsilon$.

10.2.5   Prove that the Cantor set K is separable if it is considered as a metric space under the absolute-value norm.

*10.2.6   For this problem, it is helpful to review Section 2.1 for the construction of the Cantor set K.

(a) Prove that every point of K is a limit point of K.

Hint: Show that, for every $x \in K$ and for every $\epsilon > 0$, an end point of one of the intervals that occur in the construction of K lies in the interval $(x - \epsilon, x + \epsilon)$.

(b) Prove that K is closed.

(c) A closed set P, every element of which is a limit point of P, is called **perfect**. Therefore K is a perfect set. Give another example of a perfect set.

10.2.7   Find the support of the function f such that

$$f(x) = \begin{cases} \exp\{-[x(1-x)]^{-1}\} & \text{if } 0 < x < 1 \\ 0 & \text{if } x \geq 1 \text{ or } x \leq 0 \end{cases} \tag{10.107}$$

10.2.8   Find the support of the function χ such that

$$\chi(x) = \begin{cases} 1 & \text{if } 0 < x < 1 \text{ and } x \in \mathbb{Q} \\ 0 & \text{otherwise} \end{cases} \tag{10.108}$$

in which $\mathbb{Q}$ is the set of all rational numbers.

10.2.9   Let S be an arbitrary set, and let ρ be the metric defined in Eq. (10.71).

(a) Show that $\forall x \in S$, $\{x\}$ is closed. Compare your result with the result of Exercise 10.1.8(b).

(b) Show that $S_1(x) = \{x\}$ and $\overline{S_1(x)} = \{x\}$, but $\overline{S}_1(x) = S$. Conclude that $\overline{S_1(x)} \neq \overline{S}_1(x)$.

10.3 CONVERGENCE OF SEQUENCES AND SERIES

10.3.1 CONVERGENCE OF SEQUENCES

So far we have defined the concepts of limit points and closed sets and have derived a few of their most basic properties. We have shown that a necessary condition for a subset L of a metric space S to have a limit point is that L must be infinite. Because every infinite set includes a countable set, we begin a more detailed investigation of limit points by assuming that L is countable, that is, that the elements of L are the terms of a sequence.

Right away we establish an important property of any subset of a metric space that possesses a limit point. Let S be a metric space, let

$$Z = (z_1, z_2, \ldots) \tag{10.109}$$

be a countable subset of S that is indexed as a sequence, and let $z \in S$ be any limit point of Z. Choose any $\epsilon > 0$ and let

$$r_n = \frac{\epsilon}{n}. \tag{10.110}$$

For every n, $S_{r_n}(z)$ includes an infinite subset of Z. Therefore there exists a point z_{i_n} such that z_{i_n} belongs both to $S_{r_n}(z)$ and to Z, and $i_n > i_{n-1}$. Then the subsequence

$$Z' = (z_{i_1}, z_{i_2}, \ldots) \tag{10.111}$$

has the property that

$$\forall n \in \mathbb{Z}^+ : \rho(z, z_{i_n}) < \frac{\epsilon}{n}. \tag{10.112}$$

We have "sifted" the original sequence in such a way as to obtain another sequence whose elements approach more and more closely to the limit point z. We use similar "sifting" methods repeatedly in what follows. This example also teaches one how to define convergence in terms of a metric.

Convergence to a Limit
A sequence

$$X = (x_1, x_2, \ldots) \tag{10.113}$$

in a metric space S is said to **converge to a point** $x \in S$ (or to **converge pointwise** to x) if and only if

$$\forall \epsilon > 0 : \exists M(\epsilon) \in \mathbb{Z}^+ :\ni: \forall n > M(\epsilon) : \rho(x, x_n) < \epsilon. \tag{10.114}$$

A sequence X that converges to a point x of a metric space S is called **convergent** (in S), and x is called the **limit** of the sequence:

$$x = \lim_{n \to \infty} x_n. \tag{10.115}$$

One also writes

$$x_n \to x \tag{10.116}$$

or

$$x_n \xrightarrow[n \to \infty]{} x. \tag{10.117}$$

For example, the sequence

$$X = (x_1, x_2, \ldots, x_n, \ldots) \quad \text{in which} \quad x_n = \frac{1}{2^n} \tag{10.118}$$

converges to the limit $x = 0$.

If a point $x \in S$ is the limit of a sequence X then x is a limit point of X, for the definition of convergence implies that for a given ϵ, the infinite subset $\{x_n \mid n > M(\epsilon)\}$ is included in the neighborhood $S_\epsilon(x)$. For the sequence Eq. (10.118), for example, a simple calculation shows that

$$n > M(\epsilon) = \left\lfloor -\frac{\log \epsilon}{\log 2} \right\rfloor \Rightarrow 0 < x_n < \epsilon. \tag{10.119}$$

Then the infinite subset $\{x_{M(\epsilon)+1}, x_{M(\epsilon)+2}, \ldots\} \subset X$ is included in the neighborhood $S_\epsilon(0) = (-\epsilon, \epsilon)$.

A sequence can converge to at most one limit. If $x_n \to x$ and $x_n \to y$, then (for every $\epsilon > 0$) there exist integers M, N such that for every $n > M$, $\rho(x, x_n) < \epsilon/2$, and for every $n > N$, $\rho(y, x_n) < \epsilon/2$. Then, for every n greater than both M and N, the triangle inequality implies that $\rho(x, y) \leq \rho(x, x_n) + \rho(x_n, y) < \epsilon$. Therefore $x = y$. It follows that a sequence converges if and only if it has a *unique* limit point.

If a sequence X is not convergent, it is called **divergent**. Note that a sequence diverges if and only if it has either no limit point or two or more limit points. For example, the sequence of positive integers, $\mathbb{Z}^+ = (1, 2, \ldots)$ diverges because it has no limit point. It is *not* correct, however, to conclude that the distance between a fixed term x_k and a general term x_n of a divergent sequence must increase without bound as n increases. (In fact, under some metrics – such as the Fréchet metric, Eq. (10.26) – the distance between two points is never greater than some least upper bound.) For example, the sequence R of all the rational numbers in the interval $[0, 1]$ diverges because every real number r such that $0 \leq r \leq 1$ is a limit point of R; hence R has uncountably many limit points. The terms of R are all bounded by 1.

Subsequences

If X (Eq. [10.113]) is any sequence and if $X' = (x_{i_1}, x_{i_2}, \ldots)$ (in which $i_1 < i_2 < \cdots$) is a sequence each of whose terms is a term of X, then X' is called a **subsequence** of X. From the definition of a limit point we have already shown that *if S is a metric space and if Z is a sequence in S, then z is a limit point of Z if and only if it is possible to extract a subsequence Z' of Z that converges to z.* For example, let R be the sequence of all the rationals in $[0, 1]$. Then every real number x in $[0, 1]$ can be approximated with arbitrary accuracy by some subsequence of R, such as a sequence of increasingly accurate rational approximations to x.

Let L_X be the set of all the limit points of a sequence $X = (x_1, x_2, \ldots)$, in which every $x_k \in \mathbb{R}$:

$$L_X := \{\alpha \in \mathbb{R} \mid \exists \{x_{n_k}\} \subseteq X :\ni: x_{n_k} \to \alpha\}. \tag{10.120}$$

We show that the set L_X is closed. Let λ be a limit point of L_X. Either there exists an integer n_1 such that $x_{n_1} \neq \lambda$, or not. If not, then we are through, because L_X is just the one-element set $\{\lambda\}$. If there exists an element x_{n_1} of X such that $x_{n_1} \neq \lambda$, let $\epsilon := |\lambda - x_{n_1}|$. Because λ is a limit point of L_X, there exists an element $\alpha_1 \in L_X$ such that $|\lambda - \alpha_1| < \epsilon/4$. Because α_1 is a limit point of X, there exists an element $x_{n_2} \in X$ such that $|\alpha_1 - x_{n_2}| < \epsilon/4$. Then

$$|\lambda - x_{n_2}| \leq |\lambda - \alpha_1| + |\alpha_1 - x_{n_2}| < \frac{\epsilon}{4} + \frac{\epsilon}{4} = \frac{\epsilon}{2}. \tag{10.121}$$

At the mth step, one finds $\alpha_m \in L_X$ such that $|\lambda - \alpha_m| < \epsilon/2^{m+1}$ and $x_{n_{m+1}} \in X$ such that $|\alpha_m - x_{n_{m+1}}| < \epsilon/2^{m+1}$. Then $|\lambda - x_{n_{m+1}}| < \epsilon/2^m$. Therefore one can construct a subsequence $\{x_{n_m}\} \subseteq X$ such that

$$x_{n_m} \to \lambda. \tag{10.122}$$

It follows that λ is a limit point of X. Then λ belongs to L_X. The same proof can be adapted to apply to the set of limit points of a sequence in any metric space (see Exercise 10.3.2).

If X is a sequence of rational or real numbers, then the least upper bound and greatest lower bound of the set of its limit points, L_X, are called the **upper limit** and **lower limit** of X, respectively:

$$\limsup x_n := \sup L_X \tag{10.123}$$
$$\liminf x_n := \inf L_X. \tag{10.124}$$

Because L_X is closed, the upper and lower limits of X both belong to L_X. For example, the upper limit of the sequence R that contains every rational number in the interval $(0, 1)$ is 1. The lower limit is 0. Both 0 and 1 belong to L_X.

Compactness

A subset C of a metric space S is called **sequentially** (or **countably**) **compact** if and only if every infinite subset of C has a limit point in C. If C is sequentially compact, and if $z \in S$ is a limit point of the countable subset $Z = \{z_1, z_2, \ldots\}$ of C, then (by Eqs. [10.109] through

[10.112]) Z includes a subsequence $Z' = (z_{i_1}, z_{i_2}, \ldots)$ which converges to z. Therefore *a subset C of a metric space S is sequentially compact if and only if every sequence in C contains a subsequence that converges in C.*

One says that a function $f : S \to \mathbb{F}$, in which S is a metric space and $\mathbb{F}$ is either $\mathbb{R}$ or $\mathbb{C}$, has **compact support** if and only if supp$[f]$ is a compact subset of S. The space

$$\mathcal{D}([a, b]; \mathbb{R}) := \{f \in \mathcal{C}^\infty([a, b]; \mathbb{R}) \mid \text{supp}[f] \text{ is compact}\} \tag{10.125}$$

of real-valued, infinitely differentiable functions whose support is contained in the interval $[a, b]$ is important in physics and engineering because it is one of the spaces to which test functions belong in the theory of distributions such as the Dirac delta "function."

It is not instantly obvious that the requirement of infinite differentiability is consistent with the requirement that f must vanish outside the finite interval $[a, b]$. One can verify straightforwardly that the function

$$f(x) := \begin{cases} \exp\{[-(x - a)(b - x)]^{-1}\}, & \text{if } a < x < b; \\ 0, & \text{if } x \leq a \text{ or } x \geq b \end{cases} \tag{10.126}$$

is infinitely differentiable on the interval $[a, b]$ and vanishes, along with all of its derivatives, at $x = a$, at $x = b$, and at all points outside the interval.

Cauchy Sequences

We derive a fundamental necessary condition for convergence. Let a sequence X (Eq. [10.113]) converge to x. Choose $N(\epsilon)$ such that

$$\forall n > N(\epsilon) : \rho(x, x_n) < \frac{\epsilon}{2}. \tag{10.127}$$

Then for every pair (m, n) of positive integers such that $m > N(\epsilon)$ and $n > N(\epsilon)$,

$$\rho(x_m, x_n) \leq \rho(x_m, x) + \rho(x, x_n) < \epsilon. \tag{10.128}$$

We have shown that if a sequence X in a metric space S is convergent, then the terms of X satisfy the **Bolzano-Cauchy condition**:

$$\forall \epsilon > 0 : \exists N(\epsilon) \in \mathbb{Z}^+ :\ni: \forall m, n > N(\epsilon) : \rho(x_m, x_n) < \epsilon. \tag{10.129}$$

A sequence whose terms satisfy this condition is called a **Cauchy sequence**. Evidently every subsequence of a Cauchy sequence is a Cauchy sequence.

In a metric space S, every convergent sequence is a Cauchy sequence. To see whether the converse is true, suppose that X is a Cauchy sequence in S. Choose $\epsilon > 0$ and let m be a fixed integer such that $m > N(\epsilon)$. Then the sequence $Y = (\rho(x_m, x_{m+1}), \rho(x_m, x_{m+2}), \ldots)$ is a bounded sequence of real numbers (because each term is bounded between 0 and ϵ). By the Bolzano-Weierstraß theorem (see Appendix E) Y has a limit point. In fact, Y has a unique limit point, because the triangle inequality implies that for all $p > n > m$,

$$|\rho(x_m, x_n) - \rho(x_m, x_p)| \leq \rho(x_n, x_p). \tag{10.130}$$

As $n \to \infty$ (with $p > n$) the right-hand side approaches zero because X is a Cauchy sequence. It follows that Y is a Cauchy sequence of real numbers. Appendix E shows that every Cauchy sequence of real numbers converges to a unique real number.

One is tempted, then, to speak of the limit point of a Cauchy sequence X. Unfortunately there are metric spaces S such that some Cauchy sequences in S do not converge to any point of S. For example, let S be the metric space of rational numbers $\mathbb{Q}$ under the absolute-value norm, and consider the sequence $(3.1, 3.14, 3.142, \ldots, x_n, \ldots)$ in which x_n is an $(n+1)$–significant-figure, base-10 approximation to π, rounded to even. The difference $|x_n - x_m|$ between an $(n+1)$–significant-figure approximation and an $(m+1)$–significant-figure approximation to π, in which $m < n$, is less than 10^{-m}. Then $(x_1, x_2, \ldots)$ is a Cauchy sequence in $\mathbb{Q}$. However, $\lim_{n \to \infty} x_n = \pi$, which is irrational, hence not in $\mathbb{Q}$.

In Section 10.7 we define *complete* metric spaces S, which have the property that every Cauchy sequence in S approaches a limit that belongs to S.

Bounded Sequences

A subset B of a metric space S is called **bounded** if and only if there exist a real number r and a point $x \in S$ such that

$$B \subseteq S_r(x). \tag{10.131}$$

A sequence X is called a **bounded sequence** if and only if all of its terms belong to some fixed neighborhood $S_r(x)$.

We show that if $X = (x_1, x_2, \ldots)$ is a Cauchy sequence, then X is bounded. Assume for the sake of obtaining a contradiction that X is not bounded. We show that we can extract a non-Cauchy subsequence of X. Choose any $y_1 \in X$ and let

$$r_1 = 1. \tag{10.132}$$

Because X is not bounded, it is not a subset of $S_{r_1}(y_1)$. Then there exists an element $y_2 \in X$ such that $y_2 \notin S_{r_1}(y_1)$, that is, such that

$$\rho(y_1, y_2) \geq r_1. \tag{10.133}$$

Let

$$r_2 = \rho(y_1, y_2) + 1. \tag{10.134}$$

Because X is not a subset of $S_{r_2}(y_1)$, we can choose a point $y_3 \in X$ such that $y_3 \notin S_{r_2}(y_1)$, and therefore $\rho(y_3, y_1) \geq r_2$ (see Fig. 10.5). Continuing this process, one constructs a sequence

$$Y = (y_1, y_2, \ldots) \tag{10.135}$$

such that

$$\rho(y_1, y_n) = r_n - 1 \geq r_{n-1}. \tag{10.136}$$

FIGURE 10.5

Illustration of the construction of y_1, y_2, and y_3. The sequence $(y_1, y_2, \ldots)$ is unbounded.

Let $n > m \geq 2$. By construction,

$$\rho(y_1, y_m) = r_m - 1. \tag{10.137}$$

But the triangle inequality implies that

$$\rho(y_1, y_n) \leq \rho(y_1, y_m) + \rho(y_m, y_n). \tag{10.138}$$

Then

$$r_m \leq \rho(y_1, y_n) \leq (r_m - 1) + \rho(y_m, y_n), \tag{10.139}$$

which implies that for all $n > m \geq 2$,

$$\rho(y_m, y_n) \geq 1. \tag{10.140}$$

Then the subsequence Y of the Cauchy sequence X violates the Bolzano-Cauchy condition (a contradiction). It follows that X is bounded.

10.3.2 NUMERICAL SEQUENCES

Monotonic Sequences

A sequence $(x_1, x_2, \ldots)$ such that every $x_n \in \mathbb{R}$ is called **monotonically increasing** if and only if

$$\forall n \in \mathbb{Z}^+ : x_n \leq x_{n+1} \tag{10.141}$$

or **monotonically decreasing** if and only if

$$\forall n \in \mathbb{Z}^+ : x_n \geq x_{n+1}. \tag{10.142}$$

For example, the sequence whose n-th term is $1/n$ is monotonically decreasing. A sequence is called **monotonic** if it is either monotonically increasing or monotonically decreasing.

An extremely simple criterion determines the convergence or divergence of a monotonic sequence: A monotonic sequence X converges if and only if it is bounded, that is, if and only if

$$b_+ := \sup X < \infty \tag{10.143}$$

if X is monotonically increasing, or

$$b_- := \inf X > -\infty \tag{10.144}$$

if X is monotonically decreasing. For simplicity's sake, we consider only the case in which X is monotonically increasing. If X is bounded, then for every $\epsilon > 0$, there exists some integer N such that $b_+ - \epsilon < x_N \le b_+$, for if there were no such term x_n then $b_+ - \epsilon$ would be an upper bound of X, in contradiction to our assumption that b_+ is the least upper bound of X. It follows that for every $n \ge N$, $x_n \in S_\epsilon(b_+) = (b_+ - \epsilon, b_+ + \epsilon)$. Then $S_\epsilon(b_+)$ includes an infinite subset of X. Therefore b_+ is a limit point of X. Conversely, if a monotonically increasing sequence X converges to a limit x, then x is the least upper bound of X (Exercise 10.3.2).

Common Numerical Sequences

Let us give some examples of the application of the definition of convergence, Eq. (10.114), by way of a review of a few useful numerical sequences. We begin by noting that the limit

$$\forall p > 0 : \lim_{n \to \infty} \frac{1}{n^p} = 0 \tag{10.145}$$

follows at once from Eq. (10.114) if one lets

$$M(\epsilon) = \left\lfloor \left(\frac{1}{\epsilon}\right)^{1/p} \right\rfloor, \tag{10.146}$$

in which $\lfloor x \rfloor$ denotes the largest integer not greater than x, for then $\forall n > M(\epsilon) : n^{-p} < \epsilon$.

For the next example we need the formula

$$\forall x \in \mathbb{R} : \ni: x \ge 0 : \forall k \in \mathbb{Z}^+ : \forall n \in \mathbb{Z}^+ : \ni: n \ge k : (1+x)^n \ge 1 + \binom{n}{k} x^k, \tag{10.147}$$

which follows directly from the binomial theorem (Eq. [2.164]).

With Eq. (10.147) (for the case $k = 1$) it is easy to establish the limit of the sequence whose n-th term is $x_n = a^{1/n}$, in which a is real and positive:

$$\forall a > 0 : \lim_{n \to \infty} a^{1/n} = 1. \tag{10.148}$$

For the case in which $a > 1$, one has $a^{1/n} > 1$. Then $a^{1/n} = 1 + y_n$, in which $y_n > 0$. It follows that $a = (1 + y_n)^n > 1 + ny_n$, and therefore that

$$0 < y_n < \frac{a-1}{n}. \tag{10.149}$$

Choose any $\epsilon > 0$, and let

$$M(\epsilon) = \left\lfloor \frac{a-1}{\epsilon} \right\rfloor. \tag{10.150}$$

Then Eq. (10.149) implies that for every $n > M(\epsilon)$,

$$0 < y_n < \frac{a-1}{n} < \epsilon. \tag{10.151}$$

According to Eq. (10.114) the limit of the sequence $\{y_n\}$ is therefore zero, implying that the limit of the sequence whose nth term is $x_n = 1 + y_n$ is 1. The case in which $0 < a < 1$ can be handled similarly by setting $a^{1/n} = 1/(1 + y_n)$, in which $y_n > 0$.

We next establish the important theorem that

$$\forall x > 0 : \forall \alpha \in \mathbb{R} : \lim_{n\to\infty} \frac{n^\alpha}{(1+x)^n} = 0, \tag{10.152}$$

one corollary of which is that the exponential function grows more rapidly than any power as the argument becomes infinite:

$$\forall \alpha \in \mathbb{R} : \lim_{n\to\infty} n^\alpha e^{-n} = 0. \tag{10.153}$$

To prove Eq. (10.152), choose an integer k such that $k \in \mathbb{Z}^+$ and $k > \alpha$. Eq. (10.147) implies that

$$(1+x)^n \geq 1 + \binom{n}{k} x^k$$
$$> \binom{n}{k} x^k. \tag{10.154}$$

The technical problem here is to make a useful estimate of the binomial coefficient as n becomes large. For every n such that $n > 2k - 2$ one has

$$\binom{n}{k} = \frac{n(n-1)\cdots(n-k+1)}{k!} > \left(\frac{n}{2}\right)^k \frac{1}{k!} \tag{10.155}$$

because $n > n/2, \ldots, (n - k + 1) > n/2$. It follows that $(1 + x)^n > n^k x^k / (2^k k!)$, and therefore that

$$0 < \frac{n^\alpha}{(1+x)^n} < \frac{2^k k!}{x^k} n^{\alpha-k} \xrightarrow[n\to\infty]{} 0 \tag{10.156}$$

because $\alpha - k < 0$ by assumption.

10.3.3 NUMERICAL SERIES

Definitions
If the terms s_n of a sequence obey a recurrence relation

$$s_{n+1} = s_n + x_{n+1}, \tag{10.157}$$

in which $s_0 = 0$ and $(x_1, x_2, \ldots)$ is a sequence, then s_n is the sum of $x_1, \ldots, x_n$:

$$s_n = \sum_{k=1}^{n} x_k. \tag{10.158}$$

For the moment we consider only the special case in which the terms x_k are real numbers. In Section 10.4 we discuss a more general case in which the terms x_k belong to a normed vector space.

The sum s_n is called the **n-th partial sum** of the **series** (or **infinite series**)

$$\sum_{k=1}^{\infty} x_k := \lim_{n \to \infty} \sum_{k=1}^{n} x_k. \qquad (10.159)$$

If the limit $s = \lim_{n \to \infty} \sum_{k=1}^{n} x_k$ exists and $|s| < \infty$, then one says that Eq. (10.159) is a **convergent series** and that s is the **sum** of the series. If the sequence of partial sums is divergent, then Eq. (10.159) is called a **divergent series**. Although it is convenient to use an infinite upper limit of summation for both convergent and divergent series, one should understand that for a divergent series the infinite sum is undefined because the limit does not exist.

For example, it follows from Eq. (2.125) for the sum of a geometric progression $1 + r + \cdots + r^n$ that the **geometric series** converges to $(1-r)^{-1}$ if $|r| < 1$ and diverges if $r \geq 1$:

$$\sum_{k=0}^{\infty} r^k = \lim_{n \to \infty} \frac{1 - r^{n+1}}{1-r} = \begin{cases} (1-r)^{-1}, & \text{if } |r| < 1; \\ \infty, & \text{if } |r| \geq 1. \end{cases} \qquad (10.160)$$

(Note that the formula fails if $r = 1$, but that in that case the n-th partial sum is just n. Hence the sequence of partial sums is $\mathbb{Z}^+$, which diverges.)

Tests for Convergence of Numerical Series

The simplest method of determining whether a series converges is the **comparison test for convergence**: If $(c_1, c_2, \ldots)$ is a sequence of positive terms such that the series

$$\sum_{k=1}^{\infty} c_k \qquad (10.161)$$

converges, and if there exists an integer M such that

$$\forall k \in \mathbb{Z}^+ :\ni: k > M : |x_k| \leq c_k, \qquad (10.162)$$

then $\sum_{k=1}^{\infty} x_k$ converges. By hypothesis, the sequence of partial sums

$$a_n = \sum_{k=1}^{n} c_k \qquad (10.163)$$

converges and is therefore a Cauchy sequence. Then for every $\epsilon > 0$ there exists an integer $N(\epsilon)$ such that

$$\forall m, n > N(\epsilon) : \sum_{k=m+1}^{n} c_k = |a_n - a_m| < \epsilon. \qquad (10.164)$$

Hence the sequence of partial sums

$$s_n = \sum_{k=1}^{n} x_n \tag{10.165}$$

also is a Cauchy sequence, for

$$|s_n - s_m| \leq \sum_{k=m+1}^{n} |x_k| \leq \sum_{k=m+1}^{n} c_k = |a_n - a_m| < \epsilon \tag{10.166}$$

for all m and n such that $\max\{M, N(\epsilon)\} < m < n$. Then the series $\sum_{k=1}^{\infty} x_k$ converges to a limit in $\mathbb{R}$.

In a similar way one establishes the **comparison test for divergence**: If a series of positive terms $\sum_{k=1}^{\infty} d_k$ is known to diverge, and if $(x_1, x_2, \ldots)$ is a sequence of *positive* terms such that

$$0 < d_k \leq x_k \tag{10.167}$$

then the series $\sum_{k=1}^{\infty} x_k$ diverges.

For example, consider the base-β fixed-point expansion of a real number,

$$r = \pm \left[\sum_{j=0}^{n} b_j \beta^j + \sum_{k=1}^{\infty} \frac{f_k}{\beta^k} \right]. \tag{10.168}$$

The fractional part of r is a series which converges by comparison with the geometric series. One has $x_k = \pm f_k/\beta^k$; therefore $|x_k| < \beta\beta^{-k}$, because $0 \leq f_k < \beta$. The series for the fractional part of r converges because the series $\beta \sum_{k=1}^{\infty} \beta^{-k}$ converges.

Comparison with the geometric series gives rise to other useful tests. For example, if for every k greater than some integer M one has

$$|x_k| \leq c_k = r^k, \tag{10.169}$$

in which $0 < r < 1$, then the series $\sum_{k=1}^{\infty} x_k$ converges. This test can be restated as the **root test**:

$$\begin{cases} \limsup\limits_{k \to \infty} |x_k|^{1/k} < 1 & \Rightarrow \sum\limits_{k=1}^{\infty} x_k \text{ converges;} \\[2ex] \limsup\limits_{k \to \infty} |x_k|^{1/k} > 1 & \Rightarrow \sum\limits_{k=1}^{\infty} x_k \text{ diverges;} \\[2ex] \limsup\limits_{k \to \infty} |x_k|^{1/k} = 1 & \text{is inconclusive.} \end{cases} \tag{10.170}$$

In the first case there exist an integer M and a constant r such that

$$\forall k > M : |x_k|^{1/k} \leq r < 1, \tag{10.171}$$

which implies that the series $\sum_{k=1}^{\infty} x_k$ converges. In the second case one has $|x_k| > 1$ for all $k > M$, which implies divergence. The fact that $\limsup_{k\to\infty} |x_k|^{1/k} = 1$ for the series $\sum_{k=1}^{\infty} k^{-1}$, which diverges, and for the series $\sum_{k=1}^{\infty} k^{-2}$, which converges, proves the third case.

For example, consider the series

$$r + s + r^2 + s^2 + \cdots + r^k + s^k + \cdots \quad 0 < r < s < 1, \tag{10.172}$$

which we created by interleaving the geometric series $\sum_{k=1}^{\infty} r^k$ and $\sum_{k=1}^{\infty} s^k$. The general terms of the interleaved series are $x_{2l} = s^l$ and $x_{2l-1} = r^l, l \in \mathbb{Z}^+$. Therefore

$$\liminf_{k\to\infty} |x_k|^{1/k} = \lim_{l\to\infty} (r^l)^{1/(2l-1)} = r^{1/2} < 1 \tag{10.173}$$

and

$$\limsup_{k\to\infty} |x_k|^{1/k} = \lim_{l\to\infty} (s^l)^{1/2l} = s^{1/2} < 1. \tag{10.174}$$

It follows that the interleaved series converges.

Comparison with the geometric series leads also to the **ratio test**:

$$\begin{cases} \limsup\limits_{k\to\infty} \left| \dfrac{x_{k+1}}{x_k} \right| < 1 & \Rightarrow \sum\limits_{k=1}^{\infty} x_k \text{ converges;} \\[4mm] \forall k > M : \left| \dfrac{x_{k+1}}{x_k} \right| > 1 & \Rightarrow \sum\limits_{k=1}^{\infty} x_k \text{ diverges.} \end{cases} \tag{10.175}$$

In the first case there exist an integer M and a constant $r < 1$ such that for all $k > M$, $|x_{k+1}/x_k| < r$, which implies that for all $N > M$,

$$\left| \sum_{k=1}^{N} x_k \right| \le \sum_{k=1}^{N} |x_k| < \sum_{k=1}^{M} |x_k| + |x_M| \sum_{k=M+1}^{N} r^k. \tag{10.176}$$

Therefore the series converges by comparison with the geometric series.

For example, consider the Bessel series

$$J_m(x) = \left(\frac{x}{2} \right)^m \sum_{k=0}^{\infty} \frac{(-1)^k}{k!(k+m)!} \left(\frac{x}{2} \right)^{2k}. \tag{10.177}$$

Here $|x_{k+1}/x_k| = |x|^2/(4(k+1)(k+m+1))$, which tends to zero as $k \to \infty$ for any finite x. Therefore the Bessel series converges for all x.

Note that the ratio test fails for the interleaved series Eq. (10.172), because $\limsup_{k\to\infty} |x_{k+1}/x_k| = \lim_{k\to\infty} (s/r)^k = \infty$. In general the root test is stronger than the ratio test in the sense that if a series converges according to the ratio test, then the root test also implies convergence, while if the root test is inconclusive, so is the ratio test.

Absolute Convergence

In many of the series that one encounters in physics and engineering, the terms are real and alternate in sign, that is,

$$x_n = (-1)^n |x_n|. \tag{10.178}$$

One says that a numerical series $\sum_{k=1}^{\infty} x_k$ **converges absolutely** if and only if the series $\sum_{k=1}^{\infty} |x_k|$ converges. If a series $\sum_{k=1}^{\infty} x_k$ converges absolutely, then $\sum_{k=1}^{\infty} x_k$ converges, by the comparison test with $c_k = |x_k|$.

If a series $\sum_{k=1}^{\infty} x_k$ converges but the series of absolute values, $\sum_{k=1}^{\infty} |x_k|$, diverges, then one says that the series converges **nonabsolutely** or **conditionally**. Any conditionally convergent series has the interesting property that it cannot be rearranged without running the risk of making the rearranged series divergent, or making it converge to a different limit than the original series. If $\phi : \mathbb{Z}^+ \to \mathbb{Z}^+$ is a bijective (i.e., one-to-one and onto) mapping of the positive integers, then one calls

$$\sum_{k=1}^{\infty} x_{\phi(k)} \tag{10.179}$$

a **rearrangement** of the series $\sum_{k=1}^{\infty} x_k$. A theorem proved by Riemann states that if $\sum_{k=1}^{\infty} x_k$ converges nonabsolutely, then for every pair l_1, l_2 of real numbers such that

$$-\infty \le l_1 \le l_2 \le \infty \tag{10.180}$$

there exists a rearrangement such that

$$\begin{aligned} \liminf s_n' &= l_1 \\ \limsup s_n' &= l_2, \end{aligned} \tag{10.181}$$

in which

$$s_n' := \sum_{k=1}^{n} x_{\phi(k)} \tag{10.182}$$

is the n-th partial sum of the rearranged series.

However, if $\sum_{k=1}^{\infty} x_k$ converges absolutely to x, then every rearrangement of $\sum_{k=1}^{\infty} x_k$ also converges absolutely to the same limit, x. Rudins (1976) provides readable proofs of these theorems.

Cauchy's Theorem and the Harmonic Series

One of the more remarkable results concerning the convergence of numerical series is **Cauchy's theorem**: Let $(x_1, x_2, \ldots)$ be a monotonically decreasing sequence of nonnegative terms, that is, $x_1 \ge x_2 \ge \cdots \ge x_k \ge \cdots \ge 0$. Then the series $\sum_{k=1}^{\infty} x_k$ converges if and only if the series

$$\sum_{k=0}^{\infty} 2^k x_{2^k}$$ (10.183)

converges. The m-th partial sum of Eq. (10.183) is

$$t_m = x_1 + 2x_2 + \cdots + 2^m x_{2^m}.$$ (10.184)

For every $n < 2^m$, one has

$$s_n \leq x_1 + (x_2 + x_3) + \cdots + (x_{2^m} + \cdots + x_{2^{m+1}-1})$$
$$\leq x_1 + 2x_2 + \cdots + 2^m x_{2^m}$$ (10.185)
$$= t_m.$$

If the series Eq. (10.183) converges, then the sequence $(t_1, t_2, \ldots)$ is bounded. We have just shown that the monotonically increasing sequence $(s_1, s_2, \ldots)$ is bounded; therefore $\sum_{k=1}^{\infty} x_k$ converges. This establishes the "if" part of the theorem. One shows similarly that if $n > 2^m$, then

$$s_n \geq \tfrac{1}{2} t_m.$$ (10.186)

It follows that if the series Eq. (10.183) does not converge, then the series $\sum_{k=1}^{\infty} x_k$ does not converge. This establishes the "only if" part of Cauchy's theorem.

We illustrate the use of Cauchy's theorem by applying it to the **harmonic series**

$$\sum_{k=1}^{\infty} \frac{1}{k} = 1 + \frac{1}{2} + \frac{1}{3} + \cdots.$$ (10.187)

In this case the series Eq. (10.183) is the series $\sum_{k=1}^{\infty} 2^k / 2^k = \sum_{k=1}^{\infty} 1$, which obviously diverges. Cauchy's theorem shows also that the series that defines the **zeta function** of real argument,

$$\zeta(x) := \sum_{k=1}^{\infty} \frac{1}{k^x},$$ (10.188)

is convergent provided that $x > 1$. For the zeta function, the test series Eq. (10.183) is $\sum_{k=1}^{\infty} 2^k / 2^{kx} = \sum_{k=1}^{\infty} 2^{(1-x)k}$, which converges by comparison with the geometric series if $x > 1$.

10.3.4 EXERCISES FOR SECTION 10.3

10.3.1   Prove that

$$\forall \alpha > 0 : \lim_{n \to \infty} \frac{1}{n^\alpha} = 0.$$ (10.189)

10.3.2   Prove that if a monotonically increasing sequence X converges to a limit x, then x is the least upper bound of X.

10.3.3   Prove that

$$\lim_{n \to \infty} n^{1/n} = 1. \tag{10.190}$$

Hint: Let $x_n = n^{1/n} - 1$. Prove that $x_n \geq 0$. Use Eq. (10.147) for $k = 2$ to prove that

$$(1 + x_n)^n \geq \frac{n(n-1)}{2} x_n^2. \tag{10.191}$$

Then obtain an inequality on x_n.

10.3.4   Show by comparison with the geometric series with $r = \frac{1}{2}$ that the series

$$e - 1 = \sum_{k=1}^{\infty} \frac{1}{k!} \tag{10.192}$$

converges. Can you devise a similar method to prove that the series for e^x converges?

10.3.5   Prove that the series

$$\sum_{k=1}^{\infty} \frac{1}{k^x} \tag{10.193}$$

converges for all real $x > 1$ *without* making use of Cauchy's theorem.

10.3.6   Let f belong to the class of rapidly decreasing functions on $\mathbb{R}$. Prove that the series

$$f(0) + \sum_{n=1}^{\infty} [f(n) + f(-n)] \tag{10.194}$$

converges absolutely.

10.4 STRONG AND POINTWISE CONVERGENCE

10.4.1 STRONG CONVERGENCE

Definitions and Basic Results

If the metric space S considered in Section 10.3.1 is a normed vector space $\mathcal{V}$ (and therefore the distance function ρ is a norm, $\rho(\bar{x}, \bar{y}) = \|\bar{x} - \bar{y}\|$) then Eq. (10.114) defines **strong convergence** for a sequence $(\bar{x}_1, \bar{x}_2, \ldots)$ of vectors in $\mathcal{V}$. One says, then, that a sequence $X = (\bar{x}_1, \bar{x}_2, \ldots)$ **converges strongly** to the vector $\bar{x}$ if and only if

$$\forall \epsilon > 0 : \exists N(\epsilon) \in \mathbb{Z}^+ :\ni: \forall n \in \mathbb{Z}^+ :\ni: n > N(\epsilon) : \|\bar{x}_n - \bar{x}\| < \epsilon. \tag{10.195}$$

For example, let

$$\mathbf{x}_n = \begin{pmatrix} 2^{-n} \\ 1 - 2^{-n} \end{pmatrix}, \quad \mathbf{x} = \begin{pmatrix} 0 \\ 1 \end{pmatrix}. \tag{10.196}$$

Then the sequence $(\mathbf{x}_1, \mathbf{x}_2, \ldots)$ converges strongly to $\mathbf{x}$ under the ∞-norm (and, in fact, under every p-norm for $1 \le p \le \infty$), because

$$\|\mathbf{x} - \mathbf{x}_n\|_\infty = \frac{1}{2^n}. \tag{10.197}$$

To give another example, convergence as it is defined for sequences of real numbers is strong convergence under the absolute-value norm. Note that it is not necessary for the normed vector space $\mathcal{V}$ to be finite-dimensional. We give several important examples of convergence in infinite-dimensional normed vector spaces in this section.

A normed vector space is a metric space, and strong convergence is simply convergence under the metric defined by the norm, Eq. (10.15). Therefore, all of the results that we have already obtained for convergent sequences in a metric space are valid for strongly convergent sequences in a normed vector space. In particular, a strongly convergent sequence $X = (\overline{x}_1, \overline{x}_2, \ldots)$ obeys the Bolzano-Cauchy condition, which reads

$$\forall \epsilon > 0 : \exists N(\epsilon) \in \mathbb{Z}^+ :\ni: \forall n > m > N(\epsilon) : \|\overline{x}_n - \overline{x}_m\| < \epsilon \tag{10.198}$$

in terms of a vector norm.

One says that the **vector series**

$$\sum_{k=1}^{\infty} \overline{x}_k \tag{10.199}$$

converges strongly to a vector $\overline{x}$ if and only if the sequence

$$(\overline{s}_1, \overline{s}_2, \ldots) \tag{10.200}$$

of partial sums

$$\overline{s}_n = \sum_{k=1}^{n} \overline{x}_k \tag{10.201}$$

converges strongly to $\overline{x}$. The Bolzano-Cauchy condition implies that a vector series converges strongly in a normed vector space $\mathcal{V}$ only if

$$\forall \epsilon > 0 : \exists N(\epsilon) \in \mathbb{Z}^+ :\ni: \forall n > m > N(\epsilon) : \left\| \sum_{k=m+1}^{n} \overline{x}_k \right\| = \|\overline{s}_m - \overline{s}_n\| < \epsilon. \tag{10.202}$$

It follows that a necessary condition for the strong convergence of a vector series such as Eq. (10.199) is that

$$\|\bar{x}_k\| \to 0. \tag{10.203}$$

Exactly as is the case for numerical series, this condition is not sufficient to ensure convergence.

The **comparison test for vector series** is: If

$$(c_1, c_2, \ldots) \tag{10.204}$$

is a sequence of positive terms such that the series

$$\sum_{k=1}^{\infty} c_k \tag{10.205}$$

converges, and if there exists an integer M such that

$$\forall k \in \mathbb{Z}^+ :\ni: k > M : \|\bar{x}_k\| \leq c_k, \tag{10.206}$$

then

$$\sum_{k=1}^{\infty} \bar{x}_k \tag{10.207}$$

converges strongly. (To prove this theorem, replace the absolute value or modulus with the norm in the proof of the comparison test for numerical series.) A vector series such as Eq. (10.199) is called **absolutely convergent** if and only if the numerical series of norms,

$$\sum_{k=1}^{\infty} \|\bar{x}\|, \tag{10.208}$$

converges. As for numerical series, absolute convergence implies strong convergence, by the comparison test.

The concept of strong convergence is important in physics and engineering because it permits one to define useful kinds of convergence for series of vectors, functions, matrices, and operators. For example, in the partial-wave series for the scattering amplitude f in potential scattering theory, the terms of the series are functions:

$$f(k, \theta) = \frac{1}{k} \sum_{l=0}^{\infty} (2l + 1) e^{i\delta_l} \sin \delta_l P_l(\cos \theta). \tag{10.209}$$

Here P_l is a Legendre polynomial, θ is the scattering angle, and δ_l is a real number called the *phase shift*. The useful norm for this problem is defined by the inner product

$$\langle f, g \rangle = \int_0^{\pi} f(\theta)^* g(\theta) \sin \theta \, d\theta. \tag{10.210}$$

The Bolzano-Cauchy condition implies that the partial-wave series converges in the space of square-integrable functions on $[-\pi, \pi]$ if and only if for every $\epsilon > 0$, there exists a positive integer $N(\epsilon)$ such that for every pair of integers m, n such that $n > m > N(\epsilon)$,

$$\left\langle \sum_{l=m+1}^{n} (2l + 1)e^{i\delta_l} \sin \delta_l \, P_l(\cos \theta), \sum_{l'=m+1}^{n} (2l' + 1)e^{i\delta_{l'}} \sin \delta_{l'} \, P_{l'}(\cos \theta) \right\rangle < \epsilon. \tag{10.211}$$

Using the orthogonality properties of the Legendre polynomials, one obtains the equivalent condition

$$\sum_{l=m+1}^{n} (2l + 1) \sin^2 \delta_l < \frac{1}{2}\epsilon. \tag{10.212}$$

In other words, the partial-wave series converges strongly in the inner-product norm if and only if the numerical series

$$\sum_{l=0}^{\infty} (2l + 1) \sin^2 \delta_l \tag{10.213}$$

converges.

Uniform Convergence

Let $\mathcal{V}$ be a normed vector space and let S be a set. One says that a sequence $(f_1, f_2, \ldots)$ of functions $f_n : S \to \mathcal{V}$ **converges uniformly** to a function $f : S \to \mathcal{V}$ if and only if

$$\forall \epsilon > 0: \ \exists N(\epsilon) \in \mathbb{Z}^+ :\ni: \forall n > N(\epsilon) : \forall x \in S : \ \|f(x) - f_n(x)\| < \epsilon. \tag{10.214}$$

In this definition, $N(\epsilon)$ is independent of the point $x \in S$. For example, S could be a subset of $\mathbb{R}$, and $\mathcal{V}$ could be $\mathbb{R}$ under the absolute-value norm; see Section 10.4.3. For another example, let S be a subset of $\mathbb{R}$, and let $\mathcal{V}$ be $\mathbb{C}$ under the modulus norm. For a third example, let S be a subset of $\mathbb{C}$, and let $\mathcal{V}$ be $\mathbb{C}$ under the modulus norm.

Equivalent Norms

Many different norms can be defined on the same vector space. For example, on $\mathbb{R}^n$ one has all of the p-norms, in which $1 \le p \le \infty$. In principle, for each different norm that is defined on a normed vector space $\mathcal{V}$ there exists a different kind of strong convergence. Two norms $\| \cdot \|_\alpha$ and $\| \cdot \|_\beta$ defined on the same metric vector space $\mathcal{V}$ are called **equivalent** if and only if, for every $\bar{x} \in \mathcal{V}$, every sequence $\{\bar{x}_n\}$ that converges to $\bar{x}$ under $\| \cdot \|_\alpha$ also converges to $\bar{x}$ under $\| \cdot \|_\beta$, and conversely:

$$\|\bar{x}_n - \bar{x}\|_\alpha \to 0 \Leftrightarrow \|\bar{x}_n - \bar{x}\|_\beta \to 0. \tag{10.215}$$

On a finite-dimensional vector space all norms are equivalent. For example, if the p-norm of $\mathbf{x}_k$, $[\sum_{j-1}^{n} |x_k^j|^p]^{1/p}$, converges to zero for one value of p, then $x_k^j \to 0$ as $k \to \infty$, implying that $\|\mathbf{x}_k\|_p \to 0$ for all values of p such that $1 \le p \le \infty$. However, we give an example in

Section 10.4.3 that shows that the absolute-value and supremum norms are not equivalent in the infinite-dimensional space of continuous functions on [0, 1].

10.4.2 OPERATORS

Definition

Let S and T be metric spaces with metrics ρ_S and ρ_T, respectively. A mapping A that carries its domain $D \subseteq S$ into T is called an **operator** (on S). Clearly an operator A is a function whose argument is in D and whose values are in $R = \text{range}[A] \subseteq T$. If D and R are subsets of $\mathbb{R}$, then A is just a real-valued function. If the range R is a subset of either $\mathbb{R}$ or $\mathbb{C}$, then A is called a **functional**. *Linear* functionals defined on a vector space V are the subject of a previous chapter.

Bounded Linear Operators

If $A : V \to X$ is a linear operator, then A is called **bounded** if and only if there exists a positive real number B such that for every vector $\bar{x}$,

$$\|A\bar{x}\|_x \leq B\|\bar{x}\|_v. \tag{10.216}$$

If A is bounded, then the supremum (least upper bound) of the set of constants B for which Eq. (10.216) is satisfied is called the **norm** of the operator A and is written

$$\|A\| = \sup_{\bar{x} \neq \bar{0}} \frac{\|A\bar{x}\|_x}{\|\bar{x}\|_v}, \tag{10.217}$$

exactly as for a linear mapping of a finite-dimensional vector space [Eq. (8.432)].

Occasionally one needs the following formulation of the definition of the norm of a linear operator:

$$\|A\| = \sup_{\|\bar{x}\| \leq 1} \|A\bar{x}\|. \tag{10.218}$$

Certainly

$$\|A\bar{x}\| \leq \|A\|\|\bar{x}\| \leq \|A\| \tag{10.219}$$

for all $\bar{x}$ such that $\|\bar{x}\| \leq 1$. Conversely, by the definition of the least upper bound, for every $\epsilon > 0$ there exists a vector $\bar{y}_\epsilon$ such that

$$\|A\bar{y}_\epsilon\| > (\|A\| - \epsilon)\|\bar{y}_\epsilon\|. \tag{10.220}$$

Define the unit vector

$$\bar{u} := \frac{\bar{y}_\epsilon}{\|\bar{y}_\epsilon\|}. \tag{10.221}$$

Then, by the homogeneity of A and the scaling property of the norm, one has

$$\|A\bar{u}\| = \frac{\|A\bar{y}_\epsilon\|}{\|\bar{y}_\epsilon\|} \tag{10.222}$$

$$> \frac{(\|A\| - \epsilon)\|\bar{y}_\epsilon\|}{\|\bar{y}_\epsilon\|} = \|A\| - \epsilon.$$

Therefore

$$\forall \epsilon > 0 : \sup_{\|\bar{x}\| \leq 1} \|A\bar{x}\| \geq \|A\bar{u}\| > \|A\| - \epsilon. \tag{10.223}$$

It follows that

$$\sup_{\|\bar{x}\| \leq 1} \|A\bar{x}\| \geq \|A\|. \tag{10.224}$$

Together with the inequality

$$\|A\bar{u}\| \leq \|A\| \tag{10.225}$$

[which follows from the definition of the operator norm, Eq. (10.216)], this proves equality in Eq. (10.218).

Pointwise Operator Convergence

One says that an operator sequence $\{A_n\}$ **converges pointwise** at $x \in S$ to the operator $A : S \to T$ if and only if $\{A_n x\}$ converges to Ax, that is, if and only if

$$\forall \epsilon > 0 : \exists N(\epsilon, x) \in \mathbb{Z}^+ : \ni : \forall n > N(\epsilon, x) : \ \rho_T(A_n x, Ax) < \epsilon. \tag{10.226}$$

If S and T are normed vector spaces $\mathcal{V}_{(1)}$, $\mathcal{V}_{(2)}$ and A_n, A are bounded linear operators, then pointwise convergence for every point in S as defined in Eq. (10.226) implies that

$$\forall \bar{x} \in \mathcal{V}_{(1)} : \forall \epsilon > 0 : \ \exists N(\epsilon, \bar{x}) \in \mathbb{Z}^+ : \ni : \forall n > N(\epsilon, \bar{x}) : \ \|A_n \bar{x} - A\bar{x}\| < \epsilon. \tag{10.227}$$

[Sometimes pointwise convergence is called "strong" convergence for operators. We do not use the term "strong" in this way because of potential confusion with strong convergence in the vector space of all bounded linear operators, where strong convergence would mean convergence in the operator norm.] A special case of the definition of pointwise operator convergence is the definition of pointwise convergence given in Section 10.4.3 for a sequence of real-valued functions.

Uniform Operator Convergence

One says that an operator sequence $\{A_n\}$ **converges uniformly** to the operator A on a subset X of a metric space S if and only if

$$\forall \epsilon > 0 : \ \exists N(\epsilon) \in \mathbb{Z}^+ : \ni : \forall n > N(\epsilon) : \forall x \in X : \ \rho_T(A_n x, Ax) < \epsilon. \tag{10.228}$$

Note that $N(\epsilon)$ is the same for all points $x \in X$. Clearly uniform convergence on X implies pointwise convergence on X. The converse is not true.

The order of quantifiers is different in the definitions of pointwise and uniform convergence. For pointwise convergence the order is

$$\forall x : \forall \epsilon : \exists N(\epsilon, x) : \forall n. \tag{10.229}$$

For uniform convergence the order is

$$\forall \epsilon : \exists N(\epsilon) : \forall n : \forall x. \tag{10.230}$$

10.4.3 SEQUENCES OF REAL-VALUED FUNCTIONS

Pointwise Convergence

Let $(f_1, f_2, \ldots)$ be a sequence of real-valued functions $f_n : [a, b] \to \mathbb{R}$. If, for every x in the interval $[a, b]$, the numerical sequence $(f_1(x), f_2(x), \ldots)$ converges, then one can define a limit function f, the value of which at x is

$$f(x) := \lim_{n \to \infty} f_n(x). \tag{10.231}$$

One says that the sequence $(f_1, f_2, \ldots)$ **converges pointwise** to f if and only if

$$\forall x \in [a, b] : \forall \epsilon > 0 : \exists N(\epsilon, x) \in \mathbb{Z}^+ : \ni : \forall n > N(\epsilon, x) : |f_n(x) - f(x)| < \epsilon. \tag{10.232}$$

There is absolutely no guarantee that the pointwise limit function f has any desirable properties such as continuity or differentiability, even if each of the functions f_n has these properties. For example, let

$$\forall x \in [0, 1] : f_n(x) := \frac{nx}{1 + nx}. \tag{10.233}$$

The sequence $(f_1, f_2, \ldots)$ converges pointwise for every $x \in [0, 1]$. For $x = 0$ one has

$$\lim_{n \to \infty} f_n(0) = 0; \tag{10.234}$$

for $x \neq 0$ one has

$$\forall x \in (0, 1] : \lim_{n \to \infty} f_n(x) = 1. \tag{10.235}$$

Then the limit function

$$f = \lim_{n \to \infty} f_n \tag{10.236}$$

is discontinuous at $x = 0$, for $f(0) = 0$ and $f(x) = 1$ for all x such that $0 < x \leq 1$. In this example, if one chooses ϵ such that $0 < \epsilon < 1$, then

$$N(\epsilon, x) = \left\lfloor \frac{1 - \epsilon}{\epsilon x} \right\rfloor \qquad (10.237)$$

in which $\lfloor \alpha \rfloor$ means the largest integer which is less than or equal to α.

Uniform Convergence

As in the preceding example, consider a sequence $(f_1, f_2, \ldots)$ of real-valued functions $f_n : [a, b] \rightarrow \mathbb{R}$ on the interval $[a, b]$. If $(f_1, f_2, \ldots)$ converges strongly to a function f under the supremum norm (Eq. [10.31]), then one says that f_n **converges uniformly** to f:

$$\forall \epsilon > 0 : \exists N(\epsilon) \in \mathbb{Z}^+ :\ni: \forall n > N(\epsilon) : \| f_n - f \|_C < \epsilon. \qquad (10.238)$$

This sort of convergence is called *uniform* because a single value of N applies to the entire interval $[a, b]$, as one may see from the more elementary definition:

$$\forall \epsilon > 0 : \exists N(\epsilon) \in \mathbb{Z}^+ :\ni: \forall n > N(\epsilon) : \forall x \in [a, b] : |f_n(x) - f(x)| < \epsilon. \qquad (10.239)$$

Uniform convergence implies pointwise convergence, but *not* conversely.

One can see from the definition of uniform convergence that a pointwise-convergent sequence $(f_1, f_2, \ldots)$ converges uniformly if and only if for every $\epsilon > 0$ there exists a finite integer $N(\epsilon)$ such that

$$\forall x \in [a, b] : \ N(\epsilon, x) \leq N(\epsilon), \qquad (10.240)$$

in which $N(\epsilon, x)$ is defined in Eq. (10.232). One can say, equivalently, that if a sequence $(f_1, f_2, \ldots)$ converges pointwise, then it converges uniformly if and only if the set $\{N(\epsilon, x) | x \in [a, b]\}$ is bounded above. For example, the convergence of the sequence $(f_1, f_2, \ldots)$ defined in Eq. (10.233) is not uniform. An upper bound $N(\epsilon)$ does not exist in this case, because the integer $N(\epsilon, x)$ given in Eq. (10.237) grows without bound as x approaches zero.

This example and the examples of pointwise convergence above illustrate the important point that, in an infinite-dimensional space, the convergence or divergence of a sequence may depend on the norm that one chooses, because all norms are not equivalent. This example shows that the absolute-value norm and the supremum norm are not equivalent on the infinite-dimensional space $\mathcal{C}^0([a, b]; \mathbb{R})$.

An important consequence of the definition of uniform convergence of a sequence of real-valued functions is that if every f_k is continuous,

$$\forall k \in \mathbb{Z}^+ : f_k \in \mathcal{C}^0([a, b]; \mathbb{R}), \qquad (10.241)$$

and if the sequence $(f_1, f_2, \ldots)$ converges *uniformly* to f in $[a, b]$, then the limit function f is continuous on $[a, b]$:

$$f \in C^0([a, b]; \mathbb{R}).\tag{10.242}$$

The proof makes use of the so-called $\epsilon/3$ method: Choose $\epsilon > 0$ and let $n > N(\epsilon/3)$, for which the integer-valued function N is defined in Eq. (10.238). The assumed continuity of f_n guarantees that for every $\epsilon > 0$ and for any point $x \in [a, b]$, there exists a real number $\delta(x, \epsilon/3) > 0$ such that $|f_n(x) - f_n(y)| < \epsilon/3$ for all points y such that $|x - y| < \delta(x, \epsilon/3)$. Then for all such points y,

$$\begin{aligned}
|f(x) - f(y)| &= |f(x) - f_n(x) + f_n(x) - f_n(y) + f_n(y) - f(y)| \\
&\leq |f(x) - f_n(x)| + |f_n(x) - f_n(y)| + |f_n(y) - f(y)| \\
&< \frac{\epsilon}{3} + \frac{\epsilon}{3} + \frac{\epsilon}{3} = \epsilon,
\end{aligned}\tag{10.243}$$

which implies that f is continuous on $[a, b]$.

Even more is true: Under the assumptions made in the preceding paragraph, the integral of the limit function f is equal to the limit of the integral of f_n. Choose any $n > N(\epsilon/|b-a|)$. Then

$$\begin{aligned}
\left| \int_a^x f(x')\, dx' - \int_a^x f_n(x')\, dx' \right| &= \left| \int_a^x [f(x') - f_n(x')]\, dx' \right| \\
&\leq \int_a^x |f(x') - f_n(x')|\, dx' \\
&\leq |x - a| \sup_{x' \in [a,b]} |f(x') - f_n(x')| \\
&\leq |b - a| \|f - f_n\|_C \\
&< \epsilon,
\end{aligned}\tag{10.244}$$

which establishes that for every $x \in [a, b]$,

$$\int_a^x f_n(x')\, dx' \rightarrow \int_a^x f(x')\, dx'.\tag{10.245}$$

The sequence must be uniformly convergent in order for one to proceed legitimately from the fourth to the fifth line of Eq. (10.244).

Quite often one needs to know whether a limit function is differentiable, that is, whether the function $f' = df/dx$ exists, and if so, whether $f_n' \rightarrow f'$. Once again, uniform convergence provides a sufficient condition: If every $f_n \in C^1([a, b]; \mathbb{R})$, if $f_n \rightarrow f$, and if the sequence $(f_1', f_2', \ldots)$ converges *uniformly*, then

$$f_n' \rightarrow f'.\tag{10.246}$$

Let g be the limit function defined by the sequence $(f_1', f_2', \ldots)$:

$$\forall x \in [a, b] : g(x) := \lim_{n \to \infty} f_n'(x). \tag{10.247}$$

Because every $f_n' \in \mathcal{C}^0([a, b]; \mathbb{R})$ and because the sequence $(f_1', f_2', \ldots)$ converges uniformly, it follows that

$$g \in \mathcal{C}^0([a, b]; \mathbb{R}). \tag{10.248}$$

Now one can apply Eq. (10.244) (replacing f with g and f_n with f_n') to show that

$$\begin{aligned}
\int_a^x g(x') \, dx' &= \lim_{n \to \infty} \int_a^x f_n'(x') \, dx' \\
&= \lim_{n \to \infty} [f_n(x) - f_n(a)] \\
&= f(x) - f(a).
\end{aligned} \tag{10.249}$$

It follows by taking the limit of the difference quotient of both sides of this equation that the derivative of f exists and that $f' = g$ everywhere in the closed interval $[a, b]$.

For an example in which a sequence of functions fails to converge uniformly and in which the sequence of derivatives fails to converge to f', we return to the functions f_n defined in Eq. (10.233). The sequence of derivatives, $\{f_n'(x) = n/(1 + nx)^2\}$, converges to 0 pointwise (but not uniformly) for every $x \neq 0$. The limit function f defined in Eqs. (10.234) and (10.236) is differentiable, and $f'(x) = 0$, for all x except for $x = 0$. In any interval that does *not* include 0, the sequence $\{f_n\}$ converges uniformly and $f_n' \to f'$.

10.4.4 SERIES OF REAL-VALUED FUNCTIONS

Pointwise Convergence of Series

If $(u_1, u_2, \ldots)$ is a sequence of functions $f_n : [a, b] \to \mathbb{R}$, then one says that the series

$$\sum_{k=1}^{\infty} u_k \tag{10.250}$$

converges pointwise to a function f at the point x if and only if the numerical series $\sum_{k=1}^{\infty} u_k(x)$ converges to $f(x)$. The series **converges absolutely** to f at x if and only if it converges pointwise to f at x and the numerical series

$$\sum_{k=1}^{\infty} |u_k(x)| \tag{10.251}$$

converges.

Uniform Convergence of Series

A series $\sum_{k=1}^{\infty} u_k$ **converges uniformly** to f on $[a, b]$ if and only if the sequence of partial sums converges uniformly to f. It follows from Eq. (10.245) that a uniformly convergent series can be integrated term-by-term. Eq. (10.243) implies that if every $u_k \in \mathcal{C}^0([a, b]; \mathbb{R})$ and if $\sum_{k=1}^{\infty} u_k$ converges uniformly to f on $[a, b]$, then the limit function is continuous, that is, $f \in \mathcal{C}^0([a, b]; \mathbb{R})$. Finally, if every term u_k is differentiable,

$$u_k \in \mathcal{C}^1([a, b]; \mathbb{R}), \tag{10.252}$$

if $\sum_{k=1}^{\infty} u_k$ converges pointwise to f on $[a, b]$, and if $\sum_{k=1}^{\infty} u'_k$ converges uniformly on $[a, b]$, then the series $\sum_{k=1}^{\infty} u'_k$ converges to f', that is, the series may be differentiated term-by-term.

The comparison test for uniform convergence of a series

$$\sum_{k=1}^{\infty} u_k \tag{10.253}$$

follows from the comparison test for uniform convergence of a sequence: If there exist an integer N and a convergent test series

$$\sum_{k=1}^{\infty} M_k \tag{10.254}$$

such that every $M_k > 0$ and such that on the interval $[a, b]$,

$$\forall k > N : \|u_k\|_C \le M_k, \tag{10.255}$$

then the series $\sum_{k=1}^{\infty} u_k$ converges uniformly on $[a, b]$. This case of the comparison test is often called the **Weierstraß M-test**.

10.4.5 EXERCISES FOR SECTION 10.4

10.4.1　Test the following series for uniform convergence on the interval indicated:

(a)

$$\sum_{k=1}^{\infty} (-1)^{k-1} \frac{\sin(2k + 1)x}{(2k + 1)^2}, \tag{10.256}$$

on $[-\pi, \pi]$.

(b)

$$\sum_{k=1}^{\infty} (-1)^k \frac{\cos(2k + 1)x}{(2k + 1)^3}, \tag{10.257}$$

on $[-\pi, \pi]$.

(c)

$$\sum_{k=0}^{\infty} \frac{x^k}{k!}, \tag{10.258}$$

on $[-R, R]$, in which R is any finite positive real number.

(d)

$$\sum_{k=1}^{\infty} \frac{\sin(2k+1)x}{2k+1}, \tag{10.259}$$

on $[-\pi, \pi]$.

10.4.2 Prove that the series in Eq. (10.257) may be differentiated term-by-term.

10.4.3 Derive Eq. (10.237).

10.4.4 Prove that if S is the subset of the set of real numbers such that $\forall\, x \in S : |x| \le R$, in which $0 < R < \infty$, then the exponential series converges uniformly on S to the exponential function, $f(x) = e^x$.

10.4.5 Prove that if S is the subset of the set of complex numbers such that $\forall\, z \in S :$ $|z| \le R$, in which $0 < R < \infty$ and $|z|$ denotes the modulus of $z \in \mathbb{C}$, then the sequence $(f_0, f_1, \dots)$ of partial sums $f_n = \sum_{m=0}^{n} z^m/m!$ of the exponential series converges uniformly on S to the exponential function, $f(z) = e^z$.

10.5 CONTINUITY

Continuous has nearly the same colloquial meaning as *uninterrupted*. The mathematical term that is closest to *continuous* in the everyday sense is *dense*. In mathematics the concept of continuity was not clearly separated from that of density until the works of Karl Weierstraß and others in the mid-nineteenth century. We shall see below that continuity, as it is defined in modern mathematics, is a property of mappings. A continuous mapping has the property that nearby points or an open set of points in the range of the mapping are always the images of nearby points or of an open set of points in the mapping's domain.

One may ask whether it is worthwhile to learn about an abstract property such as continuity in an age of digital, and therefore discrete, computation. Two answers are that a continuous function is a useful abstraction, much like an irrational number, and that one would hardly be able to analyze the convergence of successive numerical approximations or study nonlinear dynamical systems without using the concept of a continuous mapping.

10.5.1 POINTWISE CONTINUITY

In undergraduate calculus one learns that a function f is called continuous at a point x if the statement "$x \to a$" implies the statement "$f(x) \to f(a)$". Equivalently, one can say that f is continuous if and only if f preserves nearness. Whether a function is continuous depends upon the manner in which one measures the nearness of points and their images under the function. For example, some real-valued functions converge pointwise (in the absolute-value norm) but not uniformly (in the supremum norm).

In a metric space one can describe nearness in terms of distances, sequences or membership in open sets. We discuss continuity from each of these equivalent points of view.

Definition of Pointwise Operator Continuity

An operator A on a metric space S is called **pointwise continuous** at $y \in S$ if and only if

$$\forall \epsilon > 0 : \exists \delta = \delta(\epsilon, y) > 0 :\ni: \forall x \in S_\delta(y) : \ Ax \in S_\epsilon(Ay). \tag{10.260}$$

It is clearly equivalent to say that A is pointwise continuous at y if and only if

$$\forall \epsilon > 0 : \exists \delta = \delta(\epsilon, y) > 0 :\ni: AS_\delta(y) \subseteq S_\epsilon(Ay), \tag{10.261}$$

or that A is pointwise continuous at y if and only if

$$\forall \epsilon > 0 : \exists \delta = \delta(\epsilon, y) > 0 :\ni: \forall x \in S :\ni: \rho_S(x, y) < \delta : \rho_T(Ax, Ay) < \epsilon. \tag{10.262}$$

An operator A is called **continuous on a set** $S' \subseteq S$ if and only if A is pointwise continuous at every point of S'.

Continuity at a point depends explicitly upon the spaces S, T and their metrics ρ_S and ρ_T. If S and T are both the set of real numbers $\mathbb{R}$, then an operator $A : \mathbb{R} \to \mathbb{R}$ is simply a real-valued function of a real variable. If ρ_S and ρ_T are both the absolute-value norm in $\mathbb{R}$, then Eq. (10.262) reduces to the freshman-calculus definition of continuity.

The notation $\delta(\epsilon, y)$ reminds one that δ depends upon both the chosen ϵ and the point y. In essence, $\delta(\epsilon, y)$ measures the sensitivity of the distance between two image points Ax and Ay to the distance between the preimages x and y. Note that y must belong to D in order for the definition of continuity at y to make sense. Note, too, that continuity at a single point does not imply continuity at any other point of S, no matter how close to y.

Continuous Images of Convergent Sequences

A continuous operator preserves the convergence of sequences. Let $A : S \to T$ be an operator with domain D and range $R \subseteq T$. We shall prove the following theorem: A *is pointwise continuous at $s \in D$ if and only if*

$$\forall \{s_m \mid s_m \in D\} :\ni: \lim_{m \to \infty} s_m = s : \lim_{m \to \infty} As_m = As, \tag{10.263}$$

that is, if and only if for every sequence $\{s_m\}$ in D that converges to $s \in D$, the sequence $\{As_m\}$ converges to $As \in R$. Eq. (10.263) is, in effect, a definition of continuity that is equivalent to our original definition.

To prove the "only if" part, assume that A is pointwise continuous at s and that there exists a sequence $\{s_m\}$ such that

$$\lim_{m\to\infty} s_m = s, \tag{10.264}$$

but that $\{As_m\}$ does not converge to As. Then there exists a real number $\zeta > 0$ such that the neighborhood $S_\zeta(As)$ includes at most a finite subset of $\{As_m\}$. [If there is no such neighborhood, then every neighborhood of As includes an infinite subset of $\{As_m\}$. Then As is a limit point of $\{As_m\}$, contradicting the assumption that $\{As_m\}$ does not converge to As.] Given ζ, there exists a real number $\eta > 0$ such that

$$\forall x :\ni: \rho_S(x, s) < \eta : \rho_T(Ax, As) < \zeta \tag{10.265}$$

by the assumed continuity of A. Since s is a limit point of $\{s_m\}$, the neighborhood $S_\zeta(As)$ therefore contains every As_m such that s_m is within a distance η of s. Thus continuity implies that $S_\zeta(As)$ includes all but a finite subset of $\{As_m\}$. This establishes a contradiction and proves the "only if" implication.

To prove the "if" part of the theorem, suppose that Eq. (10.263) holds and that A is not continuous at s. Then there exists no real number $\eta > 0$ such that Eq. (10.265) holds for all $x \in S_\eta(s)$. Therefore, in each neighborhood $S_{1/n}(s)$ there exists some $t_n \in S_{1/n}(s)$ such that

$$At_n \notin S_\zeta(As). \tag{10.266}$$

By the definition of a limit and the fact that $t_n \in S_{1/n}(s)$,

$$\lim_{n\to\infty} t_n = s. \tag{10.267}$$

By construction, $\{At_n\}$ does not converge to As, because $S_\zeta(As)$ contains no point of $\{At_n\}$. This contradicts Eq. (10.263) and establishes the theorem.

An immediate consequence is that if an operator A is continuous on a sequentially (countably) compact subset X of a metric space S, then the image $AX \subseteq T$ is also sequentially compact. Consider any sequence $\{As_m\}$ in AX. Because X is sequentially compact, there exists a subsequence $\{s_{m_k}\}$ of $\{s_m\}$ that converges to $s \in X$. Then the continuity of A implies that the subsequence $\{As_{m_k}\}$ converges to As. In other words, *a continuous image of a compact set is compact*.

Continuity in Terms of the Preimage of an Open Set

We can rephrase the definition of continuity given in Eq. (10.260) in a way that lends itself to generalization. Let us temporarily replace A with f in order to conform to the usual notation for functions. Then, according to the definition of continuity, for every $\epsilon > 0$ there exists $\delta(y) > 0$ such that $f(S_\delta(y)) \subseteq S_\epsilon(z)$, where $z = f(y)$. This statement is equivalent to saying that given any open sphere $S_\epsilon(z)$ in the range of f, the open sphere $S_\delta(y)$ is a subset of the preimage $f^{-1}(S_\epsilon(z))$.

Generalizing from open spheres to open sets, we observe that [by the definition of an open set] for every open subset O of range $[f]$ and for every point $f(y) \in O$, there exists some open sphere $S_\epsilon(f(y)) \subseteq O$. By the continuity of f, there exists a real number $\delta(y) > 0$ such that $f(S_{\delta(y)}(y)) \subseteq S_\epsilon(f(y))$. In other words, for every $f(y) \in O$ [hence for every $y \in f^{-1}(O)$], there exists an open sphere in $f^{-1}(O)$ that contains y, $S_{\delta(y)}(y) \subseteq f^{-1}(O)$. It follows that $f^{-1}(O)$ is open. Conversely, if for every open subset $O \subseteq$ range $[f]$ it follows that $f^{-1}(O)$ is open, then f is continuous, because for every $f(y) \in O$ there exists an open sphere $S_{\delta(y)}(y) \subseteq f^{-1}(S_\epsilon(f(y)))$. Therefore we can restate the definition of continuity as follows: *If S and T are metric spaces, then a function $f : S \to T$ is continuous on S if and only if $f^{-1}(O)$ is open for every open subset $O \subseteq f(S)$.* This definition is general enough to apply when S and T are topological spaces, not metric spaces.

10.5.2 UNIFORM CONTINUITY

Continuity as we defined it in Eq. (10.260) is a point property, because $\delta(\epsilon, y)$ depends upon the point y. If an operator A is pointwise continuous on an infinite subset X of a metric space S, then the set $\{\delta(\epsilon, y) \mid y \in X\}$ of values of δ for a given ϵ does not necessarily have a lower bound that is greater than zero.

An operator $A : S \to T$, where S and T are metric spaces, is called **uniformly continuous on the set** $X \subseteq S$ if and only if

$$\forall \epsilon > 0 : \exists \delta = \delta(\epsilon) > 0 :\ni: \forall x, y \in X :\ni: \rho_S(x, y) < \delta : \rho_T(Ax, Ay) < \epsilon. \quad (10.268)$$

Obviously, if A is uniformly continuous on X, then A is continuous on X. The converse is *not* true; counterexamples abound. For instance, let $S := (0, 1]$ and let $Ax := 1/x$. From the identity

$$\frac{1}{x} - \frac{1}{x \pm \delta} = \frac{\pm \delta}{x(x \pm \delta)} \quad (10.269)$$

it follows that

$$|Ax - A(x \pm \delta)| < \epsilon \Leftrightarrow \delta(\epsilon, x) < \epsilon \min\{|x(x - \delta)|, |x(x + \delta)|\}. \quad (10.270)$$

Therefore the greatest lower bound of $\delta(\epsilon, x)$ for $x \in (0, 1]$ is zero. Clearly A is continuous, but not uniformly continuous, on $(0, 1]$.

The theorem that a real-valued function that is continuous on a bounded, closed interval is uniformly continuous on that interval is well-known. The generalization to operators on metric spaces runs as follows: *Let S and T be metric spaces and let $A : S \to T$ be continuous on S. If S is sequentially compact, then A is uniformly continuous on S.* Let ρ_S and ρ_T be the distance functions in S and T, respectively, and let ϵ be any positive real number. For the sake of developing a contradiction, we assume that A is continuous on S but is not uniformly continuous. Then for every integer k there exists a pair of points in S which are closer than

k^{-1} but whose images under A are not closer than ϵ:

$$\forall k \in \mathbb{Z}^+ : \exists x_k, x_k' \in S :\ni: \rho_S(x_k, x_k') < \frac{1}{k} \quad \text{and} \quad \rho_T(Ax_k, Ax_k') \geq \epsilon. \tag{10.271}$$

Because S is sequentially compact, the sequence $(x_1, x_2, \ldots)$ includes a subsequence $(x_{i_1}, x_{i_2}, \ldots)$ which converges to a point $x \in S$:

$$x_{i_n} \to x. \tag{10.272}$$

For any $\eta > 0$, there exists an integer i_n such that

$$\rho_S(x_{i_n}, x) < \frac{\eta}{2} \quad \text{and} \quad \frac{1}{i_n} < \frac{\eta}{2}. \tag{10.273}$$

Because $\rho_S(x_{i_n}, x_{i_n}') < 1/i_n$, the sequence $(x_{i_1}', x_{i_2}', \ldots)$ converges to x:

$$\begin{aligned}
\rho_S(x_{i_n}', x) &\leq \rho_S(x_{i_n}', x_{i_n}) + \rho_S(x_{i_n}, x) \\
&< \frac{\eta}{2} + \frac{\eta}{2} = \eta.
\end{aligned} \tag{10.274}$$

But A is continuous, which implies that

$$Ax_{i_n} \to Ax \quad \text{and} \quad Ax_{i_n}' \to Ax. \tag{10.275}$$

Therefore there exists an integer i_N such that for every $i_n > i_N$,

$$\rho_T(Ax_{i_n}, Ax) < \frac{\epsilon}{2} \quad \text{and} \quad \rho_T(Ax_{i_n}', Ax) < \frac{\epsilon}{2}. \tag{10.276}$$

Then

$$\begin{aligned}
\rho_T(Ax_{i_n}, Ax_{i_n}') &\leq \rho_T(Ax_{i_n}, Ax) + \rho_T(Ax, Ax_{i_n}') \\
&< \frac{\epsilon}{2} + \frac{\epsilon}{2} = \epsilon.
\end{aligned} \tag{10.277}$$

This contradiction with the assertion that $\rho_T(Ax_{i_n}, Ax_{i_n}') \geq \epsilon$ establishes the theorem without forcing one to give an algorithm for computing a number $\delta(\epsilon)$ such that $\rho_S(y, x) < \delta(\epsilon)$ implies that $\rho_T(Ay, Ax) < \epsilon$.

10.6 BEST APPROXIMATIONS IN THE MAXIMUM AND SUPREMUM NORMS

Suppose that one approximates a continuous function $f : S \to \mathbb{R}$ by projection methods using a family of orthogonal functions, as described in Section 8.3. Section 8.4 shows that the result is a least-squares approximation, i.e., an approximation that minimizes the inner-product norm $\|f - p\|_{ip}$ of the difference between f and the approximating function p. Section 10.7.3 shows that for a *complete* orthonormal family of functions (such as the Fourier functions), it is possible to construct an infinite series that converges strongly to f, provided only that f is square-integrable. However, the value of a least-squares approximating function p at any given point $x \in S$ is not necessarily required to be numerically close to the value of f. In more formal terms, the **maximum deviation** δ of the approximating function p from f is not bounded, in which δ is defined as the Chebyshev norm of the error,

$$\delta := \|f - p\|_C. \tag{10.278}$$

The assumption that f is continuous is essential for the purpose of constructing a sequence $(p_1, p_2, \ldots)$ of approximating functions p_n such that $\|f - p_n\|_C \to 0$ as $n \to \infty$. One can rather easily construct examples of functions and a sequence of approximations such that the Chebyshev norm of the error is a constant, no matter how small one is able to make the inner-product norm of the error, $\|f - p\|_2$. For example, Section 10.7.4 shows that the orthogonal (Fourier) series for a square wave does not converge uniformly because of Gibbs' phenomenon (Figures 10.9 and 10.10). For a square wave of unit amplitude, the supremum norm of the error, $\|f - p\|_C$, approaches the constant value 0.17898... (Eq. 10.401), although the inner-product (least-squares) norm of the error approaches zero. In other words, the Chebyshev and inner-product norms are inequivalent.

Section 8.4 shows that, without assuming that f is continuous, one can still find a sequence of approximations to f that converges to f in the inner-product norm, but a sequence of functions that converges to f in the inner-product norm may fail to converge in the supremum norm (i.e., uniformly). However, there are many applications in physics and engineering for which one needs to minimize the supremum norm of the error, $\max_{x \in S} |f(x) - p(x)| = \|f - p\|_C$, instead of the inner-product norm of the error. It turns out that the polynomials obtained in the Chebyshev approximations defined in Eqs. (8.305) and (8.308), which came about by minimizing $\|f - p\|_{ip}$, also are very close to the polynomials that minimize the supremum or maximum norm of the error.

The functions that must be approximated in practical applications usually are, or are believed to be, continuous. If one specifies in advance an acceptable maximum pointwise error $\delta > 0$ and if one wishes to approximate a continuous function $f : [a, b] \to \mathbb{R}$, where $[a, b]$ is a finite, closed interval, then the Weierstraß approximation theorem (Appendix F) implies that, without loss of generality, one can assume that the approximating function p such that $\|f - p\|_C < \delta$ is a polynomial.

The Weierstraß theorem guarantees only the existence of a polynomial p such that $\|f - p\|_C < \delta$. The theorem gives no information about a computationally useful method for constructing p. From a practical point of view, it makes sense to turn the problem around and look for a polynomial p of a fixed degree, n, that minimizes $\|f - p\|_C$. In other words, one temporarily sets aside the problem of the convergence of a sequence of approximations to f in favor of the more practical problem of finding an approximation that, once optimized, requires only a certain finite number of computational steps to evaluate.

A degree-n polynomial p is called a **polynomial of best approximation** to a continuous function $f : S \to \mathbb{R}$ if and only if the maximum deviation of p from f is equal to the greatest lower bound (the infimum) of the maximum deviations from f of all polynomials of degree n.

This is a **minimax** property: The polynomial of best approximation is such that its maximum deviation from f is a minimum.

Because physical data usually are available only for a finite set of sampling points (or times), we begin with the problem of finding a polynomial of best approximation at points $x_0 = a < x_1 < \cdots < x_{N-1} = b$. In this case, a function $f : S \to \mathbb{R}$ that is defined on a finite set $S = \{x_0, x_1, \ldots, x_{N-1}\}$ is to be approximated on S by a polynomial of degree n. The supremum norm of the error is replaced by the maximum norm of the error, $\max_{k \in (0:N-1)} |f(x_k) - p(x_k)|$. Although convenient, this approach does not address the question of the approximation of f by a polynomial when S is a closed interval $[a, b]$. We discuss approximations in the supremum norm on a closed interval in Section 10.6.2.

10.6.1 BEST APPROXIMATIONS IN THE MAXIMUM NORM

Suppose that a function f is sampled at N points $x_0 = a, x_1, \ldots, x_{N-1} = b$ and is approximated by a degree-n polynomial

$$p(n, \mathbf{a}, x) := a_0 + a_1 x + \cdots + a_n x^n. \tag{10.279}$$

The sampled values of f are the components of the vector

$$\mathbf{f} = \begin{pmatrix} f(x_0) \\ f(x_1) \\ \vdots \\ f(x_{N-1}) \end{pmatrix}, \tag{10.280}$$

and the sampling points are the components of the vector

$$\mathbf{x} = \begin{pmatrix} x_0 \\ x_1 \\ \vdots \\ x_{N-1} \end{pmatrix}. \tag{10.281}$$

Both vectors belong to $\mathbb{R}^N$. The coefficients a_k of the approximating polynomial $p(n, \mathbf{a}, x)$ can be regarded as the components of a vector $\mathbf{a} \in \mathbb{R}^{n+1}$.

Polynomials of Best Approximation

Let $\delta(\mathbf{a}, f, \mathbf{x})$ be the maximum deviation of the polynomial with coefficients $\mathbf{a}$ from the function f on the points $\mathbf{x}$:

$$\delta(\mathbf{a}, f, \mathbf{x}) := \max_{k \in (0:N-1)} |f(x_k) - p(n, \mathbf{a}, x_k)|. \tag{10.282}$$

The set of values of $\delta(\mathbf{a}, f, \mathbf{x})$ for all possible coefficient vectors $\mathbf{a}$ is a subset of $\mathbb{R}$ that is bounded from below by zero. Therefore there exists a greatest lower bound

$$\rho(n, f, \mathbf{x}) := \inf_{\mathbf{a}} \delta(\mathbf{a}, f, \mathbf{x}). \tag{10.283}$$

As defined above, a polynomial of best approximation $p(n, \mathbf{a}^{(0)}, x)$ is such that

$$\delta(\mathbf{a}^{(0)}, f, \mathbf{x}) = \rho(n, f, \mathbf{x}). \tag{10.284}$$

If the number of sampling points is one greater than the degree of the polynomial ($N = n+1$), then according to Section 7.6.1 there exists a unique interpolating polynomial, which one can (in principle) compute by solving the $n + 1$ linear equations

$$\forall k \in (0:n): \ a_0 + a_1 x_k + \cdots + a_n x_k^n = f(x_k) \tag{10.285}$$

for the $n + 1$ unknown coefficients $a_0, a_1, \ldots, a_n$. In this case, $\rho(n, f, \mathbf{x}) = 0$, because, on the sampling points, the values of the interpolating polynomial are equal to the values of the function being approximated. Therefore the interpolating polynomial is the unique polynomial of best approximation.

Let us turn to the case in which $N = n + 2$. There are still $n + 1$ unknown coefficients $a_0, a_1, \ldots, a_n$, but now there are $n + 2$ values of f to be approximated. One expects that if the magnitude of the deviation $|f(x_k) - p(x_k)|$ of the value of the approximating polynomial p from the value of the function at one point x_k were much greater than the magnitudes of the deviations at the other points, then it would be possible to decrease the maximum deviation by choosing different coefficients $a_0, a_1, \ldots, a_n$ to make the deviations $|f(x_k) - p(n, \mathbf{a}, x_k)|$ more nearly equal. Further, if all of the deviations were of the same sign, then it would be possible to subtract a constant (i.e., change the value of a_0) to make some deviations positive and others negative, thereby decreasing the magnitude of the maximum deviation. It turns out that there is a unique polynomial of best approximation, $p(n, \mathbf{a}^{(0)}, x)$, and that its deviations from the sampled function values, $f(x_k) - p(n, \mathbf{a}^{(0)}, x_k)$, are of the same magnitude and alternate in sign as k runs from 0 to $n + 1$.

One says that a polynomial p has the **equioscillation property** with respect to a function f if and only if there exists a real number h such that

$$\forall k \in (0:n+1): \ f(x_k) - p(x_k) = (-1)^k h. \tag{10.286}$$

For a polynomial that has the equioscillation property, the deviations (i.e., errors) $f(x_k) - p(x_k)$ alternate in sign and are of exactly the same magnitude. If such a polynomial exists, then the maximum deviation of p from f is $|h|$.

We prove that if the number of sampling points is two greater than the degree of the approximating polynomial ($N = n + 2$), and if some polynomial $p(n, \mathbf{a}^{(0)}, x)$ has the equioscillation property with respect to f, then $p(n, \mathbf{a}^{(0)}, x)$ is a polynomial of best approximation. This result is part of a more general theorem known as **Chebyshev's equioscillation theorem**.

To establish the result using proof by contradiction, let us assume that there exists a real number h such that Eq. (10.286) holds, and that the greatest lower bound of all maximum deviations is less than $|h|$:

$$\rho(n, f, \mathbf{x}) < |h| = \max_{k \in (0:n+1)} |f(x_k) - p(n, \mathbf{a}^{(0)}, x_k)|. \tag{10.287}$$

Then there exists a degree-n polynomial $p(n, \mathbf{a}^{(1)}, x)$ with coefficients $\mathbf{a}^{(1)}$ such that, for some $k = k_1$,

$$\left| f\left(x_{k_1}\right) - p\left(n, \mathbf{a}^{(1)}, x_{k_1}\right) \right| < |h|. \tag{10.288}$$

The difference between the two approximating polynomials,

$$q(x) := p\left(n, \mathbf{a}^{(1)}, x\right) - p\left(n, \mathbf{a}^{(0)}, x\right), \tag{10.289}$$

is a polynomial of degree n. The values of $q(x)$ at the sampling points are

$$\begin{aligned}
q(x_k) &= p\left(n, \mathbf{a}^{(1)}, x_k\right) - p\left(n, \mathbf{a}^{(0)}, x_k\right) \\
&= \left[f(x_k) - p\left(n, \mathbf{a}^{(0)}, x_k\right)\right] + \left[p\left(n, \mathbf{a}^{(1)}, x_k\right) - f(x_k)\right] \\
&= (-1)^k h - \left[f(x_k) - p\left(n, \mathbf{a}^{(1)}, x_k\right)\right].
\end{aligned} \tag{10.290}$$

According to Eq. (10.288), the error of the approximating polynomial $p(n, \mathbf{a}^{(1)}, x)$ is not greater than $|h|$ at any sampling point. Therefore, Eq. (10.290) implies that, at each sampling point, $q(x_k)$ has the same sign as $(-1)^k h$.

Because the polynomial $q(x)$ changes sign $n + 1$ times in the interval $[a, b]$, it must have at least $n + 1$ roots. But this is impossible for a polynomial of degree n. This contradiction establishes that $p(n, \mathbf{a}^{(0)}, x)$ is a polynomial of best approximation.

At this point it is clear that, if one regards the maximum deviation as another unknown, one obtains a system of linear equations that determine both the coefficients of the polynomial of best approximation and the maximum deviation from the function being approximated. The $n + 2$ equations

$$\forall k \in (0 : n + 1) : \quad (-1)^k h + p\left(n, \mathbf{a}^{(0)}, x_k\right) = f(x_k), \tag{10.291}$$

which are linear in the $n + 2$ unknowns $h, a_0^{(0)}, a_1^{(0)}, \ldots, a_n^{(0)}$, have a unique solution if and only if the determinant of the coefficients is nonzero:

$$\Delta := \det \begin{pmatrix} 1 & 1 & x_0 & \cdots & x_0^n \\ -1 & 1 & x_1 & \cdots & x_1^n \\ \vdots & 1 & \vdots & \ddots & \vdots \\ (-1)^{n+1} & 1 & x_{n+1} & \cdots & x_{n+1}^n \end{pmatrix} \neq 0. \tag{10.292}$$

Expanding in minors of the first column, one finds that

$$\Delta = \sum_{l=0}^{n+1} v_n(x_0, \ldots, x_{l-1}, x_{l+1}, \ldots, x_{n+1}). \tag{10.293}$$

where v_n is a Vandermonde determinant, as defined in Eq. (6.350). According to Eq. (6.356), a Vandermonde determinant is strictly positive if its arguments are arranged in numerical order. Therefore each Vandermonde determinant in the sum given in Eq. (10.293) is strictly positive. It follows that $\Delta > 0$, and therefore that the linear equations (10.291) have a unique solution. In principle, at least, this solves the problem of determining the polynomial of best approximation when the number of sampling points is two greater than the degree of the approximating polynomial.

Fitting at the Chebyshev Extrema

For a given degree n of the approximating polynomial, some choices of the values of the sampling points will produce a smaller maximum deviation, $|h|$, than others. A polynomial of degree $N - 1$ can be made to pass through the $N = n + 2$ points (x_k, d_k), where $d_k = (-1)^k h$ is the deviation from $f(x_k)$ of the polynomial of best approximation. Therefore it makes sense to assume that the deviations are the values of an error polynomial at the sampling points, and look for a polynomial that is known to have small extremal values in a specified interval. For simplicity we choose the interval $[a, b]$ to be $[-1, 1]$; we will discard this assumption after we have found the optimal values of x_k in the interval $[-1, 1]$, because we can translate and scale $[-1, 1]$ to coincide with any finite closed interval.

Chebyshev proved that, in the interval $[-1, 1]$, the polynomial $2^{1-m} T_m(x)$, where $T_m(x)$ is the Chebyshev polynomial of degree m, has the smallest upper bound of its absolute value, $|2^{1-m} T_m(x)|$, of any polynomial $p(x)$ of degree m with leading coefficient $a_m = 1$. The proof, again by contradiction, begins with the assumption that, for all values of x in the interval $[-1, 1]$,

$$|p(x)| < \max_{z \in [-1,1]} |2^{1-m} T_m(z)|. \tag{10.294}$$

The difference between the two polynomials,

$$q(x) := 2^{1-m} T_m(x) - p(x), \tag{10.295}$$

is a polynomial of degree $m - 1$ since the leading terms of $2^{1-m} T_m(x)$ and $p(x)$ cancel. Because $T_m(x) = \cos(m \cos^{-1} x)$, the extremal values of $2^{1-m} T_m(x)$ are ± 1, and they occur at the $m + 1$ points

$$\forall k \in (0 : m) : \quad x'_{m,k} := \cos\left(\frac{k\pi}{m}\right). \tag{10.296}$$

The extremal value of $|2^{1-m} T_m(x'_{m,k})|$ is 1. Because we have assumed that $|p(x'_{m,k})| < 1$, the sign of $q(x'_{m,k})$ is the same as the sign of $2^{1-m} T_m(x'_{m,k})$, which is alternately positive and negative. Therefore $q(x)$ has m zeros in $[-1, 1]$, which is impossible for a polynomial of degree $m - 1$. This contradiction with Eq. (10.294) proves the theorem.

It follows that, if one can choose the sampling points, one should choose the points given in Eq. (10.296), at which the Chebyshev polynomial of degree $m = N - 1$ achieves its extremal values. Recalling the definition of the Chebyshev polynomials in Eq. (8.196) and the definition of the Chebyshev extremal abscissae in Eq. (10.296), one sees that

$$(-1)^k = \cos(k\pi) = \cos((N-1) \cos^{-1} x'_{N-1,k}) = T_{N-1}(x'_{N-1,k}). \tag{10.297}$$

Because $x'_{N-1,k}$ decreases from 1 to -1 as k goes from 0 to $N-1$, and because we want x_k to increase (not decrease) as k increases, we choose as the sampling points

$$x_k = x'_{N-1,N-k-1} = \cos\left(\frac{(N-k-1)\pi}{N-1}\right). \tag{10.298}$$

If the sampling points x_k are chosen in this way, the extrema of the polynomial $2^{-N} T_{N-1}(x)$ occur at the sampling points. Noting that $T_{N-1}(x_k) = \cos((N-k-1)\pi) = (-1)^{N-k-1} = (-1)^{N-1}(-1)^{-k} = (-1)^{N-1}(-1)^k$, one finds that, at the sampling points,

$$\forall k \in (0 : N-1) : \quad f(x_k) - p\left(n, \mathbf{a}^{(0)}, x_k\right) = (-1)^{N-1} h\, T_{N-1}(x_k). \tag{10.299}$$

If the interval on which one needs to approximate f is not $[-1, 1]$, one translates and stretches the interval $[-1, 1]$ to the interval $[a, b]$ as follows:

$$\forall k \in (0 : N-1) : \quad x_k = \frac{b+a}{2} + \left(\frac{b-a}{2}\right) x'_{N-k-1}$$

$$= \frac{b+a}{2} + \left(\frac{b-a}{2}\right) \cos\left(\frac{(N-k-1)\pi}{N-1}\right). \tag{10.300}$$

Solving the linear system (10.299) determines the value of h and the coefficients of the polynomial of best approximation.

Table 10.1 gives the Chebyshev polynomials for orders 0 through 10.

Computational Considerations

The linear equations (10.299) determine h by subtracting two quantities, $f(x_k)$ and $p(n, \mathbf{a}^{(0)}, x_k)$, which are generally much larger than h. From Section 1.4.3 one knows that the

TABLE 10.1 Chebyshev polynomials for orders 0 through 10

Order (n)	Chebyshev polynomial $T_n(x)$
0	1
1	x
2	$2x^2 - 1$
3	$4x^3 - 3x$
4	$8x^4 - 8x^2 + 1$
5	$16x^5 - 20x^3 + 5x$
6	$32x^6 - 48x^4 + 18x^2 - 1$
7	$64x^7 - 112x^5 + 56x^3 - 7x$
8	$128x^8 - 256x^6 + 160x^4 - 32x^2 + 1$
9	$256x^9 - 576x^7 + 432x^5 - 120x^3 + 9x$
10	$512x^{10} - 1280x^8 + 1120x^6 - 400x^4 + 50x^2 - 1$

TABLE 10.2 Powers x^0 through x^{10} in terms of Chebyshev polynomials

Power (x^n)	Expression in terms of Chebyshev polynomials
x^0	T_0
x^1	T_1
x^2	$\frac{1}{2}(T_2 + 1)$
x^3	$\frac{1}{4}(T_3 + 3T_1)$
x^4	$\frac{1}{8}(T_4 + 4T_2 + 3)$
x^5	$\frac{1}{16}(T_5 + 5T_3 + 10T_1)$
x^6	$\frac{1}{32}(T_6 + 6T_4 + 15T_2 + 10T_0)$
x^7	$\frac{1}{64}(T_7 + 7T_5 + 21T_3 + 35T_1)$
x^8	$\frac{1}{128}(T_8 + 8T_6 + 28T_4 + 56T_2 + 35T_0)$
x^9	$\frac{1}{256}(T_9 + 9T_7 + 36T_5 + 84T_3 + 126T_1)$
x^{10}	$\frac{1}{512}(T_{10} + 10T_8 + 45T_6 + 120T_4 + 210T_2 + 126T_0)$

subtraction of two numbers of comparable magnitudes can result in a catastrophic cancellation of significant digits. It follows that, in order to be sure that the value of h one computes is not simply a result of the pathologies of floating-point arithmetic, one should carry out the computation with many more digits (or bits) than one needs in the answer. For example, if one believes that single-precision accuracy is adequate for h, then one should compute h and the coefficients of the polynomial of best approximation in double precision.

Even if one has found a numerical approximation to the polynomial of best approximation, one's computational difficulties are not over. Section 3.2.2 shows that the numerical evaluation of a polynomial may be subject to severe errors as a result of rounding and the properties of floating-point representations. Another way of making the same point is to observe that, given a tolerance δ, there exist a positive integer M and a linear combination of the monomials 1, $x, \ldots, x^{M-2}$ such that the difference between x^M and the linear combination of lower-degree monomials is less than δ on the interval $[-1, 1]$. In other words, for purposes of computation with a desired absolute precision δ, x^M is linearly dependent on $1, x, \ldots, x^{M-2}$.

To prove this assertion, let us begin with the expression for x^M in terms of Chebyshev polynomials given in Eq. (8.199) (see also Table 10.2). The leading term in this expression is $2^{1-M}T_M(x)$; the next term is $2^{1-M}M\,T_{M-2}(x)$, which is a polynomial of degree $M-2$. Because the supremum of $T_M(x)$ on the interval $[-1, 1]$ is 1, the supremum of $2^{1-M}|T_M(x)|$ is 2^{1-M}. Given $\delta > 0$, choose M such that $2^{1-M} < \delta$. From the exact expression in Eq. (8.199) one sees that the error incurred by omitting the term $2^{1-M}T_M(x)$ is less than δ. Therefore,

$$\forall x \in [-1, 1]: \left| x^M - 2^{1-n}\left[\binom{M}{1} T_{M-2}(x) + \cdots + \binom{M}{k} T_1(x) \right] \right| < \delta \qquad (10.301)$$

if $M = 2k + 1$, or

$$\forall x \in [-1, 1]: \left| x^M - 2^{1-n} \left[\binom{M}{1} T_{M-2}(x) + \cdots + \binom{M}{k} \frac{1}{2} T_0(x) \right] \right| < \delta \qquad (10.302)$$

if $M = 2k$. For example, if $\delta = 10^{-3}$, then $2^{1-M} < \delta$ if $M - 1 \geq 10$. Hence, for $M \geq 11$, x^M is, for purposes of computation with an accuracy δ, linearly dependent on $1, x, \ldots, x^{M-2}$ in the interval $[-1, 1]$.

It follows from the approximate linear dependence of monomials of high degree on the monomials of lower degrees that one should avoid computing a polynomial as a sum of monomials. Instead, one should recall that orthogonality implies linear independence (Section 8.1.5), and approximate f using a sum of orthogonal polynomials, such as the Chebyshev polynomials. Unfortunately the superficial simplicity of a sum of monomials and the time-honored tradition of obtaining an approximation by truncating a power series seem likely to ensure a long life for the computationally dangerous operation of trying to approximate a function with a polynomial of the form of Eq. (10.279).

We discuss the use of Chebyshev expansions to obtain polynomial approximations in Section 10.6.2.

10.6.2 BEST APPROXIMATIONS IN THE SUPREMUM NORM

There are several approaches to finding a polynomial that is close to the polynomial of best approximation to a continuous function $f : [-1, 1] \to \mathbb{R}$. All of these approaches make use of the **equal-ripple** property of the Chebyshev polynomials, which means simply that the extremal values of a Chebyshev polynomial in the interval $[-1, 1]$ are equal in absolute value (since the extremal values of $\cos \theta$ are ± 1). We proved above that if the number of sampling points is two greater than the degree of the approximating polynomial ($N = n + 2$), and if some polynomial $p(n, \mathbf{a}^{(0)}, x)$ has the equioscillation property with respect to f, then $p(n, \mathbf{a}^{(0)}, x)$ is a polynomial of best approximation. Thus it is sufficient to find a set of points on which to evaluate f and an approximating polynomial that has the equioscillation property with respect to f. In other words, it is sufficient to find an approximating polynomial p such that the error $f - p$ is (to an accuracy that is adequate for the approximation that one seeks) proportional to the Chebyshev polynomial of order $N - 1$, in order to conclude that p is close to the polynomial of best approximation.

The first approach is to choose (or guess) a degree, n, of the approximating polynomial, and then to carry out Lagrangian interpolation (Section 7.6) using the Chebyshev abscissae (zeros of the Chebyshev polynomial of degree $n + 1$; see Eq. (8.204)) as the points at which the value of the polynomial equals the value of f. The reason why this approach may be useful is that, according to Eq. (D.2), the error in Lagrangian interpolation is

$$f(x) - p(x) = \frac{f^{(n+1)}(\xi(x))}{(n+1)!}(x - x_1) \cdots (x - x_{n+1}) \qquad (10.303)$$

where $\xi(x)$ is an interior point of the smallest interval that contains the points $x, x_1, \ldots, x_{n+1}$.

Note that ξ depends on x. If $x_1, \ldots, x_{n+1}$ are chosen as the zeros of T_{n+1}, then the remainder is equal to

$$f(x) - p(x) = \frac{f^{(n+1)}(\xi)}{2^n(n+1)!} T_{n+1}(x). \tag{10.304}$$

If ξ is approximately constant for values of $x \in [a, b]$, then the remainder is proportional to an equal-ripple function. Therefore the equioscillation condition is (approximately) satisfied, and p is (approximately) a polynomial of best approximation. The obvious weakness in this approach is that, in order to evaluate the accuracy of approximating f by p, one must evaluate ξ for (at least) several values of x.

A second approach is to choose (or guess) a degree, n, of the approximating polynomial, and then fit f at the Chebyshev extremal abscissae defined in Eq. (10.296) by solving the system of linear equations given in Eq. (10.299). The problem with this approach is that in the continuous case (unlike the discrete case discussed above) there is absolutely no reason to think that the error is, in fact, proportional to a Chebyshev polynomial, or is a polynomial at all. However, this approach is useful as the first step in an iterative approach known as Remes' method [Dem'yanov 1974] that leads eventually to a modified set of sampling points and a polynomial of best approximation. In general, the error $f - p$ that results from this iterative process is not proportional to a Chebyshev polynomial.

A third approach is to choose a degree, n, of the approximating polynomial p, and then use Eq. (8.305) to project the function f onto the subspace of $\mathbb{C}^0([-1, 1]; \mathbb{R})$ that is spanned by the Chebyshev polynomials of degrees 0 through n. One then sets p equal to the resulting linear combination of Chebyshev polynomials:

$$\forall x \in [-1, 1] : \quad p(x) = \sum_{m=0}^{n} \frac{1}{\langle T_m, T_m \rangle} \langle T_m, f \rangle T_m(x). \tag{10.305}$$

Since the inner product

$$\langle T_m, f \rangle = \int_{-1}^{1} T_m(x) f(x) \frac{1}{[1 - x^2]^{1/2}} \, dx = \int_0^{\pi} \cos(m\theta) f(\cos\theta) \, d\theta \tag{10.306}$$

is proportional to a Fourier coefficient, one can use well-known techniques such as the fast Fourier transform to compute it. If the coefficients in the linear combination (10.305) decrease rapidly enough that the error committed in truncating the expansion at degree n is dominated by a single Chebyshev polynomial, then one can conclude that the approximating polynomial is close to the polynomial of best approximation. A virtue of this approach is that one can find out by inspecting the coefficients $\langle T_m, f \rangle$ in Eq. (8.305) whether the error can be approximated adequately by a single Chebyshev polynomial.

A fourth approach is useful if one already has a truncated power series for f,

$$f(x) = \sum_{k=0}^{M} f_k x^k + O(x^{M+1}). \tag{10.307}$$

Suppose that the final result must be accurate within a certain tolerance δ, and that the error of the truncated power series is not larger than $\delta/2$:

$$\left| f(x) - \sum_{k=0}^{M} f_k \, x^k \right| < \frac{\delta}{2}. \tag{10.308}$$

Let us express x^k in terms of the Chebyshev polynomials $T_k(x), \ldots, T_1(x)$ or $T_0(x)$ using Eq. (8.199) or Table 10.2 and then rearrange the summation in Eq. (10.307) to obtain a linear combination of Chebyshev polynomials,

$$f(x) = \sum_{k=0}^{M} t_k \, T_k(x) + O(x^{M+1}). \tag{10.309}$$

The error if one discards terms of order x^{M+1} is unchanged (and is less than $\delta/2$).

One sees directly from Eqs. (10.307) and (10.309) that the coefficient of T_M is

$$t_M = \frac{f_M}{2^{M-1}}. \tag{10.310}$$

We would like to simplify the Chebyshev sum (10.309) by discarding the highest term, $t_M \, T_M(x)$. (It may be possible, in other cases, to discard more than just the highest-order Chebyshev polynomial.) If the coefficient f_M of the highest term in the truncated power series obeys the inequality

$$f_M < 2^{M-2} \delta, \tag{10.311}$$

then

$$t_M < \frac{\delta}{2} \Rightarrow |t_M \, T_M| < \frac{\delta}{2}. \tag{10.312}$$

The error if one keeps only $T_{M-2}(x), \ldots, T_1(x)$ or $T_0(x)$ is still less than δ:

$$\left| f(x) - \sum_{k=0}^{M-2} t_k \, T_k(x) \right| < \left| f(x) - \sum_{k=0}^{M} f_k \, x^k \right| + |t_M \, T_M(x)| < \frac{\delta}{2} + \frac{\delta}{2} = \delta. \tag{10.313}$$

The result is that, after discarding all Chebyshev polynomial terms that can be omitted without causing the error to exceed δ, one has succeeded in **economizing** the original truncated power series, Eq. (10.307).

For example, let $f(x) = \sin(x)$, $M = 5$, and $\delta = 1.25 \times 10^{-3}$. Then $f_5 = 1/120 < 2^3 \delta = 10^{-2}$, from which it follows that the inequality (10.311) is satisfied. Therefore we truncate the power series after the term $x^5/5!$, obtaining

$$\sin(x) = x - \frac{x^3}{3!} + \frac{x^5}{5!} + O\left(\frac{x^7}{7!}\right). \tag{10.314}$$

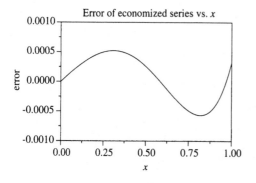

FIGURE 10.6

Error in the economized sine series, Eq. (10.317).

Since the sine series is alternating, the error in the truncated power series is less than the first neglected term, which is $x^7/7! < 2 \times 10^{-4} < \delta/2$ on $[-1, 1]$. The expansion of the truncated power series in terms of Chebyshev polynomials is

$$\sin(x) = \frac{1}{1920} T_5(x) - \frac{15}{384} T_3(x) + \frac{338}{384} T_1(x) + O\left(\frac{x^7}{7!}\right). \tag{10.315}$$

We can drop the term $T_5(x)/1920$ because the error committed in so doing is $1/1920 = 5.21 \times 10^{-4} < \delta/2 = 6.25 \times 10^{-4}$. Thus

$$\sin(x) \approx \frac{338}{384} T_1(x) - \frac{15}{384} T_3(x) \tag{10.316}$$

with an error less than 1.25×10^{-3}.

Although Eq. (10.316) is in the best form for computation (because the Chebyshev polynomials are orthogonal), it is interesting to re-express the economized Chebyshev approximation (10.316) in terms of the monomials x and x^3. One obtains as the economized power series for the sine

$$\sin(x) \approx p(x) = \frac{383}{384} x - \frac{5}{32} x^3, \tag{10.317}$$

which is close to, but significantly different from, the power series truncated after the cubic term $(x - x^3/6)$. Figure 10.6 shows the error $\sin(x) - p(x)$ as a function of x for the interval $[0, 1]$. Because the sine is an odd function, the curve for the interval $[-1, 0]$ can be obtained from the curve shown in Figure 10.6 by reflection in the horizontal and vertical axes. The approximation given in Eq. (10.317) is not quite equal-ripple, the maximum deviations being -5.67×10^{-4} and $+5.21 \times 10^{-4}$.

10.6.3 EXERCISES FOR SECTION 10.6

10.6.1 Use Eq. (10.302) to approximate x^{10} in terms of the Chebyshev polynomials $T_0(x), \ldots, T_8(x)$ on the interval $[-1, 1]$. What is the *maximum* absolute value of the error of this approximation?

10.6.2 The purpose of this problem is to explore a little more deeply why x^{10} can be approximated to 2 parts in 1000 by a polynomial of order 8 on the interval

$[-1, 1]$. For simplicity, assume that the weight function is

$$w(x) = 1, \qquad (10.318)$$

and let

$$\phi_j(x) = N_j\, x^j \qquad (10.319)$$

where $x \in [-1, 1]$ and N_j is a normalization constant.

(a) Find N_j such that $\|\phi_j\| = 1$.

(b) Calculate $\langle \phi_i, \phi_j \rangle$.

(c) For $j > 1$, let

$$\phi_j = \chi_j + \psi_j, \qquad (10.320)$$

where χ_j is the orthogonal projection of ϕ_j on ϕ_{j-2} and $\psi_j \perp \phi_{j-2}$. Write down a simple general expression for χ_j. Use the Pythagorean theorem to compute $\|\psi_j\|^2$.

(d) Show from the results of parts (a)–(c) that

$$\lim_{j \to \infty} \|\psi_j\|^2 = 0. \qquad (10.321)$$

(e) Repeat (a)–(d) for the weight function

$$w(x) = \frac{1}{[1 - x^2]^{1/2}}. \qquad (10.322)$$

10.6.3   This problem concerns polynomial approximations of the function

$$f(x) = \frac{1}{3 + x}$$

where $x \in [-1, 1]$.

(a) What first-degree polynomial $(n = 1)$ does one get by fitting at the Chebyshev extrema? What is the *maximum* absolute value of the error?

Hint: You will need to find the maxima in the interior of the interval and compare them with the errors at $x = \pm 1$.

(b) Now let $n = 9$. By fitting at the Chebyshev extrema using Eq. (10.299), obtain the magnitude of the maximum deviation, h.

10.6.4   Obtain an economized Chebyshev series, and then an economized power series, for the function $\sin(x)$ on the interval $[-1, 1]$, assuming a tolerance $\delta = 10^{-5}$.

10.6.5   Obtain an economized Chebyshev series, and then an economized power series, for the function $\sin(x)$ on the interval $[-\frac{1}{2}\pi, \frac{1}{2}\pi]$, assuming a tolerance $\delta = 10^{-2}$.

10.6.6   Obtain an economized Chebyshev series, and then an economized power series, for the function $\cos(x)$ on the interval $[-1, 1]$, assuming a tolerance $\delta = 1.25 \times 10^{-3}$.

10.7 HILBERT AND BANACH SPACES

10.7.1 SURVEY OF COMPLETE METRIC VECTOR SPACES

We show in Section 10.3.1 that if a sequence X in a metric space S is convergent, then X is a Cauchy sequence. The converse, that every Cauchy sequence is convergent, fails if S does not contain the limit of one or more of its Cauchy sequences. A metric space S is called **complete** if and only if every Cauchy sequence in S converges to a point in S. If S is not complete, then it is called **incomplete**. The set of rational numbers $\mathbb{Q}$ is incomplete with respect to the absolute-value norm, as one can see by considering any sequence of decimal fractions that successively approximate an irrational number. We show in Appendix E that $\mathbb{R}$ and $\mathbb{C}$ are complete with respect to the absolute-value and modulus norms, respectively. *In a complete metric space, a sequence converges if and only if it is a Cauchy sequence.*

Every metric space can be "completed" by a formal construction that is a generalization of the construction that Cantor used to obtain the system of real numbers, starting from the rationals (Kolmogorov and Fomin [1975]).

Fréchet Spaces

A complete metric vector space $\mathcal{F}$ in which the metric is translation-invariant,

$$\rho(\overline{x} + \overline{z}, \overline{y} + \overline{z}) = \rho(\overline{x}, \overline{y}), \tag{10.323}$$

and in which scalar multiplication is continuous,

$$\forall \lambda \in \mathbb{F} : \forall \overline{x} \in \mathcal{F} : \forall \epsilon > 0 : \exists \delta > 0 :\ni: \forall \overline{x}' :\ni: \rho(\overline{x}', \overline{x}) < \delta : \rho(\lambda \overline{x}', \lambda \overline{x}) < \epsilon, \tag{10.324}$$

is called a **Fréchet space**. The requirement that scalar multiplication must be continuous ensures that if $\overline{x}$ and $\overline{x}'$ are close under the metric of $\mathcal{F}$, then $\lambda \overline{x}$ and $\lambda \overline{x}'$ are also close. The continuity of vector addition is a consequence of the triangle inequality (See Exercise 10.7.2).

For example, the spaces $\mathcal{D}([a, b]; \mathbb{R})$ and $\mathcal{S}$ are Fréchet spaces. We know already that these spaces possess translation-invariant metrics. To prove completeness, we note that if $\{f_n\}$ is a Cauchy sequence in $\mathcal{D}([a, b]; \mathbb{R})$ under the supremum norm, then for every $m = 0, 1, \ldots$ the sequence $\{f_n^{(m)}\}$ of m-th derivatives converges uniformly on $[a, b]$. We proved previously that if a sequence of real-valued functions converges uniformly to a function f and the sequence of their derivatives converges uniformly to a function g, then $g = f'$. Induction then shows that the function $f := \lim f_n$ possesses derivatives of all orders and vanishes outside the open interval (a, b); therefore $f \in \mathcal{D}([a, b]; \mathbb{R})$. The proof of completeness for $\mathcal{S}$ is similar.

Banach Spaces

A complete normed vector space is called a **Banach space** in honor of Stefan Banach, who studied such spaces in great depth. For example, $L^p([a, b]; \mathbb{C})$ is a Banach space. Every Banach space is a Fréchet space, but the converse is not true.

Finite-dimensional Banach spaces are useful in computational mathematics, in part because a vector space $\mathcal{V}$ and the vector space of linear mappings from $\mathcal{V}$ into $\mathcal{V}$ are both Banach spaces under any p-norm such that $1 < p < \infty$. Infinite-dimensional Banach spaces are of great importance in pure mathematics.

Hilbert Spaces

A special case of a Banach space is a **Hilbert space**, which is a Euclidean or unitary space that is complete with respect to the inner-product norm Eq. (10.17). *Separable* Hilbert spaces are extremely important in physics and engineering, because the inner-product norm is often directly related to energy or probability, and because the state space of a quantum-mechanical system is a Hilbert space. Because inseparable Hilbert spaces are less tractable than separable Hilbert spaces, physicists frequently use stratagems that reduce an inseparable to a separable Hilbert space, such as conceptually enclosing a spatially unbounded system of particles and fields in a finite box.

A proof of the completeness of a particular space depends on the nature of that space. Appendix E proves the completeness of $\mathbb{R}$ and of all finite-dimensional Euclidean or unitary spaces. We prove the completeness of $l^2(\mathbb{F})$ subsequently. For a proof of the completeness of $L^2([a, b]; \mathbb{F})$ (in which $\mathbb{F} = \mathbb{R}$ or $\mathbb{C}$), see Kolmogorov and Fomin (1975).

A vector subspace of a Hilbert space is often called a **linear manifold**. A **Hilbert subspace** is a closed linear manifold, in keeping with the convention that a named substructure must obey the same axioms as its parent.

Space of Complex Numbers

$\mathbb{C}$ is a Hilbert space over $\mathbb{R}$. We prove the completeness of $\mathbb{C}$ as part of the Bolzano-Weierstraß theorem in Appendix E.

Spaces of Column Vectors

Both $\mathbb{R}^n$ and $\mathbb{C}^n$ are Hilbert spaces if

$$\rho(\mathbf{x}, \mathbf{y}) = \|\mathbf{x} - \mathbf{y}\|_{ip} = \langle \mathbf{x} - \mathbf{y}, \mathbf{x} - \mathbf{y} \rangle^{1/2}. \tag{10.325}$$

If, on the other hand,

$$\rho(\mathbf{x}, \mathbf{y}) = \|\mathbf{x} - \mathbf{y}\|_p \tag{10.326}$$

in which $p \neq 2$, then $\mathbb{R}^n$ and $\mathbb{C}^n$ are Banach spaces but not Hilbert spaces. Once again, see Appendix E for a proof of completeness.

Matrix Spaces

Both $\mathbb{R}^{n \times n}$ and $\mathbb{C}^{n \times n}$ are Hilbert spaces over their respective fields under the inner product

$$\langle \mathbf{A}, \mathbf{B} \rangle = \text{trace}[\mathbf{A}^\dagger \mathbf{B}]. \tag{10.327}$$

More generally, we have shown that the set hom $[\mathcal{V}]$ of all linear mappings of a finite-dimensional Euclidean or unitary space is itself a finite-dimensional Euclidean or unitary space. We show in Appendix E that every finite-dimensional Euclidean or unitary space is complete. Therefore hom $[\mathcal{V}]$ is complete.

Sequence Spaces

The space of finite complex sequences $\phi(\mathbb{F})$ is *not* a Hilbert space over $\mathbb{F}$, because the limit of a sequence of elements of $\phi(\mathbb{F})$ may be an infinite sequence, hence not in $\phi(\mathbb{F})$. Therefore $\phi(\mathbb{F})$ is not complete. It can be shown that both $m(\mathbb{F})$ and $c(\mathbb{F})$ are complete (see Exercises 10.7.3 and 10.7.4).

l^p Spaces

We show that $l^p(\mathbb{F})$ (in which $\mathbb{F} = \mathbb{R}$ or $\mathbb{C}$) is complete. Let $\bar{x}_n = (x_n^1, x_n^2, \ldots)$ and let $(\bar{x}_1, \bar{x}_2, \ldots)$ be a Cauchy sequence in some p-norm. Then, for every $\epsilon > 0$, there exists an integer $M(\epsilon)$ such that

$$\forall m, n > M(\epsilon): \ \|\bar{x}_m - \bar{x}_n\|_p = \left[\sum_{j=1}^{\infty} |x_m^j - x_n^j|^p \right]^{1/p} < \epsilon. \tag{10.328}$$

Then, for every j and for every $m, n > M(\epsilon)$,

$$|x_m^j - x_n^j| < \epsilon. \tag{10.329}$$

By the completeness of $\mathbb{F}$ (which we prove in Appendix E), for each j there exists a number $x^j \in \mathbb{F}$ such that

$$x^j = \lim_{n \to \infty} x_n^j. \tag{10.330}$$

Then each coordinate of $\{\bar{x}_n\}$ converges to the corresponding coordinate of

$$\bar{x} := (x^1, x^2, \ldots). \tag{10.331}$$

Equation (10.328), which is true for every n, is also true in the limit $n \to \infty$. Then

$$\forall m > M(\epsilon): \ \|\bar{x}_m - \bar{x}\|_p \leq \epsilon, \tag{10.332}$$

which implies that $\{\bar{x}_n\}$ converges strongly to $\bar{x}$.

Finally, we must show that $\bar{x}$ belongs to $l^p(\mathbb{F})$. By the triangle inequality,

$$\forall \epsilon > 0 : \forall m > M(\epsilon): \ \|\bar{x}\|_p = \|\bar{x} - \bar{x}_m + \bar{x}_m\|_p \leq \|\bar{x} - \bar{x}_m\|_p + \|\bar{x}_m\|_p. \tag{10.333}$$

Then

$$\forall \epsilon > 0 : \forall m > M(\epsilon): \ \|\bar{x}\|_p < \epsilon + \|\bar{x}_m\|_p. \tag{10.334}$$

Because the Cauchy sequence $(\bar{x}_1, \bar{x}_2, \ldots)$ is bounded, it follows that for every ϵ, $\|\bar{x}\|_p$ cannot be greater than $\epsilon + \limsup_{m \to \infty} \|\bar{x}_m\|_p$ and is therefore finite. Then the limit vector $\bar{x}$ belongs to $l^p(\mathbb{F})$. It follows that $l^p(\mathbb{F})$ is complete.

Space of Continuous Real-Valued Functions on $[a, b]$

We show in Section 10.4.3 that every uniformly convergent sequence of continuous real-valued functions on $[a, b]$ converges to a continuous function. For real-valued functions on $\mathbb{R}$, uniform convergence is convergence in the supremum norm. Therefore $\mathcal{C}^0([a, b]; \mathbb{R})$ is complete under the supremum norm.

L^2 Spaces

$L^2([a, b]; \mathbb{C})$ is a Hilbert space over $\mathbb{C}$, and $L^p([a, b]; \mathbb{C})$ is a Banach space over $\mathbb{C}$. For a proof of completeness, see Kolmogorov and Fomin (1975).

10.7.2 COMPLETE ORTHONORMAL SETS

Let $\mathcal{H}$ be an infinite-dimensional Hilbert space that is separable under its inner-product norm. The definition of separability guarantees that there exists a sequence of vectors $(\bar{v}_1, \bar{v}_2, \ldots)$ that is dense in $\mathcal{H}$. However, for most practical computations involving a vector space a dense sequence is much less convenient than a basis. We show that every separable Hilbert space has a basis by constructing one from a dense sequence. This result is the foundation for expansions in series of orthogonal functions, which are very widely used in physics and engineering.

We show in Section 8.2.2 that if a set of vectors spans a finite-dimensional Euclidean or Hermitian space, then the Gram-Schmidt procedure produces an orthonormal basis. The Gram-Schmidt procedure can also be applied to a *countable* set of vectors because the procedure is inductive. Each new vector $\bar{u}_{n+1}$ is constructed in a way that does not change any of the vectors $\bar{u}_1, \ldots, \bar{u}_n$ that have already been calculated.

If one applies the Gram-Schmidt procedure to a dense sequence $(\bar{v}_1, \bar{v}_2, \ldots)$, discarding along the way any vector $\bar{v}_m$ that is linearly dependent on its predecessors, then one obtains a countable orthonormal set

$$B := \{\bar{u}_1, \ldots, \bar{u}_n, \ldots\}. \tag{10.335}$$

A countable orthonormal set such as $\{\bar{u}_n\}$ is called a **complete orthonormal set** in $\mathcal{H}$ if and only if its finite linear combinations are dense in $\mathcal{H}$. We show that B is complete in $\mathcal{H}$.

To prove that $\{\bar{u}_n\}$ is a complete orthonormal set, one must show that for every $\epsilon > 0$, and for every $\bar{x} \in \mathcal{H}$, there exists a *finite* linear combination $\bar{s}$ of the $\{\bar{u}_n\}$ such that $\|\bar{x} - \bar{s}\|_{\text{ip}} < \epsilon$. (As usual, $\| \cdot \|_{\text{ip}}$ stands for the inner-product norm in $\mathcal{H}$.) Because the set $\{\bar{v}_n\}$ is dense in $\mathcal{H}$, there exist elements $\bar{v}_q$ such that

$$\|\bar{x} - \bar{v}_q\|_{\text{ip}} < \epsilon. \tag{10.336}$$

By the well-ordering principle, the set of positive integers $\{q\}$ for which this inequality holds has a least element, k. Let l be the dimension of the subspace $\mathcal{W}_l$ spanned by $\{\bar{v}_1, \ldots, \bar{v}_k\}$:

$$\begin{aligned} \mathcal{W}_l &:= \text{span}[\bar{v}_1, \ldots, \bar{v}_k] \\ l &:= \dim[\mathcal{W}_l]. \end{aligned} \tag{10.337}$$

$\mathcal{W}_l$ is spanned by the first l orthonormal vectors produced by the Gram-Schmidt process, $\{\bar{u}_1, \ldots, \bar{u}_l\}$. Certainly

$$l \leq k; \tag{10.338}$$

l is less than k if the set $\{\bar{v}_1, \ldots, \bar{v}_k\}$ is linearly dependent. Then

$$\bar{v}_k = \sum_{j=1}^{l} \langle \bar{u}_j, \bar{v}_k \rangle \bar{u}_j. \tag{10.339}$$

Let P_l be the projector on $\mathcal{W}_l$:

$$\mathsf{P}_l \bar{x} = \sum_{j=1}^{l} \langle \bar{u}_j, \bar{x} \rangle \bar{u}_j. \tag{10.340}$$

From the Pythagorean theorem one has

$$\|\bar{x} - \bar{v}_k\|_{\text{ip}}^2 = \|\bar{x} - \mathsf{P}_l \bar{x}\|_{\text{ip}}^2 + \|\mathsf{P}_l \bar{x} - \bar{v}_k\|_{\text{ip}}^2 \tag{10.341}$$

because $\mathsf{P}_l \bar{x} - \bar{v}_k \in \mathcal{W}_l$ and because $\bar{x} - \mathsf{P}_l \bar{x}$ is orthogonal to $\mathcal{W}_l$. Then

$$\|\bar{x} - \mathsf{P}_l \bar{x}\|_{\text{ip}}^2 = \|\bar{x} - \bar{v}_k\|_{\text{ip}}^2 - \|\mathsf{P}_l \bar{x} - \bar{v}_k\|_{\text{ip}}^2 \leq \|\bar{x} - \bar{v}_k\|_{\text{ip}}^2 < \epsilon^2, \tag{10.342}$$

which implies that

$$\forall \bar{x} \in \mathcal{H} : \forall \epsilon > 0 : \exists l \in \mathbb{Z}^+ :\ni: \left\| \bar{x} - \sum_{j=1}^{l} \langle \bar{u}_j, \bar{x} \rangle \bar{u}_j \right\|_{\text{ip}} < \epsilon. \tag{10.343}$$

That is, there exists a finite linear combination $\sum_{j=1}^{l} \langle \bar{u}_j, \bar{x} \rangle \bar{u}_j$ that approximates the vector $\bar{x}$ within the specified error ϵ. This establishes that $\{\bar{u}_1, \bar{u}_2, \ldots\}$ is a complete orthonormal set.

For example, the set of polynomials with rational coefficients, $\mathcal{P}(\mathbb{Q})$, which is countable and dense in the space of all continuous functions on a finite interval, $\mathcal{C}^0((a, b); \mathbb{R})$, can be taken as the set $\{\bar{v}_l\}$. The monomials $1, x, x^2, \ldots$ occur in the sequence of polynomials with rational coefficients. We have already shown that if the interval is $(-1, 1)$ and the weight function is unity, and if one applies the Gram-Schmidt procedure to the sequence $1, x, x^2, \ldots$, the result is the Legendre polynomials. Therefore the Legendre polynomials are an example of a complete orthonormal set $\{u_1, u_2, \ldots\}$. We give an example of a Legendre series in Section 10.7.5.

We have shown that if $\mathcal{H}$ is separable, then it includes a complete orthonormal set. Conversely, if $\mathcal{H}$ includes a complete orthonormal subset $\{\bar{u}_n\}$ then the countable set that consists of the finite linear combinations of the $\bar{u}_n$ with rational (or rational complex) coefficients is dense in $\mathcal{H}$, implying that $\mathcal{H}$ is separable. Therefore *a Euclidean or unitary space $\mathcal{H}$ has a complete orthonormal set if and only if $\mathcal{H}$ is separable.*

If $\mathcal{H}$ is a separable complex Hilbert space, then $l^2(\mathbb{C})$ and $\mathcal{H}$ are isomorphic as Hilbert spaces under the mapping

$$\forall n \in \mathbb{Z}^+ : \bar{u}_n \mapsto \bar{e}_n \in l^2(\mathbb{C}), \tag{10.344}$$

in which $\{\bar{u}_n\}$ is a complete orthonormal set in $\mathcal{H}$. Therefore, just as there is (up to isomorphism) only one n-dimensional unitary space, there is (up to isomorphism) only one infinite-dimensional, separable complex Hilbert space.

10.7.3 ORTHOGONAL SERIES

Strong Convergence of Orthogonal Series
We show in Section 8.4 that the finite sum

$$\mathsf{P}_l\bar{x} = \sum_{k=1}^{l} \langle \bar{u}_k, \bar{x} \rangle \, \bar{u}_k \tag{10.345}$$

is the linear combination of orthonormal vectors $\{\bar{u}_1, \ldots, \bar{u}_l\}$ that best approximates a vector $\bar{x}$ under the inner-product norm. We show here that if an orthonormal sequence $(\bar{u}_1, \bar{u}_2, \ldots)$ is constructed from a dense sequence $(\bar{v}_1, \bar{v}_2, \ldots)$ as described in Section 10.7.2, then

$$\lim_{l \to \infty} \|\bar{x} - \mathsf{P}_l\bar{x}\|_{\mathrm{ip}} = 0, \tag{10.346}$$

that is, the sequence of sums $\sum_{k=1}^{l} \langle \bar{u}_k, \bar{x} \rangle \, \bar{u}_k$ converges strongly to $\bar{x}$ under the inner-product norm.

A series

$$\sum_{k=1}^{\infty} \langle \bar{u}_k, \bar{x} \rangle \, \bar{u}_k, \tag{10.347}$$

in which $(\bar{u}_1, \bar{u}_2, \ldots)$ is a complete orthonormal set, is called an **orthogonal series**. Fourier series, Legendre series, and other orthogonal series are used constantly in theoretical physics and in many branches of engineering. Before describing specific orthogonal series, we establish some general properties of all orthogonal series.

Let us express $\bar{x}$ and $\bar{x} - \mathsf{P}_l\bar{x}$ as sums of orthogonal vectors:

$$\bar{x} = \bar{x} - \mathsf{P}_1\bar{x} + \mathsf{P}_1\bar{x} = (\bar{x} - \mathsf{P}_1\bar{x}) + \langle \bar{u}_1, \bar{x} \rangle \, \bar{u}_1 \tag{10.348}$$

for $l = 1$, or

$$\bar{x} - \mathsf{P}_l\bar{x} = (\bar{x} - \mathsf{P}_{l+1}\bar{x}) + (\mathsf{P}_{l+1}\bar{x} - \mathsf{P}_l\bar{x}) = (\bar{x} - \mathsf{P}_{l+1}\bar{x}) + \langle \bar{u}_{l+1}, \bar{x} \rangle \, \bar{u}_{l+1} \tag{10.349}$$

for every $l > 1$. Because $\bar{u}_{l+1}$ is orthogonal to $\bar{x} - P_{l+1}\bar{x}$ and $\|\bar{u}_{l+1}\|_{ip}^2 = 1$ by construction, Pythagoras's theorem implies that

$$\|\bar{x} - P_l\bar{x}\|_{ip}^2 = \|\bar{x} - P_{l+1}\bar{x}\|_{ip}^2 + |\langle \bar{u}_{l+1}, \bar{x}\rangle|^2. \tag{10.350}$$

Then

$$\|\bar{x} - P_{l+1}\bar{x}\|_{ip}^2 = \|\bar{x} - P_l\bar{x}\|_{ip}^2 - |\langle \bar{u}_{l+1}, \bar{x}\rangle|^2, \tag{10.351}$$

which shows that $(\|\bar{x} - P_1\bar{x}\|_{ip}^2, \|\bar{x} - P_2\bar{x}\|_{ip}^2, \ldots)$ is a monotonically decreasing sequence that is bounded from below by zero. Let

$$\epsilon_{min}^2 := \inf_l \|\bar{x} - P_l\bar{x}\|_{ip}^2. \tag{10.352}$$

The infimum is either zero or positive and finite. If $\epsilon_{min}^2 > 0$, one can choose the positive square root to make $\epsilon_{min} > 0$. For every positive value of ϵ_{min}, there exists a vector $\bar{v}_m$ in the dense sequence $(\bar{v}_1, \bar{v}_2, \ldots)$ such that

$$\|\bar{x} - \bar{v}_m\|_{ip} < \epsilon_{min}. \tag{10.353}$$

It follows that there exists an integer $q \leq m$ such that

$$\|\bar{x} - P_q\bar{x}\|_{ip} < \epsilon_{min}. \tag{10.354}$$

This contradiction establishes that $\epsilon_{min} = 0$. Therefore

$$\lim_{l\to\infty} \left\| \bar{x} - \sum_{j=1}^{l}\langle \bar{u}_j, \bar{x}\rangle \bar{u}_j \right\|_{ip} = 0 \tag{10.355}$$

for *every* vector $\bar{x} \in \mathcal{H}$. To put it differently, an orthogonal series for a vector $\bar{x} \in \mathcal{H}$ *always* converges with respect to the inner-product norm, provided that the expansion is carried out with respect to a complete orthonormal set. Usually one writes the orthogonal expansion of $\bar{x}$ as a series,

$$\bar{x} = \sum_{j=1}^{\infty}\langle \bar{u}_j, \bar{x}\rangle \bar{u}_j, \tag{10.356}$$

instead of the more cumbersome form Eq. (10.355). However, one should always bear in mind that what is meant is the strong limit under the inner-product norm, and not some other sort of limit such as pointwise convergence.

The parent of all orthogonal series is the Fourier series. In the nineteenth century, mathematicians devoted a great deal of effort to attempts to understand the sense in which one can say that, for example,

$$f(x) = \sum_{m=-\infty}^{\infty}\left(\frac{1}{2\pi}\int_{-\pi}^{\pi}e^{-imx'}f(x')\,dx'\right)e^{imx}, \tag{10.357}$$

for what class of functions the sequence of partial sums converges pointwise to f, and how other methods of summing the series extend the class of functions for which the series converges pointwise to f. Equation (10.355) is almost the only relatively easily proved assertion in the entire subject of the convergence of orthogonal series. From the point of view of many applications, Eq. (10.355) is also the most important statement about orthogonal series.

We give an example below that shows that Eq. (10.355) does not hold in general for norms other than the inner-product norm. This fact should not be terribly surprising, for one knows that different norms are not necessarily equivalent on an infinite-dimensional vector space.

Parseval's Equality

From Eqs. (10.348) and Eq. (10.350) it follows that

$$\forall l \in \mathbb{Z}^+ : \|\bar{x}\|_{\text{ip}}^2 = \sum_{j=1}^{l} |\langle \bar{u}_j, \bar{x} \rangle|^2 + \|\bar{x} - P_l \bar{x}\|_{\text{ip}}^2. \tag{10.358}$$

Therefore

$$\|\bar{x}\|_{\text{ip}}^2 = \lim_{l \to \infty} \sum_{j=1}^{l} |\langle \bar{u}_j, \bar{x} \rangle|^2 \tag{10.359}$$

by Eq. (10.355). This relation, which is called **Parseval's equality**, is the infinite-dimensional analog of the expression for the 2-norm in a finite-dimensional space.

Parseval's equality has physical significance in many applications. For example, if $\{\bar{u}_1, \bar{u}_2, \ldots\}$ is a basis of energy eigenfunctions of a discrete quantum system, then Parseval's equality says that the probability of the state that corresponds to the Hilbert vector $\bar{x}$ is equal to the sum of the probabilities of occupying the energy eigenstates.

Equation (10.359) also shows that *the orthogonal complement of the vector space* $\mathbb{W}$ *spanned by the finite linear combinations of the elements of B is the zero vector subspace* $\mathbb{O}$, for if $\bar{z}$ is orthogonal to every $\bar{u}_n$, then $\|\bar{z}\|^2 = 0$ and therefore $\bar{z} = \bar{0}$. This property justifies calling $\{\bar{u}_n\}$ a *complete* orthonormal set.

The Riemann-Lebesgue Lemma

It follows from Bessel's inequality for a complete orthogonal set $\{\bar{u}_j\}$,

$$\forall l \in \mathbb{Z}^+ : \sum_{j=1}^{l} |\langle \bar{u}_j, \bar{x} \rangle|^2 \le \|\bar{x}\|_{\text{ip}}^2, \tag{10.360}$$

that for every $\bar{x}$, the coefficient $\langle \bar{u}_j, \bar{x} \rangle$ of $\bar{u}_j$ approaches zero as j becomes large:

$$\lim_{j \to \infty} \langle \bar{u}_j, \bar{x} \rangle = 0. \tag{10.361}$$

For example, substituting the Fourier functions $(2\pi)^{-1/2}e^{imx}$ for the $\bar{u}_j$ gives a special case of the **Riemann-Lebesgue lemma**:

$$\lim_{|m|\to\infty} \int_{-\pi}^{\pi} e^{-imx} f(x)\,dx = 0. \tag{10.362}$$

In the full Riemann-Lebesgue lemma, the integer m may be replaced with any real number t, the interval $[-\pi, \pi]$ may be replaced with any interval of the real line, and f may be any Lebesgue-integrable function.

Uniqueness of Orthogonal Series

We show that the orthogonal expansion Eq. (10.356) is unique, that is, if for every $\bar{x} \in \mathcal{H}$ there exist scalars c_i such that

$$\lim_{l\to\infty} \left\| \bar{x} - \sum_{j=1}^{l} c_j \bar{u}_j \right\|_{ip} = 0, \tag{10.363}$$

then

$$c_j = \langle \bar{u}_j, \bar{x} \rangle. \tag{10.364}$$

This result rests on the triangle inequality and the Pythagorean theorem. The assumption Eq. (10.363) and Eq. (10.355) imply that for every $\epsilon > 0$, there exist integers l_1 and l_2 such that

$$\left\| \bar{x} - \sum_{j=1}^{l_1} \langle \bar{u}_j, \bar{x} \rangle \bar{u}_j \right\|_{ip} < \frac{\sqrt{\epsilon}}{2} \quad \text{and} \quad \left\| \bar{x} - \sum_{j=1}^{l_2} c_j \bar{u}_j \right\|_{ip} < \frac{\sqrt{\epsilon}}{2}. \tag{10.365}$$

Let $l = \max\{l_1, l_2\}$. Then the triangle inequality implies that

$$\left\| \sum_{j=1}^{l} [c_j - \langle \bar{u}_j, \bar{x} \rangle] \bar{u}_j \right\|_{ip} \leq \left\| \bar{x} - \sum_{j=1}^{l} \langle \bar{u}_j, \bar{x} \rangle \bar{u}_j \right\|_{ip} + \left\| \sum_{j=1}^{l} c_j \bar{u}_j - \bar{x} \right\|_{ip}$$

$$< \frac{\sqrt{\epsilon}}{2} + \frac{\sqrt{\epsilon}}{2} = \sqrt{\epsilon}. \tag{10.366}$$

By Pythagoras's theorem,

$$\left\| \sum_{j=1}^{l} [c_j - \langle \bar{u}_j, \bar{x} \rangle] \bar{u}_j \right\|_{ip}^2 = \sum_{j=1}^{l} |c_j - \langle \bar{u}_j, \bar{x} \rangle|^2 < \epsilon. \tag{10.367}$$

Then

$$\forall \epsilon \in \mathbb{R}^+ : \forall j \in (1:l) : |c_j - \langle \bar{u}_j, \bar{x} \rangle|^2 < \epsilon, \tag{10.368}$$

which implies Eq. (10.364).

10.7.4 PRACTICAL ASPECTS OF FOURIER SERIES

In this section we consider in some detail the exponential Fourier series

$$f(x) = \sum_{m=-\infty}^{\infty} c_m e^{imx} \tag{10.369}$$

of a function $f : [-\pi, \pi] \to \mathbb{C}$, in which

$$c_m = \frac{1}{2\pi} \int_{-\pi}^{\pi} e^{-imx'} f(x') \, dx' \tag{10.370}$$

are the Fourier coefficients. Important special cases include Fourier series for odd or even functions, as well as sine and cosine Fourier series of the restriction of f to the interval $[0, \pi]$. Because the convergence of Fourier series in the inner-product norm follows from the discussion in Section 10.7.3, we concentrate here on pointwise convergence.

Special Cases
The complex conjugate of a Fourier coefficient of a real-valued function f is

$$c_m^* = \frac{1}{2\pi} \int_{-\pi}^{\pi} e^{imx} f(x) \, dx = c_{-m}. \tag{10.371}$$

If f takes on purely imaginary values, then

$$c_m^* = -c_{-m}. \tag{10.372}$$

If f is an **even function**, that is, if

$$f(-x) = f(x), \tag{10.373}$$

then

$$\begin{aligned}
c_{-m} &= \frac{1}{2\pi} \int_{-\pi}^{\pi} e^{imx} f(x) \, dx \\
&= \frac{1}{2\pi} \int_{-\pi}^{\pi} e^{imx} f(-x) \, dx \\
&= \frac{1}{2\pi} \int_{-\pi}^{\pi} e^{-imx} f(x) \, dx \\
&= c_m.
\end{aligned} \tag{10.374}$$

The Fourier series for an even function therefore is a **cosine series**,

$$
\begin{aligned}
f(x) &= c_0 + \sum_{m=-\infty}^{-1} c_m e^{imx} + \sum_{1}^{\infty} c_m e^{imx} \\
&= c_0 + \sum_{m=1}^{\infty} c_m (e^{imx} + e^{-imx}) \\
&= c_0 + 2 \sum_{m=1}^{\infty} c_m \cos mx.
\end{aligned}
\tag{10.375}
$$

If f is both real-valued and even, then $c_m^* = c_{-m} = c_m$; hence every term in Eq. (10.375) is real. For a Fourier cosine series it sometimes is convenient to define new coefficients

$$
\begin{aligned}
a_m &:= c_m + c_{-m} \\
&= \frac{1}{\pi} \int_{-\pi}^{\pi} \cos mx \, f(x) \, dx,
\end{aligned}
\tag{10.376}
$$

in terms of which the Fourier expansion of an even, real-valued function is

$$
f(x) = \frac{a_0}{2} + \sum_{m=1}^{\infty} a_m \cos mx.
\tag{10.377}
$$

If one defines the function $g : [0, \pi] \to \mathbb{R}$ as the restriction of $f : [-\pi, \pi] \to \mathbb{R}$ to the interval $[0, \pi]$, then g has the cosine series Eq. (10.377). Conversely, given a function $g : [0, \pi] \to \mathbb{R}$, one can extend g to an even function on the larger interval $[-\pi, \pi]$ by defining $g(-x) := g(x)$. Hence g can be represented with a Fourier cosine series.

Similarly, if f is an **odd function**, that is, if

$$
f(-x) = -f(x),
\tag{10.378}
$$

then

$$
c_{-m} = -c_m,
\tag{10.379}
$$

and the Fourier series of f is a **sine series**,

$$
f(x) = 2i \sum_{m=1}^{\infty} c_m \sin mx.
\tag{10.380}
$$

If f is both real-valued and odd, then $c_m^* = c_{-m} = -c_m$; hence every Fourier coefficient is purely imaginary. In this case it sometimes is convenient to define the new coefficients

$$
\begin{aligned}
b_m &:= i(c_m - c_{-m}) \\
&= \frac{1}{\pi} \int_{-\pi}^{\pi} \sin mx \, f(x) \, dx,
\end{aligned}
\tag{10.381}
$$

in terms of which the Fourier expansion of an odd, real-valued function is

$$f(x) = \sum_{m=1}^{\infty} b_m \sin mx.$$
(10.382)

Again, if one defines the function $g : [0, \pi] \to \mathbb{R}$ as the restriction of $f : [-\pi, \pi] \to \mathbb{R}$ to the interval $[0, \pi]$, then g has the sine series Eq. (10.382). Conversely, given a function $g : [0, \pi] \to \mathbb{R}$, one can extend g to an odd function on the larger interval $[-\pi, \pi]$ by *defining* $g(-x) := -g(x)$. Hence g can be represented with a Fourier sine series.

It may at first sight appear puzzling that a function $g : [0, \pi] \to \mathbb{R}$ can have both a Fourier sine series and a Fourier cosine series. For example, if g does not vanish at $x = 0$, then how is it possible to represent g with a series of functions that vanish at 0? There are two partial answers to this question. First, the Fourier series converges in the 2-norm, not pointwise. One can change the value of a function at an isolated point without affecting the 2-norm of the difference between the function and its Fourier series, because the 2-norm is an integral. Second, if one extends a function $g : [0, \pi] \to \mathbb{R}$ that does not vanish at $x = 0$ as an odd function on $[-\pi, \pi]$, then $x = 0$ is a point of discontinuity. If $g(0) \neq 0$, then at $x = 0$ the Fourier sine series converges to 0, which is the mean of the limit of g as x approaches 0 from above, $g(0+) := \lim_{\epsilon \to 0} g(0 + \epsilon)$, and the limit of g as x approaches 0 from below, $g(0-) := \lim_{\epsilon \to 0} g(0 + \epsilon) = -g(0+)$. In this section we give other examples in which the Fourier series of a function with a jump discontinuity converges to the mean of the limits of the function values at the left and right of the discontinuity.

Fourier series exist for functions that are square-integrable on intervals other than $[-\pi, \pi]$. For example, the Fourier series of a function $f \in L^2([a, b]; \mathbb{C})$ can be derived by making the substitution

$$x = (b - a)\frac{x'}{2\pi} + \frac{1}{2}(a + b)$$
(10.383)

in the general Fourier series Eq. (10.369). The new variable,

$$x' := \left(\frac{2\pi}{b - a}\right)\left[x - \frac{1}{2}(a + b)\right],$$
(10.384)

ranges from $-\pi$ to π. Thus a Fourier series exists for any signal (voltage, optical intensity, etc.) that is observed only over a finite interval of time, $t \in [a, b]$.

Any square-integrable function $f \in L^2([a, b]; \mathbb{R})$ has a Fourier cosine series, which can be derived by defining a function g that agrees with f on $[a, b]$ and is symmetric in the interval $[2a - b, b]$, that is, by defining $g(2a - x) = g(x) = f(x)$ for all $x \in [a, b]$. (Note that a is the center of the interval $[2a - b, b]$.) Substituting

$$x = a + \left(\frac{b - a}{\pi}\right)x'',$$
(10.385)

in Eq. (10.375), in which $x'' \in [-\pi, \pi]$, yields a cosine series that converges to g under an inner-product norm on the interval $[2a - b, b]$ and therefore converges to f under the inner-product norm on $[a, b]$.

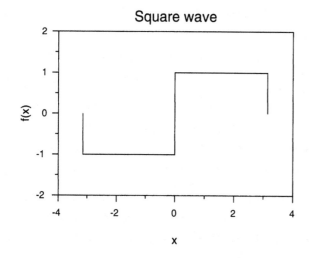

FIGURE 10.7

Periodic square wave of unit amplitude.

Examples

The periodic square wave

$$f(x) = \begin{cases} 1, & \text{if } -\pi \leq x < 0, \\ 0, & \text{if } x = 0, \text{ or} \\ -1, & \text{if } 0 < x < \pi, \end{cases} \qquad (10.386)$$

which is shown in Fig. 10.7, is an odd function and therefore has a sine series. The Fourier coefficients for $m > 0$ vanish if m is even:

$$c_m = \frac{1 - \cos \pi m}{i \pi m} = \begin{cases} 0, & \text{if } m = 2k; \\ \dfrac{2}{i \pi (2k + 1)}, & \text{if } m = 2k + 1. \end{cases} \qquad (10.387)$$

Therefore, by Eq. (10.380), the Fourier series for the function Eq. (10.386) is

$$f(x) = \frac{4}{\pi} \sum_{k=0}^{\infty} \frac{\sin(2k + 1)x}{2k + 1}. \qquad (10.388)$$

This series converges pointwise conditionally, not absolutely. We show in our discussion of Gibbs' phenomenon that the series does not converge uniformly.

The square wave Eq. (10.386) is an example of a **piecewise continuous** function, which is defined as a function that is continuous except for a finite set of jump discontinuities. In turn, a **jump discontinuity** of a function f occurs at a point a for which the right and left limits of f do not agree, that is, such that

$$\lim_{x \downarrow a} f(x) \neq \lim_{x \uparrow a} f(x). \qquad (10.389)$$

The absolute difference

$$| \lim_{x \downarrow a} f(x) - \lim_{x \uparrow a} f(x)| \qquad (10.390)$$

is called the **jump** or **saltus** of f. Likewise, a function f is called **piecewise smooth** if and only if f and its derivative $f' = df/dx$ are piecewise continuous.

It can be shown that if f is a piecewise smooth function such that $\int_{-\pi}^{\pi} |f(x)| \, dx$ exists, then the Fourier series of f converges pointwise to $f(a)$ at every point a at which f is continuous and converges to $\frac{1}{2}[\lim_{x \downarrow a} f(x) + \lim_{x \uparrow a} f(x)]$ at every point a at which f is discontinuous. If f is continuous on the interval $[-\pi, \pi]$, then the Fourier series of f converges absolutely and uniformly on $[-\pi, \pi]$. For proofs, see Tolstov (1976).

For example, the function

$$\forall m \in \mathbb{Z} : \forall x \in ((2m-1)\pi, (2m+1)\pi] : \quad f(x) = (x - 2m\pi)^n, \qquad (10.391)$$

in which $n \in \mathbb{Z}^+$, is piecewise smooth. One sees that if n is odd, then at $x = \pm \pi$ f has jump discontinuities and f' is continuous. If n is even, then Eq. (10.391) defines a function that is continuous for all $x \in [-\pi, \pi]$. However, the derivative f' has jump discontinuities at $x = \pm \pi$. Figure 10.8 shows three periods of the function defined in Eq. (10.391) for $n = 3$. The Fourier coefficient of Eq. (10.391) for $m = 0$ is

$$c_0 = \frac{1}{2\pi} \int_{-\pi}^{\pi} x^n \, dx = \begin{cases} 0, & \text{if } n \text{ is odd}; \\ \dfrac{\pi^n}{n+1}, & \text{if } n \text{ is even}. \end{cases} \qquad (10.392)$$

The Fourier coefficients c_m for $m \neq 0$ are equal to $(2\pi)^{-1} \int_{-\pi}^{\pi} e^{-imx} x^n \, dx$. The integral can be calculated either by performing n successive integrations by parts or by differentiating the integral of e^{-imx} with respect to m. We choose the latter course because it is closely related to the method of characteristic functions, which is widely used in statistical theory in both

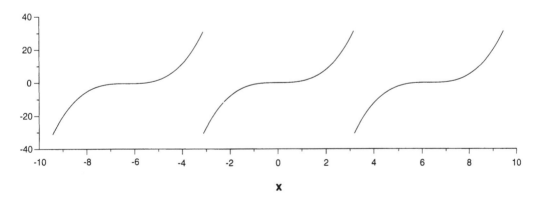

FIGURE 10.8

Three periods of the function (10.391) for $n = 3$, showing that at odd multiples of π the function has jump discontinuities, but its derivatives are continuous.

physics and mathematics. Evidently for every $m \neq 0$,

$$
\begin{aligned}
c_m &= \frac{1}{2\pi} i^n \left(\frac{d^n}{dm^n}\right) \int_{-\theta}^{\theta} e^{-imx} \, dx \bigg|_{\theta=\pi} \\
&= i^n \left(\frac{d^n}{dm^n}\right) \left(\frac{\sin m\theta}{m\pi}\right) \bigg|_{\theta=\pi}.
\end{aligned}
\tag{10.393}
$$

If n is odd, one obtains a sine series; if n is even, one obtains a cosine series. For example, the series that follow from Eqs. (10.392) and (10.394) for $n = 1$ and $n = 2$ are

$$
x = 2 \sum_{m=1}^{\infty} \frac{(-1)^{m+1}}{m} \sin mx,
\tag{10.394}
$$

which converges pointwise for all $x \in (-\pi, \pi)$, and

$$
x^2 = \frac{\pi^2}{3} + 4 \sum_{m=1}^{\infty} \frac{(-1)^m}{m^2} \cos mx,
\tag{10.395}
$$

which converges pointwise for all $x \in [-\pi, \pi]$. For values of x outside of the interval $[-\pi, \pi]$, these series give the values of the periodic functions that agree with x or x^2, respectively, if $x \in (-\pi, \pi)$.

Often one can obtain useful numerical series by evaluating a Fourier series for special values of the argument. For example, if one evaluates Eq. (10.388) at $x = \pi/2$, in which $\sin(2k + 1)x = (-1)^k$ and $f(x) = 1$, one obtains

$$
\frac{\pi}{4} = \sum_{k=0}^{\infty} \frac{(-1)^k}{2k + 1} = 1 - \frac{1}{3} + \frac{1}{5} - \cdots
\tag{10.396}
$$

The same series follows from Eq. (10.394) if one sets $x = \pi/2$. If one takes $x = \pi$ in Eq. (10.395), one obtains the series

$$
\frac{\pi^2}{6} = \sum_{m=1}^{\infty} \frac{1}{m^2} = 1 + \frac{1}{2^2} + \frac{1}{3^2} + \cdots
\tag{10.397}
$$

Gibbs' Phenomenon

The partial sums of the series Eq. (10.388) for $k = 5$ and $k = 20$ are shown in Figs. 10.9 and 10.10, respectively. (Similar curves are familiar to all who have used an oscilloscope.) Perhaps surprisingly, the maxima nearest to the points of discontinuity at $x = 0$ and $x = \pi$ show no sign of approaching 1, which is the value of f for $0 < x < \pi$. Likewise, the minima nearest to the discontinuities at $x = 0$ and $x = -\pi$ do not seem to approach -1. We show that, in fact, the value of f at the maximum nearest to $x = 0$ approaches a constant value that differs from 1 as $m \to \infty$. The property of the Fourier series of a piecewise continuous function f that the first maximum of the partial sums near a point a at which the function has a jump discontinuity tends to a limit other than $\lim_{x \downarrow a} f(x)$ or $\lim_{x \uparrow a} f(x)$ is called **Gibbs' phenomenon**.

Fourier sum (11 terms)

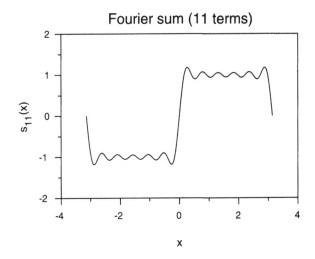

FIGURE 10.9

Partial sum of the Fourier series (10.388) for $k = 0$ through $k = 5$.

Fourier sum (41 terms)

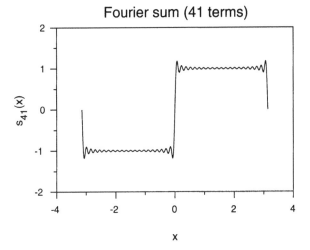

FIGURE 10.10

Partial sum of the Fourier series (10.388) for $k = 0$ through $k = 20$.

The partial sum of $n = 2K + 1$ terms of the square-wave Fourier series Eq. (10.388) is equal to

$$
\begin{aligned}
s_K(x) &= \frac{4}{\pi} \sum_{k=0}^{K} \frac{\sin(2k+1)x}{2k+1} \\[2mm]
&= \frac{4}{\pi} \sum_{k=0}^{K} \int_0^x \cos(2k+1)u \, du \\[2mm]
&= \frac{4}{\pi} \int_0^x \sum_{k=0}^{K} \cos(2k+1)u \, du \\[2mm]
&= \frac{2}{\pi} \int_0^x \frac{\sin 2(K+1)u}{\sin u} \, du,
\end{aligned}
\tag{10.398}
$$

in which we have made use of Eq. (2.167) in the last step.

The maxima and minima of s_K occur at the points x at which $s'_K(x) = ds_K(x)/dx = 0$, and therefore at the zeros of the function $\sin[2(K + 1)u]/\sin u$. The first maximum such that $x > 0$ occurs at the point

$$x_1 = \frac{\pi}{2(K + 1)}. \tag{10.399}$$

The value of s_K at $x = x_1$ is

$$\begin{aligned}
s_K(x_1) &= \frac{2}{\pi} \int_0^{x_1} \frac{\sin 2(K + 1)u}{\sin u} \, du \\
&= \frac{2}{\pi} \int_0^{\pi} \left(\frac{\sin v}{v} \right) \left(\frac{v/[2(K + 1)]}{\sin[v/[2(K + 1)]]} \right) dv,
\end{aligned} \tag{10.400}$$

in which $v = 2(K + 1)u$. Now that the upper limit of the integral has become a constant, one can easily evaluate the limit as $K \to \infty$:

$$\lim_{K \to \infty} s_K(x_1) = \frac{2}{\pi} \int_0^{\pi} \frac{\sin v}{v} \, dv = \frac{2}{\pi} \mathrm{Si}(\pi) = 1.17898 \cdots, \tag{10.401}$$

in which Si stands for the **sine integral**,

$$\mathrm{Si}(x) := \int_0^x \frac{\sin v}{v} \, dv. \tag{10.402}$$

One says that the Fourier series of the square wave "overshoots" by almost 18% at x_1.

It can be shown that at a jump discontinuity, the Fourier series of a piecewise continuous function f always overshoots and undershoots such that the total jump of the function defined as the limit of the Fourier series is greater than the jump of f by the factor $1.17898 \cdots$. It follows that the Fourier series of a piecewise continuous function never converges uniformly in an interval that contains a jump discontinuity of the function.

10.7.5 ORTHOGONAL-POLYNOMIAL EXPANSIONS

Legendre Series

The Weierstraß approximation theorem implies that the set $\mathcal{P}(\mathbb{Q})$ of polynomial functions with rational coefficients is dense in the space of continuous functions on an interval $[a, b]$. It can be shown that the latter space, $\mathcal{C}^0((a, b); \mathbb{R})$, is dense in the Hilbert space $\mathcal{H} = L^2([a, b]; \mathbb{R})$ under the norm defined by the inner product

$$\langle f, g \rangle := \int_{-1}^1 f(x) g(x) \, dx. \tag{10.403}$$

The Euclidean space $\mathcal{P}(\mathbb{Q})$ is the set of finite, rational linear combinations of the monomials $1, x, x^2, \ldots$. Therefore one can obtain a complete orthonormal set by applying the Gram-Schmidt procedure to the sequence $(1, x, x^2, \ldots)$. The result, as one knows from

Exercise 8.2.1, is a set of orthonormal polynomials p_l that are, apart from a normalization constant, equal to the Legendre polynomials P_l:

$$\forall x \in [-1, 1]: \quad p_l(x) := \sqrt{\frac{2l + 1}{2}} \, P_l(x), \quad \text{where } \langle p_l, p_m \rangle = \delta_{mn}. \tag{10.404}$$

Then Eq. (10.355) asserts that for *any* function $f \in L^2([-1, 1]; \mathbb{R})$, the inner-product norm of the difference $f - \sum_{j=1}^{l} \langle p_l, f \rangle \, p_l$ approaches zero as l becomes large without bound. Thus one obtains the **Legendre series**

$$f(x) = \sum_{l=0}^{\infty} \frac{2l + 1}{2} \, c_l P_l(x) \tag{10.405}$$

in which

$$c_l := \int_{-1}^{1} f(x) \, P_l(x) \, dx \tag{10.406}$$

and in which the infinite sum in Eq. (10.405) is understood as the limit in the inner-product norm of a sequence of finite partial sums.

Example of Legendre Series

We derive the Legendre series of the step function

$$s(x) = \begin{cases} -1, & \text{if } -1 \le x < 0; \\ 0, & \text{if } x = 0; \\ 1, & \text{if } 0 < x \le 1. \end{cases} \tag{10.407}$$

Because $P_l(-x) = (-1)^l P_l(x)$, and because s is an odd function, the coefficients in the Legendre series Eq. (10.405) are

$$c_{2k} = 0 \tag{10.408}$$

if l is even, $l = 2k$, and

$$c_{2k+1} = 2 \int_{0}^{1} P_{2k+1}(x) \, dx \tag{10.409}$$

if l is odd, $l = 2k + 1$. Therefore we need to consider only $l > 0$, l odd.

It can be shown that the Legendre polynomials obey the recurrence relation

$$P'_{l+1}(x) - P'_{l-1}(x) = (2l + 1) P_l(x), \tag{10.410}$$

with which one can express the integral of a Legendre polynomial in terms of the values of other Legendre polynomials. It can be shown also that

$$\forall l \in \mathbb{Z}^+ : P_l(1) = 1, \tag{10.411}$$

that

$$P_{2k+1}(0) = 0, \tag{10.412}$$

and that

$$P_{2k}(0) = (-1)^k \frac{(2k-1)!!}{2^k k!}, \tag{10.413}$$

in which the definition of the double factorial function of an odd integer is

$$(2k-1)!! := (2k-1)(2k-3)\cdots 1. \tag{10.414}$$

Because l is odd, we let $l = 2k + 1$ and calculate

$$\begin{aligned}
\int_0^1 P_l(x)\, dx &= (2l+1)^{-1}[P_{l+1}(x) - P_{l-1}(x)]|_0^1 \\
&= (2l+1)^{-1}[P_{l-1}(0) - P_{l+1}(0)] \\
&= (4k+3)^{-1}[P_{2k}(0) - P_{2k+2}(0)] \\
&= \frac{(-1)^k(2k-1)!!}{2^{k+1}(k+1)!}.
\end{aligned} \tag{10.415}$$

Noting that $\frac{2l+1}{2}c_l = (4k+3)\int_{-1}^1 P_{2k+1}(x)\, dx$, one finds from Eqs. (10.405), Eq. (10.409), and Eq. (10.415) that the Legendre series of the step function Eq. (10.407) is

$$s(x) = \sum_{k=0}^{\infty} \frac{(-1)^k(4k+3)(2k-1)!!}{2^{k+1}(k+1)!} P_{2k+1}(x). \tag{10.416}$$

Chebyshev Series

Series similar to the Legendre series, Eq. (10.405), exist for the other classical orthogonal polynomials. For example, the **Chebyshev series** of a function $f : [-1, 1] \to \mathbb{R}$,

$$f(x) = \sum_{n=0}^{\infty} c_n T_n(x), \tag{10.417}$$

in which

$$c_n = \frac{1}{\|T_n\|_{ip}^2} \int_{-1}^1 \frac{T_n(x)\, f(x)}{[1-x^2]^{1/2}}\, dx, \tag{10.418}$$

converges to f in the norm defined by the inner product

$$\langle f, g \rangle = \int_{-1}^1 \frac{f(x)\, g(x)}{[1-x^2]^{1/2}}\, dx. \tag{10.419}$$

Because $T_n(x) = \cos(n \cos^{-1} x)$, one can consider a Chebyshev series as a disguised Fourier cosine series. For example, the Fourier series of a square wave given in Eq. (10.388) can easily be converted to the Chebyshev series of the shifted square wave c such that

$$\forall x \in [-\pi, 0]: \quad c(x) := f(x + \pi/2). \tag{10.420}$$

Note that the interval $[-\pi, 0]$ maps to the interval $[-1, 1]$ under the mapping $x \mapsto z = \cos x$. With this definition of c, the graph of the step function to be expanded in a Chebyshev series is the portion of Fig. 10.7 that lies between $-\pi/2$ and $\pi/2$. Therefore the function

$$\forall z \in [-1, 1]: \quad s(z) = c(\cos^{-1} z) \tag{10.421}$$

is identical to the step function s defined in Eq. (10.407).

From Eq. (10.388), the definition of c, and the formula

$$\cos(\theta + k\pi) = (-1)^k \cos \theta \tag{10.422}$$

it follows that

$$c(x) = \frac{4}{\pi} \sum_{k=0}^{\infty} (-1)^k \frac{\cos(2k+1)x}{2k+1} = \frac{4}{\pi} \sum_{k=0}^{\infty} \frac{(-1)^k}{2k+1} T_{2k+1}(\cos x). \tag{10.423}$$

Then the Chebyshev series of the step function Eq. (10.407) is

$$s(z) = \frac{4}{\pi} \sum_{k=0}^{\infty} \frac{(-1)^k}{2k+1} T_{2k+1}(z). \tag{10.424}$$

Convergence of Orthogonal Series at a Jump Discontinuity

Because the Chebyshev series Eq. (10.424) was obtained from the Fourier series Eq. (10.388) through a change of the independent variable, the Chebyshev series of the piecewise continuous step function s exhibits Gibbs' phenomenon. The overshoot and undershoot obey Eq. (10.401).

As for Fourier series, the Legendre series of a piecewise continuous function displays Gibbs's phenomenon at a jump discontinuity. In Figs. 10.11 and 10.12 we show graphs of the partial sums of the Legendre series, Eq. (10.416), of the step function s defined in Eq. (10.407). It is plausible from Figs. 10.11 and 10.12, and it can be proven (see Tolstov [1976]), that Gibbs's phenomenon occurs for any orthogonal series of a piecewise continuous function at a jump discontinuity, and that the overshoot and undershoot obey Eq. (10.401).

10.7.6 EXERCISES FOR SECTION 10.7

10.7.1 Show that in $\mathbb{R}^{n \times n}$ and $\mathbb{C}^{n \times n}$,

$$\|\mathbf{A}_m - \mathbf{A}\|_p \to 0 \Leftrightarrow \forall i \in (1:n): \ \forall j \in (1:n): \ a_{mi}^j \to a_i^j. \tag{10.425}$$

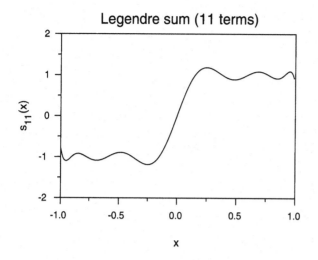

FIGURE 10.11

Partial sum of the Legendre series (10.416) for $k = 0$ through $k = 5$.

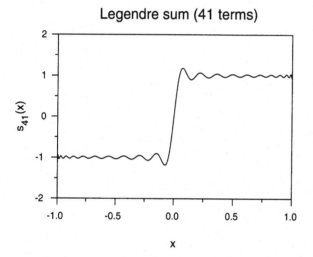

FIGURE 10.12

Partial sum of the Legendre series (10.416) for $k = 0$ through $k = 20$.

10.7.2 Prove that vector addition is continuous in a Fréchet space.

10.7.3 Prove that scalar multiplication is continuous in a Banach space.

10.7.4 Prove that $m(\mathbb{C})$ is complete.

10.7.5 Prove that $c(\mathbb{C})$ is complete.

10.7.6 Obtain the Fourier cosine series of the function f such that

$$\forall x \in [-\pi, \pi] : f(x) = |x|. \tag{10.426}$$

10.7.7 Obtain the Fourier cosine series of the function f such that

$$\forall x \in [-\pi, \pi] : f(x) = |\sin x|. \tag{10.427}$$

10.7.8   Obtain the exponential Fourier series of the function f such that

$$\forall x \in [-\pi, \pi] : f(x) = e^{-\alpha|x|}, \tag{10.428}$$

in which α is a real constant.

10.7.9   Obtain the Fourier cosine series of the function f such that

$$\forall x \in [-\pi, \pi] : f(x) = \cos \alpha x, \tag{10.429}$$

in which α is a real, non-integral constant.

10.7.10   Using the result of the preceding problem, show that

$$\frac{1}{\sin z} = \frac{1}{z} + 2 \sum_{k=1}^{\infty} (-1)^k \frac{z}{z^2 - (k\pi)^2} \tag{10.430}$$

provided that z is not an integral multiple of π.

10.7.11   Obtain the Chebyshev series that converges to the function $|z|$ for all $z \in [-1, 1]$.

10.8 BIBLIOGRAPHY

Halmos, Paul R. *Introduction to Hilbert Space and the Theory of Spectral Multiplicity*. 2 ed. New York: Chelsea, 1957.

Kolmogorov, Andrei Nikolaevich, and Sergei Vasilevich Fomin. *Introductory Real Analysis*. New York: Dover, 1975.

Loomis, Lynn H., and Shlomo Sternberg. *Advanced Calculus*. Reading: Addison-Wesley, 1968.

Pervin, William J. *Foundations of General Topology*. New York: Academic Press, 1964.

Rudin, Walter. *Principles of Mathematical Analysis*, 2 ed. New York: McGraw-Hill, 1976.

Tolstov, Georgi P. *Fourier Series*. New York: Dover, 1976.

CHAPTER 11

GROUP REPRESENTATIONS

11.1 PRELIMINARIES

11.1.1 BACKGROUND

Definitions

We recall from Chapter 4 that if S is a set of some physical interest (such as two- or three-dimensional space $\mathbb{E}^2$ or $\mathbb{E}^3$, or the surface of the unit sphere in $\mathbb{E}^3$); if G is a group, the elements of which are mappings $g : S \to S$ (such as rotations); and if $\{\Gamma(g)\}$ is a family of nonsingular linear mappings $\Gamma(g) : \mathcal{V} \to \mathcal{V}$ of a vector space $\mathcal{V}$ that obey the homomorphism law

$$\forall g_1, g_2 \in G: \; \Gamma(g_1 g_2) = \Gamma(g_1)\Gamma(g_2), \tag{11.1}$$

then the mapping

$$\forall g \in G: \; g \mapsto \Gamma(g) \tag{11.2}$$

is called a **representation** of the group G. The set S is called a G-set. One says that the vector space $\mathcal{V}$ **carries** the representation Γ, or that $\mathcal{V}$ is a **carrier space** for Γ. The dimension of $\mathcal{V}$ is called the **dimension** of Γ. If Γ is a group isomorphism, then Γ is called a **faithful** representation of G.

For example, Eq. (4.186) defines a matrix representation Γ_E of the dihedral group $G = D_3$. The carrier space is $\mathcal{V} = \mathbb{E}^2$, of dimension two. Because the (matrix) multiplication table of the matrices Γ_E is the same as the group multiplication table of D_3, Γ_E is a faithful representation of D_3.

Basic Properties of Group Representations

If Γ is a representation of a group G, then, by Eqs. (4.165) and (4.170),

$$\Gamma(e) = \mathbf{1}, \tag{11.3}$$

in which $\mathbf{1}$ is the identity linear mapping, and

$$\forall a \in G: \; [\Gamma(a)]^{-1} = \Gamma(a^{-1}). \tag{11.4}$$

The image of G under Γ,

$$\Gamma(G) := \{\Gamma(a) \mid a \in G\}, \tag{11.5}$$

is a subgroup of the general linear group of $\mathcal{V}$,

$$GL(\mathcal{V}) \cong GL(n,\ \mathbb{F}), \tag{11.6}$$

which we define in Section 6.2.4.

Matrix Representations

Suppose that Γ is a representation of a group G, and that $\mathcal{V}$ is the carrier space. Assume that the dimension of $\mathcal{V}$ is $n < \infty$. One can assume that a basis $\{\bar{e}_i \mid i \in (1 : n)\}$ has been chosen in $\mathcal{V}$. The set of matrices $\Gamma(a)$ (relative to the basis $\{\bar{e}_i \mid i \in (1 : n)\}$) that realize the linear mappings $\Gamma(a)$ is a group under matrix multiplication and is a homomorphic image of G. To prove this assertion, let

$$\Gamma(a) = M\Gamma(a)M^{-1} \tag{11.7}$$

in which $M : \mathcal{V} \to \mathbb{F}^n$ is the vector isomorphism $\bar{e}_i \mapsto \mathbf{e}_i$ that maps the basis vectors of $\mathcal{V}$ to the canonical basis vectors of $\mathbb{F}^n$. (See Eq. [5.202] for the definition of M.) It follows from Eq. (11.7) and the homomorphism property for Γ, Eq. (11.2), that the matrices $\Gamma(a)$ obey the homomorphism law,

$$\begin{aligned}
\Gamma(ba) &= M\Gamma(ba)M^{-1} \\
&= M\Gamma(b)M^{-1}M\Gamma(a)M^{-1} \\
&= \Gamma(b)\Gamma(a),
\end{aligned} \tag{11.8}$$

and therefore are the elements of a group that is homomorphic with G under the mapping

$$a \xrightarrow{\ \Gamma\ } \Gamma(a) \xrightarrow{\ M\ } \Gamma(a). \tag{11.9}$$

The matrix group

$$\Gamma(G) := \{\Gamma(a) \mid a \in G\} \tag{11.10}$$

is a matrix representation of G as defined in Section 4.3.1.

A representation matrix $\Gamma(a)$ acts on one of the canonical basis vectors $\mathbf{e}_i$ according to the usual rules of matrix-vector multiplication:

$$\Gamma(a)\mathbf{e}_i = \sum_{j=1}^{n} \gamma_{ji}(a)\mathbf{e}_j. \tag{11.11}$$

For example, if $i = 1$ one has

$$\begin{pmatrix} \gamma_{11} & \cdots & \gamma_{1n} \\ \gamma_{21} & \cdots & \gamma_{2n} \\ \vdots & \ddots & \vdots \\ \gamma_{n1} & \cdots & \gamma_{nn} \end{pmatrix} \begin{pmatrix} 1 \\ 0 \\ \vdots \\ 0 \end{pmatrix} = \begin{pmatrix} \gamma_{11} \\ \gamma_{21} \\ \vdots \\ \gamma_{n1} \end{pmatrix} = \sum_{j=1}^{n} \gamma_{j1}(a)\mathbf{e}_j. \tag{11.12}$$

Note that the sum is on the rows of the representation matrix $\Gamma(a)$ because the formula applies to basis vectors, not to the components of a vector. This formula generalizes to vector spaces in which the elements are functions instead of n-tuples.

11.1.2 SYMMETRY-ADAPTED FUNCTIONS

We show in Section 4.3.1 that if G is a group, S is a G-set, and $f : S \to \mathbb{C}$ is a complex-valued function defined on S, then the operators P_g that are defined by the equation

$$\forall x \in S: \ \forall g \in G : \ (P_g f)(x) := f(g^{-1}x) \tag{11.13}$$

preserve the group operation in the sense that they obey the homomorphism law

$$P_{g_1 g_2} = P_{g_1} P_{g_2}. \tag{11.14}$$

Therefore the mapping of group elements to operators defined by

$$g \mapsto P_g \tag{11.15}$$

is a homomorphism of G, and the set

$$\{P_g \mid g \in G\} \tag{11.16}$$

is an operator representation of the group G.

A set of vectors or functions $\{f_i : S \to \mathbb{C}\}$ carries a matrix representation Γ of a group G if and only if

$$\forall x \in S: \ \forall g \in G : \ P_g f_i = f_i(g^{-1}x) = \sum_j \gamma_{ji}(g) \, f_j(x) \tag{11.17}$$

in which $\gamma_{ji}(g)$ is the element of the matrix $\Gamma(g)$ in row j and column i. One says in this case that f_j transforms according to the j-th row of Γ. Note that Eq. (11.17) (for functions) is of exactly the same form as Eq. (11.11) (for basis vectors).

The group-representation property,

$$\Gamma(\underbrace{g_m g_l}_{\text{group product}}) = \underbrace{\Gamma(g_m)\Gamma(g_l)}_{\text{matrix product}}, \tag{11.18}$$

which reads

$$\gamma_{ki}(g_m g_l) = \sum_j \gamma_{kj}(g_m)\gamma_{ji}(g_l) \tag{11.19}$$

in terms of matrix elements, follows directly from Eq. (11.17) if one calculates $P_{g_m g_l} f_i = P_{g_m} P_{g_l} f_i$.

In a finite group, the matrix elements $\gamma_{ji}(g_l)$ are complex numbers. If one indexes the group elements as $\{g_l \mid l \in (1:|G|)\}$, then each matrix element γ_{ji} is a complex-valued function on $(1 : l)$. In a Lie group, the parameters of a group product $g = g_m g_l$ are analytic functions of the parameters of g_m and g_l. For example, in the group of rotations about a point in the two-dimensional Euclidean plane, $SO(2)$, each group element corresponds to a unique angle of rotation $\theta \in [0, 2\pi)$. Thus the elements of $SO(2)$ can be specified uniquely by a single parameter, θ. The product of a rotation through θ_1 and a rotation through θ_2 is a rotation through $\theta(\theta_1, \theta_2) = \theta_1 + \theta_2$. Obviously θ is an analytic function of θ_1 and θ_2. If the expression for the parameters of g in terms of those of g_m and g_l is sufficiently simple, as is the case for $SO(2)$, then it is easy to determine the matrix elements $\{\gamma_{ji}(g)\}$ directly from Eq. (11.19).

The functions $\{f_i\}$ that appear in Eq. (11.17) are called **symmetry-adapted functions**, or **G-adapted functions** if it is necessary to make clear which symmetry group G is meant. If the representation matrix elements $\{\gamma_{ji}(g)\}$ are known, then it is possible to determine the symmetry-adapted functions from Eq. (11.17) by holding x fixed and letting g vary over the group G.

If G is a group that leaves the Hamiltonian or Lagrangian of a physical system invariant, then the functions that carry irreducible representations of G have physical significance. For example, the quantum-mechanical angular-momentum state vectors $|jm\rangle$ are eigenfunctions of the Hamiltonian (energy eigenfunctions) for a system that is invariant under rotations in $\mathbb{E}^3$. If the spatial coordinates of the system are subjected to a three-dimensional rotation $\mathbf{R}$, the angular-momentum eigenfunctions $|jm\rangle$ transform according to the irreducible representation $\mathbf{D}^{(j)}$ of the group $SU(2)$:

$$P_\mathbf{R}|jm\rangle = \sum_{m'=-j}^{j} D_{m'm}^{(j)}(\mathbf{R})\,|jm'\rangle. \tag{11.20}$$

$SU(2)$ is called the two-dimensional **special unitary group** because its elements are 2×2 unitary matrices $\mathbf{U}$ that obey the special condition

$$\det[\mathbf{U}] = 1. \tag{11.21}$$

Applications of $SU(2)$ range from the quantum theory of angular momentum to the characterization of partially polarized light.

$SU(2)$ and other special unitary groups $(SU(n))$ are important in the physics of elementary particles. For example, the state vector of a quark in the static model belongs to a row of the fundamental representation $\mathbf{D}^{(10)}$ of the group $SU(3)$.

11.1.3 PARTNER FUNCTIONS

If G is a group, S is a G-set, and $f : S \to \mathbb{C}$ is a complex-valued function on S, then a **partner function** of f is any function $P_g f$ that can be obtained from f by applying one of the operators P_g. The **partner space** of a function $f : S \to \mathbb{C}$ is the vector space $\mathcal{P}_f$ that is spanned by the orbit of f:

$$\mathcal{P}_f := \mathrm{span}[\{P_g f \mid g \in G\}]. \tag{11.22}$$

The proof that the partner space obeys the vector-space axioms is left as Exercise 11.1.3. Because the partner space is a vector space, it carries a representation of G.

For example, let

$$f(x, y, z) = z \tag{11.23}$$

and let $\mathsf{P_R}$ be the operator on f and its partner functions induced by transforming the vector $\mathbf{r} = (x, y, z)^T$ by an element $\mathsf{R} \in D_3$ according to the usual rule,

$$(\mathsf{P_R}f)(\mathbf{r}) = f(\mathsf{R}^{-1}\mathbf{r}). \tag{11.24}$$

Because the π rotations in D_3 send $z \mapsto -z$, but the $\pm 2\pi/3$ rotations and the identity send $z \mapsto z$, it follows that for this example the partner space is spanned by the single function f and therefore is one-dimensional.

A useful method for calculating representation matrices is to compute the matrix elements of the operators P_g relative to a basis of partner functions. In the example of Eq. (11.23), the matrix elements of the three π rotations are -1; the matrix elements of the $\pm 2\pi/3$ rotations and the identity are all $+1$. Therefore the function f defined in Eq. (11.23) carries the A_2 representation of D_3.

Consider now the three functions

$$\begin{aligned}
f_1(\mathbf{r}) &= x = \mathbf{e}_1 \cdot \mathbf{r}, \\
f_2(\mathbf{r}) &= y = \mathbf{e}_2 \cdot \mathbf{r}, \\
f_3(\mathbf{r}) &= z = \mathbf{e}_3 \cdot \mathbf{r},
\end{aligned} \tag{11.25}$$

in which $\mathbf{r} \in \mathbb{R}^3$ and x, y, z are the first, second, and third coordinates of the (vector) argument of f. Because

$$\mathbf{e}_j \cdot (\mathsf{R}^{-1}\mathbf{r}) = ((\mathsf{R}^T)^{-1}\mathbf{e}_j) \cdot \mathbf{r} = (\mathsf{R}\mathbf{e}_j) \cdot \mathbf{r}, \tag{11.26}$$

the functions f_1, f_2, f_3 transform like the basis vectors $\mathbf{e}_1, \mathbf{e}_2, \mathbf{e}_3$ under rotations. Therefore it is convenient to regard f_1, f_2, f_3 as the components of a linear functional, which one can write as a *row* vector

$$\mathbf{f}^T = (f_1, f_2, f_3). \tag{11.27}$$

If applying a linear mapping to a linear functional, one puts the row vector that represents the functional on the left of the transformation matrix:

$$\mathsf{P_R}(f_1, f_2, f_3) = \mathbf{f}^T \mathsf{R}, \tag{11.28}$$

in which R is the matrix of the one-to-one linear mapping R relative to the basis $\{\mathbf{e}_j\}$. Note that the row vector that results after the application of two group operations (say, a, then b) is $\mathbf{f}^T \Gamma(b)\Gamma(a)$. If one carries out a transformation by a product of group operations, the *last* operation applied to a column vector is the first operation applied to a row vector. This is not an arbitrary rule; it is a straightforward consequence of Eq. (11.24).

If one uses Eqs. (11.24) through (11.28) to calculate $P_R f_j$ for $R \in D_3$, one obtains the following matrix representation of D_3:

$$\Gamma(e) = \begin{pmatrix} 1 & 0 & 0 \\ 0 & 1 & 0 \\ 0 & 0 & 1 \end{pmatrix}, \quad \Gamma(a) = \begin{pmatrix} -1 & 0 & 0 \\ 0 & 1 & 0 \\ 0 & 0 & -1 \end{pmatrix},$$

$$\Gamma(b) = \begin{pmatrix} \frac{1}{2} & \frac{1}{2}\sqrt{3} & 0 \\ \frac{1}{2}\sqrt{3} & -\frac{1}{2} & 0 \\ 0 & 0 & -1 \end{pmatrix}, \quad \Gamma(c) = \begin{pmatrix} \frac{1}{2} & -\frac{1}{2}\sqrt{3} & 0 \\ -\frac{1}{2}\sqrt{3} & -\frac{1}{2} & 0 \\ 0 & 0 & -1 \end{pmatrix}, \quad (11.29)$$

$$\Gamma(d) = \begin{pmatrix} -\frac{1}{2} & -\frac{1}{2}\sqrt{3} & 0 \\ \frac{1}{2}\sqrt{3} & -\frac{1}{2} & 0 \\ 0 & 0 & 1 \end{pmatrix}, \quad \Gamma(f) = \begin{pmatrix} -\frac{1}{2} & \frac{1}{2}\sqrt{3} & 0 \\ -\frac{1}{2}\sqrt{3} & -\frac{1}{2} & 0 \\ 0 & 0 & 1 \end{pmatrix}.$$

In this example, the linearly independent functions f_1, f_2, and f_3 carry a three-dimensional representation of D_3. Let V be the vector space spanned by f_1, f_2, and f_3. The partner space of f_1 is the span of f_1 and f_2, because an operation of D_3 never mixes f_3 with f_1 or f_2. If one lets W_E be the span of f_1 and f_2, and lets W_{A_1} be the span of f_3, then

$$V = W_E \oplus W_{A_1}, \quad (11.30)$$

that is, V is the direct sum of W_E and W_{A_1} (see Section 4.3.3 for the definition of the direct sum of additive groups). Also, if one lets Γ_E be the representation of D_3 carried by f_1 and f_2, and lets Γ_{A_1} be the representation carried by f_3, then each matrix $\Gamma(a)$ in the representation Γ in Eq. (11.29) is the direct sum

$$\Gamma(a) = \Gamma_E(a) \oplus \Gamma_{A_1}(a) \quad (11.31)$$

of the matrices $\Gamma_E(a)$ and $\Gamma_{A_1}(a)$. (The direct sum of matrices is defined in Section 6.1.4.)

11.1.4 EXERCISES FOR SECTION 11.1

11.1.1 Find the matrices of the representation of the dihedral group D_4 that is carried by the functions f_1, f_2, and f_3 defined in Eq. (11.25).

11.1.2 Find the matrices of the representation of the orthogonal group O that is carried by the functions f_1, f_2, and f_3 defined in Eq. (11.25).

11.1.3 Show that the partner space of a function obeys the axioms for a vector space.

11.2 REDUCIBILITY OF REPRESENTATIONS

At first sight, there appears to be a bewildering variety of representations of a group G. For example, any nontrivial permutation of the basis vectors of the carrier space V produces a different set of representation matrices. Our strategy for characterizing the representations of G as simply as possible is to characterize the carrier space up to isomorphism. This, in turn, defines an equivalence relation for representations.

A fundamentally important property of the carrier space is the presence or absence of subspaces that are imaged completely into themselves under the linear mappings induced by the elements of the group. Carrier spaces that have no such subspaces have an elementary, or "irreducible," nature. We show that any carrier space is a direct sum of "irreducible" carrier spaces, and thus that any representation (of a finite group) can be built up from representations that are carried by "irreducible" subspaces, as in Eq. (11.29).

11.2.1 INVARIANT SUBSPACES AND IRREDUCIBILITY

Recall from Section 9.3 that a subspace $\mathcal{W}$ of a vector space $\mathcal{V}$ is called *invariant* under a linear mapping $A : \mathcal{V} \to \mathcal{V}$ if and only if

$$A\mathcal{W} \subseteq \mathcal{W}, \tag{11.32}$$

that is, if and only if the image of $\mathcal{W}$ under A is a subspace of $\mathcal{W}$. If $\Gamma : G \to GL(\mathcal{V})$ is a representation of a group G, and if a subspace $\mathcal{W} \subseteq \mathcal{V}$ is invariant under every linear mapping $\Gamma(a) \in \Gamma(G)$,

$$\forall a \in G : \ \Gamma(a)\mathcal{W} \subseteq \mathcal{W}, \tag{11.33}$$

then $\mathcal{W}$ is called **G-invariant** under Γ. A **proper G-invariant subspace** $\mathcal{W}$ is a proper subspace of $\mathcal{V}$ that is G-invariant under Γ.

Every G-invariant subspace carries a representation of G. Let

$$\forall a \in G : \ \Gamma'(a) = \Gamma(a) \upharpoonright \mathcal{W} \tag{11.34}$$

be the restriction of the representation Γ to the G-invariant subspace $\mathcal{W}$. Because Γ is a representation of G and $\mathcal{W}$ is a G-invariant subspace of Γ,

$$\forall a \in G : \ \forall \overline{w} \in \mathcal{W} : \Gamma'(a)\overline{w} = \Gamma(a)\overline{w} \in \mathcal{W}. \tag{11.35}$$

Then

$$\begin{aligned} \forall a \in G : \ \forall \overline{w} \in \mathcal{W} : \ \Gamma'(a)\Gamma'(b)\overline{w} &= \Gamma(a)\Gamma(b)\overline{w} \\ &= \Gamma(ab)\overline{w} \\ &= \Gamma'(ab)\overline{w} \in \mathcal{W}. \end{aligned} \tag{11.36}$$

It follows that Γ' is a representation of G and that the carrier space of Γ' is $\mathcal{W}$.

A representation Γ is called **irreducible** if and only if there exists no proper G-invariant subspace of $\mathcal{V}$ under Γ other than the null subspace $\mathcal{O}$. A representation that is not irreducible is called **reducible**. Irreducible representations of a group G are fundamental in the sense that every representation of G can be expressed in terms of irreducible representations (see

Eq. [11.142]). By studying the irreducible representations of a group G, one studies all representations of G.

Evidently every one-dimensional representation, such as the $\Gamma^{(A_2)}$ representation of D_3, is irreducible. For example, consider the representation Γ of D_3 defined in Eq. (11.29). Evidently the subspace generated by $\mathbf{e}_3 = (0, 0, 1)^T$ is G-invariant. Therefore Γ is reducible.

It is easy to verify that the two-dimensional matrix representation $\Gamma^{(E)}$ of the group D_3 defined in Eq. (4.186) is irreducible. Because the mappings $\Gamma^{(E)}(k)$ are 2×2 matrices, a proper G-invariant subspace (other than $\mathcal{O}$) must be *one*-dimensional. Then the matrix representation $\Gamma^{(E)}$ is reducible if and only if there exists a one-dimensional G-invariant subspace, that is, if and only if there exists a vector $\mathbf{x} \in \mathbb{C}^2$ such that

$$\forall k \in D_3 : \exists \alpha(k) \in \mathbb{C} :\ni: \Gamma^{(E)}(k)\mathbf{x} = \alpha(k)\mathbf{x}. \tag{11.37}$$

In other words, some vector must be an eigenvector of every representation matrix simultaneously in order for $\Gamma^{(E)}$ to be reducible. But the eigenvectors of $\Gamma^{(E)}(a)$ are (scalar multiples of)

$$\begin{pmatrix} 1 \\ 0 \end{pmatrix} \text{ (for } \alpha(a) = -1) \text{ or } \begin{pmatrix} 0 \\ 1 \end{pmatrix} \text{ (for } \alpha(a) = 1), \tag{11.38}$$

and the eigenvectors of $\Gamma^{(E)}(b)$ are (scalar multiples of)

$$\begin{pmatrix} -\sqrt{3} \\ 1 \end{pmatrix} \text{ (for } \alpha(b) = 1) \text{ and } \begin{pmatrix} 1 \\ \sqrt{3} \end{pmatrix} \text{ (for } \alpha(b) = -1). \tag{11.39}$$

If a common eigenvector $\mathbf{x}$ of all of the representation matrices $\{\Gamma^{(E)}(k) \mid k \in D_3\}$ exists, then $\mathbf{x}$ must be a scalar multiple of *each* of the eigenvectors above (a contradiction). Therefore the representation $\Gamma^{(E)}$ of D_3 is irreducible.

More powerful methods for investigating the reducibility of a representation depend on the character relations derived in Section 11.3.5.

11.2.2 SCHUR'S LEMMA

The main result of this section, Schur's lemma, is used at several points in the derivation of the fundamental orthogonality relations for matrix elements of irreducible representations. The orthogonality theorems lead directly to explicit formulas with which one can find out whether a representation is reducible, and, if so, which irreducible representations occur.

Equivalent Representations

Two representations $\Gamma^{(1)} : G \to GL(\mathcal{V}_{(1)})$ and $\Gamma^{(2)} : G \to GL(\mathcal{V}_{(2)})$ of a group G are called **equivalent** if and only if there exists a vector isomorphism of the carrier spaces, $S : \mathcal{V}_{(1)} \to \mathcal{V}_{(2)}$, that commutes with the linear mappings that represent group elements in the sense that

$$\forall a \in G : S\Gamma^{(1)}(a) = \Gamma^{(2)}(a)S. \tag{11.40}$$

It is straightforward to show that equivalence of representations is an equivalence relation in the sense that it is reflexive, symmetric, and transitive (recall Section 2.2.5). We use the notation

$$\Gamma^{(2)} \cong \Gamma^{(1)} \tag{11.41}$$

to indicate that $\Gamma^{(2)}$ is equivalent to $\Gamma^{(1)}$. For every linear mapping $A_{(1)} : \mathcal{V}_{(1)} \to \mathcal{V}_{(1)}$ and for every vector isomorphism $S : \mathcal{V}_{(1)} \to \mathcal{V}_{(2)}$, the mapping M_S such that

$$M_S A_{(1)} := S A_{(1)} S^{-1} \tag{11.42}$$

is a linear mapping of the vector space $\mathcal{V}_{(2)}$ into itself, and, in fact, is a vector isomorphism of $GL(\mathcal{V}_{(1)})$ with $GL(\mathcal{V}_{(2)})$. The commutative diagram

$$
\begin{array}{ccc}
GL(\mathcal{V}_{(1)}) & \xrightarrow{\ M_S\ } & GL(\mathcal{V}_{(2)}) \\[1em]
\Big\uparrow{\scriptstyle\Gamma^{(1)}} & & \Big\uparrow{\scriptstyle\Gamma^{(2)}} \\[1em]
\mathcal{V}_{(1)} & \xrightarrow[\ S\ ]{} & \mathcal{V}_{(2)}
\end{array}
\tag{11.43}
$$

summarizes the relations among representations and vector isomorphisms.

It follows that two matrix representations $\Gamma^{(1)}$ and $\Gamma^{(2)}$ of the same group are equivalent if and only if they are related by a (constant) similarity transformation:

$$
\begin{aligned}
\forall a \in G: \ &\mathbf{S}\Gamma^{(1)}(a) = \Gamma^{(2)}(a)\mathbf{S} \\
&\Rightarrow \Gamma^{(2)}(a) = \mathbf{S}\Gamma^{(1)}(a)\mathbf{S}^{-1}.
\end{aligned}
\tag{11.44}
$$

We emphasize once again that the same matrix $\mathbf{S}$ must work for every group element a.

Group Characters

The **character** of a representation Γ of a group G is the mapping $\chi : G \to \mathbb{C}$ such that

$$\chi(a) := \text{trace}[\Gamma(a)]. \tag{11.45}$$

If Γ is irreducible, then χ is called an **irreducible character**.

Because the matrix representative of the group identity element is always the $n \times n$ unit matrix,

$$\chi(e) = \text{trace}[\Gamma(e)] = n \tag{11.46}$$

in which n is the dimension of the representation Γ.

By Eq. (6.301), the trace of a matrix is invariant under a similarity transformation:

$$\text{trace}[\mathbf{S}\Gamma(a)\mathbf{S}^{-1}] = \text{trace}[\Gamma(a)]. \tag{11.47}$$

It follows at once that if two matrix representations $\Gamma^{(1)}$ and $\Gamma^{(2)}$ of the same group G are equivalent (see Eq. [11.44]), then

$$
\begin{aligned}
\forall a \in G : \chi^{(1)}(a) &= \text{trace}\big[\Gamma^{(1)}(a)\big] \\
&= \text{trace}\big[\Gamma^{(2)}(a)\big] = \chi^{(2)}(a).
\end{aligned}
\tag{11.48}
$$

TABLE 11.1 Characters of the
Irreducible Representations of D_3

D_3	$\|\mathcal{C}\|, \mathcal{C}$		
	E	$3C_2$	$2C_3$
A_1	1	1	1
A_2	1	-1	1
E	2	0	-1

If one lets

$$S = \Gamma(b) \tag{11.49}$$

in which b is any element of G, then one sees that the matrices that represent conjugate elements of G have equal traces:

$$\chi(bab^{-1}) = \text{trace}[\Gamma(b)\Gamma(a)\Gamma^{-1}(b)]$$
$$= \text{trace}[\Gamma(a)]. \tag{11.50}$$

Then for all $a, b \in G$:

$$\boxed{\chi(bab^{-1}) = \chi(a).} \tag{11.51}$$

Therefore, the characters of equivalent representations are equal, and the values of a character on group elements that belong to the same class are equal. For example, Table 11.1 gives the characters of the irreducible representations of the group D_3 (see Eqs. [4.174] and [4.186], and recall that Eqs. [11.38] and [11.39] establish that $\Gamma^{(E)}$ is irreducible).

We show in Section 11.3.5 that the converse is true (if the characters of two representations are equal, then the representations are equivalent). It follows that *two representations of a finite group are equivalent if and only if their characters are equal*:

$$\boxed{\Gamma^{(2)} \cong \Gamma^{(1)} \Leftrightarrow \forall a \in G : \chi^{(2)} = \chi^{(1)}.} \tag{11.52}$$

Schur's Lemma

Let $\Gamma^{(1)}$ and $\Gamma^{(2)}$ be *irreducible* representations of the same group G carried by vector spaces $V_{(1)}$ and $V_{(2)}$, respectively. **Schur's lemma** states that if there exists a linear mapping $A : V_{(1)} \to V_{(2)}$ such that

$$\forall a \in G : A\Gamma^{(1)}(a) = \Gamma^{(2)}(a)A, \tag{11.53}$$

then

- Either A is nonsingular or $A = \mathbf{0}$, and
- If A is nonsingular, then $\mathcal{V}_{(2)}$ and $\mathcal{V}_{(1)}$ are isomorphic.

The key to proving the first assertion is to show that null[A] is a G-invariant subspace of $\mathcal{V}_{(1)}$. Then irreducibility comes into play to show that either null[A] $= \mathcal{O}$ or null[A] $= \mathcal{V}_{(1)}$. But null[A] is invariant under every $\Gamma^{(1)}(a)$ because

$$
\begin{aligned}
A\bar{x} = \bar{0}' \in \mathcal{V}_{(2)} &\Rightarrow \forall a \in G : A\Gamma^{(1)}(a)\bar{x} = \Gamma^{(2)}(a)A\bar{x} = \bar{0}' \\
&\Rightarrow \Gamma^{(1)}(a)\bar{x} \in \text{null}[A].
\end{aligned}
\tag{11.54}
$$

Because $\Gamma^{(1)}$ is irreducible and null[A] is G-invariant, than, by the definition of irreducibility,

$$
\text{null}[A] = \begin{cases} \mathcal{O}, & \text{implying that A is nonsingular, or} \\ \mathcal{V}_{(1)}, & \text{implying that } A = \mathbf{0}. \end{cases}
\tag{11.55}
$$

This establishes the first part of Schur's Lemma. Now assume that A is nonsingular; therefore range[A] $\cong \mathcal{V}_{(1)}$. But range[A] is G-invariant under $\Gamma^{(2)}$, for

$$
\begin{aligned}
\bar{y}' = A\bar{x} &\Rightarrow \forall a \in G : \Gamma^{(2)}(a)\bar{y}' = \Gamma^{(2)}(a)A\bar{x} \\
&= A\Gamma^{(1)}(a)\bar{x} \in \text{range}[A].
\end{aligned}
\tag{11.56}
$$

Because $\Gamma^{(2)}$ is irreducible and range[A] is G-invariant, either

$$
\text{range}[A] = \begin{cases} \mathcal{O}' & \text{(contradiction), or} \\ \mathcal{V}_{(2)}, & \text{implying that } \mathcal{V}_{(1)} \cong \mathcal{V}_{(2)}. \end{cases}
\tag{11.57}
$$

This completes the proof.

If $\mathcal{V}_{(1)}$ and $\mathcal{V}_{(2)}$ are the same vector space,

$$
\mathcal{V}_{(1)} = \mathcal{V}_{(2)} = \mathcal{V}, \; .
\tag{11.58}
$$

and if A is nonsingular, then Eq. (11.53) implies that $\Gamma^{(1)}$ and $\Gamma^{(2)}$ are equivalent. If $\Gamma^{(1)} = \Gamma^{(2)} = \Gamma$ and if the vector space $\mathcal{V}$ is complex, then Eq. (11.53) implies that

$$
\forall \lambda \in \mathbb{C} : \forall a \in G : [A - \lambda\mathbf{1}]\Gamma(a) = \Gamma(a)[A - \lambda\mathbf{1}],
\tag{11.59}
$$

which is the starting assumption of Schur's lemma applied to all of the linear mappings $A - \lambda\mathbf{1}$. Therefore, either $A - \lambda\mathbf{1}$ is nonsingular for all λ, or $A - \lambda\mathbf{1} = \mathbf{0}$ for some λ. In fact, if λ is one of the roots of the polynomial equation $\det[A - \lambda\mathbf{1}] = 0$, then $A - \lambda\mathbf{1}$ is singular. Therefore, by Schur's lemma,

$$
A - \lambda\mathbf{1} = \mathbf{0}.
\tag{11.60}
$$

It follows that if a linear mapping A commutes with the linear mappings that belong to an irreducible representation of a group, then A is equal to a multiple of the identity mapping.

Matrix Form of Schur's Lemma

We soon need the matrix form of Schur's lemma. If one introduces bases in the vector spaces $\mathcal{V}_{(1)}$ (of dimension n_1) and $\mathcal{V}_{(2)}$ (of dimension n_2), then one obtains matrices $\Gamma^{(1)}(a)$ of the irreducible representation $\Gamma^{(1)}$ and matrices $\Gamma^{(2)}(a)$ of the irreducible representation $\Gamma^{(2)}$. Schur's lemma implies that if there exists an $n_2 \times n_1$ matrix $\mathbf{A}$ such that

$$\forall a \in G : \mathbf{A}\Gamma^{(1)}(a) = \Gamma^{(2)}(a)\mathbf{A}, \tag{11.61}$$

then

- Either $\mathbf{A}$ is nonsingular or $\mathbf{A} = \mathbf{0}$, and
- If $\mathbf{A}$ is nonsingular, then (obviously) $n_2 = n_1$.

Further, if for every $a \in G$, $\Gamma^{(2)}(a) = \Gamma^{(1)}(a) = \Gamma(a)$ and if the vector space $\mathcal{V}$ is complex, then $\mathbf{A}$ is a multiple of the unit matrix,

$$\mathbf{A} = \lambda\mathbf{1}. \tag{11.62}$$

11.2.3 EIGENVECTORS OF INVARIANT OPERATORS

The origin of the wide applicability of group theory in physics and engineering is the direct relation that exists between representations of a group G and eigenvectors of operators A that are invariant under the operations of G. We show that if A is unchanged under the linear mappings $\Gamma(g)$ that are induced by the elements $g \in G$, then the eigenvectors of A carry a representation of G. We show also that the eigenvectors of A that carry an irreducible representation of G are degenerate (have the same eigenvalue).

In the contexts of signal processing and numerical computation, for example, the operator A is a shift-invariant digital filter, and G is the group of discrete translations. Because G is Abelian in this example, its irreducible representations are one-dimensional. A function f can carry an irreducible representation of an Abelian group only if f is an eigenfunction of the operators $\Gamma(g)$. The values of an eigenfunction f of discrete translations at uniformly spaced sample points, $f(t_n)$, turn out to be proportional to z^n, in which z is a complex number. The result that f is also an eigenfunction of a shift-invariant digital filter leads immediately to the definition of the transfer function, which is fundamental in such apparently disparate fields as numerical computation and electrical engineering.

For a quantum-mechanical system, transformations such as permutations, spatial rotations, or translations, or rotations in isospin space, induce linear mappings of the state vectors ψ and the Hamiltonian operator H. The profoundly important result that symmetry implies that certain energy (or mass) eigenvalues must be degenerate has determined the evolution of elementary-particle physics over the past forty years. In particular, particle physicists have postulated that if an energy (or mass) degeneracy exists, then there must be an underlying group, of which the degenerate state vectors carry an irreducible representation.

Suppose, then, that G is a group, $\mathcal{V}$ is a carrier space for G, Γ is an operator representation of G on $\mathcal{V}$, and A is a linear operator on $\mathcal{V}$ that is invariant under G in the sense that

$$\forall g \in G : \Gamma(g)A = A\Gamma(g). \tag{11.63}$$

Choose a nonzero vector $\bar{x} \in \mathcal{V}$, and let $\mathcal{W}_{\bar{x}}$ be the partner space of $\bar{x}$. Evidently $\mathcal{W}_{\bar{x}}$ carries a representation of G; let this representation be Γ. There are two possibilities: Either Γ is irreducible, or Γ is reducible. If Γ is irreducible, then Eq. (11.63) and Schur's lemma imply that the restriction of A to $\mathcal{W}_{\bar{x}}$ is a multiple of the identity operator:

$$A \upharpoonright \mathcal{W}_{\bar{x}} = \lambda \mathbf{1}. \tag{11.64}$$

Therefore the partner space of $\bar{x}$ is an eigensubspace of the operator A. This shows that if A is invariant under G, then all of the partner vectors (or partner functions) of $\bar{x}$ are eigenvectors (or eigenfunctions) of A belonging to the same eigenvalue.

Let us start from a slightly different point, assuming that $\mathcal{W}_\lambda$ is the eigensubspace (of a carrier space $\mathcal{V}$) that belongs to the eigenvalue λ of an operator A on $\mathcal{V}$. If A is invariant under a group G in the sense that Eq. (11.63) is obeyed, then, for every group element $g \in G$, and for every vector $\overline{w} \in \mathcal{W}_\lambda$,

$$A[\Gamma(g)\overline{w}] = \Gamma(g)A\overline{w} = \lambda\Gamma(g)\overline{w}. \tag{11.65}$$

Therefore every vector of the form $\Gamma(g)\overline{w}$ is an eigenvector of A with the same eigenvalue, λ. It follows that

$$\forall g \in G : \Gamma(g)\mathcal{W}_\lambda \subseteq \mathcal{W}_\lambda. \tag{11.66}$$

Therefore $\mathcal{W}_\lambda$ is an invariant subspace under G and is a carrier space of the representation Γ of G. Moreover, if Γ is finite-dimensional, then the dimension of $\mathcal{W}_\lambda$ cannot be less than the dimension of Γ. Thus the minimum dimension of an eigensubspace of A is determined by symmetry.

To summarize: If an operator obeys Eq. (11.63), then its eigensubspaces carry representations of the symmetry group G. The dimensions of the irreducible representations of G determine the minimum dimension (the degeneracy) of each eigensubspace. There are as many possible "kinds" of eigensubspaces of A as there are inequivalent irreducible representations of G.

For example, consider a system of three electrons moving in a known and given potential. Let the coordinates of the electrons be $\mathbf{r}_1$, $\mathbf{r}_2$, and $\mathbf{r}_3$. The Hamiltonian operator is

$$H = -\frac{\hbar^2}{2m}[\nabla_1^2 + \nabla_2^2 + \nabla_3^2] + V \tag{11.67}$$

in which the potential operator V is symmetric in $\mathbf{r}_1$, $\mathbf{r}_2$, and $\mathbf{r}_3$. If one neglects all relativistic effects, the energy eigenfunctions are solutions of the partial differential eigenvalue equation

$$H\psi(\mathbf{r}_1, \mathbf{r}_2, \mathbf{r}_3) = E\psi(\mathbf{r}_1, \mathbf{r}_2, \mathbf{r}_3). \tag{11.68}$$

Because H is invariant under permutations of the set of electron labels, $S = \{1, 2, 3\}$, a natural symmetry group is $G = S_3$, the symmetric group of three objects. Let ψ be a solution of Eq. (11.68). The partner functions of ψ are the transforms under permutations:

$$\psi_{213}(\mathbf{r}_1, \mathbf{r}_2, \mathbf{r}_3) := \psi(\mathbf{r}_2, \mathbf{r}_1, \mathbf{r}_3), \quad \psi_{321}(\mathbf{r}_1, \mathbf{r}_2, \mathbf{r}_3) := \psi(\mathbf{r}_3, \mathbf{r}_2, \mathbf{r}_1),$$
$$\psi_{132}(\mathbf{r}_1, \mathbf{r}_2, \mathbf{r}_3) := \psi(\mathbf{r}_1, \mathbf{r}_3, \mathbf{r}_2), \quad \psi_{231}(\mathbf{r}_1, \mathbf{r}_2, \mathbf{r}_3) := \psi(\mathbf{r}_2, \mathbf{r}_3, \mathbf{r}_1), \tag{11.69}$$
$$\psi_{312}(\mathbf{r}_1, \mathbf{r}_2, \mathbf{r}_3) := \psi(\mathbf{r}_3, \mathbf{r}_1, \mathbf{r}_2).$$

The partner space $\mathcal{W}_\psi$ is the vector space spanned by these partner functions. It follows from Eq. (11.64) that each of the partner functions (and every other function that belongs to the partner space) is an eigenfunction of H with the same eigenvalue as ψ.

The energy eigenfunctions that belong to a particular eigenvalue E carry a representation of the symmetric group of three objects, $G = S_3$. Because $S_3 \cong D_3$, the irreducible representations of S_3 are equivalent to the A_1, A_2, and E representations of D_3. Thus there are three kinds of enegy eigenfunctions, with respect to behavior under permutations of the set of electron labels, $S = \{1, 2, 3\}$:

- Totally symmetric eigenfunctions (every partner function is equal to ψ):

$$P_\pi \psi(\mathbf{r}_1, \mathbf{r}_2, \mathbf{r}_3) = \psi(\mathbf{r}_{\pi^{-1}1}, \mathbf{r}_{\pi^{-1}2}, \mathbf{r}_{\pi^{-1}3}) = \psi(\mathbf{r}_1, \mathbf{r}_2, \mathbf{r}_3). \tag{11.70}$$

The minimum dimension of a symmetric eigensubspace implied by group representation theory is equal to 1.
- Totally antisymmetric eigenfunctions (every partner function is equal to $\pm\psi$):

$$P_\pi \psi(\mathbf{r}_1, \mathbf{r}_2, \mathbf{r}_3) = \psi(\mathbf{r}_{\pi^{-1}1}, \mathbf{r}_{\pi^{-1}2}, \mathbf{r}_{\pi^{-1}3}) = \sigma(\pi)\psi(\mathbf{r}_1, \mathbf{r}_2, \mathbf{r}_3), \tag{11.71}$$

in which the value of $\sigma(\pi)$ is $+1$ if π is an even permutation, and -1 if π is an odd permutation. The minimum dimension of an antisymmetric eigensubspace implied by group representation theory is equal to 1.
- Eigenfunctions that belong to the E representation, and hence span an eigensubspace of minimum dimension equal to 2.

In quantum physics only the first two possibilities are known experimentally.

11.2.4 EXERCISES FOR SECTION 11.2

11.2.1 Demonstrate that the matrix form of Schur's lemma applies to the representations $\Gamma^{(A_2)}$ and $\Gamma^{(E)}$ of D_3. (Recall that Eqs. [11.38] and [11.39] establish that $\Gamma^{(E)}$ is irreducible.)

11.2.2 Let Γ be the matrix representation of the dihedral group D_4 found in Exercise 11.1.1. Demonstrate that Γ is reducible, and that

$$\Gamma = \Gamma^{(E)} \oplus \Gamma^{(A_2)} \tag{11.72}$$

in which $\Gamma^{(E)}$ is two-dimensional and $\Gamma^{(A_2)}$ is one-dimensional.

11.3 UNITARITY AND ORTHOGONALITY

We show in this section that every representation of a finite group is equivalent to a unitary representation. In other words, if G is finite, it is always possible to find a unitary inner product and a basis that is orthonormal with respect to the inner product, such that

$$\Gamma(g^{-1}) = [\Gamma(g)]^{\dagger}. \tag{11.73}$$

With the group-representation property, this equation implies that for every $g \in G$, $\Gamma(g)$ is a unitary matrix:

$$[\Gamma(g)]^{\dagger}\Gamma(g) = \mathbf{1}. \tag{11.74}$$

This property holds also for some infinite groups.

It can be shown that every finite-dimensional representation of a compact Lie group is equivalent to a unitary representation. In other words, for carrier spaces of finite-dimensional representations of compact Lie groups G such as $SU(n)$, $O(n)$, or $Sp(n)$, it is always possible to find a unitary inner product and a basis that is orthonormal with respect to the inner product, such that representation matrices relative to this basis are unitary. Moreover, all irreducible representations of compact Lie groups are finite-dimensional. However, for a noncompact Lie group such as the Lorentz group or the group of translations in $\mathbb{E}^{n}$, there are *no* unitary finite-dimensional representations. For a noncompact group, then, it is *not* true that every representation is equivalent to a unitary representation. Moreover, a noncompact Lie group may have infinite-dimensional irreducible representations. In Chapter 12 we exhibit an important infinite-dimensional representation of the Euclidean group in the plane.

11.3.1 CONSEQUENCES OF THE REARRANGEMENT THEOREM

In this section we consider only representations of finite groups. The reason for making this apparently arbitrary restriction is that we must make use of the following corollary of the rearrangement theorem (see Exercise 4.4.8): Let $f : G \to S$ be a function defined on a group G with values in a set S on which an associative addition and a distributive scalar multiplication are defined. (The elements of S may be real or complex numbers, vectors or matrices, for example.) The sum of the values $f(a)$ over all elements of the group is invariant under left-multiplication, right-multiplication, and conjugation by elements of the group:

$$\forall b \in G : \sum_{a \in G} f(a) = \sum_{a \in G} f(ab) = \sum_{a \in G} f(ba)$$
$$= \sum_{a \in G} f(ab^{-1}) = \sum_{a \in G} f(b^{-1}a) \tag{11.75}$$
$$= \sum_{a \in G} f(bab^{-1}).$$

These results follow immediately from the fact that the mappings $a \mapsto ab$, $a \mapsto ab^{-1}$, $a \mapsto bab^{-1}$, and so forth are one-to-one and onto and therefore are permutations of G (see

Exercise 4.4.8). A continuum analog of Eq. (11.75) exists only for compact Lie groups. In that case the sum is replaced by an integral.

Equation (11.75) is the key to showing that every representation of a finite group or a compact Lie group is equivalent to a unitary representation, and to deriving important orthogonality relations for representation matrix elements, group characters, and partner functions.

11.3.2 UNITARY REPRESENTATIONS

Let G be a finite group, and let Γ be an n-dimensional matrix representation of G relative to an orthonormal basis of the carrier space. We can assume without loss of generality that the carrier space is $\mathcal{V} = \mathbb{U}^n$, in which case the inner product of two vectors in $\mathcal{V}$ is

$$\langle \mathbf{x}, \mathbf{y} \rangle = \mathbf{x}^\dagger \mathbf{y}. \tag{11.76}$$

The matrix

$$\mathbf{H} := |G|^{-1} \sum_{a \in G} \Gamma(a)\Gamma^\dagger(a) \tag{11.77}$$

is Hermitian (by inspection) and, for every $b \in G$, has the property of being invariant under the transformation

$$\mathbf{H} \mapsto \Gamma(b)\mathbf{H}\Gamma^\dagger(b) = |G|^{-1} \sum_{a \in G} \Gamma(b)\Gamma(a)\Gamma^\dagger(a)\Gamma^\dagger(b)$$

$$= |G|^{-1} \sum_{c \in G} \Gamma(c)\Gamma^\dagger(c) \tag{11.78}$$

$$= \mathbf{H}.$$

The step from the first line to the second is nontrivial. It depends on the fact that if a runs over the group, then $c = ba$ also runs over the group (by the rearrangement theorem).

We show that the matrix $\mathbf{H}$ can be considered as the Gram matrix of a positive-definite inner product on $\mathcal{V}$. This inner product motivates the definition of a most useful kind of orthogonality. Let

$$\langle\langle \mathbf{x}, \mathbf{y} \rangle\rangle := \mathbf{x}^\dagger \mathbf{H} \mathbf{y}. \tag{11.79}$$

This inner product is Hermitian,

$$\langle\langle \mathbf{y}, \mathbf{x} \rangle\rangle = \langle\langle \mathbf{x}, \mathbf{y} \rangle\rangle^*, \tag{11.80}$$

because $\mathbf{H}$ is Hermitian. From the definition of $\mathbf{H}$ one has

$$\mathbf{x}^\dagger \mathbf{H} \mathbf{y} = |G|^{-1} \sum_{a \in G} \mathbf{x}^\dagger \Gamma(a)\Gamma^\dagger(a)\mathbf{y}$$

$$= |G|^{-1} \sum_{a \in G} [\Gamma^\dagger(a)\mathbf{x}]^\dagger [\Gamma^\dagger(a)\mathbf{y}]. \tag{11.81}$$

Then

$$\langle\langle\mathbf{x}, \mathbf{x}\rangle\rangle = |G|^{-1} \sum_{a\in G} \|\mathbf{\Gamma}^\dagger(a)\mathbf{x}\|_{\text{ip}}^2 \geq 0, \tag{11.82}$$

equality holding if and only if $\mathbf{x} = \mathbf{0}$. Therefore the inner product defined in Eq. (11.79) is positive-definite as well as Hermitian. It follows that the carrier space is a unitary space under this inner product.

Let b be any element of the group G. It is straightforward to verify that the matrix $\mathbf{\Gamma}^\dagger(b)$ is unitary with respect to the inner product $\langle\langle\mathbf{x}, \mathbf{y}\rangle\rangle$:

$$\langle\langle\mathbf{\Gamma}^\dagger(b)\mathbf{x}, \mathbf{\Gamma}^\dagger(b)\mathbf{y}\rangle\rangle = \mathbf{x}^\dagger \left[|G|^{-1} \sum_{a\in G} \mathbf{\Gamma}(b)\mathbf{\Gamma}(a)\mathbf{\Gamma}^\dagger(a)\mathbf{\Gamma}^\dagger(b) \right] \mathbf{y}$$

$$= \mathbf{x}^\dagger \left[|G|^{-1} \sum_{a\in G} \mathbf{\Gamma}(ba)\mathbf{\Gamma}^\dagger(ba) \right] \mathbf{y} \tag{11.83}$$

$$= \mathbf{x}^\dagger \mathbf{H} \mathbf{y} = \langle\langle\mathbf{x}, \mathbf{y}\rangle\rangle.$$

The third line follows from the second by virtue of the rearrangement theorem; see Eq. (11.75).

It is clear now that the representation matrices are unitary relative to a properly chosen basis. There are two ways to proceed: Transform to a basis of eigenvectors of $\mathbf{H}$, which can be orthonormalized because $\mathbf{H}$ is Hermitian; or use the Cholesky decomposition Eq. (9.363),

$$\mathbf{H} = \mathbf{C}\mathbf{C}^\dagger, \tag{11.84}$$

in which $\mathbf{C}$ is nonsingular and the columns of $\mathbf{C}^\dagger$ are the basis vectors for which $\mathbf{H}$ is the Gram matrix; see Eq. (8.30).

Taking the latter course, we use the Cholesky decomposition to transform the inner product $\langle\langle\mathbf{\Gamma}^\dagger(b)\mathbf{x}, \mathbf{\Gamma}^\dagger(b)\mathbf{y}\rangle\rangle$ to an ordinary inner product. First, note that in terms of the Cholesky matrix $\mathbf{C}$, the inner product $\langle\langle\mathbf{x}, \mathbf{y}\rangle\rangle$ becomes the ordinary inner product of $\mathbf{C}^\dagger\mathbf{x}$ and $\mathbf{C}^\dagger\mathbf{y}$:

$$\langle\langle\mathbf{x}, \mathbf{y}\rangle\rangle = [\mathbf{C}^\dagger\mathbf{x}]^\dagger \mathbf{C}^\dagger\mathbf{y} = \langle\mathbf{C}^\dagger\mathbf{x}, \mathbf{C}^\dagger\mathbf{y}\rangle. \tag{11.85}$$

Next, one finds easily that

$$\langle\langle\mathbf{\Gamma}^\dagger(b)\mathbf{x}, \mathbf{\Gamma}^\dagger(b)\mathbf{y}\rangle\rangle = [\mathbf{U}^\dagger(b)\mathbf{C}^\dagger\mathbf{x}]^\dagger \mathbf{U}^\dagger(b)\mathbf{C}^\dagger\mathbf{y}$$
$$= \langle\mathbf{U}^\dagger(b)\mathbf{C}^\dagger\mathbf{x}, \mathbf{U}^\dagger(b)\mathbf{C}^\dagger\mathbf{y}\rangle \tag{11.86}$$

in which

$$\mathbf{U}^\dagger(b) := \mathbf{C}^\dagger\mathbf{\Gamma}^\dagger(b)[\mathbf{C}^\dagger]^{-1}. \tag{11.87}$$

Comparing Eqs. (11.83), (11.85), and (11.86), one sees that

$$\langle\mathbf{U}^\dagger(b)\mathbf{C}^\dagger\mathbf{x}, \mathbf{U}^\dagger(b)\mathbf{C}^\dagger\mathbf{y}\rangle = \langle\mathbf{C}^\dagger\mathbf{x}, \mathbf{C}^\dagger\mathbf{y}\rangle. \tag{11.88}$$

Because $\mathbf{C}^\dagger$ is nonsingular, it follows that $\mathbf{x}' = \mathbf{C}^\dagger \mathbf{x}$ and $\mathbf{y}' = \mathbf{C}^\dagger \mathbf{y}$ may be any vectors in the carrier space. Then

$$\langle \mathbf{U}^\dagger(b)\mathbf{x}', \mathbf{U}^\dagger(b)\mathbf{y}' \rangle = \langle \mathbf{x}', \mathbf{y}' \rangle, \tag{11.89}$$

which shows that the matrix $\mathbf{U}^\dagger(b)$ is unitary with respect to the ordinary inner product $\langle \mathbf{x}, \mathbf{y} \rangle = \mathbf{x}^\dagger \mathbf{y}$.

Because $\mathbf{U}^\dagger(b)$ is unitary, it follows that

$$\mathbf{U}(b) = [\mathbf{U}^\dagger(b)]^\dagger = \mathbf{C}^{-1}\mathbf{\Gamma}(b)\mathbf{C} \tag{11.90}$$

is also unitary, for every $b \in G$. Also, $\mathbf{U}$ is a matrix representation of G, for

$$\mathbf{U}(a)\mathbf{U}(b) = \mathbf{C}^{-1}\mathbf{\Gamma}(a)\mathbf{C}\mathbf{C}^{-1}\mathbf{\Gamma}(b)\mathbf{C} = \mathbf{C}^{-1}\mathbf{\Gamma}(ab)\mathbf{C} = \mathbf{U}(ab). \tag{11.91}$$

The representations $\mathbf{U}$ and $\mathbf{\Gamma}$ are equivalent, according to the definition in Eq. (11.44). It follows that *every representation of a finite group is equivalent to a unitary representation.*

11.3.3 ORTHOGONALITY THEOREMS

Both the matrix elements of a representation $\mathbf{\Gamma}$ of a finite group G and the character χ are complex-valued functions defined on G. The set of such functions is a vector space over the complex numbers $\mathbb{C}$, for if $f : G \to \mathbb{C}$ and $g : G \to \mathbb{C}$ are complex-valued functions on G, then so are αf and $f + g$ (in which $\alpha \in \mathbb{C}$). Evidently one can consider any function $f : G \to \mathbb{C}$ as an element of the complex vector space $\mathbb{C}^{|G|}$, for f is defined uniquely by the vector

$$\mathbf{f} = \begin{pmatrix} f(a_1) \\ \vdots \\ f(a_{|G|}) \end{pmatrix} \tag{11.92}$$

of its values on the elements $a_1 = e, \ldots, a_{|G|}$ of G.

As soon as one has defined a vector space, one wants to find a basis. It turns out to be very useful for this purpose to define the obviously Hermitian **group inner product**

$$\langle f, g \rangle_G := |G|^{-1} \sum_{a \in G} f(a)^* g(a), \tag{11.93}$$

because orthogonality with respect to an inner product automatically implies linear independence.

Matrix Elements of Irreducible Representations

Let $\mathbf{\Gamma}^{(i)}$ and $\mathbf{\Gamma}^{(j)}$ be irreducible representations of a finite group G, such that $\mathbf{\Gamma}^{(i)}$ and $\mathbf{\Gamma}^{(j)}$ are inequivalent for all $j \neq i$. Let n_i and n_j be the dimensions of $\mathbf{\Gamma}^{(i)}$ and $\mathbf{\Gamma}^{(j)}$, respectively. By Section 11.3.2, one can assume that $\mathbf{\Gamma}^{(i)}$ and $\mathbf{\Gamma}^{(j)}$ are unitary with respect to the canonical inner products in $\mathbb{U}^{n_i}$ and $\mathbb{U}^{n_j}$, respectively. Therefore we write matrix elements as a_{mn} instead of a_n^m.

Let

$$\mathbf{A}(i, j, \mathbf{X}) := \sum_{a \in G} \mathbf{\Gamma}^{(j)}(a) \mathbf{X} \mathbf{\Gamma}^{(i)}(a^{-1}) \tag{11.94}$$

in which $\mathbf{X}$ is any $n_i \times n_j$ matrix. Because $\mathbf{\Gamma}^{(j)}(a) = \mathbf{\Gamma}^{(j)}(b)\mathbf{\Gamma}^{(j)}(b^{-1}a)$, one has

$$\forall b \in G: \quad \mathbf{A}(i, j, \mathbf{X})\mathbf{\Gamma}^{(i)}(b) = \sum_{a \in G} \mathbf{\Gamma}^{(j)}(a) \mathbf{X} \mathbf{\Gamma}^{(i)}(a^{-1}b)$$

$$= \mathbf{\Gamma}^{(j)}(b) \sum_{a \in G} \mathbf{\Gamma}^{(j)}(b^{-1}a) \mathbf{X} \mathbf{\Gamma}^{(i)}(a^{-1}b) \tag{11.95}$$

$$= \mathbf{\Gamma}^{(j)}(b) \mathbf{A}(i, j, \mathbf{X}).$$

By Schur's lemma, $\mathbf{A}(i, j, \mathbf{X}) = \mathbf{0}$ for all $j \neq i$, and there exists a complex constant $\lambda(i, \mathbf{X})$ such that $\mathbf{A}(i, i, \mathbf{X}) = \lambda(i, \mathbf{X})\mathbf{1}$. One can summarize both of these statements in the matrix equation

$$\mathbf{A}(i, j, \mathbf{X}) = \delta_{ij}\lambda(i, \mathbf{X})\mathbf{1}, \tag{11.96}$$

which is equivalent to the component equations

$$a_{kl}(i, j, \mathbf{X}) = \sum_{a \in G} \sum_{r,s} \gamma_{kr}^{(j)}(a) x_{rs} \gamma_{sl}^{(i)}(a^{-1}) = \delta_{ij}\delta_{kl}\lambda(i, \mathbf{X}). \tag{11.97}$$

If

$$x_{rs} = \delta_{rm}\delta_{sn} \tag{11.98}$$

then the matrix $\mathbf{X}$ has a 1 in at the intersection of row m and column n, and 0 elsewhere. Taking $j = i$ and $l = k$, one obtains

$$\lambda(i, \mathbf{E}_m^n) = \sum_{a \in G} \gamma_{km}^{(i)}(a) \gamma_{nk}^{(i)}(a^{-1}) \tag{11.99}$$

for every $k \in (1 : n_i)$, because $\mathbf{A}(i, j, \mathbf{X})$ is a multiple of the identity matrix. By virtue of the rearrangement theorem one can replace a with a^{-1} and a^{-1} with a, obtaining

$$\lambda(i, \mathbf{E}_m^n) = \sum_{a \in G} \gamma_{km}^{(i)}(a^{-1}) \gamma_{nk}^{(i)}(a)$$

$$= \sum_{a \in G} \gamma_{nk}^{(i)}(a) \gamma_{km}^{(i)}(a^{-1}) \tag{11.100}$$

$$= \delta_{mn}\lambda(i, \mathbf{E}_k^k),$$

independently of the value of k. It follows that

$$a_{kl}(i, j, \mathbf{E}_m^n) = \delta_{kl}\delta_{mn}\mu_i, \tag{11.101}$$

in which μ_i is independent of k. To evaluate μ_i, we take $l = k, n = m$ and sum on $k \in (1 : n_i)$:

$$\sum_{k=1}^{n_i} a_{kk}\left(i, i, \mathbf{E}_m^m\right) = \sum_{a \in G} \sum_{k=1}^{n_i} \gamma_{km}^{(i)}(a^{-1}) \gamma_{mk}^{(i)}(a)$$

$$= n_i \mu_i$$

$$= \sum_{a \in G} \gamma_{mm}^{(i)}(e) \tag{11.102}$$

$$= \sum_{a \in G} 1 = |G|.$$

Then

$$\mu_i = \frac{|G|}{n_i}. \tag{11.103}$$

Finally, because the representations $\boldsymbol{\Gamma}^{(i)}$ and $\boldsymbol{\Gamma}^{(j)}$ are unitary relative to an orthonormal basis, one can set

$$\boldsymbol{\Gamma}^{(i)}(a^{-1}) = \boldsymbol{\Gamma}^{(i)}(a)^{\dagger} \tag{11.104}$$

in the preceding equations.

These results can be summarized in the **great orthogonality theorem**

$$\boxed{\sum_{a \in G} \gamma_{kl}^{(j)}(a) \gamma_{mn}^{(i)*}(a) = \delta_{ij} \delta_{km} \delta_{ln} \frac{|G|}{n_i}} \tag{11.105}$$

for the matrix elements $\gamma_{kl}^{(j)}(a)$ of inequivalent, irreducible, unitary representations of a finite group. In terms of the group inner product, the theorem reads

$$\left\langle \gamma_{mn}^{(i)}, \gamma_{kl}^{(j)} \right\rangle_G = \frac{\delta_{ij} \delta_{km} \delta_{ln}}{n_i}. \tag{11.106}$$

Equations (11.105) and (11.106) turn out to be the group-theoretical foundation of many important orthogonality relations, including selection rules in physics.

For Lie groups, the matrix elements of irreducible representations are functions of the parameters that label the group elements (such as the axis and angle of rotation in the case of the group $SO(3)$ of proper rotations in $\mathbb{E}^3$). The matrix elements turn out to be among the special functions of mathematical physics, and the analog of Eq. (11.105) for compact Lie groups turns out to be an orthogonality relation for special functions.

Orthogonality Relations for Characters

Orthogonality relations for irreducible characters follow immediately from Eq. (11.105). Setting $l = k$ and $n = m$ to select the diagonal matrix elements and then summing on k and

m, one obtains

$$\langle \chi^{(i)}, \chi^{(j)}\rangle_G = |G|^{-1} \sum_{a \in G} \chi^{(j)}(a)\chi^{(i)*}(a) = \delta_{ij}. \tag{11.107}$$

Because the values of a group character on elements of the same conjugacy class are equal, one can state this relation somewhat more succinctly in the form of a sum over classes:

$$|G|^{-1} \sum_{\mathcal{C} \subset G} |\mathcal{C}|\, \chi^{(j)}(\mathcal{C})\, \chi^{(i)*}(\mathcal{C}) = \delta_{ij}. \tag{11.108}$$

In this equation, $|\mathcal{C}|$ is the order of $\mathcal{C}$ (that is, the number of elements of $\mathcal{C}$).

One can interpret Eq. (11.108) as an orthogonality relation between vectors

$$\chi^{(i)} = \begin{pmatrix} \chi^{(i)}(\mathcal{C}_1) \\ \vdots \\ \chi^{(i)}(\mathcal{C}_{K(G)}) \end{pmatrix} \quad \text{and} \quad \chi^{(j)} = \begin{pmatrix} \chi^{(j)}(\mathcal{C}_1) \\ \vdots \\ \chi^{(j)}(\mathcal{C}_{K(G)}) \end{pmatrix} \tag{11.109}$$

in a complex vector space $\mathbb{C}^{K(G)}$. The dimension of the space is $K(G)$, the number of classes of G. The Gram matrix $\mathbf{G}$, or metric tensor, is diagonal,

$$\mathbf{G} = |G|^{-1}\mathrm{diag}\big[|\mathcal{C}_1|, \dots, |\mathcal{C}_{K(G)}|\big], \tag{11.110}$$

the diagonal elements being the orders of the classes. Therefore the inner product

$$(\!(\chi, \chi')\!) := |G|^{-1} \sum_{\mathcal{C} \subset G} |\mathcal{C}|\, \chi^*(\mathcal{C})\, \chi'(\mathcal{C}) \tag{11.111}$$

is positive-definite. We denote as $\mathcal{K}_G$ the space $\mathbb{C}^{K(G)}$, equipped with the inner product Eq. (11.111).

Equation (11.108) implies that the irreducible characters of a finite group G are linearly independent in $\mathcal{K}_G$. Therefore the number of irreducible representations of a finite group cannot exceed the number of classes (the dimension of $\mathcal{K}_G$). It can be shown that, in fact, the number of irreducible representations is equal to the number of classes.

Let

$$\mathbf{y}^{(i)} := |G|^{-1/2} \begin{pmatrix} \chi^{(i)}(\mathcal{C}_1)|\mathcal{C}_1|^{1/2} \\ \vdots \\ \chi^{(i)}(\mathcal{C}_{K(G)})|\mathcal{C}_{K(G)}|^{1/2} \end{pmatrix}, \tag{11.112}$$

in which $i \in (1 : K(G))$, be the vector of weighted values of the character of the irreducible representation $\Gamma^{(i)}$. The orthogonality relation Eq. (11.108) is equivalent to the canonical inner products

$$\langle \mathbf{y}^{(i)}, \mathbf{y}^{(j)} \rangle = \mathbf{y}^{(i)\dagger}\mathbf{y}^{(j)} = \delta_{ij} \tag{11.113}$$

in $\mathbb{C}^{K(G)}$. In turn, these vector relations can be summarized in the matrix equation

$$\mathbf{Y}^{\dagger}\mathbf{Y} = \mathbf{1} \tag{11.114}$$

in which

$$\mathbf{Y} = \left[\mathbf{y}^{(1)}, \dots, \mathbf{y}^{(K(G))}\right] \tag{11.115}$$

is the $K(G) \times K(G)$ matrix of which the vectors $\mathbf{y}^{(i)}$ are the columns. It follows at once that $\mathbf{Y}$ is unitary. Then

$$\mathbf{Y}\mathbf{Y}^{\dagger} = \mathbf{1}, \tag{11.116}$$

implying the new character orthogonality relations

$$|G|^{-1}\sum_{i=1}^{K(G)} \chi^{(i)*}(\mathcal{C}_m)\chi^{(i)}(\mathcal{C}_n)[|\mathcal{C}_m||\mathcal{C}_n|]^{1/2} = \delta_{mn}. \tag{11.117}$$

One can take advantage of the Kronecker delta to rewrite these relations in the more convenient form

$$\boxed{\sum_{i=1}^{K(G)} \chi^{(i)*}(\mathcal{C}_m)\chi^{(i)}(\mathcal{C}_n) = \frac{|G|}{|\mathcal{C}_m|}\delta_{mn}.} \tag{11.118}$$

In this orthogonality relation the sum runs over representations (that is, over the columns of $\mathbf{Y}$); in Eq. (11.108) the sum runs over classes (that is, over the rows of $\mathbf{Y}$).

One of the more useful special cases of Eq. (11.118) occurs for the class of the identity. Because the identity element is in a class by itself, $|\mathcal{C}_e| = 1$. Also, the matrix that represents the group identity is always the identity matrix, the character of which is equal to the dimension of the representation:

$$\chi^{(i)}(\mathcal{C}_e) = n_i. \tag{11.119}$$

Substituting in Eq. (11.118), one finds that the sum of the squares of the dimensions of the irreducible representations of a finite group G is equal to the order of G:

$$\boxed{\sum_{i=1}^{K(G)} (n_i)^2 = |G|.} \tag{11.120}$$

For example, if one knew only that the order of D_3 is 6 and that D_3 has three classes, one would instantly know that the dimensions of the irreducible representations are 1, 1, and 2, for $1^2 + 1^2 + 2^2 = 6$, and there are no other triplets of integers whose squares add to 6.

If one applies Eq. (11.120) to a finite Abelian group, one sees instantly that all of the irreducible representations are one-dimensional, because each element is in a class by itself. For example, the irreducible representations of any cyclic group (such as C_2, C_3, etc.) are one-dimensional. The irreducible representations of the infinite cyclic group defined in Eq. (4.81) are also one-dimensional.

In general, Eq. (11.118) implies that the vectors

$$
\chi(\mathcal{C}_m) := \begin{pmatrix} \chi^{(1)}(\mathcal{C}_m) \\ \chi^{(2)}(\mathcal{C}_m) \\ \vdots \\ \chi^{(K(G))}(\mathcal{C}_m) \end{pmatrix}
\tag{11.121}
$$

are mutually orthogonal.

11.3.4 PRODUCT RELATION FOR CHARACTERS

In order to calculate the characters of the irreducible representations of a finite group, one needs not only the orthogonality relations derived above, but also a relation for the product of the irreducible characters of two classes.

We begin by defining the matrix that one obtains by summing the matrices of a particular irreducible representation $\Gamma^{(i)}$ over the elements of a particular conjugacy class, $\mathcal{C}_k$:

$$
\mathbf{C}_k^{(i)} := \sum_{g \in \mathcal{C}_k} \Gamma^{(i)}(g)
\tag{11.122}
$$

From Exercise 4.4.11, one sees that

$$
\forall a \in G: \ \Gamma^{(i)}(a)\mathbf{C}_k^{(i)}\Gamma^{(i)}(a^{-1}) = \sum_{g \in \mathcal{C}_k} \Gamma^{(i)}(aga^{-1}) = \mathbf{C}_k^{(i)}
\tag{11.123}
$$

because the conjugacy mapping $g \in \mathcal{C}_k \mapsto aga^{-1} \in \mathcal{C}_k$ is simply a permutation of the elements of the class $\mathcal{C}_k$. It follows at once that the product of two class-sum matrices is also invariant under all conjugacy mappings:

$$
\forall a \in G: \ \Gamma^{(i)}(a)\mathbf{C}_j^{(i)}\mathbf{C}_k^{(i)}\Gamma^{(i)}(a^{-1}) = \mathbf{C}_j^{(i)}\mathbf{C}_k^{(i)}.
\tag{11.124}
$$

We show that this implies that the product $\mathbf{C}_j^{(i)}\mathbf{C}_k^{(i)}$ is equal to a linear combination of class-sum matrices with integral coefficients.

From the representation property, $\Gamma(b)\Gamma(a) = \Gamma(ba)$, one sees that the product of two class-sum matrices is a sum over representation matrices of the elements of the group, with non-negative, integral coefficients:

$$\mathbf{C}_j^{(i)}\mathbf{C}_k^{(i)} = \sum_{a \in G} c_{jk;a}\boldsymbol{\Gamma}^{(i)}(a).$$ (11.125)

If $c_{jk;a} \neq 0$ for some $a \in G$ in Eq. (11.125), then, by Eq. (11.124), for every element xax^{-1} of $\mathcal{C}_a$ the matrix $\boldsymbol{\Gamma}^{(i)}(xax^{-1})$ has the coefficient $c_{jk;a}$ in Eq. (11.125). Then there exist nonnegative integers c_{jkl} such that

$$\mathbf{C}_j^{(i)}\mathbf{C}_k^{(i)} = \sum_{l=1}^{K(G)} c_{jkl}\mathbf{C}_l^{(i)}.$$ (11.126)

Equation (11.123) is equivalent to the hypothesis of Schur's lemma:

$$\forall a \in G: \boldsymbol{\Gamma}^{(i)}(a)\mathbf{C}_k^{(i)} = \mathbf{C}_k^{(i)}\boldsymbol{\Gamma}^{(i)}(a).$$ (11.127)

Therefore $\mathbf{C}_k^{(i)}$ is a multiple of the unit matrix:

$$\mathbf{C}_k^{(i)} = \eta_k^{(i)}\mathbf{1}.$$ (11.128)

Substituting this equation in Eq. (11.126), one finds that

$$\eta_j^{(i)}\eta_k^{(i)} = \sum_{l=1}^{K(G)} c_{jkl}\eta_l^{(i)}.$$ (11.129)

In order to make this relation useful, we must evaluate the constants $\eta_k^{(i)}$. We do this by calculating the trace of the matrix $\mathbf{C}_k^{(i)}$ in two different ways. From Eq. (11.128), one gets

$$\text{trace}\left[\mathbf{C}_k^{(i)}\right] = n_i\eta_k^{(i)}.$$ (11.130)

Taking the trace of both sides of Eq. (11.122), one gets

$$\text{trace}\left[\mathbf{C}_k^{(i)}\right] = \sum_{g \in \mathcal{C}_k} \chi^{(i)}(\mathcal{C}_k) = |\mathcal{C}_k|\chi^{(i)}(\mathcal{C}_k).$$ (11.131)

It follows that

$$\eta_k^{(i)} = \frac{|\mathcal{C}_k|\chi^{(i)}(\mathcal{C}_k)}{n_i}.$$ (11.132)

Substituting this relation into Eq. (11.129), one obtains

$$|\mathcal{C}_j|\chi^{(i)}(\mathcal{C}_j)|\mathcal{C}_k|\chi^{(i)}(\mathcal{C}_k) = n_i \sum_{l=1}^{K(G)} c_{jkl}|\mathcal{C}_l|\chi^{(i)}(\mathcal{C}_l)$$ (11.133)

TABLE 11.2 Nonzero Constants c_{jkl} for the Group D_3, in which $\mathcal{C}_1 = \{e\}$, $\mathcal{C}_2 = \{a, b, c\}$, and $\mathcal{C}_3 = \{d, f\}$

j	k	l	c_{jkl}
1	2	2	1
1	3	3	1
2	2	1	3
2	2	3	3
2	3	2	2

Table 11.2 gives the constants c_{jkl} for the group D_3. From the table one finds, for example, that

$$\mathbf{C}_2 \mathbf{C}_2 = 3\mathbf{C}_1 + 3\mathbf{C}_3. \tag{11.134}$$

11.3.5 REDUCTION OF UNITARY REPRESENTATIONS

Let V be a unitary space that carries a representation Γ of a finite group G. By Section 11.3.2, one can assume that Γ is unitary. Let Γ be reducible, and let W be a proper G-invariant subspace of V.

It follows that the orthogonal complement $W^\perp$ is also G-invariant. Because Γ is unitary, one has

$$\forall a \in G: \ [\Gamma(a^{-1})]^H = [\Gamma(a^{-1})]^{-1} = \Gamma(a). \tag{11.135}$$

Then

$$\forall \overline{w} \in W: \ \forall \overline{z} \in W^\perp: \ \forall a \in G: \ \langle \overline{z}, \Gamma(a^{-1})\overline{w} \rangle = 0$$
$$= \langle \Gamma(a)\overline{z}, \overline{w} \rangle \tag{11.136}$$

in which the first line follows because W is G-invariant. Therefore $\Gamma(a)\overline{z} \in W^\perp$, which shows that $W^\perp$ is G-invariant.

Let the restrictions of Γ to W and $W^\perp$ be $\Gamma^{(1)}$ and $\Gamma^{(2)}$, respectively:

$$\forall a \in G: \ \Gamma^{(1)}(a) = \Gamma(a) \restriction W \quad \text{and} \quad \Gamma^{(2)}(a) = \Gamma(a) \restriction W^\perp. \tag{11.137}$$

Because V is the direct sum of the G-invariant subspaces W and $W^\perp$,

$$V = W \oplus W^\perp, \tag{11.138}$$

it follows from Section 6.1.4 that every linear mapping $\Gamma(a)$ in the representation Γ is equal to the direct sum of $\Gamma^{(1)}(a)$ and $\Gamma^{(2)}(a)$:

$$\forall a \in G: \ \Gamma(a) = \Gamma^{(1)}(a) \oplus \Gamma^{(2)}(a). \tag{11.139}$$

The canonical realization of a direct sum of two linear mappings is a block-diagonal matrix (see Eqs. [5.272] and [6.84]).

Therefore there exists at least one basis (in fact, there are many) relative to which the matrix of every $\Gamma(a)$ is block-diagonal, meaning that there exists a nonsingular matrix $\mathbf{T}$ such that

$$\forall a \in G: \ \mathbf{T}\Gamma(a)\mathbf{T}^{-1} = \begin{pmatrix} \Gamma^{(1)}(a) & \mathbf{0} \\ \mathbf{0} & \Gamma^{(2)}(a) \end{pmatrix} \tag{11.140}$$

(see Eq. [6.84]). If such a basis exists (or, in basis-free terms, if the carrier space $\mathcal{V}$ is the direct sum of at least two G-invariant subspaces) then one says that the representation Γ is **fully reducible**. We have just shown that every reducible unitary representation is fully reducible. Because every representation of a finite group is equivalent to a unitary representation, it follows that every reducible representation of a finite group is fully reducible.

For example, the reducible representation of $G = D_3$ given in Eq. (11.29) is in fully reduced form. In this example,

$$\Gamma = \Gamma_{A_2} \oplus \Gamma_E, \tag{11.141}$$

as one can verify from Eqs. (4.174) and (4.186).

If, in turn, $\mathcal{W}'$ is equal to the direct sum of two other G-invariant subspaces, then one can fully reduce $\Gamma^{(2)}$ to the direct sum of the representations carried by the G-invariant subspaces, and so on. One continues until one reaches irreducible representations. It follows that any reducible representation of a finite group G is equivalent to a direct sum of irreducible representations $\Gamma^{(1)}, \ldots, \Gamma^{(r)}$:

$$\forall a \in G: \ \Gamma(a) \cong \Gamma^{(1)}(a) \oplus \cdots \oplus \Gamma^{(r)}(a). \tag{11.142}$$

This direct sum is equivalent to the matrix equation

$$\forall a \in G: \ \mathbf{T}\Gamma(a)\mathbf{T}^{-1} = \begin{pmatrix} \Gamma^{(1)}(a) & & \mathbf{0} \\ & \ddots & \\ \mathbf{0} & & \Gamma^{(r)}(a) \end{pmatrix} \tag{11.143}$$

in which $\mathbf{T}$ is a nonsingular matrix such that $\mathbf{T}^{-1}$ transforms the original basis to a basis in which the matrix of every $\Gamma(a)$ is block-diagonal. The irreducible representations $\Gamma^{(1)}, \ldots, \Gamma^{(r)}$ are called the **irreducible components** of the representation Γ.

It follows that the character of a reducible representation Γ is equal to the sum of the characters of its irreducible components:

$$\forall a \in G: \chi(a) = \sum_{j=1}^{K(G)} r_j \chi^{(j)}(a) \tag{11.144}$$

in which

$$\chi^{(j)}(a) := \text{trace } \Gamma^{(j)}(a) \tag{11.145}$$

and r_j is the number of times that the irreducible representation $\Gamma^{(j)}$ occurs in the direct sum Eq. (11.142). The integer r_j is zero if $\Gamma^{(j)}$ does not occur in the reduction of Γ. There is no upper bound on r_j, apart from the dimension of Γ, because a given reducible representation may be equivalent to a direct sum that contains an arbitrary number of copies of $\Gamma^{(j)}$.

Given a group character χ, one can use the orthogonality relation given in Eq. (11.108) to find the number of times each irreducible component occurs if a representation (of which χ is the character) is reduced:

$$r_j = |G|^{-1} \sum_{\mathcal{C} \subset G} |\mathcal{C}| \chi^{(j)*}(\mathcal{C}) \chi(\mathcal{C}). \tag{11.146}$$

For example, consider the dihedral group $G = D_3$ and the three-dimensional matrix representation Γ defined in Eq. (11.29). Table 11.3 gives the character of this representation. From Table 11.1 and Eq. (11.146), one finds

$$r_{A_1} = \tfrac{1}{6}[(1)(1)(3) + (3)(1)(-1) + (2)(1)(0)] = 0,$$
$$r_{A_2} = \tfrac{1}{6}[(1)(1)(3) + (3)(-1)(-1) + (2)(1)(0)] = 1, \tag{11.147}$$
$$r_E = \tfrac{1}{6}[(1)(2)(3) + (3)(0)(-1) + (2)(-1)(0)] = 1,$$

in agreement with the reduction of Γ demonstrated in Eq. (11.29). The above results for r_{A_1},

TABLE 11.3 Character of the
Reducible Representation of D_3
Defined in Eq. (11.29)

D_3	$|\mathcal{C}|, \mathcal{C}$		
	E	$3C_2$	$2C_3$
χ	3	-1	0

TABLE 11.4 Starting Point in Finding the Characters of the Irreducible Representations of D_3 from the Orthogonality Relations and the Product Relation

D_3	$|\mathcal{C}|, \mathcal{C}$		
	E	$3C_2$	$2C_3$
A_1	1	1	1
A_2	1	α	β
E	2	γ	δ

and so forth, often are written in the form

$$\Gamma \cong \Gamma^{(A_2)} \oplus \Gamma^{(E)}, \tag{11.148}$$

which indicates that Γ is equivalent to the direct sum of the representations $\Gamma^{(A_2)}$ and $\Gamma^{(E)}$.

11.3.6 CONSTRUCTION OF CHARACTER TABLES

Although it is not at all obvious, Eqs. (11.108), (11.118), and (11.133) determine the characters of the irreducible representations of a finite group. For example, if one knew only that the group D_3 has three classes, $\mathcal{C}_e$, $\mathcal{C}_a$, and $\mathcal{C}_f$, with 1, 3, and 2 elements, respectively, one would be able to calculate that $6 = \sum n_i^2 = 1^2 + 1^2 + 2^2$, and to fill in the unknown characters α, β, γ and δ in Table 11.4 using the orthogonality of the rows and columns of the Table.

One knows the first row of Table 11.4 because the identity (A_1) representation "matrices" are all equal to the number 1. One knows the first column because the character of the class of the group identity element is equal to the dimension of the representation. From the mutual orthogonality of $\chi^{(A_1)}$, $\chi^{(A_2)}$, and $\chi^{(E)}$ one gets (summing over classes)

$$\begin{aligned}
1 + 3\alpha + 2\beta &= 0 \\
2 + 3\gamma + 2\delta &= 0 \\
2 + 3\alpha\gamma + 2\beta\delta &= 0.
\end{aligned} \tag{11.149}$$

From the mutual orthogonality of $\chi(E)$, $\chi(3C_2)$, and $\chi(2C_3)$ one gets (summing over representations)

$$\begin{aligned}
1 + \alpha + 2\gamma &= 0 \\
1 + \beta + 2\delta &= 0 \\
1 + \alpha\beta + \gamma\delta &= 0.
\end{aligned} \tag{11.150}$$

To solve these equations, we eliminate all but one variable. From the first and second part of

Eq. (11.150) and the second part of Eq. (11.149) it follows that

$$\alpha = -1 - 2\gamma$$
$$\beta = -1 - 2\delta \tag{11.151}$$
$$\delta = -1 - \tfrac{3}{2}\gamma.$$

Observing that $\beta = 1 + 3\gamma$, and substituting these into the last part of Eq. (11.150), one obtains a quadratic equation for γ:

$$\gamma \left(-6 - \tfrac{15}{2}\gamma\right) = 0. \tag{11.152}$$

Thus *two* sets of characters are consistent with the orthogonality relations for the group D_3:

$$\alpha = -1, \quad \beta = 1,$$
$$\gamma = 0, \quad \delta = -1, \tag{11.153}$$

and

$$\alpha = \tfrac{3}{5}, \quad \beta = -\tfrac{7}{5},$$
$$\gamma = -\tfrac{4}{5}, \quad \delta = \tfrac{1}{5}. \tag{11.154}$$

However, only one of these two sets is correct.

The product relation, Eq. (11.134), implies that for the group D_3,

$$3\alpha^2 = 1 + 2\beta. \tag{11.155}$$

The characters in Eq. (11.154) do not satisfy this relation, but the characters in Eq. (11.153) do. This confirms that Table 11.1 gives the unique correct set of irreducible characters for the group D_3.

11.3.7 CHARACTERS OF KRONECKER PRODUCTS

In Section 7.7.5 we define the Kronecker product

$$\mathbf{C} = \mathbf{A} \otimes \mathbf{B} \tag{11.156}$$

of two matrices $\mathbf{A}$ and $\mathbf{B}$ as the "supermatrix" with elements

$$c_{(k,l)}^{(i,j)} = a_k^i \, b_l^j. \tag{11.157}$$

The Kronecker product matrix acts on the tensor product of vector spaces,

$$\mathbf{C} : \mathbb{F}^m \otimes \mathbb{F}^n \to \mathbb{F}^m \otimes \mathbb{F}^n. \tag{11.158}$$

Kronecker product matrices arise naturally in quantum physics, in which one constructs eigenfunctions of a Hamiltonian operator that acts on several different particles, or on different degrees of freedom of one particle, by taking tensor products of functions and vectors that belong to different vector spaces.

For example, if a particle has both orbital and spin angular momenta, the state vector belongs to the tensor product $\mathcal{V}_O \otimes \mathcal{V}_S$ of the space $\mathcal{V}_O$ of orbital angular momentum eigenfunctions and the space $\mathcal{V}_S$ of spin angular momentum eigenvectors. If $\mathcal{V}_O$ carries the representation $\Gamma^{(O)}$ of the rotation group $SO(3, \mathbb{R})$, and $\mathcal{V}_S$ carries the representation $\Gamma^{(S)}$ of $SO(3, \mathbb{R})$, then a spin-orbital state vector such as $\psi_j^{(O)} \otimes \psi_k^{(S)}$ transforms according to the Kronecker product representation $\Gamma^{(O)} \otimes \Gamma^{(S)}$:

$$P_R\left(\psi_j^{(O)} \otimes \psi_k^{(S)}\right) = \sum_{j=1}^{m} \sum_{l=1}^{n} \gamma_{j'j}^{(O)} \gamma_{k'k}^{(S)} \, \psi_{j'}^{(O)} \otimes \psi_{k'}^{(S)}. \tag{11.159}$$

In general, one obtains a Kronecker product representation whenever the same group operation is applied on all of the vector spaces that make up a tensor product.

It is convenient to write the Kronecker product of two representations as

$$\Gamma^{(O \otimes S)} := \Gamma^{(O)} \otimes \Gamma^{(S)}. \tag{11.160}$$

The corresponding notation for matrix elements is

$$\gamma_{(j',k'),(j,k)}^{(O \otimes S)} := \gamma_{j'j}^{(O)} \gamma_{k'k}^{(S)}. \tag{11.161}$$

To find the character of $\Gamma^{(O \otimes S)}$, one sets the row index (j', k') equal to the column index (j, k) and sums on the row index,

$$\begin{aligned} \chi^{(O \otimes S)}(\mathcal{C}_l) &:= \operatorname{trace}\left[\Gamma^{(O \otimes S)}(\mathcal{C}_l)\right] \\ &= \sum_{j=1}^{m} \sum_{k=1}^{n} \gamma_{(j,k),(j,k)}^{(O \otimes S)}(a) \\ &= \sum_{j=1}^{m} \sum_{k=1}^{n} \gamma_{jj}^{(O)}(a) \gamma_{kk}^{(S)}(a) \\ &= \chi^{(O)}(\mathcal{C}_l)\, \chi^{(S)}(\mathcal{C}_l), \end{aligned} \tag{11.162}$$

in which a is a group element in the class $\mathcal{C}_l$. Summarizing the result, one has

$$\boxed{\chi^{(O \otimes S)}(\mathcal{C}_l) = \chi^{(O)}(\mathcal{C}_l)\, \chi^{(S)}(\mathcal{C}_l).} \tag{11.163}$$

In other words, the character of a Kronecker product representation $\Gamma^{(O \otimes S)}$ is the product of the characters of the components $\Gamma^{(O)}$ and $\Gamma^{(S)}$.

For example, Table 11.5 gives the character of the representation $\Gamma^{(E \otimes E)} = \Gamma^{(E)} \otimes \Gamma^{(E)}$ of the group D_3. A glance at Table 11.1, which gives the irreducible characters of D_3, shows that $\Gamma^{(E \otimes E)}$ is reducible. A straightforward application of Eq. (11.146) yields the result that

TABLE 11.5 Character of the Reducible
Representation of D_3 Defined by the
Representation $\Gamma^{(E)} \otimes \Gamma^{(E)}$

	$|\mathcal{C}|, \mathcal{C}$		
D_3	E	$3C_2$	$2C_3$
$\chi^{(E \otimes E)}$	4	0	1

$\Gamma^{(E \otimes E)}$ reduces to the direct sum of all three of the irreducible representations of D_3:

$$\Gamma^{(E \otimes E)} \cong \Gamma^{(A_1)} \oplus \Gamma^{(A_2)} \oplus \Gamma^{(E)}. \tag{11.164}$$

11.3.8 EXERCISES FOR SECTION 11.3

11.3.1   Obtain the table of irreducible characters for the group D_4.

11.3.2   Obtain the table of irreducible characters for the group C_3.

   Hint: Some of the characters are complex.

11.3.3   Obtain the table of irreducible characters for the tetrahedral group T.

   Hint: There are twelve elements, arranged in the four classes E, $3C_2$, $4C_3$, and $4C_3'$. Some of the characters are complex.

11.3.4   Obtain the table of irreducible characters for the octahedral group O, as shown in Table 11.6

11.3.5   Verify Eq. (11.134) using the representation property, Eq. (11.18), and the multiplication table for the group D_3, Table 4.1

11.3.6   Obtain the character of the representation of the group D_4 found in Exercise 11.1.1, and find the irreducible components of this representation using Eq. (11.146).

TABLE 11.6 Characters of the Irreducible Representations
of the Octahedral Group O

	$|\mathcal{C}|, \mathcal{C}$				
O	E	$6C_2$	$3C_2 = C_4^2$	$8C_3$	$6C_4$
A_1	1	1	1	1	1
A_2	1	-1	1	1	-1
E	2	0	2	-1	0
F_1	3	-1	1	0	1
F_2	3	1	-1	0	-1

TABLE 11.7 Character of the Reducible Representation of D_3 Defined by the F_1 Representation of O

D_3	$\lvert \mathcal{C} \rvert, \mathcal{C}$		
	E	$3C_2$	$2C_3$
χ	3	-1	0

11.3.7 Find the irreducible components of the representation of D_3 whose character is given in Table 11.7. The result of this exercise gives the splitting of an F_1 energy level expected if the symmetry of an originally cubic crystal field is lowered to D_3 by stretching or compressing along one of the octahedron's three-fold axes.

11.3.8 Find the irreducible components of the representation of D_4 whose character is given in Table 11.8. The result of this exercise gives the splitting of an F_1 energy level expected if the symmetry of an originally cubic crystal field is lowered to D_4 by stretching or compressing along one of the octahedron's four-fold axes.

11.3.9 Derive Eq. (11.164).

11.4 TWO-DIMENSIONAL ROTATION GROUP

Certain important equations, such as Laplace's equation in two dimensions,

$$\left(\frac{\partial^2}{\partial x^2} + \frac{\partial^2}{\partial y^2} \right) \phi(\mathbf{r}) = 0, \tag{11.165}$$

are invariant under the group of rotations in two-dimensional Euclidean space, $SO(2)$ (see Exercise 11.4.1). In the equation above, the radius $r = (\mathbf{r} \cdot \mathbf{r})^{1/2}$ is invariant under any rotation $\mathbf{R}$,

$$r = (\mathbf{r} \cdot \mathbf{r})^{1/2}$$
$$= (\mathbf{R}\mathbf{r} \cdot \mathbf{R}\mathbf{r})^{1/2}, \tag{11.166}$$

because $\mathbf{R}$ is an orthogonal transformation on $\mathbb{E}^2$.

TABLE 11.8 Character of the Reducible Representation of D_4 Defined by the F_1 Representation of O

D_4	$\lvert \mathcal{C} \rvert, \mathcal{C}$				
	E	$C_2 = C_4^2$	$2C_4$	$2C_2'$	$2C_2''$
χ	3	-1	1	-1	-1

The usual realization of a rotation (proper orthogonal transformation) in two dimensions, $\mathbf{R}(\theta) : \mathbb{E}^2 \to \mathbb{E}^2$, is a 2×2 matrix

$$\mathbf{R}(\theta) = \begin{pmatrix} \cos\theta & -\sin\theta \\ \sin\theta & \cos\theta \end{pmatrix}. \tag{11.167}$$

According to Eq. (11.13), a rotation $\mathbf{R}(\theta)$ maps a function $f : \mathbb{E}^2 \to \mathbb{C}$ to another function $\mathsf{P}_{\mathbf{R}(\theta)} f$ such that

$$\left(\mathsf{P}_{\mathbf{R}(\theta)} f\right)(\mathbf{r}) = f(\mathbf{R}(\theta)^{-1}\mathbf{r}). \tag{11.168}$$

In terms of the plane polar coordinates

$$\rho = (x^2 + y^2)^{1/2}, \quad \phi = \tan^{-1}(y/x) \tag{11.169}$$

in which $\rho \in [0, \infty)$ and $\phi \in (-\pi, \pi]$, Eq. (11.168) reads

$$\left(\mathsf{P}_{\mathbf{R}(\theta)} f\right)(\rho, \phi) = f(\rho, \phi - \theta) \tag{11.170}$$

in view of Eq. (11.166). Therefore, in looking for functions that carry irreducible representations of $SO(2)$ one can safely ignore the dependence of g on ρ and consider only functions $f : (-\pi, \pi] \to \mathbb{C}$, for which

$$\left(\mathsf{P}_{\mathbf{R}(\theta)} f\right)(\phi) = f(\phi - \theta). \tag{11.171}$$

This definition realizes a rotation in $\mathbb{E}^2$ as a mapping of the circumference of the unit circle (realized as the interval $(-\pi, \pi]$) onto itself.

Because single-valuedness is part of the definition of a function (see Section 2.2.2), f must be single-valued. This requirement implies that the only acceptable definition of f outside the interval $(-\pi, \pi]$ is

$$\forall \phi \in (-\pi, \pi] : \forall n \in \mathbb{Z} : f(\phi + 2n\pi) := f(\phi). \tag{11.172}$$

In other words, f must be a periodic function of ϕ. One can say more formally that f must be invariant under the operations of the discrete translation group $\{\mathsf{T}_{2n\pi} \mid n \in \mathbb{Z}\}$; see Section 11.6.

11.4.1 REPRESENTATION SPACE FOR SO(2)

In order to characterize the Hilbert space $\mathcal{H}$ to which f should belong, we refer again to physics. Let $\psi : \mathbb{E}^2 \to \mathbb{C}$ be the wave function of a quantum-mechanical particle that is confined to move in two dimensions x, y. The canonical inner product in $\mathcal{H}$ is

$$\begin{aligned} \langle \psi, \psi' \rangle &= \int_{\mathbb{E}^2} \psi(\mathbf{r})^* \, \psi'(\mathbf{r}) \, d^2r \\ &= \int_0^{2\pi} \int_0^{\infty} \psi(\rho, \phi)^* \, \psi'(\rho, \phi) \, \rho \, d\rho \, d\phi. \end{aligned} \tag{11.173}$$

The most natural choice for the coordinate of a particle that is constrained to move on a circular ring is ϕ, the angle of rotation from a fixed reference direction. Therefore the appropriate inner product for the Hilbert space of a particle on a ring of unit radius follows from Eq. (11.173) by neglecting ρ:

$$\langle f, g \rangle = \frac{1}{2\pi} \int_{-\pi}^{\pi} f(\phi)^* g(\phi) \, d\phi. \tag{11.174}$$

It follows that the natural choice for the representation space of the two-dimensional rotation group is the Hilbert space $\mathcal{H} = \mathcal{L}^2((-\pi, \pi]; \, \mathbb{C})$.

Relative to the inner product Eq. (11.174), the representation

$$R(\theta) \mapsto P_{R(\theta)} \tag{11.175}$$

of $SO(2)$ is unitary, for in this case

$$\left(P_{R(\theta)} f, P_{R(\theta)} g \right) = \langle f, g \rangle. \tag{11.176}$$

The group density function that enters into the inner product Eq. (11.174) is $(2\pi)^{-1}$, which is trivially invariant under the mapping Eq. (11.171).

11.4.2 REPRESENTATIONS OF $SO(2)$

In Chapter 8 we showed that the Fourier functions

$$u_m(\phi) := \frac{1}{\sqrt{2\pi}} e^{im\phi} \tag{11.177}$$

are orthonormal on the unit circle under the inner product Eq. (11.174). In Chapter 10 we showed that the set $\{u_m\}$ is complete in the Hilbert space $\mathcal{H} = \mathcal{L}^2((-\pi, \pi]; \, \mathbb{C})$. If one takes $f = u_m$ in Eq. (11.171), then one has

$$\left(P_{R(\theta)} u_m \right)(\phi) = u_m(\phi - \theta) = e^{-im\theta} u_m(\phi). \tag{11.178}$$

Because

$$e^{-im\theta_1} e^{-im\theta_2} = e^{-im(\theta_1 + \theta_2)}, \tag{11.179}$$

and so forth, it follows that the mapping

$$R(\theta) \mapsto e^{-im\theta} \tag{11.180}$$

is a representation of $SO(2)$ and that the functions u_m carry the representation $e^{-im\theta}$. Clearly each of the representations $\{e^{-im\theta}\}$ is unitary. It can be shown that these are the only unitary irreducible representations of $SO(2)$.

The infinitesimal generator of rotations about the z-axis is the operator $-i\partial/\partial\phi$. The $SO(2)$-adapted function u_m is an angular-momentum eigenfunction with eigenvalue m, for

$$-i\frac{\partial}{\partial\phi}e^{im\phi} = m\,e^{im\phi}. \tag{11.181}$$

The $SO(2)$-adapted basis is also the angular-momentum basis.

Every square-integrable function $f \in \mathcal{H}$ possesses an angular-momentum–eigenfunction expansion

$$f(\phi) = \sum_{m=-\infty}^{\infty} c_m\,u_m(\phi), \tag{11.182}$$

in terms of which one can compute the result of applying the operator $\mathsf{P}_{\mathsf{R}(\theta)}$ as follows:

$$\left(\mathsf{P}_{\mathsf{R}(\theta)}f\right)(\phi) = \sum_{m=-\infty}^{\infty} e^{-im\theta}\,c_m\,u_m(\phi). \tag{11.183}$$

(We get $e^{-im\theta}$ because a rotation maps $e^{im\phi} \mapsto e^{im(\phi-\theta)}$.) Relative to the basis $\{u_m\}$, the representation $\mathsf{P}_{\mathsf{R}(\theta)}$ of $SO(2)$ therefore reduces to the direct sum of the representations $\Gamma_m(\theta) = e^{-im\theta}$. In other words, the matrix of $\mathsf{P}_{\mathsf{R}(\theta)}$ is equivalent to an infinite-dimensional diagonal matrix with elements $e^{-im\theta}$,

$$\mathsf{P}_{\mathsf{R}(\theta)} \cong \begin{pmatrix} \ddots & & & & \\ & e^{-im\theta} & & & \\ & & e^{-i(m-1)\theta} & & \\ & & & e^{-i(m-2)\theta} & \\ & & & & \ddots \end{pmatrix}. \tag{11.184}$$

This reduction occurs in the basis that carries the unitary irreducible representations of $SO(2)$, with respect to which the angular-momentum operator is diagonal.

11.4.3 COMPLETENESS RELATION FOR $\{e^{-im\theta}\}$

The mapping $m \mapsto e^{-im\theta}$ defines a one-to-one correspondence between the set of integers $\mathbb{Z}$ and the set of irreducible representations $\{e^{-im\theta}\}$ of $SO(2)$. The set of all such representations is, apart from the sign of the exponents and the trivial normalization factor $1/\sqrt{2\pi}$, identical with the set of Fourier functions, which one knows to be complete. Therefore the irreducible representations $\{e^{-im\phi} \mid m \in \mathbb{Z}\}$ of $SO(2)$ define a complete set, illustrating the following assertion: *Every classical special function can be expressed as a matrix element of an irreducible representation of some Lie group G. The set of matrix elements is complete in some infinite-dimensional space, and carries the irreducible representations of G.*

Completeness relations can usefully be expressed in terms of Dirac delta distributions (which are better known as delta "functions"). The completeness relation for the $SO(2)$-

adapted functions is

$$f(\theta) = \frac{1}{2\pi} \sum_{m=-\infty}^{\infty} \int_{-\pi}^{\pi} e^{im(\theta-\theta')} f(\theta') d\theta'$$

$$= \int_{-\pi}^{\pi} \left(\frac{1}{2\pi} \sum_{m=-\infty}^{\infty} e^{im(\theta-\theta')} \right) f(\theta') d\theta'.$$

(11.185)

The object set off in large parentheses in the second line of this equation is clearly not a function in the ordinary sense. In fact, it is the distribution

$$d(\theta - \theta') := \frac{1}{2\pi} \sum_{m=-\infty}^{\infty} e^{im(\theta-\theta')},$$

(11.186)

the action of which on a suitable test function f is defined by Eq. (11.185). If f belongs to $\mathcal{D}[-\pi, \pi]$, the space of infinitely differentiable functions of compact support in the interval $[-\pi, \pi]$, then by the definition of the delta distribution one has

$$\forall \theta, \theta' \in (-\pi, \pi) : f(\theta) = \int_{-\pi}^{\pi} \delta(\theta - \theta') f(\theta') d\theta'.$$

(11.187)

Therefore

$$d(\theta - \theta') = \delta(\theta - \theta')$$

(11.188)

for all $\theta, \theta' \in (-\pi, \pi)$.

Equation (11.188) is not correct as it stands if θ and θ' lie outside the interval $[-\pi, \pi]$. Let f belong to $\mathcal{S}(\mathbb{R})$, the space of rapidly decreasing functions on the real line. Because $e^{im(x+2n\pi-x')} = e^{im(x-x')}$ for any integer n, one has

$$\int_{-\infty}^{\infty} d(x + 2n\pi - x') f(x') dx' = \int_{-\infty}^{\infty} d(x - x') f(x') dx'$$

(11.189)

for every test function $f \in \mathcal{S}(\mathbb{R})$. Heuristically, therefore, $d(x - x')$ is periodic:

$$\forall n \in \mathbb{Z} : d(x + 2n\pi - x') = d(x - x').$$

(11.190)

It follows that $d(x - x')$ is equal to an infinite formal sum of delta functions,

$$\frac{1}{2\pi} \sum_{m=-\infty}^{\infty} e^{im(x-x')} = \sum_{n=-\infty}^{\infty} \delta(x + 2n\pi - x').$$

(11.191)

Both sides of this equation are well defined as distributions on the space of rapidly decreasing test functions (see Exercise 10.3.6), because the infinite formal sum results in a convergent series after multiplying by a test function and integrating. The right-hand side is sometimes called "Dirac's comb."

11.4.4 EXERCISE FOR SECTION 11.4

11.4.1   Demonstrate that the two-dimensional Laplacian,

$$\nabla^2 = \frac{\partial^2}{\partial x^2} + \frac{\partial^2}{\partial y^2},\tag{11.192}$$

is invariant in form under the two-dimensional rotation defined in Eq. (6.125). In other words, demonstrate that $\nabla'^2 = \nabla^2$.

11.5 SYMMETRY AND THE ONE-DIMENSIONAL WAVE EQUATION

11.5.1 BOUNDARY CONDITIONS AND SYMMETRY

When an ordinary or partial differential equation is invariant under the operations of a Lie group G, one says that the equation is **symmetric under G**. For example, we show that the Helmholtz equation in $\mathbb{E}^n$,

$$(\nabla^2 + k^2)f = 0,\tag{11.193}$$

is invariant under all of the operations of the group of translations and rotations in n-dimensional Euclidean space.

In order to describe a specific physical problem, one needs not only a partial differential equation to specify the dynamics of the system, but also a set of boundary conditions or initial conditions (or both) that ensure that the dynamical equation has one and only one solution. Boundary conditions often restrict the symmetry group of the physical problem to a subgroup of the full symmetry group of the differential equation. For example, if the problem is to solve for an acoustic wave in a cylindrical enclosure subject to the boundary condition that the acoustic displacement must vanish on the walls of the cylinder, then the problem is invariant under rotations about the axis of the cylinder but not under translations. The obvious symmetry group in this example is $SO(2)$. It comes as no surprise, then, that the amplitude of the acoustic wave can be expressed as a series in the functions that carry the irreducible representations of $SO(2)$. Not accidentally, these functions are the solutions that one obtains using the method of separation of variables (see Section 12.1.1).

However, the usefulness of the $SO(2)$-adapted functions $e^{im\phi}$ extends far beyond systems that obviously possess rotational symmetry. Any square-integrable function f that is periodic, such that

$$f(x + L) = f(x)\tag{11.194}$$

for all $x \in \mathbb{R}$ and for some fixed $L \in \mathbb{R}$, can be mapped to a function on the unit circle by the transformation

$$\phi = 2\pi(x - a)/L.\tag{11.195}$$

Therefore f possesses an expansion in terms of $\{e^{im\phi}\}$, that is, a Fourier series. We give now a classic example of a partial differential equation, the solution of which is periodic *if* certain boundary conditions are imposed.

11.5.2 WAVE EQUATION FOR A VIBRATING STRING

Let u be the x-component of the displacement of a string that is stretched along the z-axis. If the slope dx/dz of the displaced string is always small, then u obeys the **one-dimensional wave equation**

$$\left(\frac{\partial^2}{\partial z^2} - \frac{1}{c^2} \frac{\partial^2}{\partial t^2} \right) u(z, t) = 0. \tag{11.196}$$

The **phase velocity** of small-amplitude waves on the string is $c = [\tau_0/\lambda_0]^{1/2}$, in which τ_0 is the tension and λ_0 is the mass per unit length. Equation (11.196) applies to many other wave-propagation problems for which a one-dimensional model is useful, such as the propagation of light waves in free space or in a laser cavity.

11.5.3 BOUNDARY CONDITIONS FOR THE ONE-DIMENSIONAL WAVE EQUATION

We assume that the phase velocity c is independent of both z and t. It follows from this assumption and from the form of the wave equation (11.196) that if $u(z, t)$ is a solution, then so is $u(z - \zeta, t - \tau)$. In this sense the partial differential equation (11.196) is invariant under arbitrary translations in both space and time.

If the string is clamped at $z = 0$ and $z = L$, then the solution u must obey the homogeneous Dirichlet boundary conditions

$$\forall t \in \mathbb{R} : u(0, t) = u(L, t) = 0 \tag{11.197}$$

as well as an initial condition at time $t = 0$ that we specify in greater detail in Section 11.5.8. Although the Dirichlet boundary conditions in Eq. (11.197) spoil the translation invariance only in z, and not in t, we show that Eqs. (11.196) and (11.197) imply that u is periodic in both z and t. We show also for a string of infinite length for which no boundary conditions are imposed in z that u is generally not periodic and that two initial conditions are necessary in order to determine a unique solution.

11.5.4 FORM INVARIANCE OF THE WAVE EQUATION

It is helpful to state the invariance of an operator under coordinate transformations in abstract terms before considering the invariance of the wave equation under translations and Lorentz transformations. We are especially interested in the operator W such that

$$(Wf)(z, t) := \left(\frac{\partial^2}{\partial z^2} - \frac{1}{c^2} \frac{\partial^2}{\partial t^2} \right) f(z, t) \tag{11.198}$$

in which z and t are the coordinates of a vector $\mathbf{x} = z\mathbf{f}_1 + t\mathbf{f}_0$ in two-dimensional space-time $\mathbb{R}^2$, and $(\mathbf{f}_0, \mathbf{f}_1)$ is a basis of $\mathbb{R}^2$ such that $\langle \mathbf{f}_0, \mathbf{f}_1 \rangle = 0$, $\langle \mathbf{f}_0, \mathbf{f}_0 \rangle = c^2$, and $\langle \mathbf{f}_1, \mathbf{f}_1 \rangle = -1$. We call W the **wave operator** in two-dimensional space-time.

Let $\mathcal{S}$ and $\mathcal{T}$ be vector spaces whose elements are functions $f : \mathbb{R}^n \to \mathbb{C}$. We temporarily use f to denote an element of $\mathcal{S}$ and g to denote an element of $\mathcal{T}$. Let $A : \mathbb{R}^n \to \mathbb{R}^n$ be a linear mapping such that

$$\mathbf{x} \mapsto \mathbf{x}' = A\mathbf{x} \in \mathbb{R}^n. \tag{11.199}$$

The linear mapping A induces mappings $\mathsf{P_A} : \mathcal{S} \to \mathcal{S}$ and $\mathsf{P_A} : \mathcal{T} \to \mathcal{T}$ of functions $f : \mathbb{R}^n \to \mathbb{C}$ such that

$$(\mathsf{P_A} f)(\mathbf{x}) := f(A^{-1}\mathbf{x}). \tag{11.200}$$

See Section 4.3 and Eq. (4.177) to review the implications of this definition.

Now let $\mathsf{K} : \mathcal{S} \to \mathcal{T}$ be a linear mapping; we eventually set $\mathsf{K} = \mathsf{W}$. The operator on coordinate-transformed functions $\mathsf{P_A} f$ that produces the same result as K is $\mathsf{P_A K P_{A^{-1}}}$, as the following commutative diagram shows:

$$
\begin{array}{ccc}
\mathcal{S} & \xrightarrow{\ \mathsf{P_A K P_A^{-1}}\ } & \mathcal{T} \\
{\scriptstyle \mathsf{P_A}}\big\uparrow & & \big\uparrow{\scriptstyle \mathsf{P_A}} \\
\mathcal{S} & \xrightarrow[\ \mathsf{K}\ ]{} & \mathcal{T}
\end{array}
\tag{11.201}
$$

One says that the operator K is **form-invariant** or **covariant** under a coordinate transformation $A : \mathbf{x} \mapsto \mathbf{x}'$ of $\mathbb{R}^n$ if and only if

$$\mathsf{P_A K P_{A^{-1}}} = \mathsf{K}, \tag{11.202}$$

that is, if and only if the operator on coordinate-transformed functions is the *same* as the original operator. Equation (11.202) is equivalent to the requirement that K and $\mathsf{P_A}$ must commute,

$$[\mathsf{P_A}, \mathsf{K}] = 0. \tag{11.203}$$

In other words, a law of physics that is expressed in terms of an equation involving an operator K (such as $\mathsf{K} f = 0$) is the same in coordinate systems that are related by a linear mapping A if and only if $[\mathsf{P_A}, \mathsf{K}] = 0$.

The requirement for form invariance, Eq. (11.202), is equivalent to the requirement that

$$\mathsf{K P_{A^{-1}}} = \mathsf{P_{A^{-1}} K}. \tag{11.204}$$

Let us see how this equation is realized if

$$\mathsf{K} = \mathsf{W} \tag{11.205}$$

and

$$A : \mathbf{x} = (z, t)^T \mapsto \mathbf{x}' = (z', t')^T \tag{11.206}$$

is a nonsingular linear mapping of $\mathbb{R}^2$. To exhibit the operator $\mathsf{P_A}$, let $u : \mathbb{R}^2 \to \mathbb{C}$ be a smooth function, and let $v(z, t)$ be the value of u at the transformed coordinates z', t':

$$v(z, t) := u(z', t'). \tag{11.207}$$

Then

$$v = \mathsf{P_{A^{-1}}} u \quad \text{and} \quad u = \mathsf{P_A} v. \tag{11.208}$$

It follows that

$$\mathsf{W} \mathsf{P_{A^{-1}}} u = \mathsf{W} v = \left(\frac{\partial^2}{\partial z^2} - \frac{1}{c^2} \frac{\partial^2}{\partial t^2} \right) v. \tag{11.209}$$

Because $\mathsf{P_{A^{-1}}} f(\mathbf{x}) = f(\mathbf{Ax}) = f(\mathbf{x}')$ for any function f, one has

$$\mathsf{P_{A^{-1}}} \mathsf{W} u(z, t) = \mathsf{P_{A^{-1}}} \left(\frac{\partial^2}{\partial z^2} - \frac{1}{c^2} \frac{\partial^2}{\partial t^2} \right) u(z, t)$$

$$= \left(\frac{\partial^2}{\partial z'^2} - \frac{1}{c^2} \frac{\partial^2}{\partial t'^2} \right) u(z', t'). \tag{11.210}$$

By Eq. (11.204), the wave operator W is form-invariant under the coordinate transformation A if and only if the right-hand sides of Eqs. (11.209) and (11.211) are equal:

$$\left(\frac{\partial^2}{\partial z^2} - \frac{1}{c^2} \frac{\partial^2}{\partial t^2} \right) v(z, t) = \left(\frac{\partial^2}{\partial z'^2} - \frac{1}{c^2} \frac{\partial^2}{\partial t'^2} \right) u(z', t') \tag{11.211}$$

for all smooth functions $u : \mathbb{R}^2 \to \mathbb{R}$.

To obtain the conditions on A that follow from Eq. (11.211), it is necessary to transform partial derivatives with respect to z and t into partial derivatives with respect to z' and t'. To set up the calculation, let $u : \mathbb{R}^2 \to \mathbb{R}$ be a smooth function, and let $u_{,1}$ be the function obtained by differentiating u with respect to its first argument,

$$u_{,1}(z, t) := \frac{\partial u(z, t)}{\partial z}. \tag{11.212}$$

Likewise, let $u_{,12}$ be the mixed second partial derivative with respect to the first and second arguments of u,

$$u_{,12}(z, t) := \frac{\partial^2 u(z, t)}{\partial z \partial t} = \frac{\partial^2 u(z, t)}{\partial t \partial z}, \tag{11.213}$$

and so on. (The mixed second partial derivatives are equal because we have assumed that u is smooth.)

Using the chain rule, one finds, for example, that

$$\frac{\partial v}{\partial z} = u_{,1} \frac{\partial z'}{\partial z} + u_{,2} \frac{\partial t'}{\partial z}. \tag{11.214}$$

The linearity of A implies that the partial derivatives of the new coordinates with respect to the old, such as $\partial t'/\partial z$, are constants. Then (see Exercise 11.5.7)

$$
\begin{aligned}
Wv = u_{,11} \left(\frac{\partial z'}{\partial z}\right)^2 &+ u_{,22} \left(\frac{\partial t'}{\partial z}\right)^2 + 2u_{,12}\frac{\partial z' }{\partial z}\frac{\partial t'}{\partial z} \\
&- \frac{1}{c^2}\left[u_{,11}\left(\frac{\partial z'}{\partial t}\right)^2 + u_{,22}\left(\frac{\partial t'}{\partial t}\right)^2 + 2u_{,12}\frac{\partial z'}{\partial t}\frac{\partial t'}{\partial t}\right].
\end{aligned}
\tag{11.215}
$$

If the wave operator is form-invariant under the coordinate transformation A, then the right-hand side of Eq. (11.215) reduces to $Wu = u_{,11} - c^{-2}u_{,22}$, and conversely.

11.5.5 INVARIANCE OF THE WAVE EQUATION UNDER TRANSLATIONS

The space-time translation operator $T_{\mathbf{a}}$, in which $\mathbf{a} = \zeta\mathbf{f}_1 + \tau\mathbf{f}_0$, maps $\mathbf{x} = z\mathbf{f}_1 + t\mathbf{f}_0$ to

$$
T_{\mathbf{a}}\mathbf{x} = \mathbf{x} + \mathbf{a} = (z + \zeta)\mathbf{f}_1 + (t + \tau)\mathbf{f}_0.
\tag{11.216}
$$

The operator $P_{\mathbf{a}}$ maps functions according to the usual rule,

$$
(P_{\mathbf{a}}u)(\mathbf{x}) = u\left(T_{\mathbf{a}}^{-1}\mathbf{x}\right) = u(z - \zeta, t - \tau).
\tag{11.217}
$$

For this example, then, one has

$$
z' = z - \zeta, \quad t' = t - \tau
\tag{11.218}
$$

in Eq. (11.207). One sees immediately from Eq. (11.215) or Eqs. (11.250) through (11.252) (in Exercise 11.5.5) that the right-hand side of Eq. (11.215) reduces to $Wu(z', t')$, implying that Eq. (11.211) holds and therefore that the wave operator is form-invariant under translations.

11.5.6 INVARIANCE OF THE WAVE EQUATION UNDER LORENTZ TRANSFORMATIONS

Let Λ be a two-dimensional Lorentz transformation, that is, a linear mapping $\Lambda : \mathbb{R}^2 \to \mathbb{R}^2$ that is unitary with respect to the inner product: $\langle \Lambda\overline{x}, \Lambda\overline{y}\rangle = \langle\overline{x}, \overline{y}\rangle = c^2 x^0 y^0 - x^1 y^1$. The coordinates z', t' in Eq. (11.207) are

$$
\begin{aligned}
z' &= z \cosh\chi - ct \sinh\chi, \\
t' &= -\frac{z}{c}\sinh\chi + t\cosh\chi
\end{aligned}
\tag{11.219}
$$

where $\chi \in (-\infty, \infty)$. One finds for this example that Eqs. (11.250) and (11.251) reduce to the identity $\cosh^2\chi - \sinh^2\chi = 1$ and that Eq. (11.252) is satisfied identically (see Exercise 11.5.5). Therefore the wave equation (11.196) is invariant under Lorentz transformations.

A physically important consequence of the Lorentz invariance of Eq. (11.196) is that the phase velocity of the waves that that equation describes is the same in every set of coordinates z', t' that is related to the z, t coordinates by a Lorentz transformation.

11.5.7 D'ALEMBERT'S SOLUTION OF THE WAVE EQUATION

One way to solve Eqs. (11.196) and (11.197) is to transform them to new independent variables, in terms of which the solution is obvious. Let

$$\xi = z + ct, \quad \eta = z - ct. \tag{11.220}$$

The chain rule for partial differentiation implies that

$$\frac{\partial}{\partial z} = \frac{\partial}{\partial \xi} + \frac{\partial}{\partial \eta}, \quad \frac{1}{c}\frac{\partial}{\partial t} = \frac{\partial}{\partial \xi} - \frac{\partial}{\partial \eta}. \tag{11.221}$$

Then

$$\left(\frac{\partial^2}{\partial z^2} - \frac{1}{c^2}\frac{\partial^2}{\partial t^2}\right) u(z, t) = \left(\frac{\partial}{\partial z} - \frac{1}{c}\frac{\partial}{\partial t}\right)\left(\frac{\partial}{\partial z} + \frac{1}{c}\frac{\partial}{\partial t}\right) u(z, t), \tag{11.222}$$

which implies that

$$4\frac{\partial^2}{\partial \xi \partial \eta} w(\xi, \eta) = 0 \tag{11.223}$$

in which

$$w(\xi, \eta) := u(z, t). \tag{11.224}$$

Equation (11.223) is called the **canonical form** of the one-dimensional wave equation, and ξ, η are called the **canonical variables**.

One can see at once that the general solution of Eq. (11.223) is

$$w(\xi, \eta) = f(\eta) + g(\xi) \tag{11.225}$$

in which f and g are arbitrary continuously twice-differentiable functions. Therefore the general solution of Eq. (11.196) is

$$u(z, t) = f(z - ct) + g(z + ct). \tag{11.226}$$

To interpret this equation, let

$$u_+(z, t) = f(z - ct) \quad \text{and} \quad u_-(z, t) = g(z + ct). \tag{11.227}$$

The wave represented by u_+ travels in the $+z$ direction because u_+ takes on the same value, $f(\xi_0)$, for every space-time point (z, t) such that $z - ct = \xi_0$. The function u_+ obeys the "one-way wave equation"

$$\frac{\partial u_+}{\partial z} + \frac{1}{c}\frac{\partial u_+}{\partial t} = 0, \tag{11.228}$$

which describes a unidirectional wave traveling in the $+z$ direction. Similarly, the wave represented by u_- satisfies the equation

$$\frac{\partial u_-}{\partial z} - \frac{1}{c}\frac{\partial u_-}{\partial t} = 0,$$

(11.229)

and travels in the $-z$ direction.

The solution Eq. (11.226), and its derivation as we have given it, were first obtained by Jean Le Rond d'Alembert in 1746. The connection to the group-theoretical development of the preceding sections is that the monomials in ξ and η carry irreducible representations of the Lorentz group in two-dimensional space-time. Under the mapping $z \mapsto z', t \mapsto t'$ defined in Eq. (11.219), in fact,

$$\xi \mapsto \xi' = e^{\chi}\xi, \quad \eta \mapsto \eta' = e^{-\chi}\eta.$$

(11.230)

Therefore ξ^m and η^n carry the irreducible representations $e^{m\chi}$ and $e^{-n\chi}$, respectively. If one assumes that f and g are analytic functions, then Eq. (11.226) is simply an expansion of the solution of the wave equation in terms of functions that carry the irreducible one-dimensional representations of the Lorentz group in $\mathbb{R}^2$.

D'Alembert's solution makes no use of the boundary conditions Eq. (11.197) and is therefore valid for a string of arbitrary length. On a string of finite length, reflections at the ends couple the two functions f and g. We incorporate the boundary conditions after discussing the initial conditions that must be imposed in order to determine unique functions f and g for an infinitely long string.

11.5.8 SOLUTION FOR A STRING OF INFINITE LENGTH

For a string of infinite length two initial conditions are necessary in order to determine the two independent functions f and g in Eq. (11.226). Let

$$d(z) := u(z, 0), \quad s(z) := \frac{1}{c}\frac{\partial u}{\partial t}(z, t)\Big|_{t=0}.$$

(11.231)

The function d specifies the initial displacement profile of the wave, the function s specifies the x-component of the velocity of the string itself, expressed as a fraction of the phase velocity of waves on the string. It follows that

$$d(z) = f(z) + g(z)$$

(11.232)

and that

$$s(z) = -f'(z) + g'(z) \Rightarrow -f(z) + g(z) = c\int_0^z s(z')\, dz'.$$

(11.233)

From Eqs. (11.226) through (11.233) one finds that the general solution of the wave equation for an infinitely long string is

$$u(x, t) = \frac{1}{2}\left[d(z - ct) + d(z + ct) + c\int_{z-ct}^{z+ct} s(z')\, dz'\right].$$

(11.234)

Figure 11.1 illustrates the fact that the value of u at the spacetime point (z, t) depends on the

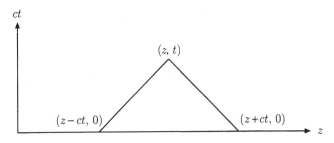

FIGURE 11.1

The domain of dependence for the wave equation on an infinitely long string.

values of d at the points $(z \pm ct, 0)$ and the values of $\partial u/\partial t$ on the interval $[z - ct, z + ct]$ along the z-axis. The straight lines with slopes ± 1 that intersect at (z, t) are examples of *characteristic curves* for the wave equation (11.196). The triangular spacetime region defined by the points (z, t) and $(z \pm ct, 0)$ is called the **domain of dependence**, because it can be shown that the value of u at (z, t) depends on $\partial u/\partial t$ along any curve that traverses this domain and intersects both characteristic curves.

Because of its very general form, the d'Alembert solution imposes no restrictions on the frequencies that may be present in a one-dimensional wave.

11.5.9 SOLUTION FOR A STRING OF FINITE LENGTH

We show that the boundary conditions for a finite string given in Eq. (11.197) imply that there is only one unknown function to be determined, instead of the two functions that appear in the d'Alembert solution Eq. (11.226), and that that function is periodic with a period equal to twice the length of the string. The boundary condition $u(0, t) = 0$ implies that

$$\forall t \in \mathbb{R} : f(0 - ct) + g(0 + ct) = 0. \tag{11.235}$$

Therefore g is simply f, reflected in the origin and reversed in sign:

$$\forall z \in \mathbb{R} : g(z) = -f(-z). \tag{11.236}$$

From a physical point of view it is obvious that the backward-traveling wave $g(z + ct)$ can exactly cancel the forward-traveling wave $f(z - ct)$ at $z = 0$ for all times t only if Eq. (11.236) is obeyed.

Equation (11.236) and the boundary condition $u(L, t) = 0$ imply that

$$\forall t \in \mathbb{R} : f(L - ct) - f(-L - ct) = 0. \tag{11.237}$$

After substituting $ct = -L - z$ one sees that f is periodic with period $2L$:

$$\forall z \in \mathbb{R} : f(z + 2L) = f(z). \tag{11.238}$$

Therefore f possesses a Fourier series on an interval of length $2L$:

$$f(z) = \sum_{m=-\infty}^{\infty} f_m e^{2\pi i m z/2L}. \tag{11.239}$$

From Eq. (11.236) it follows that

$$g(z) = -\sum_{m=-\infty}^{\infty} f_m e^{-im\pi z/L}. \tag{11.240}$$

Then Eq. (11.226) implies that

$$u(z, t) = \sum_{m=-\infty}^{\infty} f_m \left(e^{im\pi(z-ct)/L} - e^{-im\pi(z+ct)/L} \right)$$

$$= 2i \sum_{m=-\infty}^{\infty} f_m e^{-im\pi ct/L} \sin(m\pi z/L). \tag{11.241}$$

The sine function guarantees that u vanishes at $z = 0$ and $z = L$.

The condition that u must be real-valued, $u(z, t)^* = u(z, t)$, implies that

$$f_m^* = f_{-m}. \tag{11.242}$$

Then

$$u(z, t) = 2i \sum_{m=1}^{\infty} (f_m e^{-im\pi ct/L} - f_{-m} e^{im\pi ct/L}) \sin(m\pi z/L). \tag{11.243}$$

The sum runs from 1 to ∞ because $\sin(m\pi z/L) = 0$ if $m = 0$. Letting

$$f_m = |f_m| e^{-i\phi_m}, \tag{11.244}$$

one finds that the general solution to the wave equation for a string of length L is

$$u(z, t) = 4 \sum_{m=1}^{\infty} |f_m| \sin(m\pi z/L) \sin(m\pi ct/L + \phi_m). \tag{11.245}$$

The mutually orthogonal functions $\sin(m\pi z/L)$ are called the **normal modes** of the vibrating string. Each mode is a standing wave, that is, a wave such that the zeros (or nodes) occur at time-independent positions.

It is noteworthy that because the Dirichlet boundary conditions in Eq. (11.197) imply that f must be periodic, the only frequencies that can appear in the solution are the frequencies of the normal modes. In this example the normal-mode frequencies are integer multiples of the **fundamental frequency** $c/2L$. Different boundary conditions give different normal-mode frequencies. For example, the mixed boundary conditions

$$u(0, t) = 0, \quad \frac{\partial u}{\partial z}(L, t) = 0 \tag{11.246}$$

give normal-mode frequencies that are odd multiples of $c/4L$ (see Exercise 11.5.1).

The initial displacement of the string is

$$u(z, 0) = 4 \sum_{m=1}^{\infty} |f_m| \sin \phi_m \sin(m \pi z / L).$$

(11.247)

The coefficients of this Fourier sine series, $b_m = 4|f_m| \sin \phi_m$, can be determined uniquely by making use of the orthogonality of the sines of different multiples of the same angle (see Eq. [10.303]). Therefore specifying the initial displacement determines a unique solution of the wave equation for all future times on a string of finite length.

11.5.10 EXERCISES FOR SECTION 11.5

11.5.1 Solve the one-dimensional wave equation (11.196) given the mixed boundary conditions in Eq. (11.246) (i.e., a Dirichlet condition at $z = 0$ and a Neumann condition at $z = L$).

11.5.2 Solve the one-dimensional wave equation given an initial displacement

$$d(z) = d_0 e^{-z^2 / z_0^2}$$

(11.248)

and an initial speed $s(0) = 0$. Plot u versus z for times $t = 0$, $t = .5z_0/c$, $t = z_0/c$, $t = 2z_0/c$, and $t = 4z_0/c$.

11.5.3 Solve the one-dimensional wave equation (11.196) subject to the **periodic boundary conditions**

$$\forall x, t \in \mathbb{R} : u(x + L, t) = u(x, t),$$

(11.249)

in which the length L is a constant.

11.5.4 Prove that the boundary conditions given in Eq. (11.197) are equivalent to the boundary conditions given in Eq. (11.224), that w must vanish along the lines defined by the equations $\xi + \eta = 0$ and $\xi + \eta = 2L$, in which ξ and η are the canonical variables defined in Eq. (11.220).

11.5.5 Show from Eqs. (11.215) and (11.211) that the wave operator $W_{\bar{x}}$ is form-invariant under a coordinate transformation A if and only if

$$\left(\frac{\partial z'}{\partial z} \right)^2 - \frac{1}{c^2} \left(\frac{\partial z'}{\partial t} \right)^2 = 1,$$

(11.250)

$$\left(\frac{\partial t'}{\partial z} \right)^2 - \frac{1}{c^2} \left(\frac{\partial t'}{\partial t} \right)^2 = -\frac{1}{c^2},$$

(11.251)

and

$$\frac{\partial z'}{\partial z} \frac{\partial t'}{\partial z} - \frac{1}{c^2} \frac{\partial z'}{\partial t} \frac{\partial t'}{\partial t} = 0.$$

(11.252)

11.5.6   Verify that for the Lorentz transformation defined in Eq. (11.219), Eq. (11.250) reduces to the identity

$$\cosh^2 \chi - \sinh^2 \chi = 1,$$
(11.253)

and that Eqs. (11.251) and (11.252) are also satisfied identically.

11.5.7   Derive Eq. (11.215).

11.6 DISCRETE TRANSLATION GROUPS

11.6.1 MOTIVATION

The group of discrete translations in one dimension, $T(\mathbb{Z})$, is important because of the use, in digital communications and numerical computation, of uniform time sampling as in Eq. (6.223), and because of the discrete translation invariance of natural and artificial lattices. In this section we establish the basic properties of shift-invariant digital filters and of waves in periodic structures.

In digital signal processing and in finite-difference methods for differential equations, the times t_n at which a signal u is sampled are usually uniformly spaced, $t_n = t_0 + nh$, as are the points of the one-dimensional crystal lattice discussed in Section 4.2.4. Both the set of sampling times, $\{t_n\}$, and the points of a one-dimensional lattice are invariant under translations through integral multiples of the lattice period (or step size) h.

The transfer function of a digital filter, such as a finite-difference method for solving a differential equation, is the ratio of the output of the filter to the input at a single frequency. A shift-invariant digital filter can be characterized completely in terms of its transfer function. Analysis of the transfer functions is a powerful technique for understanding the global behavior of digital filters. For this reason, transfer functions are a most useful supplement, and even an alternative, to traditional, Taylor-series methods of analyzing the errors in computational methods.

11.6.2 INVARIANCE UNDER THE DISCRETE TRANSLATION GROUP

A typical element of $T(\mathbb{Z})$ is T_{n-m}, in which

$$\mathsf{T}_{n-m}t := t + (n - m)h$$
(11.254)

takes one from the sampling time (or lattice point) t_m to the sampling time (or lattice point) t_n:

$$\mathsf{T}_{n-m}t_m = t_n.$$
(11.255)

The set of all sequences of time-sampled values $u_n = u(t_n)$ of functions u is the vector space $\mathbb{F}^{\mathbb{Z}}$, the elements of which are

$$\bar{u} = (\ldots, u_{-m}, \ldots, u_0, u_1, \ldots, u_n, \ldots).$$
(11.256)

It is convenient to assume that u can be any continuous square-integrable function,

$$u \in \mathcal{C}(\mathbb{R}; \mathbb{F}) \cap \mathcal{L}^2(\mathbb{R}; \mathbb{F}). \tag{11.257}$$

Equation (4.85) defines the operator representation of $T(\mathbb{Z})$ that is carried by u:

$$(\mathsf{P}_m u)(t) = u\left(\mathsf{T}_m^{-1} t\right) = u(t - mh). \tag{11.258}$$

This property is significant both for sequences of discrete-time sampled values, and for waves in periodic structures.

The Hamiltonian operator H of an electron moving in a one-dimensional periodic potential is invariant under discrete translations, because both the kinetic energy operator $\mathsf{T} = -(\hbar^2/2m)d^2/dx^2$ and the potential energy function V are unchanged under discrete translations $x \mapsto x - mh$. Hence

$$\mathsf{H}\mathsf{P}_m = \mathsf{P}_m\mathsf{H}. \tag{11.259}$$

It follows that the eigenfunctions ψ_E of H belong to irreducible representations of $T(\mathbb{Z})$.

11.6.3 DISCRETE-SHIFT-INVARIANT DIGITAL FILTERS

Returning to signal processing, and specializing Eq. (11.258) to a sampling time $t = t_n$ and to $m = 1$, one sees that P_1 induces a right-shift operation on a sequence of sampled values u_n, because

$$\mathsf{P}_1 u(t_n) = u(t_n - h) = u(t_{n-1}) = u_{n-1}. \tag{11.260}$$

Therefore

$$\begin{aligned}
(\ldots, \mathsf{P}_1 u_{-m}, &\ldots, \mathsf{P}_1 u_0, \mathsf{P}_1 u_1, \ldots, \mathsf{P}_1 u_n, \ldots) \\
&= (\ldots, u_{-m-1}, \ldots, u_{-1}, u_0, \ldots, u_{n-1}, \ldots) \\
&= \mathsf{S}\bar{u}
\end{aligned} \tag{11.261}$$

in which the right-shift operator S is defined in Eq. (6.205).

A linear digital filter Φ is **shift-invariant** if and only if

$$\Phi\mathsf{S} = \mathsf{S}\Phi. \tag{11.262}$$

We show now that digital filters with constant coefficients, such as the nonrecursive filter $\bar{y} = \mathsf{C}\bar{u}$ defined in Eq. (6.224) or the recursive filter $\bar{y} = \mathsf{C}\bar{u} + \mathsf{D}\bar{y}$ defined in Eq. (6.236), are shift-invariant. It suffices to show that

$$\mathsf{C}\mathsf{S} = \mathsf{S}\mathsf{C} \quad \text{and} \quad \mathsf{D}\mathsf{S} = \mathsf{S}\mathsf{D} \tag{11.263}$$

for then one has the same relation between $S\bar{u}$ and $S\bar{y}$ as one has between $\bar{u}$ and $\bar{y}$:

$$S\bar{y} = C(S\bar{u}) \tag{11.264}$$

for a nonrecursive filter, or

$$S\bar{y} = C(S\bar{u}) + D(S\bar{y}) \tag{11.265}$$

for a recursive filter. To show that $SC\bar{u} = CS\bar{u}$, for example, one calculates the n-th terms of the sequences $SC\bar{u}$ and $CS\bar{u}$:

$$(S\bar{u})_n = u_{n-1} \Rightarrow (CS\bar{u})_n = \sum_{k=-N}^{N} c_k u_{n-k-1} = y_{n-1}$$

$$(SC\bar{u})_n = (S\bar{y})_n = y_{n-1}. \tag{11.266}$$

This establishes the shift invariance asserted in Eq. (11.263).

11.6.4 REPRESENTATIONS OF THE DISCRETE TRANSLATION GROUP

Because $T(\mathbb{Z})$ is Abelian, its irreducible representations are one-dimensional. The easy way to construct these representations is to find a family of carrier functions. If a function u carries an irreducible representation of $T(\mathbb{Z})$, then there exists a mapping $\gamma : \mathbb{Z} \to \mathbb{C}$ such that

$$P_n u = \gamma_n u. \tag{11.267}$$

The representation property $P_m P_n = P_{m+n}$ implies that

$$\gamma_m \gamma_n = \gamma_{m+n}, \tag{11.268}$$

which yields the two-term recurrence relation

$$\gamma_{m+1} = \gamma_1 \gamma_m. \tag{11.269}$$

Let

$$z := \gamma_1; \tag{11.270}$$

clearly $z \in \mathbb{C}$. Then

$$\gamma_m = z^m \Rightarrow P_m u(t) = z^m u(t). \tag{11.271}$$

For a *unitary* one-dimensional representation, one must have

$$|z| = 1 \Rightarrow z = e^{i\omega h} \tag{11.272}$$

in which $\omega \in \mathbb{R}$ *labels the representation*. Thus a function $u^{(\omega)}$ carries the irreducible representation of the discrete translation group $T(\mathbb{Z})$ labeled by the real number ω if and only if

$$P_m u^{(\omega)}(t) = u^{(\omega)}(t - mh) = e^{im\omega h} u^{(\omega)}(t). \qquad (11.273)$$

To complete the characterization of functions that carry irreducible representations of $T(\mathbb{Z})$, we observe that if one lets

$$\phi^{(\omega)}(t) := e^{i\omega t} u^{(\omega)}(t) \qquad (11.274)$$

then

$$u^{(\omega)}(t) = e^{-i\omega t} \phi^{(\omega)}(t), \qquad (11.275)$$

in which $\phi^{(\omega)}$ is periodic with the period of the lattice (or the sampling time interval), h:

$$\phi^{(\omega)}(t - mh) = \phi^{(\omega)}(t). \qquad (11.276)$$

Therefore every function that carries an irreducible representation of the one-dimensional discrete translation group $T(\mathbb{Z})$ is equal to the product of the exponential function $e^{-i\omega t}$ and a periodic function, $\phi^{(\omega)}$. This result is known in solid-state physics as **Bloch's theorem**.

11.6.5 DISCRETE-TIME TRANSFER FUNCTION

For both recursive and nonrecursive shift-invariant digital filters, we have shown that if the input sequence $\bar{u}$ carries the irreducible representation of $T(\mathbb{Z})$ that is labeled by ω, then so must the output sequence $\bar{y} = \Phi\bar{u}$. In this case, there exists a constant, which we write for consistency as $\psi^{(\omega)}(t_0)$, such that

$$y_m^{(\omega)} = \psi^{(\omega)}(t_0) e^{-im\omega h}. \qquad (11.277)$$

For discrete-time sequences of sampled values, the constants $\phi^{(\omega)}(t_0)$ and $\psi^{(\omega)}(t_0)$ are the amplitudes of the input and output signals, respectively.

The results obtained so far illustrate both the power and the limitations of a group-theoretical approach. Group theory tells one the general form of waves in a periodic structure, as in Eq. (11.275), but tells nothing about the periodic function $\phi^{(\omega)}$. The proportionality of u to the exponential $e^{-i\omega t}$ is enforced by symmetry, but $\phi^{(\omega)}$ is different for each specific wave-propagation problem.

For shift-invariant digital filters, an immediate consequence of Eq. (11.275) is the existence of a transfer function. Because $\phi^{(\omega)}$ is periodic,

$$u_m^{(\omega)} = u^{(\omega)}(t_m) = e^{-im\omega h} \phi^{(\omega)}(t_0). \tag{11.278}$$

Substituting this equation for $u_m^{(\omega)}$ into the definition of a shift-invariant, nonrecursive digital filter, Eq. (6.224), one obtains

$$
\begin{aligned}
y_m^{(\omega)} &= \psi^{(\omega)}(t_0) e^{-im\omega h} \\
&= \left(C \bar{u}^{(\omega)} \right)_m \\
&= \sum_{k=-N}^{N} c_k \phi^{(\omega)}(t_0) e^{-i(m-k)\omega h} \\
&= H(\omega h) \phi^{(\omega)}(t_0) e^{-im\omega h} \\
&= H(\omega h) u_m^{(\omega)},
\end{aligned}
\tag{11.279}
$$

in which

$$\boxed{H(\omega h) := \sum_{k=-N}^{N} c_k e^{ik\omega h}} \tag{11.280}$$

is called the **transfer function** of the nonrecursive digital filter. The result that

$$H(\omega h) = \frac{\psi^{(\omega)}(t_0)}{\phi^{(\omega)}(t_0)} \tag{11.281}$$

illustrates the more general definition of a transfer function as the ratio of the output amplitude to the input amplitude.

Because the transfer function H depends on the frequency ω only through the periodic function $e^{i\omega h}$, it follows that H is itself periodic. The period, $2\pi/h$, is equal to the Nyquist (or sampling) frequency ω_N defined in Eq. (8.360). Therefore one needs to calculate the transfer function on only one of its periods, say from $-\frac{1}{2}\omega_N$ to $\frac{1}{2}\omega_N$.

For example, the transfer function of the centered–moving-average filter defined in Eq. (6.229), for which $c_k = 1/(2N + 1)$, is

$$
\begin{aligned}
H(\omega h) &= \frac{1}{2N + 1} \sum_{k=-N}^{N} e^{ik\omega h} \\
&= \frac{1}{2N + 1} g_N(\omega h),
\end{aligned}
\tag{11.282}
$$

in which the grating function g_N is defined in Eq. (7.70). Figure 7.5 shows one period of this function, from $\omega h = 0$ to $\omega h = 2\pi$. Because the centered moving average is a symmetric filter with real coefficients, $H(-\omega h) = \overline{H(\omega h)}$; hence one really needs only the portion of Fig. 7.5 between 0 and π.

An ideal smoothing filter would pass low frequencies and reject high frequencies. Although Fig. 7.5 verifies that the centered moving average is a low-pass filter, the figure also reveals a tendency to introduce oscillations at the frequencies for which the grating function g_M has a local maximum, $\omega h = [(2l + 1)/(2N + 1)]\pi$. Thus an uncritical analysis of data that has been smoothed with a centered moving average might "discover" spurious oscillations, which in fact are no more than numerical artifacts. In general, one expects the transfer function to oscillate whenever the window is "hard," that is, if the coefficients c_k go from a large value to zero at $k = \pm N$.

For the shift-invariant recursive digital filter defined in Eq. (6.235), Eq. (11.278) implies that

$$
\begin{aligned}
y_m^{(\omega)} &= \left(C\overline{u}^{(\omega)}\right)_m + \left(D\overline{y}^{(\omega)}\right)_m \\
&= \sum_{k=-N}^{N} c_k \, \phi^{(\omega)}(t_0) e^{-i(m-k)\omega h} + \sum_{k=1}^{M} d_k \psi^{(\omega)}(t_0) e^{-i(m-k)\omega h} \\
&= u_m^{(\omega)} \sum_{k=-N}^{N} c_k \, e^{ik\omega h} + y_m^{(\omega)} \sum_{k=1}^{M} d_k \, e^{ik\omega h}.
\end{aligned}
\tag{11.283}
$$

Collecting the terms in $y_m^{(\omega)}$, one finds the transfer function, $\psi^{(\omega)}(t_0)/\phi^{(\omega)}(t_0)$, to be

$$
H(\omega h) = \frac{\displaystyle\sum_{k=-N}^{N} c_k \, e^{ik\omega h}}{1 - \displaystyle\sum_{k=1}^{M} d_k \, e^{ik\omega h}}.
\tag{11.284}
$$

For example, Euler's method for solving the ordinary differential equation $dy/dt = f$, Eq. (3.97), is a recursive digital filter with coefficients

$$
c_1 = h, \quad d_1 = 1.
\tag{11.285}
$$

Therefore the transfer function of Euler's method is

$$
\begin{aligned}
H_{\text{Euler}}(\omega h) &= \frac{h e^{e^{i\omega h}}}{1 - e^{i\omega h}} \\
&= \frac{h e^{e^{i\omega h/2}}}{e^{i\omega h/2} - e^{i\omega h/2}} \\
&= -\frac{h e^{e^{i\omega h/2}}}{2i \, \sin(\omega h/2)}.
\end{aligned}
\tag{11.286}
$$

Transfer-function ratio for Euler's method

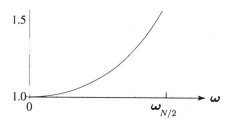

FIGURE 11.2

The ratio of the transfer function of Euler's method to the transfer function of an ideal integrator.

In order to interpret this equation, let us recall that Euler's method yields an approximate solution of the first-order differential equation $dy/dt = f(t)$, and that the transfer function gives the ratio of output $[y(t)]$ to input $[f(t)]$ if the input is $f(t) = e^{-i\omega t}$. The analytical solution of the differential equation $dy/dt = e^{-i\omega t}$ is the integral of the exponential,

$$y(t) = -\frac{e^{-i\omega t}}{i\omega}.$$ (11.287)

Then the transfer function of an ideal integrator is

$$H_{\text{ideal}}(\omega h) = \frac{y(t)}{f(t)} = -\frac{h}{i\omega h}.$$ (11.288)

To compare the frequency response of Euler's method with that of an ideal integrator, we form the **transfer-function ratio**

$$P(\omega h) := \frac{H_{\text{Euler}}(\omega h)}{H_{\text{ideal}}(\omega h)}$$

$$= \frac{\omega h}{2\sin(\omega h/2)}.$$ (11.289)

Figure 11.2 shows $P(\omega h)$ for $\omega h \in [0, \pi]$. By inspection,

$$\lim_{\omega h \to 0} P(\omega h) = 1.$$ (11.290)

Therefore Euler's method gives the correct frequency response in the limit of small step size, neglecting rounding errors (see Section 3.5.2). However, for step sizes or frequencies such that ωh is larger than roughly $\omega_N/6$, the Euler transfer function differs from the ideal transfer function by more than 10% – a large error.

11.6.6 EXERCISES FOR SECTION 11.6

11.6.1 Calculate and plot the transfer function of the straight-line differentiator defined in Eq. (6.230).

11.6.2 Calculate and plot the ratio of the transfer function of Simpson's rule, Eq. (3.82), to the transfer function of an ideal integrator. Note that the transfer-function ratio diverges as $\omega \to \omega_N/2$.

11.6.3 Calculate and plot the ratio of the transfer function for the midpoint method, Eq. (3.114), to the transfer function of an ideal integrator.

11.6.4 Calculate and plot the ratio of the transfer function for the backward Euler method, Eq. (3.126), to the transfer function of an ideal integrator.

11.6.5 Calculate and plot the ratio of the transfer function for the trapezoidal method, Eq. (3.131), to the transfer function of an ideal integrator.

11.6.6 Calculate and plot the ratio of the transfer function for the midpoint-trapezoidal method, Eqs. (3.136) and (3.137), to the transfer function of an ideal integrator.

11.7 CONTINUOUS TRANSLATION GROUPS

11.7.1 TRANSLATION GROUP OF THE REAL LINE

The most important symmetry group in engineering and classical physics may be the continuous translation group of the real line, $T(\mathbb{R})$, because many models of physical systems, and the equations that describe them, are invariant under arbitrary translations in time. We have more to say in Section 11.9 about these so-called continuous-shift-invariant systems.

The elements of $T(\mathbb{R})$ are the translations T_a such that

$$\forall x \in \mathbb{R} : \mathsf{T}_a x := x + a. \tag{11.291}$$

$T(\mathbb{R})$ is an infinite Abelian group. The mapping

$$\mathsf{T}_a \mapsto \mathsf{P}_a, \tag{11.292}$$

in which P_a is the operation of translating a function,

$$(\mathsf{P}_a f)(x) = f\left(\mathsf{T}_a^{-1} x\right) = f(x - a), \tag{11.293}$$

is an isomorphism and defines an operator representation of $T(\mathbb{R})$.

11.7.2 IRREDUCIBLE REPRESENTATIONS OF $T(\mathbb{R})$

Because $T(\mathbb{R})$ is an Abelian group, its irreducible representations are one-dimensional. It follows that if a function f carries an irreducible representation of $T(\mathbb{R})$, then for every $a \in \mathbb{R}$ there exists a complex number $\gamma(a)$ such that

$$(\mathsf{P}_a f)(x) = f(x - a) = \gamma(a) f(x). \tag{11.294}$$

The representation property

$$\mathsf{P}_{a+b} = \mathsf{P}_a \mathsf{P}_b \tag{11.295}$$

implies that

$$\gamma(a + b) = \gamma(a)\,\gamma(b). \tag{11.296}$$

We assume that f and the function $\gamma : \mathbb{R} \to \mathbb{C}$ are differentiable. Letting γ' represent the derivative of γ, one obtains

$$\gamma'(a + b) = \gamma(a)\,\gamma'(b) \tag{11.297}$$

on differentiating with respect to b. At the identity translation ($\mathsf{T}_b = \mathbf{1} \Rightarrow b = 0$) this becomes a differential equation for γ,

$$\gamma'(a) = \alpha\,\gamma(a), \tag{11.298}$$

in which

$$\alpha = \frac{d\gamma}{db}(b)\bigg|_{b=0}. \tag{11.299}$$

Evidently α may be any complex number. The irreducible representation determined by any given $\alpha \in \mathbb{C}$ is inequivalent to the irreducible representation determined by any $\alpha' \neq \alpha$.

Then the general form of an irreducible representation of $T(\mathbb{R})$ is

$$\gamma(a) = \kappa\,e^{\alpha a} \tag{11.300}$$

in which κ is a constant. The theorem that the homomorphic image of the group identity must be the identity element in the new group implies that

$$\gamma(0) = 1. \tag{11.301}$$

Then $\kappa = 1$.

If one replaces a with $-a$ in Eq. (11.294) and then evaluates both sides at $x = 0$, one sees that the general form of a function f that carries an irreducible representation of $T(\mathbb{R})$ is

$$f(a) = f(0)\,\gamma(-a) = f(0)\,e^{-\alpha x}. \tag{11.302}$$

Clearly $f(0)$ is an arbitrary constant that determines the normalization of the $T(\mathbb{R})$-adapted functions.

By definition, the unitary irreducible representations of $T(\mathbb{R})$ are such that

$$\gamma(a)^*\,\gamma(a) = 1. \tag{11.303}$$

Then an irreducible representation $\gamma(a) = e^{\alpha a}$ is unitary if and only if α is purely imaginary,

$$\alpha = -ik, \tag{11.304}$$

in which $k \in \mathbb{R}$. Therefore it is the functions

$$\psi_k(x) := \frac{1}{\sqrt{2\pi}} e^{ikx} \tag{11.305}$$

that carry the unitary irreducible representations of $T(\mathbb{R})$. The choice $f(0) = (2\pi)^{-1/2}$ that we have made in Eq. (11.305) is common in physics and engineering.

Two features distinguish the unitary irreducible representations e^{-ika} of $T(\mathbb{R})$ from the unitary irreducible representations $e^{-im\theta}$ of $SO(2)$. First, the index k that labels the irreducible representation $\gamma^{(k)}(a) = e^{-ika}$ is continuous, while the index m which labels the irreducible representations of $SO(2)$ is discrete. Second, the functions ψ_k are defined on the entire real line but are not square-integrable. Lighthill (1958) shows that the ψ_k belong to a larger vector space than $\mathcal{L}^2(\mathbb{R})$.

11.7.3 {ψ_k} AS MOMENTUM EIGENFUNCTIONS

By construction, every f that carries an irreducible representation of $T(\mathbb{R})$ is an eigenfunction of the momentum operator

$$\mathsf{p} := -i\frac{d}{dx}, \tag{11.306}$$

for one has $\mathsf{p}f = i\alpha f$. In particular, the functions $\{\psi_k\}$ defined in Eq. (11.305), which carry the unitary irreducible representations of $T(\mathbb{R})$, belong to real eigenvalues of p:

$$\mathsf{p}\psi_k = k\,\psi_k. \tag{11.307}$$

The momentum operator used in nonrelativistic quantum mechanics is equal to $\hbar\mathsf{p}$; we omit the factor $\hbar = h/2\pi$ because we are concerned here with purely geometrical properties and not with quantum dynamics.

It should come as no surprise that functions that are translation eigenfunctions are also momentum eigenfunctions; and converse also is true. Let $f \in \mathcal{C}^\infty(\mathbb{R}; \mathbb{C})$. By definition,

$$(\mathsf{P}_a f)(x) = f(x - a). \tag{11.308}$$

The Taylor series

$$f(x - a) = f(x) - af'(x) + \frac{1}{2!}a^2 f''(x) - \cdots \tag{11.309}$$

converges absolutely and uniformly to f in any finite interval. The series Eq. (11.309) can be written formally as

$$f(x - a) = \exp\left\{-a\frac{d}{dx}\right\} f(x)$$
$$= \exp\{-ia\mathsf{p}\}\, f(x), \tag{11.310}$$

in which the exponential is defined as the power series $e^x = 1 + x + x^2/2! + \cdots$. Therefore

$$P_a = \exp\{-iap\} \tag{11.311}$$

if P_a acts on infinitely differentiable functions.

It follows from Eq. (11.310) that an infinitely differentiable function that is a momentum eigenfunction is also a translation eigenfunction. Conversely, let f be an eigenfunction of P_a, and let

$$g(a) = f(x - a). \tag{11.312}$$

Then

$$\begin{aligned} g(a) &= P_a f(x) \\ &= \gamma(a) f(x). \end{aligned} \tag{11.313}$$

The functions g and γ are continuously differentiable. Hence

$$g'(0) = -ipf(x) = \gamma'(0)f(x), \tag{11.314}$$

which shows that f is an eigenfunction of p.

11.7.4 REPRESENTATION OF $T(\mathbb{E}^2)$ CARRIED BY ψ_k

We show that the plane wave

$$\forall \mathbf{r} \in \mathbb{E}^2 : \ \psi_k(\mathbf{r}) := \frac{1}{2\pi} e^{i\mathbf{k}\cdot\mathbf{r}} \tag{11.315}$$

carries an irreducible representation of the translation group $T(\mathbb{E}^2)$ and is an eigenfunction of the momentum operator $p = -i\nabla$. Let P_a be the mapping of the set of functions $f : \mathbb{E}^2 \to \mathbb{C}$ induced by a translation $T_a \in T(\mathbb{E}^2)$:

$$\begin{aligned} (P_a f)(\mathbf{r}) &:= f\left(T_a^{-1}\mathbf{r}\right) \\ &= f(\mathbf{r} - \mathbf{a}). \end{aligned} \tag{11.316}$$

Then

$$P_a \psi_k(\mathbf{r}) = \psi_k(\mathbf{r} - \mathbf{a}) = e^{-i\mathbf{k}\cdot\mathbf{a}}\psi_k(\mathbf{r}). \tag{11.317}$$

The mapping

$$T_a \mapsto e^{-i\mathbf{k}\cdot\mathbf{a}} \tag{11.318}$$

is a representation of the group $T(\mathbb{E}^2)$, because

$$T_{a_2} T_{a_1} \mapsto e^{-i\mathbf{k}\cdot\mathbf{a}_2} e^{-i\mathbf{k}\cdot\mathbf{a}_1} = e^{-i\mathbf{k}\cdot(\mathbf{a}_1 + \mathbf{a}_2)}, \tag{11.319}$$

and $e^{-i\mathbf{k}\cdot(\mathbf{a}_1+\mathbf{a}_2)}$ is the image of $T_{\mathbf{a}_1+\mathbf{a}_2} = T_{\mathbf{a}_2}T_{\mathbf{a}_1}$. This representation is one-dimensional and is therefore irreducible. The vector $\mathbf{k}$ labels the representation.

The property that $\psi_{\mathbf{k}}$ is an eigenfunction of the momentum operator p,

$$\mathsf{p}\psi_{\mathbf{k}} = \mathbf{k}\psi_{\mathbf{k}}, \tag{11.320}$$

which follows from differentiating Eq. (11.315), is another way of expressing the representation property that we have already expressed in Eq. (11.317). To demonstrate this assertion we derive Eq. (11.320) directly from Eq. (11.317). Certainly

$$\nabla_{\mathbf{a}}\psi_{\mathbf{k}}(\mathbf{r} - \mathbf{a}) = -\nabla_{\mathbf{r}}\psi_{\mathbf{k}}(\mathbf{r} - \mathbf{a}) \tag{11.321}$$

in which $\nabla_{\mathbf{a}}$ is the gradient operator with respect to the components of $\mathbf{a}$. But Eq. (11.317) implies that

$$\nabla_{\mathbf{a}}\psi_{\mathbf{k}}(\mathbf{r} - \mathbf{a}) = \nabla_{\mathbf{a}}e^{-i\mathbf{k}\cdot\mathbf{a}}\psi_{\mathbf{k}}(\mathbf{r}) = -i\mathbf{k}\psi_{\mathbf{k}}(\mathbf{r} - \mathbf{a}). \tag{11.322}$$

Multiplying this equation by i and evaluating at the identity translation ($\mathbf{a} = \mathbf{0}$) in order to obtain an element of the Lie algebra of $T(\mathbb{E}^2)$ give Eq. (11.317).

11.7.5 TRANSLATION GROUP OF EUCLIDEAN n-SPACE

In $\mathbb{R}^n$ the operation of translation through a vector $\mathbf{a}$,

$$\forall \mathbf{r} \in \mathbb{E}^n : T_{\mathbf{a}}\mathbf{r} := \mathbf{r} + \mathbf{a} \tag{11.323}$$

induces the transformation

$$(\mathsf{P}_{\mathbf{a}}f)(\mathbf{r}) = f(T_{\mathbf{a}}^{-1}\mathbf{r}) = f(\mathbf{r} - \mathbf{a}) \tag{11.324}$$

of functions $f : \mathbb{R}^n \to \mathbb{C}$. The collection $T(\mathbb{R}^n)$ of all the translations $T_{\mathbf{a}}$ is an Abelian group. Therefore each of its irreducible representations is one-dimensional. Because we have not yet defined an inner product on $\mathbb{R}^n$, this conclusion applies to spaces with a nontrivial signature, such as Minkowski space $\mathcal{M}$, as well as to n-dimensional Euclidean space $\mathbb{E}^n$.

We derive the irreducible representations of $T(\mathbb{E}^n)$ before considering more general cases such as $T(\mathcal{M})$. Every function f that carries an irreducible representation of $T(\mathbb{E}^n)$ must satisfy the equation

$$(\mathsf{P}_{\mathbf{a}}f)(\mathbf{r}) = f(\mathbf{r} - \mathbf{a}) = \gamma(\mathbf{a})\,f(\mathbf{r}). \tag{11.325}$$

As in the one-dimensional case, the representation property

$$\mathsf{P}_{\mathbf{a}+\mathbf{b}} = \mathsf{P}_{\mathbf{a}}\mathsf{P}_{\mathbf{b}} \tag{11.326}$$

implies that

$$\gamma(\mathbf{a} + \mathbf{b}) = \gamma(\mathbf{a})\,\gamma(\mathbf{b}). \tag{11.327}$$

If γ is differentiable, then

$$\nabla_b \gamma(a + b) = (\nabla_b \gamma(b)) \gamma(a) \tag{11.328}$$

in which ∇_b is the gradient operator with respect to the components of b. It follows that at $b = 0$,

$$\nabla_a \gamma(a) = \alpha \gamma(a) \tag{11.329}$$

in which

$$\alpha = \nabla_a \gamma(a)|_{a=0} \tag{11.330}$$

is a fixed vector in $\mathbb{C}^n$. The solution of the differential equation (11.329) subject to the initial condition $\gamma(0) = 1$,

$$\gamma(a) = e^{\alpha \cdot a}, \tag{11.331}$$

in which $a \cdot b$ is the canonical inner product of vectors in $\mathbb{E}^n$ defined in Eq. (8.15), gives the general form of an irreducible representation of $T(\mathbb{E}^n)$.

More generally, if one endows $\mathbb{R}^n$ with a non-positive-definite inner product such as the Minkowski inner product on $\mathbb{R}^4$, Eq. (11.331) holds if one replaces the Euclidean inner product with a more general inner product. For example, in Minkowski space the solution for γ is

$$\gamma(a) = \exp\{-\alpha^0 a^0 + \alpha^1 a^1 + \alpha^2 a^2 + \alpha^3 a^3\}. \tag{11.332}$$

In order for the irreducible representation $\gamma(a) = e^{\langle \alpha, a \rangle}$ to be unitary, $\gamma(a)$ must be a complex number of modulus one. Then α must be equal to i times a vector $-k \in \mathbb{E}^n$. In Euclidean n-space, this implies that γ is a plane wave,

$$\gamma(a) = e^{-ik \cdot a}, \tag{11.333}$$

which is a solution of the n-dimensional Helmholtz equation

$$(\nabla_a^2 + k \cdot k)\gamma = 0. \tag{11.334}$$

In Minkowski space, one has

$$\gamma(a) = e^{-i(k^0 a^0 - k^1 a^1 - k^2 a^2 - k^3 a^3)}, \tag{11.335}$$

which is a solution of the wave equation in three space dimensions and one time dimension,

$$\left(\frac{\partial^2}{\partial (a^0)^2} - \frac{\partial^2}{\partial (a^1)^2} - \frac{\partial^2}{\partial (a^2)^2} - \frac{\partial^2}{\partial (a^3)^2} \right) \gamma = 0. \tag{11.336}$$

One can find a function that carries the irreducible representation $e^{\langle \alpha, a \rangle}$ by replacing a

with $-\mathbf{a}$ in Eq. (11.325) and then evaluating at $\mathbf{r} = \mathbf{0}$:

$$f(\mathbf{a}) = f(\mathbf{0})\,\gamma(-\mathbf{a}) = f(\mathbf{0})\,e^{\alpha \cdot \mathbf{a}}. \tag{11.337}$$

The choice $f(\mathbf{0}) = (2\pi)^{-n/2}$ ensures that the resulting functions are delta-orthonormalized; see Eq. (11.366) below. Thus the functions

$$\forall \mathbf{r} \in \mathbb{E}^n : \psi_{\mathbf{k}}(\mathbf{r}) := \frac{1}{(2\pi)^{n/2}} e^{i \mathbf{k} \cdot \mathbf{r}} \tag{11.338}$$

carry the unitary irreducible representations

$$\gamma^{(\mathbf{k})}(\mathbf{a}) = e^{-i \mathbf{k} \cdot \mathbf{a}} \tag{11.339}$$

of $T(\mathbb{E}^n)$.

Equations (11.329) and (11.337) imply that every function f that carries an irreducible representation of $T(\mathbb{E}^n)$ is an eigenfunction of the momentum operator

$$\mathbf{p} := -i \nabla_{\mathbf{r}}. \tag{11.340}$$

The plane wave $\psi_{\mathbf{k}}$ is a momentum eigenfunction with "eigenvalue" $\mathbf{k}$:

$$\mathbf{p}\,\psi_{\mathbf{k}} = \mathbf{k}\,\psi_{\mathbf{k}}. \tag{11.341}$$

This equation means, of course, that $-i \partial \psi_{\mathbf{k}} / \partial x^j = k^j \psi_{\mathbf{k}}$, in which k^j is the j-th Cartesian component of $\mathbf{k}$.

Also as in the one-dimensional case, one can write the translation of an infinitely differentiable function,

$$(\mathsf{P}_{\mathbf{a}} f)(\mathbf{r}) = f(\mathbf{r} - \mathbf{a}), \tag{11.342}$$

in terms of the Taylor series

$$f(\mathbf{r} - \mathbf{a}) = f(\mathbf{r}) - \mathbf{a} \cdot \nabla_{\mathbf{r}} f(\mathbf{r}) + \frac{1}{2!}(\mathbf{a} \cdot \nabla_{\mathbf{r}})^2 f(\mathbf{r}) - \cdots, \tag{11.343}$$

which is equivalent to the exponential relation

$$\begin{aligned} f(\mathbf{r} - \mathbf{a}) &= \exp\{-\mathbf{a} \cdot \nabla_{\mathbf{r}}\} f(\mathbf{r}) \\ &= \exp\{-i \mathbf{a} \cdot \mathbf{p}\} f(\mathbf{r}). \end{aligned} \tag{11.344}$$

It follows that an infinitely differentiable function that is an eigenfunction of the translation operator is also an eigenfunction of the momentum operator, and that the converse is true.

11.7.6 EXERCISES FOR SECTION 11.7

11.7.1   Prove that the mapping

$$T_a \mapsto e^{-ika} \tag{11.345}$$

is a representation of $T(\mathbb{R})$.

11.7.2   The goal of this problem is to establish the form of the irreducible representations of $T(\mathbb{R})$ assuming only continuity, not differentiability.

(a) Using the representation property Eq. (11.296), prove that for every $\rho \in \mathbb{Q}$,

$$\gamma(\rho) = \exp[\rho \ln \gamma(1)] = [\gamma(1)]^\rho. \tag{11.346}$$

Hint: Let $\sigma(a) = \ln[\gamma(a)]$ and prove that $\sigma(m/n) = (m/n)\sigma(1)$.

(b) Assume that γ is continuous and prove that for every $a \in \mathbb{R}$,

$$\gamma(a) = \exp[a \ln \gamma(1)]. \tag{11.347}$$

11.8   FOURIER TRANSFORMS

The purpose of discussing Fourier transforms here is to point out their fundamental relation to the translation group of $\mathbb{E}^n$. For a more complete discussion of Fourier transforms, see Lighthill (1958).

11.8.1   FOURIER TRANSFORM IN ONE DIMENSION

We have already shown that the $SO(2)$-adapted functions

$$u_m(\phi) = (2\pi)^{-1/2} e^{im\phi} \tag{11.348}$$

are a complete orthonormal set, and that function $f \in \mathcal{L}^2((-\pi, \pi]; \mathbb{C})$ that is square-integrable on the finite interval $(-\pi, \pi]$ possesses a unique expansion

$$f = \sum_{m=-\infty}^{\infty} f_m u_m, \tag{11.349}$$

in which m labels an irreducible representation of $SO(2)$, and the expansion (Fourier) coefficients are

$$f_m = \frac{1}{\sqrt{2\pi}} \int_{-\pi}^{\pi} u_m(\phi)^* f(\phi) \, d\phi. \tag{11.350}$$

Lighthill (1958) shows that every rapidly decreasing function $f \in \mathcal{S}$ possesses a unique expansion in terms of the functions

$$\psi_k(x) = \frac{1}{\sqrt{2\pi}} e^{ikx}, \tag{11.351}$$

namely,

$$f(x) = \int_{-\infty}^{\infty} \psi_k(x)\, \tilde{f}(k)\, dk = \frac{1}{\sqrt{2\pi}} \int_{-\infty}^{\infty} \tilde{f}(k)\, e^{ikx}\, dk. \tag{11.352}$$

The function $\tilde{f}$, which is called the **Fourier transform** of f, is the analog of the Fourier coefficient f_m. The Fourier transform, Eq. (11.352), is the analog of the Fourier series for rapidly decreasing functions, which are defined on the entire real line, not merely on a finite interval. The Fourier transform $\tilde{f}$ is given by the **Fourier inversion formula**

$$\tilde{f}(k) = \frac{1}{\sqrt{2\pi}} \int_{-\infty}^{\infty} f(x)\, e^{-ikx}\, dx, \tag{11.353}$$

which is the analog of the formula for the Fourier coefficient f_m.

For example, the Fourier transform of a Gaussian is a Gaussian:

$$f(x) = e^{-x^2/(2\sigma^2)} \Rightarrow \tilde{f} = \sigma\, e^{-\sigma^2 k^2/2} \tag{11.354}$$

(see Lighthill [1958]). We refer to the vector space of Fourier transforms $\tilde{f}$, which is isomorphic to $\mathcal{S}$, as **Fourier-transform space**. The Fourier-transform variable k has the physical significance that

$$k = 2\pi/\lambda, \tag{11.355}$$

in which λ is the wavelength of the wave that e^{ikx} represents.

Because the real and imaginary parts of the function $e^{i(kx-\omega t)}$ represent waves that travel in the $+x$ direction, physicists often reverse the signs in the exponents of Eqs. (11.356) and (11.357) if the argument of f is interpreted as time rather than position:

$$f(t) = \frac{1}{\sqrt{2\pi}} \int_{-\infty}^{\infty} \tilde{f}(\omega)\, e^{-i\omega t}\, d\omega, \tag{11.356}$$

$$\tilde{f}(\omega) = \frac{1}{\sqrt{2\pi}} \int_{-\infty}^{\infty} f(t)\, e^{i\omega t}\, dt. \tag{11.357}$$

In other words, one expands in terms of the functions ψ_ω^*, in which $\psi_\omega^*(t) = (2\pi)^{-1/2} e^{-i\omega t}$. In this case the Fourier-transform variable ω, which also labels the irreducible representation that is carried by ψ_ω^*, is a temporal frequency.

Lighthill (1958) establishes that the Fourier transform $\tilde{f}$ of a rapidly decreasing function f is also rapidly decreasing and is infinitely differentiable. In fact, $\tilde{f}$ is analytic for complex arguments ω with positive imaginary part, $\text{Im}[\omega] > 0$ (i.e., the upper half-plane), and perhaps in some strip $\text{Im}[\omega] > -|c|$ below the real axis.

11.8.2 COMPLETENESS RELATION FOR THE $\{\psi_k\}$

By the definition of the Dirac delta functional,

$$\forall f \in \mathcal{S}(\mathbb{R}) : \forall x \in \mathbb{R} : f(x) = \int_{-\infty}^{\infty} \delta(x - x')\, f(x')\, dx'. \tag{11.358}$$

Therefore the formula

$$\forall f \in \mathcal{S}(\mathbb{R}): \ \forall x \in \mathbb{R}: \ f(x) = \frac{1}{2\pi} \int_{-\infty}^{\infty} \int_{-\infty}^{\infty} e^{ik(x-x')} f(x') \, dk \, dx' \tag{11.359}$$

(which follows from Eqs. [11.352] and [11.353]) becomes

$$\frac{1}{2\pi} \int_{-\infty}^{\infty} e^{ik(x-x')} \, dk = \delta(x - x'). \tag{11.360}$$

This is the completeness relation for the functions $\{\psi_k\}$ that carry the unitary irreducible representations of $T(\mathbb{R})$. Note that unlike the completeness relation for the functions $\{u_m\}$ that carry the unitary irreducible representations of $SO(2)$, the completeness relation for the $\{\psi_k\}$ is an integral, not a sum, because for each $k \in \mathbb{R}$ there is an irreducible representation of $T(\mathbb{R})$.

If one replaces the variables x and x' in Eq. (11.360) with k and k' and replaces k with x, one obtains the equivalent formula

$$\frac{1}{2\pi} \int_{-\infty}^{\infty} e^{i(k-k')x} \, dx = \delta(k - k'). \tag{11.361}$$

The left-hand side of this equation is equal to the formal inner product $\langle \psi_{k'}, \psi_k \rangle$. Obviously $\psi_{k'}$ and ψ_k are "orthogonal" if $k' \neq k$. In such a case one speaks loosely of "delta-function orthonormalization," even though the integral (11.361) has no meaning outside of the context of distribution theory.

11.8.3 FOURIER TRANSFORMS IN n-DIMENSIONAL EUCLIDEAN SPACE

The symmetry-adapted functions for the translation group in n-dimensional Euclidean space are the plane waves $e^{i\langle \mathbf{k}, \mathbf{r} \rangle}$. The Fourier transform $\tilde{f}$ of a rapidly decreasing function $f : \mathbb{E}^n \to \mathbb{C}$ is defined as

$$\tilde{f}(\mathbf{k}) := \frac{1}{(2\pi)^{n/2}} \int_{\mathbb{E}^n} e^{-i\langle \mathbf{k}, \mathbf{r} \rangle} f(\mathbf{r}) \, d^n r. \tag{11.362}$$

The inner product in this equation is the canonical inner product in $\mathbb{E}^n$. (For the translation group in Minkowski space, one must use the Minkowski inner product in Eq. [11.362].)

In $\mathbb{E}^n$ the Fourier inversion theorem reads

$$f(\mathbf{r}) = \frac{1}{(2\pi)^{n/2}} \int_{\mathbb{E}^n} e^{i\langle \mathbf{k}, \mathbf{r} \rangle} \tilde{f}(\mathbf{k}) \, d^n k. \tag{11.363}$$

Then the completeness relation for the $T(\mathbb{E}^n)$-adapted functions $\{\psi_\mathbf{k}\}$ is

$$f(\mathbf{r}) = \frac{1}{(2\pi)^n} \int \int_{\mathbb{E}^n} e^{i\langle \mathbf{k}, \mathbf{r} - \mathbf{r}' \rangle} f(\mathbf{r}') \, d^n r' \, d^n k. \tag{11.364}$$

In terms of the delta distribution in n dimensions, the completeness relation reads

$$\frac{1}{(2\pi)^n} \int_{\mathbb{E}^n} e^{i(\mathbf{k},\mathbf{r}-\mathbf{r}')} d^n k = \delta(\mathbf{r} - \mathbf{r}') \tag{11.365}$$

in the sense that Eq. (11.364) holds for all rapidly decreasing functions on $\mathbb{E}^n$.

As for one-dimensional Fourier transforms, Eq. (11.365) can be reexpressed as the completeness relation in Fourier-transform space

$$\frac{1}{(2\pi)^n} \int_{\mathbb{E}^n} e^{i(\mathbf{k}-\mathbf{k}',\mathbf{r})} d^n r = \delta(\mathbf{k} - \mathbf{k}'). \tag{11.366}$$

The left-hand side of this equation has the form of an inner product $\langle \psi_{\mathbf{k}'}, \psi_{\mathbf{k}} \rangle$. As in one dimension, one often says that the functions $\{\psi_{\mathbf{k}}\}$ that carry the unitary irreducible representations of $T(\mathbb{E}^n)$ are "delta-orthonormalized," with the understanding that both sides of Eq. (11.366) are distributions.

11.8.4 POISSON SUM FORMULA

Equation (11.191) for "Dirac's comb" implies a formula that is sometimes useful for the purpose of transforming a slowly-convergent series into a rapidly convergent series. Multiplying both sides of Eq. (11.191) by $f \in \mathcal{S}(\mathbb{R})$ and integrating, one obtains the **Poisson sum formula**

$$\frac{1}{\sqrt{2\pi}} \sum_{m=-\infty}^{\infty} \tilde{f}(m) = \sum_{n=-\infty}^{\infty} f(2n\pi). \tag{11.367}$$

See Jackson (1975) for the application of this formula to the computation of the spectrum of synchrotron radiation.

11.8.5 EXERCISES FOR SECTION 11.8

11.8.1 Assume that if $f \in \mathcal{S}(\mathbb{R})$, then $\tilde{f} \in \mathcal{S}(\mathbb{R})$. Prove that the series

$$\sum_{m=-\infty}^{\infty} \tilde{f}(m) \tag{11.368}$$

converges absolutely.

11.8.2 Show that

$$\frac{1}{\sigma\sqrt{2\pi}} \sum_{n=-\infty}^{\infty} e^{-n^2/(2\sigma^2)} = \sum_{m=-\infty}^{\infty} e^{-2\pi^2\sigma^2 m^2}. \tag{11.369}$$

Comment on the rates of convergence of the two sides of this equation if $\sigma \gg 1$.

11.9　LINEAR, SHIFT-INVARIANT SYSTEMS

11.9.1　CONTINUOUS-TIME-SHIFT INVARIANCE

An operator A is called **continous-time-shift invariant** if and only if

$$\forall \tau \in \mathbb{R} : P_\tau A = A P_\tau \tag{11.370}$$

in which P_τ is the operation of translating a function in time,

$$(P_\tau f)(t) = f(t - \tau). \tag{11.371}$$

For example, the operator d^2/dt^2 is time-shift–invariant on functions $f \in \mathcal{C}^2(\mathbb{R}; \mathbb{R})$, because $(P_\tau d^2 f/dt^2)(t) = (d^2/dt^2)(P_\tau f)(t) = f''(t - \tau)$.

Time-shift invariance is important in physics and engineering because many of the operators that enter into basic laws are time-shift invariant. Newton's second law, $\mathbf{f} = d\mathbf{p}/dt$, is time-shift–invariant if the force $\mathbf{f}$ is a constant vector, because d/dt is a time-shift–invariant operator. The time-dependent Schrödinger equation, $H\psi = i\hbar \partial \psi/\partial t$, is time-shift–invariant if the Hamiltonian operator H commutes with the time-translation operator,

$$[H, P_\tau] = H P_\tau - P_\tau H = \mathbf{0}. \tag{11.372}$$

Engineering systems such as a linear circuit in which the components have time-independent values of resistance, capacitance, or inductance are also described in terms of time-shift–invariant operators.

In the two-dimensional rotation group $SO(2)$, the operator $P_{R(\theta)}$ that implements a finite rotation, $(P_{R(\theta)} f)(\phi) = f(\phi - \theta)$, is diagonal relative to the symmetry-adapted basis u_m according to Eq. (11.178). In the one-dimensional translation group, it is still the case that the operator P_τ, which implements a finite translation, is diagonal in Fourier-transform space,

$$(P_\tau f)(t) = f(t - \tau) = \int_{-\infty}^{\infty} e^{i\omega\tau}\, \tilde{f}(\omega)\, e^{-i\omega t}\, d\omega, \tag{11.373}$$

even though the $T(\mathbb{R})$-adapted functions ψ_ω are not a basis of $\mathcal{L}^2(\mathbb{R})$. In the language of group theory, the representation P_τ of the one-dimensional translation group $T(\mathbb{R})$ reduces in Fourier-transform space to a sum (or an integral, in this case) over the irreducible representations $e^{i\omega\tau}$.

If one interprets the independent variable as a spatial coordinate, not as time, then the function $\psi_k(x) = (2\pi)^{-1/2} e^{ikx}$ is a momentum eigenfunction: $p\psi_k = -i d\psi/dx = k\,\psi_k$. Thus the Fourier-integral representation of f, Eq. (11.352), expresses f as a superposition of momentum eigenfunctions, which also carry irreducible representations of the translation group $T(\mathbb{R})$.

Suppose now that a linear operator A is time-shift–invariant and that $f : \mathbb{R} \to \mathbb{C}$ is a function such that

$$P_\tau f = \chi f \tag{11.374}$$

in which $\chi \in \mathbb{C}$. (One knows, of course, that if $|\chi| = 1$, then $\chi = e^{i\omega\tau}$ for some ω.) It follows that

$$AP_\tau f = A(\chi f) = \chi Af = P_\tau Af, \tag{11.375}$$

that is, that Af is an eigenfunction of P_τ with the same eigenvalue χ. If the eigenvalue χ is nondegenerate, then there exists a number $\lambda \in \mathbb{C}$ such that

$$Af = \lambda f. \tag{11.376}$$

That is, f is also an eigenfunction of A. One expects, therefore, that time-shift–invariant operators will be "diagonal" in Fourier-transform space. We devote the next several paragraphs to an especially important example.

11.9.2 CONTINUOUS-TIME TRANSFER FUNCTION

In many models of physical and engineering systems one encounters an equation of the form

$$A\bar{x} = \bar{b} \tag{11.377}$$

in which $\bar{x}$ and $\bar{b}$ are functions that depend upon time, and in which the differential or integral operator A is time-shift–invariant. In finite-dimensional models A is realized by a matrix. An integral operator of the form

$$(Af)(t) = \int_{-\infty}^{\infty} \alpha(t, t') \, f(t') \, dt' \tag{11.378}$$

implements the infinite-dimensional analog of a finite-dimensional matrix-vector product, $x'^j = \sum_{k=1}^{n} a_k^j x^k$. The "matrix element" function α is called the **kernel**. We assume for the purposes of this discussion that every f belongs to S, the space of rapidly decreasing real-valued functions, and that the kernel α is absolutely integrable.

If A is time-shift–invariant, then

$$
\begin{aligned}
(AP_\tau f)(t) &= \int_{-\infty}^{\infty} \alpha(t, t') \, f(t' - \tau) \, dt' \\
&= \int_{-\infty}^{\infty} \alpha(t, t' + \tau) \, f(t') \, dt' \\
&= (P_\tau Af)(t) = \int_{-\infty}^{\infty} \alpha(t - \tau, t') \, f(t') \, dt'
\end{aligned} \tag{11.379}
$$

for all $t, \tau \in \mathbb{R}$. Then the kernel is a function of $t - t'$ alone, for one has (for all t, t', $\tau \in \mathbb{R}$)

$$\alpha(t, t' + \tau) = \alpha(t - \tau, t') \Rightarrow \alpha(t, \tau) = \alpha(t - \tau, 0). \tag{11.380}$$

Therefore the general form of a linear, time-shift–invariant integral operator A is

$$g(t) = (Af)(t) = \int_{-\infty}^{\infty} h(t - t') \, f(t') \, dt'. \tag{11.381}$$

This equation can be used to describe the output $g(t)$ of a linear, time-shift–invariant system (such as an ideal dielectric) given an external time-varying input $f(t)$ (such as an electric field). If $f(t') = \delta(t' - t_0)$, then $g(t) = h(t - t_0)$; thus h is the **impulse response**.

As Eq. (11.376) implies, the input-output relation Eq. (11.381) for a linear, time-shift–invariant system simplifies if one introduces the Fourier transforms of f, g, and h as in Eq. (11.357). The convolution theorem implies that

$$g(t) = \int_{-\infty}^{\infty} e^{-i\omega t} \, \tilde{g}(\omega) \, d\omega = 2\pi \int_{-\infty}^{\infty} e^{-i\omega t} \, \tilde{h}(\omega) \, \tilde{f}(\omega) \, d\omega. \tag{11.382}$$

Therefore the Fourier-space version of the input-output relation Eq. (11.381) is "diagonal,"

$$\tilde{g}(\omega) = 2\pi \, \tilde{h}(\omega) \, \tilde{f}(\omega). \tag{11.383}$$

In other words, the spectral amplitude of the output function g is equal to the spectral amplitude of the input function f, multiplied by the spectral amplitude of the response function h. Usually one calls $\tilde{h}$ the **transfer function** of the system.

11.9.3 EXERCISES FOR SECTION 11.9

11.9.1   Interpreting Eq. (11.382) as the $\mathcal{L}^2(\mathbb{R})$ inner product of the functions $\tilde{f}^* e^{i\omega t}$ and $\tilde{h}$, apply the Cauchy-Schwarz-Bunyakovsky inequality to show that the maximum output amplitude occurs if one chooses the transfer (or filter) function so that, for a given input signal f,

$$\tilde{h}(\omega) = K \, e^{i\omega t} [\tilde{f}(\omega)]^*. \tag{11.384}$$

Conclude that the impulse response of such a **matched filter** is proportional to the time-reversed image of the input signal f for which one wants to optimize the output:

$$h(t') = K \, f(t - t'). \tag{11.385}$$

11.10 TWO-DIMENSIONAL EUCLIDEAN GROUP $E(2)$

The group of translations and rotations in two-dimensional Euclidean space $\mathbf{r} \in \mathbb{E}^2$ is called the **Euclidean group** in two dimensions. Let us write a translation-rotation as an ordered pair

$$F = (\mathbf{a}, \mathbf{R}(\theta)). \tag{11.386}$$

A translation-rotation acts on a vector $\mathbf{r} \in \mathbb{R}^2$ as follows:

$$\mathbf{r} \mapsto \mathbf{r}' = (\mathbf{a}, \mathbf{R}(\theta))\mathbf{r} := \mathbf{a} + \mathbf{R}(\theta)\mathbf{r}. \tag{11.387}$$

From this definition one obtains the law of composition

$$F_2 F_1 = (\mathbf{a}_2, \mathbf{R}(\theta_2))(\mathbf{a}_1, \mathbf{R}(\theta_1)) = (\mathbf{R}(\theta_2)\mathbf{a}_1 + \mathbf{a}_2, \mathbf{R}(\theta_2)\mathbf{R}(\theta_1)). \tag{11.388}$$

It follows from this law of composition that the set

$$E(2) := \{(\mathbf{a}, \mathbf{R}(\theta)) \mid \mathbf{a} \in \mathbb{R}^2, \mathbf{R}(\theta) \in SO(2, \mathbb{R})\} \tag{11.389}$$

is a group. Equation (11.388) immediately implies closure. Exercise 11.10.2 establishes associativity. The identity element is $(\mathbf{0}, \mathbf{R}(0))$. The inverse of $(\mathbf{a}, \mathbf{R}(\theta))$ is

$$(\mathbf{a}, \mathbf{R}(\theta))^{-1} = (-\mathbf{R}(-\theta)\mathbf{a}, \mathbf{R}(-\theta)) \tag{11.390}$$

in which

$$\mathbf{R}(-\theta) = [\mathbf{R}(\theta)]^{-1}. \tag{11.391}$$

This completes the proof that $E(2)$ is a group. We call $E(2)$ the **Euclidean group** of $\mathbb{E}^2$.

It follows from Appendix B that $(\mathbf{a}, \mathbf{R}(\theta))$ is an affine transformation of $\mathbb{E}^2$. The translations and Lorentz transformations on Minkowski space $\mathcal{M}$ are also affine transformations. The group of space-time translations and Lorentz transformations is called the **Poincaré group** or the **inhomogeneous Lorentz group**.

Two subgroups of $E(2)$ are especially important: the **translational subgroup**

$$\{(\mathbf{a}, \mathbf{R}(0)) \mid \mathbf{a} \in \mathbb{R}^2, \mathbf{R}(0) = \mathbf{1} \in SO(2, \mathbb{R})\} \cong T(\mathbb{E}^2) \tag{11.392}$$

and the **rotational subgroup**

$$\{(\mathbf{0}, \mathbf{R}(\theta)) \mid \mathbf{0} \in \mathbb{R}^2, \mathbf{R}(\theta) \in SO(2, \mathbb{R})\} \cong SO(2). \tag{11.393}$$

The translational subgroup is normal in $E(2)$, but the rotational subgroup is not (see Exercise 11.10.4).

The mapping that a translation-rotation $F \in E(2)$ induces on functions $f : \mathbb{E}^2 \to \mathbb{C}$ is

$$\begin{aligned} (P_F f)(\mathbf{r}) &:= f(F^{-1}\mathbf{r}) \\ &= f(\mathbf{R}(\theta)^{-1}(\mathbf{r} - \mathbf{a})). \end{aligned} \tag{11.394}$$

To show that the set of mappings $\{P_F \mid F \in E(2)\}$ is a representation of $E(2)$, one must show that this definition ensures that

$$P_{F_2 F_1} = P_{F_2} P_{F_1}. \tag{11.395}$$

The key is the equation that gives the product of two elements of $E(2)$. The goal of Exercise 11.10.1 is to work out the details involved in establishing Eqs. (11.395) and (11.388).

11.10.1 REPRESENTATION OF $E(2)$ CARRIED BY $\{\psi_k\}$

The plane wave ψ_k also carries a representation of the full two-dimensional Euclidean group $E(2)$. From Eqs. (11.386) and (11.394) one has

$$\left(P_{(\mathbf{a}, \mathbf{R})} \psi_\mathbf{k}\right)(\mathbf{r}) = \psi_\mathbf{k}(\mathbf{R}^{-1}(\mathbf{r} - \mathbf{a})) = \frac{1}{2\pi} e^{i\mathbf{k} \cdot (\mathbf{R}^{-1}(\mathbf{r} - \mathbf{a}))}. \tag{11.396}$$

Because $\mathbf{R}$ is an orthogonal matrix,

$$
\begin{aligned}
\mathbf{k} \cdot (\mathbf{R}^{-1}(\mathbf{r} - \mathbf{a})) &= \mathbf{k}^T (\mathbf{R}^T (\mathbf{r} - \mathbf{a})) \\
&= (\mathbf{R}\mathbf{k})^T (\mathbf{r} - \mathbf{a}) \\
&= (\mathbf{R}\mathbf{k}) \cdot (\mathbf{r} - \mathbf{a}).
\end{aligned}
\tag{11.397}
$$

Then

$$\left(P_{(\mathbf{a}, \mathbf{R})} \psi_\mathbf{k}\right)(\mathbf{r}) = e^{-i\mathbf{R}\mathbf{k} \cdot \mathbf{a}} \psi_{\mathbf{R}\mathbf{k}}(\mathbf{r}) \tag{11.398}$$

defines the image of $\psi_\mathbf{k}$ under the operator $P_{(\mathbf{a}, \mathbf{R})}$, which represents a general element $(\mathbf{a}, \mathbf{R})$ of $E(2)$.

The mapping

$$(\mathbf{a}, \mathbf{R}) \mapsto P_{(\mathbf{a}, \mathbf{R})} \tag{11.399}$$

is a representation of $E(2)$ (see Exercise 11.10.3). However, this representation is *not* reduced with respect to the rotational subgroup of $E(2)$, because, under a pure rotation,

$$\left(P_{(\mathbf{0}, \mathbf{R})} \psi_\mathbf{k}\right)(\mathbf{r}) = \psi_{\mathbf{R}\mathbf{k}}(\mathbf{r}). \tag{11.400}$$

However, a function that carries an irreducible representation of the rotation group $SO(2)$ is simply multiplied by $e^{-im\theta}$ as the result of a rotation. We show in Section 12.2 that if one reduces the representation Eq. (11.399) with respect to the rotational subgroup, one immediately obtains the Bessel functions of the first kind of integer order.

11.10.2 DISCUSSION

Exercise B.7 shows that the two-dimensional translation group $T(\mathbb{E}^2)$ is a normal subgroup of $E(2)$. However, $E(2)$ is not what is called a semisimple group, because it has a nontrivial Abelian invariant subgroup. What this means in practice is that one must be cautious about applying intuitions based on a semisimple group (such as the rotation group in three dimensions) to $E(2)$. We show in Chapter 12 that what breaks in the case of $E(2)$ is the possibility of indexing irreducible representations with an integer (or any other discrete index). It turns out that the irreducible representations of $E(2)$ must be indexed with a positive real number.

The general principle that the discussion of Eqs. (11.320), (11.321), and (11.322) illustrates is that if the irreducible representations of a Lie group G must be continuously indexed, that is, indexed by one or more real numbers (rather than by integers), the functions that carry the irreducible representations of G are not square-integrable. This principle follows from the fact that irreducible representations of noncompact Lie groups, for which the group volume is infinite, must be indexed by real number(s).

In Section 12.2 we express the function ψ_k, which carries the representation Eq. (11.318) of the subgroup $T(2)$ of $E(2)$, in terms of the angular-momentum eigenfunctions $e^{im\theta}$, which carry the irreducible representations of the rotational subgroup $SO(2)$ of $E(2)$. We then show that this construction enables one to derive all of the classical formulas involving Bessel functions and also permits one to construct irreducible representations of the full two-dimensional Euclidean group $E(2)$.

11.10.3 EXERCISES FOR SECTION 11.10

11.10.1 Derive Eqs. (11.395) and (11.388).

Hint: Note that in order to establish Eq. (11.395), you must use the inverse of Eq. (11.388).

11.10.2 Prove that the law of composition derived in the preceding problem is associative.

11.10.3 Prove that the mapping defined in Eq. (11.399) is a representation of $E(2)$.

11.10.4 Show that the translational subgroup defined in Eq. (11.392) is normal in $E(2)$.

11.11 BIBLIOGRAPHY

Hall, Marshall. *The Theory of Groups.* New York: Macmillan, 1959.

Hamermesh, Morton. *Group Theory and Its Application to Physical Problems.* New York: Dover, 1990.

Jackson, J. D. *Classical Electrodynamics.* 2 ed. New York: Wiley, 1975, Exercise 14.6.

Lighthill, M. J. *Introduction to Fourier Analysis and Generalised Functions.* Cambridge: University Press, 1958.

Weyl, Hermann. *The Theory of Groups and Quantum Mechanics* (H. P. Robertson; trans.). New York: Dover, 1950 (first German edition: 1928).

Wigner, Eugene P. *Group Theory and Its Applications to the Quantum Mechanics of Atomic Spectra.* New York: Academic Press, 1959, Chapters 7, 8.

CHAPTER 12

SPECIAL FUNCTIONS

12.1 GROUP THEORY AND SPECIAL FUNCTIONS

12.1.1 SEPARATION OF VARIABLES

In a traditional presentation, the so-called *special functions of mathematical physics* such as the Bessel functions or Legendre polynomials usually appear as the solutions of certain ordinary differential equations. In turn, these differential equations arise if one tries to solve an important partial differential equation such as Laplace's or Helmholtz's equation by the method of separation of variables. For example, given the two-dimensional Helmholtz equation in plane polar coordinates ρ, ϕ,

$$\left[\frac{1}{\rho} \frac{\partial}{\partial \rho} \left(\rho \frac{\partial}{\partial \rho} \right) + \frac{1}{\rho^2} \frac{\partial^2}{\partial \phi^2} + k^2 \right] f(\mathbf{r}) = 0, \tag{12.1}$$

one substitutes the trial solution

$$f(\mathbf{r}) = P(\rho)\Phi(\phi) \tag{12.2}$$

and derives the equation

$$\frac{1}{\rho P} \frac{d}{d\rho} \left(\rho \frac{dP}{d\rho} \right) + k^2 = -\frac{1}{\rho^2 \Phi} \frac{d^2 \Phi}{d\phi^2}. \tag{12.3}$$

In this equation the left-hand side is a function of ρ alone and the right-hand side is a function of ϕ alone; therefore each side is equal to a constant m^2, the so-called separation constant. Then the functions P and Φ must obey the **separated equations**

$$\frac{1}{\rho} \frac{d}{d\rho} \left(\rho \frac{dP}{d\rho} \right) + \left(k^2 - \frac{m^2}{\rho^2} \right) P = 0 \tag{12.4}$$

and

$$\frac{d^2 \Phi}{d\phi^2} = -m^2 \Phi. \tag{12.5}$$

It follows that the *angular function* is $\Phi(\phi) = e^{im\phi}$; the boundary condition $\Phi(\phi + 2\pi) = \Phi(\phi)$ implies that m must be an integer. The *radial function* P obeys Bessel's differential equation. The boundary conditions in ρ determine whether the solution is a Bessel function of the first kind (J_m) or of the second kind (the Neumann function, Y_m), or some linear combination of J_m and Y_m.

In this chapter we present special functions using the language of modern physics – group theory – as much as possible, because the traditional separation-of-variables approach to special functions leaves some important questions unanswered. For example, one simply assumes that every function that satisfies both the Helmholtz equation and a set of consistent boundary conditions can be expanded in a double series, each term of which is proportional to $e^{im\phi} Z_m(k\rho)$, in which the *cylinder function* Z_m is J_m, Y_m or a linear combination of the two. From a more fundamental point of view, the traditional approaches give no clue to why certain special functions tend to occur in pairs such as $J_m(k\rho)\, e^{im\phi}$ or the spherical Bessel–spherical harmonic pair $j_m(kr)\, Y_{lm}(\theta, \phi)$, apart from the fact that these functions turn up on opposite sides of a separation equation such as Eq. (12.3).

We show how, in some important special cases, both the separability of a partial differential equation, and the properties of the special functions that are the solutions of the separated equations, follow from the properties of the irreducible representations of the symmetry group of the partial differential equation and its boundary conditions. Although group theory can provide complete developments of both the properties of the special functions and the method of separation of variables, a general treatment is beyond the scope of this book. Interested readers should consult the books by Miller, Talman and Wigner, and Vilenkin listed among the references at the end of this chapter.

12.1.2 SPECIAL FUNCTIONS AS MATRIX ELEMENTS

Many of this chapter's derivations of properties of the special functions are based upon the discovery by Wigner, Vilenkin and Miller that if Γ is the matrix of an irreducible representation of a Lie group G, then a matrix element $\gamma_{ji}(g)$, considered as a function of the parameters that designate a specific group element g (such as the angle of rotation if G is the rotation group in the plane, or the Euler angles in the case of the three-dimensional rotation group), is proportional to one of the classical special functions or to a product of special functions. In fact, the matrix elements $\{\gamma_{ji}(g)\}$ furnish a complete set of symmetry-adapted functions. (Completeness follows from the theorem that the set of vectors that carry an irreducible vector representation of a finite group or a compact Lie group G span the representation space.) The representation property Eq. (11.19) then becomes an *addition theorem* for the family of special functions $\{\gamma_{ji}(g)\}$. Readers who are familiar with quantum theory will recognize that a modern theory of the special functions strongly overlaps the modern theories of the fundamental particles and of the energy eigenstates of atoms, molecules, and solids.

12.1.3 SYMMETRIES OF THE HELMHOLTZ EQUATION

The solutions of the homogeneous Helmholtz equation

$$(\nabla^2 + k^2)\, \psi(\mathbf{r}) = 0 \tag{12.6}$$

are waves; for example, an idealized solution of particular importance is a plane wave, $e^{i\mathbf{k}\cdot\mathbf{r}}$. In the language of modern physics, the homogeneous Helmholtz equation describes the dynamics of a free (that is, noninteracting) scalar field ψ. The inhomogeneous Helmholtz equation

$$(\nabla^2 + k^2)\psi(\mathbf{r}) = s(\mathbf{r}) \tag{12.7}$$

describes the interaction of the field ψ with another system, because one can regard s as a "current" that radiates the waves that ψ represents.

Following an approach that has become standard in modern physics, we ask, "What is the global continuous symmetry group of the Helmholtz equation?" We expect that the answer will give useful information about the kinds of waves of which ψ is made up, exactly as the irreducible representations of $SO(2)$ provide functions in terms of which every function $f \in \mathcal{L}^2((-\pi, \pi]; \mathbb{C})$ can be expressed in a Fourier series.

We recall that a system possesses a *global* symmetry if the system's Lagrangian or Hamiltonian, or its governing equation, is invariant under a group in which the same transformation is applied at every spatial point or at every space-time point. A group is called *continuous* if, for any group elements g_1 and g_2, the parameters that label the element $g_2 g_1$ are continuous functions of the parameters that label g_1 and g_2. A group element g is *continuously connected to the identity element* if g can be "shrunk" back to the group's identity element e by continuously changing the parameters that label g. Let $\mathbf{x} = (x^1, x^2)^T$ be a vector in two-dimensional Euclidean space, $\mathbb{E}^2$. By the continuous symmetries of the Helmholtz equation $(\nabla^2 + k^2)\psi = 0$ we mean the set of coordinate transformations

$$x'^1 = f^1(\mathbf{x}), \quad x'^2 = f^2(\mathbf{x}) \tag{12.8}$$

that are invertible and infinitely differentiable and for which the two-dimensional Laplacian is invariant in form,

$$\frac{\partial^2}{\partial (x'^1)^2} + \frac{\partial^2}{\partial (x'^2)^2} = \frac{\partial^2}{\partial (x^1)^2} + \frac{\partial^2}{\partial (x^2)^2}. \tag{12.9}$$

The set of such transformations is a group.

The coordinate transformation Eq. (12.8) induces a linear mapping in the tangent space of vectors $(dx^1, dx^2)^T$,

$$\begin{pmatrix} dx'^1 \\ dx'^2 \end{pmatrix} = \mathbf{J} \begin{pmatrix} dx^1 \\ dx^2 \end{pmatrix}, \tag{12.10}$$

in which $\mathbf{J}$ is the Jacobian matrix,

$$\mathbf{J} = \begin{pmatrix} \dfrac{\partial x'^1}{\partial x^1} & \dfrac{\partial x'^1}{\partial x^2} \\ \dfrac{\partial x'^2}{\partial x^1} & \dfrac{\partial x'^2}{\partial x^2} \end{pmatrix}. \tag{12.11}$$

For a global symmetry $\mathbf{J}$ is independent of the position $\mathbf{x}$.

From Section 9.4 it follows that the singular-value decomposition of the real 2×2 matrix **J** is

$$\mathbf{J} = \mathbf{U\Sigma V}^T \tag{12.12}$$

in which **U** and **V** are real, orthogonal 2×2 matrices,

$$\Sigma = \begin{pmatrix} \sigma_1 & 0 \\ 0 & \sigma_2 \end{pmatrix}, \tag{12.13}$$

and $\sigma_1 \geq \sigma_2 \geq 0$. If $\sigma_2 = 0$ then the Jacobian determinant $\det[\mathbf{J}]$ vanishes, implying that a region of nonzero volume is mapped onto a region of zero volume. We assume for obvious reasons that this is not the case; then $\sigma_1 \geq \sigma_2 > 0$.

For the moment let us disregard the orthogonal transformations **U** and **V**. Then $dx'^1 = \sigma_1 dx^1$, $dx'^2 = \sigma_2 dx^2$, and therefore

$$\nabla^2 = \frac{\partial^2}{\partial(x^1)^2} + \frac{\partial^2}{\partial(x^2)^2} = \sigma_1^2 \frac{\partial^2}{\partial(x'^1)^2} + \sigma_2^2 \frac{\partial^2}{\partial(x'^2)^2}. \tag{12.14}$$

The right-hand side is equal to the Laplacian in the new coordinates x'^1, x'^2 if and only if $\sigma_1^2 = \sigma_2^2 = 1$, which (with $\sigma_1 \geq \sigma_2 > 0$) implies that

$$\sigma_1 = \sigma_2 = 1. \tag{12.15}$$

The only effect of the orthogonal transformations **U** and **V** is to change bases in the tangent space. Because we show in Chapter 9 that the Laplacian in $\mathbb{E}^n$ is invariant under orthogonal transformations, Eq. (12.15) holds in general.

We have shown that Σ is the 2×2 identity matrix. Therefore

$$\mathbf{J} = \mathbf{UV}^T \tag{12.16}$$

is an orthogonal matrix:

$$\mathbf{J}^T \mathbf{J} = \mathbf{VU}^T \mathbf{UV}^T = \mathbf{1}. \tag{12.17}$$

It follows that $(\det[\mathbf{J}])^2 = 1$, and therefore that the Jacobian is $\det[\mathbf{J}] = \pm 1$. We exclude the case in which $\det[\mathbf{J}] = -1$ because the determinant is a continuous function of the matrix elements, which are continuous functions of the angle of rotation; therefore only an orthogonal matrix with determinant equal to $+1$ can be obtained continuously from the 2×2 identity matrix. It follows that **J** is a rotation matrix:

$$\mathbf{J} = \mathbf{R}(\theta). \tag{12.18}$$

Then

$$d\mathbf{r}' = \mathbf{R}(\theta) \, d\mathbf{r} \tag{12.19}$$

in which the rotation angle θ is the same for all $\mathbf{r}$ because the transformation is global, not local.

A translation $\mathbf{r} \mapsto \mathbf{r} + \mathbf{a}$ also leaves the Helmholtz operator invariant and commutes with the mapping defined in Eq. (12.10). Therefore the most general global coordinate transformation, continuously connected to the identity, that leaves the Helmholtz operator $\nabla^2 + k^2$ invariant is a rotation, followed by a translation.

12.1.4 EXERCISE FOR SECTION 12.1

12.1.1   Show that the set of transformations of the form given in Eq. (12.8) that are continuous and invertible, and that obey Eq. (12.9), is a group under the composition operation defined in Eq. (2.47).

12.2 DEFINITION OF THE BESSEL FUNCTIONS

12.2.1 FOURIER EXPANSION OF A PLANE WAVE

Let us expand the momentum eigenfunction

$$e^{i\mathbf{k}\cdot\mathbf{r}} = e^{ik\rho \cos\phi} \tag{12.20}$$

in terms of the angular-momentum eigenfunctions $e^{im\phi}$. In Eq. (12.20) we choose the direction of $\mathbf{k}$ as the x-axis and use the plane polar coordinates defined in Eq. (11.169),

$$x = \rho \cos\phi, \quad y = \rho \sin\phi. \tag{12.21}$$

Because $\{e^{im\phi}\}$ is a complete set on the interval $(-\pi, \pi]$, the infinitely differentiable periodic function $e^{ik\rho \cos\phi} = e^{ik\rho \cos(\phi+2\pi)}$ has a uniformly convergent Fourier series. We define the Fourier coefficients as proportional to the Bessel functions $J_m(k\rho)$.

In order to expand $e^{ik\rho \cos\phi}$, $\mathrm{Re}[e^{ik\rho \cos\phi}] = \cos(k\rho \cos\phi)$, and $\mathrm{Im}[e^{ik\rho \cos\phi}] = \sin(k\rho \cos\phi)$ in Fourier series as simply as possible, we temporarily introduce the angle

$$\phi' := \phi + \frac{\pi}{2} \Rightarrow \cos\phi = \sin\phi', \tag{12.22}$$

as shown in Fig. 12.1. This substitution makes the real part even and the imaginary part odd

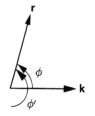

FIGURE 12.1

The position vector $\mathbf{r}$ and the wave vector $\mathbf{k}$ used in the Fourier expansion of a plane wave.

in ϕ':

$$e^{ik\rho \cos \phi} = e^{ik\rho \sin \phi'}$$

$$= \underbrace{\cos(k\rho \sin \phi')}_{\text{even}} + i \underbrace{\sin(k\rho \sin \phi')}_{\text{odd}}. \tag{12.23}$$

The even function $\cos(k\rho \sin \phi')$ has the Fourier cosine series

$$\cos(k\rho \sin \phi') = \frac{1}{2}a_0(k\rho) + \sum_{m=1}^{\infty} a_m(k\rho) \cos m\phi' \tag{12.24}$$

and the odd function $\sin(k\rho \sin \phi')$ has the Fourier sine series

$$\sin(k\rho \sin \phi') = \sum_{m=1}^{\infty} b_m(k\rho) \sin m\phi'. \tag{12.25}$$

The Fourier coefficients are

$$a_m(k\rho) = \frac{1}{\pi} \int_{-\pi}^{\pi} \cos(k\rho \sin \phi') \cos m\phi' \, d\phi' \tag{12.26}$$

and

$$b_m(k\rho) = \frac{1}{\pi} \int_{-\pi}^{\pi} \sin(k\rho \sin \phi') \sin m\phi' \, d\phi'. \tag{12.27}$$

The integrals work out most simply in terms of ϕ rather than ϕ'. One has

$$\cos m\phi' = \cos\left(m\frac{\pi}{2} + m\phi\right), \quad \sin m\phi' = \sin\left(m\frac{\pi}{2} + m\phi\right). \tag{12.28}$$

If m is even, $m = 2n$, then

$$\cos m\phi' = \cos(n\pi + m\phi) = (-1)^n \cos m\phi$$
$$\sin m\phi' = \sin(n\pi + m\phi) = (-1)^n \sin m\phi. \tag{12.29}$$

Therefore

$$b_{2n}(k\rho) = \frac{(-1)^n}{\pi} \int_{-\pi}^{\pi} \sin(k\rho \cos \phi) \sin 2n\phi \, d\phi = 0 \tag{12.30}$$

because $\sin(k\rho \cos \phi)$ is even but $\sin 2n\phi$ is odd. On the other hand, if m is odd, $m = 2n + 1$, then

$$\cos m\phi' = \cos\left(n\pi + \frac{\pi}{2} + m\phi\right) = (-1)^{n+1} \sin m\phi$$

$$\sin m\phi' = \sin\left(n\pi + \frac{\pi}{2} + m\phi\right) = (-1)^n \cos m\phi. \tag{12.31}$$

Therefore

$$a_{2n+1}(k\rho) = \frac{(-1)^{n+1}}{\pi} \int_{-\pi}^{\pi} \cos(k\rho \cos \phi) \sin m\phi \, d\phi = 0 \tag{12.32}$$

because $\cos(k\rho \cos \phi)$ is even but $\sin m\phi$ is odd.

12.2.2 DEFINITION OF THE BESSEL FUNCTION J_m OF INTEGER ORDER

Because the even Fourier-sine coefficients b_{2n} and the odd Fourier-cosine coefficients a_{2n+1} vanish, only one Fourier coefficient, either $a_m(k\rho)$ or $b_m(k\rho)$, is nonzero for any given m. Then one can define the single family of coefficients

$$2J_m(k\rho) := \begin{cases} a_m(k\rho), & \text{if } m \text{ is even}; \\ b_m(k\rho), & \text{if } m \text{ is odd}. \end{cases} \tag{12.33}$$

J_m is called the **Bessel function of the first kind of order** m.

So far we have defined the Bessel functions only for nonnegative, integral orders. We show that the definition can be extended to negative integral orders. From Eqs. (12.26) and (12.27) one sees that if m is even

$$2J_m(k\rho) = \frac{1}{\pi} \int_{-\pi}^{\pi} \cos(k\rho \sin \phi') \cos m\phi' \, d\phi' = 2J_{-m}(k\rho) \tag{12.34}$$

but that if m is odd

$$2J_m(k\rho) = \frac{1}{\pi} \int_{-\pi}^{\pi} \sin(k\rho \sin \phi') \sin m\phi' \, d\phi' = -2 J_{-m}(k\rho). \tag{12.35}$$

Therefore

$$J_{-m}(k\rho) = (-1)^m J_m(k\rho) \tag{12.36}$$

for any integer order m.

12.2.3 JACOBI-ANGER EXPANSION

Let us see how the Fourier series in Eqs. (12.24) and (12.25) look in terms of the Bessel functions. In view of Eqs. (12.30), (12.32), and (12.36), one can write

$$\begin{aligned} a_m(k\rho) \cos m\phi' &= J_m(k\rho) \cos m\phi' + J_{-m}(k\rho) \cos(-m\phi') \\ b_m(k\rho) \sin m\phi' &= J_m(k\rho) \sin m\phi' + J_{-m}(k\rho) \sin(-m\phi') \end{aligned} \tag{12.37}$$

regardless of whether the order m is even or odd. (For example, if m is odd then the right-hand side vanishes on the first line.) Then

$$e^{ik\rho \sin \phi'} = \cos(ik\rho \sin \phi') + i \sin(ik\rho \sin \phi')$$

$$= J_0(k\rho) + \sum_{m=1}^{\infty}(J_m(k\rho) \cos m\phi' + J_{-m}(k\rho) \cos(-m\phi'))$$

$$+ i \sum_{m=1}^{\infty}(J_m(k\rho) \sin m\phi' + J_{-m}(k\rho) \sin(-m\phi'))$$

$$= \sum_{m=-\infty}^{\infty} J_m(k\rho)(\cos m\phi' + i \sin m\phi'),$$

(12.38)

which shows that the Fourier series for $e^{ik\rho \sin \phi'}$ is

$$e^{ik\rho \sin \phi'} = \sum_{m=-\infty}^{\infty} J_m(k\rho) e^{im\phi'}. \qquad (12.39)$$

The formula

$$J_m(z) = \frac{1}{2\pi} \int_{-\pi}^{\pi} e^{iz \sin \phi} e^{-im\phi} \, d\phi, \qquad (12.40)$$

which follows directly from Eqs. (12.26) and (12.27), gives the Fourier coefficients. This equation can easily be transformed into **Bessel's integral**,

$$J_m(z) = \frac{1}{2\pi} \int_{0}^{2\pi} \cos(m\phi - z \sin \phi) \, d\phi, \qquad (12.41)$$

which Bessel[1] took as the definition of the Bessel functions, and from which he derived their other properties (see Exercise 12.2.3).

One can see also from Eq. (12.40) or Eq. (12.41) that the Bessel functions of the first kind take on the special values

$$J_m(0) = \begin{cases} 1, & \text{if } m = 0; \\ 0, & \text{if } m \neq 0 \end{cases} \qquad (12.42)$$

(see Exercise 12.2.2). The series Eq. (12.39) is our primary tool in deriving the Bessel recurrence relations, from which we shall obtain the Bessel raising and lowering operators.

---

[1] Bessel calculated the orbit of Halley's comet at the age of 20 while he was employed in the service sector (as a warehouse worker). He later became Professor of Astronomy at the University of Königsberg (now Kaliningrad).

In terms of the angle between $\mathbf{k}$ and $\mathbf{r}$, $\phi = \phi' - \pi/2$, the Fourier series Eq. (12.39) becomes the **Jacobi-Anger expansion**

$$
e^{ik\rho \cos \phi} = \sum_{m=-\infty}^{\infty} i^m J_m(k\rho) e^{im\phi}.
\tag{12.43}
$$

12.2.4 FREQUENCY CONTENT OF AN FM SIGNAL

A significant application of the Fourier expansion of the Bessel function of the first kind occurs in the mathematical description of frequency modulation of a radio signal. One can define the instantaneous frequency of a signal as i times the time derivative of the phase. Suppose that the instantaneous frequency $\omega(t)$ of a constant-amplitude voltage e is

$$
\omega(t) = \omega_0 + M\omega \sin(\omega t),
\tag{12.44}
$$

in which M is a real constant and ω is the frequency at which the phase of the signal is modulated. (For example, ω could be an audio frequency.) Then

$$
e(t) = A \operatorname{Re} \left(e^{-i(\omega_0 t - M \cos(\omega t) - \psi)} \right)
\tag{12.45}
$$

in which A and ψ are real constants. By Eq. (12.43),

$$
e(t) = A \operatorname{Re} \left(e^{-i(\omega_0 t - \psi)} \sum_{m=-\infty}^{\infty} i^m J_m(M) e^{im\omega t} \right).
\tag{12.46}
$$

This result implies that a signal whose frequency is modulated with a single-frequency sinusoid has an infinite set of sidebands at frequencies $\omega_0 \pm |m|\omega$. Therefore the frequency bandwidth required for accurate reception of a frequency-modulated signal can be much larger than for an amplitude-modulated signal, because amplitude modulation,

$$
\begin{aligned}
e(t) &= (A + M \cos \omega t) \cos(\omega_0 t) \\
&= A \cos(\omega_0 t) + \tfrac{1}{2} M (\cos((\omega_0 + \omega)t) + \cos((\omega_0 - \omega)t)),
\end{aligned}
\tag{12.47}
$$

produces only two sidebands at the frequencies $\omega_0 \pm \omega$.

12.2.5 EXERCISES FOR SECTION 12.2

12.2.1   Derive the usual Bessel generating function,

$$
\exp \left[\frac{z}{2} \left(t - \frac{1}{t} \right) \right] = \sum_{m=-\infty}^{\infty} t^m J_m(z),
\tag{12.48}
$$

from Eq. (12.39).

12.2.2 Using Eq. (12.40), derive Eq. (12.42).

12.2.3 From Eq. (12.40), derive Bessel's integral, Eq. (12.41).

12.2.4 Use Eq. (12.39) to establish the following identities:

(a)

$$\sum_{m=-\infty}^{\infty} J_m(z) = 1 \qquad (12.49)$$

(b)

$$\sum_{m=-\infty}^{\infty} i^m J_m(z) = e^{iz} \qquad (12.50)$$

(c)

$$\sum_{m=-\infty}^{\infty} i^{-m} J_m(z) = e^{-iz} \qquad (12.51)$$

(d)

$$\sum_{m=-\infty}^{\infty} (-1)^m J_m(z) = 1 \qquad (12.52)$$

12.3 BESSEL-FUNCTION ADDITION FORMULAS

All special functions have *addition formulas* that relate a single function to the product of a pair of functions of the same family, or to a sum of pair products. For the functions $\psi_k(\mathbf{r}) = e^{i\mathbf{k}\cdot\mathbf{r}}$, the addition formula is

$$e^{i\mathbf{k}\cdot(\mathbf{r}_1+\mathbf{r}_2)} = e^{i\mathbf{k}\cdot\mathbf{r}_1} e^{i\mathbf{k}\cdot\mathbf{r}_2}. \qquad (12.53)$$

This and other special-function addition formulas are directly related to representations of groups.

One can look on $e^{i\mathbf{k}\cdot\mathbf{r}}$ either as the function that carries the irreducible representation of $T(\mathbb{E}^2)$ labeled by the wave vector $\mathbf{k}$ or as the representative of the translation $T_{-\mathbf{r}}$. Under the latter interpretation, Eq. (12.53) expresses the representation property (11.18) for the representation of $T(\mathbb{E}^2)$ labeled by $\mathbf{k}$. We obtain the Bessel addition formulas from Eqs. (12.53) and (12.43).

Figure 12.2 illustrates the geometry for this derivation. The x-axis is the direction of $\mathbf{k}$. The polar coordinates of $\mathbf{r}_1$, $\mathbf{r}_2$, and $\mathbf{r}$ are (ρ_1, ϕ_1), $(\rho_2, \phi_1 + \phi_2)$, and $(\rho, \phi_1 + \phi)$, respectively. Then Eq. (12.53) becomes

$$e^{ik\rho \cos(\phi_1+\phi)} = e^{ik\rho_1 \cos\phi_1} e^{ik\rho_2 \cos(\phi_1+\phi_2)}, \qquad (12.54)$$

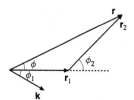

FIGURE 12.2

Geometry for the derivation of Graf's addition theorem for Bessel functions.

which, with the Jacobi-Anger expansion Eq. (12.39), implies that

$$\sum_{n=-\infty}^{\infty} i^n J_n(k\rho) e^{in(\phi_1+\phi)} = \sum_{l=-\infty}^{\infty} i^l J_l(k\rho_1) e^{il\phi_1} \sum_{m=-\infty}^{\infty} i^m J_m(k\rho_2) e^{im(\phi_1+\phi_2)}. \tag{12.55}$$

One can interchange the order of the summations on the right-hand side because the Fourier series of a continuous function converges absolutely. Let $n' = l + m$; then

$$\sum_{n=-\infty}^{\infty} i^n J_n(k\rho) e^{in(\phi_1+\phi)} = \sum_{n'=-\infty}^{\infty} i^{n'} e^{in'\phi_1} \sum_{m=-\infty}^{\infty} e^{im\phi_2} J_{n'-m}(k\rho_1) J_m(k\rho_2). \tag{12.56}$$

By the uniqueness theorem for orthogonal series, one can equate the coefficients of $e^{in\phi_1}$, thereby obtaining **Graf's addition theorem**

$$J_n(k\rho) e^{in\phi} = \sum_{m=-\infty}^{\infty} e^{im\phi_2} J_{n-m}(k\rho_1) J_m(k\rho_2). \tag{12.57}$$

Several special cases of this formula are useful in practice; see Exercises 12.3.1 through 12.3.5.

Graf's addition theorem must be recast in order to show its significance in terms of the Euclidean group $E(2)$. We begin with the translation subgroup of $E(2)$, which consists of all rotation-translations $(\mathbf{a}, \mathbf{R}(\theta))$ such that $\mathbf{R}(\theta) = \mathbf{1}$, the identity mapping. The functions f_i that carry a representation of this subgroup satisfy the equation

$$(P_{(\mathbf{a},1)} f_i)(\mathbf{r}) = f_i(\mathbf{r} - \mathbf{a}) = \sum_j \gamma_{ji}(\mathbf{a}, 1) f_j(\mathbf{r}). \tag{12.58}$$

In order to put Eq. (12.57) into this form, we use the geometry shown in Fig. 12.3, with which one has

$$J_{n'}(k\rho') e^{in'(\phi'-\phi)} = \sum_{m=-\infty}^{\infty} e^{im(\alpha-\phi-\pi)} J_{n'-m}(k\rho) J_m(ka). \tag{12.59}$$

(Note that in Fig. 12.3, the azimuthal angle of $\mathbf{a}$ is equal to the angle through which $\mathbf{k}$ must

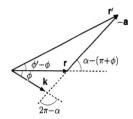

FIGURE 12.3

Geometry to illustrate that the Bessel functions carry representations of the translation subgroup of the Euclidean group in the plane. Because the x-axis is along $\mathbf{k}$, the angle ϕ' is the azimuthal angle of $\mathbf{r}' = \mathbf{r} - \mathbf{a}$, and α is the azimuthal angle of $\mathbf{a}$.

be rotated in a positive sense in order to be aligned with $\mathbf{a}$.) The substitution $n := n' - m$ converts Eq. (12.59) into

$$J_{n'}(k\rho') e^{in'\phi'} = \sum_{n=-\infty}^{\infty} (-1)^{n'-n} J_{n'-n}(ka) e^{i(n'-n)\alpha} J_n(k\rho) e^{in\phi}. \tag{12.60}$$

In terms of the functions

$$H_m(z, \phi) := J_m(z) e^{im\phi} \tag{12.61}$$

and the matrix elements

$$\gamma_{nn'}^{(k)}(\mathbf{a}, \mathbf{1}) := (-1)^{n'-n} H_{n'-n}(ka, \alpha), \tag{12.62}$$

Eq. (12.60) becomes

$$\boxed{H_{n'}(k\rho', \phi') = \sum_{n=-\infty}^{\infty} \gamma_{nn'}^{(k)}(\mathbf{a}, \mathbf{1}) H_n(k\rho, \phi),} \tag{12.63}$$

which is of the same form as Eq. (12.58). Evidently the functions $H_m(k\cdot, \cdot)$ (in which the dots denote arguments) carry a representation $\Gamma^{(k)}$ of the translation subgroup. Because we show next that $\Gamma^{(k)}$ is an irreducible representation of $E(2)$, and because $E(2)$ is the symmetry group of the Helmholtz equation, we call H_m a **Helmholtz function**.

Let us calculate now the effect of a rotation-translation $(\mathbf{a}, \mathbf{R}(\theta)) \in E(2)$ on the Helmholtz function $H_{n'}(k\cdot, \cdot)$. The transformation of functions on $\mathbb{E}^2$ that represents the rotation-translation $(\mathbf{a}, \mathbf{R}(\theta))$ is

$$\left(P_{(\mathbf{a}, \mathbf{R}(\theta))} f_i\right)(\mathbf{r}) = f_i(\mathbf{R}(\theta)(\mathbf{r} - \mathbf{a})). \tag{12.64}$$

The length of the vector $\mathbf{r}' = \mathbf{R}(\theta)(\mathbf{r} - \mathbf{a})$ is equal to ρ', the length of $\mathbf{r} - \mathbf{a}$ in Eq. (12.63) and Fig. 12.3. Under $\mathbf{R}(\theta)$, one has $\alpha \mapsto \alpha - \theta$ and $\phi \mapsto \phi - \theta$. Then

$$\left(P_{(\mathbf{a}, \mathbf{R}(\theta))} H_{n'}\right)(k\rho, \phi)$$

$$= \sum_{n=-\infty}^{\infty} (-1)^{n'-n} J_{n'-n}(ka) e^{i(n'-n)(\alpha-\theta)} H_n(k\rho, \phi - \theta) \tag{12.65}$$

$$= \sum_{n=-\infty}^{\infty} (-1)^{n'-n} e^{-in\alpha} J_{n'-n}(ka) e^{in'(\alpha-\theta)} H_n(k\rho, \phi).$$

It follows that the (n, n') element of the infinite-dimensional matrix $\Gamma^{(k)}(\mathbf{a}, \mathbf{R}(\theta))$ is equal to

$$\gamma_{nn'}^{(k)}(\mathbf{a}, \mathbf{R}(\theta)) = (-1)^{n'-n} e^{-in\alpha} J_{n'-n}(ka) e^{in'(\alpha-\theta)} \tag{12.66}$$

for an arbitrary rotation-translation $(\mathbf{a}, \mathbf{R}(\theta))$ belonging to $E(2)$. The most general Bessel addition theorem, which follows from Eq. (11.19), adds nothing that goes essentially beyond the results presented in Eqs. (12.57) and (12.65)

It is significant that the representation $\Gamma^{(k)}$ must be indexed by a positive real number k, which belongs to an uncountably infinite set, rather than by an index such as an integer that belongs to a countably infinite set. It can be shown that all *compact* Lie groups, that is, Lie groups in which the volume of the group's parameter space is finite, have irreducible representations that can be indexed by a set that is equivalent to the integers. All *noncompact* Lie groups, for which the volume of parameter space is infinite, have representations that must be indexed by a set that is equivalent to the set of real numbers, $\mathbb{R}$. Besides the Euclidean groups $E(n)$, the most important noncompact Lie groups from the point of view of physics are the four-dimensional Lorentz and Poincaré groups.

12.3.1 EXERCISES FOR SECTION 12.3

12.3.1   Show that with the substitutions $k\rho = w, k\rho_1 = u, k\rho_2 = v, \phi = \chi$, and $\phi_2 = \pi - \alpha$, Graf's addition theorem as given in Eq. (12.57) reduces to a special case of formula 9.1.79 of Abramowitz and Stegun.

12.3.2   Prove that

$$J_n(k(\rho_1 + \rho_2)) = \sum_{m=-\infty}^{\infty} J_{n-m}(k\rho_1) J_m(k\rho_2). \tag{12.67}$$

12.3.3   Prove that

$$J_n(k(\rho_1 - \rho_2)) = \sum_{m=-\infty}^{\infty} J_{n+m}(k\rho_1) J_m(k\rho_2). \tag{12.68}$$

12.3.4   Prove that

$$\delta_{n0} = J_n(0) = \sum_{m=-\infty}^{\infty} J_{n+m}(k\rho_1) J_m(k\rho_1). \tag{12.69}$$

12.3.5   Prove that

$$1 = J_0(0) = [J_0(k\rho_1)]^2 + 2\sum_{m=1}^{\infty} [J_m(k\rho_1)]^2. \tag{12.70}$$

12.4   BESSEL RAISING AND LOWERING OPERATORS

12.4.1   RECURRENCE RELATIONS

We derive recurrence relations for the Bessel functions by manipulating Eq. (12.40) for the Fourier coefficients of the series Eq. (12.39). The integral Eq. (12.40) converges absolutely, and so does the integral obtained by differentiating with respect to z. Then

$$
\begin{aligned}
J'_m(z) &= \frac{1}{2\pi} \int_{-\pi}^{\pi} \left(\frac{d}{dz} e^{iz\sin\phi} \right) e^{-im\phi} \, d\phi \\
&= \frac{1}{2\pi} \int_{-\pi}^{\pi} i\sin\phi \, e^{iz\sin\phi} e^{-im\phi} \, d\phi \\
&= \frac{1}{4\pi} \int_{-\pi}^{\pi} e^{iz\sin\phi} \left(e^{-i(m-1)\phi} - e^{-i(m+1)\phi} \right) \, d\phi \\
&= \frac{1}{2}(J_{m-1}(z) - J_{m+1}(z)).
\end{aligned}
\tag{12.71}
$$

Therefore the Bessel functions J_m obey the recurrence relation

$$
J_{m-1}(z) - J_{m+1}(z) = 2J'_m(z). \tag{12.72}
$$

For future reference it is interesting to observe that the factor of two and the minus sign in the above Bessel recurrence relation come from the definition of the sine in terms of exponentials.

Another recurrence relation, independent of Eq. (12.72), would permit one to eliminate either J_{m-1} or J_{m+1}, thereby obtaining a relation involving only two orders instead of three. Let us see what what happens if one forms the sum, rather than the difference, of $J_{m-1}(z)$ and $J_{m+1}(z)$, using the Fourier-coefficient formula Eq. (12.40):

$$
\begin{aligned}
J_{m-1}(z) + J_{m+1}(z) &= \frac{1}{2\pi} \int_{-\pi}^{\pi} e^{iz\sin\phi} \left(e^{-i(m-1)\phi} + e^{-i(m+1)\phi} \right) \, d\phi \\
&= \frac{2}{2\pi} \int_{-\pi}^{\pi} e^{iz\sin\phi} \cos\phi \, e^{-im\phi} \, d\phi \\
&= \frac{1}{i\pi z} \int_{-\pi}^{\pi} \left(\frac{\partial}{\partial\phi} e^{iz\sin\phi} \right) e^{-im\phi} \, d\phi \\
&= \frac{1}{i\pi z} \left[e^{iz\sin\phi} e^{-im\phi} \Big|_{-\pi}^{\pi} + im \int_{-\pi}^{\pi} e^{iz\sin\phi} e^{-im\phi} \, d\phi \right].
\end{aligned}
\tag{12.73}
$$

The last equation follows through integration by parts. The boundary term gives zero; the integral is simply $2\pi J_m(z)$. Then one has a second Bessel recurrence relation:

$$
J_{m-1}(z) + J_{m+1}(z) = \frac{2m}{z} J_m(z). \tag{12.74}
$$

Note that the factor of two in Eq. (12.74) comes directly from the factor of $\frac{1}{2}$ in the cosine function that results from computing $J_{m-1} + J_{m+1}$. The resemblance of the patterns of the Bessel recurrence relations in Eqs. (12.72) and (12.74) to the definitions of the sine and cosine in terms of exponentials is a consequence of the "higher order" that symmetry under the group $E(2)$ imposes.

12.4.2 RAISING AND LOWERING OPERATORS FOR J_m

From a physicist's point of view it is much more natural to use raising and lowering operators than recurrence relations, because one is already familiar with raising and lowering operators for the harmonic-oscillator and angular-momentum eigenfunctions. (Recall that an operator R is a *raising operator* for a family of functions $\{f_m\}$ if for every m, $f_{m+1} = Rf_m$.) Raising and lowering operators are a common feature of Lie algebras.

If one adds Eqs. (12.72) and (12.74) to eliminate J_{m+1}, one obtains

$$B_-^{(m)} J_m = J_{m-1} \tag{12.75}$$

in which

$$B_-^{(m)} := \frac{m}{z} + \frac{d}{dz} \tag{12.76}$$

is the **Bessel lowering operator**. Subtracting Eq. (12.74) from Eq. (12.72) in order to eliminate J_{m-1}, one obtains

$$B_+^{(m)} J_m = J_{m+1} \tag{12.77}$$

in which

$$B_+^{(m)} := \frac{m}{z} - \frac{d}{dz} \tag{12.78}$$

is the **Bessel raising operator**.

12.4.3 RAISING AND LOWERING OPERATORS FOR THE HELMHOLTZ FUNCTIONS

That the order m appears explicitly in the Bessel raising and lowering operators is surprising. A possible point of view is that the angular-momentum and harmonic-oscillator ladder operators, in which the order does not appear explicitly, are especially simple cases. One has every right to find this point of view unsatisfying, particularly in view of the fact that we derived the Bessel recurrence relations very simply by exploiting their relationship with the functions $e^{im\phi}$. If one notices that m is the "order" of $e^{im\phi}$ as well as of J_m, then it makes sense to try to find raising and lowering operators for the Helmholtz functions $H_m(z, \phi) = J_m(z) e^{im\phi}$ defined in Section 12.3. One finds with the help of the chain rule for partial differentiation that

$$\begin{aligned} B_- H_m &= H_{m-1} \\ B_+ H_m &= H_{m+1} \end{aligned} \tag{12.79}$$

in which

$$B_- := e^{-i\phi}\left(-\frac{i}{z}\frac{\partial}{\partial\phi} + \frac{\partial}{\partial z}\right)$$

$$B_+ := e^{i\phi}\left(-\frac{i}{z}\frac{\partial}{\partial\phi} - \frac{\partial}{\partial z}\right)$$

(12.80)

are the raising and lowering operators for the family $\{H_m\}$. We call $B_\pm$ the **Helmholtz raising and lowering operators**.

12.4.4 EXERCISES FOR SECTION 12.4

12.4.1   Prove that

$$B_-^{(m+1)}\, B_+^{(m)} = B_+^{(m-1)}\, B_-^{(m)}.$$

(12.81)

12.4.2   Show that the Helmholtz raising and lowering operators commute:

$$[B_-, B_+] = 0.$$

(12.82)

12.4.3   Verify Eq. (12.79) using Eq. (12.80).

12.5 BESSEL DIFFERENTIAL EQUATIONS

12.5.1 BESSEL'S DIFFERENTIAL EQUATION

If one applies first a raising operator and then a lowering operator to J_m, one returns to J_m again. Thus, from Eqs. (12.75) and (12.77), one has

$$B_-^{(m+1)}\, B_+^{(m)}\, J_m = J_m.$$

(12.83)

If one substitutes the definitions of the raising and lowering operators in Eq. (12.83), one obtains **Bessel's differential equation,**

$$\left[\frac{d^2}{dz^2} + \frac{1}{z}\frac{d}{dz} + \left(1 - \frac{m^2}{z^2}\right)\right] J_m = 0.$$

(12.84)

J_m can be defined as the solution of Bessel's equation that is regular at the origin. Because Bessel's equation is of second order and is of Sturm-Liouville form (see Exercise 5.2.21), one expects two linearly independent solutions. A second solution can be constructed using Eq. (5.176), but this approach does not yield the standard form. It can be shown that, if $J_\nu(z)$ is defined for non-integral values of ν as the power series occurring in Eq. (12.161), then $J_{-\nu}(z)$ is linearly independent of $J_\nu(z)$; see [Watson 1966]. It can also be shown that $Y_n(z) := \lim_{\nu\to n}\{[J_\nu(z) - (-1)^n J_{-\nu}(z)]/(\nu - n)\}$ (in which n is an integer) exists and defines a second solution of Bessel's equation that is linearly independent of J_n.

12.5.2 HELMHOLTZ EQUATION IN TWO DIMENSIONS

The equation

$$(B_- B_+ - 1) H_m = 0 \tag{12.85}$$

turns out to be even more interesting than Bessel's differential equation. Substituting the definitions Eq. (12.80), one finds that H_m is a solution of the partial differential equation

$$\left(\frac{\partial^2}{\partial z^2} + \frac{1}{z} \frac{\partial}{\partial z} + \frac{1}{z^2} \frac{\partial^2}{\partial \phi^2} + 1 \right) H_m = 0. \tag{12.86}$$

Recognizing the two-dimensional Laplacian in polar coordinates,

$$\nabla^2 = \frac{\partial^2}{\partial \rho^2} + \frac{1}{\rho} \frac{\partial}{\partial \rho} + \frac{1}{\rho^2} \frac{\partial^2}{\partial \phi^2}, \tag{12.87}$$

and recalling that $z = k\rho$, one sees that each function H_m satisfies the two-dimensional Helmholtz equation:

$$(\nabla^2 + k^2) H_m(k\rho, \phi) = 0. \tag{12.88}$$

To understand the operator $B_- B_+$ in the group $E(2)$, it may be helpful to recall the properties of the total–angular-momentum operator $L^2 = \frac{1}{2}(L_+ L_- + L_- L_+) + L_z^2$ for the group $SO(3)$. Just as the angular-momentum eigenfunctions $|lm\rangle$ satisfy the equation $L^2 |lm\rangle = l(l+1)|lm\rangle$ (which is independent of the row m to which $|lm\rangle$ belongs), the functions $H(k\cdot, \cdot)$ satisfy the equation $B_- B_+ H_m = H_m$, which is also independent of the row to which H_m belongs.

12.5.3 QUALITATIVE BEHAVIOR OF J_m

Bessel's differential equation can be manipulated to give useful qualitative information about the behavior of $J_m(z)$ if $z \ll m$ or $z \gg m$. In order to make the picture as clear as possible, we transform away the first-derivative term in Eq. (12.84). Let

$$J_m(z) = \frac{u(z)}{z^{1/2}}. \tag{12.89}$$

Then

$$\frac{dJ_m}{dz} = -\frac{1}{2} z^{-3/2} u + z^{-1/2} \frac{du}{dz} \tag{12.90}$$

and

$$\frac{d^2 J_m}{dz^2} = \frac{3}{4} z^{-5/2} u - z^{-3/2} \frac{du}{dz} + z^{-1/2} \frac{d^2 u}{dz^2}. \tag{12.91}$$

The resulting differential equation for u,

$$\frac{d^2u}{dz^2} + \left(1 - \frac{m^2 - \frac{1}{4}}{z^2}\right)u = 0, \tag{12.92}$$

is completely equivalent to Bessel's equation.

The point at which the coefficient of the zero-order term in Eq. (12.92) vanishes,

$$z_{\text{turn}} = \left(m^2 - \frac{1}{4}\right)^{1/2}, \tag{12.93}$$

is called the **turning point** of Eq. (12.92). As z increases from zero, J_m grows in a locally exponential manner with a local growth rate equal to $((z_{\text{turn}}/z)^2 - 1)^{1/2}$ for as long as z is less than z_{turn}. For all $z > z_{\text{turn}}$, $u(z)$ oscillates with a local frequency equal to $(1 - (z_{\text{turn}}/z)^2)^{1/2}$. If z is much larger than z_{turn}, the frequency of oscillation is approximately equal to 1 and the amplitude of oscillation of u is constant. The qualitative behavior of J_m therefore consists of rapid growth near the origin, slower growth as z approaches z_{turn} from below, and oscillation with an amplitude proportional to $z^{-1/2}$ for $z > z_{\text{turn}}$. The only exception to this picture occurs for $m = 0$, at which the behavior of u and J_0 is entirely oscillatory. Figure 12.4 illustrates the initial growth, turning point, and oscillatory decay of J_0 through J_9. Note that the turning point occurs when the argument of the Bessel function is approximately equal to the order.

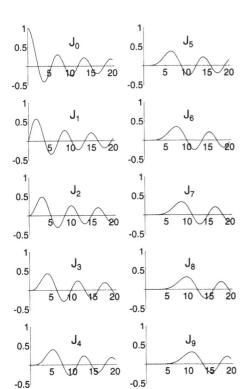

FIGURE 12.4

The values of the Bessel functions $J_0(x)$ through $J_9(x)$ (*vertical axes*) plotted as functions of x (*horizontal axes*).

Evidently there exist an infinite set of zeros of J_m and another infinite set of zeros of the first derivative, J_m'. For future reference we define

$$j_{m,n} := n\text{-th zero of } J_m \tag{12.94}$$

and

$$j_{m,n}' := n\text{-th zero of } J_m'. \tag{12.95}$$

Unlike the zeros of $\sin z$ and its derivative, $\cos z$, the Bessel and Bessel-derivative zeros are not equally spaced.

12.5.4 EXERCISES FOR SECTION 12.5

12.5.1   Verify Eq. (12.84) using the definitions given in Eqs. (12.76) and (12.78).

12.5.2   Derive Eq. (12.86) from Eq. (12.85).

12.6 ORTHOGONAL SERIES IN J_n

12.6.1 BOUNDARY CONDITIONS THAT ENSURE SELF-ADJOINTNESS

Rewritten as

$$\left[\frac{1}{\rho} \frac{d}{d\rho} \left(\rho \frac{d}{d\rho} \right) - \frac{n^2}{\rho^2} \right] J_n(k\rho) = -k^2 J_n(k\rho), \tag{12.96}$$

Bessel's differential equation Eq. (12.84) has the form of an eigenvalue-eigenvector problem $A\bar{x} = \lambda\bar{x}$ in which the operator A is the **Bessel operator of order n**,

$$B_n := \frac{1}{\rho} \frac{d}{d\rho} \left(\rho \frac{d}{d\rho} \right) - \frac{n^2}{\rho^2}, \tag{12.97}$$

and in which the eigenvalue is $\lambda = -k^2$. Therefore the Bessel function $J_n(k\cdot)$ is an eigenfunction of B_n. The spectrum of B_n is continuous, for k^2 can be any positive real number. However, one may guess by analogy with the one-dimensional wave equation that boundary conditions may force k^2 to take on only certain discrete values.

One can see by inspection that the Bessel operator of order n is formally self-adjoint with respect to the weight function $w(\rho) = \rho$ (see Eq. [9.115] and Section 9.2.5). The inner product with this weight function,

$$\langle f, g \rangle = \int_a^b f(\rho) g(\rho) \, \rho \, d\rho, \tag{12.98}$$

is positive-definite because ρ must be nonnegative, and therefore one can safely assume that $0 \le a < b$. The space of square-integrable functions with respect to this inner product,

$\mathcal{H} = \mathcal{L}^2_w([a, b]; \mathbb{R})$, is a Hilbert space. Let f and g be any elements of the space $\mathcal{C}^\infty([a, b]; \mathbb{R})$ of infinitely differentiable functions, which is dense in $\mathcal{H}$. Given the inner product Eq. (12.98), Eqs. (9.115) and (9.116), or integration by parts, imply that

$$\langle f, B_n g \rangle = \underline{\beta}[f, g] + \langle B_n f, g \rangle \tag{12.99}$$

in which

$$\underline{\beta}[f, g] = \rho \left[f \frac{dg}{d\rho} - \frac{df}{d\rho} g \right] \Big|_a^b \tag{12.100}$$

is the boundary functional for the Bessel operator on the interval $[a, b]$. Therefore the Bessel operator is self-adjoint if and only if every f and every g obey boundary conditions that make the boundary functional vanish.

The integrated terms vanish identically at $\rho = 0$:

$$\rho \left[f \frac{dg}{d\rho} - \frac{df}{d\rho} g \right] \Big|_{\rho=0} = 0. \tag{12.101}$$

This property turns out to be essential for the purpose of finding sets of boundary conditions that can be satisfied if f and g are both Bessel functions of the first kind, $f(\rho) = J_m(k\rho)$ and $g(\rho) = J_m(k'\rho)$.

Because the idea of self-adjointness makes sense only for an operator that acts on some vector space, one can make the following stronger statement: *The Bessel boundary functional vanishes, and the Bessel operator of order n is self-adjoint, if and only if (1) $\underline{\beta}[f, g] = 0$ for all infinitely differentiable functions f, g that obey chosen boundary conditions, and (2) the set of all infinitely differentiable functions that obey the chosen boundary conditions is a linear manifold $\mathcal{M} \subset \mathcal{H}$.* In other words, one must ensure not only that all admissible functions f and g satisfy physically meaningful boundary conditions, but also that the set of functions that satisfy these boundary conditions is a Hilbert space, that is, a linear manifold in $\mathcal{H}$. Because the null space of a linear functional is a linear manifold, one can meet both of these requirements with boundary conditions that take the form $\underline{\phi}[f] = 0$, in which $\underline{\phi}$ is a linear functional on $\mathcal{H}$ (see Chapter 7).

Boundary conditions that satisfy these criteria include the following *homogeneous* boundary conditions:

1. **Dirichlet boundary conditions:**

$$\forall f \in \mathcal{M} : \underline{\delta}_a[f] = \underline{\delta}_b[f] = 0 \tag{12.102}$$
$$\Rightarrow f(a) = f(b) = 0,$$

in which

$$\underline{\delta}_a[f] := f(a) \tag{12.103}$$

and $0 \leq a < b$. Then the linear manifold on which the Bessel operator is self-adjoint is

$$\mathcal{M} = \text{null}[\delta_a] \cap \text{null}[\delta_b] \cap \mathcal{C}^\infty([0, b]; \mathbb{R}). \tag{12.104}$$

If one of the functions f, g is a Bessel function of the first kind, $f(\rho) = J_m(k\rho)$, then the Dirichlet boundary condition $J_m(ka) = 0$ implies that there exists some integer n such that

$$ka = j_{m,n} \Rightarrow k = \frac{j_{m,n}}{a}, \tag{12.105}$$

in which $j_{m,n}$ is the n-th zero of J_m. Likewise, $J_m(k\cdot)$ satisfies a Dirichlet boundary condition at $\rho = b$ if and only if

$$k = \frac{j_{m,n'}}{b} \tag{12.106}$$

for some integer n'. Because the zeros $\{j_{m,n}\}$ of J_m are unique real numbers and are not equally spaced, it is generally impossible to choose boundary conditions such that J_m vanishes at both $\rho = a$ and $\rho = b$ except in the special case in which $a = 0$. In that case the integrated terms vanish at the origin (because they are proportional to ρ) unless one of the functions f, g, f', g' is singular at the origin.

The linear manifold $\mathcal{M}$ on which the Bessel operator is self-adjoint given a Dirichlet boundary condition

$$J_m(kb) = 0 \tag{12.107}$$

at $\rho = b$ and a condition of continuity and differentiability at the origin is, therefore,

$$\mathcal{M} = \text{null}[\delta_b] \cap \mathcal{C}^\infty([0, b]; \mathbb{R}). \tag{12.108}$$

A solution of Bessel's equation that satisfies Dirichlet conditions at $\rho = a > 0$ and $\rho = b$ is a linear combination of two linearly independent solutions of Bessel's equation. In this case Dirichlet boundary conditions at a and b determine k as the solution of a particular eigenvalue problem. For example, let $Y_m(k\cdot)$ be an eigenfunction of the Bessel operator for order m and eigenvalue k that is linearly independent of $J_m(k\cdot)$. If

$$f(\rho) = \alpha J_m(k\rho) + \beta Y_m(k\rho) \tag{12.109}$$

then the equations

$$f(a) = \alpha J_m(ka) + \beta Y_m(ka) = 0$$
$$f(b) = \alpha J_m(kb) + \beta Y_m(kb) = 0 \tag{12.110}$$

have a nontrivial solution for α and β if and only if

$$\begin{vmatrix} J_m(ka) & Y_m(ka) \\ J_m(kb) & Y_m(kb) \end{vmatrix} = J_m(ka)Y_m(kb) - J_m(kb)Y_m(ka)$$

$$= 0 \tag{12.111}$$

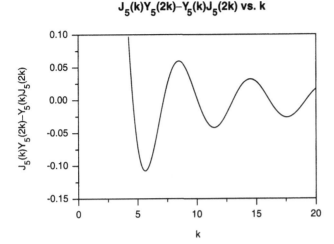

FIGURE 12.5

The value of the determinant defined in Eq. (12.111), plotted versus k for $m = 5$, $a = 1$, and $b = 2$.

such that k is real. Figure 12.5 shows a plot of the left-hand side of this equation for $m = 5$, $a = 1$, and $b = 2$. Let $k_{m,n}(a, b)$ be the n-th solution of Eq. (12.111). Let

$$f_{m,n}(\rho) := \alpha J_m(k_{m,n}(a, b)\rho) + \beta Y_m(k_{m,n}(a, b)\rho). \tag{12.112}$$

Then, by construction, $f_{m,n}$ satisfies Dirichlet boundary conditions at $\rho = a$ and $\rho = b$.

2. **Neumann boundary conditions:**

$$\forall f \in \mathcal{M} : \underline{\delta}'_a[f] = \underline{\delta}'_b[f] = 0$$
$$\Rightarrow f'(a) = f'(b) = 0, \tag{12.113}$$

in which

$$\underline{\delta}'_a[f] := -f'(a). \tag{12.114}$$

If $a > 0$, then the linear manifold on which the Bessel operator is self-adjoint is

$$\mathcal{M} = \text{null}[\underline{\delta}'_a] \cap \text{null}[\underline{\delta}'_b] \cap \mathcal{C}^\infty([0, b]; \mathbb{R}). \tag{12.115}$$

If a solution $\alpha J_m(k\rho) + \beta Y_m(k\rho)$ of Bessel's equation satisfies Eq. (12.113), then k is a real solution of the eigenvalue problem

$$\begin{vmatrix} J'_m(ka) & Y'_m(ka) \\ J'_m(kb) & Y'_m(kb) \end{vmatrix} = 0. \tag{12.116}$$

If $a = 0$, then the linear manifold on which the Bessel operator is self-adjoint, given a Neumann boundary condition

$$J'_m(kb) = 0 \tag{12.117}$$

at $\rho = b$ and a condition of continuity and differentiability at the origin, is

$$\mathcal{M} = \text{null}[\underline{\delta}'_b] \cap \mathcal{C}^\infty([0, b]; \mathbb{R}). \tag{12.118}$$

If a Bessel function of the first kind, $J_m(k\cdot)$, satisfies a Neumann boundary condition at $\rho = b$, then

$$k = \frac{j'_{m,n}}{b} \tag{12.119}$$

in which $j'_{m,n}$ is the n-th zero of J'_m and is real for every n.

3. **Mixed boundary conditions:** In this case the linear functional that must vanish at the endpoints is a linear combination of the function and its first derivative. Let

$$\begin{aligned} \forall f \in \mathcal{M} : \underline{\phi}_a[f] &= -\underline{\delta}'_a[f] + \eta \underline{\delta}_a[f] \\ &= f'(a) + \eta f(a) = 0, \\ \text{and} \quad \underline{\phi}_b[f] &= -\underline{\delta}'_b[f] + \eta \underline{\delta}_b[f] \\ &= f'(b) + \eta f(b) = 0, \end{aligned} \tag{12.120}$$

in which η is a real constant. If $a > 0$, then the linear manifold on which the Bessel operator is self-adjoint is

$$\mathcal{M} = \text{null}[\underline{\phi}_a] \cap \text{null}[\underline{\phi}_b] \cap \mathcal{C}^\infty([a, b]; \mathbb{R}). \tag{12.121}$$

If $a = 0$, then the appropriate linear manifold is

$$\mathcal{M} = \text{null}[\underline{\phi}_b] \cap \mathcal{C}^\infty([a, b]; \mathbb{R}). \tag{12.122}$$

The set of values of k such that

$$J'_m(kb) + \eta J_m(kb) = 0 \tag{12.123}$$

is countable. Unlike Neumann or Dirichlet conditions at $\rho = b$, however, the mixed boundary condition Eq. (12.123) may lead to complex values of k.

12.6.2 ORTHOGONALITY RELATIONS

Let $J_m(k\cdot)$ and $J_m(k'\cdot)$ satisfy the same boundary condition at $\rho = b$, and let that boundary condition be Eq. (12.107), Eq. (12.117), or Eq. (12.123). Because $J_m(k\cdot)$ and $J_m(k'\cdot)$ are eigenfunctions of a self-adjoint linear mapping, they are orthogonal if $k' \neq k$:

$$k' \neq k \Rightarrow \int_0^b J_m(k\rho) \, J_m(k'\rho) \, \rho \, d\rho = 0. \tag{12.124}$$

In particular, this equation holds in the case of Dirichlet boundary conditions at $\rho = b$, for which

$$k = \frac{j_{m,n}}{b}, \quad k' = \frac{j_{m,n'}}{b}, \tag{12.125}$$

and in the case of Neumann boundary conditions at $\rho = b$, for which

$$k = \frac{j'_{m,n}}{b}, \quad k' = \frac{j'_{m,n'}}{b}. \tag{12.126}$$

Orthogonality relations also hold for linear combinations of Bessel functions which obey Dirichlet, Neumann, or mixed boundary conditions at the end points a, b of a finite interval $[a, b]$ if $0 < a < b$. For example,

$$n' \neq n \Rightarrow \int_a^b f_{m,n}(\rho) \, f_{m,n'}(\rho) \, \rho \, d\rho = 0 \tag{12.127}$$

in which $f_{m,n}$ is defined in Eq. (12.112). The key to orthogonality among solutions of Bessel's equation on a finite interval $[a, b]$ is that different solutions are orthogonal if they all obey the same boundary conditions and if those boundary conditions imply that the Bessel operator is self-adjoint.

12.6.3 FOURIER-BESSEL SERIES

It can be shown that the orthonormal set

$$\left\{ \frac{1}{K_{m,n}(b)} J_m(j_{m,n}\rho/b) \,\middle|\, n \in \mathbb{Z}^+ \right\}, \tag{12.128}$$

in which

$$[K_{m,n}(b)]^2 := \int_0^b [J_m(j_{m,n}\rho/b)]^2 \, \rho \, d\rho, \tag{12.129}$$

is complete in the Hilbert space $\mathcal{L}_\rho^2([0, b]; \mathbb{R})$.[2] Note that the Bessel order m is fixed. Orthogonal eigenfunctions are indexed by different values of n, and therefore have different numbers of zeros in the interval $[0, b]$. By the definition of the Bessel zeros $j_{m,n}$, every function in the complete set Eq. (12.128) vanishes at $\rho = b$. It can be shown that the value of the

---

[2] For a proof, see [Watson 1966], Chapter XVIII.

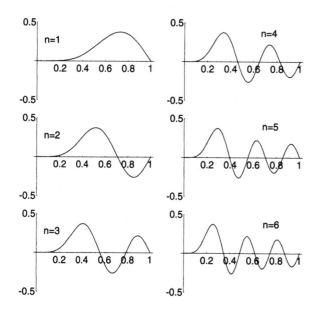

FIGURE 12.6

The orthogonal functions $J_5(j_{5,n}x)$ (*vertical axes*) as functions of x (*horizontal axes*) for $n = 1$ through $n = 6$. The integer n is equal to the number of zeros of $J_5(j_{5,n}x)$ for $x \neq 0$ in the interval $[0, 1]$.

normalization constant is given by the formula

$$[K_{m,n}(b)]^2 = \frac{b^2}{2}[J_{m+1}(j_{m,n})]^2. \tag{12.130}$$

Figure 12.6 shows the orthogonal functions $J_5(j_{5,n}x)$ as functions of $x \in [0, 1]$ for $n = 1$ through $n = 6$.

The orthonormal set

$$\left\{ \frac{1}{K'_{m,n}(b)} J_m(j'_{m,n}\rho/b) \,\middle|\, n \in \mathbb{Z}^+ \right\}, \tag{12.131}$$

in which

$$[K'_{m,n}(b)]^2 := \int_0^b [J_m(j'_{m,n}\rho/b)]^2 \, \rho \, d\rho, \tag{12.132}$$

is also complete in $\mathcal{L}_\rho^2([0, b]; \mathbb{R})$. Every element of the latter set has a vanishing derivative at $\rho = b$.

Orthogonal series using either of the complete sets Eq. (12.128) or Eq. (12.131) are called **Fourier-Bessel series**. Any square-integrable function on the interval $[0, b]$ can be expanded in the Fourier-Bessel series

$$f(\rho) = \sum_{n=1}^{\infty} f_n \, J_m(j_{m,n}\rho/b) \tag{12.133}$$

or

$$f(\rho) = \sum_{n=1}^{\infty} f'_n \, J_m(j'_{m,n}\rho/b) \tag{12.134}$$

in which the coefficients are

$$f_n = \frac{1}{K_{m,n}} \int_0^b J_m(j_{m,n}\rho/b)\, f(\rho)\, \rho\, d\rho \tag{12.135}$$

and

$$f_n' = \frac{1}{K_{m,n}'} \int_0^b J_m(j_{m,n}'\rho/b)\, f(\rho)\, \rho\, d\rho. \tag{12.136}$$

The series in Eqs. (12.133) and (12.134) converge under the inner-product norm defined by Eq. (12.98) for the special case in which $a = 0$. Of course, convergence under an inner-product norm does not imply pointwise or uniform convergence. It can be shown that if f is continuous, if the variation of f is bounded, and if $f(b) = 0$, then the convergence of the Fourier-Bessel series in Eqs. (12.133) and (12.134) is uniform (for a proof, see Tolstov [1962, chapter 3]).

12.6.4 EXERCISE FOR SECTION 12.6

12.6.1 Prove that

$$\rho \left[J_m(j_{m,n}\rho/b) \frac{d}{d\rho} J_m(j_{m,n'}\rho/b) - \frac{d}{d\rho} J_m(j_{m,n}\rho/b) J_m(j_{m,n'}\rho/b) \right]\Big|_0^b$$

$$= \frac{j_{m,n}^2 - j_{m,n'}^2}{b^2} \int_0^b J_m(j_{m,n}\rho/b) J_m(j_{m,n'}\rho/b)\, \rho\, d\rho. \tag{12.137}$$

This formula can be used as the basis for an alternative proof of the orthogonality of Bessel functions of the first kind and of the same order, all of which satisfy a Dirichlet boundary condition at $\rho = b$.

12.7 VIBRATIONS OF A DRUMHEAD

The problem of finding the normal modes of a uniformly stretched, vibrating drumhead illustrates how the results derived in this chapter can be used in practice. A simple application of Newton's second law shows that the vertical displacement u of the drumhead satisfies the two-dimensional wave equation

$$\left(\nabla^2 - \frac{1}{c_d^2} \frac{\partial^2}{\partial t^2} \right) u(\mathbf{r}, t) = 0 \tag{12.138}$$

in the limit $u \ll b$, in which b is the radius of the drumhead, $c_d = (\sigma/\mu)^{1/2}$ is the velocity of elastic waves, σ is the surface tension, and μ is the mass per unit area.

By definition, the time-dependence of a normal mode of a linear system is sinusoidal. Assuming that u can be expressed as either a sum or an integral over functions of the form

$\text{Re}[e^{-i\omega t}\,\psi(\mathbf{r})]$, one obtains the two-dimensional Helmholtz equation

$$(\nabla^2 + k^2)\psi(\mathbf{r}) = 0 \tag{12.139}$$

for the normal-mode amplitude ψ at frequency ω, in which $k = \omega^2/c_d^2$. (One can obtain this equation also by noticing that the wave equation is invariant under time translations, which suggests an expansion in terms of the functions $e^{-i\omega t}$.)

We assume that the drumhead is clamped rigidly at $\rho = b$. Then ψ must obey the Dirichlet boundary condition

$$\psi(b, \phi) = 0 \tag{12.140}$$

for all $\phi \in [0, 2\pi]$.

On physical grounds one can say that $\int_0^b \int_0^{2\pi} |\psi(\rho, \phi)|^2 \rho\,d\phi\,d\rho < \infty$, because the elastic energy of the drumhead is proportional to the integral of $|\psi|^2$. Again on physical grounds, ψ must be at least twice-differentiable for all $\rho < b$ and for all ϕ, because the elastic force per unit area is proportional to $\nabla^2 \psi$. Also, $\psi(\rho, \phi + 2\pi) = \psi(\rho, \phi)$ because ψ must be single-valued. Therefore at each ρ,

$$\psi(\rho, \phi) = \sum_{m=-\infty}^{\infty} \psi_m(\rho)\, e^{im\phi} \tag{12.141}$$

in which

$$\psi_m(\rho) = \frac{1}{2\pi} \int_0^{2\pi} e^{-im\phi}\, \psi(\rho, \phi)\,d\phi \tag{12.142}$$

is a function of ρ alone. Because ψ_m is continuous and $\psi_m(b) = 0$, it follows from Eq. (12.133) that each ψ_m has a uniformly convergent Fourier-Bessel series

$$\psi_m(\rho) = \sum_{n=1}^{\infty} f_{m,n}\, J_m(j_{m,n}\rho/b). \tag{12.143}$$

To find the coefficients $f_{m,n}$, one uses by Eq. (12.135), replacing f with ψ_m.

The requirement that ψ must satisfy the Helmholtz equation implies that only one term of the Fourier and Fourier-Bessel expansions can be nonzero, because each normal mode $e^{im\phi} J_m(j_{m,n}\rho/b)$ satisfies a Helmholtz equation with a unique propagation constant

$$k_{m,n} = j_{m,n}/b. \tag{12.144}$$

It follows that the solution of the wave equation for the vibrations of a drumhead, including the harmonic time dependence of each mode, is

$$u(\rho, \phi, t) = \sum_{m=-\infty}^{\infty} \sum_{n=1}^{\infty} f_{m,n} e^{im\phi} J_m(j_{m,n}\rho/b) e^{-i\omega_{m,n}t}, \tag{12.145}$$

in which

$$\omega_{m,n}^2 = c_d^2 k_{m,n}^2. \tag{12.146}$$

The coefficients $f_{m,n}$ must be determined from an initial condition such as

$$u(\rho, \phi, 0) = f(\rho, \phi), \tag{12.147}$$

for which the Fourier-Bessel coefficients are

$$f_{m,n} = \frac{1}{2\pi K_{m,n}} \int_0^{2\pi} \int_0^b e^{-im\phi} J_m(j_{m,n}\rho/b) f(\rho, \phi) \rho \, d\rho \, d\phi. \tag{12.148}$$

12.7.1 EXERCISE FOR SECTION 12.7

12.7.1 Plot the displacement, Eq. (12.145), as a three-dimensional surface plot for selected values of t. If possible, make a movie showing how the displacement changes in time.

12.8 POWER SERIES FOR J_m

12.8.1 DERIVATION USING RAISING AND LOWERING OPERATORS

For some purposes it is useful to have a power series for J_m. One can derive a series expansion by exploiting the close relationship of the Bessel functions with the group $E(2)$. Because the translation group $T(\mathbb{E}^2)$ is a subgroup of $E(2)$ and because we obtained the Bessel functions $J_m(k\rho)$ from the representation $e^{ik\cdot r}$ of $T(\mathbb{E}^2)$, it makes sense to begin by considering the translation operator $\mathsf{T_a}$. By definition,

$$\forall r \in \mathbb{E}^2 : \mathsf{T_a} r = r + a. \tag{12.149}$$

We show in Chapter 4 that $\mathsf{T_a}$ induces a mapping of functions on $\mathbb{E}^2$, as follows:

$$\forall r \in \mathbb{E}^2 : (\mathsf{P_a} f)(r) := f(\mathsf{T_a}^{-1})(r) = f(r - a). \tag{12.150}$$

One can choose a representation space that contains solely functions f that are infinitely differentiable, for the space of such functions is dense in the Hilbert space $\mathcal{L}^2(\mathbb{E}^2)$.

From Taylor's theorem in two dimensions, one has

$$f(r + a) = e^{a \cdot \nabla} f(r). \tag{12.151}$$

Then

$$P_{-a} = e^{a \cdot \nabla}.$$

(12.152)

If one takes $f = H_m$, then it makes sense to try to expand the translation operator $e^{a \cdot \nabla}$ in an exponential series involving the Bessel raising and lowering operators. One sees now that it is possible to determine the functional form of H_m from its transformation properties:[3]

$$\begin{aligned}
H_m(\mathbf{a}) &= (P_{-a} H_m)(\mathbf{r})|_{\mathbf{r} = 0} \\
&= e^{a \cdot \nabla} H_m(\mathbf{r})|_{\mathbf{r} = 0}.
\end{aligned}$$

(12.153)

The remaining steps, then, are to transform

$$\mathbf{a} \cdot \nabla = a_x \frac{\partial}{\partial x} + a_y \frac{\partial}{\partial y}$$

(12.154)

into plane polar coordinates, to recognize the Helmholtz raising and lowering operators $B_{\pm}$, and to evaluate $e^{a \cdot \nabla}$.

Using the chain rule for partial differentiation, one finds (see Exercise 12.8.1) that

$$\begin{aligned}
\frac{\partial f}{\partial x} &= \frac{k}{2}(-B_+ + B_-) \\
\frac{\partial f}{\partial y} &= \frac{ik}{2}(B_+ + B_-).
\end{aligned}$$

(12.155)

Then (see Exercise 12.8.2)

$$\mathbf{a} \cdot \nabla = \frac{ka}{2}(-e^{-i\alpha} B_+ + e^{i\alpha} B_-)$$

(12.156)

in which $\alpha = \tan^{-1}(a_y/a_x)$ is the polar angle of $\mathbf{a}$. Putting this result together with Eq. (12.152) for P_{-a}, one obtains

$$P_{-a} = \exp\left[\left(-\frac{ka}{2}\right) e^{-i\alpha} B_+ + \left(\frac{ka}{2}\right) e^{i\alpha} B_-\right].$$

(12.157)

Because B_+ and B_- commute, and therefore obey the same laws of algebra as ordinary numbers, the law of exponents is valid for the above exponential. Separating the exponential factors, and expanding each one in an infinite series, one obtains

$$\begin{aligned}
P_{-a} &= \exp\left[\left(-\frac{ka}{2}\right) e^{-i\alpha} B_+\right] \exp\left[\left(\frac{ka}{2}\right) e^{i\alpha} B_-\right] \\
&= \sum_{j=0}^{\infty} \frac{(-1)^j}{j!} \left(\frac{ka}{2}\right)^j e^{-ij\alpha} B_+^j \sum_{l=0}^{\infty} \frac{1}{l!} \left(\frac{ka}{2}\right)^l e^{il\alpha} B_-^l.
\end{aligned}$$

(12.158)

---

[3] The insight that transformation properties under a group can completely or partly determine the wave function of a quantum system was one of Wigner's achievements. See [Wigner 1959] Chapter 19.

Applying the right-hand side of this equation to $J_m(k\rho)\, e^{im\phi}$, one encounters $(\mathsf{B}_-)^l\, H_m = H_{m-l}$ and $(\mathsf{B}_+)^j\, H_{m-l} = H_{m+j-l}$. From Eq. (12.42) and Exercise 12.2.2,

$$H_{m+j-l}(0, \phi) = \delta_{m+j-l,0} \tag{12.159}$$

because $J_n(0) = \delta_{n,0}$. Then

$$\begin{aligned}
H_m(\mathbf{a}) &= J_m(ka)\, e^{im\alpha} \\
&= (\mathsf{P}_{-\mathbf{a}} H_m)(0) \\
&= \sum_{j=0}^{\infty} \frac{(-1)^j}{j!\,(j+m)!} \left(\frac{ka}{2}\right)^{2j+m} e^{i(j+m-j)\alpha}
\end{aligned} \tag{12.160}$$

from which it follows at once that

$$J_m(z) = \sum_{j=0}^{\infty} \frac{(-1)^j}{j!\,(j+m)!} \left(\frac{z}{2}\right)^{2j+m}. \tag{12.161}$$

Thus the power series for J_m is a straightforward consequence of symmetry and can be derived from the exponential series for the translation operator in two-dimensional Euclidean space.

One shows straightforwardly that the radius of convergence of the power series Eq. (12.161) for J_m is $R = \infty$. It follows that the series converges for all finite complex values of z and therefore that J_m is an entire analytic function. For every finite z, the power series converges absolutely and uniformly.

The power series in Eq. (12.161) can be taken as the definition of the Bessel function of the first kind of integer order m.

12.8.2 PROPERTIES DEDUCED FROM THE POWER SERIES

The partial sums of the series in Eq. (12.161) are alternating and are therefore subject to catastrophic cancellation of significant digits in any fixed-precision floating-point representation. The power series is therefore unsuitable for numerical evaluation of $J_m(z)$ unless $|z| \ll 1$, in which case one has the **asymptotic form for small argument**,

$$J_m(z) \approx \frac{z^m}{2^m m!} = \frac{z^m}{(2m)!!}, \tag{12.162}$$

which simply approximates $J_m(z)$ by the first term of the series.

12.8.3 EXERCISES FOR SECTION 12.8

12.8.1   Derive Eqs. (12.155).

12.8.2   Verify Eq. (12.156).

12.8.3   Prove that the radius of convergence of the power series for J_m given in Eq. (12.161) is $R = \infty$.

12.9   COMPLETENESS RELATIONS USING J_n

Often one must expand a solution of the Helmholtz equation over all of two-dimensional space, rather than in a finite disk as in Section 12.7. In this section we develop analogs of Fourier-Bessel expansions such as Eq. (12.133) and of normal-mode expansions such as Eq. (12.145), that converge over the entire two-dimensional plane, and in which different terms are automatically orthogonal. For the sake of simplicity, we restrict the functions to be expanded to the class of rapidly decreasing, complex-valued functions on $\mathbb{E}^2$.

Our basic tool is the Fourier transform. If applied in two-dimensional Euclidean space, Eq. (11.362) for the Fourier transform of a rapidly decreasing function $f : \mathbb{E}^2 \to \mathbb{C}$ becomes

$$\tilde{f}(\mathbf{k}) = \frac{1}{2\pi} \int_{\mathbb{E}^2} e^{-i\mathbf{k} \cdot \mathbf{r}} \, f(\mathbf{r}) \, d^2 r. \tag{12.163}$$

Let the plane polar coordinates of $\mathbf{k}$ and $\mathbf{r}$ be (k, θ) and (r, ϕ), respectively. We express $e^{-i\mathbf{k} \cdot \mathbf{r}}$ in terms of Bessel functions using the Jacobi-Anger expansion (Eq. [12.43]) to obtain

$$\begin{aligned}
\tilde{f}(k, \theta) &= \frac{1}{2\pi} \int_0^\infty \int_0^{2\pi} e^{-ik\rho \cos(\theta - \phi)} \, f(\rho, \phi) \, \rho \, d\phi \, d\rho \\
&= \frac{1}{2\pi} \sum_{n=-\infty}^{\infty} (-i)^n \, e^{-in\theta} \int_0^\infty \int_0^{2\pi} J_n(k\rho) \, e^{in\phi} \, f(\rho, \phi) \, \rho \, d\phi \, d\rho,
\end{aligned} \tag{12.164}$$

which holds for a general rapidly decreasing function on $\mathbb{E}^2$.

If f is the solution of a problem with rotational symmetry, then f is equal to a (possibly infinite) sum of functions that transform according to irreducible representations of $SO(2)$. Let

$$f(\rho, \phi) = g(\rho) \, e^{-im\phi}, \tag{12.165}$$

in which g is a rapidly decreasing function on $\mathbb{R}^+$. Clearly f carries the irreducible representation $e^{im\phi}$ of $SO(2)$. Then we use the orthogonality relation

$$\int_0^{2\pi} e^{i(n-m)\phi} \, d\phi = 2\pi \, \delta_{n,m} \tag{12.166}$$

in Eq. (12.164) to obtain

$$\tilde{f}(k, \theta) = \frac{1}{2\pi} (-i)^m \, e^{-im\theta} \int_0^\infty J_m(k\rho) \, g(\rho) \, \rho \, d\rho. \tag{12.167}$$

It follows that in this case the Fourier transform $\tilde{f}$ carries the same representation of $SO(2)$ as f, and that

$$\tilde{f}(k, \theta) = (-i)^m e^{-im\theta} g^{[H,m]}(k),\tag{12.168}$$

in which $g^{[H,m]}$ is the **Hankel transform** of g of order m,

$$g^{[H,m]}(k) := \int_0^\infty J_m(k\rho) g(\rho) \rho \, d\rho.\tag{12.169}$$

A Hankel transform is simply the radial part of the two-dimensional Fourier transform of a function with definite symmetry under $SO(2)$.

We now obtain the completeness relations for the Helmholtz and Bessel functions, as well as the inversion formula for Hankel transforms. Our starting point is the Fourier inversion formula Eq. (11.364) in $\mathbb{E}^2$,

$$f(\mathbf{r}) = \frac{1}{(2\pi)^2} \int\int_{\mathbb{E}^2} e^{i\mathbf{k}\cdot(\mathbf{r}-\mathbf{r}')} f(\mathbf{r}') d^2r' \, d^2k.\tag{12.170}$$

Expanding $\exp i\mathbf{k} \cdot (\mathbf{r} - \mathbf{r}') = e^{i\mathbf{k}\cdot\mathbf{r}} e^{-i\mathbf{k}\cdot\mathbf{r}'}$ using the Jacobi-Anger expansion Eq. (12.43), one obtains

$$
\begin{aligned}
f(\rho, \phi) &= \frac{1}{(2\pi)^2} \int_0^\infty \int_0^\infty \int_0^{2\pi} \int_0^{2\pi} e^{ik\rho\cos(\theta-\phi)} e^{-ik\rho'\cos(\theta-\phi')} f(\rho', \phi') k\,\rho'\, d\theta\, d\phi'\, dk\, d\rho' \\
&= \frac{1}{(2\pi)^2} \sum_{m,n=-\infty}^{\infty} i^{m-n} \int_0^\infty \int_0^\infty \int_0^{2\pi} \int_0^{2\pi} e^{i[m(\theta-\phi)+n(\phi'-\theta)]} \\
&\qquad\qquad\qquad \times J_m(k\rho) J_n(k\rho') f(\rho', \phi') k\,\rho'\, d\theta\, d\phi'\, dk\, d\rho' \\
&= \frac{1}{2\pi} \sum_{n=-\infty}^{\infty} \int_0^\infty \int_0^\infty \int_0^{2\pi} e^{in(\phi'-\phi)} J_n(k\rho) J_n(k\rho') f(\rho', \phi') k\,\rho'\, d\phi'\, dk\, d\rho'
\end{aligned}\tag{12.171}
$$

after using Eq. (12.166). Because f is any rapidly decreasing function, it follows that

$$
\begin{aligned}
&\frac{1}{2\pi} \sum_{n=-\infty}^{\infty} \int_0^\infty e^{in(\phi'-\phi)} J_n(k\rho) J_n(k\rho') k \, dk \\
&= \frac{1}{2\pi} \sum_{n=-\infty}^{\infty} \int_0^\infty H_n(k\rho, \phi)^* H_n(k\rho', \phi') k \, dk \\
&= \frac{1}{\rho'}\delta(\rho' - \rho)\delta(\phi' - \phi).
\end{aligned}\tag{12.172}
$$

This equation is the completeness relation for the Helmholtz functions on $\mathbb{E}^2$. The completeness relation holds also with complex conjugation applied to $H_n(k\rho', \phi')$ instead of to $H_n(k\rho, \phi)$.

In the special case in which f belongs to an irreducible representation of $SO(2)$, Eq. (12.165), one finds that for every $g \in \mathcal{S}(\mathbb{R}^+)$,

$$g(\rho) = \int_0^\infty \int_0^\infty J_m(k\rho) J_m(k\rho') g(\rho') k \, \rho' \, dk \, d\rho', \tag{12.173}$$

which implies that

$$\boxed{\int_0^\infty J_m(k\rho) J_m(k\rho') k \, dk = \frac{1}{\rho'} \delta(\rho' - \rho).} \tag{12.174}$$

This equation is the completeness relation for J_m on the positive real axis.

From Eq. (12.173) one obtains the **Hankel-transform inversion formula**

$$g(\rho) = \int_0^\infty J_m(k\rho) g^{[H,m]}(k) k \, dk, \tag{12.175}$$

which holds for all integral orders m and (it turns out) also for all real orders ν such that $\nu > -\frac{1}{2}$.

12.10 BIBLIOGRAPHY

Lighthill, M. J. *Introduction to Fourier Analysis and Generalised Functions.* Cambridge: University Press, 1958.

Talman, James D. *Special Functions: A Group Theoretic Approach.* New York: Benjamin, 1968.

Tolstov, Georgi P. *Fourier Series.* Englewood Cliffs, NJ: Prentice-Hall, 1962.

Vilenkin, N. Ja. *Special Functions and the Theory of Group Representations* (*Translations of Mathematical Monographs*, vol. 22). Providence, RI: American Mathematical Society, 1968.

Watson, G. N. *A Treatise on the Theory of Bessel Functions*, Cambridge: University Press, 1966.

Wigner, E. P. *Group Theory and Its Applications to the Quantum Mechanics of Atomic Spectra*, New York: Academic Press, 1959.

APPENDIX A

INDEX OF NOTATION

A.1 QUANTIFIERS AND OTHER LOGICAL SYMBOLS

$\exists$, there exists
$\forall$, for all
$\ni$, such that
$\neg$, negation
$:=$, is defined as

A.2 SETS AND MAPPINGS

$\in$, belongs to (a set)
$\notin$, does not belong to (a set)
$\cap$, intersection of sets
$\cup$, union of sets
$\triangle$, symmetric difference of sets ($S \triangle T := (S \backslash T) \cup (T \backslash S)$)
$\subset$, is a proper subset of
$\subseteq$, is a subset of
$\emptyset$, the empty set (i.e., the null set)
$\mathfrak{P}(S)$, the power set of the set S (i.e., the set of all subsets of S)
$\prec$, precedes
inf S, infimum (greatest lower bound) of the partially ordered set $S \subset \mathbb{R}$
sup S, supremum (least upper bound) of the partially ordered set $S \subset \mathbb{R}$
T°, interior of subset T of a metric space
$\overline{T}$, closure of the subset T of a metric space
∂T, boundary of subset T of a metric space
$\{x \in S \mid s(x)\}$, set of elements x of the set S such that the specification $s(x)$ is true
$(x_1, \ldots, x_n)$, ordered n-tuple
$(n_1 : n_2)$ (in which $n_1 \leq n_2$), set containing the integers from n_1 to n_2
$S \backslash T$, complement of T relative to S ($S \backslash T := \{x \in S \mid x \notin T\}$)
$\phi : S \to T$, mapping ϕ under which the image of the set S is a subset of T
$\phi : x \mapsto y$, mapping ϕ under which the image of x is y
$f(S)$, image of set S under mapping f
$f^{-1}(T)$, inverse image of T under f ($f^{-1}(T) := \{x \mid f(x) \in T\}$)

supp[f], support of function $f : S \to \mathbb{F}$ (supp[f] := $\overline{\{x \in S \mid f(x) \neq 0\}}$)
(a, b), ordered pair; $(x_1, x_2, \ldots)$, sequence with terms $x_1, x_2, \ldots$
(α, β), open interval $\{x \in \mathbb{R} \mid \alpha < x < \beta\}$
$[\alpha, \beta]$, closed interval $\{x \in \mathbb{R} \mid \alpha \leq x \leq \beta\}$

A.3 VECTOR SPACES, LINEAR MAPPINGS AND MATRICES

A, linear mapping
A^T, transposed linear mapping
A^H, Hermitian adjoint of A
A, matrix of linear mapping A
$\mathbf{A}^T$, transpose of matrix **A**
$\mathbf{A}^\dagger$, transpose, complex conjugate of matrix **A**
$\mathbf{A}^H$, Hermitian adjoint of matrix **A**
[A, B], commutator of operators A and B ([A, B] := AB − BA)
$\mathcal{C}^n([a, b]; \mathbb{R})$, space of n-times continuously differentiable functions from $[a, b]$ to $\mathbb{R}$
$\mathbb{C}$, set of complex numbers
$\mathbb{C}^n$, space of column vectors with n complex components
$\mathbb{C}_n$, space of row vectors with n complex components
$\mathbb{C}^{m \times n}$, space of $m \times n$ matrices with complex elements
$\mathcal{D}(\mathbb{R}^n)$, space of infinitely differentiable functions with compact support in $\mathbb{R}^n$
$\mathcal{D}'(\mathbb{R}^n)$, topological dual of $\mathcal{D}(\mathbb{R}^n)$; space of distributions
$\mathbb{F}$, number field (usually $\mathbb{R}$ or $\mathbb{C}$)
$\mathbb{F}^{\mathbb{Z}}$, space of infinite sequences with terms in $\mathbb{F}$, labeled by elements of $\mathbb{Z}$
$\mathcal{H}$, Hilbert space
$\mathcal{L}^2([a, b]; \mathbb{C})$, space of complex-valued, Lebesgue–square-integrable functions on $[a, b]$
$\mathcal{M}$, linear manifold
$\mathcal{M}$, Minkowski space (four-dimensional space-time)
$\mathcal{O}$, vector space consisting solely of the zero vector
$\mathbb{Q}$, set of rational numbers
$\mathbb{R}$, set of real numbers; $\mathbb{R}^+$, set of positive real numbers
$\mathbb{R}^n$, space of column vectors with n real components
$\mathbb{R}_n$, space of row vectors with n real components
$\mathbb{R}^{m \times n}$, space of $m \times n$ matrices with real elements
$\mathcal{V}$, abstract vector space
$\mathcal{V}^*$, algebraic dual of $\mathcal{V}$ (space of linear functionals on $\mathcal{V}$)
$\mathcal{V}'$, topological dual of $\mathcal{V}$ (space of continuous linear functionals on $\mathcal{V}$)
$\mathcal{W}, \mathcal{X}, \mathcal{Y}, \mathcal{Z}$, vector subspaces
$\mathcal{W}^\perp$, orthogonal complement of $\mathcal{W}$
$\mathbb{Z}$, set of integers
$\mathbb{Z}^+$, set of positive integers
$\mathbb{N}$, set of nonnegative integers (natural numbers)

a^i_j, element of matrix $\mathbf{A}$ in row i and column j

$c(\mathbb{F})$, vector space of convergent sequences with terms in $\mathbb{F}$

$m(\mathbb{F})$, vector space of bounded sequences with terms in $\mathbb{F}$

$\ell^2(\mathbb{Z}; \mathbb{F})$, vector space of sequences with terms in $\mathbb{F} = \mathbb{R}$ or $\mathbb{F} = \mathbb{C}$, indexed by the elements of $\mathbb{Z}$, such that $\sum_{\mathbb{Z}} |c_m|^2 < \infty$

$\mathbf{x}$, column vector

$\mathbf{x}^T$, row vector

$\overline{x}$, abstract vector

$\underline{x}$, linear functional such that $\underline{x}[\overline{y}] = \langle \overline{x}, \overline{y} \rangle$

x^i, i-th component of $\mathbf{x}$; i-th contravariant component of $\overline{x}$

x_i, i-th covariant component of $\overline{x}$

α, scalar

ϕ, abstract linear functional

$\phi[\overline{x}]$, value of linear functional ϕ on vector $\overline{x}$

0, zero scalar in $\mathbb{R}$ or $\mathbb{C}$

$\overline{0}$, zero vector

$\underline{0}$, null functional (for all vectors $\overline{x}$, $\underline{0}[\overline{x}] := 0$)

$\mathbf{0}$, null mapping or null matrix (matrix with all elements equal to zero)

$\mathbf{1}$, identity mapping or unit matrix

A.4 NORMS AND INNER PRODUCTS

$\|\mathsf{A}\|$, norm of linear mapping A

$\|\mathbf{A}\|_F$, Frobenius norm of matrix $\mathbf{A}$

$\|\mathbf{A}\|_p$, p-norm of matrix $\mathbf{A}$

$\|f\|_C$, Chebyshev (supremum) norm of real-valued function f

$|r|$, absolute value of $r \in \mathbb{R}$

$\langle \overline{x}, \overline{y} \rangle$, inner product of vectors $\overline{x}$ and $\overline{y}$

$\|\overline{x}\|$, norm of abstract vector $\overline{x}$

$\|\overline{x}\|_{ip}$, inner-product norm of abstract vector $\overline{x}$

$\|\mathbf{x}\|_p$, p-norm of column vector $\mathbf{x}$

$|z|$, modulus of complex number z ($|x + iy| := [x^2 + y^2]^{1/2}$ if $x, y \in \mathbb{R}$)

A.5 FUNCTIONS

A.5.1 GENERAL NOTATION FOR FUNCTIONS

$f(\cdot)$, "place-holder" notation for a function f

$f(x)$, value of the function f at x (the image of x under f)

$\text{sign}(x)$, function with value $+1$ if $x > 0$ and -1 if $x < 0$ (value undefined if $x = 0$)

A.5.2 SPECIAL FUNCTIONS

$J_\nu(z)$, value at z of the Bessel function of the first kind of order ν

$P_l(z)$, value at z of the Legendre polynomial of order l

$T_n(z)$, value at z of the Chebyshev polynomial of the first kind of order n

$Y_\nu(z)$, value at z of the Neumann function of order ν

$Y_{lm}(\theta, \phi)$, value at θ, ϕ of the spherical harmonic of order l and index m

$j_l(z)$, value at z of the spherical Bessel function of the first kind of order l

$y_l(z)$, value at z of the spherical Neumann function of order l

$\Gamma(z)$, value of the gamma function at z

A.6 PROBABILITY

$E[f]$, mean value (expectation value) of function f

APPENDIX B

AFFINE MAPPINGS

B.1 AFFINE GROUP OF A VECTOR SPACE

The linear mappings defined in Chapter 6 are not the only mappings of a vector space that preserve important structures. In this appendix we discuss a more general class of invertible mappings, affine transformations, that preserve the parallelism of affine subspaces but are not additive in the sense of Eq. (6.1). In order to preserve the parallelism of affine subspaces, a mapping must map parallel lines to parallel lines, parallel planes to parallel planes, and so forth.

For example, a Lorentz transformation is an affine transformation. Einstein postulated that physics is the same in any two frames that are connected by a Lorentz transformation. A necessary condition for physics to be the same is that the transformation must preserve essential geometrical features such as dimensionality and parallelism.

A translation, which we defined in Section 5.2, is also a parallelism-preserving mapping. It may be helpful to keep this example in mind while studying the following definition.

An **affine mapping** in a real or complex vector space V is a mapping $F : V \to V$ that has the following properties:

1. F is bijective.
2. The image under F of an affine subspace $\bar{x} + W$ is an affine subspace through $F\bar{x}$ parallel to a unique vector subspace W' that is determined by F and W:

$$F(\bar{x} + W) = F\bar{x} + W'. \tag{B.1}$$

The property that W' is independent of $\bar{x}$ implies that F preserves parallelism. For example, $\bar{x} + W$ and $\bar{y} + W$ are parallel, and so are the images under F, $F\bar{x} + W'$ and $F\bar{y} + W'$.

3. The image under F of any affine subspace $\bar{y} + W = T_{\bar{y}-\bar{x}}(\bar{x} + W)$ is the affine subspace of V that results from applying the translation $T_{F\bar{y}-F\bar{x}}$ to $F(\bar{x} + W)$:

$$\begin{aligned}
\forall \bar{y} \in V : F(\bar{y} + W) &= F\bar{y} + W' \\
&= F[T_{\bar{y}-\bar{x}}(\bar{x} + W)] \\
&= T_{F\bar{y}-F\bar{x}} F(\bar{x} + W) \\
&= T_{F\bar{y}-F\bar{x}}(F\bar{x} + W')
\end{aligned} \tag{B.2}$$

The first line of (B.2) follows from (B.1); the second line follows from the definition of a translation. The third line is a transcription of assumption 3. The fourth line follows from the third line and Eq. (B.1).

4. The mapping $A : \mathcal{V} \to \mathcal{V}$ defined by

$$A\bar{x} := F\bar{x} - \bar{a}, \tag{B.3}$$

in which

$$\bar{a} := F\bar{0}, \tag{B.4}$$

is homogeneous:

$$\forall \alpha \in \mathbb{F} : \forall \bar{x} \in \mathcal{V} : A(\alpha\bar{x}) = \alpha A\bar{x}. \tag{B.5}$$

In essence, this definition says that an affine transformation not only preserves the parallelism of affine subspaces but also preserves the relationship of the transformed affine subspaces to the underlying vector space.

The motivation for part 4 of the definition requires a brief explanation. We define the mapping A because it is necessary to understand how F maps vector subspaces. It turns out that A maps the vector subspace $\mathcal{W}$ onto a vector subspace $\mathcal{W}'$. Using only the first three properties, one can show that A is homogeneous with respect to *rational* scalars (Exercise B.1.9). Equation (B.5) simply ensures that what works for rational scalars also works for irrational or complex scalars (see the comments on the completeness of the real numbers in Section B.2). If $\mathcal{V}$ is a metric space (as every inner-product space is), then one can show that it is possible to replace (B.5) with the requirement that A must be continuous with respect to the metric of $\mathcal{V}$.

If one takes $\alpha = 0$ in the homogeneity equation (B.5), one obtains

$$A\bar{0} = \bar{0}. \tag{B.6}$$

It follows from Eqs. (B.1) through (B.4) that $A(-\bar{x}) = -A\bar{x}$ (see Exercise B.1.8).

Equations (B.3) and (B.4) are consistent with properties 1 through 3. One sees from (B.2) that

$$\forall \bar{y} \in \mathcal{V} : \forall \bar{w} \in \mathcal{W} : \exists \bar{w}' \in \mathcal{W}' : \ni : F(\bar{y} + \bar{w}) = T_{F\bar{y} - F\bar{x}}(F\bar{x} + \bar{w}') = F\bar{y} + \bar{w}'. \tag{B.7}$$

Take $\bar{y} = \bar{0}$; then

$$\forall \bar{w} \in \mathcal{W} : F\bar{w} = F\bar{0} + \bar{w}' = \bar{a} + \bar{w}'. \tag{B.8}$$

Then

$$\bar{w}' = F\bar{w} - \bar{a} = A\bar{w}. \tag{B.9}$$

Therefore $F\bar{w} = \bar{a} + A\bar{w}$. Of course, $\bar{w}$ can be any vector in $\mathcal{V}$, because $\mathcal{W}$ can be any subspace in $\mathcal{V}$.

The result that if $F : \mathcal{V} \to \mathcal{V}$ is an affine mapping, then there exist a vector $\bar{a} \in \mathcal{V}$ and a linear mapping $A : \mathcal{V} \to \mathcal{V}$ such that

$$\boxed{\forall \bar{x} \in \mathcal{V} : F\bar{x} = A\bar{x} + \bar{a}} \tag{B.10}$$

is called the **fundamental theorem of affine geometry**.

Property 1 implies that F is invertible. However, it does *not* follow from this definition that F is an isomorphism of $\mathcal{V}$. Affine transformations do not necessarily have the additivity or homogeneity properties defined in Eqs. (5.180) and (5.181). In fact, the simplest example of an affine transformation, a translation by a fixed vector $\bar{a}$, is neither additive nor homogeneous.

Other examples of affine transformations that are important in physics and engineering include deformations (Exercise B.1.1), shears (Exercise B.1.2), rotations (Exercise B.1.3), Lorentz transformations (Exercise B.1.4), Galilean transformations (Exercise B.1.5), and Poincaré transformations (Exercise B.1.6).

Because F is one-to-one, so is A. We can see that A maps $\mathcal{V}$ *onto* $\mathcal{V}$ as follows: Choose any $\bar{x} \in \mathcal{V}$. The vector $\bar{y}$ such that $\bar{x} = A\bar{y}$ is

$$\bar{y} = F^{-1}(\bar{x} + \bar{a}) \tag{B.11}$$

because $A\bar{y} = F\bar{y} - \bar{a} = \bar{x}$.

Because A is one-to-one and onto, the inverse mapping A^{-1} exists. We show that A is an automorphism of $\mathcal{V}$. Because

$$F\bar{x} = A\bar{x} + \bar{a}, \tag{B.12}$$

it follows that every affine transformation is an automorphism of $\mathcal{V}$ followed by a translation.

The first step is to show that A is additive, unlike F. From Eqs. (B.7) and (B.10) one has

$$\forall \bar{x}, \bar{y} \in \mathcal{V} : F(\bar{x} + \bar{y}) = F\bar{x} + A\bar{y} = \bar{a} + A\bar{x} + A\bar{y}. \tag{B.13}$$

But

$$F(\bar{x} + \bar{y}) = F(\bar{0} + \bar{x} + \bar{y}) = F\bar{0} + A(\bar{x} + \bar{y}). \tag{B.14}$$

Then A is additive:

$$\forall \bar{x}, \bar{y} \in \mathcal{V} : A(\bar{x} + \bar{y}) = A\bar{x} + A\bar{y}. \tag{B.15}$$

We have already defined A to be homogeneous. Because A is one-to-one, onto, additive, and homogeneous, it is an isomorphism of $\mathcal{V}$. In particular, A is an isomorphism of vector subspaces:

$$\mathcal{W}' = A\mathcal{W} \cong \mathcal{W}. \tag{B.16}$$

If $\mathcal{W}$ is finite-dimensional, this establishes that $\dim[\mathcal{W}'] = \dim[\mathcal{W}]$, establishing that F maps points to points, lines to lines, and so forth.

We have established that every affine transformation of $\mathcal{V}$ is uniquely determined by (and uniquely determines) a vector automorphism A and a translation $T_{\bar{a}}$. From now on we write affine transformations in the form

$$(\bar{a}, A)\bar{x} := F\bar{x} = \bar{a} + A\bar{x}. \tag{B.17}$$

The **product** or **composition** of two affine transformations is another affine transformation:

$$(\bar{a}_2, A_2)(\bar{a}_1, A_1)\bar{x} = (\bar{a}_2, A_2)[\bar{a}_1 + A_1\bar{x}] = A_2 A_1 \bar{x} + (A_2 \bar{a}_1 + \bar{a}_2). \tag{B.18}$$

Therefore the product law for affine transformations is

$$(\bar{a}_2, A_2)(\bar{a}_1, A_1) = (A_2\bar{a}_1 + \bar{a}_2, A_2 A_1). \tag{B.19}$$

Let $A(\mathcal{V})$ be the set of all affine transformations of $\mathcal{V}$. Evidently $A(\mathcal{V})$ is closed under composition. It is straightforward to check that the product of any three affine transformations is associative. The identity affine transformation is $(\bar{0}, \mathbf{1})$. The inverse of the affine transformation $(\bar{a}, A)$ is

$$(\bar{a}, A)^{-1} = (-A^{-1}\bar{a}, A^{-1}). \tag{B.20}$$

Therefore one calls $A(\mathcal{V})$ the **affine group of** $\mathcal{V}$.

Certain subgroups of $A(\mathcal{V})$ are especially important. Clearly every affine transformation of the form $(\bar{a}, \mathbf{1})$ is a translation, and every affine transformation of the form $(\bar{0}, A)$ is an automorphism of $\mathcal{V}$. Therefore both aut$[\mathcal{V}]$ and $T(\mathcal{V})$ are subgroups of $A(\mathcal{V})$. It is an exercise (Exercise B.1.7) to show that $T(\mathcal{V})$ is a *normal* subgroup of $A(\mathcal{V})$.

The subgroup $E(n)$ of $A(\mathbb{R}^n)$ whose elements $(\mathbf{a}, R)$ are such that R is a rotation is important in physics and engineering because it is the group that leaves invariant the n-**dimensional Helmholtz equation,**

$$\sum_{j=1}^{n} \frac{\partial^2 \psi}{\partial x^{j2}} + k^2 \psi = 0, \tag{B.21}$$

in which k is a real constant. $E(n)$ is called the n-**dimensional Euclidean group.** It can be shown that the matrix elements of the irreducible group representations of $E(n)$ are proportional to a Bessel function times an n-dimensional spherical harmonic; see Chapter 12 for the case $n = 2$.

B.2 COORDINATE TRANSFORMATIONS

Up to this point we have discussed vector isomorphisms and automorphisms abstractly as one-to-one mappings of one vector space onto another (or onto itself) that preserve the vector-space structure. For many practical purposes one needs concrete realizations of a vector automorphism A in terms of coordinates relative to specific bases. In this subsection we show how to express a vector automorphism in terms of a transformation of coordinates.

Two points of view are possible with respect to coordinate transformations in a vector space: In the **active** point of view, one regards A as a transformation of vectors $\bar{x} \mapsto \bar{x}'$. The transformation of coordinates induced by A can be found by making coordinate expansions of $\bar{x}$ and $\bar{x}'$ relative to a fixed basis. In the **passive** point of view, A transforms the basis $\{\bar{e}_i\}$ to a new basis $\{\bar{e}'_j\}$, but the vectors $\bar{x}$ remain fixed. The transformation of coordinates induced by A can be found by making coordinate expansions of $\bar{x}$ relative to the old and new bases.

B.2.1 ACTIVE TRANSFORMATIONS

Using the active viewpoint, let us make coordinate expansions of $\bar{x} \in \mathcal{V}$ and its image under an affine transformation,

$$\bar{x}' := (\bar{a}, A)\bar{x} = A\bar{x} + \bar{a}, \tag{B.22}$$

relative to a fixed basis $\{\bar{e}_i\}$. The coordinate expansion of $A\bar{x}$ is

$$A\bar{x} = A(x^i \bar{e}_i) = x^i (A\bar{e}_i). \tag{B.23}$$

Because $A\bar{e}_i$ belongs to $\mathcal{V}$, it has a coordinate expansion (rel$\{\bar{e}_i\}$):

$$A\bar{e}_i = a_i^j \bar{e}_j. \tag{B.24}$$

Therefore the coordinate expansion of the image of $\bar{x}$ under the affine transformation $(\bar{a}, A)$,

$$\bar{x}' = x^{j'} \bar{e}_j, \tag{B.25}$$

is

$$x^{j'} = a_i^j x^i + a^j. \tag{B.26}$$

B.2.2 PASSIVE TRANSFORMATIONS

A transformation made using the passive viewpoint consists of the following steps:

1. Shift the origin of coordinates from $\bar{0}$ to $\bar{b}$ by attaching $\mathcal{V}$ at $\bar{b}$:

$$\bar{x} \mapsto \bar{x} - \bar{b}. \tag{B.27}$$

2. Make coordinate expansions of $\bar{x} - \bar{b}$ relative to the old basis $\{\bar{e}_i\}$ and the new basis $\{\bar{e}'_j\}$:

$$\bar{x} - \bar{b} = (x^i - b^i)\bar{e}_i = x^{j'} \bar{e}'_j. \tag{B.28}$$

One can regard this step equivalently as defining coordinate expansions in the "copy" of $\mathcal{V}$ that is attached at $\bar{b}$.

3. By (5.200), one can define a vector automorphism B such that

$$\bar{e}'_j = B\bar{e}_j. \tag{B.29}$$

B.2.3 RELATION BETWEEN ACTIVE AND PASSIVE TRANSFORMATIONS

These steps imply the existence of an important relationship between A and B. It follows from Eqs. (B.28) and (B.29) and the additivity of B that

$$\bar{x} - \bar{b} = \bar{x}^{j'} B\bar{e}_j = B(x^{j'}\bar{e}_j). \tag{B.30}$$

With the definition (B.25) (which is consistent with either point of view), one obtains

$$\bar{x} - \bar{b} = B\bar{x}'. \tag{B.31}$$

Because B is a vector automorphism, its inverse exists:

$$\bar{x}' = B^{-1}(\bar{x} - \bar{b}) = (\bar{b}, B)^{-1}\bar{x}. \tag{B.32}$$

In the last step, we have used (B.20). We see from Eqs. (B.22) and (B.32) that the affine mappings by which $\bar{x}'$ is obtained in the active and passive viewpoints are inverses of one another:

$$B = A^{-1}, \quad -B^{-1}\bar{b} = \bar{a}. \tag{B.33}$$

Then the coordinate transformation (B.26) holds in both the active and the passive viewpoints. However, it is still important to remember which viewpoint one is using in a practical computation, because forgetting that the vector automorphism by which basis vectors are transformed in the passive viewpoint is the inverse of the vector automorphism that transforms vectors in the active viewpoint can cause major problems in interpreting one's results.

B.3 EXERCISES

B.3.1.   Demonstrate that the deformation defined in Eq. (6.123) is an affine mapping.

B.3.2.   Demonstrate that the shear defined in Eq. (6.124) is an affine mapping.

B.3.3.   Demonstrate that the rotation defined in (6.125) is an affine mapping of $\mathbb{R}^2$.

B.3.4.   Demonstrate that the Lorentz transformation defined in Eq. (6.126) is an affine mapping in $\mathbb{R}^2$.

B.3.5.   Let $G : \mathbb{R}^2 \to \mathbb{R}^2$ map the contravariant components of a vector, relative to the basis $\{e_1, e_2\}$, as follows:

$$\begin{aligned} x^1 &\mapsto x'^1 = x^1 - vx^2 \\ x^2 &\mapsto x'^2 = x^2 - \tau \end{aligned} \tag{B.34}$$

in which v and τ are real and nonzero. Show that **G** is an affine transformation in $\mathbb{R}^2$. (With the identifications

$$\begin{aligned} x^1 &= x \\ x^2 &= t \end{aligned} \tag{B.35}$$

G is an example of a **Galilean transformation** in nonrelativistic two-dimensional space-time.)

B.3.6. Establish the following results concerning the translation subgroup $T(\mathcal{V})$ of the affine group $A(\mathcal{V})$ of a vector space $\mathcal{V}$:

(a) Derive the formula

$$(\bar{a}, A)(\bar{b}, B)(\bar{a}, A)^{-1} = (A\bar{b}, \mathbf{1}), \tag{B.36}$$

thereby proving that $T(\mathcal{V})$ is a normal subgroup of $A(\mathcal{V})$.

(b) Prove that

$$A(\mathcal{V})/T(\mathcal{V}) \cong \mathsf{aut}[\mathcal{V}]. \tag{B.37}$$

B.3.7. Show that if A is defined as in (B.3) and (B.4), then

$$A(-\bar{x}) = -A\bar{x}. \tag{B.38}$$

Hint: Use the additivity property $A(\bar{x} + \bar{y}) = A\bar{x} + A\bar{y}$, as well as Eq. (B.6).

B.3.8. Let A be an affine mapping. Show, using only the additivity of A (Eq. [B.15]), and without making any use of the homogeneity property (B.5), that

$$\forall m, n \in \mathbb{Z} \setminus \{0\} : \; A\left(\frac{m}{n}\bar{x}\right) = \frac{m}{n}A\bar{x}. \tag{B.39}$$

Conclude that the homogeneity property (B.5) for *rational* scalars α follows from the other parts of the definition of an affine mapping.

Hint: If $m > 0$, then

$$\frac{m}{n}\bar{x} = \underbrace{\frac{1}{n}\bar{x} + \cdots + \frac{1}{n}\bar{x}}_{m \text{ terms}}. \tag{B.40}$$

APPENDIX C

PSEUDO-UNITARY SPACES

Recall that an inner-product space is called *pseudo-Euclidean* or *pseudo-unitary* if the self–inner product of a nonzero vector can be negative or zero as well as positive. A physically important example of a pseudo-Euclidean space is Minkowski space $\mathcal{M}$. In this appendix we prove that a pseudoorthonormal basis (such that $\langle \bar{e}_i, \bar{e}_i \rangle = \pm 1$) exists in every finite-dimensional pseudo-Euclidean or pseudo-unitary space $\mathcal{V}$.

Because $\mathcal{V}$ is only pseudo-Euclidean or pseudo-unitary, one has no guarantee that any given vector of a particular basis has a nonzero self–inner product. But there exists at least one vector $\bar{a}_1 \in \mathcal{V}$ such that $\bar{a}_1 \neq \bar{0}$ and $\langle \bar{a}_1, \bar{a}_1 \rangle \neq 0$. (If not, then whichever of Eqs. [8.141] and [8.145] applies to $\mathcal{V}$ implies that $\langle \bar{x}, \bar{y} \rangle$ vanishes identically for all $\bar{x}, \bar{y} \in \mathcal{V}$. This conclusion violates Eq. [8.59] and implies that $\mathcal{V}$ is not pseudo-Euclidean or pseudo-unitary after all.) Define the vector $\bar{q}_1$ as in (8.157), and let

$$d_1 := |\langle \bar{q}_1, \bar{q}_1 \rangle|^{1/2},$$
$$\bar{u}_1 := \frac{1}{d_1} \bar{q}_1. \tag{C.1}$$

By construction, $\bar{u}_1$ is a pseudo-unit vector:

$$\sigma_1 := \langle \bar{u}_1, \bar{u}_1 \rangle = \pm 1. \tag{C.2}$$

Let $\mathcal{W}_1$ be the subspace of $\mathcal{V}$ spanned by $\bar{u}_1$. Let $\mathcal{X}_1$ be any complement of $\mathcal{W}$. (A complement of $\mathcal{W}$ exists, because one can extend $\bar{u}_1$ to a basis of $\mathcal{V}$; see Eqs. [5.158] and [5.159].) By (5.255), $\dim[\mathcal{X}_1] = n - 1$. Let $\{\bar{a}_2^{(1)}, \ldots, \bar{a}_n^{(1)}\}$ be a basis of $\mathcal{X}_1$, and define

$$\forall j \in (2 : n) : \bar{a}_j^{(2)} := \bar{a}_j^{(1)} - \sigma_1 r_{1j} \bar{q}_1$$
$$= \bar{a}_j^{(1)} - \sigma_1 \langle \bar{u}_1, \bar{a}_j^{(1)} \rangle \bar{u}_1 \tag{C.3}$$
$$\Rightarrow \langle \bar{u}_1, \bar{a}_j^{(2)} \rangle = 0.$$

No $\bar{a}_j^{(2)}$ vanishes, because $\bar{u}_1 \notin \mathcal{X}_1$, which implies that $\bar{u}_1$ is linearly independent of $\{\bar{a}_2^{(1)}, \ldots, \bar{a}_n^{(1)}\}$. As for a Euclidean or unitary space, every $\bar{a}_j^{(2)} \perp \bar{u}_1$, implying that $\{\bar{u}_1, \bar{a}_2^{(2)}, \ldots, \bar{a}_n^{(2)}\}$ is linearly independent (see Section 8.1). Then every $\bar{a}_j^{(2)}$ belongs to $\mathcal{W}_1^{\perp}$, the orthogonal complement of $\mathcal{W}_1$.

We show that $\{\bar{u}_1, \bar{a}_2^{(2)}, \ldots, \bar{a}_n^{(2)}\}$ spans $\mathcal{V}$ and is therefore a basis of $\mathcal{V}$. By (5.257) and (5.258), every vector $\bar{x} \in \mathcal{V}$ has a unique decomposition

$$\bar{x} = \xi^1 \bar{u}_1 + \bar{w}' \tag{C.4}$$

in which $\overline{w}' \in X_1$. Let

$$\overline{w}' = \sum_{j=2}^{n} \xi^j \overline{a}_j^{(1)}. \tag{C.5}$$

By (C.3),

$$\overline{w}' = \sum_{j=2}^{n} \xi^j \left(\overline{a}_j^{(2)} + \sigma_1 \langle \overline{u}_1, \overline{a}_j^{(1)} \rangle \overline{u}_1 \right). \tag{C.6}$$

Therefore

$$\overline{x} = x^1 \overline{u}_1 + \sum_{j=2}^{n} \xi^j \overline{a}_j^{(2)} \tag{C.7}$$

in which

$$x^1 = \xi^1 + \sigma_1 \sum_{j=2}^{n} \langle \overline{u}_1, \overline{a}_j^{(1)} \rangle. \tag{C.8}$$

Equations (C.7) and (C.8) establish that, for every vector $\overline{x}$ in a finite-dimensional pseudo-Euclidean or pseudo-unitary vector space, there exists a decomposition with respect to W_1 and $W_1^{\perp}$ of the form (8.250). Because W_1 and $W_1^{\perp}$ are complementary subspaces, the decomposition (C.8) is unique, and V is equal to the direct sum of W_1 and $W_1^{\perp}$ (Eq. [8.249]). Therefore $\{\overline{u}_1, \overline{a}_2^{(2)}, \ldots, \overline{a}_n^{(2)}\}$ is a basis of V, and $\mathrm{span}[\overline{a}_2^{(2)}, \ldots, \overline{a}_n^{(2)}]$ is a complement of W_1.

Let us verify that the list $\{\overline{a}_2^{(2)}, \ldots, \overline{a}_n^{(2)}\}$ is a basis of $W_1^{\perp}$. Suppose that there exists a vector $\overline{w}' \in W_1^{\perp}$ such that the n-element list $\{\overline{w}', \overline{a}_2^{(2)}, \ldots, \overline{a}_n^{(2)}\}$ is linearly independent. Then $\overline{u}_1$ is linearly dependent on $\{\overline{w}', \overline{a}_2^{(2)}, \ldots, \overline{a}_n^{(2)}\}$, because $\dim[V] = n$ (see Eqs. [5.155] and [5.156]). But, because $\langle \overline{u}_1, \overline{u}_1 \rangle \neq 0$, $\overline{u}_1$ cannot be equal to a linear combination of vectors to which it is orthogonal. This contradiction establishes that $\{\overline{a}_2^{(2)}, \ldots, \overline{a}_n^{(2)}\}$ is a basis of $W_1^{\perp}$.

It turns out that $W_1^{\perp}$ is itself a pseudo-Euclidean or pseudo-unitary space. For, let $\overline{z} \in W_1^{\perp}$, $\overline{z} \neq \overline{0}$, and suppose that

$$\forall \overline{y} \in W_1^{\perp} : \langle \overline{z}, \overline{y} \rangle = 0. \tag{C.9}$$

Then

$$\forall \overline{x} \in V : \langle \overline{z}, \overline{x} \rangle = \langle \overline{z}, \alpha_1 \overline{u}_1 + \overline{v} \rangle = 0, \tag{C.10}$$

which implies that the scalar product of every vector in V with the nonzero vector $\overline{z}$ vanishes identically, contradicting the hypothesis that V is pseudo-Euclidean or pseudo-unitary. We now proceed with $W_1^{\perp}$ as with V, and so forth, obtaining a basis $\{\overline{u}_1, \overline{u}_2, \ldots\}$ of V that is orthogonal,

$$\langle \overline{u}_i, \overline{u}_j \rangle = 0 \quad \text{if } i \neq j, \tag{C.11}$$

and such that each of the basis vectors has a nonzero norm,

$$\forall i = 1, \ldots, n : \langle \overline{u}_i, \overline{u}_i \rangle \neq 0. \tag{C.12}$$

The **pseudo-unit vectors**

$$\forall i \in (1 : n) : \overline{e}_i := |\langle \overline{u}_i, \overline{u}_i \rangle|^{-1/2} \overline{u}_i \tag{C.13}$$

then produce a diagonal Gram matrix of the form (8.153). If the inner product is definite (Eq. [8.42]), then every basis vector $\overline{u}_i$ has a positive self–inner product $\langle \overline{u}_i, \overline{u}_i \rangle$, and one gets only $+$ signs in (8.153).

The resolution of a vector $\overline{x}$ relative to a finite pseudoorthonormal basis, (8.153), is

$$\overline{x} = \sum_i^n \sigma_i \langle \overline{e}_i, \overline{x} \rangle \overline{e}_i \tag{C.14}$$

because $\langle \overline{e}_j, \overline{x} \rangle = \sigma_j x^j$ (no summation convention).

APPENDIX D

REMAINDER TERM

According to Eq. (7.124), for any given set of n real numbers $x_1 < \cdots < x_n$ and for any function $f : \mathbb{R} \to \mathbb{R}$, there exists a unique polynomial p of degree $n - 1$ such that

$$\forall i \in (1 : n) \: : \: p(x_i) = f(x_i). \tag{D.1}$$

We show that if $f \in C^n([a, b]; \mathbb{R})$, then the **remainder term in polynomial interpolation** is

$$
\begin{aligned}
r_n(x) &:= f(x) - p(x) \\
&= \frac{f^{(n)}(\xi)}{n!}(x - x_1) \cdots (x - x_n)
\end{aligned}
\tag{D.2}
$$

in which ξ is an interior point of the smallest interval that contains the points $x, x_1, \ldots, x_n$. The proof shows also that ξ depends on x.

The proof is a direct application of Rolle's theorem, which states that if for a given function $f : \mathbb{R} \to \mathbb{R}$ there exist two points a and b such that $f(b) = f(a)$ and $f \in C^1([a, b]; \mathbb{R})$, then there exists a point ξ such that $a < \xi < b$ such that $f'(\xi) = 0$. In order to take advantage of Rolle's theorem, we let

$$g(z) := (z - x_1) \cdots (z - x_n). \tag{D.3}$$

Choose $x \in \mathbb{R}$ and let

$$s(z) := f(z) - p(z) - c(x)g(z), \tag{D.4}$$

in which the function c remains to be defined. Because g vanishes for every x_i and because $p(x_i) = f(x_i)$, it follows that s vanishes at $z = x_1, \ldots, x_n$. We choose the function c such that

$$s(x) = 0 \tag{D.5}$$

also. Rolle's theorem implies that s' has n zeros in the interval defined by the points x, $x_1, \ldots, x_n$, s'' has $n - 1$ zeros in the same interval ; ...; and $s^{(n)}$ has two zeros. Applying Rolle's theorem one last time, one sees that for every x there exists a real number ξ in the interior of the smallest interval that contains $x, x_1, \ldots, x_n$ such that

$$s^{(n)}(\xi) = 0. \tag{D.6}$$

We calculate $s^{(n)}$ by differentiating (D.4) with respect to z, noting that c is independent of z. Because p is a polynomial of degree $n - 1$, $p^{(n)}(z) = 0$ for all real z, and $g^{(n)}(z) = n!$. Then for every z,

$$s^{(n)}(z) = f^{(n)}(z) - n!c(x). \tag{D.7}$$

For the particular value $z = \xi$ one obtains

$$c(x) = \frac{f^{(n)}(x)}{n!}. \tag{D.8}$$

Then (D.4) implies (D.2).

APPENDIX E

BOLZANO-WEIERSTRASS THEOREM

This appendix establishes one of the basic theorems of analysis, which makes rigorous the intuitively obvious fact that if a closed interval of the real line contains an infinite subset B, then the interval must contain at least one limit point, in any neighborhood of which there are infinitely many points of B. The theorem can be extended to any bounded subset B of a finite-dimensional Euclidean or unitary vector space $\mathcal{V}$. Clearly it is enough to show that every bounded sequence in $\mathcal{V}$ has a limit point in $\mathcal{V}$.

E.1 THE REAL NUMBERS

We turn first to the metric space $\mathbb{R}$ under the absolute-value norm. The **Bolzano-Weierstraß theorem** in $\mathbb{R}$ asserts that if L is a bounded infinite subset of $\mathbb{R}$, then L has a limit point $l \in \mathbb{R}$. The proof makes use of the least-upper-bound property of $\mathbb{R}$. Because L is bounded under the absolute-value norm, there exists a positive integer M such that $|x| < M$ for every $x \in L$. Then there exists a finite closed interval $[a, b]$ such that $L \subseteq [a, b]$. If we subdivide $[a, b]$ into two equal subintervals,

$$[a, b] = \left[a, \frac{b+a}{2} \right] \bigcup \left[\frac{b+a}{2}, b \right],$$

(E.1)

then one (or both) of the subintervals must include an infinite subset of L. Let $I_1 = [a_1, b_1]$ include an infinite subset of L and continue the process of halving the chosen interval and choosing a subinterval that includes an infinite subset of L. At the n-th step the chosen interval is $[a_n, b_n]$, for which

$$b_n - a_n = \frac{b-a}{2^n} \xrightarrow[n\to\infty]{} 0$$

(E.2)

and

$$a \leq a_n < b_n \leq b.$$

(E.3)

Define the sets

$$A = \{a_n\} \quad \text{and} \quad B = \{b_n\}.$$

(E.4)

735

A is bounded above; hence sup A exists. B is bounded below; hence inf B exists. By the definitions of supremum and infimum,

$$a_n < \sup A \le \inf B < b_n. \tag{E.5}$$

Take the limit $n \to \infty$. Then

$$l := \sup A = \inf B \tag{E.6}$$

defines a unique real number l. We claim that l is a limit point of L. For, let $r > 0$. For every n that is sufficiently large that $b_n - a_n < r/2$,

$$[a_n, b_n] \subseteq S_r(l). \tag{E.7}$$

But $[a_n, b_n]$ includes an infinite subset of L. Then for every $r > 0$, $S_r(l)$ contains a point of L that is different from l. Therefore l is a limit point of L.

It is important to notice that it does not follow that $l \in L$. For example, if

$$L = \left\{ 1 - \frac{1}{2^n} \,\middle|\, n \in \mathbb{Z}^+ \right\}, \tag{E.8}$$

then $l = 1$, but l does not belong to L.

E.2 FINITE-DIMENSIONAL HILBERT SPACES

We generalize now to the case in which the metric space S is a finite-dimensional Hilbert space $\mathcal{H}$ over a field $\mathbb{F} = \mathbb{R}$ or $\mathbb{F} = \mathbb{C}$. We make use of the Bolzano-Weierstraß theorem in $\mathbb{R}$ (and, provisionally, in $\mathbb{C}$) to demonstrate the **generalized Bolzano-Weierstraß theorem**: *Every bounded subset B of a finite-dimensional Hilbert space $\mathcal{H}$ over $\mathbb{R}$ or $\mathbb{C}$ is sequentially compact.* Along the way, we show that *every finite-dimensional Euclidean or unitary space is complete and is therefore a Hilbert space.*

Let $\mathcal{V}$ be a finite-dimensional Euclidean or unitary space. Then we can choose an orthonormal basis $\{\bar{e}_1, \bar{e}_2, \ldots, \bar{e}_n\}$, in which $\dim[\mathcal{V}] = n$. Let $B \subseteq \mathcal{V}$ be bounded. Then there exist a vector $\bar{x} \in B$ and a real number $r > 0$ such that $B \subseteq S_r(\bar{x})$. Let

$$Y = \{\bar{y}_1, \bar{y}_2, \ldots\} \tag{E.9}$$

be an infinite sequence in B, and let

$$\bar{x} = \sum_{i=1}^{n} x^i \bar{e}_i, \tag{E.10}$$

$$\bar{y}_k = \sum_{i=1}^{n} y_k^i \bar{e}_i, \tag{E.11}$$

$$\bar{z}_k = \bar{x} - \bar{y}_k = \sum_{i=1}^{n} z_k^i \bar{e}_i. \tag{E.12}$$

The basic idea of what comes next is to extract a vector subsequence that converges componentwise. Because $B \subseteq S_r(\overline{x})$,

$$\rho(\overline{x}, \overline{y}_k) = \left[\sum_{i=1}^{n} |z_k^i|^p \right]^{1/p} < r. \tag{E.13}$$

Then

$$\forall i = 1, \ldots, n : \forall k \in \mathbb{Z}^+ : |z_k^i| < r. \tag{E.14}$$

Because the sequence of first components $\{z_1^1, z_2^1, \ldots\}$ is bounded in $\mathbb{F}$, the Bolzano-Weierstraß theorem in $\mathbb{F}$ implies that $\{z_1^1, z_2^1, \ldots\}$ has a limit point in $\mathbb{F}$. Then we can extract a subsequence

$$Z^1 = \{z_{i_1}^1, z_{i_2}^1, \ldots\} \tag{E.15}$$

that converges to $z^1 \in \mathbb{F}$. The next step is to extract, from the bounded sequence $\{z_{i_1}^2, z_{i_2}^2, \ldots\}$ of second components of the vector subsequence $\{\overline{z}_{i_1}, \overline{z}_{i_2}, \ldots\}$, a subsequence

$$Z^2 = \{z_{j_1}^2, z_{j_2}^2, \ldots\} \tag{E.16}$$

that converges to $z^2 \in \mathbb{F}$. So far we have constructed a *vector* subsequence $\{\overline{z}_{j_1}, \overline{z}_{j_2}, \ldots\}$ in which the first and second components of the (vector) terms converge to z^1 and z^2, respectively. We continue to extract vector subsequences such that successively larger numbers of components of each vector converge in $\mathbb{F}$ until we arrive at a vector subsequence

$$Z = \{\overline{z}_{k_1}, \overline{z}_{k_2}, \ldots\} \tag{E.17}$$

in which

$$\forall i = 1, \ldots, n : \lim_{k_l \to \infty} z_{k_l}^i = z^i. \tag{E.18}$$

Now let

$$\overline{z} = \sum_{i=1}^{n} z^i \overline{e}_i. \tag{E.19}$$

Because componentwise convergence implies vector (strong) convergence in a finite-dimensional normed vector space (see section CC.3), the vector subsequence Z converges to $\overline{z}$:

$$\overline{z} = \lim_{k_l \to \infty} \overline{z}_{k_l}. \tag{E.20}$$

Therefore $\overline{x} - \overline{z}$ is a limit point of the bounded sequence Y. This establishes the theorem.

The generalized Bolzano-Weierstraß theorem holds in $\mathbb{C}$, because $\mathbb{C}$ is a two-dimensional Euclidean inner-product space over $\mathbb{R}$. In particular, $\mathbb{C}$ is complete with respect to the modulus norm. Therefore the generalized Bolzano-Weierstraß theorem holds in all finite-dimensional unitary spaces.

Clearly we have also established that *every finite-dimensional Euclidean or unitary space $\mathcal{V}$ is complete under any p-norm such that $1 \leq p \leq \infty$ and therefore that $\mathcal{V}$ is a Banach space or a Hilbert space, depending on the value of p.*

APPENDIX F

WEIERSTRAß APPROXIMATION THEOREM

The **Weierstraß approximation theorem** is: *For every function $f \in \mathcal{C}^0([a, b]; \mathbb{C})$, there exists a sequence of polynomials $(p_0, p_1, \ldots)$ that converges uniformly to f on $[a, b]$:*

$$\forall \epsilon > 0 : \exists L(\epsilon) : \forall l > L(\epsilon) : \forall x \in [a, b] : |f(x) - p_l(x)| < \epsilon. \tag{F.1}$$

Note that the theorem applies to complex-valued continuous functions. It turns out that the polynomials that we construct to prove the theorem are real-valued if f is real-valued.

The Weierstraß approximation theorem is an extremely important part of the foundations of both mathematical and computational physics because it guarantees the existence of useful polynomial approximations. When applied to trigonometric polynomials, the Weierstraß theorem implies that for any continuous complex-valued function $f : [0, 2\pi] \to \mathbb{C}$, there exists a sequence of linear combinations of Fourier functions

$$u_m(\phi) := \frac{1}{\sqrt{2\pi}} e^{im\phi}. \tag{F.2}$$

that approximates f uniformly on $[0, 2\pi]$. Because $\mathcal{C}^0([0, 2\pi]; \mathbb{C})$ is dense in $\mathcal{L}^2([0, 2\pi]; \mathbb{C})$, the Weierstraß theorem therefore implies the completeness of the Fourier functions.

We simplify the uniform-approximation problem as stated above in two ways before proceeding to a proof. First, an arbitrary finite interval $[a, b]$ can be scaled to the interval $[0, 1]$ through the change of variables

$$x \mapsto x' = \frac{x - a}{b - a}. \tag{F.3}$$

Second, we can, restrict our attention to functions that vanish at $x = 0$ and at $x = 1$. Given any $f \in \mathcal{C}^0([0, 1]; \mathbb{C})$, let

$$g(x) = f(x) - f(0) - x[f(1) - f(0)]. \tag{F.4}$$

Because $f - g$ is the polynomial $h(x) = f(0) + x[f(1) - f(0)]$, if the Weierstraß theorem is true for g, then it is true also for $f = g + h$. Therefore we assume in the following that

$$f(0) = f(1) = 0. \tag{F.5}$$

We assume also that f vanishes outside the interval $[0, 1]$; hence $[0, 1]$ is the support of f,

$$\text{supp}[f] = [0, 1]. \tag{F.6}$$

The assumption that f is continuous on the closed interval $[0, 1]$ implies that f is uniformly continuous on $[0, 1]$. This facts and our assumptions that $f(0) = f(1) = 0$ and that f vanishes outside the interval $[0, 1]$, imply that f is uniformly continuous on $\mathbb{R}$.

Let

$$\varphi_l(x) := \kappa_l(1 - x^2)^l \tag{F.7}$$

in which κ_l is a constant such that

$$\int_{-1}^{1} \varphi_l(x)\,dx = 1. \tag{F.8}$$

Obviously φ_l is positive on the open interval $(-1, 1)$. It can be shown that

$$\kappa_l = \frac{(2l + 1)!!}{2(2^l l!)^2} < l + 1. \tag{F.9}$$

The exact bound is not important; what matters is that κ_l increases no faster than some polynomial in l.

We show that the functions p_l such that

$$p_l(x) := \int_{-1}^{1} f(x + y)\varphi_l(y)\,dy \tag{F.10}$$

are polynomials that satisfy Eq. (F.1). Note that the integral with respect to y must run from -1 to 1, and not from 0 to 1, because the extreme values of x are -1 and 1. In order to make $x + y = 0$ with $x = 1$, one must have $y = -1$; in order to make $x + y = 1$ with $x = 0$, one must have $y = 1$. To show that p_l is a polynomial, let $x' = x + y$ in Eq. (F.10):

$$\begin{aligned} p_l(x) &= \int_{-1+x}^{1+x} f(x')\varphi_l(x' - x)\,dx' \\ &= \int_{0}^{1} f(x')\varphi_l(x' - x)\,dx' \end{aligned} \tag{F.11}$$

where the second line follows for any $x \in [0, 1]$ because $\mathrm{supp}[f] = [0, 1]$. The second line makes it clear that p_l is a polynomial, because φ_l is a polynomial. The interpretation of the convolution of f with φ_l in Eq. (F.11) is that p_l is the function f, "filtered" by the polynomial φ_l.

The fact that $1 - x^2 < 1$ if $0 < |x| \le 1$ implies that the functions $(1 - x^2)^l$ decrease exponentially as l increases for every $x \neq 0$. Because κ_l is bounded by a polynomial in l, $\varphi_l(x)$ approaches zero as $l \to \infty$ for every nonzero value of x:

$$\forall x \in [-1, 0) \cup (0, 1] : \lim_{l \to \infty} \varphi_l(x) = 0. \tag{F.12}$$

In particular, if $0 < \delta < 1$,

$$\forall x \ni: 1 \ge |x| \ge \delta : (1 - x^2)^l \le (1 - \delta^2)^l \tag{F.13}$$

and therefore

$$\forall x :\ni: 1 \geq |x| \geq \delta :\ 0 \leq \varphi_l(x) \leq \kappa_l(1 - \delta^2)^l < (l + 1)(1 - \delta^2)^l. \tag{F.14}$$

Let F be the least upper bound of the modulus of $f(x)$:

$$F = \sup_{x \in [0,1]} |f(x)|. \tag{F.15}$$

Then

$$\forall \epsilon > 0 : \exists L(\epsilon) > 0 :\ni: \forall l > L(\epsilon) : (l + 1)(1 - \delta^2)^l < \frac{\epsilon}{8F}, \tag{F.16}$$

from which it follows at once that

$$\forall \epsilon > 0 : \exists L(\epsilon) > 0 :\ni: \forall l > L(\epsilon) : \varphi_l(x) < \frac{\epsilon}{8F}. \tag{F.17}$$

Therefore

$$\int_\delta^1 \varphi_l(y)\, dy < (1 - \delta)\frac{\epsilon}{8F} < \frac{\epsilon}{8F} \tag{F.18}$$

and

$$\int_{-1}^{-\delta} \varphi_l(y)\, dy < (1 - \delta)\frac{\epsilon}{8F} < \frac{\epsilon}{8F}. \tag{F.19}$$

Choose $\epsilon > 0$. Because f is uniformly continuous on $\mathbb{R}$,

$$\forall \epsilon > 0 : \exists \delta \in (0, 1) :\ni: \forall x, y \in \mathbb{R} :\ni: |x - y| < \delta : |f(x) - f(y)| < \frac{\epsilon}{2}. \tag{F.20}$$

Choose l as in Eq. (F.17). Then, for all $x \in [0, 1]$:

$$
\begin{aligned}
|f(x) - p_l(x)| &= \left| \int_{-1}^1 [f(x) - f(x + y)]\varphi_l(y)\, dy \right| \\
&\leq \int_{-1}^1 |f(x) - f(x + y)|\varphi_l(y)\, dy \\
&\leq 2F \int_{-1}^{-\delta} \varphi_l(y)\, dy + \frac{\epsilon}{2} \int_{-\delta}^{\delta} \varphi_l(y)\, dy + 2F \int_\delta^1 \varphi_l(y)\, dy \\
&< \frac{\epsilon}{4} + \frac{\epsilon}{2} + \frac{\epsilon}{4} = \epsilon.
\end{aligned}
\tag{F.21}
$$

It follows that $p_l \to f$ uniformly on $[0, 1]$ as $l \to \infty$.

The extension of the Weierstraß approximation theorem to functions of, and polynomials in, more than one variable is straightforward. For example, consider a continuous function $f : [0, 1] \times [0, 1] \to \mathbb{C}$, which maps the unit square into the field of complex numbers. Each boundary of the unit square is defined by one of the four equations $x = 0$, $x = 1$, $y = 0$, or

$y = 1$. We can assume, without loss of generality, that f vanishes on the boundaries of the square, *i.e.*, that

$$\forall x \in [0, 1] : \quad f(x, 0) = 0 \quad \text{and} \quad f(x, 1) = 0 \tag{F.22}$$

and

$$\forall y \in [0, 1] : \quad f(0, y) = 0 \quad \text{and} \quad f(1, y) = 0. \tag{F.23}$$

To see this, note that $f(x, 0)$, and so forth, are continuous functions on the unit closed interval, and therefore can be approximated uniformly by (different) polynomials. The function

$$
\begin{aligned}
g(x, y) = {}& f(x, y) - \left[f(x, 0)(1 - y) + f(x, 1)y + f(0, y)(1 - x) + f(1, y)x \right] \\
& + \left[f(0, 0)(1 - x)(1 - y) + f(1, 0)x(1 - y) \right. \\
& \left. + f(0, 1)(1 - x)y + f(1, 1)xy \right]
\end{aligned}
\tag{F.24}
$$

vanishes on all four boundaries, and the terms in square brackets can be approximated uniformly by polynomials. We assume from now on that f vanishes on the boundaries of the unit square, and that

$$\text{supp}[f] = [0, 1] \times [0, 1]. \tag{F.25}$$

It follows that f is uniformly continuous on $\mathbb{R}^2$.

The approximating polynomial is

$$p_l(x, y) := \int_{-1}^{1} \int_{-1}^{1} f(x + x', y + y') \varphi_l(x') \varphi_l(y') \, dx' dy'. \tag{F.26}$$

The modulus of the difference between f and p_l,

$$
\begin{aligned}
& |f(x, y) - p_l(x, y)| \\
& = \int_{-1}^{1} \int_{-1}^{1} [f(x, y) - f(x + x', y + y')] \varphi_l(x') \varphi_l(y') \, dx' dy',
\end{aligned}
\tag{F.27}
$$

can be estimated by dividing the unit square into nine different regions, as follows (beginning at the upper left corner, and proceeding to the right and downwards):

$$
\begin{array}{ccc}
[-1, -\delta] \times [\delta, 1], & [-\delta, \delta] \times [\delta, 1], & [\delta, 1] \times [\delta, 1], \\
[-1, -\delta] \times [-\delta, \delta], & [-\delta, \delta] \times [-\delta, \delta], & [\delta, 1] \times [-\delta, \delta], \\
[-1, -\delta] \times [-\delta, -1], & [-\delta, \delta] \times [-\delta, -1], & [\delta, 1] \times [-\delta, -1].
\end{array}
\tag{F.28}
$$

By Eq. (F.12),

$$\forall \epsilon > 0 : \exists L(\epsilon) > 0 : \ni: \forall l > L(\epsilon) : \quad \varphi_l(x) < \frac{\epsilon}{16F}. \tag{F.29}$$

Because f is uniformly continuous on $\mathbb{R}^2$,

$$\forall \epsilon > 0 : \exists \delta \in (0, 1) :\ni: \forall x, y, x', y' \in \mathbb{R} :\ni: |x - x'| < \delta \quad \text{and}$$

$$|y - y'| < \delta : |f(x, y) - f(x', y')| < \frac{\epsilon}{2}. \tag{F.30}$$

The contribution of each of the four square regions at the corners (such as $[-1, -\delta] \times [\delta, 1]$) is less than

$$2F \left(\int_{-1}^{-\delta} \varphi_l(x') \, dx' \right) \left(\int_{\delta}^{1} \varphi_l(y') \, dy' \right) < 2F \left(\frac{\epsilon}{16F} \right)^2 = \frac{\epsilon^2}{32F} \tag{F.31}$$

and can be neglected. The contribution of each of the four strips (regions such as $[-1, -\delta] \times [-\delta, \delta]$) is less than

$$2F \left(\int_{-1}^{-\delta} \varphi_l(x') \, dx' \right) \left(\int_{-\delta}^{\delta} \varphi_l(y') \, dy' \right) < \epsilon/8. \tag{F.32}$$

The contribution of the center region, $[-\delta, \delta] \times [-\delta, \delta]$, is less than

$$\int_{-\delta}^{\delta} \int_{-\delta}^{\delta} \left(\frac{\epsilon}{2} \right) \varphi_l(x') \varphi_l(y') \, dx' \, dy' < \epsilon/2. \tag{F.33}$$

Therefore the right-hand side of Eq. (F.27) is less than the sum $\epsilon/4 + \epsilon/4 + \epsilon/4 + \epsilon/4 + \epsilon/2 = \epsilon$. It follows that f can be approximated uniformly to any desired accuracy by a polynomial on the unit square. The extension to arbitrary rectangles follows by translating and stretching the unit square, using transformations such as Eq. (F.3) in both x and y.

We show, finally, that any continuous, complex-valued function on the interval $[0, 2\pi]$ such that $f(2\pi) = f(0)$ can be approximated uniformly to any desired accuracy ϵ by a polynomial of degree $l = l(\epsilon)$ in $\cos\theta$ and $\sin\theta$, where $\theta = \tan^{-1}(y/x)$. We note that f is defined on the circumference of the unit circle, and extend f to a function g on the square $[-1, 1] \times [-1, 1]$ with the definition (valid for all x and y in the square)

$$g(x, y) := \begin{cases} (x^2 + y^2)^\mu f(\theta), & \text{if } x \text{ or } y, \text{ or both, are nonzero;} \\ 0, & \text{if } x = y = 0, \end{cases} \tag{F.34}$$

where $\mu > 0$ and

$$\theta = \tan^{-1}\left(\frac{y}{x}\right). \tag{F.35}$$

Evidently

$$\xi = \frac{x}{\left(x^2 + y^2\right)^{1/2}} = \cos\theta \quad \text{and} \quad \eta = \frac{y}{\left(x^2 + y^2\right)^{1/2}} = \sin\theta \tag{F.36}$$

are the x and y coordinates of a point on the circumference of the unit circle.

By the Weierstraß approximation theorem, for every $\epsilon > 0$ there exists a polynomial $p_l(x, y)$ of degree l in both x and y such that, for all x and y in the square $[-1, 1] \times [-1, 1]$,

$$|g(x, y) - p_l(x, y)| < \epsilon. \tag{F.37}$$

For any point (ξ, η) on the circumference of the unit circle, then,

$$|g(\xi, \eta) - p_l(\xi, \eta)| = |f(\theta) - p_l(\cos\theta, \sin\theta)| < \epsilon. \tag{F.38}$$

With the help of the usual relations between the trigonometric functions and $e^{i\theta}$, p_l can be expressed as a linear combination of the exponentials $e^{-il\theta}, \ldots, e^{il\theta}$, in other words, as a linear combination of the Fourier basis functions. It follows that the linear combinations of the Fourier functions are dense in $\mathcal{C}^0([0, 2\pi]; \mathbb{C})$. Because the trigonometric polynomial $p_l(\cos\theta, \sin\theta)$ approximates f uniformly for $\theta \in [0, 2\pi]$, it follows also that the sequence $(p_0, p_1, \ldots)$ of approximating trigonometric polynomials converges uniformly to f, if f is continuous on $[0, 2\pi]$.

INDEX

Printed in the United States
By Bookmasters